The Physics of Solar Energy Conversion

The Physics of Solar Energy Conversion

Juan Bisquert
Universitat Jaume I, Castelló, Spain

CRC Press
Taylor & Francis Group
Boca Raton London New York

CRC Press is an imprint of the
Taylor & Francis Group, an **informa** business

About the Cover Image: Transparent and flexible perovskite solar cell from Saule Technologies, reprinted with permission from Saule Technologies.

First edition published 2020
by CRC Press
6000 Broken Sound Parkway NW, Suite 300, Boca Raton, FL 33487-2742

and by CRC Press
2 Park Square, Milton Park, Abingdon, Oxon, OX14 4RN

ISBN: 978-1-138-58464-8 (hbk)
ISBN: 978-0-367-49602-9 (pbk)
ISBN: 978-0-429-50587-4 (ebk)

Typeset in Times
by Deanta Global Publishing Services, Chennai, India

A Joan i Joel

Contents

Preface xv
Acknowledgments xvii
Author xix

Chapter 1 Introduction to Energy Devices 1

References 9

PART I Equilibrium Concepts and Kinetics

Chapter 2 Electrostatic and Thermodynamic Potentials of Electrons in Materials 13

2.1 Electrostatic Potential 13
2.2 Energies of Free Electrons and Holes 14
2.3 Potential Energy of the Electrons in the Semiconductor 17
2.4 The Vacuum Level 17
2.5 The Fermi Level and the Work Function 20
2.6 The Chemical Potential of Electrons 21
2.7 Potential Step of a Dipole Layer or a Double Layer 23
2.8 Origin of Surface Dipoles 24
2.9 The Volta Potential 25
2.10 Equalization of Fermi Levels of Two Electronic Conductors in Contact 27
2.11 Equilibration of Metal Junctions and the Contact Potential Difference 28
2.12 Equilibrium across the Semiconductor Junction 29
General References 31
References 31

Chapter 3 Voltage, Capacitors, and Batteries 33

3.1 The Voltage in the Device 33
3.2 Anode and Cathode 34
3.3 Applied Voltage and Potential Difference 35
3.4 The Capacitor 37
3.5 Measurement of the Capacitance 38
3.6 Energy Storage in the Capacitor 40
3.7 Electrochemical Systems: Structure of the Metal/Solution Interface 40
3.8 Electrode Potential and Reference Electrodes 42
3.9 Redox Potential in Electrochemical Cells 44
3.10 Electrochemical and Physical Scales of Electron Energy in Material Systems 45
3.11 Changes of Electrolyte Levels with pH 46
3.12 Principles of Electrochemical Batteries 47
3.13 Capacity and Energy Content 50
3.14 Practical Electrochemical Batteries 51
3.14.1 Zinc-Silver Battery 51
3.14.2 Sodium-Sulfur Battery 52
3.15 Li-Ion Battery 53
General References 57
References 57

Chapter 4 Work Functions and Injection Barriers 59

4.1 Injection to Vacuum in Thermionic Emission 59
4.2 Richardson–Dushman Equation 60
4.3 Kelvin Probe Method 61
4.4 Photoelectron Emission Spectroscopy 63
4.5 Injection Barriers 66
4.6 Pinning of the Fermi Level and Charge-Neutrality Level 69
General References 73
References 73

Chapter 5 Thermal Distribution of Electrons, Holes, and Ions in Solids 75

5.1 Equilibration of the Electrochemical Potential of Electrons 75
5.2 Configurational Entropy of Weakly Interacting Particles 76
5.3 Equilibrium Occupancy of Conduction Band and Valence Band States 76
5.4 Equilibrium Fermi Level and the Carrier Number in Semiconductors 79
5.5 Transparent Conducting Oxides 81
5.6 Hot Electrons 82
5.7 Screening 84
5.8 The Rectifier at Forward and Reverse Voltage 85
5.9 Semiconductor Devices as Thermal Machines that Realize Useful Work 88
5.10 Cell Potential in the Lithium Ion Battery 90
5.11 Insertion of Ions: The Lattice Gas Model 94
General References 98
References 98

Chapter 6 Interfacial Kinetics and Hopping Transitions 101

6.1 Principle of Detailed Balance 101
6.2 Form of the Transition Rates 104
6.3 Kinetics of Localized States: Shockley-Read-Hall Recombination Model 106
6.4 Reorganization Effects in Charge Transfer: The Marcus Model 107
6.5 Polaron Hopping 112
6.6 Rate of Electrode Reaction: Butler-Volmer Equation 115
6.6.1 Availability of Electronic Species 116
6.6.2 Availability of Redox Species 116
6.6.3 The Kinetic Constant for Charge Transfer 117
6.7 Electron Transfer at Metal-Semiconductor Contact 120
6.8 Electron Transfer at the Semiconductor/Electrolyte Interface 121
General References 126
References 127

Chapter 7 The Chemical Capacitance 131

7.1 Carrier Accumulation and Energy Storage in the Chemical Capacitance 131
7.2 Localized Electronic States in Disordered Materials and Surface States 133
7.3 Chemical Capacitance of a Single State 135
7.4 Chemical Capacitance of a Broad DOS 136
7.5 Filling a DOS with Carriers: The Voltage and the Conductivity 138
7.6 Chemical Capacitance of Li Intercalation Materials 139
7.7 Chemical Capacitance of Graphene 140
General References 142
References 143

Chapter 8 The Density of States in Disordered Inorganic and Organic Conductors 145

8.1 Capacitive and Reactive Current in Cyclic Voltammetry 145
8.2 Kinetic Effects in CV Response 149
8.3 The Exponential DOS in Amorphous Semiconductors 150
8.4 The Exponential DOS in Nanocrystalline Metal Oxides 152
8.5 Basic Properties of Organic Layers 156
8.6 The Gaussian DOS 160
General References 162
References 163

Chapter 9 Planar and Nanostructured Semiconductor Junctions 167

9.1 Structure of the Schottky Barrier at a Metal/Semiconductor Contacts 167
9.2 Changes of the Schottky Barrier by the Applied Voltage 168
9.3 Properties of the Planar Depletion Layer 170
9.4 Mott–Schottky Plots 171
9.5 Capacitance Response of Defect Levels and Surface States 172
9.6 Semiconductor Electrodes and the Flatband Potential 173
9.7 Changes of Redox Level and Band Unpinning 176
9.8 Inversion and Accumulation Layer 180
9.9 Heterojunctions 181
9.10 Effect of Voltage on Highly Doped Nanocrystalline Semiconductors 183
9.11 Homogeneous Carrier Accumulation in Low-Doped Nanocrystalline Semiconductors 188
General References 192
References 192

PART II Foundations of Carrier Transport

Chapter 10 Carrier Injection and Drift Transport 197

10.1 Transport by Drift in the Electrical Field 197
10.2 Injection at Contacts 198
10.3 The Metal-Insulator-Metal Model 202
10.4 The Time-of-Flight Method 205
General References 206
References 206

Chapter 11 Diffusion Transport 209

11.1 Diffusion in the Random Walk Model 209
11.2 Macroscopic Diffusion Equation 211
11.3 The Diffusion Length 212
11.4 Chemical Diffusion Coefficient and the Thermodynamic Factor 213
General References 215
References 215

Chapter 12 Drift-Diffusion Transport 217

12.1 General Transport Equation in Terms of Electrochemical Potential 217
12.2 The Transport Resistance 217
12.3 The Einstein Relation 219
12.4 Drift-Diffusion Equations 220

12.5 Ambipolar Diffusion Transport 221
12.6 Relaxation of Injected Charge 222
12.7 Transient Current in Insulator Layers 223
12.8 Modeling Transport Problems 224
General References 227
References 227

Chapter 13 Transport in Disordered Media 229

13.1 Multiple Trapping and Hopping Transport 229
13.2 Transport by Hopping in a Single Level 231
13.3 Trapping Factors in the Kinetic Constants 233
13.4 Two-Level (Single-Trap) Model 235
13.5 Multiple Trapping in Exponential DOS 237
13.6 Activated Transport in a Gaussian DOS 237
13.7 Multiple Trapping in the Time Domain 239
13.8 Hopping Conductivity 241
13.9 The Transport Energy 242
13.10 Variable Range Hopping 243
General References 245
References 245

Chapter 14 Thin Film Transistors 249

14.1 Organic Thin Film Transistors 249
14.2 Carrier Density in the Channel 250
14.3 Determination of the DOS in Thin Film Transistor Configuration 252
14.4 Current-Voltage Characteristics 255
14.5 The Mobility in Disordered Semiconductors 257
14.6 Electrochemical Transistor 258
General References 259
References 259

Chapter 15 Space-Charge-Limited Transport 263

15.1 Space-Charge-Limited Current 263
15.2 Injected Carrier Capacitance in SCLC 265
15.3 Space Charge in Double Injection 267
General References 269
References 269

Chapter 16 Impedance and Capacitance Spectroscopies 271

16.1 Frequency Domain Measurements 271
16.2 Dielectric Relaxation Functions 272
16.3 Resistance and Capacitance in Equivalent Circuit Models 274
16.4 Relaxation in Time Domain 279
16.5 Universal Properties of the Frequency-Dependent Conductivity 281
16.6 Electrode Polarization 283
General References 284
References 284

PART III Radiation, Light, and Semiconductors

Chapter 17 Blackbody Radiation and Light ... 289

17.1 Photons and Light ... 289
17.2 Spread and Direction of Radiation ... 289
17.3 Color and Photometry ... 291
17.4 Blackbody Radiation ... 293
17.5 The Planck Spectrum ... 294
17.6 The Energy Density of The Distribution of Photons in Blackbody Radiation ... 295
17.7 The Photon and Energy Fluxes in Blackbody Radiation ... 297
17.8 The Solar Spectrum ... 299
General References ... 302
References ... 302

Chapter 18 Light Absorption, Carrier Recombination, and Luminescence ... 305

18.1 Absorption of Incident Radiation ... 305
18.2 Luminescence and Energy Transfer ... 307
18.3 The Quantum Efficiency ... 310
18.4 The Recombination of Carriers in Semiconductors ... 311
18.5 Recombination Lifetime ... 314
General References ... 316
References ... 316

Chapter 19 Optical Transitions in Organic and Inorganic Semiconductors ... 319

19.1 Light Absorption in Inorganic Solids ... 319
19.2 Free Carrier Phenomena ... 323
19.3 Excitons ... 325
19.4 Quantum Dots ... 328
19.5 Organic Molecules and Materials ... 330
19.6 The CT Band in Organic Blends and Heterojunctions ... 333
General References ... 336
References ... 336

PART IV Photovoltaic Principles and Solar Energy Conversion

Chapter 20 Fundamental Model of a Solar Cell ... 343

20.1 Majority Carrier Injection Mechanisms ... 343
20.2 Majority Carrier Devices ... 344
20.3 Minority Carrier Devices ... 345
20.4 Fundamental Properties of a Solar Cell ... 346
20.5 Physical Properties of Selective Contacts in Solar Cells ... 348
General References ... 351
References ... 351

Chapter 21 Recombination Current in the Semiconductor Diode 353

21.1 Dark Equilibrium of Absorption and Emission of Radiation 353
21.2 Recombination Current 355
21.3 Dark Characteristics of Diode Equation 356
21.4 Light-Emitting Diodes 357
21.5 Dye Sensitization and Molecular Diodes 360
General References 363
References 363

Chapter 22 Radiative Equilibrium in a Semiconductor 365

22.1 Utilization of Solar Photons 365
22.2 Fundamental Radiative Carrier Lifetime 368
22.3 Radiative Emission of a Semiconductor Layer 369
22.4 Photons at Nonzero Chemical Potential 370
General References 373
References 373

Chapter 23 Reciprocity Relations in Solar Cells and Fundamental Limits to the Photovoltage 375

23.1 The Reciprocity between LED and Photovoltaic Performance Parameters 375
23.2 Factors Determining the Photovoltage 378
23.3 External Radiative Efficiency 382
23.4 Photon Recycling 383
23.5 Radiative Cooling in EL and Photoluminescence 386
23.6 Reciprocity of Absorption and Emission in a CT Band 387
General References 391
References 392

Chapter 24 Charge Separation and Material Limits to the Photovoltage 395

24.1 Light Absorption 395
24.2 Charge Separation 395
24.3 Materials Limits to the Photovoltage 398
General References 403
References 404

Chapter 25 Operation of Solar Cells and Fundamental Limits to Their Performance 407

25.1 Current-Voltage Characteristics 407
25.2 Power Conversion Efficiency 408
25.3 Analysis of FF 410
25.4 Shockley–Queisser Efficiency Limits 412
25.5 Practical Solar Cells Efficiency Limits 413
General References 419
References 419

Chapter 26 Charge Collection in Solar Cells 421

26.1 Introduction to Charge Collection Properties 421
26.2 Charge Collection Distance 422

26.3 General Modeling Equations 424
26.4 The Boundary Conditions 425
26.4.1 Charge Extraction Boundary Condition 426
26.4.2 Blocking Boundary Condition 427
26.4.3 Generalized Boundary Conditions 428
26.5 A Photovoltaic Model with Diffusion and Recombination 429
26.6 The Gärtner Model 433
26.7 Diffusion-Recombination and Collection in the Space-Charge Region 435
26.8 Solar Cell Simulation 436
26.9 Classification of Solar Cells 437
26.10 Measuring and Reporting Solar Cell Efficiencies 439
General References 442
References 442

Chapter 27 Spectral Harvesting and Photoelectrochemical Conversion 445

27.1 Conversion of Photon Frequencies for Solar Energy Harvesting 445
27.2 Tandem Solar Cells 448
27.3 Solar Fuel Generation 450
General References 456
References 456

Appendix 459

Index 463

Preface

The investigation of solar energy conversion materials and devices has come to the forefront of global scientific research, and it reached a state of maturity in recent decades. Nonetheless, enormous challenges lie ahead. Substantial scientific innovation will be necessary to obtain the essential contributions and steps towards a clean energy–based economy that is still only a promise.

The conversion of photons to electricity using light-absorbing semiconductors in a stable and highly efficient device that withstands operation over time is an extraordinarily complex process requiring matching physical processes in tailored materials at multiple spatial and temporal scales. The operation of solar energy conversion involves a set of concepts and tools that fluctuate among different disciplines of physics or chemistry. Examples of rather broad concepts and methods for such applications include: the synthesis of active molecules and solids that generate and transfer charge; the actuation of Fermi levels in thin films; and the detailed balance of light absorption and emission.

Following several decades of intensive investigation, a new broad landscape of candidate materials and devices were discovered and systematically studied and reported, including organic, inorganic, solid, liquid, soft matter and their combinations, and a myriad of contact materials and interfacial processes. In this evolution and expansion of the classes of materials in the past three decades, with a particular emphasis on dye-sensitized solar cells, the organic solar cells, and the metal halide perovskite solar cells, the conceptual picture of a solar cell has been extended and refined. New concepts and a powerful picture that embraces very different types of devices have been established based upon many discussions as well as conceptual clashes.

The resulting scientific consensus is the story I want to tell. The realization of this project, this volume, took one decade and came about through the consecutive publication of three original separate books, as different units of the plan were being fulfilled:

Nanostructured Energy Devices. Equilibrium Concepts and Kinetics
Nanostructured Energy Devices. Foundation of Carrier Transport
Physics of Solar Cells. Perovskites, Organics, and Fundamentals of Photovoltaics

This book has been formed by merging the three previous separate volumes and in 2020 updating several chapters to reflect the latest scientific advances of the field. The preceding trilogy was already conceived as a unity at the time of the first publication. Hence the chapters of this volume, *Physics of Solar Energy Conversion*, follow in most cases the sequence of chapters of the previous books, but are now numerated sequentially. However, cross citation between chapters, which is used frequently and emphasizes the intricate web of concepts, ideas and methods, is now much easier.

In this volume the reader will obtain a firm basis for understanding solar energy conversion devices, which may be formed by a great variety of materials, at the same time obeying general principles and constraints in order to reach the final goal of forming useful energy conversion technologies. The principal spirit of this approach is to facilitate starting researchers to get an overall picture of what needs to be known, and also hopefully support experts to achieve a summarizing scheme.

Solar cell light absorber materials can be inversely operated for light emission and also in electrochemical solar fuel conversion. A broad class of devices for such types of purposes use electrochemical setups, thin films, and organic materials, normally formed by solution-processed routes for low-cost applications. Due to the existence of these strong direct material and conceptual connections, one goal of this volume is to put energy devices of solar cells in a wider perspective, including a range of energy devices that are operationally related and often based on similar materials. From the broader standpoint, the text covers the main concepts that apply to several types of devices so that the reader gains an insight into the general view of principles of operation of the energy devices. We analyze the fundamental concepts, main properties, and key applications of energy devices including perovskite solar cells, organic solar cells, electrochemical batteries, diodes, LEDs and OLEDs, transistors, and the direct conversion of solar radiation to chemical fuels. One benefit of the broader perspective is that innovations can be shared and exported across communities producing cross-fertilization. Such principles will be discussed in connection with an array of physical, electrochemical, and optoelectronic measurement techniques and theoretical simulation methods that are widely used and constitute essential tools in the field.

The book has been organized to provide an introduction to the fundamentals of operation of very different devices that nonetheless have operational or material relationships. A unified view of the device operation is emphasized based on some general concepts such as the energy diagrams and the electrochemical potentials. We aim to pull together the views and terminologies used by several communities including the solid state science, surface science, electrochemistry, and electronic devices physics. The book does not aim to exhaustively list materials and their properties but rather takes some prototypical and representative cases of materials widely used for devices and applications, with the main focus being on laying conceptual foundations rather than summarizing innovations of the past decades.

The first part of the book, Chapters 1–9, examines fundamental principles of semiconductor energetics, interfacial charge transfer, basic concepts and methods of measurement, and the properties of important classes of materials such as metal oxides and organic semiconductors. These materials and their properties are important in the operation of organic and perovskite solar cells either as the bulk absorber or as a selective contact structure. Electrolytic and solid ionic conductor properties also play relevant roles in organic and perovskite solar cells. This part is mainly focused on equilibrium properties at open circuit condition, often represented in energy diagrams, and also on the basic aspect of carrier kinetics.

The second part, Chapters 10–16, presents a catalog of the physics of carrier transport in semiconductors with a view to establishing energy device models. This part systematically explains the diffusion-drift model that is central to solar cell operation, the different responses of band bending and electrical field distribution that occur when a voltage is applied to a device with contacts, and the central issue of injection and mechanisms of contacts. The carrier transport in disordered materials is explored, as these often appear as good candidates for easily processed solar cells. There are also excursions into other important topics such as the transistor configuration and frequency domain techniques, such as Impedance Spectroscopy, that produce central experimental tools for the characterization of these devices.

The third part, Chapters 17–19, consists of the background to essential physical and material properties about electromagnetic radiation fields, optical properties of semiconductors, and the interaction of light with the latter causing the generation of carriers that are at the heart of photovoltaic action: creating energetic electrons that go around an external circuit.

The fourth part, Chapters 20–27, addresses the main goal of the book: to provide an explanation of the operation of solar energy conversion devices, which embraces concepts from a diversity of classes of solar cells, from nanostructured and highly disordered materials to highly efficient devices such as the metal halide perovskite solar cells. The approach is to establish from the beginning a simple but very rich model of a solar cell, in order to develop and understand step-by-step the photovoltaic operation according to fundamental physical properties and constraints. The subject matter is initiated by formulating the physical basis of the recombination diode, and from the standpoint of the built-in asymmetry of the flow of electrons and holes, state the effects of the interaction of light and the semiconductors, the creation of the split of Fermi levels, and the production of output electrical power. In many textbooks, solar cell operation is related to a diode element as the starting point. Here, the physical mechanisms determining the operation of the diode in photovoltaic conversion is clarified, as this is an essential point for the understanding and design of solar cell structures. This model was developed historically from the central insights of William Shockley, Hans-Joachim Queisser and Robert T. Ross in papers published 50 years ago (see Chapter 22). It focuses on the aspects pertaining to the functioning of a solar cell and the determination of limiting efficiencies of energy conversion, by intentionally removing the many avoidable losses such as transport gradients, which are treated in the final chapters. This is then extended to the physical limitations to optimal conversion obtaining more realistic and specific effects, configurations, and shortcomings that provide a summary of the multitude of effects that may come into play at the time of experimental investigations and technological development.

Building on the knowledge that has been gathered in the book allows exploration of the physical and electrochemical properties of devices, from molecular aspects to macroscopic models, which can provide the reader with a strong foundation both for understanding solar cell operation and for exploring related materials and a broader class of possibilities for solar energy conversion devices.

Acknowledgments

I am very grateful to many colleagues who pointed out improvements on parts of this book, and especially:

Osbel Almora
Doron Aurbach
Henk Bolink
Chris Case
Germà Garcia Belmonte
Antonio Guerrero
Albert Ferrando
Tom Hamann
Andreas Klein
Volodia Kytin
José A. Manzanares
Juan P. Martínez Pastor
Rudolph Marcus
Truls Norby
Luis M. Pazos Outón
Kim Puigdollers
Andrey Rogach,
Rafael Sánchez,
Iván Mora Seró
Greg Smestad
Sergio Trasatti
Eduard Vakarin
Koen Vandewal

I am especially grateful to Mehdi Ansari-Rad, Sandheep Ravishankar, Pilar López-Varo, Thomas Kirchartz, and Luca Bertoluzzi for their revisions of some parts of the book, in the preparation of the final versions. Any remaining mistakes are solely my fault.

Author

Juan Bisquert is a professor of applied physics at the Universitat Jaume I de Castelló and the funding director of the Institute of Advanced Materials at UJI. He earned an MSc in physics in 1985 and a PhD from the Universitat de València in 1992. The research work is in perovskite solar cells, semiconductor optoelectronics, mixed ionic-electronic conductors, and solar fuel converters based on visible light and semiconductors for water splitting and CO_2 reduction. His most well-known work is about the mechanisms governing the operation of nanostructured and solution-processed thin film solar cells. He has developed insights in the electronic processes in hybrid organic–inorganic solar cells, combining the novel theory of semiconductor nanostructures, photoelectrochemistry, and systematic experimental demonstration. His contributions produced a broad range of concepts and characterization methods to analyze the operation of photovoltaic and optoelectronic devices. He is a senior editor of the *Journal of Physical Chemistry Letters*. He has been distinguished several times in the list of ISI Highly Cited Researchers. Bisquert created nanoGe Conferences and is the president of the Fundació Scito. He wrote a novel of speculative fiction, *The Canamel Conjecture*.

1 Introduction to Energy Devices

A system far removed from its condition of equilibrium is the one chosen if we wish to harness its processes for the doing of useful work.

(G. N. Lewis and M. Randall, *Thermodynamics*, 1923, p. 111)

Many technologies have been developed throughout history to provide for human needs: procuring heat, light, and mechanical work for homes, agricultural and industrial use, transportation, and in general to facilitate a broad set of activities of life. These technologies and systems share many features, and the concept that unifies them is *energy*.

In physics, energy is the capacity to do work, meaning that a force can displace an object along a distance. But this is later generalized to a large number of energies, such as thermal energy, related to heat; kinetic energy, in moving objects; chemical energy; electric energy; and many others. The connection of energy to the history of civilization has been well summarized by Smil (2017).

While we live in the midst of thermal energy that is readily available, the energy that we are interested in is one that we could utilize readily for our needs. We wish to easily direct such energy to power a machine or portable electronics that will perform the desired task. The energy flowing in the environment as heat is not useful for these purposes. Energy that is able to do work has to be produced from available sources. Usable energy also has to be stored for use at the time when it is needed, or transported to the required place of consumption.

This book is about using a combination of materials for a useful purpose, either related to energy production and storage or for some practical need that consumes energy. Such an arrangement is called a "device." It is usually operated electrically; therefore, it has contacts where we plug in the external circuit, and internally it is composed of a variety of condensed phase components. Here, we will not examine the basic properties of the materials as a wide variety of books are available on that topic, either from a chemistry or solid-state physics angle. Rather, here we will discuss the methods of analysis of the properties of combinations of the materials in a device, the contacts between the phases, the conductivity, the light absorption or luminescence, the production of chemical species, and their use in providing output power. We will discuss the generation of electricity that does not involve any macroscopic moveable parts and no other motion than the electrons around the circuit. This is achieved by using different kinds of species: photons impinging on light absorber materials and electrons and ions in the solid state and in liquid electrolytes. The final use of energy devices is to transform energy from raw sources to usable energy, to store that usable energy, and also, in some cases, to produce desired processes with substantial energy resource savings.

Industry, transportation, lighting, and communication depend on the supply and utilization of low-priced electricity and clean fuels. As a result of the rising population and increasing living standards around the world, the need for a reliable, convenient, and adaptable energy supply has become a central concern. Currently, chemical energy is the most convenient and useful form of portable energy, and this explains the dependence on fossil fuels (coal, petroleum, and natural gas). However, extensive use of fossil fuels has had a dramatic impact on the world, particularly in environmental change and effects on health. Many people believe that profound changes need to occur in the production, delivery, and utilization of energy. Another important concern is the rising demand of power storage and supply in a variety of forms for feeding portable electronics and electric cars as well as for the continuous availability of solar or wind energy. The development of new energy devices can deeply transform the ways in which energy is obtained and produced, with broad implications for the environment and for the human needs.

A formulation of the desirable system of renewable energy supply is well understood. Sunlight and water are, in effect, inexhaustible resources to provide the required diversity of energy supplies. Sunlight is the largest energy resource base currently available. Solar energy is well distributed across the most populated areas of the globe, making the resource accessible where it is most needed. Furthermore, because solar energy is carbon-neutral and environmentally benign, its conversion represents an ideal method of powering the planet.

However, the conversion of solar photons into a usable form of energy is a huge challenge. Solar energy on earth is dispersed in many ways: light arrives at low concentration (1 kW m^{-2} at midday), in an intermittent fashion, and with a wide dispersion of frequencies across the electromagnetic spectrum. Analogous to photosynthesis, storing solar energy in chemical bonds (i.e., solar fuels) would be an ideal method to provide usable energy. It is particularly attractive to use solar energy to split water into H_2 and O_2, and subsequently the H_2 could be used in a fuel cell or other energy conversion scheme. High-performance lightweight batteries can be used to propel electric vehicles. However, major technological advances are required to achieve systems that have the capability to produce electricity, break H_2O bonds, and store the energy either in batteries, as molecular hydrogen, or in carbonized fuel compounds. Large-scale reduction of CO_2 assisted by solar photons can alleviate the greenhouse effect and form the basis for useful chemistries. Currently, such systems pose formidable challenges in terms of the required materials, efficiency, cost, and design of practical devices.

For several billion years, bacteria and plants have been effectively converting solar energy into chemical fuel by means of photosynthesis. To compensate for the low concentration of solar energy, living systems spread wide area converters—the leaves—built with organic components that can self-heal when failure occurs during the photochemical operation. To counter the degradation of the organic components, the solar collectors in plants can ultimately be dropped and renewed. These systems are thought to provide a main source of inspiration for the formation of artificial systems for large-scale supply of clean energy. However, we must note that photosynthesis, on average, has a conversion efficiency of only about 1%.

The operation of energy devices is rooted in fundamental physical principles. One key concept that should be kept in mind is the unidirectional flow of all systems toward the final state of equilibrium. The flows of the carriers in the device will be regulated by the entropy as a quantity that is always increasing in all natural phenomena. This approach focuses attention on this irreversible process (Tolman and Fine, 1948). Lewis and Randall developed a conclusion in 1923: a system far removed from its condition of equilibrium is the one chosen if we wish to harness its processes for the doing of useful work. While a battery at open circuit is usually regarded as an equilibrated system (from an electrical point of view), we should look at this system as either far from equilibrium or useless. In energy systems for sunlight energy conversion, as in batteries used for energy storage, we must arrange electrons and ions in materials in a far-from-equilibrium, highly reactive state, which in order to be productive should react only in a very specific way. Thus, in a solar cell, the primary excitation created by a photon is an energetic electron-hole pair. The pair would naturally react (or recombine) at the site of creation, but instead the energy is harnessed by provoking the recombination in an external circuit. This requires a directional arrangement that funnels the potentially reacting carriers toward separate pathways.

Since solar energy is diffuse and intermittent, the power conversion efficiency of viable devices must be high. Light absorption and forward process must be optimized and backward reaction cannot be afforded. In general, the cornerstone of energy devices is to find an effective way to channel processes in a given direction and avoid reciprocal processes that constitute a waste. Another stratagem that is widely used is to deploy specific layers covering the contact that will realize the selection of one specific carrier from a homogeneous film. Such layers are termed selective contacts (Bisquert et al., 2004). In a charged lithium battery, the two electrodes contain electrons that withstand an energy distance of 3–4 eV (Goodenough and Kim, 2010). The high-energy (strongly reducing) electrons in the negative electrode tend to fall to the low-energy state in the positive electrode (the cathode), and such flow is contained by an electrolyte that conducts ions and not electrons. However, the electrons cascade if the electrolyte *is* reduced or oxidized. The electrochemical stability of the electrolyte limits the voltage that can be gained by modifying the lithium intercalation materials to high positive values in the electrochemical scale. In contrast to the solar cell and any thermal engine, a crucial feature of the battery is the noncyclic character of its operation. Once the useful energy has been degraded by using the battery, the initial (charged) state has to be restored from the outside via a flux of negative entropy toward the battery, which cannot be spontaneous.

The need for a large area for effective scale energy production leads us to search for materials combinations and systems that can combine low fabrication cost with medium-to-high conversion efficiencies. The general term nanotechnology refers to the technology of design, manufacture, and application of nanostructures and nanomaterials, while nanoscience is the study of the fundamental relationships between physical and chemical phenomena in materials dimensions on a nanometric scale. Nanoscale materials and devices for energy harvesting, storage, delivery, and for improving the energy efficiency of certain needs have been successfully developed. Nanosized materials and their combinations provided several opportunities for innovation and improvement of certain functions required in energy devices. Beneficial features are an increased surface area, better tolerance to volume changes, increased functionality at the interface between the different phases of the device, and the formation of transport pathways toward the active interfaces or the charge-accumulation phases. The small dimension of semiconductor nanostructures also permits the tailoring of physicochemical properties, such as the light absorption or thermodynamics of carriers, basically controlling the size of nanoparticles, nanowires, or nanotubes.

Nanostructured materials, based on combinations of inorganic and organic materials, emerged to efficiently convert solar energy into electricity or fuels. Large-area devices can be obtained by low-temperature solution-processed techniques. An archetypical energy device is a nanostructured electrode in which a porous layer with the required properties is deposited on a conducting substrate that allows electrical manipulation of the internal structure. Examples of metal oxide mesoporous electrodes are shown in Figure 1.1. Figure 1.2 shows an example of TiO_2 made of octahedron-like single crystal anatase particles, deposited on transparent conducting oxide (TCO) over glass. The structure of individual single crystals of anatase phase is shown in the inset of Figure 1.2, and Figure 1.3 shows the atomic structure of the nanocrystallite, which is about 10 unit cells wide. The fundamental electronic operation of this type of electrode (Wang et al., 2014) can be seen in Figure 1.4: specific electronic processes occurring in the semiconductor matrix allow the exchange of electrons with the substrate, via the transport in the nanostructured film, and such electrons enter the outer circuit where they serve to do electric work. Figure 1.4 also suggests a density of electronic states (DOS) in the TiO_2 nanoparticles, which is associated to disorder in this kind of material, although a broad DOS can be found even in the case of crystallites, due to grain boundaries, small size effects, interaction with the medium in the pores, and other effects. The main structural unit of electrical operation, however, requires two contacts so that electrons can go around the circuit. Separation between the active electrode and the second contact is also required in order to avoid a short circuit that disables the device. One way to achieve the operational architecture is by utilizing a redox electrolyte to contact the active electrode and counterelectrode, as shown in Figure 1.5. The liquid medium forms a barrier to electrons in the nanostructure, so that electrons are channeled to the substrate. However, the generation and extraction of electron requires their replenishment for a continuous process to be maintained. This can be achieved by the transport of ionic species that take or give electrons by electrochemical reaction at the interfaces. Thus, the electrochemical reaction provides a fundamental switch between electronic and ion conduction with numerous implications for energy devices such as batteries and photoelectrochemical solar cells.

The actual energy conversion process relies on a set of sequential steps, as outlined in Figure 1.6. Photon absorption creates electronic excitation of the absorbing component with concomitant electronic charge creation. The excitation can be an electron-hole pair in a semiconductor or an electronic excitation of a

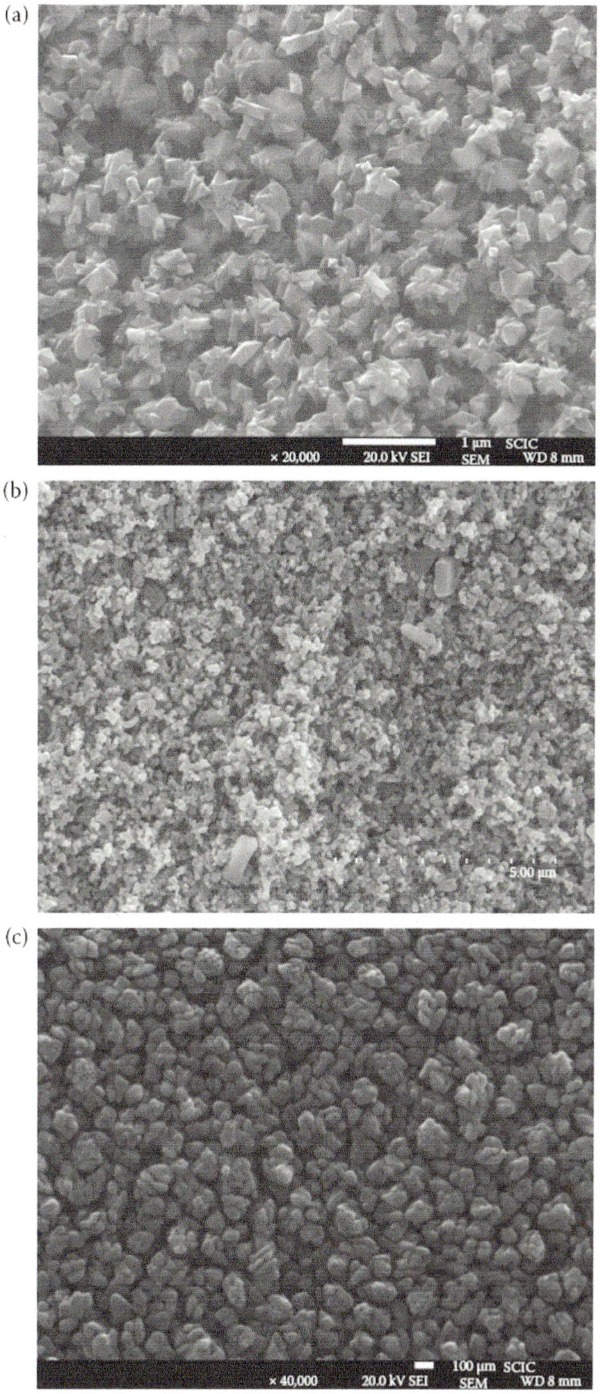

FIGURE 1.1 Top view micrographs of mesoporous films composed by different nanostructured metal oxides deposited on transparent conducting substrates: (a) TiO_2 nanoparticles, 250 nm particle size, (b) WO_3 nanoparticles processed by hydrothermal synthesis, (c) Fe_2O_3 deposited by spray pyrolysis. (Courtesy of Sixto Giménez.)

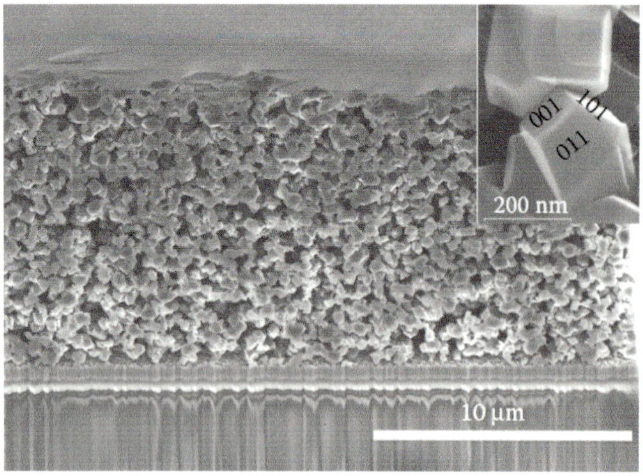

FIGURE 1.2 SEM image of the cross section of a porous electrode formed by sintered TiO_2 anatase crystals (inset), deposited on a transparent substrate that consists of a glass support and thin TCO layer. (Courtesy of Ronen Gottesman and Arie Zaban.)

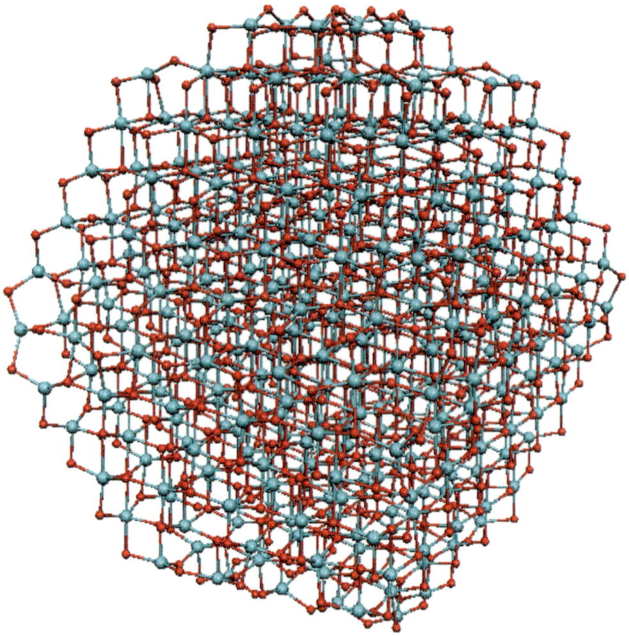

FIGURE 1.3 Atomic structure of anatase TiO_2 nanocrystal. (Adapted from Nunzi, F. et al., *Energy and Environmental Science* 2013, *6*, 1221–1229. With permission.)

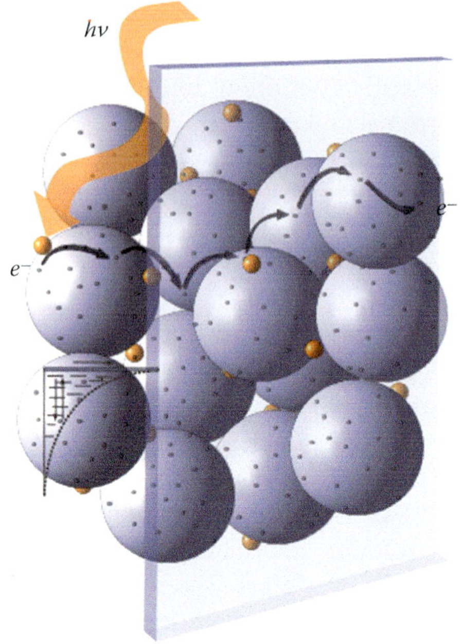

FIGURE 1.4 Scheme of electron transport in a metal oxide nanostructure, showing a photogeneration event, and the density of states in the semiconductor bandgap. (Courtesy of Juan A. Anta and José P. González.)

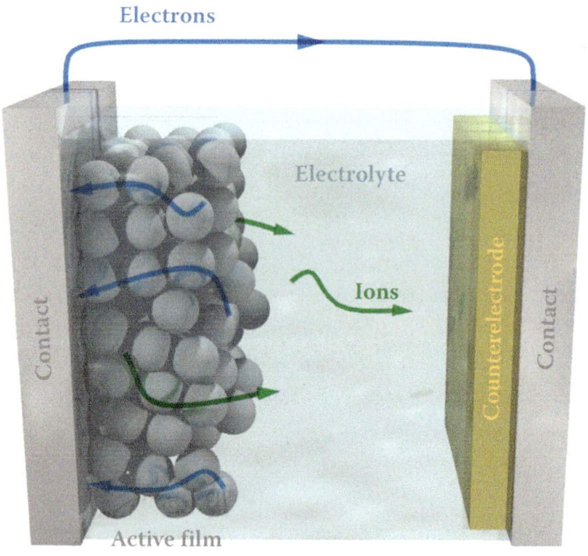

FIGURE 1.5 Scheme of an electrochemical cell consisting of a nanostructured semiconductor electrode that realizes efficient electron transport to the substrate contact, an electrolyte filling the pores, and transport of ions in the electrolyte. (Courtesy of Emilio J. Juárez-Pérez.)

molecule or polymer, usually in the form of excitons. Separation of the electronic charges before they return to the ground state (recombination) allows investing the energy of the carriers in the required energy production functionality. Basically, the carriers are transported to contacts or reactive sites, using the cell structure described earlier, so that either a photocurrent emerges as the output or a desired chemical reaction is performed. The energy converter has to satisfy a series of stringent requirements, including a narrow forbidden energy gap for photon absorption; a large light absorption coefficient; a high photon-to-charge conversion yield; good charge separation properties and long diffusion lengths (electrons and holes directed toward

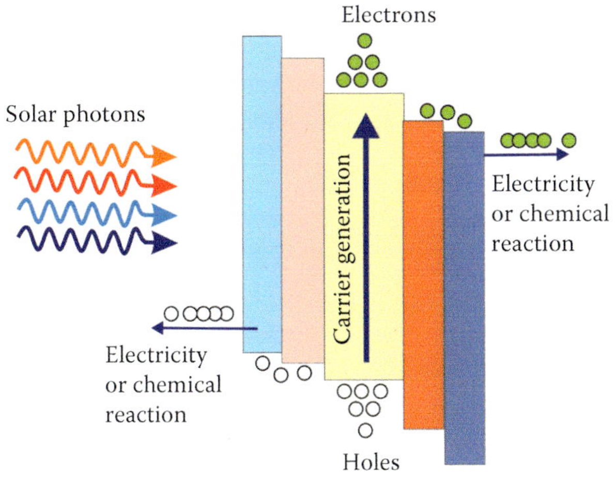

FIGURE 1.6 Scheme of solar energy conversion scheme using nanoheterostructures.

the useful function); a strong rate of product formation (electricity/chemical fuel) by good contact selectivity; and fast charge transfer kinetics of the forward reactions. The final requirements include robustness, duration, and low materials process costs.

In the second half of the twentieth century, a great deal of research effort was invested in the search for specific inorganic semiconductors that would perform *all* of the required functions in the conversion of solar photons into electricity or fuel. Success was obtained in the case of photovoltaic cells, with the development of defect-free, high-purity crystalline silicon solar cells. The quest for a semiconductor able to perform photoelectrochemical water splitting without degradation by corrosion was not successful and was abandoned in the mid-1980s. In 1990s, photovoltaic devices, pioneered by M. Grätzel, A. J. Heeger, and many others, were suggested, that operate on the basis of a combination of materials that takes advantage of the properties and phenomena occurring in nanoscale units and at interfaces (O'Regan and Grätzel, 1991; Scharber et al., 2006). Various specific photovoltaic nanostructured devices as well as some of their main physical properties are summarized in Figure 1.7. Nanoporosity was demonstrated to be a very efficient feature for effective charge separation and transport in different phases. Thus, dye-sensitized solar cells (DSC), inorganic quantum-dot sensitized cells, and organic

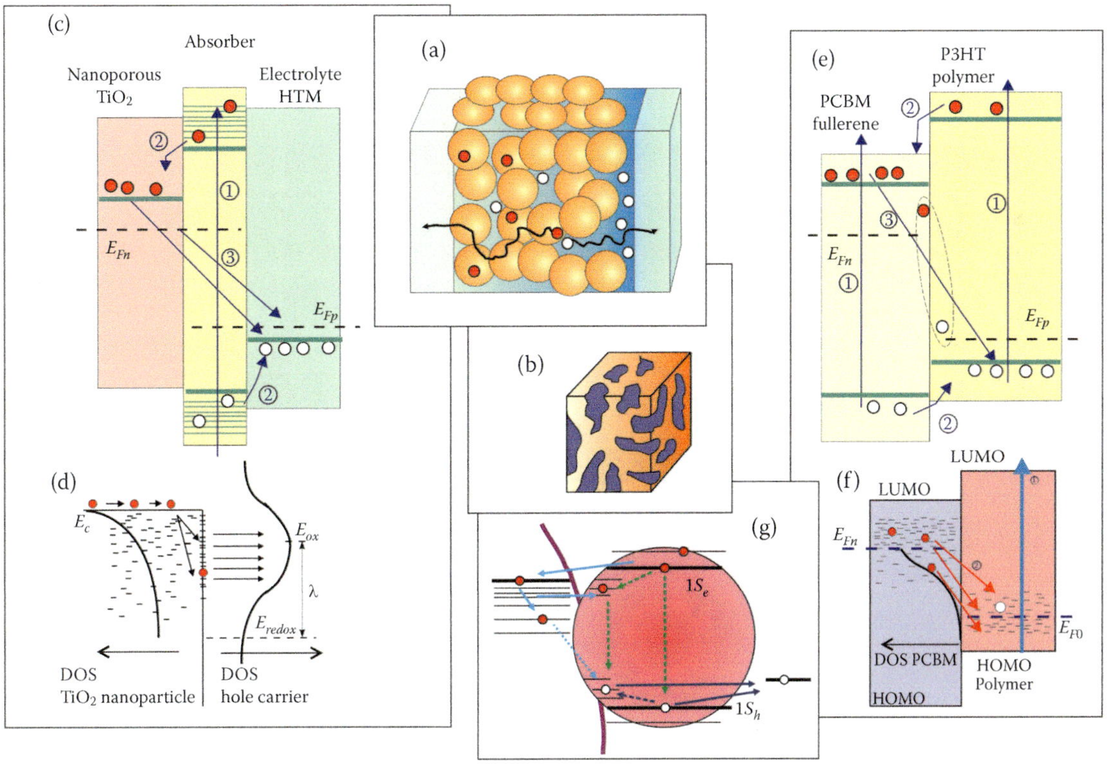

FIGURE 1.7 (a) Scheme of nanostructure that carries electrons and holes in different phases, (b) the structure of bulk heterojunction solar cells, with two organic phases closely mixed and separately interconnected, (c) energetic scheme of a dye-sensitized solar cell, showing the absorber dye molecule between electron transport material, and hole transport material, (d) the energetic disorder in recombination process, (e) energetic scheme of a bulk heterojunction solar cell, with carrier generation in both fullerene PCBM and polymer P3HT, (f) the energetic disorder in recombination process, (g) generation and charge transfer in a quantum-dot sensitizer.

solar cells (especially bulk heterojunction solar cells) became topics of major interest in research, both for their fundamental properties as well as reachable technological realizations. Semiconductor quantum dots also emerged as a very important sensitizer for photovoltaics, due to several attractive features such as the tunability of absorption by size and high extinction coefficients.

The schemes for solar energy conversion using low-cost, large-area nanostructures involve directed transfer of photogenerated carriers from the generation site to the final energy production step. These carriers use several physical phenomena that occur in nanomaterials and at their interfaces, including carrier accumulation, transport, and transfer across the interface. Insights into the physical rules that govern such phenomena in relation to specific materials and interface properties is a key element in the future design of energy conversion devices based on nanostructured, low-cost organic and inorganic materials. This will be an important topic of this book.

Research on solution-processed materials for solar cells produced significant outcomes in materials that can also operate as thin films. Semiconductor materials based on metal halide perovskite have now been extensivelly investigated because of their exceptional opto-electronic properties, which allow for applications in areas such as solar cells, light emitting diodes (LED), X-ray detectors, and so on; see Figure 1.8. Research in this area initially branched off from the field of dye solar cell research. In the quest for an efficient inorganic quantum-dot sensitized liquid cell, this topic rapidly formed a huge field of its own.

The perovskite crystal structure shown in Figure 1.8a has been adopted by a large class of materials formed by a combination of three elements or molecules placed on the corners of the lattice and inner regular octahedral configuration. The chemical formula of the perovskite can be described as ABX_3, where A = monovalent cations such as methylammonium (MA), formamidinium (FA), cesium (Cs), B = divalent cations such as lead (Pb), tin (Sn) and X = halide anions including chloride (Cl), bromide (Br), iodide (I). $MAPbX_3$ was first synthesized by Weber in 1978; see the crystals shown in Figure 1.9. The first attempt to apply methylammonium lead halide

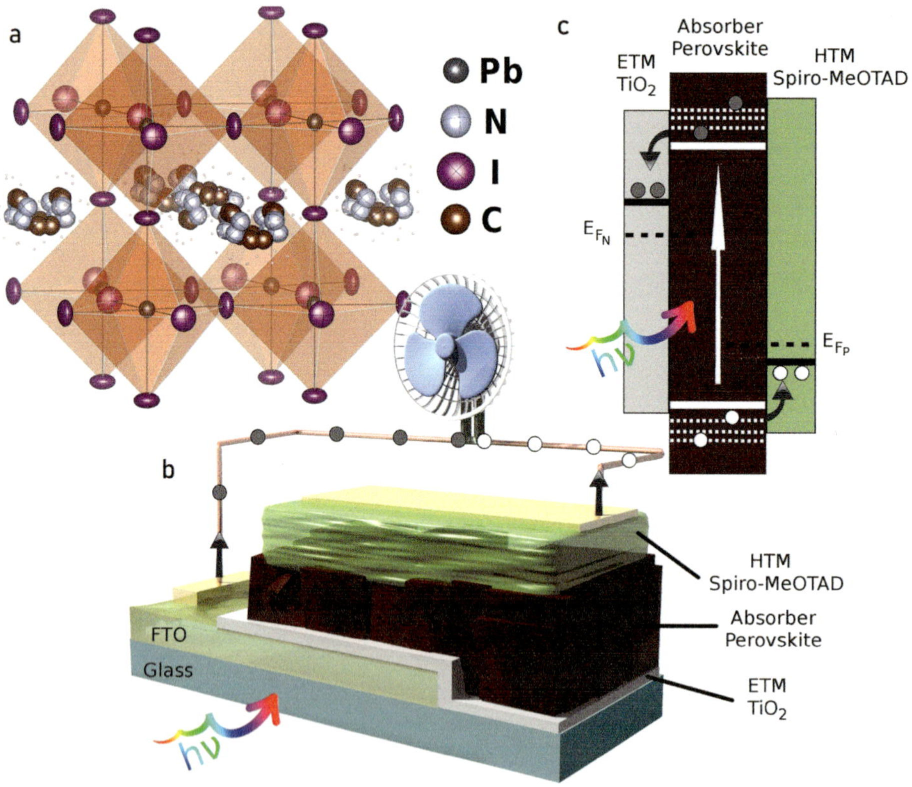

FIGURE 1.8 (a) Tetragonal crystal structure of $CH_3NH_3PbI_3$ perovskite as determined from XRD measurements in single crystal (ref no. 238610, Inorganic Crystal Structure Database), (b) perovskite solar cell conventional stack section using glass, FTO, TiO_2 compact layer, perovskite absorber, and Spiro-MeOTAD layer, holes are extracted from the Spiro-MeOTAD side and electrons from TiO_2 compact layer side; both meet again on the load in this case represented by a fan closing the electric circuit, (c) simplified energy band scheme for perovskite solar cell stack showing hybrid perovskite absorber material sandwiched between electron transport material (ETM, TiO_2) and hole transport material (HTM, Spiro-MeOTAD). (Courtesy of Emilio J. Juárez-Pérez.)

FIGURE 1.9 SEM picture of $CH_3NH_3PbI_3$ crystals. (Courtesy of Nam-Gyu Park.)

FIGURE 1.10 SEM cross-sectional view of a $CH_3NH_3PbBr_3$ perovskite solar cell. Bottom contact (front contact of the solar cell) is nanostructured TiO_2 on top of compact TiO_2 deposited on FTO. The upper contact is Spiro-OMeTAD hole collector layer with a gold overlayer as metal contact. The inset shows the crystalline perovskite layer without top contact. (Courtesy of Clara Aranda and Antonio Guerrero.)

(bromide, iodide) ($MAPbBr_3$, $MAPbI_3$) in solar cells was reported by T. Miyasaka and coworkers in 2009. They introduced perovskite nanocrystals as a light absorber into a conventional sensitized solar cell using iodide/triiodide liquid electrolyte. In 2011, higher concentrated precursor solutions almost doubled the power conversion efficiency of the 2009 device. After the initial configurations, that used a liquid contact to close the circuit, the first devices that adopted a solid-state hole transport material (HTM) during 2012 appeared. Such developments were also used in DSCs but, contrary to expected behavior observed in a DSC, perovskite solar cells with the solid-state HTM reached a remarkable ~10% of power conversion efficiency and overcame the ephemeral stability reached using liquid electrolyte selective contact. In 2013, the first basic layer stacks and procedures to assemble the device were published. For a short history of the development of the field see Bisquert et al. (2017).

In the next few years the advancement in material engineering and interface optimization continuously pushed the power conversion efficiency of perovskite solar cells; see the dense film formed by large perovskite grains on a nanostructured TiO_2-based electron collector in Figure 1.10. A device formed by organic contacts is shown in Figure 1.11a. A certified record efficiency of 25.2% for single-junction solar cells was reached in 2020, which is comparable to crystalline silicon solar cells and even surpasses the efficiency of thin film CIGS and CdTe solar cells.

The most attractive features for perovskite-based photovoltaic and optoelectronic technologies include a low-temperature solution-processing method, exceptional semiconductor properties, and high efficiency achieved with very thin layer (<500 nm), as well as the abundance of raw materials. These make them viable for large-scale production for lightweight, flexible devices using cost-effective methods such as roll-to-roll, screen-printing, or slot-die to facilitate low-cost solar electricity production and lighting in the future. The facile design of the bandgap obtained by modifying the compositions of the constituents makes hybrid perovskites suitable for high-efficiency LEDs, as indicated in Figure 1.12, and for the combination with other materials for high-efficiency tandem solar cells, as shown in Figure 1.11b. Note the improvement of charge collection by the application of an antireflective coating inthe combined tandem arrangement shown in Figure 1.11c. The perovskite/Si two junction tandem arrangement has reached 29.1% record efficiency in 2020.

The rise of a succession of new classes of organic, inorganic, and hybrid materials with photovoltaic applications, the consequent realization and determination of their properties, and the optimization of materials and interfaces for better performance, are generally due to rather broad physical and chemical concepts and principles. The progress and understanding of the device operation of these materials has provided a pattern that will be systematically explained in this book. In addition, functionalities sometimes respond to unique and specific features.

In the pursuit of new photovoltaic conversion materials, the initial questions posed are, how can we extract charge at all, how does charge separation occur, and why do charges arrive at the electrical contact? Therefore, historically, the explanation of the functioning of a solar cell gave a heavy significance to the transport of carriers and the action of electrical fields. It has now been recognized that this approach may be highly misleading

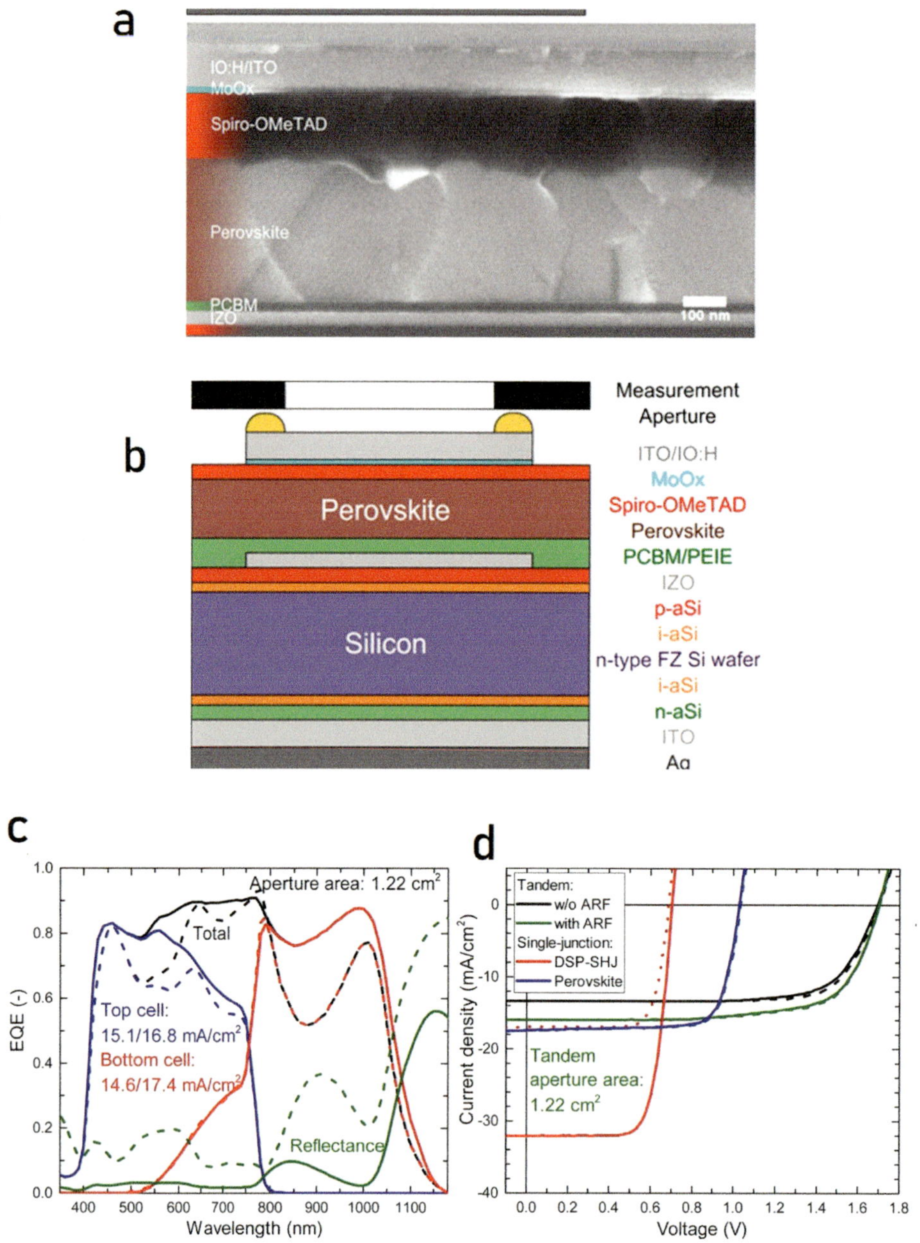

FIGURE 1.11 (a) SEM cross-sectional view of a perovskite solar cell, (b) schematic drawing of a planar monolithic perovskite/(amorphous/crystalline silicon heterojunction) tandem cell layer stack, (c) EQE spectra of a perovskite/Si monolithic tandem with (solid lines) and without (dashed lines) antireflective foil (ARF) as well as the corresponding reflectance (green curves); the integrated photocurrent for both top and bottom cells are given in the legend (without ARF/with ARF), (d) j–V measurements of the best perovskite/Si monolithic tandem with 1.22 cm^2 aperture area and of the single-junction perovskite and Si cells; reverse (solid lines) and forward (dashed lines) scans are shown for perovskite single-junction and tandem cells; the dotted red curve shows the j–V curve of the Si cell when illuminated at an intensity of 0.53 suns. (Reproduced with permission from Werner et al., 2016.)

about the core of the operation of photovoltaic conversion. When the material becomes better understood, carrier collection is usually not the main issue of photovoltaic cells. In most cases, the crucial issue is obtaining the maximal photovoltage allowed by the fundamental physical constraints; therefore, one begins to worry about reducing the recombination and finding the optimized structures for selective contacts. If the technology progresses sufficiently and reaches a level of very high conversion efficiencies, the electronic operation of the device becomes reproducible and proficient, and then it is necessary to maximize the extraction of power from every photon that comes to the device surface. The researcher is then forced to attend to the photonic

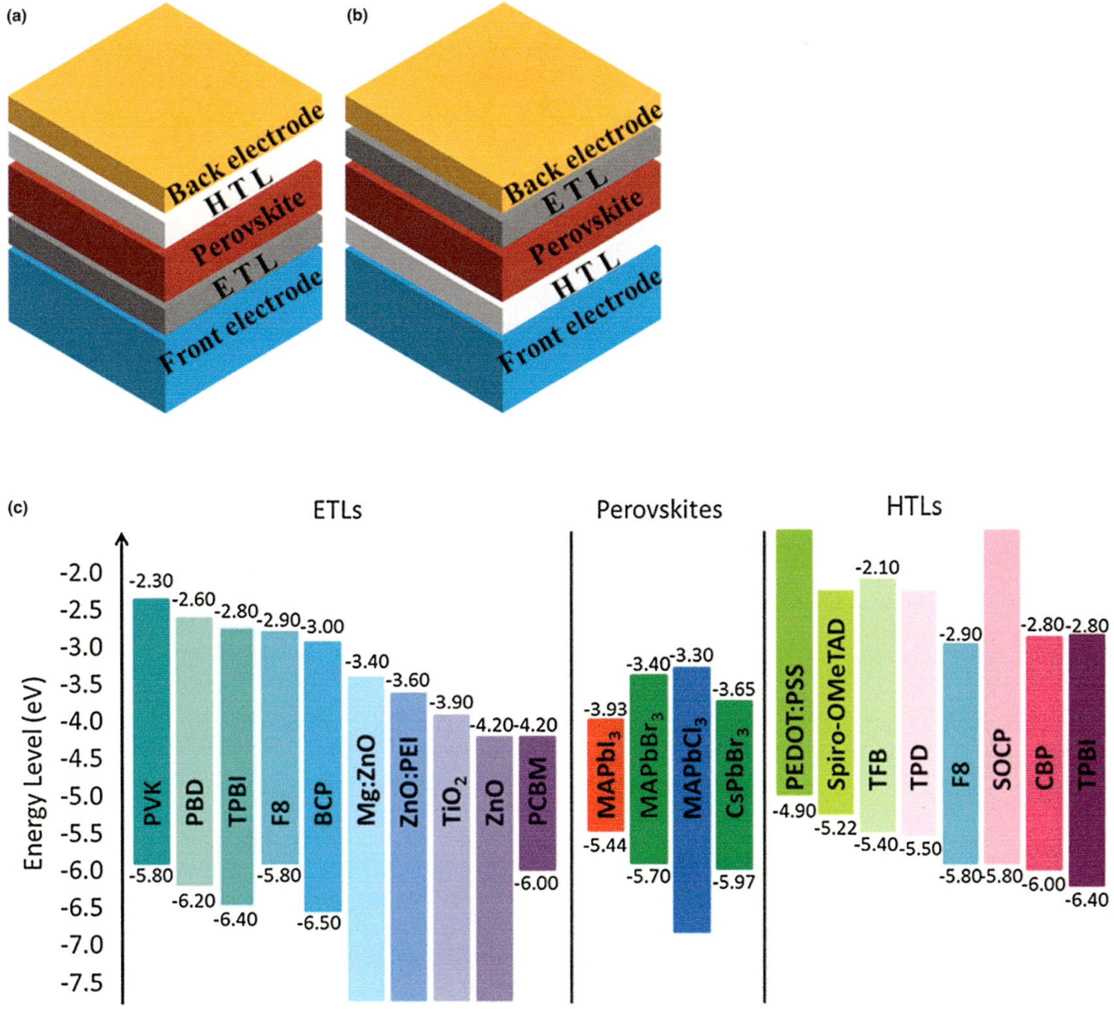

FIGURE 1.12 (a, b) Schematic diagrams of perovskite LED architectures in the conventional and inverted configuration, respectively, (c) energy level alignment of various materials used as perovskites, ETLs, and HTLs in the perovskite LEDs. (Reproduced by permission from Adjokatse et al., 2017.)

characteristics. Thereafter, schemes are developed to harvest the full solar spectrum, a task that requires a combination of absorbers. Therefore, to keep up with the pace of research, one needs to go through a wide variety of problems.

In general, this book is about understanding solar cell device operation and related devices. Device understanding usually consists of a picture of how the device operates internally, sufficient for rationalizing the outcome of different manipulations and materials variations. These tests and studies can be performed using the established array of experimental tools, as well as the new techniques that can be designed for specific problems that appear during the analysis of a new class of materials/devices. Such understanding is based on a broad range of physical principles, on the mastering of chemical features of inorganic lattices and organic molecules and their aggregates, and on a comprehension of different classes of materials and their interfaces.

By understanding, we obtain a tool for explaining device operation that guides us in developing modifications and improvements, and opens doors for better designs of materials, structures, and interfaces. Achieving such a tool is the purpose of the forthcoming chapters.

REFERENCES

Adjokatse, S.; Fang, H.-H.; Loi, M. A. Broadly tunable metal halide perovskites for solid-state light-emission applications. *Materials Today* 2017, *20*, 413–424.

Bisquert, J.; Cahen, D.; Rühle, S.; Hodes, G.; Zaban, A. Physical chemical principles of photovoltaic conversion with nanoparticulate, mesoporous dye-sensitized solar cells. *The Journal of Physical Chemistry B* 2004, *108*, 8106–8118.

Bisquert, J.; Juárez-Pérez, E. J.; Kamat, P. V. *Hybrid Perovskite Solar Cells. The Genesis and Early Developments, 2009–2014*; Fundació Scito, València, 2017.

Goodenough, J. B.; Kim, Y. Challenges for rechargeable Li batteries. *Chemistry of Materials* 2010, *22*, 587–603.

Kojima, A.; Teshima, K.; Shirai, Y.; Miyasaka, T. Organometal halide perovskites as visible-light sensitizers for photovoltaic cells. *Journal of the American Chemical Society* 2009, *131*, 6050–6051.

Nunzi, F.; Mosconi, E.; Storchi, L.; Ronca, E.; Selloni, A.; Grätzel, M.; De Angelis, F. Inherent electronic trap states in TiO_2 nanocrystals: Effect of saturation and sintering. *Energy and Environmental Science* 2013, *6*, 1221–1229.

O'Regan, B.; Grätzel, M. A low-cost high-efficiency solar cell based on dye-sensitized colloidal TiO_2 films. *Nature* 1991, *353*, 737–740.

Scharber, M. C.; Mühlbacher, D.; Koppe, M.; Denk, P.; Waldauf, C.; Heeger, A. J.; Brabec, C. J. Design rules for donors in bulk-heterojunction solar cells—Towards 10% energy-conversion efficiency. *Advanced Materials* 2006, *18*, 789–794.

Smil, V. *Energy and Civilization*; MIT Press, Cambridge, MA, 2017.

Tolman, R. C.; Fine, P. C. On the irreversible production of entropy. *Reviews of Modern Physics* 1948, *20*, 51–77.

Wang, P.; Bai, Y.; Mora-Sero, I.; De Angelis, F.; Bisquert, J. Titanium dioxide nanomaterials for photovoltaic applications. *Chemical Reviews* 2014, *114*, 10095–10130.

Weber, D. $CH_3NH_3PBX_3$, a PB(II)-system with cubic perovskite structure. *Zeitschrift Fur Naturforschung Section B* 1978, *33*, 1443–1445.

Werner, J.; Weng, C.-H.; Walter, A.; Fesquet, L.; Seif, J. P.; De Wolf, S.; Niesen, B.; et al. Efficient monolithic perovskite/silicon tandem solar cell with cell area >1 cm^2. *The Journal of Physical Chemistry Letters* 2016, *7*, 161–166.

Part I

Equilibrium Concepts and Kinetics

2 Electrostatic and Thermodynamic Potentials of Electrons in Materials

The devices for energy production and storage are formed with different types of materials and interfaces. We start our discussion with metals and semiconductors that are at the heart of modern electronic and photovoltaic systems. Some basic properties of the semiconductors are reviewed, and several elements of the energy diagrams are built that will be used to describe and understand the operational function of devices. Electric potential energies are determined by electrostatic conditions that depend on the distribution of electric charges. This method is sufficient to describe energetics of electrons if the electrons reside in a material or phase that is chemically homogeneous. But in order to make junctions in working devices we consider different phases, in contact, and the electron in transit across the interface will experience a combination of chemical and electric forces which cannot be simply separated. Therefore, electrostatics is generally limited to describe energetics of electrons, and one must treat the electron system in terms of equilibrium states of thermodynamics. In this chapter, fundamental concepts such as the Fermi level, the surface dipole, the chemical potential, and the Volta and Galvani potentials will be defined and discussed with examples. In this chapter, these concepts are described in total equilibrium conditions, but they form the basis for the description of nonequilibrium phenomena such as the diffusion flux.

2.1 ELECTROSTATIC POTENTIAL

The unit of charge is the coulomb (C). The elementary charge is $q = 1.60 \times 10^{-19}$ C and the charge of an electron is $-q$. The number of atoms in a mole is the Avogadro constant ($N_A = 6.022 \times 10^{23}$ mol^{-1}). The term qN_A is given the name Faraday's constant, F, and it has the value of 96,485 C mol^{-1}. The charge of one mole of electrons is therefore 96,485 C.

The *electric potential difference* between two points A and B is defined from the work W to carry a test charge Q_0 from A to B against the electric field, as follows:

$$\varphi_B - \varphi_A = \frac{W_{AB}}{Q_0} \tag{2.1}$$

If Q_0 is positive and $W_{AB} > 0$, then B is at higher electrostatic potential than A. The electric potential is measured in volts, V, being 1 volt = 1 J/C^{-1}.

For simplicity, we refer to situations in which the potentials and concentrations vary only along one spatial axis x. In this book, the magnitude of the electric field is denoted in bold letter **E**, while energy is denoted as E. The electric field is given by the expression

$$\mathbf{E} = -\frac{\partial \varphi}{\partial x} \tag{2.2}$$

In a solid or liquid medium we can relate the electric field to the density of *excess charges* (with respect to the neutral ionic–electronic background charge) using Poisson equation. In a region where a charge distribution of density ρ_e per unit volume exists, we have

$$\frac{\partial \mathbf{E}}{\partial x} = \frac{\rho_e}{\varepsilon_r \varepsilon_0} \tag{2.3}$$

where ε_r is the static dielectric constant of the medium and $\varepsilon_0 = 8.85 \times 10^{-12}$ F m^{-1} is the permittivity of the vacuum. Hence,

$$\frac{\partial^2 \varphi}{\partial x^2} = -\frac{\rho_e}{\varepsilon} \tag{2.4}$$

Here, $\varepsilon = \varepsilon_r \varepsilon_0$.

Consider the semiconductor shown in Figure 2.1. In an n-type semiconductor there is a density of free electrons n (carriers/cm^3), each with charge $-q$. For simplicity we assume that all dopant ions are ionized. (In general, the ionization of dopants depends on the temperature and the electrostatic potential.) Therefore, everywhere in the semiconductor there is a background density N_D of fixed foreign cations, each with a charge $+q$. The net electric charge density at any point is

$$\rho_e = q\left(N_D - n\right) \tag{2.5}$$

Normally we refer to the bulk material as any part of the material far from the interfaces, where there is no net charge. Therefore, the electron density equals the doping density

$$n_0 = N_D \tag{2.6}$$

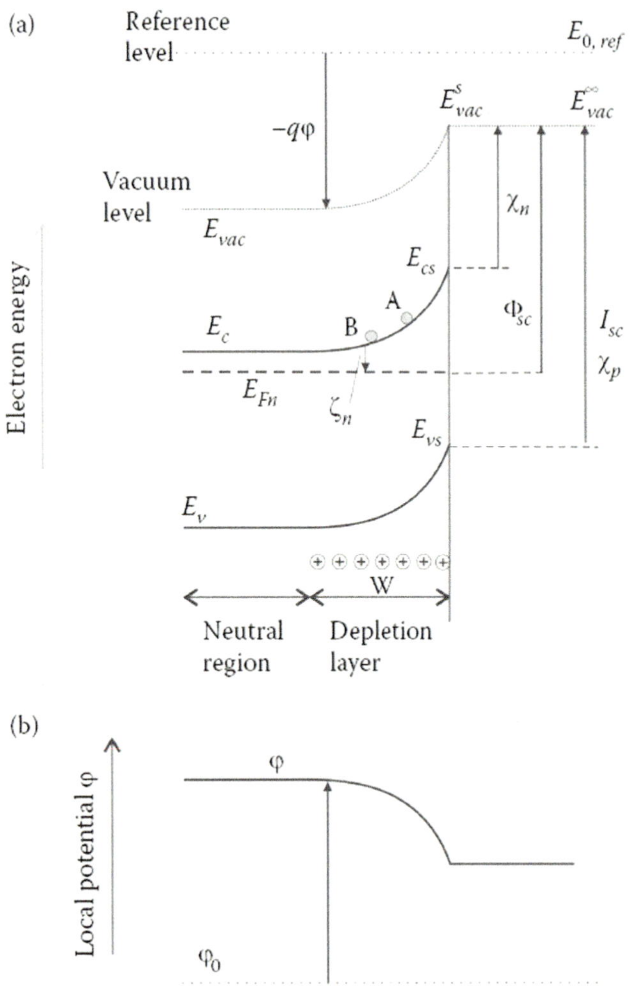

FIGURE 2.1 Energy diagram of an n-type doped semiconductor showing upward band bending at the surface: (a) Reference energy level $E_{0,ref}$, local vacuum level E_{vac}, energy band edges E_c and E_v, the Fermi level E_{Fn}, the chemical potential ζ_n, the electron affinity χ_n, work function Φ, and ionization energy I (hole affinity χ_p). The subscript s denotes the value at the surface. (b) Local electric potential.

This is denoted as the *neutral region*. We also emphasize that free charges may accumulate at surfaces or interfaces of the material, for example, electrons trapped in surface states. In Figure 2.1, close to the surface of the semiconductor, there is a region where electronic charge has been withdrawn by some other material that is in contact with this semiconductor (or surface states). This means that in a region of width w there only remains the fixed positively charged ions. This is called the *depletion region*. In general, any zone where net charge exists in a material is said to contain *space charge*.

The depletion zone in Figure 2.1 provides a very simple example of the application of Poisson equation, because in this case the spatial distribution of space charge is both homogeneous and fixed. Hence, we have the equation

$$\frac{\partial^2 \varphi}{\partial x^2} = -\frac{qN_D}{\varepsilon} \tag{2.7}$$

Equation 2.7 corresponds to the depletion approximation (see Chapter 9), which implies that the potential is parabolic as shown in Figure 2.1b. In semiconductor devices, the depletion zone at the surface controls a number of carrier injection phenomena. The depletion layer at the metal–semiconductor interface is called a Schottky barrier (this will be explained in more detail in Chapter 9).

2.2 ENERGIES OF FREE ELECTRONS AND HOLES

The work, and energy changes, in the displacement of electronic charges, is usually given in electron volts, 1 eV = q × V. 1 eV corresponds to 96.48 kJ mol^{-1} = 23.05 kcal mol^{-1} or to the energy of a quantum of wavelength 1240 nm.

The electronic states in a periodic lattice described by a weak periodic potential are formed by strongly correlated quasiwave functions denoted as Bloch functions that are determined by wavenumber **k**. The electron (crystal) quasimomentum $\mathbf{p}_e$ is $\mathbf{p}_e = \hbar\mathbf{k}$ where $\hbar = h/2\pi$ and $h = 4.136 \times 10^{-15}$ eV s is Planck's constant. Hereafter, the prefix quasi in the word quasimomentum will be omitted. The dispersion relationship between electronic energy E and **k** that is obtained solving Schrödinger's equation has the consequence that in a semiconductor crystal some intervals of energy, the gaps, do not have allowed quantum states. The upper occupied band in a semiconductor is called the valence band, and the lowest empty band (at zero Kelvin) is the conduction band. The energy of the bottom of the conduction band, E_c, and the energy at the top of the valence band, E_v, are separated by the energy gap (see Figure 2.2a)

$$E_g = E_c - E_v \tag{2.8}$$

Electrons in the valence band can be thermally or optically excited to the conduction band, leaving a hole with charge $+q$ in the valence band. The kinetic energy of an electron of energy E is given by $E_k = E - E_c$. We note that the presence of an electric field will destroy the periodicity of the total potential of the system, but our discussion is limited to the case of a low-intensity electric field that can be considered a small perturbation

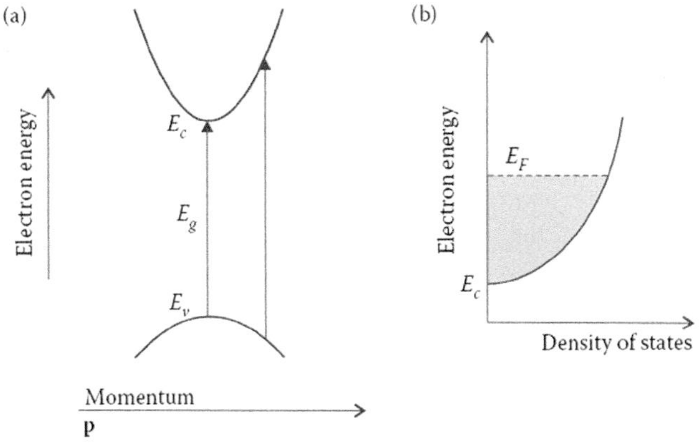

FIGURE 2.2 (a) Energy of the conduction band (edge energy E_c) and valence band (edge energy E_v) electronic states, as a function of electron momentum, in a semiconductor of bandgap energy E_g. The vertical arrows indicate photoexcitation events that promote electrons from the valence band to the conduction band. (b) The density of states in the conduction band of a metal, showing the states filled with electrons up to the Fermi level at $T = 0$.

of the total potential. In the conduction process, the carriers must gain energy in the electric field. So conductivity is not simply controlled by the number of carriers and the mobility; it is also necessary for the carriers to have free available states in order to gain kinetic energy. Therefore, conductivity is realized by carriers in a partially filled band, and those carriers are termed *free carriers*, as opposed to carriers in localized states or deeper atomic levels.

The relationship $E(\mathbf{p}_e)$ is symmetric to momentum inversion, otherwise an electronic current would be generated in equilibrium. Therefore, in a first approximation, the bottom of the conduction band shows a quadratic dependence on the momentum

$$E - E_c = \frac{\mathbf{p}_e^2}{2m_e^*} \tag{2.9}$$

Here, m_e^* is the *effective mass of the electron*, that is defined as $m^* = \left(d^2E/d|\mathbf{p}|^2\right)_{p=0}^{-1}$ and may be written $m_e^* = m_e m_0$, in terms of the mass of the free, noninteracting electron, m_0 and a normalized mass m_e. Similarly, for holes in the valence band, $m_h^* = m_h m_0$. If Equation 2.9 is satisfied far from the critical point we say that the band is parabolic, that is, the effective mass is a constant.

The electrons in the semiconductor crystal will be distributed in different states labeled by momentum and spin. According to Pauli principle, two electrons, which have spin 1/2, cannot occupy the same quantum state. We will see in Chapter 5 that the distribution of electrons in thermal equilibrium is determined only by the *energy* of the electrons, and not by their momentum. When there are many quantum states lumped together in the energy axis, the states are better described by a continuous function, the *density of states* (DOS) rather than by discrete levels. If the number of states per unit volume, as a function of energy, is $N(E)$, then the DOS $D(E)$ defined as the number of states per energy interval is

$$D(E) = \frac{dN}{dE} \tag{2.10}$$

When the momentum of the electron increases, the possibilities to occupy a unit volume in momentum space increase. Hence, the DOS increases with increasing electron energy, that is, from the conduction band edge upward, as shown in Figure 2.2b, and the DOS for holes increases from the valence band edge downward. For electrons in a parabolic band with effective mass m_e^*, it can be shown that the DOS has the form

$$D_c(E)dE = \frac{\left(2m_e^*\right)^{3/2}}{2\pi^2\hbar^3}\left(E - E_c\right)^{1/2} dE \tag{2.11}$$

The occupancy probability of a level at energy E is given by the Fermi–Dirac distribution function $f(E - E_F)$, that is described in detail in Chapter 5. The distribution of carriers at equilibrium depends both on the Fermi level, E_F (defined below), and the temperature of the semiconductor lattice, T. Figure 2.2b shows the sharp distribution of the band in a metal, up to the Fermi level, at $T = 0$ K.

In a semiconductor, the Fermi level usually lies in the gap, so that the conduction band will be empty at $T = 0$ K, but electrons can be excited thermally. The Fermi–Dirac distribution function has a tail at large energies, described by the *Boltzmann factor* $f(E - E_F) = e^{(E_F - E)/k_BT}$, where $k_B = 1.3806 \times 10^{-23}$ J K^{-1} = 8.617×10^{-5} eV K^{-1} is Boltzmann's constant. In equilibrium, electrons occupy the bottom of the available DOS in the conduction band, and the holes occupy the top of the valence band states. The number of electrons corresponding to an energy interval dE in the DOS is

$$dn(E) = f(E - E_F)D(E)dE \tag{2.12}$$

The thermal distributions are shown later in Figures 5.5 and 5.9.

The average energy of the electrons in the conduction band is given by the integral of Equation 2.12, considering the Boltzmann distribution that provides the result (Würfel, 2005)

$$E = \int_{E_c}^{\infty} Ef(E - E_F)D(E)dE = E_c + \frac{3}{2}k_B T \lim_{x\to\infty} \quad (2.13)$$

Equation 2.13 implies that most of the electrons are situated in an energy interval of the order 3/2 k_BT, from the edge of the conduction band upward. At room temperature, the thermal energy is $k_BT = 0.026$ eV.

To simplify the calculation of electron density in the conduction band, the occupation of the true DOS is equated to the thermal occupation of a single level with the *effective density of states* N_c and similarly for the valence band, with an effective density of states N_v. The integration of the DOS provides the following value:

$$N_c = 2\left(\frac{2\pi m_e^* k_B T}{h^2}\right)^{3/2} \quad (2.14)$$

For $m_e = 1$

$$N_c = 2\times 10^{19}\,\text{cm}^{-3} \quad (2.15)$$

Values of these parameters for representative crystalline semiconductors are given in Table 2.1. In summary, we usually say that a free electron has energy E_c and a free hole has the energy E_v.

A photon absorbed by the semiconductor creates an electron–hole pair. An electron is excited from a state in the valence band, to a state in the conduction band, as shown in Figure 2.2a, leaving a hole in the valence band. The distributions of photogenerated carriers in the DOS are affected by different relaxation and recombination processes. Immediately after excitation, the electrons may be widely distributed in the DOS. For example, photons with energy quite in excess of the bandgap can be absorbed. Since the electrons have much higher energies than the lowest states in the conduction band, they are considered *hot carriers*. In a timescale of about 10^{-11} s, the electrons relax in the lattice (see Section 5.5). Another significant timescale is that for the return of excited electron to the valence band in a recombination process that characteristically occurs in a time of $10^{-6} - 10^{-3}$ s.

In between carrier relaxation and recombination timescales, an important time span occurs in which carriers can be considered relaxed to equilibrium distribution into the respective band. Therefore, the excess population of each type of carrier spreads over the respective energy levels according to the equilibrium functions. Since both sets of carriers—electrons and holes—can

TABLE 2.1
Characteristics of Crystalline Semiconductors

Name	Symbol	Germanium	Silicon G	Gallium Arsenide
Bandgap energy at 300 K	E_g (eV)	0.66	1.14	1.424
Density	(g/cm^3)	5.33	2.33	5.32
Effective density of states in the conduction band at 300 K	N_c (cm^{-3})	1.02 × 10^{19}	2.82 × 10^{19}	4.35 × 10^{17}
Effective density of states in the valence band at 300 K	N_v (cm^{-3})	5.65 × 10^{18}	1.83 × 10^{19}	7.57 × 10^{18}
	Density of states Effective Mass			
Electrons	m_n^*/m_0	0.55	1.08	0.067
Holes	m_p^*/m_0	0.37	0.81	0.45
Electron affinity	χ_n(eV)	4.0	4.05	4.07
	Mobility at 300 K (Undoped)			
Electrons	u_n (cm^2/V^{-1}s^{-1})	3900	1400	8800
Holes	u_p (cm^2/V^{-1}s^{-1})	1900	450	400
Relative dielectric constant	ε	16	11.9	13.1

Source: Zeghbroeck, B. V. *Principles of Semiconductor Devices;* http://ecee.colorado.edu/~bart/book/, 2011.

be separately equilibrated by the respective relaxation processes, one introduces the separate Fermi levels of electrons, E_{Fn}, and holes, E_{Fp}. This topic will be further discussed in Chapter 5.

2.3 POTENTIAL ENERGY OF THE ELECTRONS IN THE SEMICONDUCTOR

The definition of the electrostatic potential φ in the solid in terms of the density of free carriers requires a spatial average over a region much larger than the spacing of the atoms in the material, but still small on a macroscopic scale. The shape of the potential φ inside the semiconductor is established by the presence of space charge through Poisson equation, from an applied electric field, or by a combination of both. This potential can be treated with respect to excess charge as indicated in Equation 2.4 (Lonergan, 2004). The electric potential φ imparts potential energy to the electron. φ is converted into electron energy by-product with $-q$, and it is converted into hole energy by product with $+q$.

Figure 2.1 illustrates the relationship of the local potential φ with the energy bands. The edge of the band follows the shape of the variation of the potential, though inverted. If the local potential becomes more positive, the energy of an electron decreases. When there is no space charge in the semiconductor the band is straight, according to Equation 2.4, but it can be tilted, corresponding to a constant electric field in the semiconductor, just by charge at the boundaries. This is the structure of a dielectric capacitor discussed in Section 3.4. However, the curvature of the band, which is called *band bending*, indicates the presence of space charge inside the semiconductor.

In Figure 2.1a we show an example in which the energy of the electron decreases in going from A to B due to the increase of electric potential between the two points, shown in Figure 2.1b, which is reflected in the lower value of E_c at point B. If the electron changes its position from A to B in the interior of a semiconductor, the change in *energy* is related to the variation of the electrostatic potential as

$$E_{cB} - E_{cA} = -q\left(\varphi_B - \varphi_A\right) \tag{2.16}$$

2.4 THE VACUUM LEVEL

The description of energy states of electrons in different materials, which allow us to account for properties of semiconductor junctions, requires one to take care of a combination of electric and chemical forces. We will see that the concept of an electric potential difference between points inside *different* phases cannot be based on the standard measurement with a voltmeter. The analysis of changes of energies of electrons across different phases needs to be based on thermodynamic states of the electrons. The *free energy* is the magnitude that provides the work that the electron can do in the process of transference between materials (see Chapter 5). We are interested in the calculation of the free energy difference between specific electronic states of the system.

Let us start by establishing a useful *reference* for the free energy of electrons in a material (solid or liquid). It turns out that electrons at rest and not interacting with anything except the local electrostatic field are a good thermodynamic state for the ensemble of electrons. We may think of a gas of electrons outside the material, which defines the energy of the *vacuum level* (VL) that we denote as E_{vac}.

The electron inside a metal or semiconductor interacts with all the atoms and free electrons, both by Coulomb attraction and repulsion and by the exchange correlation forces (Pauli principle). Interactions at the atomic level are electric forces to a great extent, but they must be treated quantum mechanically. The crystal forms a quantum well so that the energy of the electrons is lower inside the material than when extracted outside the surface. By *lower energy* we mean more negative in the electron energy scale, but it must be noted that "lower" energy means a stronger binding energy of the electron. A detailed microscopic view of the crystal potential is shown in Figure 2.3 for a stack of five-layer gold atoms. However, we normally simplify the energy diagram of a metal as shown in Figure 2.4 that illustrates the crystal potential φ as well as the local electrostatic potential φ (Gerischer and Ekardt, 1983). The horizontal part outside of the solid is the VL. If the electron gains enough energy to overcome the VL, it can escape from the solid.

The energy diagram of a semiconductor illustrating the VL is shown in Figure 2.5. The *electron affinity* χ_n is the energy required to excite an electron from the conduction band to the VL:

$$\chi_n = E_{vac} - E_c \tag{2.17}$$

The *ionization energy* (I_{sc}) *or hole affinity* (χ_p) is the energy difference between the valence band and the surface vacuum level,

$$\chi_p = I_{sc} = E_{vac} - E_c \tag{2.18}$$

and we have

$$\chi_p - \chi_n = E_g \tag{2.19}$$

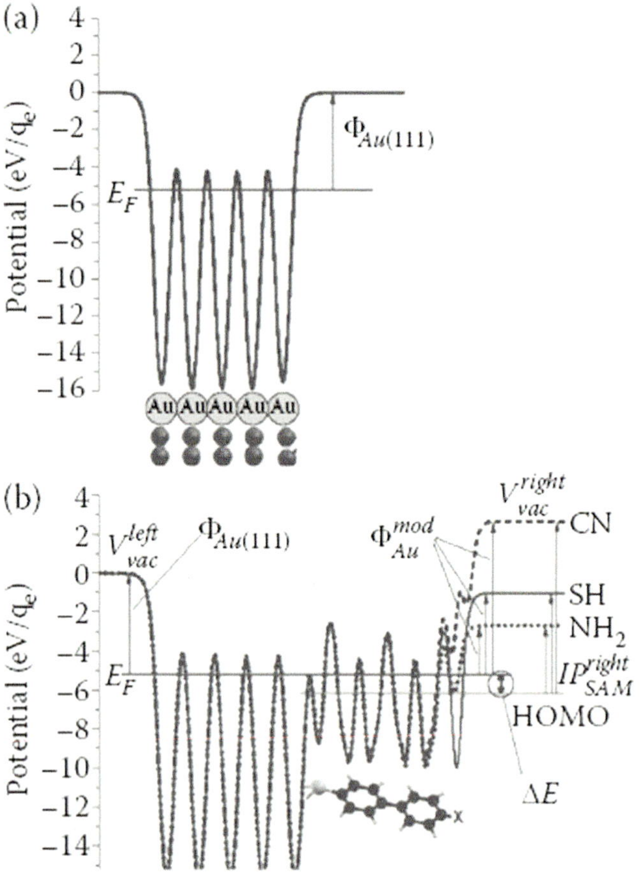

FIGURE 2.3 (a) The distribution of crystal potential of a five-layer Au(111) slab. Also indicated is the metal Fermi energy E_F, and its work function. (b) Potential of the self-assembled organic monolayers (SAM) adsorbed onto Au(111). The HOMO is located at ΔE below E_F. Also indicated is the vacuum level, and the modified work function of Au depending on the end group at the SAM. (Adapted with permission from Heimel, G. et al. *Physical Review Letters* 2006, *96*, 196806.)

Electron energies will be referred to the VL in many applications, such as photoelectron spectroscopies and barrier modeling in devices. However, it is important to distinguish the VL in three specific locations as shown in Figure 2.5b and defined as follows:

a. The *surface vacuum level* (SVL), E_{vac}^{s}, is the energy of an electron at rest "just outside the surface," which means that the electron is free of the binding to the crystal lattice, but is still within the range of the crystal's electrostatic potential (although outside the reach of image force effect). This point is typically taken at a distance 10^{-4} cm outside the solid and is also called the "near surface VL." It is the main reference for measurement of energetics of electron in solid materials.

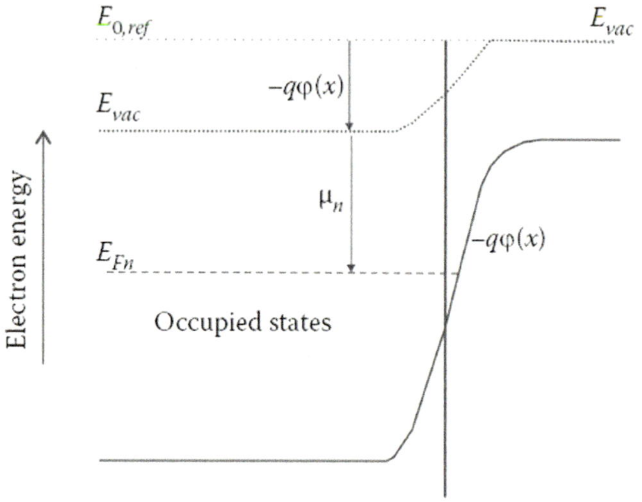

FIGURE 2.4 Energy diagram of a metal, showing the crystal potential φ(*x*), and the local (Galvani) potential due to surface dipole and unscreened charge, φ(*x*), that changes the local vacuum level.

b. The vacuum level away from the surface, E_{vac}^{∞}, means a distance far beyond the reach of electrostatic forces due to surface dipoles or net electric charge of the material.
c. The *local vacuum level* (LVL) is a prolongation of the vacuum level *inside* the crystal. The LVL represents the energy of an electron at a given point if it were at rest (no kinetic energy) and free from the influence of the crystal potential (Marshak, 1989). Thus, by definition, the LVL represents the potential energy other than that due to the influence of the crystal.

In addition to the VL, it is useful to introduce a reference level $E_{0,ref}$ for electrostatic potential φ (see Figure 2.6). Thus, we can write

$$E_{vac}(x) = E_{0,ref} - q\varphi(x) \tag{2.20}$$

E_{vac} is free from the crystal macroscopic potential but it follows the macroscopic potentials caused by space charge regions. Any change of electrostatic potential changes the potential energy of the electron and appears as an equal change (of opposite sign) in the LVL. Then, in the picture of a semiconductor device, E_{vac} follows the band bending at all points in the device, as shown in Figure 2.1, and, as a consequence, Equation 2.20 is consistent with Equation 2.16. Furthermore, the level E_{vac} allows us to account for the structure of interfacial potential between materials of different composition, displaying any dipole layer that produces a change of

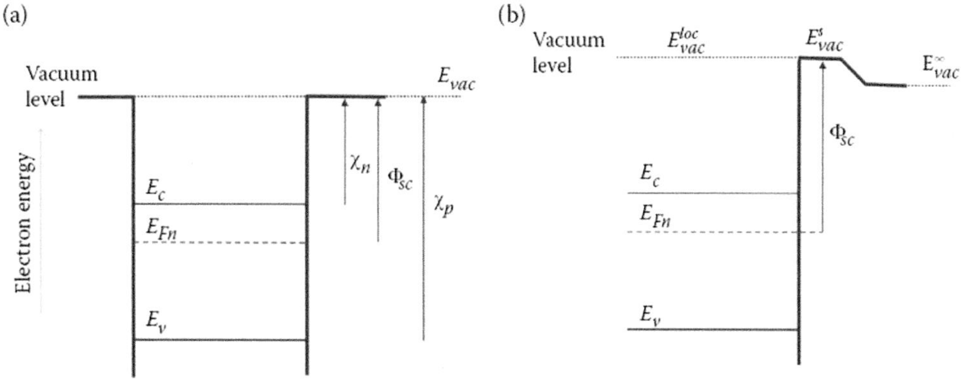

FIGURE 2.5 (a) Energy diagram of a semiconductor indicating the energy band edges E_c and E_v, vacuum level E_{vac}, the Fermi level E_{Fn}, the electron affinity χ_n, work function Φ_{sc}, and ionization energy χ_p. (b) The local, surface, and infinite vacuum level.

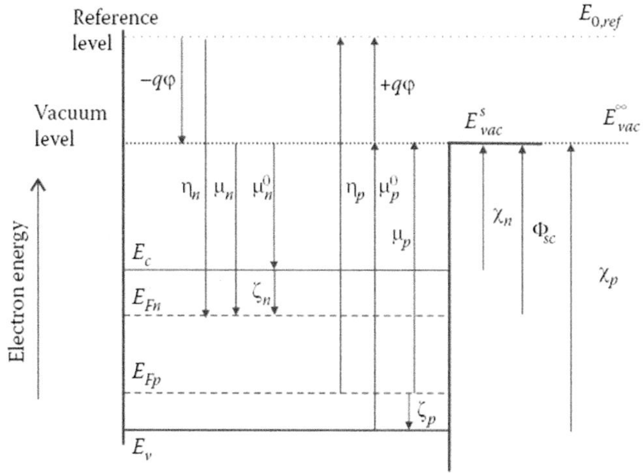

FIGURE 2.6 Energy diagram of a semiconductor indicating the reference energy level $E_{0,ref}$ the vacuum level E_{vac}, energy band edges E_c and E_v, and the different statistical quantities defined in the text.

electrostatic energy at the surface, as will be discussed later.

In calculations of electrostatics in elementary textbooks, the reference of potential is the electron at rest at infinite, as shown in Figure 2.4, so that

$$E_{0,ref} = E_{vac}^{\infty} = 0 \text{ for } \varphi = 0 \left(\text{particular choice}\right) \quad (2.21)$$

However, electronic states in materials are not directly measurable with respect to the vacuum level at infinite. Therefore, in many applications the zero of energy scale is placed at E_{vac}^{s}

$$E_{0,ref} = E_{vac}^{s} = 0 \ \left(\text{particular choice}\right) \quad (2.22)$$

We will make extensive use of energy diagrams as the one shown in Figure 2.6. In the scheme, the vertical direction indicates increasing electron energy. Energy levels in the diagram are *absolute electron energies* and *absolute electron free energies* such as E_{vac}, E_c, E_F, which are measured with respect to a reference level, as indicated in Equation 2.22. On the contrary, *energy differences* and *free energy differences*, as χ, Φ, $-q\varphi$, $h\nu$, are represented by an arrow in the energy diagram. The arrow for A, joining levels *B* (base of the arrow) and *C* (tip of the arrow), means that the defined quantity is $A = C - B$. An arrow pointing downward is a negative energy difference.

In Figure 2.6 we observe that

$$E_c = -\chi_n - q\varphi \quad (2.23)$$

Therefore, the total potential energy of the conduction band electrons, E_c, consists of two parts: the energy $-\chi_n$ due to the crystal potential, and energy $-q\varphi$ due to the electrostatic field. For electrons in the valence band, we have

$$E_v = -q\varphi - E_g - \chi_n \quad (2.24)$$

By Equation 2.19, we can write

$$E_v = -\chi_p - q\varphi \quad (2.25)$$

In a semiconductor crystal that becomes of the size of a few nm, the electronic properties (excited states, ionization potential, and electron affinity) depend strongly on the quantum confinement of the electron and hole in the reduced space. Such small crystal is called a *quantum dot* (QD).

In organic crystals, and in organic solids in general, the concept of the conduction band is replaced by the lowest unoccupied molecular orbital (LUMO), and the valence band edge is denominated highest occupied molecular orbital (HOMO). Small molecular systems show reorganization effects and the electron affinity and ionization energy often differ from the one electron LUMO and HOMO levels (Savoie et al., 2013).

2.5 THE FERMI LEVEL AND THE WORK FUNCTION

To determine the amount of *work* necessary to bring a set of Δn electrons into the semiconductor, we need to introduce a quantity, termed the *Fermi level* of the electrons, E_{Fn}, which is the same as the *electrochemical potential* of the electrons, η_n,

$$\eta_n = E_{Fn} \tag{2.26}$$

More rigorously, the Fermi level is defined as $\eta_n(T = 0)$, but the difference at finite temperature can usually be neglected.

The concept of the Fermi level is based on the difference of energies for the N-electron system in two different states (Lang and Kohn, 1971). We take as an example a neutral metal slab. E_N is the ground state energy and E'_N is the energy of an excited state in which $(N - 1)$ electrons reside in the lowest possible state in the metal, with energy E_{N-1}, and one electron is at rest in the vacuum at infinite. Thus, we have that the difference of energy for a one electron state between the electron inside the material at the Fermi level and the electron at rest in vacuum is

$$E_{Fn} - E_{vac}^{\infty} = E_{N-1} - E_N \tag{2.27}$$

In Chapters 3 and 5, the Fermi level will be connected to the voltage in a device and also to the occupation of electronic states according to the energy level of such state. In metals, the Fermi energy is situated inside the continuum of states of the conduction band and divides occupied and unoccupied electron energy levels, as shown in Figure 2.2b. A similar concept applies to a semiconductor, although usually there are no electrons with the energy E_{Fn} due to the fact that the Fermi level lies in the energy gap. In general, the work associated to bring Δn electrons from infinity and add them to a solid or liquid material is equal to $E_{Fn}\Delta n$ minus the free energy that these electrons Δn had as a vapor at infinity, which is $E_{vac}^{\infty}\Delta n$.

The *work function*, Φ, is defined as the difference in energy of an electron at rest in vacuum *just outside the surface* and an electron at the Fermi energy into the solid (see Figure 2.5b)

$$\Phi = E_{vac}^{s} - E_{Fn} \tag{2.28}$$

We have already stated earlier that the electron at rest in vacuum is a convenient thermodynamic state of reference, only determined by the local electrostatic potential (the Volta potential; see later in this chapter). Therefore, the work function is defined as a difference of free energies, and it can be related to other thermodynamic quantities such as electrode potentials and free energies of reaction. From definition (2.28), the work function is the onset energy for photoelectric emission for a metal as will be discussed in Chapter 4. The work function of a metal may be considered its ionization energy as well as its electron affinity, $\Phi = I = \chi$.

Having introduced the energy of electron at rest in vacuum just outside the surface, as a standard reference for electron energies in materials, we can establish the *absolute energy scale*, with a zero as $E_{0,ref} = E_{vac}^{s} = 0$, which allows us to place energies and Fermi levels of electrons in materials, by measurement of work functions that will be outlined in Chapter 4. Work functions of several metals are indicated in Table 2.2. Energies for a range of metals and organic and inorganic semiconductors and redox electrolytes used in energy devices are shown in Figure 2.7. It will be commented later that the changes in the surface created by deposition conditions or adsorption layers produce strong modifications of the energies of electrons in the material with respect to the surrounding. Therefore, values in Figure 2.7 must be taken as indicative. The right scale is the electrochemical scale for solid/electrolyte interfaces that has the zero

TABLE 2.2
Free Electron Densities, Electric Conductivities, and Work Functions[a] of Selected Metal Elements at 300 K

Ion	$n(10^{22}$ cm$^{-3})$	$\sigma(10^6\ \Omega^{-1}$ cm$^{-1})$	ΦeV)
Li	4.70 (78 K)	0.12	2.38–2.93
Na	2.65 (5 K)	0.23	2.35
Ba	3.15	0.017	2.49–2.7
Ca	4.61	0.30	2.80–2.87
Zn	13.2	0.18	3.63–4.9
Al	18.1	0.41	4.06–4.28
Cu	8.47	0.64	4.53–5.10
Ag	5.86	0.66	4.52–4.74
Au	5.90	0.49	5.1 (4.7–5.8)

Source: Ashcroft, N. W.; Mermin, N. D. *Solid State Physics;* Thomson Learning: Florence, 1976.

[a] Work functions can change based on the crystal face orientation and surface conditions.

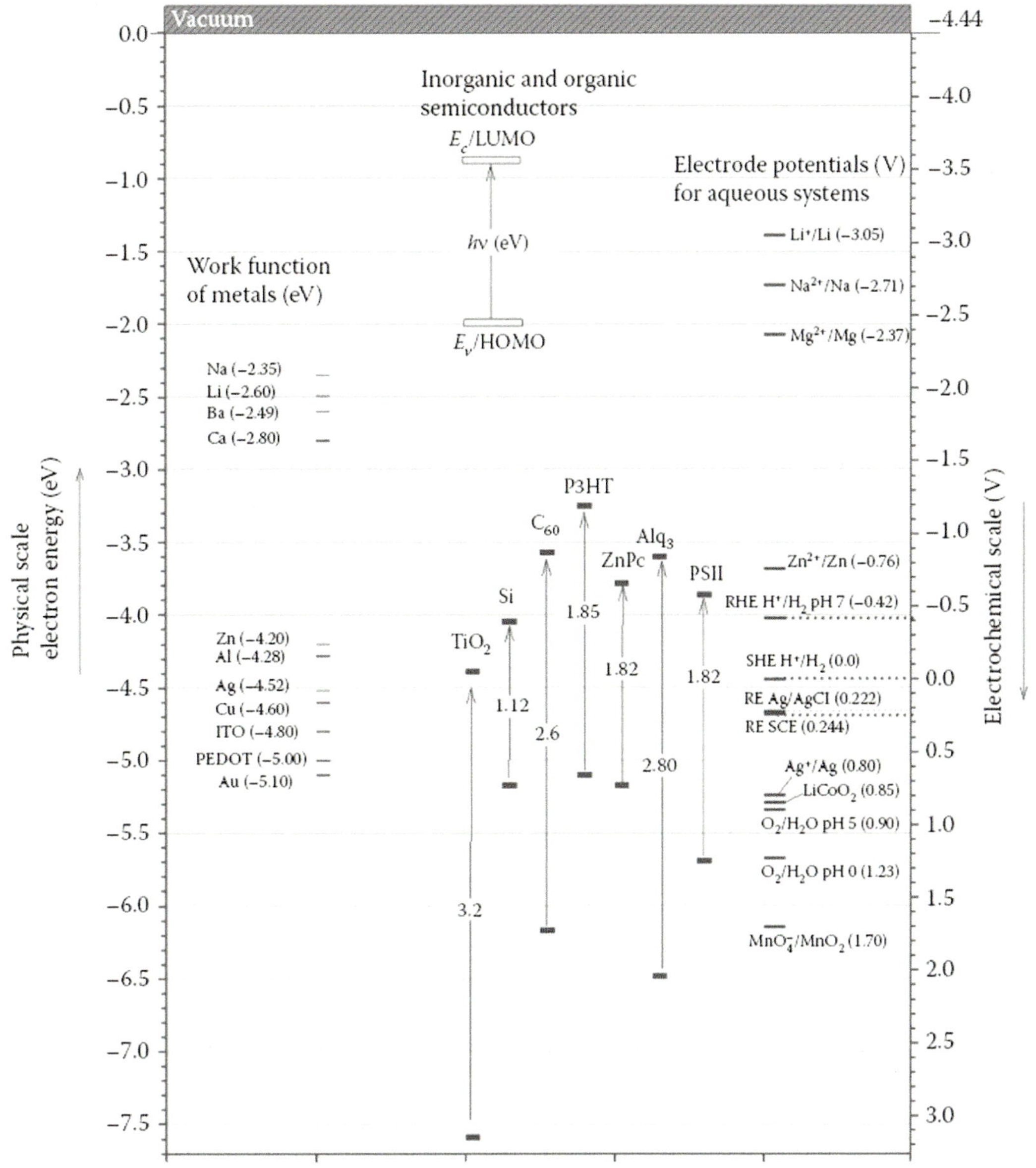

FIGURE 2.7 Work functions of metals and contact materials, energy levels of inorganic and organic semiconductors and photoactive molecules, redox level of electrode materials, reactions in aqueous solution, and reference electrodes.

at the potential of standard hydrogen electrode; this will be introduced in Chapter 3.

2.6 THE CHEMICAL POTENTIAL OF ELECTRONS

The electrochemical potential η_i of a species of charge $z_i q$ ($z = \pm 1, 2\ldots$) is composed of the electrostatic energy, $z_i q\varphi$, and a chemical potential, μ_i, which includes the binding energy of the medium and an entropic contribution further discussed in Chapter 5. Therefore, we have

$$\eta_i = z_i q\varphi + \mu_i \tag{2.29}$$

These quantities are represented in Figure 2.6. For the electron by Equation 2.26, we can write

$$E_{Fn} = -q\varphi + \mu_n \tag{2.30}$$

For the analysis of semiconductor devices, it is useful to separate the chemical potential into a constant term and concentration-dependent term (Parrott, 1996):

$$\mu_n = \mu_n^0 + \zeta_n \tag{2.31}$$

where

$$\zeta_n = k_B T \ln \frac{n}{n^0} \tag{2.32}$$

Here, n^0 is a reference concentration associated to the value of chemical potential μ_n^0. In rigorous terms, Equations 2.31 and 2.32 are valid only under applicability of the one-electron model to n electrons, for parabolic energy bands over the range of occupied energies. For metals, μ_n does not depend on electron concentration, which is a constant.

Here, we adopt as the standard reference state $n^0 = N_c$ so that the expression of the chemical potential is

$$\mu_n = \mu_n^0 + k_B T \ln \frac{n}{N_c} \quad (2.33)$$

and furthermore,

$$\mu_n^0 = -\chi_n \quad (2.34)$$

$\zeta_n = E_{Fn} - E_c$ indicates the distance of the Fermi level to the conduction band and hence the extent of occupation of the conduction band, as shown in Figure 2.8.

Taking into account Equation 2.23, we obtain

$$E_{Fn} = -\chi_n - q\varphi + \zeta_n = E_c + \zeta_n = E_c + k_B T \ln \frac{n}{N_c} \quad (2.35)$$

Thus, the electron concentration can be expressed as

$$n = N_c e^{(E_{Fn}-E_c)/k_BT} \quad (2.36)$$

This expression is valid provided that the Fermi level is far from the band edge (see Equation 5.17).

For holes in the valence band with concentration p, and with the reference state $p^0 = N_v$ we obtain the following expressions for the electrochemical potential,

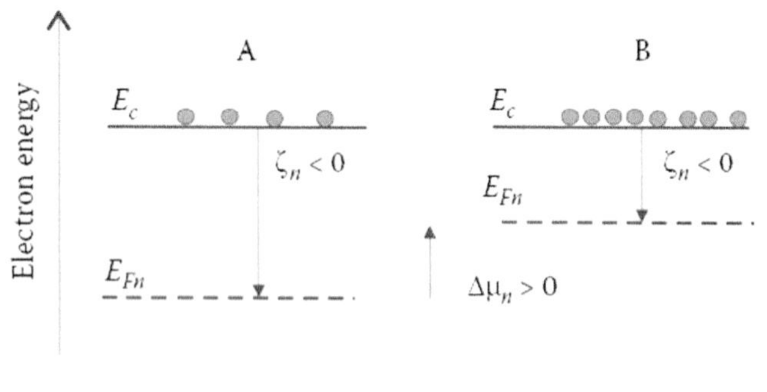

FIGURE 2.8 Energy diagram semiconductor with different values of the chemical potential μ_n. E_c is the conduction band edge and E_{Fn} is the Fermi level. The increase of the chemical potential $\Delta\mu_n > 0$ produces a larger concentration of electrons in the conduction band.

chemical potential, reference term, and concentration term:

$$\eta_p = +q\varphi + \mu_p \quad (2.37)$$

$$\mu_p = \mu_p^0 + \zeta_p \quad (2.38)$$

$$\mu_p^0 = \chi_p \quad (2.39)$$

$$\zeta_p = k_B T \ln \frac{p}{N_v} \quad (2.40)$$

Since the hole Fermi level is drawn in the electron energy scale, we must write

$$E_{Fp} = -\eta_p = -q\varphi - \mu_p \quad (2.41)$$

Using Equation 2.25, we have $\zeta_p = E_v - E_{Fp}$ and

$$E_{Fp} = -\chi_p - q\varphi - \zeta_p = E_v - \zeta_p = E_v - k_B T \ln \frac{p}{N_v} \quad (2.42)$$

The hole concentration dependence on hole Fermi level takes the form

$$p = N_v e^{-(E_{Fp}-E_v)/k_BT} \quad (2.43)$$

The different relationships outlined in this section are shown in Figure 2.6.

We can express the work function Equation 2.28 in the following way:

$$\Phi_{sc} = E_{vac}^s - E_{Fn} = \left(E_{vac}^s - E_c\right) + \left(E_c - E_{Fn}\right) \quad (2.44)$$

The first term in Equation 2.44 is the electron affinity of the solid, and the second term represents the Fermi energy with relation to the bottom of the conduction band. Therefore, we can write

$$\Phi_{sc} = \chi_n - \zeta_n \quad (2.45)$$

Finally, we consider the total electrochemical energy of an electron–hole pair

$$\eta_n + \eta_p = E_{Fn} - E_{Fp} = -q\varphi + \mu_n + q\varphi + \mu_p = \mu_n + \mu_p \quad (2.46)$$

Since the pair is uncharged, the total electrochemical energy is independent of the local electrostatic potential and equals the total chemical potential. Equation 2.46 has great significance for energy conversion devices, and it coincides with the voltage that can be extracted from a semiconductor, as will be seen in Equation 3.1.

2.7 POTENTIAL STEP OF A DIPOLE LAYER OR A DOUBLE LAYER

One important effect occurring at the surface of a material, or at the interface between different media, is a local separation of charges over a *very narrow distance* (1 nm or less) that is described in a first approximation as a sheet dipole. The example shown in Figure 2.9a consists of a positive sheet situated just at the surface of the material, and the compensating negative sheet of charge at the vacuum side of the interface, separated at a distance d. The charge separation may arise due to several reasons discussed in Section 2.8, or from a double layer in electrochemical interfaces (see Section 3.7). From Equation 2.2, the potential drop across the layer relates to the electric field as

$$\varphi_2 - \varphi_1 = -\int \mathbf{E}\,dx \tag{2.47}$$

In principle, we neglect the extension of the surface and take the charge sheets as infinitely large. Since d can be a fraction of a nanometer, this is a good approximation provided that we do not move the test charge a macroscopic distance away from the surface. Adopting this restriction, the electric field **E** is normal to the surface as shown in Figure 2.9a. Gauss law provides the connection of the electric field to the amount of surface charge. We consider a surface that encloses a total quantity of charge Q in the positive sheet. The area of the test surface that cuts the lines of the electric field is A. Therefore,

$$\mathbf{E} = \frac{\sigma_q}{\varepsilon} \tag{2.48}$$

where $\sigma_q = Q/A$ is the surface charge. The electric field has the constant value in Equation 2.48 between the two sheets of charge. The field is zero outside of the charged sheets (the separate contribution of each sheet of charge to the field is $\sigma_q/2\varepsilon$). Consequently, the potential rises linearly from one charge plane to the other one. The step of the energy is

$$\Delta_d = q(\varphi_1 - \varphi_2) = \frac{qd\sigma_q}{\varepsilon} \tag{2.49}$$

Therefore, the step of the vacuum energy is

$$E_{vac}^{v} - E_{vac}^{m} = -q(\varphi_v - \varphi_m) = +\Delta_d \tag{2.50}$$

as shown in Figure 2.9a. Δ_d is positive in Figure 2.9a and negative in Figure 2.9b. This choice follows the custom in the literature of work functions of the metals, in which Δ_d is taken to be positive when the vacuum energy is raised *outside* of the material (i.e., when the negative side of the dipole is pointing outward).

If the surface charge density is N_d (m^{-2}) ions of charge $+zq$, Equation 2.49 can be written as

$$\Delta_d = \frac{qN_d m_d}{\varepsilon} \tag{2.51}$$

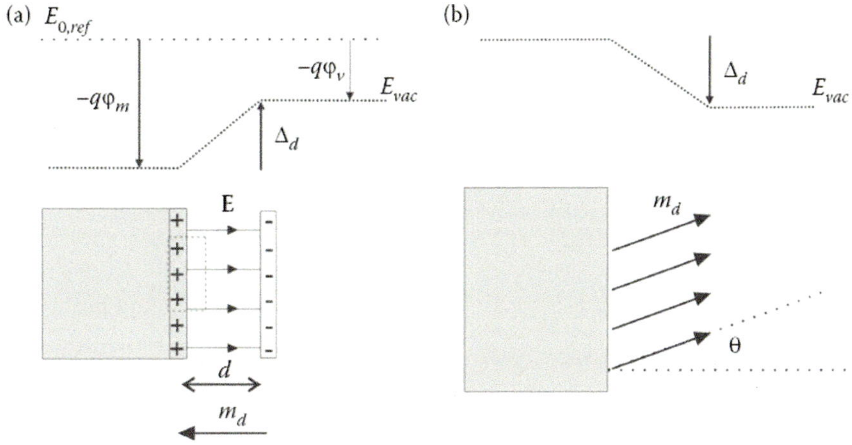

FIGURE 2.9 (a) Energy diagram of a dipole layer consisting of two sheets of negative and positive charge separated a distance d. φ is the electrostatic potential, **E** is the electric field, and Δ_d is the step of the vacuum energy across the dipole layer. (b) Energy diagram (vacuum level) of a dipole layer of polar molecules with dipole moment m_d that are tilted an angle θ with respect to the normal.

where

$$m_d = zqd \tag{2.52}$$

is the electric dipole moment (C m) associated to one ion and the negative countercharge. m_d is often reported in debye units (1D = 3.33×10^{-30} C m). If d = 0.1 nm and z = 1 then m_d = 4.8 D.

In addition to the intrinsic dipole layer, polar adsorbed molecules may contribute a dipole m_d to the step of VL as shown in Figure 2.9b. If the molecules are tilted on average an angle θ with respect to the normal, then

$$\Delta_d = \frac{qm_d\cos\theta}{A_d\varepsilon} \tag{2.53}$$

Here, A_d is the area per molecule (m²). ε is the dipole layer's dielectric constant, derivable from the molecular polarizabilities (Fröhlich, 1958; Iwamoto et al., 1996). Equation 2.53 is called *the Helmholtz equation*.

The constant potential due to the dipole layer sheet cannot be maintained at large distances, when the layer extension L cannot be neglected. The potential at the surface remains constant for a distance $x \ll L$ in which the dipole layer can be considered semi-infinite. But when removing the electron a distance $x \gg L$, the dipole layer can be considered a point dipole and the electrostatic potential decreases rapidly (see Figure 2.10) (Ishii et al., 1999). The potential of a rigid (nonpolarizable) dipole with moment m_d in an infinite medium of dielectric constant ε at distance r on the direction ϑ is given by

$$\varphi(r,\theta) = \frac{m_d\cos\vartheta}{\varepsilon r^2} \tag{2.54}$$

When the distance from the dipole layer is larger than the layer extension L, the potential decays as $1/r^2$. Such decay of the surface dipole is one reason why there may be a difference between the vacuum levels "just outside the surface" E_{vac}^s and the value at a larger distance E_{vac}^{∞} as seen in Figure 2.5b, even if the solid is electrically neutral as a whole. This difference is called the Volta potential, as described in Section 2.9.

2.8 ORIGIN OF SURFACE DIPOLES

Conducting or semiconducting materials are electrically neutral in the bulk, but charge separation occurs at the surface. If we consider a piece of material that is uncharged, the total number of positive and negative charges is balanced; however, a separation of charge at the interface occurs. A surface dipole exists at every surface due to any inhomogeneous distribution of the charge at the surface. The surface dipole is indicated by a step change of the local electrostatic potential φ and the corresponding step of the VL as shown in Figure 2.4.

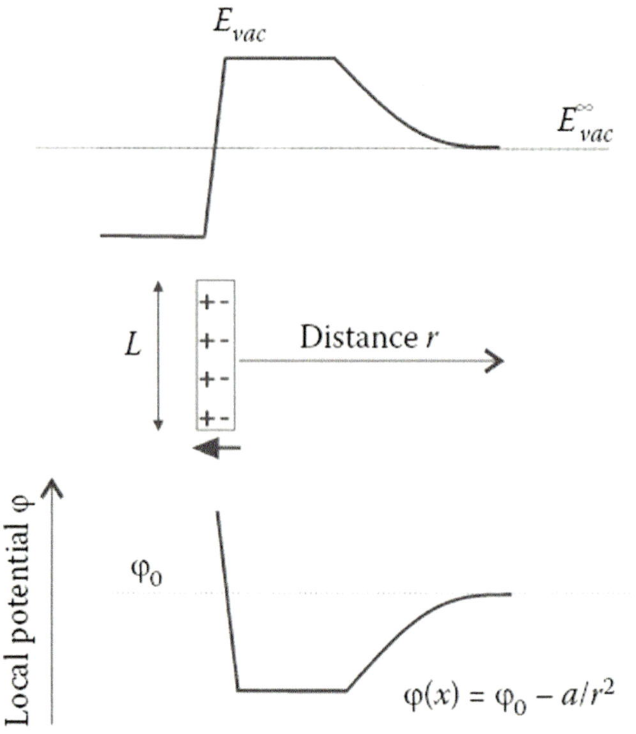

FIGURE 2.10 Energy diagram (vacuum level) of a dipole layer of finite extension.

The dipole at the surface may have diverse physicochemical origins but the result is a change of the internal energy levels with respect to the surface vacuum level that influences the work function and the formation of barriers and junctions. The work function of a metal is very sensitive to the microscopic structure of the surface, and is influenced by the specific crystallographic orientation, and the chemical or structural defects. It is important to remark that even for a single crystal, the magnitude of this dipole layer differs for different crystallographic faces, as shown in Figure 2.11, as a result of the different arrangements of the atoms on these faces. For a clean metal surface, the values are well established and should depend only on orientation, but in practice Φ may vary about 1 eV, depending on surface conditions.

In the metal bulk structure, because electronic conductivity is large, the electron density is homogeneously distributed throughout space to cancel the positive charge of ions in fixed lattice sites. But such compensation is not exact at the surface, and the electronic density decays over a distance of d = 0.1 nm out of the outer layer of positive ions. The tailing of the negatively charged electron cloud "spills out" of the atomic plane into vacuum and makes the potential at the vacuum side negative,

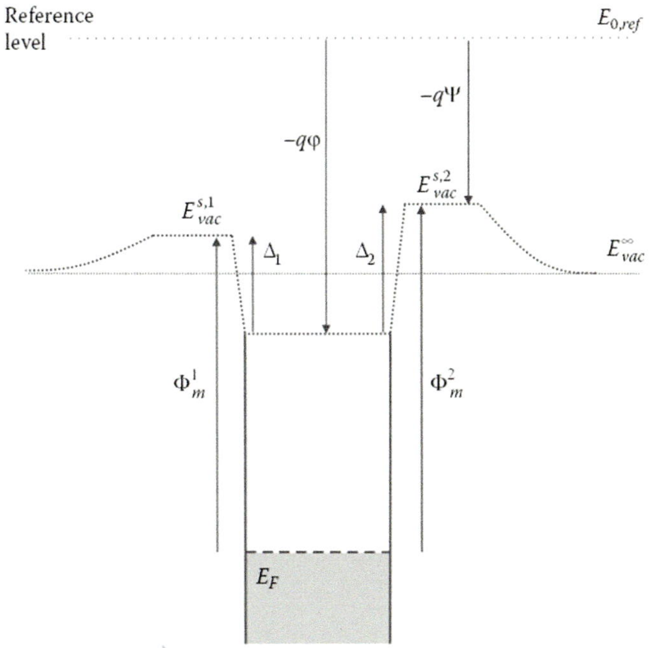

FIGURE 2.11 Energy diagram of an electron in and out of a metal crystal. The different crystal facets produce different surface dipole and change the work function Φ. (Adapted from Ishii, H. et al. *Advanced Materials* 1999, *11*, 605.)

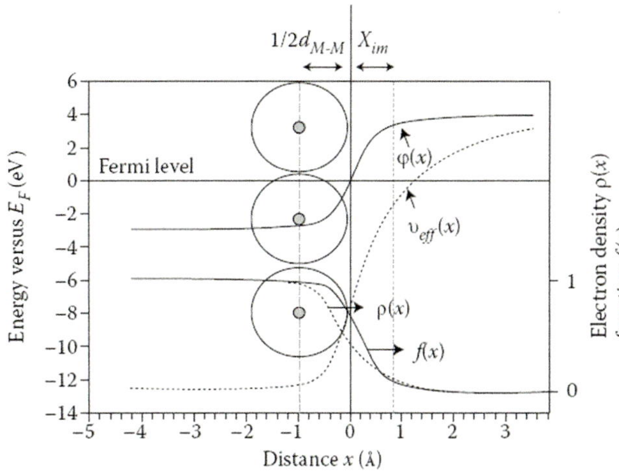

FIGURE 2.12 DFT calculations within the jellium model give the evolution of the electron density ρ(x) across the metal/vacuum interface (bottom dotted curve), the effective potential $v_{eff}(x)$ (top dotted curve), and the electric potential φ(x) (top solid line). The large circles depict the atoms on the metal surface; the zero on the x axis is positioned at one-half the interatomic layer distance from the surface nuclei plane ($x > 0$ indicates the metal bulk). f(x) is a fitting sigmoid function. (Reproduced with permission from Crispin, X. et al. *Journal of the American Chemical Society*. 2002, *124*, 8131–8141.)

while the lack of electrons inside the surface makes the bulk side positive, as shown in Figure 2.12.

In a semiconductor crystal the periodicity of the lattice is maintained throughout the crystal, except at impurities or defects. But the atoms in the surface lack the neighboring atoms they would have on the bulk, and a different configuration is realized to obtain a minimum of the free energy. The surface is said to be *reconstructed* and attains a different symmetry configuration. The surface reconstruction involves the outward or inward displacement of surface atoms, and electrostatic dipole layers are formed, which change the measured affinity.

In addition to the intrinsic properties of the exposed surface, extrinsic species attached or adsorbed to the surface produce changes of surface dipole value. An example is shown in Figure 2.3b that shows that the end group of the molecular layer is crucial to determine the Au slab work function. Chemisorption and physisorption of atoms or molecules on the surface produce a significant redistribution of the surface charge and consequently modify the surface dipole. The formation of a metal/organic interface involves establishing chemical bonds of the organic molecules on the metal plane, which leads to charge transfer between the organic molecules and localized states of the metal. Charge transfer between molecular layers in contact at the interface implies the formation of a potential step.

If the molecule that is attached to the surface is polar, it can contribute a dipole layer. Covering the surface with a tiny, well-oriented molecular layer can therefore substantially modify the effective work function of the semiconductor. The Helmholtz Equation 2.53 is useful for the estimation of the change of work function, Δ, caused by an organic monolayer on metal or semiconductor substrate. Figure 2.13 shows the change of work function of thin films of titanylphthalocyanine (OTiPc), vacuum deposited on graphite.

A liquid or electrolyte also contains electronic states with energy levels that are affected by the specific dipole layer at the liquid surface. While the solid surface has a limited ability for reconstruction, at the liquid/vapor interface, polar molecules have a large freedom of rearrangement in order to minimize the free energy. The molecules in the surface will tend to orient so as to place their electric fields as much as possible in the high-dielectric constant liquid, rather than the low-dielectric-constant vapor. The orientation causes a net surface dipole (Stillinger and Ben-Naim, 1967).

2.9 THE VOLTA POTENTIAL

As discussed in the previous sections, a variety of factors determine the actual value of the work function of a metal or semiconductor surface. The crystal potential is

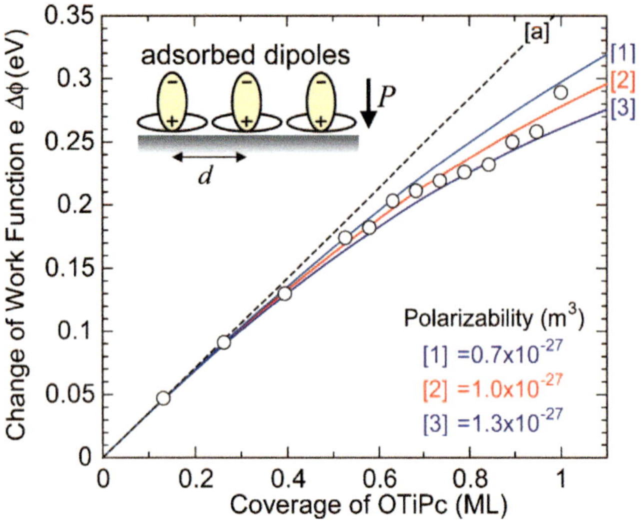

FIGURE 2.13 Dependence of work function on OTiPc coverage. This molecule has an electric dipole moment perpendicular to the molecular plane. The schematic representation of the molecular orientation and the direction of the dipole in the monolayer of OTiPc are indicated. In the low coverage region (≈ 0.3 ML), the work function changes along the dashed line (a). The thin curves represent fitting curves to different models. (Reproduced with permission from Fukagawa, H. et al. *Physical Review B* 2006, 73, 041302.)

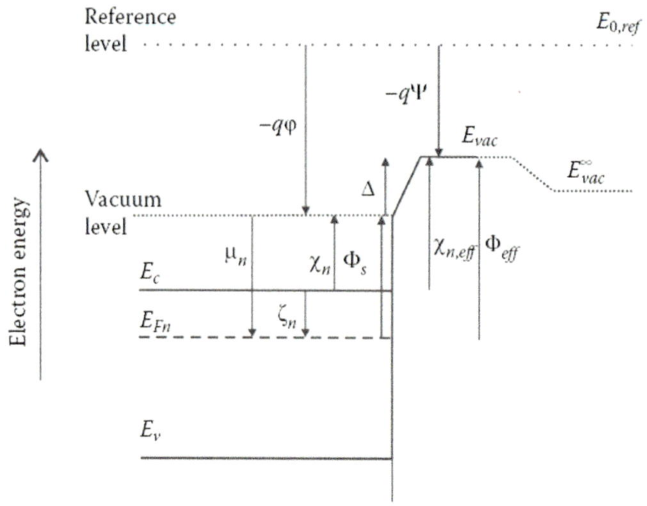

FIGURE 2.14 Energy diagram of an electron in and out of a semiconductor crystal with a surface dipole.

a material property but the step of the VL at the surface depends on specific properties of the surface or interface, both intrinsic and due to extrinsic agents, that establish a surface dipole. The presence of the surface dipole modifies the energy difference of the electron between the conducting medium and the vacuum outside the surface, as shown in Figure 2.11. It is, therefore, useful to consider that the work function, which is the work to extract an electron from the material, consists of two parts: the specific binding energy of the electron in the bulk material, and the work required to move the electron across the surface dipole layer (Bardeen, 1936). The effective (measurable) work function can be written as

$$\Phi_{eff} = E_{vac}^{s} - E_{Fn} = \left(E_{vac}^{b} - E_{Fn}\right) + \left(E_{vac}^{s} - E_{vac}^{b}\right) \quad (2.55)$$

so that it consists of an intrinsic work function Φ and the surface polarization term, Δ,

$$\Phi_{eff} = \Phi + \Delta \quad (2.56)$$

This last expression is illustrated in Figure 2.14. Similarly, one may introduce a partition of the electron affinity, into the bulk binding energy χ_n and the surface dipole (see Figure 2.14)

$$\chi_{n,eff} = \chi_n + \Delta \quad (2.57)$$

We have observed that the VL *outside* the surface, which is the energy of electron at rest, has a special importance as the thermodynamic reference state for the free energies inside the crystal such as the work function. When the electrons are considered outside the surface, they do not interact with the chemical forces inside the crystal or liquid, and they neither interact among themselves. The total energy of the electrons at rest just outside the surface is determined only by the local electrostatic potential.

The electrostatic potential, at the surface, receives a special denomination, and it is called *the Volta potential* ψ, and also the *outer potential* (see Figures 2.11 and 2.14). The Fermi level can now be expressed as the potential energy just outside the surface, minus the effective work function:

$$E_{Fn} = -q\psi - \Phi_{eff} \quad (2.58)$$

It is important to remark that Equation 2.58 relates quantities that *can be measured*: the outer potential, the effective work function, and the Fermi level. Methods of measurement will be discussed in Chapter 4.

The Volta potential is determined by electrostatic forces of the crystal surface, so that there is a difference between the outer potential and the electrostatic potential at infinite, as indicated in Figure 2.11. The Volta potential at the surface of a charged material follows from Equation 2.48:

$$\frac{\partial \psi}{\partial x} = -\frac{\sigma_q}{\varepsilon} \quad (2.59)$$

If a material is effectively charged, then the electrostatic potential decays with distance as $1/r$, that is, more slowly than the dipole potential in Equation 2.54. Therefore, the Volta potential will decrease with distance so that a transition occurs between E^s_{vac} and E^{∞}_{vac}.

The Volta potential also has the importance that in any fragment of a real material one has to account for the lateral transference of surface charge between different surface planes. Thus, outer potential ψ changes in zones of low work function with respect to those of high work function, as indicated in Figure 2.11, so that there will be no difference of the total binding energy of electrons as they are transferred from E_F inside the material to E^{∞}_{vac}.

The electric potential inside the solid or liquid phase is called the *Galvani potential* φ_G, or the *inner potential*. The Galvani potential difference is an electric potential difference between points in the bulk of two phases. Due to the presence of surface dipoles, the Galvani potential difference is measurable with a voltmeter only when the two phases have identical composition, as explained in Chapter 3. The only measurable potential difference is that between two regions immediately outside the two phases (in a contact between metalsemiconductor, or electrolyte), which is the Volta potential difference or the contact potential difference. The inner or Galvani potential difference between two phases takes a major role in the derivation of the voltage of electrochemical cells, including different electronic and ionic species (see, for example, Section 5.10).

We must also emphasize that the potential φ has a definite variation across the interface, including the potential step at dipole layer, as indicated in Figure 2.4. The electrons are in equilibrium across the interface due to the compensation of the chemical forces and electrostatic forces. The electrostatic field at the interface is real but the situation is somewhat confusing as we cannot apply the criterion of Equation 2.1 using an electron as a test charge, since, in addition to the electrostatic force due to the dipole sheet at the interface, the electron is subjected to the Pauli exclusion principle and chemical forces in general. The reality of the electrostatic potential difference and the vacuum step at the interface is better established using a positron as a test charge (Makinen et al., 1991). The positron will experience a force in a unique direction, of size $q\,|\nabla\varphi|$.

In the remainder of this book, we will use the same letter φ for the inner (Galvani) *electric* potential, and for the outer (Volta) *electrostatic* potential, unless they need to be distinguished from each other.

2.10 EQUALIZATION OF FERMI LEVELS OF TWO ELECTRONIC CONDUCTORS IN CONTACT

The knowledge of the work functions of metals and semiconductors is an essential resource that allows us to predict and analyze the properties of the junctions made of such materials. If two electronic systems come into contact, as shown in Figure 2.15, and the electrons have different electrochemical potentials in the two systems, there will be a transference of electrons from the system with the higher electrochemical potential to that with the lower η_n. The transference is compositionally negligible and it will occur if there is no kinetic barrier impeding the effective passage of electrons, and will stop when the condition of equilibrium (that is demonstrated in Chapter 5) is met

$$\eta_1 = \eta_2 \tag{2.60}$$

or alternatively

$$E_{Fn1} = E_{Fn2} \tag{2.61}$$

The condition of equilibria (2.60) and (2.61) is akin to the usual chemical equilibrium condition.

Assuming that both materials are good electronic conductors, if they are brought into contact with a wire, before making the junction, the transference of charge will produce sheets of charges in the surface that result in a Volta potential difference. When the two materials are joined to form the junction, the Volta potential difference has to be accommodated at the interface (in metals) or across extended space charge layer in semiconductors, so that the Fermi levels remain aligned and the bulk energy levels become flat. The lineup of the Fermi levels of two conductors in electronic equilibrium implies changes of the Volta potential on the surface of each conductor, $\Delta\psi_i$. The relative amount of the change for each conductor depends on the relative sizes of the conductors (Trasatti, 1986; Hansen and Hansen, 1987a, b) and in the case of semiconductors, on the relative values of the DOS at the Fermi level (Gerischer et al., 1994).

In most cases, one medium accumulates excess electrons, and the other material contains a positive surface layer formed by the removal of electrons (although it is also possible to remove positively charged holes, leaving negative ionic space charge). The distribution of charge in each medium depends on the electronic distribution and *charge shielding* or *screening* in the medium. In a metal, either negative or positive charge remains very close to the surface (*Thomas-Fermi screening length*, less than 1 nm). In a semiconductor, the extent of the space-charge region as that in Figure 2.1 depends on the doping density: the larger the doping density, the less is the required thickness of the depletion region w, as explained in Chapter 9. If the doping density is very large, of the order of 10^{20} cm^{-3}, then a depletion can take

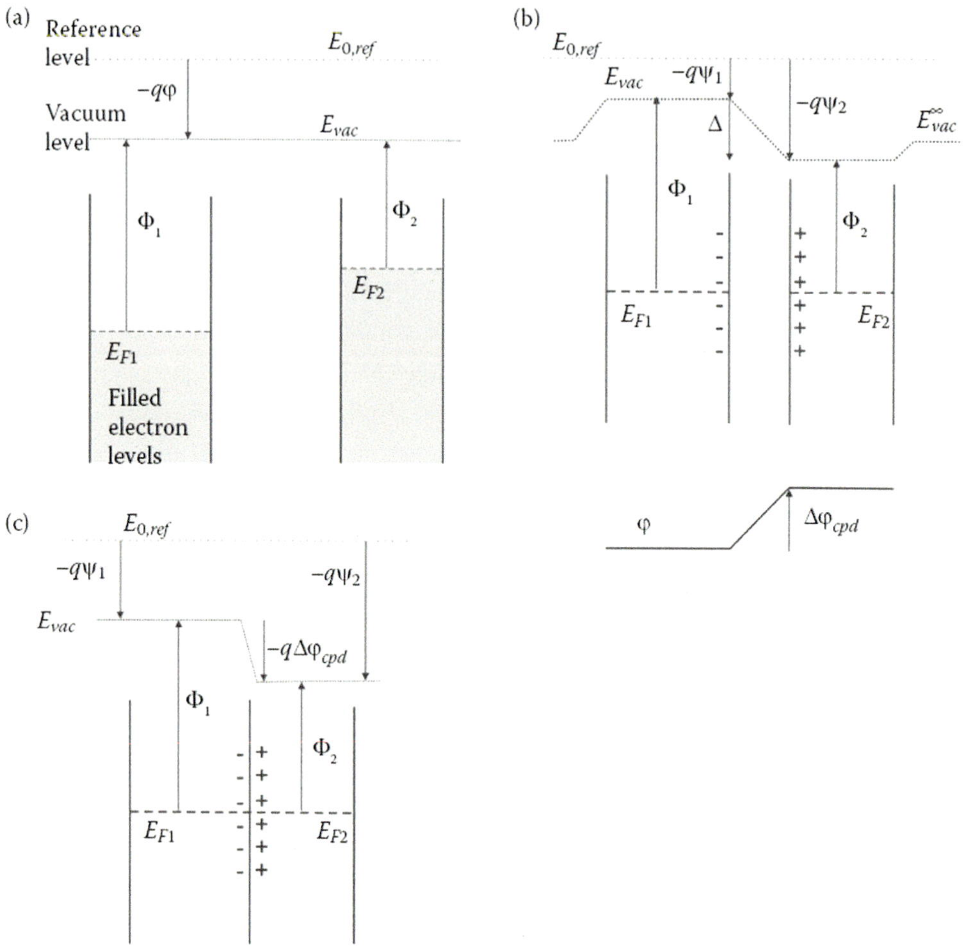

FIGURE 2.15 (a) Metal blocks of different work function. (b) The Volta potentials ψ caused by net charges in each plate and the contact potential difference $\Delta\varphi$ between the two metals when the Fermi levels are in equilibrium. (c) The metal surfaces in contact.

only about 2 nm. This space charge layer easily allows for electron tunneling, and its internal structure does not need to be described.

2.11 EQUILIBRATION OF METAL JUNCTIONS AND THE CONTACT POTENTIAL DIFFERENCE

In this section, we consider the electronic equilibration of two metals as in Figure 2.15 and consequent establishment of a dipole layer to accommodate the Fermi level difference, called the built-in potential. For simplicity, we neglect any initial dipole layers in the starting materials and assume $\psi_i = \varphi_{G,i}$ at each material, while the influence of dipole layer in the interface characteristics will be discussed in Chapter 4.

There are two ways to establish electronic equilibrium between two metal plates: one is to connect them by a conducting wire, and the other one is to bring the two metals into close proximity so that they will form an intimate junction. In the second case, once the gap between the surfaces is sufficiently small, electrons are able to cross it by tunneling and flow from the metal of higher Fermi level (lower work function) and cause a difference of potential (Simmons, 1963).

In any case, at the instant after the contact the difference of Fermi level causes transference of electrons from material 1 to 2. The flow of electrons is progressively prevented by the increase of the step of the potential, but the process continues until the potential difference has shifted the levels in 1 relative to those in 2 so that the top of the distributions of electrons inside the two metals is at the same height in the energy axis, Figure 2.15b. This effectively means the equilibration of the Fermi levels.

Initially, the Fermi levels lie at different distances from the vacuum, as shown in Figure 2.15a,

$$E_{F1} = E_{vac} - q\Phi_1 \tag{2.62}$$

$$E_{F2} = E_{vac} - q\Phi_2 \tag{2.63}$$

When the final equilibrium occurs (Figure 2.15b), there is a difference of potential between the plates that is called the *contact potential difference* (CPD).

$$V_{cpd} = \psi_2 - \psi_1 = \frac{\Phi_1 - \Phi_2}{q} \tag{2.64}$$

The dipole layer at the interface sustains the CPD maintaining the alignment of the Fermi levels. This is visualized in the step of the local vacuum level (Figure 2.15c) that we denote as before in Equation 2.50, $\Delta = -V_{cpd}$. Therefore,

$$\Delta = \Phi_2 - \Phi_1 \tag{2.65}$$

The built-in potential is the final step of the VL that corresponds to the initial difference of Volta potential, or work functions

$$V_{bi} = \Delta\psi_{CPD} = \psi_1^{eq} - \psi_2^{eq} = \frac{\Phi_2 - \Phi_1}{q} \tag{2.66}$$

Figure 2.16 illustrates important concepts about the potentials that can be measured in a combination of materials. Electrons at the surface still feel the electrostatic potential of the solid, so that the Volta potential with respect to infinite is not null. The difference of the Volta potentials $\Delta\psi = \psi_B - \psi_A$ can be measured as stated above. Often the two pieces of metals are connected by wire and one of them is taken as reference. In this case, the Fermi levels line up, and $\Delta\psi$ is measured by Kelvin probe method or photoelectron spectroscopies (Chapter 4). The Volta potential difference $\Delta\psi$ corresponds rigorously with the CPD (Trasatti, 1986). More generally, Figure 2.16 shows that the two contacts can

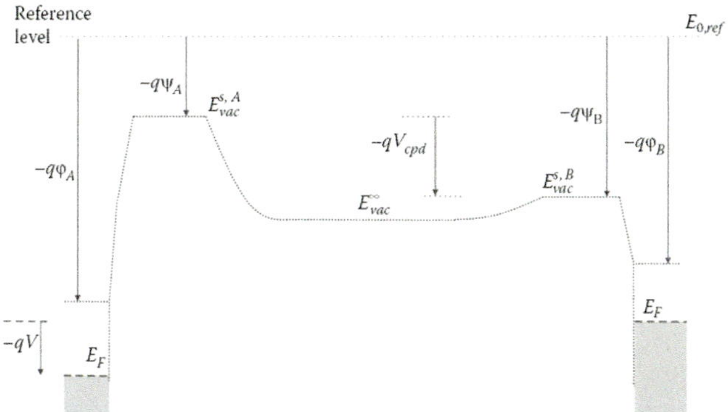

FIGURE 2.16 Energy diagram of potentials, vacuum levels, and Fermi levels, in two metal crystals.

be biased by a voltage V, showing a difference of Fermi levels, as will be further explained in Chapter 3.

2.12 EQUILIBRIUM ACROSS THE SEMICONDUCTOR JUNCTION

We now treat a situation as shown in Figure 2.17 in which two semiconductors, with different affinities and chemical potentials, come into contact. At the instant after the contact the difference of Fermi level causes electron transference (Figure 2.17a). In the absence of external influence, the system composed by the two semiconductors in good electronic contact must come to equilibrium situation defined by Equation 2.61 and shown in Figure 2.17b. Since the Fermi levels have the same value, from Equation 2.35 we obtain

$$q(\varphi_2 - \varphi_1) = (\chi_{n1} - \zeta_{n1}) - (\chi_{n2} - \zeta_{n2}) \tag{2.67}$$

Equation 2.67 establishes the condition of equilibrium. A difference of potentials exists in Figure 2.17b which changes the overall energies in the two phases in order to align the Fermi levels. The potential difference between the phases is shown in the diagram by the inclined step in the local vacuum level. If the extension of the layer containing the potential drop is of the order of a few atomic distances or less, it is called a dipole layer (or Helmholtz double layer in electrochemistry). If the interfacial barrier is an extended space-charge depletion region, well above the nm range, as in Figure 2.1, it is not considered as part of the dipole layer.

Now we can write the condition of equilibrium between two semiconductors as follows:

$$q(\varphi_2 - \varphi_1) = \Phi_1 - \Phi_2 \tag{2.68}$$

We consider in Figure 2.18 the equilibration of two separate pieces of a semiconductor, which are n-doped and p-doped, with a moderate extent of doping. When the junction is formed an interfacial barrier occurs that brings the Fermi level to alignment, much as in Figure 2.17. However, in Figure 2.18 the space charge layer takes considerable space to the interior of the bulk materials, and the band bending is visible. This structure is a p–n junction. Instead of an interfacial dipole layer, we obtain an extended barrier whose height is characterized by the built-in potential, equal to the original difference of Fermi levels. A detailed study of semiconductor junctions, including the effect of the dipole layer, will be presented in Chapter 9.

We have so far described electronic equilibration of materials in contact without a change in the properties of

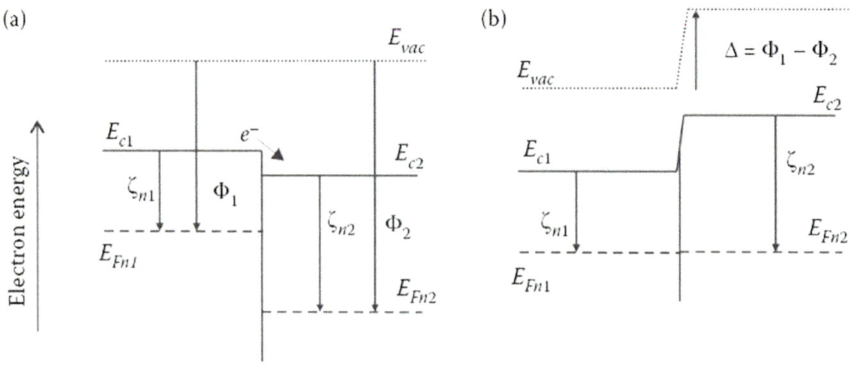

FIGURE 2.17 Energy diagram of two semiconductors in contact in different situations: (a) Nonequilibrium situation. The two materials are not in equilibrium as the electron Fermi level E_{Fn} is higher in the material at the left side. This causes electron transfer toward the material at the right side. (b) Equilibrium situation. The electron transfer causes a rearrangement of the potential difference between the materials that is reflected in the step of the vacuum energy, which equals the difference of the work functions of the materials, and a change in the conduction band offset $E_{c2}-E_{c1}$. Now electrons have the same free energy at both sides of the interface and the net flux across the interface is zero.

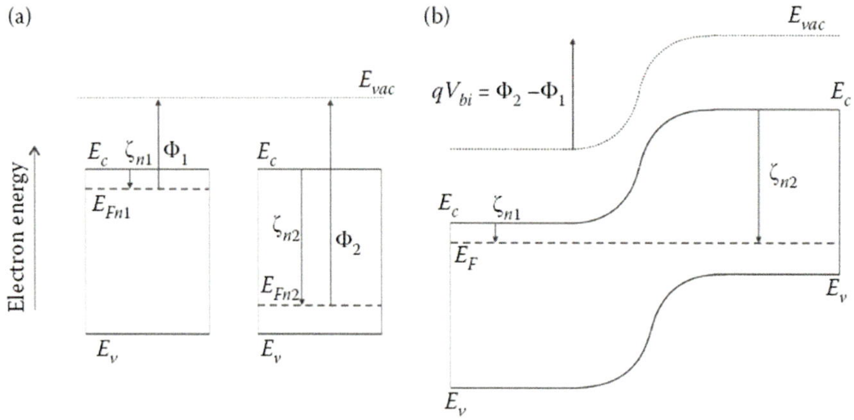

FIGURE 2.18 Energy diagram of two pieces of semiconductor with different doping that form a p–n junction. (a) The left piece is n-doped, and the right one is p-doped. (b) When the materials come to contact, electronic equilibrium is established equalizing the Fermi level by means of a space charge region that produces band bending at the junction.

the parent materials. In the simplest models, we assume that the interface is abrupt, and that bulk properties are maintained right to the interface. As mentioned earlier, the surface has different properties than the bulk. Metal or semiconductor surfaces undergo surface reconstruction. Specific electronic states exist at the surface or interface. In addition to special electronic properties at the surface, the formation of the junction may lead to chemical reaction of the parts in contact, in which case the electronic properties are changed with respect to the bulk materials. In addition, in the preparation of multilayer devices or contacts, interdiffusion may occur that will also alter the chemical and physical properties of the parent materials. Semiconductor and metal surfaces exposed to the ambient atmosphere can rapidly form an oxidized layer that drastically changes the surface and interfacial properties. Semiconductors in contact with electrolytes may corrode under applied voltage. Interfaces can be poised upon electrochemical reaction; ions may be intercalated into an open solid structure when voltage is applied. In general, a multitude of effects modify interfaces by chemical interactions and electronic or ionic transfer. Possible effects that effectively modify interface and junction properties should be carefully checked in the interpretation of specific experiments. For the purpose of general theories, we neglect such effects and assume a stable interface that abruptly separates two phases with well-defined electronic and ionic properties, but nonetheless it is very important to probe the real surface conditions.

GENERAL REFERENCES

Band calculations by methods of solid state physics: Klingshirn (1995), Kittel (2004).

Concept of vacuum level: Guggenheim (1929), Delahay (1982), Hansen and Hansen (1987a), Hansen and Hansen (1987b), Marshak (1989), Ishii et al. (1999).

Electrochemical potential of electrons: Herring and Nichols (1949), Lang and Kohn (1971), Ashcroft and Mermin (1976), Rickert (1982), Reiss (1985). Properties of the semiconductor materials: Pankove (1971), Mönch (1993), Würfel (2005).

Surface dipole: Nonnenmacher et al. (1991), Gensterblum et al. (1994), Bruening et al. (1997), Scott (2003), Flores et al. (2009), Hwang et al. (2009), Topham et al. (2011).

The chemical potential of electrons: Harvey (1962), Rickert (1982), Parrott (1996).

The Volta and Galvani potentials: Guggenheim (1929), Parsons (1974), Jaegermann (1996), Ishii et al. (1999), Fawcett (2004), Tonti et al. (2009).

REFERENCES

Ashcroft, N. W.; Mermin, N. D. *Solid State Physics*; Thomson Learning: Florence, 1976.

Bardeen, J. Theory of the work function, II: The surface double layer. *Physical Review* 1936, *49*, 653–663.

Bruening, M.; Cohen, R.; Guillemoles, J. F.; Moav, T.; Libman, J.; Shanzer, A.; Cahen, D. Simultaneous control of surface potential and wetting of solids with chemisorbed multifunctional ligands. *Journal of the American Chemical Society* 1997, *119*, 5720–5728.

Delahay, P. Photoelectron emission spectroscopy of aqueous solutions. *Accounts of Chemical Research* 1982, *15*, 40–45.

Fawcett, W. R. *Liquids, Solutions and Interfaces*; Oxford University Press: Oxford, 2004.

Flores, F.; Ortega, J.; Vazquez, H. Modelling energy level alignment at organic interfaces and density functional theory. *Physical Chemistry Chemical Physics* 2009, *11*, 8658–8675.

Fröhlich, H. *Theory of Dielectrics*, 2nd ed.; Oxford University Press: Oxford, 1958.

Gensterblum, G.; Hevesi, K.; Han, B. Y.; Yu, L. M.; Pireaux, J. J.; Thiry, P. A.; Caudano, R.; et al. Growth mode and electronic structure of the epitaxial C_{60}(111)/GeS(001) interface. *Physical Review B* 1994, *50*, 11981–11995.

Gerischer, H.; Decker, F.; Scrosati, B. The electronic and ionic contribution to the free energy of Alkali Metals in intercalation compounds. *Journal of the Electrochemical Society* 1994, *141*, 2297–2300.

Gerischer, H.; Ekardt, W. Fermi levels in electrolytes and the absolute scale of redox potentials. *Applied Physics Letters* 1983, *43*, 393–395.

Guggenheim, E. A. The conceptions of electrical potential difference between two phases and the individual activities of ions. *The Journal of Physical Chemistry* 1929, *33*, 842.

Hansen, W. N.; Hansen, G. J. Absolute half-cell potential: A simple direct measurement. *Physical Review A* 1987a, *36*, 1396–1402.

Hansen, W. N.; Hansen, G. J. Reference states for absolute half-cell potentials. *Physical Review Letters* 1987b, *59*, 1049–1052.

Harvey, W. W. The relation between the chemical potential of electrons and energy parameters of the band theory as applied to semiconductors. *Journal of Physics and Chemistry of Solids* 1962, *23*, 1545–1548.

Herring, C.; Nichols, M. H. Thermionic emission. *Reviews of Modern Physics* 1949, *21*, 185–270.

Hwang, J.; Wan, A.; Kahn, A. Energetics of metal/organic interfaces: New experiments and assessment of the field. *Materials Science and Engineering: R: Reports* 2009, *64*, 1–31.

Ishii, H.; Sugiyama, K.; Ito, E.; Seki, K. Energy level alignment and interfacial electronic structures at organic/metal and organic/organic interfaces. *Advanced Materials* 1999, *11*, 605.

Iwamoto, M.; Mizutani, Y.; Sugimura, A. Calculation of the dielectric constant of monolayer films on a material surface. *Physical Review B* 1996, *54*, 8186–8190.

Jaegermann, W. The semiconductor/electrolyte interface: A surface science approach. *Modern Aspects of Electrochemistry* 1996, *30*, 1–186.

Kittel, C. *Introduction to Solid State Physics*; Wiley: New York, 2004.

Klingshirn, C. F. *Semiconductor Optics*; Springer-Verlag: Berlin, 1995.

Lang, N. D.; Kohn, W. Theory of metal surfaces: Work function. *Physical Review B* 1971, *3*, 1215–1223.

Lonergan, M. Charge transport at conjugated polymer-inorganic semiconductor and conjugated polymer-metal interfaces. *Annual Review of Physical Chemistry* 2004, *55*, 257–298.

Makinen, J.; Corbel, C.; Hautojarvi, P.; Mathiot, D. Measurement of positron mobility in Si at 300 K. *Physical Review B* 1991, *43*, 12114–12117.

Marshak, A. H. Modeling semiconductor devices with position-dependent material parameters. *IEEE Transactions on Electron Devices* 1989, *36*, 1764–1772.

Mönch, W. *Semiconductor Surfaces and Interfaces*; Springer: Berlin, 1993.

Nonnenmacher, M.; O'Boyle, M. P.; Wickramasinghe, H. K. Kelvin probe force microscopy. *Applied Physics Letters* 1991, *58*, 2921–2923.

Pankove, J. I. *Optical Processes in Semiconductors*; Prentice-Hall: Englewood Cliffs, 1971.

Parrott, J. E. Thermodynamic theory of transport processes in semiconductors. *IEEE Transactions on Electron Devices* 1996, *43*, 809–826.

Parsons, R. Electrochemical nomenclature. *Pure and Applied Chemistry* 1974, *37*, 501.

Reiss, H. The Fermi level and the redox potential. *The Journal of Physical Chemistry* 1985, *89*, 3783–3791.

Rickert, H. *Electrochemistry of Solids*; Springer Verlag: Berlin, 1982.

Savoie, B. M.; Jackson, N. E.; Marks, T. J.; Ratner, M. A. Reassessing the use of one-electron energetics in the design and characterization of organic photovoltaics. *Physical Chemistry Chemical Physics* 2013, *15*, 4538–4547.

Scott, J. C. Metal—organic interface and charge injection in organic electronic devices. *Journal of Vacuum Science and Technology A: Vacuum, Surfaces, and Films* 2003, *21*, 521–531.

Simmons, J. G. Generalized formula for the electric tunnel effect between similar electrodes separated by a thin insulating film. *Journal of Applied Physics* 1963, *34*, 1793–1803.

Stillinger, F. H.; Ben-Naim, A. Liquid-vapor interface for water. *The Journal of Chemical Physics* 1967, *47*, 4431–4438.

Tonti, D.; Zanoni, R.; Garche, J. *Encyclopedia of Electrochemical Power Sources*; Elsevier: Amsterdam, 2009; pp. 673–695.

Topham, B. J.; Kumar, M.; Soos, Z. G. Profiles of work function shifts and collective charge transfer in submonolayer metal–organic films. *Advanced Functional Materials* 2011, *21*, 1931–1940.

Trasatti, S. The absolute electrode potential: An explanatory note. *Pure and Applied Chemistry* 1986, *58*, 955–966.

Würfel, P. *Physics of Solar Cells: From Principles to New Concepts*; Wiley: Weinheim, 2005.

Zeghbroeck, B. V. *Principles of Semiconductor Devices*; http://ecee.colorado.edu/~bart/book/, 2011.

3 Voltage, Capacitors, and Batteries

This chapter examines the meaning and implications of applied voltage on a device by taking into consideration some important examples such as the capacitor, electrochemical cells in general, and the specific case of the battery. We focus on voltage viewed as a difference of Fermi levels at the contacts, in contrast to an electric potential difference. The notion of capacitance is then introduced, and we treat some fundamental problems of its measurement, which leads to the idea of small perturbation measurement technique. Elementary notions of electrochemistry are discussed, and we describe a variety of concepts like the electrode potential, the redox potential, and the structure of the electrochemical cell that uses a reference electrode for the determination of electrode potentials. We establish the relationship between the energy scale referenced to vacuum level, and the potential scale used in electrochemistry that takes the zero at a reference electrode. The final part is a summary of the principles of electrochemical batteries, with specific examples of practical battery cells and special emphasis on the operation and materials of lithium-ion cells.

3.1 THE VOLTAGE IN THE DEVICE

The contacts to a device allow us to change the applied voltage, inject or extract electric current, control carrier distributions, and govern different device properties, which are required to operate the device. The main property desired for a contact is a very high electronic conductivity; thus, the archetype contact is a thin metal layer as indicated in Figure 3.1a. In general, devices are contacted by a high conductivity material that does not undergo chemical transformation. Such material is often termed the *current collector*. Other common denominations of the contacts are the *electrodes*, *plates*, or *terminals*. The *electrode* is sometimes extended to include active phases in the device, as well as the first layers of electrolyte in contact with the solid.

Nanostructured energy devices usually operate by the combination of several materials. Groupings of different phases are arranged to funnel the charge carriers in a desired direction or provoke required chemical reactions. A widely used type of structure is shown in Figure 3.1b in which the nanostructured semiconductor is contacted at the left electrode, while the surrounding liquid phase is contacted by the right electrode. A blocking layer on the left electrode prevents direct contact of the liquid with the metal, so that the left contact selectively takes electrons from the semiconductor but not from the ionic species in the electrolyte. This structure is realized with mesoporous semiconductors, as shown in Figures 1.1 and 1.2. The porous structure can be easily permeated with electrolyte.

In physical terms, the voltage V measures the work necessary to carry an electron from one plate of the device to the other. This work involves the measurement of an electric current with an electrometer and is given by the difference of electron Fermi levels between the two plates:

$$-qV = E_{FA} - E_{FB} \tag{3.1}$$

In Figure 3.2a, a negative potential at the left electrode indicates that the Fermi energy of the electrons in the left electrode, E_{FA}, is higher by the quantity, $|qV|$, than the Fermi level E_{FB}, at the right electrode. (Note that each Fermi level refers to the whole electron-conducting system of the electrode, including lead wires to the measuring devices.) An alternative denomination of applying a voltage is to *polarize* a device. A *potentiostat* is a voltage source that maintains the applied voltage stable independently of the circulating electric current.

As remarked in Section 2.4, the energy diagram plots electron energies, and energy differences as qV are plotted as arrows. To denote the sign of voltage in the energy diagram, we depict the quantity $-qV$ An arrow pointing upward denotes a negative voltage $V < 0$ or negative electric potential φ (positive electron energy). While we use predominantly the letter V for the voltage applied in a device, and φ (or ψ) for a local electric potential, a voltage is generally synonymous to an electric potential difference. Thus, we also use V_x for characteristic magnitudes with dimension of a potential difference such as the built-in potential, V_{bi}, the Helmholtz potential, V_H, or the band bending at space charge region, V_{sc}.

If the device is a *load* such as a resistor, it consumes electric power. The voltage has to be applied from an external power source, for example, using a battery as indicated in Figure 3.3a. If the device itself produces power such as a battery, or a solar cell, voltage exists even in the absence of any external power source (Figure 3.3b). This is sometimes termed an *electromotive force*, ε, or *emf* (Varney and Fischer, 1980).

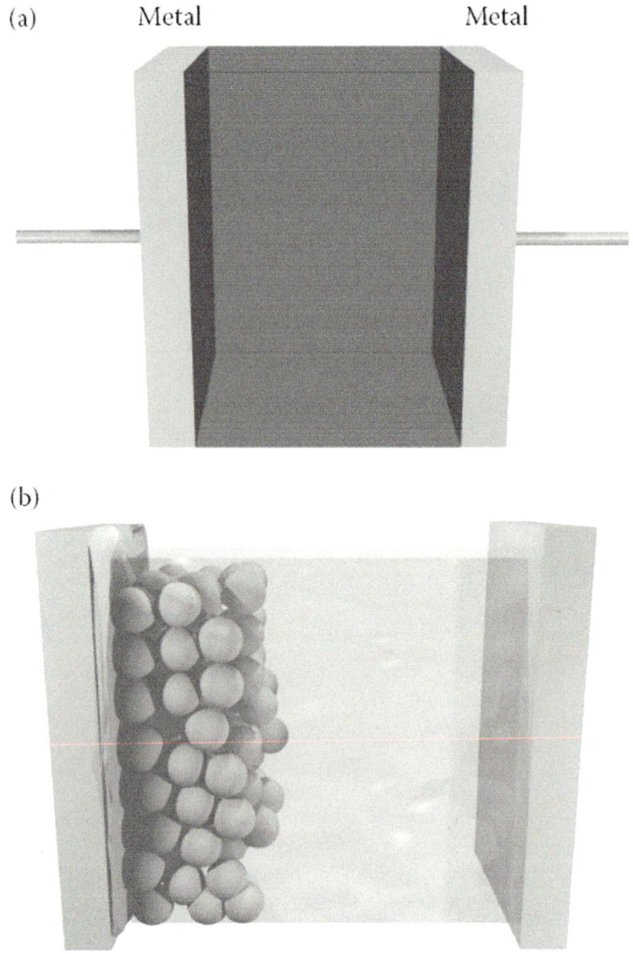

FIGURE 3.1 Basic structure of a device: (a) planar geometry, (b) nanostructured particulate semiconductor contacts a current collector (including a blocking layer) and is surrounded by liquid electrolyte that contacts the right electrode.

3.2 ANODE AND CATHODE

In Figure 3.3a we show that a negative bias voltage is applied to the left electrode, which is termed the *cathode*, and this is where electrons tend to flow inside the device. The effect of positive bias at the left electrode is shown in Figure 3.2b. The positively biased electrode is termed the *anode*. (Note that by convention the direction of a positive electric current flow is that of the positive charges.)

The terms anode and cathode are widely used in practice. However, the two electrodes become anode and cathode as we apply an external potential difference to an electrochemical cell, such as an electrolyzer. In a power source, the appropriate terms would be positive and negative electrode. Nonetheless, in the following paragraphs, we will describe these widely held terminologies.

The denomination of anode and cathode refers to the characterization of electrochemical reactions. Faraday defined anode and cathode in terms of current flow to and from the electrochemical cell, and this convention has been thereafter followed by chemists. Current enters the cell by the anode, and leaves by the cathode (Saslow, 1999). In a chemical reaction, *reduction* corresponds to the addition of electrons, and *oxidation* corresponds to the loss of electrons (since oxygen normally removes electrons when it forms a chemical bond). A *redox reaction* involves electron transfer between two species (a "redox couple"). Each oxidation and reduction reaction in the overall reaction is called a "half reaction." Since electrons flow in the opposite direction to that of current flow, at the *cathode* of the electrochemical cell electrons enter the electrode toward the solution producing a reduction reaction, and at the anode the electrode extracts electrons in *oxidation* reactions at the electrode/solution interface.

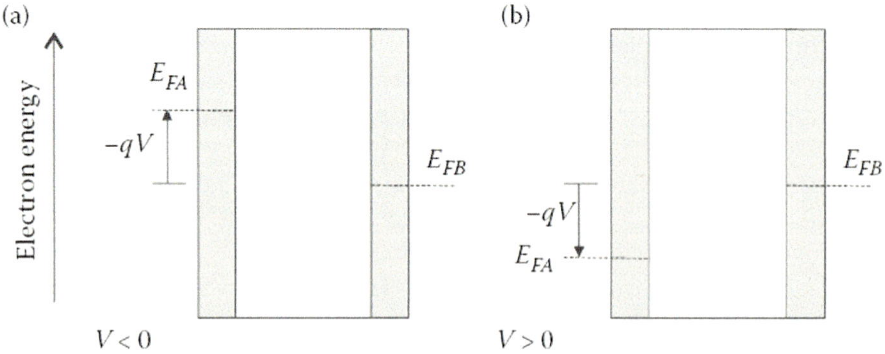

FIGURE 3.2 The voltage applied in the device, V, is proportional to the difference of the electron Fermi level in the left electrode, E_{FA}, and the electron Fermi level in the right electrode, E_{FB}, which is taken as a graphical reference. The vertical scale represents electron energy increasing upwards. (a) Negative bias voltage at the left electrode. (b) Positive bias voltage at the left electrode.

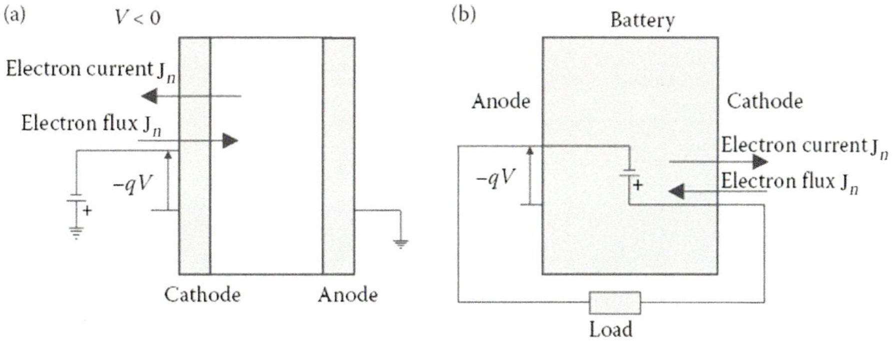

FIGURE 3.3 (a) Negative bias at the left electrode of a diode. Electrons enter the diode at the negative electrode, the cathode. (b) The battery produces negative voltage at the left electrode. Electrons leave the battery through the negative terminal, and enter the battery through the positive terminal, which is the cathode.

However, the relation between the voltage in the contacts and the sense of current flow is not unique, since the sign of current is reversed if the device either consumes or produces electric power. In general, the cathode in a polarized device is the electrode where current flows out of the device (electrons enter the device); see Figure 3.3. Thus,

1. In a device that consumes power as an electrolyzer, the cathode is at negative voltage with respect to the anode.
2. In a device that provides power, the cathode is at positive potential.

In a battery, the sign of the current changes when the battery is being charged or discharged. Irrespective of the current flow, the electrode containing the more electropositive material, that is, with lower work function (e.g., graphite in a lithium-ion cell) is termed the anode.

In the case of solar cells, terminology varies depending on the specific experimental tradition. Solid-state and also organic solar cells are considered as electronic devices and probed as diodes. In the dark, the cathode is the negative voltage terminal and electrons are injected across it to the device bulk. These devices belong to case (*1*). However, other photovoltaic technologies see solar cells as electrochemical devices. This is the case of dye-sensitized solar cells. Here, the terminology follows case (*2*), in which the negative electrode is labeled the photoanode.

3.3 APPLIED VOLTAGE AND POTENTIAL DIFFERENCE

We have emphasized that a voltage in a device is a difference of Fermi levels between the two electrodes. Let us point out the dissimilarity between a voltage between two plates and the difference of electrostatic potentials. Consider two metal plates made of different materials. If we connect them by a wire in which we place a voltmeter, in the absence of any electric power source, the voltage is zero, because the Fermi levels come to equilibrium. However, there is a difference of potentials, the contact potential difference (CPD), as discussed in Figure 2.15. Thus, the voltage in a device defined in Equation 3.1 does not coincide with a difference of potentials.

In general, however, a voltage applied to a device is widely viewed as synonymous to a potential difference between the electrodes. This terminology is perfectly acceptable provided that one is aware of the limitations of the concept of electric potential difference. In general, a difference of potentials between two contacts cannot sustain an external current flow, as mentined above for the CPD, while a difference of Fermi level does. In addition, as discussed in Chapter 2, it is not possible to measure the difference in electric potential between two points, unless those two points are located in the same phase, or in different phases of similar composition. The Galvani potential difference cannot be measured straightforwardly among different phases, since unknown dipole layers exist at the surface, while the Volta potential difference can be measured (Oldham and Myland, 1994).

Figure 3.4 clarifies the connection between measurement of voltage with an electrometer and the associated electrostatic potential difference between the electrodes. Two metal contacts to the device, made of the same material, for example, copper, are shown separately. The contacts are at different Fermi levels, due, for example, to the action of a battery or photovoltaic cell. We first discuss the measurement of the electrostatic potential difference between the two electrodes.

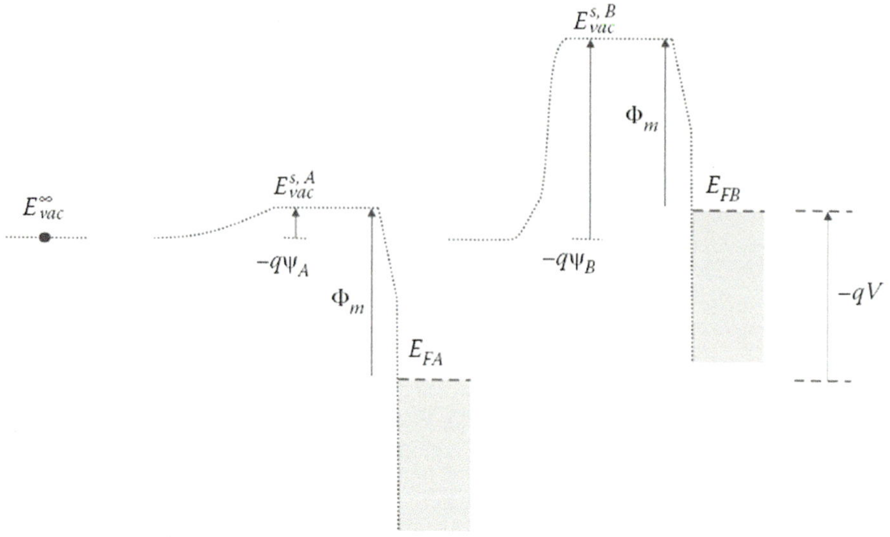

FIGURE 3.4 Energy diagram of potentials, vacuum levels, and Fermi levels, in two metal contacts of the same material.

We take as reference a point in the vacuum far away from both contacts. Consider that an electronic gas at infinity has zero energy. A quantity of electrons Δn is brought isothermally from infinite at level E^{∞}_{vac} to just outside the surface at level $E^{s,A}_{vac}$. The potential difference between the point far away and the point just outside the surface can be measured: both points are in the same medium, the vacuum, and the measured value is in fact the Volta potential ψ_A. The work required is given by $W = -q\Delta n\psi_A$. The same measurement can be done for the point at the other electrode surface $E^{s,B}_{vac}$. The work is a conservative quantity; so the work to bring Δn electrons from the surface of one contact to infinity, and to bring these electrons from infinite to the other contact surface, gives the work to take the electrons from one contact to the other. Therefore, $W = -q\Delta\psi$ can be measured. Let us show that this quantity of work coincides with Equation 3.1.

Suppose that the Δn electrons at the surface $E^{s,A}_{vac}$ level are brought inside the material in equilibrium at the Fermi level E_{FA}. The work associated with this process is given by the work function, $W = \Delta n\ \Phi$. The total work to add these electrons from the reference point at infinite, to a point inside the device, at equilibrium with the metal contact, is $W = E_{FA}\Delta n$. The total work to bring the electrons from one contact to the other is $W = (E_{FA} - E_{FB})\Delta n$, which is measured by the electrometer as indicated in Equation 3.1. The work functions in the two contacts are the same, since the contacts are made of the same material. If the contacts Fermi levels stand at different position of the energy axis, the same difference exists between the Volta potentials, that is, $\Delta E_F = -q\Delta\psi$.

On the vacuum side, $V = \Delta\psi$ is an electrostatic potential difference that corresponds to the work to take an electron from one contact to the other. Inside the device, $-qV$ is a difference of electrochemical potential of the electrons corresponding to take the electron from one contact to the other.

The previous arguments do not work if the contact materials are different, due to the disparity of Volta and Galvani potentials. As stated earlier, the voltmeter measures the difference of electrochemical potential of the two terminals. If these terminals are of the same material, say Cu, then the difference of Fermi levels in Equation 3.1 can be identified with a difference of the electric potential inside each copper wire (the Galvani potential), so that

$$V = -\frac{E_{FA} - E_{FB}}{q} = \Delta\varphi \tag{3.2}$$

However, with the voltmeter we cannot determine the potential difference between pieces of different metals, as shown in Figure 3.5. The contacts Cu/Ag and Cu/Au are separately equilibrated with respect to electrochemical potentials. Then, the voltmeter provides the difference of electrochemical potentials (Fermi levels) of electrons between Ag and Au, though not the electric potential difference $\varphi_{Ag} - \varphi_{Au}$. There are different CPD at the two interfaces Cu/Ag and Cu/Au, and we cannot determine the difference of potentials between the interior of the gold and the silver pieces, with just a voltmeter. The experimental methods that are used to determine the work function and surface dipole are described in Chapter 4.

Nevertheless, even if the contacts consist of different materials, a *modification* of the voltage applied to

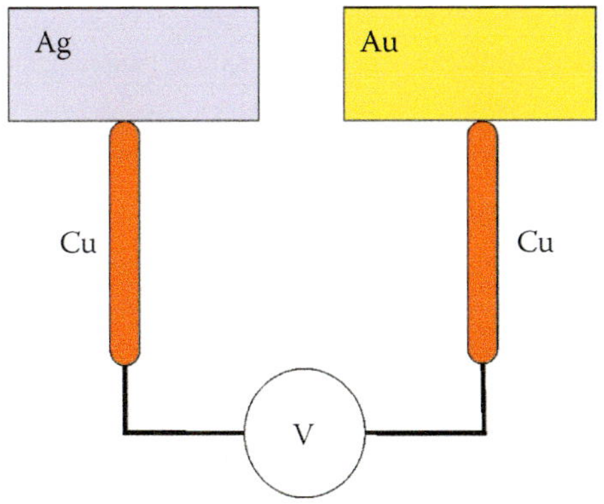

FIGURE 3.5 Measurement of voltage between pieces of silver and gold, using a voltmeter that has copper terminals.

the device (difference of Fermi levels between the electrodes) causes an equal *modification* of the electrostatic potential difference between the contacts (the Volta potential difference $\Delta\psi$). This remark follows from the fact that the contact is a metal in which the chemical potential of electrons cannot be modified. Therefore, upon application of voltage, electrochemical potential and electrostatic potential must be changed consistently at each contact. The applied voltage produces a corresponding difference of potentials *in the metal electrodes*.

3.4 THE CAPACITOR

Two metal plates will form a CPD depending on the net electric charge at each separate plate. This problem can be solved fully by electrostatics. Due to the mutual repulsion between charges of the same sign, the charge on each conductor resides in a thin surface layer. Due to the attraction of charges of opposite signs, the charges are located on the surfaces of the plates that face each other. The charge storage with respect to potential difference is measured by the dielectric capacitance.

Figure 3.6a shows two plates made of the same material separated by a dielectric medium (vacuum or air). In the absence of bias voltage, there is no CPD between the plates (Figure 3.6a). An applied voltage V_{app} changes the position of the Fermi levels and establishes a CPD between the plates (a Volta potential difference) (Figure 3.6b), of value

$$\Delta\varphi = \varphi_1 - \varphi_2 = V_{app} \tag{3.3}$$

The change of potential difference occurs by transference of electrons to the left plate that leaves behind a defect of negative charge that causes positive charging. As the metal is equipotential, and if there is no interfacial potential drop at the metal/vacuum interface, the potential difference creates an electric field between the two plates situated at a distance d, given by

$$\mathbf{E} = \frac{\Delta\varphi}{d} \tag{3.4}$$

The *capacitance* of two isolated conductors that can exchange electric charge is defined as

$$C = \frac{Q}{\Delta\varphi} \tag{3.5}$$

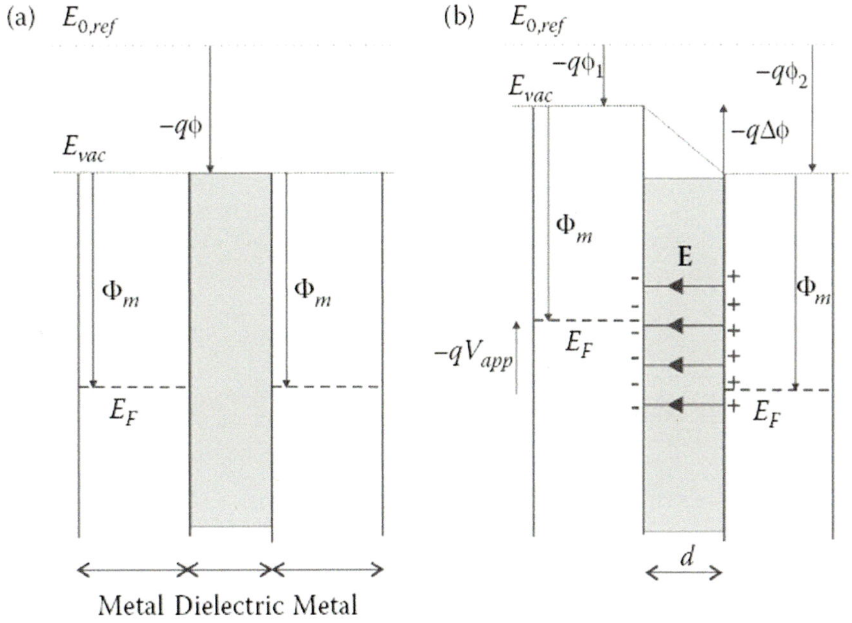

FIGURE 3.6 A capacitor formed by two plates of the same metal: (a) zero voltage, (b) the left plate in biased negatively.

By Equation 2.48, we obtain for the case of Figure 3.6

$$C = A\frac{\varepsilon}{d} \tag{3.6}$$

where $\varepsilon = \varepsilon_r\varepsilon_0$. The unit of the capacitance is called the farad (F). The plates of a capacitor of 1 F carry a charge of 1 C if their potential difference is 1 V.

We now consider two electrodes made of metals with different work functions Φ_i as shown in Figure 3.7a. We apply a voltage in this system as shown in Figure 3.7b. The Fermi levels in the two plates are given by

$$E_{F1} = -q\varphi_1 - \Phi_1 \tag{3.7}$$

$$E_{F2} = -q\varphi_2 - \Phi_2 \tag{3.8}$$

The bias voltage relates to the potentials in the plates as

$$-qV_{app} = E_{F1} - E_{F2} = -q(\varphi_1 - \varphi_2) - (\Phi_1 - \Phi_2) \tag{3.9}$$

The built-in potential is $V_{bi} = \varphi_1^{eq} - \varphi_2^{eq} = (\Phi_2 - \Phi_1)/q$. Therefore,

$$V_{app} = \varphi_1 - \varphi_2 - V_{bi} \tag{3.10}$$

where the *built-in potential*, V_{bi}, that corresponds to the contact potential difference was described in Chapter 2 in terms of the difference of Volta potentials in equilibrium existing when Fermi levels are aligned, $V_{bi} = \Delta\psi_{CPD} = \psi_1^{eq} - \psi_2^{eq}$.

In this problem we find that the capacitor is charged when the system is unbiased, which is a result of the CPD discussed earlier. The charge in the plates relates to the applied voltage as

$$V_{app} = \frac{Q}{C} - V_{bi} \tag{3.11}$$

where C is given by Equation 3.6. If the left plate is negatively biased to cancel the built-in potential, then there is no charge in the plates and no electric field across them. The potential is flat, and this voltage recovers the situation in which the materials are disconnected, as in Figure 3.7a.

3.5 MEASUREMENT OF THE CAPACITANCE

According to Equation 3.5, the measurement of the capacitance requires a determination of the total charge stored in the capacitor at a given potential difference

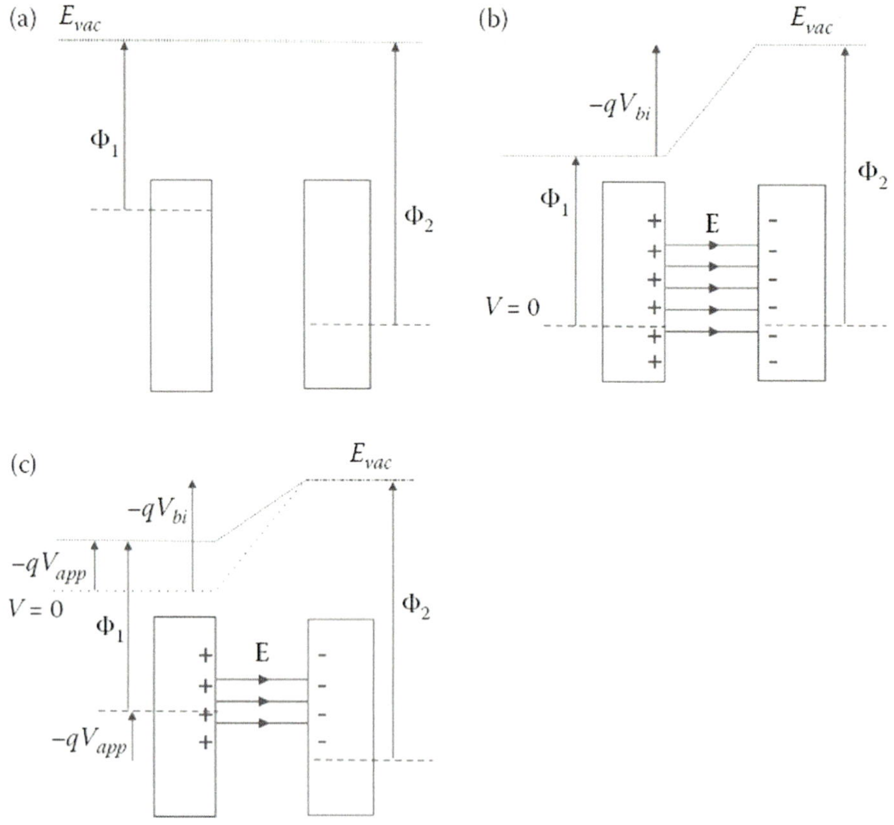

FIGURE 3.7 A capacitor formed by two plates of different material: (a) the separate materials, (b) the capacitor at zero voltage, (c) the capacitor biased negatively at the left contact.

between the plates. However, rather than measuring stored electric charge, it is easier to measure currents with an electrometer. As the current for a change of the charge Q is

$$I = \frac{dQ}{dt} \tag{3.12}$$

we can obtain the charge by integration of the current

$$Q = \int_0^t I dt \tag{3.13}$$

Note that the charge cannot be measured at steady state, in which no current flows, since the capacitor is an insulator. Therefore, a transient measurement is necessary to determine the capacitance. One needs to apply a probe voltage that induces the charging process. The time-varying voltage produces a changing electric field between the plates, and there is an electric current density that does not involve the motion of charges. It is a displacement current, which according to Maxwell equations is given by the expression

$$j = \frac{\partial D}{\partial t} \tag{3.14}$$

where the dielectric displacement D is related to the electric field as

$$D = \varepsilon \mathbf{E} \tag{3.15}$$

The displacement current only flows across the plates in transient conditions.

Capacitors involving displacement current will be here called a *dielectric capacitor*, to distinguish from a *chemical capacitance* and *electrochemical capacitance* that are charged by conduction currents (Chapter 7). The dielectric capacitor can have conduction current as well, in fact most capacitors do show some degree of current loss across the dielectric material. Still, the current across the plates, related to the charging of the plates, is a displacement current.

The battery is a device that primarily stores and delivers energy, while having a large discharge rate is not a primary consideration, except for specific applications. On the contrary, an ordinary capacitor stores a limited amount of energy but the discharge is fast. Supercapacitors are electrochemical devices that possess a very large capacitance of mF or higher, and deliver the energy at a large rate since the main finality of the device is to provide high electric power.

We continue the analysis of the ordinary dielectric capacitor. To measure the capacitance, we apply a time-dependent perturbation and analyze the response. We denote a small perturbation value of a quantity A by a hat, $\hat{A}$, and the value at steady state is denoted by an overbar, $\bar{A}$. The most convenient perturbation is a harmonic function of frequency f and angular frequency $\omega = 2\pi f$. The variable perturbation maintains the displacement current over time, so far as the frequency is larger than zero. The case $\omega = 0$ is called dc conduction and $\omega > 0$ is the ac conduction. The relationship of input/output signals is an impedance, Z.

The capacitor is set at a given voltage $\bar{V}$, which is unchanging with time (the *steady-state condition*). This voltage causes a steady polarization but has no dielectric current. Over $\bar{V}$ we apply a small oscillating voltage ΔV of amplitude $\hat{V}$ that enables us to probe the properties of the steady state at a specific $\bar{V}$. The total voltage applied by the power source is

$$V = \bar{V} + \Delta V(t) \tag{3.16}$$

where the time-dependent small voltage is

$$\Delta V(t) = \hat{V}_{in} \cos(\omega t) \tag{3.17}$$

We have shown earlier that the voltage applied to the capacitor plates changes the electrostatic potential difference between the plates. By the differential of Equation 3.10, we obtain for the small perturbation

$$\hat{V}_{in} = \Delta\hat{\varphi} \tag{3.18}$$

We observe that the background potential V_{bi} in Equation 3.11 is included in $\bar{V}$ but makes no contribution to $\hat{V}$, because V_{bi} is a constant that cannot oscillate. In conclusion, we have the value of the perturbation of the dielectric displacement

$$\hat{D} = \frac{\varepsilon}{d}\hat{V}_{in} \tag{3.19}$$

Therefore, the current resulting from the oscillating perturbation is

$$\Delta j = -\frac{\varepsilon\omega}{d}\hat{V}_{in} \sin(\omega t) \tag{3.20}$$

There are two important aspects to this result. The first is that the magnitude of the current increases with the frequency. Just by the faster variation of the electric field in time, the displacement current is larger. We also remark that the capacitor, being an insulator, blocks all electric current at zero frequency. However, it conducts very efficiently at high frequency.

The second important aspect of the current is that it is delayed a quarter period from the voltage, as we can see by writing Equation 3.20 in the form

$$\Delta j = -\frac{\varepsilon\omega}{d}\hat{V}_{in}\cos\left(\omega t + \frac{\pi}{2}\right) \quad (3.21)$$

and comparing it with Equation 3.17. This delay reflects the fact that the capacitor stores and releases energy without dissipation, as the average of the power

$$S_{el} = \hat{j}\hat{V} \quad (3.22)$$

over one cycle is zero. To avoid the use of trigonometric functions, we express the voltage perturbation as

$$\Delta V = \hat{V}_{in}e^{i\omega t} \quad (3.23)$$

Then, the current takes the form

$$\Delta j = -\frac{\varepsilon i\omega\hat{V}_{in}}{d}e^{i\omega t} = \hat{j}e^{i\omega t} \quad (3.24)$$

and the phase shift is represented in the factor i (the complex number). The impedance is defined as

$$Z = -\frac{\hat{V}}{\hat{j}} \quad (3.25)$$

For the capacitor, we obtain

$$Z = \frac{d}{i\omega\varepsilon} = \frac{1}{i\omega C} \quad (3.26)$$

where C is the capacitance per unit area. We see that the impedance is infinite at dc and decreases at large frequency, reflecting the conduction properties of the capacitor that have been explained above.

The admittance is defined as

$$Y = \frac{1}{Z} \quad (3.27)$$

therefore,

$$Y = i\omega C \quad (3.28)$$

The capacitance that is derived from the small perturbation impedance measurement only counts the change of charge in the capacitor with respect to the change of voltage. This is called a *differential capacitance*, as discussed in Section 7.1.

3.6 ENERGY STORAGE IN THE CAPACITOR

In general, the free energy F stored in a capacitor, when the amount of charge dQ is brought at the voltage V, is

$$dF = VdQ = CV\,dV \quad (3.29)$$

and the total free energy obtained by charging the capacitor is

$$F = \int_0^V CV\,dV \quad (3.30)$$

If the capacitance is independent of the voltage, we obtain

$$F = \frac{1}{2}CV^2 \quad (3.31)$$

For the capacitance in Equation 3.6, the stored energy is

$$F = \frac{1}{2}\left(\frac{A\varepsilon}{d}\right)(\mathbf{E}d)^2 = (Ad)\frac{1}{2}\varepsilon\mathbf{E}^2 \quad (3.32)$$

From electromagnetic theory, it is known that if the electric displacement is increased by an amount dD, then the influx of energy into a dielectric is (Fröhlich, 1958)

$$dU = \mathbf{E}dD \quad (3.33)$$

Integration gives the total stored energy per unit volume

$$U = \frac{1}{2}\varepsilon\mathbf{E}^2 \quad (3.34)$$

We observe that the energy in a dielectric capacitor is stored as spatial energy density in the electric field between the plates.

3.7 ELECTROCHEMICAL SYSTEMS: STRUCTURE OF THE METAL/SOLUTION INTERFACE

Electrochemistry is the study of processes at the interface between an electronic conductor and an ionic conductor (the electrolyte). The combination of these two elements is called the *electrode*. These studies involve a wide range of systems including metals, semiconductors (organic and inorganic), and even at times insulators, as the electronic conductor, and solutions, molten salts (ionic liquids), and solid electrolytes as ionic conductors.

Reactions involving the charge transfer at the electrode interface are called electrochemical reactions. The current in the electrochemical cell has the particular characteristic that the current is carried by electronic carriers in the metal (or semiconductor) part of the electrodes, and by ionic carriers in the electrolyte. Thus, the electrochemical reaction allows changing from electronic to ion conduction in the electrochemical and photoelectrochemical devices. Important systems for energy conversion such as the lithium-ion battery (LIB) are based on mixed ionic/electronic conductors that function as electrodes.

Highly concentrated electrolytes possess the ability to shield electric fields, because electroneutrality must prevail at length scales longer than the Debye screening length (see Section 5.7). Therefore, effective spatial charge separation occurs only at the boundaries of the liquid electrolyte. However, as discussed in Chapter 2, the excess charge in a metal is also located at the surface. Thus, any step of the vacuum level in the metal/electrolyte system occurs predominantly at the metal/solution interface. The contact is characterized by the buildup of charge separation and a consequent potential difference across the interface. For example, in Figure 3.8a negative electronic charge lays at the surface of the metal, and a compensating sheet of positive ionic charge lays in the solution side that is called the *Helmholtz layer.* This situation can be more complicated, especially if the concentration of ions is low, because the charge can be distributed along a few Debye lengths from the surface, and it will form a thermal distribution of anions and cations (the *Gouy–Chapman layer*). In general, all the region of space charge compensating that at the metal surface is termed the *electrical double layer.*

For highly concentrated solutions, the main voltage drop across the interface is the potential difference at the Helmholtz layer V_H. V_H is maintained by two sheets of charge separated a distance d. In the solution side, the charges originate from ions that are attracted to the surface until the distance of closest approach (the distance of the ionic center from the metal surface), the *outer Helmholtz plane*, including specifically adsorbed ions. On the solid side of the double layer, the charge arises from accumulation of free electronic charge (electrons or holes) and electronic charge trapped in surface states. V_H is defined as

$$V_H = \varphi_s - \varphi_{el} \tag{3.35}$$

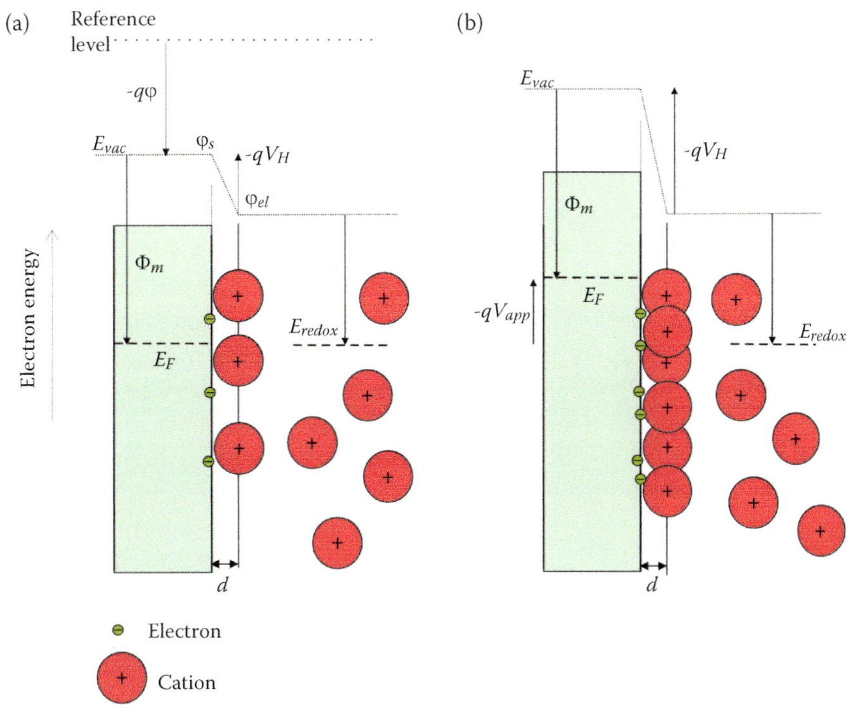

FIGURE 3.8 Structure of the Helmholtz layer at the contact between the metal electrode and an electrolyte. (a) At zero applied potential, a contact potential difference is built (that accommodates the difference between the metal work function and the redox energy of the electrolyte), consisting of a voltage drop V_H, across the interfacial dipole formed by electrons at the metal surface and ions at the solution side. (b) With a negative applied potential, the amount of positive and negative charge at the surface increases, and so does the interfacial potential drop V_H.

where φ_s is the potential at the solid surface and φ_{el} is the potential at the outer Helmholtz (see Figure 3.8a). In the example shown in the figure, V_H is negative.

The Helmholtz layer gives place to an interfacial capacitance, C_H, the Helmholtz capacitance, that is totally dielectric:

$$C_H = \frac{\varepsilon_r \varepsilon_0}{d} \tag{3.36}$$

The Helmholtz capacitance is strongly influenced by the dielectric properties of the solvent, represented by the local dielectric constant at the interface ε_r (see Equation 2.53). Usually, C_H can be assumed constant so that the applied potential changes the amount of charge at the interface, and V_H will vary linearly with the amount of charge as

$$V_H = \frac{qc_s}{C_H} \tag{3.37}$$

where c_s is the surface concentration of ionic (or electronic) charge. This situation is indicated in Figure 3.8b. The voltage at which the surface is uncharged is called the *potential of zero charge*. The value of C_H is of the order of 20 μF cm^{-2} away from the potential of zero charge, and depends strongly on the size of ions attracted to the metal surface.

There are two ways in which current is carried across the metal–electrolyte system, and according to Grahame (1952) these are called the faradaic and nonfaradaic paths. In the faradaic pathway, the current crosses the interface by virtue of an electrochemical reaction such as the reduction or oxidation of ionic species. In nonfaradaic current, charged particles do not cross the interface; the current is carried by the charging and discharging of the interface capacitor that is formed by the Helmholtz layer. It is important to remark that a process that allows a continuous current to flow must be regarded as faradaic. Conversely, capacitive current cannot give dc current, because the current finishes when the capacitor is charged at the given potential.

3.8 ELECTRODE POTENTIAL AND REFERENCE ELECTRODES

Liquid electrolytes are usually electronic insulators, so that electronic states in the electrolyte that can exchange electrons with the surrounding exist only in ionic species that may have different oxidation states. Such species can diffuse in the liquid medium and exchange electrons with the metal electrode by the electrochemical reaction consisting of electron transfer at the electrode interface. The electrolyte has a Fermi level, which is the electrochemical potential of electrons in redox species, that is denoted the redox potential E_{redox}.

The voltage between two metal electrodes immersed in electrolyte in an electrochemical cell can be measured easily, and has the same meaning as that in Equation 3.1. However, in electrochemistry it is especially important to ascertain the local difference of Fermi levels between the metal part of the electrode and the electrolyte *at a single electrode*. This difference drives the electrochemical reaction and further carrier transport. The electronic conducting part of the electrode takes a definite value of the electron Fermi level, which is known as *the electrode potential*. When an electrolyte containing reduction and oxidation species (a redox couple) comes in contact with a metal, the electrode potential corresponds to the equilibrium established between the Fermi level of the electrons in the electrode and the redox level (see Figure 3.9a). The electrode potential is denoted E or U in the literature of electrochemistry but here we use V for the potential at one electrode in electrochemical measurements, as it can be defined in terms of a difference of Fermi levels (see Equation 3.38).

The electrode potential of one electrode alone is denoted as a *half-cell potential*. The absolute electrode potential has the meaning of absolute electron energies in the ionic species in solution with respect to electron in vacuum just outside the solution/vacuum interface, which is the standard thermodynamic reference just as in the definition of the work function for solids discussed in Chapter 2 (Hansen and Hansen, 1987a, b). Indeed, with a suitable change of energy scale discussed in Section 3.10, a half-cell potential corresponds to the electron Fermi level of the electrode with respect to electron at rest in vacuum. However, in electrochemistry the absolute scale of electron energies is not generally applied. The electrolyte interface introduces a great uncertainty about the value of the surface dipole, and the absolute electron energy measurement is not usually practical for liquid systems.

A specific tool is introduced for the measurement of the voltage of an electrode with respect to a reference of potential in the electrolyte. While for solids E_F is considered an "absolute" value with respect to E_{vac}, the electrochemical potentials of electrolyte solutions (electrode potentials) are measured *relative to one particular electrode* denoted as the reference electrode (RE). The measurement of electrode potential can then be made simply with a voltmeter.

The electrochemical cell has two electrodes as shown in Figure 3.9b: the working electrode (WE), which is the

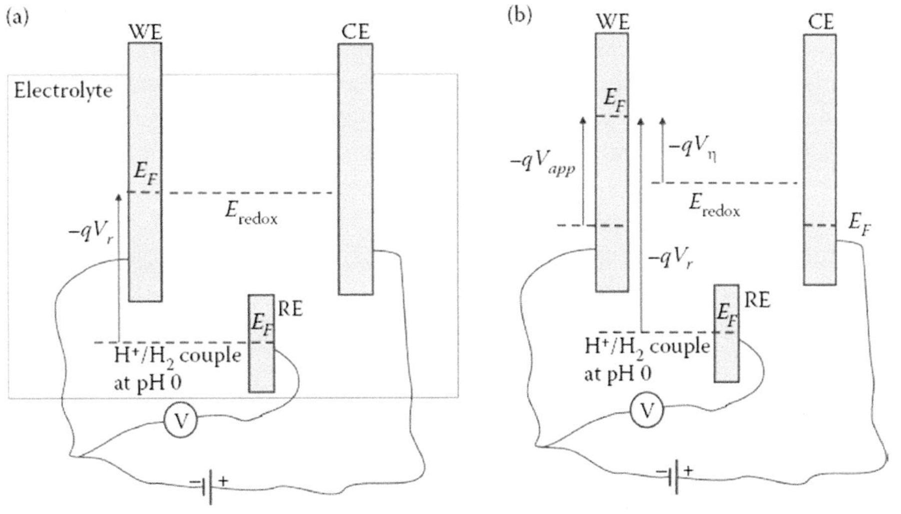

FIGURE 3.9 Scheme of an electrochemical cell, consisting of the working electrode (WE), counterelectrode (CE), and reference electrode (RE). The potential V_r is measured with respect to a reference electrode in solution, which in this case is the SHE defined in terms of the H^+/H_2 couple at pH = 0. The voltage V_r is a difference of Fermi levels between the Fermi level of electrons in the metal WE, E_F, and the Fermi level of the metallic contact to the SHE. (a) Measurement of the redox potential, $V_r = V_{redox}$. (b) Determination of the overpotential at the WE when the cell is under applied voltage. The overpotential V_η is the difference between the potential of the Fermi level in the metal WE and the equilibrium (redox) potential V_{redox} (indicated as a redox energy level $E_{redox} = -qV_{redox}$). The applied voltage V_{app} is related to the difference of Fermi levels between WE and CE.

one we wish to measure, and the counterelectrode (CE). In addition, the RE allows us to determine the electrode potential, termed here V_r, that is the voltage of the WE with respect to the RE (Figure 3.9a). It corresponds to energy difference between the Fermi level of electrons in the WE, E_F^{WE}, and the Fermi level of the metallic contact to the RE:

$$V_r = -\frac{1}{q}\left(E_F^{WE} - E_F^{RE}\right) \tag{3.38}$$

The applied potential in the electrochemical cell (Figure 3.9b) is the voltage between the WE and CE. Thus,

$$V_{app} = -\frac{1}{q}\left(E_F^{WE} - E_F^{CE}\right) \tag{3.39}$$

Note that V_{app} controls the current density j that flows across the cell. The RE is built in such a way that it takes no part in energy transference in terms of current (no current flows between WE and RE), but allows to know the potential of the working electrode (WE), that is the position of the Fermi level of electrons, E_F^{WE}, with respect to a reference that does not change during operation of the cell.

As commented by Trasatti (1986), "electrode potential" is often misinterpreted as the electric potential difference between a point in the bulk of the solid conductor and a point in the bulk of the electrolyte solution. In reality, the electrode potential also takes into account the change of chemical forces in the transference of one electron across the two phases, and therefore corresponds to a free energy difference, measured by the relative Fermi level of the electrode with respect to that of the reference electrode.

Electrode potentials in electrochemical cells are given with respect to a standard reference electrode that has been adopted by convention. The standard electrode consists of H_2 gas at a pressure of 1 atmosphere, bubbled over platinum in an electrolyte in which the activity of hydrogen *ions* (not atoms or molecules) is unity. Such specific concentration of hydrogen corresponds to water at pH 0. This electrode is known as the *standard hydrogen electrode*, or SHE. It is sometimes called the *normal hydrogen electrode* (NHE), but this denomination is obsolete, as a normal solution is not necessary in the standard state. The reaction is given by

$$H_2(g) \leftrightarrow 2H^+ + 2e^- \tag{3.40}$$

Provided that H^+ is solvated in the bulk of the solution, the free energies of $H_2(g)$ and H^+ are independent of the metal used as electrode, and the equilibrium of the reaction (3.40) completely determines the Fermi level of the metal electrode. The potential of the reaction (3.40) is zero by definition of the electrochemical scale of potentials, and all other electrode potentials are given with respect to this one. The electrochemical scale is shown at the right in Figure 2.7. Several different half-cell potentials are shown in the right part of the figure.

We have emphasized the electronic equilibration at the electrode interface for the measurement of electrode potential. It should also be pointed out that in electrochemical cells, it may be the *ions* that come to equilibrium across different phases (the electronic conductors and electrolyte). Then it is the electrochemical potential of ions that is homogeneous across the cell, as in the LIB discussed in Section 5.10.

3.9 REDOX POTENTIAL IN ELECTROCHEMICAL CELLS

For an ionic species with charge $z_k q$ in a dilute solution, the electrochemical potential can be expressed as in Equation 2.29

$$\eta_k = z_k q \varphi + \mu_k \tag{3.41}$$

where φ is the Galvani potential in the interior of the phase. The chemical potential has the form

$$\mu_n = \mu_n^0 + k_B T \ln c_k \tag{3.42}$$

Here, c_k is the volume density of the species (cm^{-3}) and μ_n^0 is a constant independent of the phase composition, called the *standard chemical potential*. For a solution the standard state is a 1M solution in the absence of particle–particle interactions, which in practice is approximated by a solution at infinite dilution (infinite solute particle–particle distance). If we express the concentration in terms of the molar concentration $\bar{c}_k = 10^3\, c_k/N_A$ (10^3 mol m^{-3}), then the chemical potential on a molar basis writes

$$\mu_n = \mu_n^{'0} + \frac{RT}{N_A} \ln \bar{c}_k \tag{3.43}$$

where $R = N_A k_B$.

If the statistics of each substance in a blend obeys the Boltzmann distribution as in Equation 3.42 it is called an "ideal solution." For concentrated and interacting systems, Equation 3.42, cannot describe well the chemical potential of one component. However, in general the chemical potential is expressed in terms of a dimensionless quantity, the *activity* a_k, as

$$\mu_n = \mu_n^0 + k_B T \ln a_k \tag{3.44}$$

The activity is a function of the concentration c_k and depends on the standard state chosen, which is taken either as the pure substance or as a state of infinite dilution. For a gaseous species, the standard state is the pure gas at a pressure $p^0 = 10^5$ Pa (1 bar). The activity is proportional to the partial pressure p_k, hence it is given by

$$a_k = \frac{p_k}{p^0} \tag{3.45}$$

For a species in the solid or liquid state, the standard state with $a_k = 1$ is the pure phase. For a solute, the activity is proportional to concentration c_k. The standard concentration is $\bar{c}^0 = 10^3$ mol m^{-3}; hence,

$$a_i = \frac{\bar{c}_i}{\bar{c}^0} \tag{3.46}$$

The departure from ideality is given in terms of the *activity coefficient* γ_i as follows:

$$a_i = \frac{\bar{c}_i}{\bar{c}^0} \gamma_i \tag{3.47}$$

An electrolyte that contains both forms of a redox couple, oxidized *ox* and reduced *red* species, can be viewed as a set of electronic states, of which *ox* are the empty levels and *red* the occupied levels. The electrochemical potential of the redox couple is called the redox potential. When a metal electrode is immersed in the electrolyte, there occurs charge transfer across the metal surface until the electrode potential equilibrates to the redox potential, as mentioned earlier. For a reaction with a number z of electrons transferred per molecule to the metal electrode

$$\mathrm{OX(S)} + Ze^-(\mathrm{m}) \rightarrow \mathrm{red(S)} \tag{3.48}$$

the electrode potential is given by the Nernst equation

$$V_{redox} = V_{redox}^0 + \frac{k_B T}{zq} \ln \frac{c_{ox}}{c_{red}} \tag{3.49}$$

where V_{redox}^0 is the *formal potential*. The redox energy of the electrolyte is

$$\begin{aligned} E_{redox} &= -qV_{redox} \\ &= E_{redox}^0 + \frac{k_B T}{z} \ln \frac{c_{red}}{c_{ox}} \end{aligned} \tag{3.50}$$

where $E_{redox}^0 = -qV_{redox}^0$ is the energy of the redox couple at $c_{ox} = c_{red}$, with respect to the RE. Redox potentials obtained from electrochemical measurement are listed in standard tables with respect to SHE, but in practice the electrode potential is often reported with respect to the reference electrode that was actually used as the reference. A silver-silver chloride electrode (Ag/AgCl), for example, has a potential $V = +0.222$ V positive with respect to SHE. For the saturated calomel electrode (SCE) $V = +0.244$ V versus SHE. A few REs are shown in Figure 2.7.

Let us consider a one-electron-transfer redox couple. At room temperature, Equation 3.50 can be also written as

$$E_{redox} = E^0_{redox} + (0.059\,\text{eV})\log\frac{c_{red}}{c_{ox}} \qquad (3.51)$$

If we increase c_{red} 10-fold, leaving c_{ox} unchanged, then the energy of the redox couple becomes more positive by 0.059 eV, and the voltage of the redox couple with respect to any RE becomes more negative by 0.059 V. Therefore, we say that the voltage (or energy) changes 0.059 V per decade of concentration.

The total concentration of redox species in solution is

$$c_{tot} = c_{ox} + c_{red} \qquad (3.52)$$

Therefore, Equation 3.50 gives

$$E_{redox} = E^0_{redox} + \frac{k_B T}{z}\ln\frac{c_{red}}{c_{tot} - c_{red}} \qquad (3.53)$$

This is exactly the form of Fermi–Dirac statistics (see Chapter 5), so that the redox energy is often called the Fermi level of the ions in solution (Gerischer and Ekardt, 1983; Reiss, 1985).

3.10 ELECTROCHEMICAL AND PHYSICAL SCALES OF ELECTRON ENERGY IN MATERIAL SYSTEMS

So far, we have introduced two different energy scales for electrons in materials and molecules: the absolute scale with respect to the energy of an electron in rest in vacuum, and the energies of electrons in redox systems, and in general the electrode potentials, given in a relative scale with respect to a reference electrode, which for protic systems is the SHE. A precise relationship between these two scales is required for many applications, for instance, to evaluate the energy levels of reactants in solution with respect to the energy levels of electrons in a metal or semiconductor.

The difference between the absolute scale, and SHE scale, is the energy E_0^{SHE} of the hydrated proton at unit activity, with respect to the electron in vacuum close to the solution surface (Trasatti, 1986), which is about $E_0^{SHE} = -4.5\,\text{V}$. Consider the measurement of a specific energy level in an electrochemical system, E, with respect to the surface vacuum level, that is taken as reference (see Figure 3.10a). The same energy level can be measured, $-qV_r$, as an electrode potential V_r with respect to an RE, as the SHE shown in Figure 3.10a. Obviously, the values of both measurements differ by E_0^{SHE}. Thus, the conversion from the one-electron absolute energy scale, to the electrochemical SHE scale is given by the expression

$$E(\text{abs,eV})E_0^{SHE} - qV_r \qquad (3.54)$$

Different values of E_0^{SHE} have been proposed, ranging from −4.44 to −4.85 eV. The origin of the electrochemical scale is usually taken at

$$E_0^{SHE} = -4.44\,\text{eV} \qquad (3.55)$$

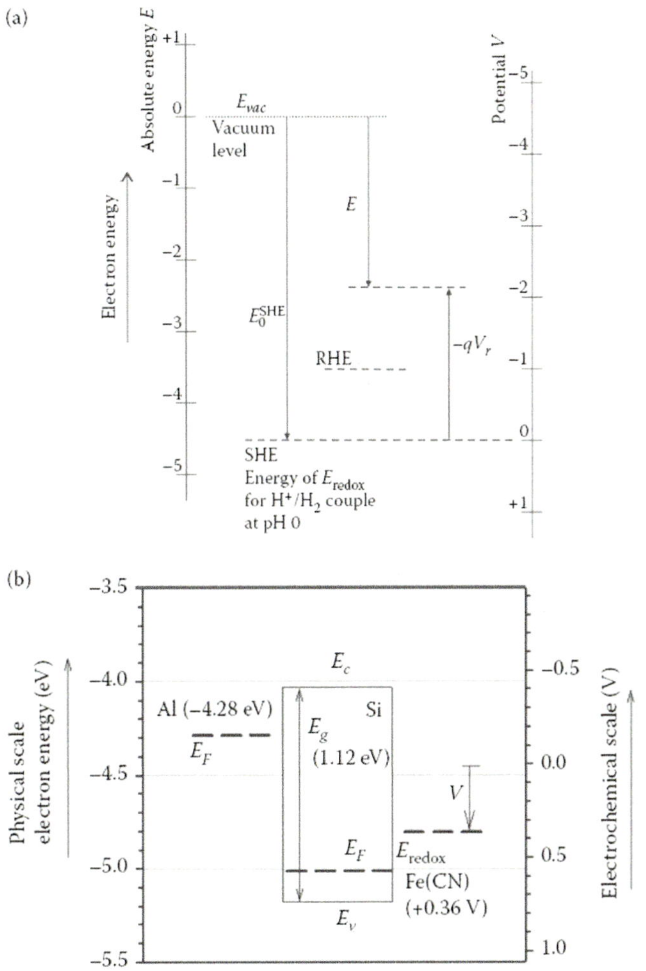

FIGURE 3.10 (a) Measurement of energy of a specific energy level, E, from the vacuum level, or as potential V_r from the reference electrode. (b) Energy level of metal/semiconductor/electrolyte system, before they reach equilibrium of equalized Fermi levels.

which is the IUPAC value (Trasatti, 1986). Recent estimates of E_0^{SHE} are in close agreement with this value (Fawcett, 2008). The potential of the vacuum level is $V_{VL} = E_0^{SHE}/q = -4.44\,\text{V}$.

The relationship between the absolute physical scale and the electrochemical scale allows referring all Fermi levels and energy levels of separate materials to a common reference value, in the situation in which the materials come into contact. An example is shown in Figure 3.10b. The (positive) potential of $Fe(CN)_6^{4-}/Fe(CN)_6^{3-}$ redox couple with respect to SHE is indicated. The diagram shows that electrons in the conduction band of Si have enough energy to reduce $Fe(CN)_6^{3-}$, and also that holes in the valence band can oxidize $Fe(CN)_6^{4-}$. The absolute energy scale is also very important for comparing electrochemical and ultrahigh vacuum experiments such as UPS (Kätz et al., 1986), and for density functional theory methods of simulation of electrolyte systems (Tripkovic et al., 2011).

3.11 CHANGES OF ELECTROLYTE LEVELS WITH PH

Water dissociates into its component ions, the proton H^+ and OH^-, by the reaction

$$H^+ + OH^- \leftrightarrow H_2O \tag{3.56}$$

The H^+ ions do not exist in aqueous solution, instead we express the hydrated proton as a hydronium ion H_3O^+ and write Equation 3.56 as

$$H_3O^+ + OH^- \leftrightarrow 2H_2O \tag{3.57}$$

The parameter pH measures the activity of H_3O^+ ions as

$$\text{pH} = -\log a_{H_3O^+} \tag{3.58}$$

The ionic concentration product is defined by the rate constant

$$K_w = \frac{a_{H_3O^+} a_{OH^-}}{a_{H_2O}} = 1.008 \times 10^{-14} \tag{3.59}$$

It is normally assumed that the activity of H^+ ions is the same as their concentration $[H^+]$. Equation 3.59 turns into

$$\left[H_3O^+\right]\left[OH^-\right] = 1.008 \times 10^{-14}\ \text{mol}^2\text{L}^{-2} \tag{3.60}$$

The constant in Equation 3.60 is the *ionic product* of water. In neutral water, the concentrations of both ionic species are the same, 10^{-7} mol L^{-1}, and the value of pH is 7.

In the definition of SHE by the reaction (3.40), the Nernst equation provides the following expression:

$$V_{redox} = +\frac{k_B T}{q} \ln \frac{a_{H^+}}{\left(p_{H_2}/p^0\right)^{1/2}} \tag{3.61}$$

Here, a_{H^+} is the activity of the hydrogen ions, p_{H2} is the partial pressure of hydrogen gas, in Pa, and p^0 is the standard pressure. In the SHE, the pressure of hydrogen gas is 1 bar, $p_{H2} = p^0$. The concentration of the protons is maintained at unity $a_{H^+} = \left[H^+\right] = 1$ and the value of the pH is 0; hence, $V_{redox} = 0$.

The change of pH is a very important instance of the modification of the redox potential by the change of concentration of the active redox species. All the reactions associated to water hydrolysis change with pH, as shown in Figure 3.11, and many semiconductors energy levels also shift with pH (see later Equation 9.51). It is therefore practical to use a reference electrode that changes with pH. The reversible hydrogen electrode (RHE) is constructed similar to the SHE but the pH is not fixed at zero, and can have any value. If the pressure of the

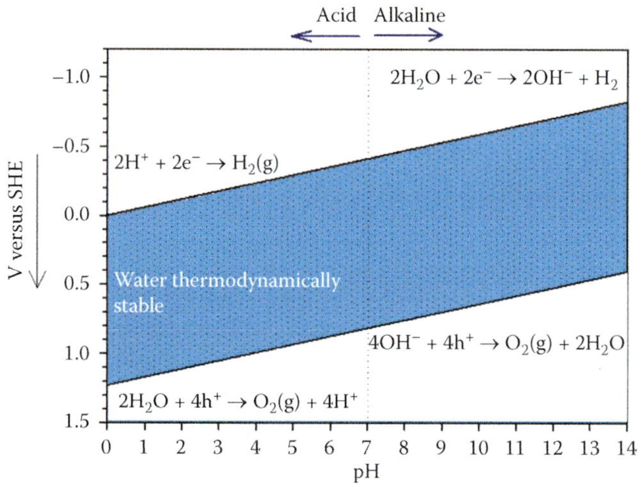

FIGURE 3.11 General form of water reduction and oxidation potentials, versus pH, or Pourbaix, diagram. The shaded region is the domain of stability of water. At more negative potentials than the upper water is reduced by electrons injected from the electrode, and at more positive potentials water is oxidized by holes injected from the electrode. At low pH proton (hydronium) ions are more abundant, while for pH corresponding to an alkaline solution the hydroxyl ion OH^- dominates, and the reduction and oxidation reactions are expressed differently.

hydrogen gas is still $p_{H2} = p^0$, the potential derived from Equation 3.61 has the value

$$V_{RHE} = V_{pH=0} - \frac{2.3k_BT}{q}(pH) \quad (3.62)$$

Therefore, the potential of the reference, RHE, changes with the pH. The potentials measured with respect to each RE are related as follows:

$$V_r^{RHE} = V_r^{SHE} + \frac{2.3k_BT}{q}(pH) \quad (3.63)$$

3.12 PRINCIPLES OF ELECTROCHEMICAL BATTERIES

The electrochemical battery produces electric energy by the transformation of chemical species. In general, a fuel is a chemical substance that releases energy when oxidized with atmospheric oxygen. The principle of the electrochemical battery is to convert chemical species into electric power by redox reactions at the electrodes. In two separate reactions, the electrons are extracted from the reductant and delivered to the oxidant via an external wire to produce electric work. The anode or negative electrode releases electrons to the external circuit and is oxidized during the discharge. The cathode or positive electrode takes electrons from the external circuit and becomes reduced during the reaction. These reactions proceed spontaneously during discharge of the battery.

The electrochemical cell is the basic unit producing the source of electric energy by direct conversion of chemical energy. A *battery* consists of several electrochemical cells arranged suitably in parallel/series connections to produce the required voltage and current for operation. In common usage the denomination of *battery* applies also for a single electrochemical cell. In a *primary battery*, the device use ends when the reactants are consumed. But a *secondary battery* can be recharged by inverting the reactions that occur in discharge, with the application of a current in the opposite direction to that of cell discharge, or an applied voltage larger than the cell voltage. A cell operating in discharge is called a galvanic cell, and a cell in recharge is called an electrolytic cell.

The amount of electric energy that the cell can deliver is expressed per unit weight (Wh kg^{-1}) or per unit volume (Wh L^{-1}). It is a function of the cell potential (V) and charge capacity (Ah kg^{-1}). Both these quantities are directly related to the chemistry of the battery. For example, the carbon–zinc system gives an

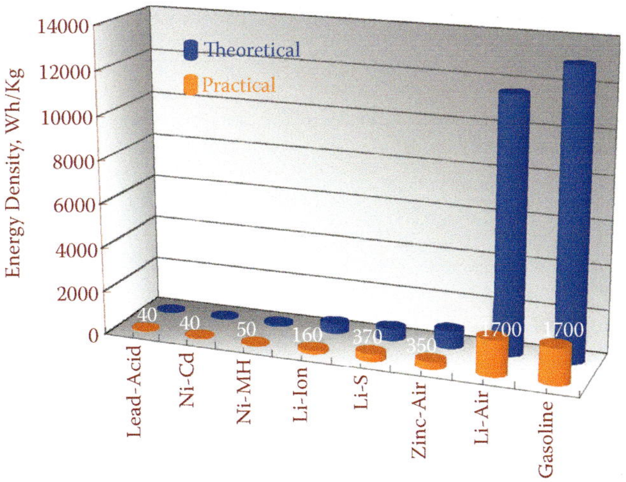

FIGURE 3.12 The gravimetric energy densities (Wh kg^{-1}) for various types of rechargeable batteries compared to gasoline. The theoretical density is based strictly on thermodynamics and is shown as the back bars while the practical achievable density is indicated by the front bars and numerical values. For Li-air, the practical value is just an estimate. For gasoline, the practical value includes the average tank-to-wheel efficiency of cars. (Reprinted with permission from Girishkumar, G. et al. *The Journal of Physical Chemistry Letters* 2010, *1*, 2193–2203.)

open-circuit voltage of 1.5 V, gravimetric energy density of 55–77 Wh kg^{-1}, and volumetric energy density of 120–152 Wh L^{-1}. Conventional lead-acid batteries store about 170 Wh kg^{-1}. The gravimetric energy densities for various types of rechargeable batteries are shown in Figure 3.12 and compared to gasoline. Another important consideration for many applications is the power density or specific power, expressed as W L^{-1} or W kg^{-1}.

The basic structure of the elementary battery electrochemical cell is shown in Figure 3.13. It is formed by two electrodes, a separator, and the electrolyte. The electrodes are electronic conductors, while the electrolyte is an ionic conductor with vanishing electronic conductivity. The reactions occur at the interface between the solid and a solution. For example, at the anode the oxidation reaction increases the state of charge of a metal M

$$M \rightarrow M^+ + e^- \qquad V_0 = V_A \quad (V) \quad (3.64)$$

and at the cathode reduction reaction decreases the state of charge of a metal oxide B,

$$B + M^+ + e^- \rightarrow BM \qquad V_0 = V_B \quad (V) \quad (3.65)$$

The requirements for anode materials are high reducing agency (negative potential in the electrochemical scale), high coulombic capacity, good conductivity, stability,

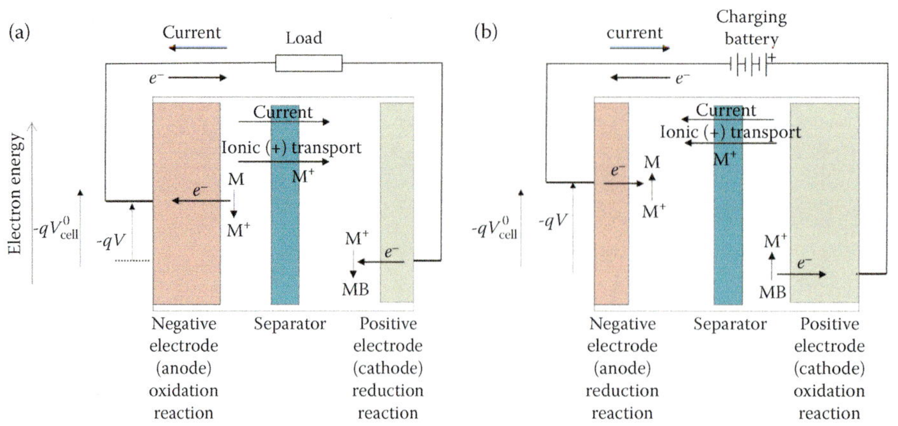

FIGURE 3.13 Principle of operation of electrochemical battery during (a) discharge and (b) charge. The arrow for electron transfer at the metal/solution interface indicates the energy level of the electron in the vertical energy scale. The arrow for ionic transport in the electrolyte indicates the flux of positive ionic charge.

and low cost. The potentials of anode materials are listed in Table 3.1. The most widely used anode materials are zinc and graphite (in LIB). The cathode material should be very positive in the electrochemical scale (i.e., efficient oxidant agent), and be stable in contact with the electrolyte. Most types of batteries use metal oxides as cathode, but other materials, such as sulfur, are also used. Oxygen can be directly used as cathode from the ambient in the metal-air battery.

The electrolyte is an aqueous solution in many cases. However, ionic liquids and nonaqueous electrolytes are used to avoid reaction of the anode with the electrolyte, and solid polymer electrolytes and gel electrolytes are used in all solid cells. During discharge of the battery as in Figure 3.13a, negative ions (anions) move toward the negative electrode, and positive ions (cations) move toward the positive electrode. The separator is necessary to avoid the electrodes from coming into contact when the battery is processed. If this happens the cell will *short circuit* and become useless because both electrodes would be at the same potential. The separator is a porous material that allows the ionic transport in the electrolyte, and sometimes selectively allows the passage of a single ionic species, as in the sodium–sulfur battery discussed later.

TABLE 3.1
Characteristics of Battery Materials

Material	Standard Reduction Potential, V	Specific Capacity, Ah kg^{-1}
Li	−3.01	3860
(Li)C_6	−2.8	370
Na	−2.71	1160
Zn	−0.76	820
H_2	0	26,590

The control of electrochemical reactions that cause solvent decomposition is a very important aspect of battery materials and performance. For example, the potential for hydrogen reduction at an anode, and that for water oxidation, leading to oxygen evolution at a cathode, change with pH but they differ by 1.23 V thermodynamically, thus the stability window of water is 1.23 V, as indicated in Figure 3.11. However, some aqueous batteries greatly exceed this limitation to the cell voltage, and this is because the reactions at anode and cathode can be kinetically inhibited by protective layers, as shown in Figure 3.14. The potential of the anode can, therefore, become more negative than the potential of reduction of the electrolyte without the reaction actually taking place and cell voltage increases (Huggins, 2009; Good-enough and Kim, 2010). For example, alkaline battery cells have an open-circuit voltage of 1.5–1.6 V. This is greater than the voltage of decomposition of water, and it is possible because the zinc negative electrode in the cell is covered by a thin layer of ionically conducting ZnO, with more negative potential for the evolution of hydrogen than in unoxidized metal electrode in contact with water. Metal hydride/nickel cells operate at 1.34 V, and oxygen evolution does not begin before reaching 1.44 V. This is due to the presence of an electronically insulating but proton-conducting layer of $NiOH_2$ on the surface in contact with the electrolyte (Colin et al., 2010).

Many active materials are poor electronic conductors; hence, the anode and cathode in practice are complex conglomerates (see examples in Figures 3.15 and 3.16) containing conductive diluents such as carbon black or carbon nanotubes and polymeric binders to hold the structure together. Nanostructured particles that

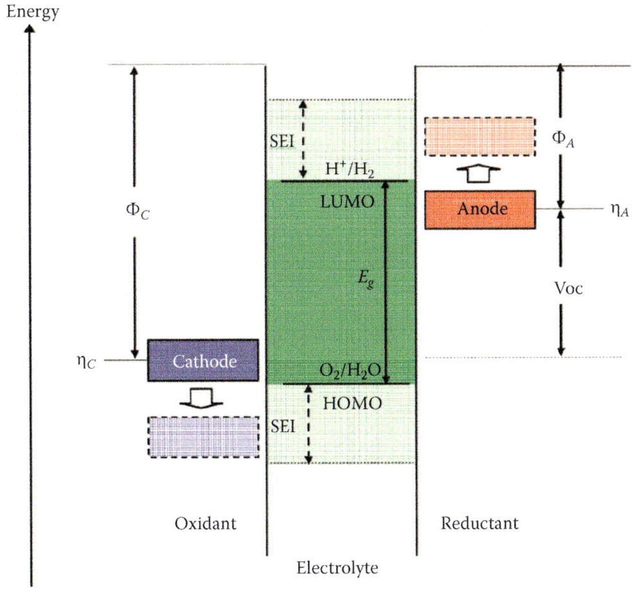

FIGURE 3.14 Schematic open-circuit energy diagram of an aqueous electrolyte. Φ_A and Φ_C are the anode and cathode work functions and η_A and η_C indicate the respective electrochemical potentials of electrons. E_g is the window of the electrolyte for thermodynamic stability. A η_A > LUMO and/or a η_C < HOMO requires a kinetic stability by the formation of an SEI layer. (Reproduced with permission from Goodenough, J. B.; Kim, Y. *Chemistry of Materials* 2010, *22*, 587–603.)

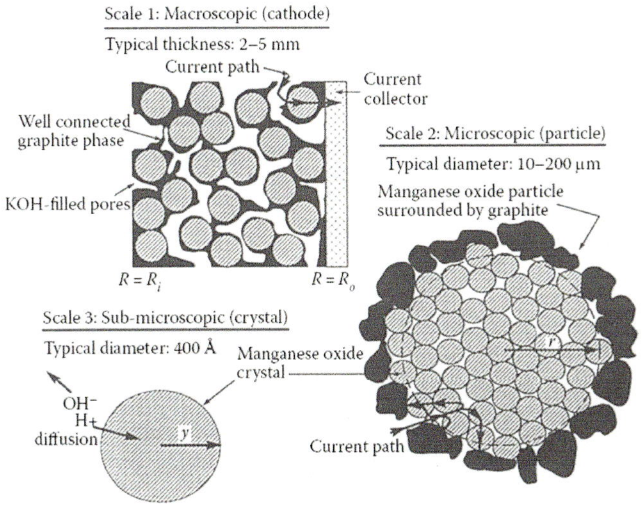

FIGURE 3.15 Different scales and morphologies of the cathode of an alkaline electrolytic manganese dioxide battery. (Reproduced with permission from Farrell, T. W. et al. *Journal of the Electrochemical Society* 2000, *147*, 4034–4044.)

possess a high capacity and adequate chemical potential for battery anode or cathode can be covered with a conducting shell of graphene to increase their electronic conductivity.

At the right of each reaction in Equations 3.64 and 3.65 is the standard potential of the reaction occurring at

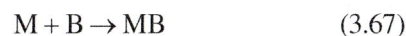

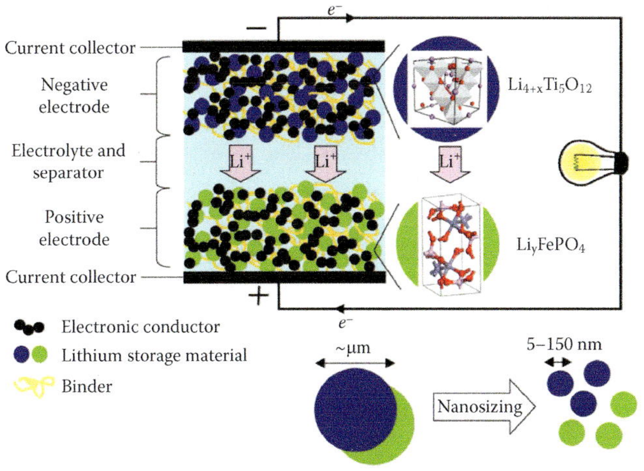

FIGURE 3.16 Basic structure of a Li-ion battery. (Reproduced with permission from Wagemaker, M.; Mulder, F. M. *Accounts of Chemical Research* 2011, *46*, 1206–1215.)

each half-cell. The equilibrium cell potential is given by the difference of potentials of cathode and anode:

$$V_{\text{cell}}^0 = V_B - V_A \tag{3.66}$$

When no current is being drawn from a reversible cell, the potential difference across its terminals, or open-circuit voltage (OCV), V_{cell}^0 is known as the *emf* of the cell. The chemical driving force across the cell is given by the change of free energy per mol of reaction (or per reacted particle), determined by the free energies of formation of products and reactants, that would occur if the *electrically neutral materials* in the two electrodes were to react chemically (Huggins, 2009). In the battery described in Figure 3.13, the free energy available is due to the transformation of the species in the cell by the net reaction:

$$\text{M} + \text{B} \rightarrow \text{MB} \tag{3.67}$$

The Gibbs-free energies turn into electrode potentials, $\Delta G = -zFV_{\text{cell}}^0$, finally represented by electrochemical potentials of electrons in the current collectors of the two electrodes (see Section 5.10). The work produced per electron that goes round the circuit is

$$W = qV_{\text{cell}}^0 = q\left(V_B - V_A\right) \tag{3.68}$$

Joining the electrodes by an external wire, we obtain a flux of electrons that can do electric work on an external load. In practice, overpotentials and *IR* drops are important aspects of the operation of practical batteries, as mentioned later in this chapter. A current *I* (A) during a time Δt (s) produces the passage of $Q = I\Delta t$ coulombs. The current in voltaic cells proceeds until the materials of the electrodes are consumed. A cell is *reversible* if the directions of the reactions (3.64) and (3.65) can be inverted by small variation of the potential around V^0_{cell}.

It is also interesting to remark that in the discharge of the battery (Figure 3.13a), the positive ion M^+ is moving from the negatively charged electrode to the positively charge electrode. This is opposite to the sense of motion that a charged particle would realize under the influence of electrostatic forces between two charged plates (Varney and Fischer, 1980). In fact, electrons and cations move through different pathways to realize the oxidation reaction at the surface of the anode and the reduction reaction at the surface of the cathode.

3.13 CAPACITY AND ENERGY CONTENT

The charge *capacity C* describes a cell's ability to deliver current. The charge capacity of the cell is the maximum amount of charge that can be supplied. We integrate the current along the discharge process to obtain the total charge passed,

$$Q_{tot} = \int_0^t I\,dt \qquad (3.69)$$

The battery capacity C is measured in terms of Ah (amp-hours) at a specific discharge rate. 1 Ah equals 3600 C. If we express the quantity of electric charge of 1 F in A h, we obtain

$$Q = 1\,\mathrm{F} = 96{,}500\,\mathrm{C\,mol^{-1}} = 26.80\ \mathrm{Ah\,g\,eq^{-1}} \qquad (3.70)$$

As example the atomic weight of lithium is 6.94 g. One atom of lithium takes part in electrochemical reactions of charge and discharge of the lithium battery; hence, the electrochemical equivalent (eq) weight of lithium is 6.94 g. The *theoretical capacity* of lithium is thus 26.80 Ah g eq^{-1}/6.94 g $\approx$ 3.86 Ah g^{-1}. Values for different materials are given in Table 3.1.

The battery is a high-energy device, and the time of discharge is in the range of hours. The *C*-rate is an expression of the discharge current as the multiple of the current that the battery can sustain in one hour. It is obtained giving the discharge current, in amps (A), as *C/t*, where t is the discharge time in hours. For example, a rate of 0.1 C means (C Ah)/(10 h), in A. Then, the total charge is exhausted in $t = 10$ h.

A cell provides a voltage given by thermodynamics of the materials as analyzed earlier, but when we draw current from the battery, kinetic phenomena, related to the speed of the reactions and the transport rates of the different species that sustain the current in the interior of the cell, come into play. The cell has an internal resistance that is originated in two main phenomena:

1. Polarization losses at the electrode, related to the overvoltage required to drive the reactions at the required current level.
2. Transport losses. At low current these usually take the form of ohmic losses, of the type *IR*. At higher current diffusion of ions in the electrolyte, and electrons and ions in the active masses, causes increasingly large transport loss.

Ideally, the discharge of the battery should occur at a constant potential until the active materials are totally utilized and the full capacity is exhausted. But when the battery is discharged at a fast rate, the internal resistance and other factors reduce the voltage of the battery, and the discharge characteristic degrades as shown in Figure 3.17. At higher currents the discharge curve shows a more slopping profile. Furthermore, the actual capacity is a function of the discharge rate. Finally, fast discharge may cause irreversible phenomena or side reactions that reduce the cyclability of the battery, and may even cause hazard damages.

Another important feature of the battery is the total energy content. If the battery supplies a current I at the voltage V, the power supply is

$$P = IV \qquad (3.71)$$

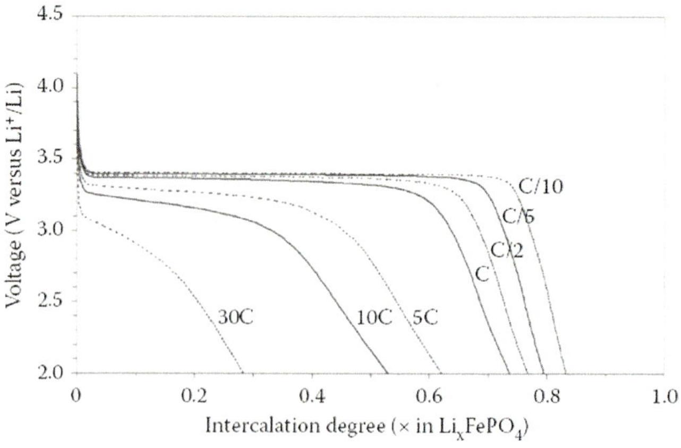

FIGURE 3.17 Discharge curves: Voltage as a function of the Li intercalation degree for different discharge rates for $LiFePO_4$ electrode (87.5% $LiFePO_4$, 0.4 mAh cm^{-2}, and 35% porosity). (Reproduced with permission from Fongy, C. et al. *Journal of the Electrochemical Society* 2010, *157*, A885–A891.)

The amount of energy content is

$$E_P = \int_0^t IV\,dt \tag{3.72}$$

where t is the time of total discharge. A battery of energy content 0.1 Wh is theoretically capable of supplying a power of 0.1 W for an hour, or 1.0 W for a tenth of an hour, although in practice the recoverable energy depends on the discharge rate.

3.14 PRACTICAL ELECTROCHEMICAL BATTERIES

The size range of batteries cover a wide range of energy supply needs. The scale extends from small button cells of energy content ≈0.1 Wh, starters for automobiles of 10^2–10^3 Wh, full traction of electric vehicles of 10^4–10^6 Wh, to load leveling batteries, used for massive energy storage, of energy content ≈10 MWh. As examples to discuss the basic concepts of battery operation, we choose the zinc-silver and sodium–sulfur rechargeable batteries.

3.14.1 Zinc-Silver Battery

The size and weight of the materials for a given amount of stored energy is of primary concerns for many applications, for example, for portable electronics. Zinc-silver battery is composed of expensive components but provides a very high-energy density. The advantages of Zn-Ag are high specific power (up to 600 W kg^{-1} continuous and 2500 W kg^{-1} for short-duration pulses), high-theoretical specific energy (up to 250 Wh kg^{-1}), and energy density (750 Wh L^{-1}). To construct the battery, a disc of metal zinc, the anode, is joined with a separator to a disc of silver oxide, the cathode. The positive electrode is made of small particle size silver powder and the negative electrode is metal zinc together with additives and binders.

The electrodes are connected by a solution of potassium hydroxide that dissociates into one K^+ ion and a hydroxide OH^- ion. The separator is a multiple layer of low-porosity films primarily based on regenerated cellulose known as battery-grade cellophane. The separator must retain silver-soluble species produced by chemical dissolution of the oxide.

At the zinc electrode the OH^- ion oxidizes zinc to form $Zn(OH)_2$, which is released into the solution. As shown in Figure 3.18, the reaction at the negative zinc electrode, the anode, is

$$Zn + 2OH^- \rightarrow Zn(OH)_2 \qquad V_0 = -1.245V \tag{3.73}$$

$$Zn(OH)_2 + 2KOH \Leftrightarrow K_2Zn(OH)_4$$

Two electrons per $Zn(OH)_2$ molecule flow through the external circuit to the silver oxide electrode, where they reduce the water molecules in the solution to form OH^-. The OH^- ions produced at the surface of the silver electrode are transported through the electrolyte toward the negative electrode to replace OH^- ions consumed in the

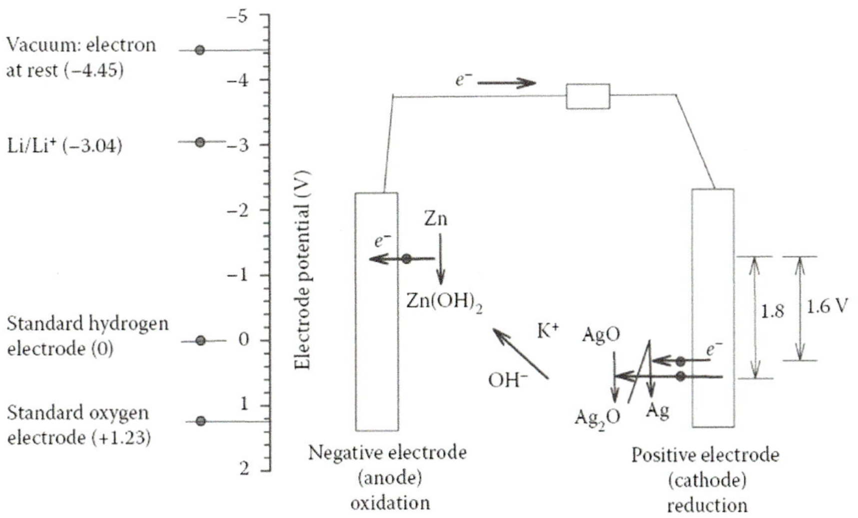

FIGURE 3.18 Processes and potentials of the reactions at the silver oxide–zinc battery. The left electrode is formed of zinc and the right electrode is AgO or AgO/Ag_2O (see the main text). The points in the arrows indicate the oxidation and reduction potentials, and the direction of electron and ion flow during battery discharge is also indicated. The points at the left of the scale indicate several reference and reaction potentials.

reaction at the zinc electrode. The reduction at the cathode occurs in two steps. The first reaction at the positive oxide electrode is

$$2AgO + H_2O + 2e^- \rightarrow Ag_2O + 2OH^- \qquad V_0 = +0.57\,V \tag{3.74}$$

From the difference of the anodic and cathodic reaction potentials, we obtain the potential of the discharge of the battery, which is 1.815 V. However, a second reaction takes place at the positive electrode at less positive potential

$$Ag_2O + H_2O + 2e^- \rightarrow 2Ag + 2OH^- \qquad V_0 = +0.344\,V \tag{3.75}$$

Therefore, in the discharge characteristic we expect a first plateau at 1.8 V until the AgO is consumed to Ag_2O, and then a second plateau at 1.6 V until the exhaustion point. These features are observed for charge and discharge in Figure 3.19a and b, respectively. In Figure 3.19b, it is observed that discharge curve is flat at 1.50 V. The change of voltage is unsuitable for many applications. Therefore, the electrode can be treated to reduce AgO in a surface layer to Ag_2O. Figure 3.19c shows an example of the discharge of a planar Zn–Ag battery constructed only with monovalent silver oxide (Ag_2O) instead of the higher density divalent form (AgO). This utilization decreases considerably the electrode capacity up to 50%.

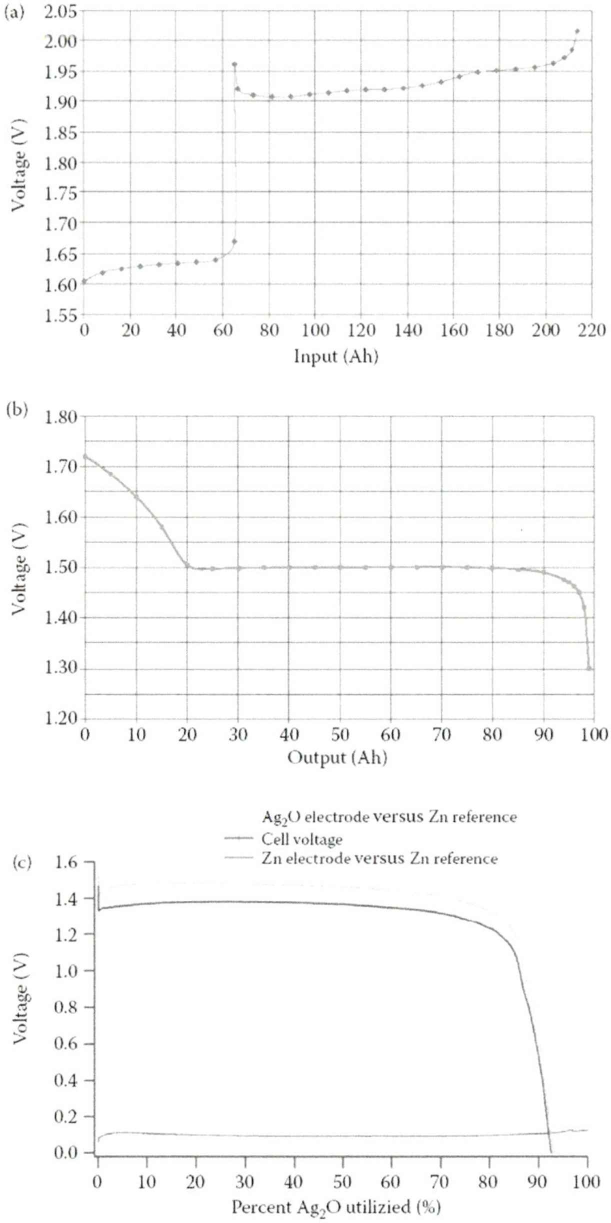

FIGURE 3.19 Voltage-charge curves at silver oxide–zinc battery. (a) Typical charge of a 200-Ah cell at 8 A. (b) Typical discharge of a 70 Ah cell at 10 A. (Reprinted from *Encyclopedia of Electrochemical Power Sources*, Salkind, A. J.; Karpinski, A. P.; Serenyi, J. R., pp. 513–523, Copyright 2009, with permission from Elsevier.) (c) A typical discharge characteristic of silver zinc battery discharged at a C/2 rate (1.408 mA). (Reprinted with permission from Braam, K. T.; Volkman, S. K.; Subramanian, V. *Journal of Power Sources* 2012, *199*, 367–372.)

3.14.2 Sodium-Sulfur Battery

High-energy applications include management of the energy supply and demand, balancing of the load curves, and peak shaving. Large battery facilities can be used as part of the main electricity supply system to store electric energy when the demand is low and so is the cost of the electricity, and then sell it when real-time electricity generation cost is high. Storing the electric energy in batteries can also solve problems associated with the intermittent nature of several renewable energy sources, that is, solar, wind, or wave power.

Sodium–sulfur (NAS) batteries present one of the best options for large energy storage systems (Bito, 2005). Because of the low cost and ready availability of its chemical components, NAS batteries are high-capacity battery systems developed for application in electric power systems. The main battery attributes are high-energy density, long cycle and shelf life, low price, and short construction interval. The theoretical energy density is very high, about 750 Wh kg^{-1}, which allows to construct a modular battery system that stores 12,000 kWh.

The battery is operated under the unusual condition that the active materials at both electrodes are *liquid* and

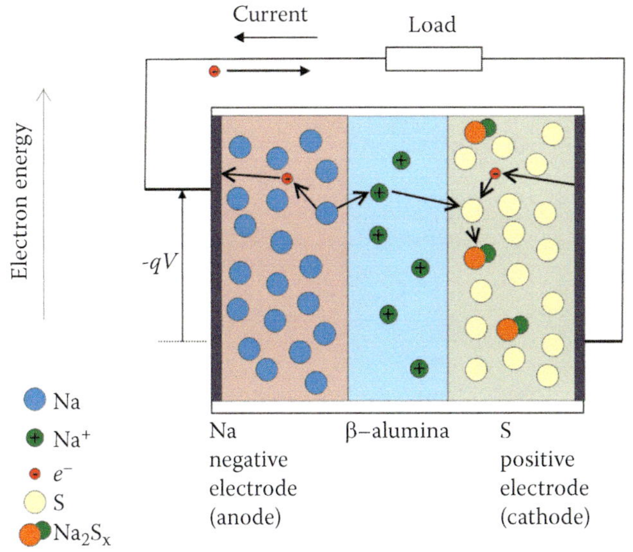

FIGURE 3.20 Principle of operation of the NAS battery during discharge. The battery works at temperatures such that anode and cathode are in a liquid state, and a solid ionic conductor functions as separator.

the electrolyte is *solid*, Figure 3.20. The cell consists of a molten sulfur positive electrode and a molten sodium negative electrode separated by a ceramic electrolyte. The separator is made of β″-alumina, a type of *fast ionic conductor* or *superionic conductor* that shows a very high ionic conductivity of 0.4 Ω^{-1} cm^{-1} at 350°C, which is specific to sodium ions as charge carriers, and a very low electronic conductivity. The cell operates at temperatures around 350°C, at which the ionic conductor allows the passage of several amps per square centimeter. During the discharge, sodium ions converted from sodium in the negative electrode pass through the solid electrolyte and reach the sulfur in the positive electrode. Sodium polysulfide is formed in the positive electrode, while sodium in the negative electrode decreases by consumption. The cell voltage derived from this reaction

$$2Na + 4S \Leftrightarrow Na_2S_4 \tag{3.76}$$

is 2.08 V. During the charge, the electric power supplied from outside forms sodium in the negative electrode and sulfur in the positive electrode.

3.15 LI-ION BATTERY

Battery technology based on Li was developed in 1970s and 1980s, motivated by the outstanding properties of Li-metal. As shown in Figure 2.7, Li is the most electropositive metal, being the Li/Li^+ redox potential at −3.05 V versus SHE. Li is also the lightest metal

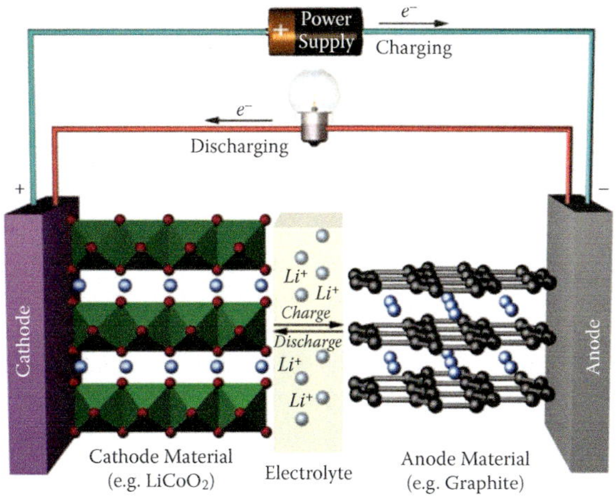

FIGURE 3.21 Schematic operation principle of a rechargeable lithium battery. (Reprinted with permission from Liu, R.; Duay, J.; Lee, S. B. *Chemical Communications* 2010, *47*, 1384–1404.)

(equivalent weight M = 6.94 g mol^{-1}, specific gravity ρ = 0.53 g cm^{-3}). These properties combined with small size of Li-ion provide the opportunity of high voltage, low weight, and fast transport, allowing for rapid kinetics for charge and discharge. Inorganic compounds as cobalt, manganese, and nickel oxides were found that allowed reversible lithiation and were later identified as intercalation compounds. These features add up to a device that provides high-energy density and power density that are required for many applications.

The LIB works by combining two materials that allow lithiation and delithiation at very different potentials. Insertion or intercalation denominates a process in which Li^+ ions enter the solid lattice without much distortion of the solid framework of the host material. Other important requisites are a high mobility of Li^+ ions, favored in layered or channeled frameworks, and sufficient electronic conductivity to assist the incorporation of cationic charge. The example shown in Figure 3.21 couples a graphite anode with very negative potential, and a cathode of Li_xCoO_2. The two electrodes of the cell are connected by a pure Li-ion conductor. The potential of the graphite anode is close to that of the pure Li/Li^+ metal potential, but the graphite is safer than the lithium metal for extended operation. However, the Li_xCoO_2 cathode is very positive on the electrochemical scale, so that the difference between anode and cathode is about 4 V. The voltage and capacity of a range of anode and cathode materials is shown in Figure 3.22.

The energy scheme shown in Figure 3.23 indicates the operation principle of the LIB cell shown in Figure 3.21. The electrochemical potential of electrons

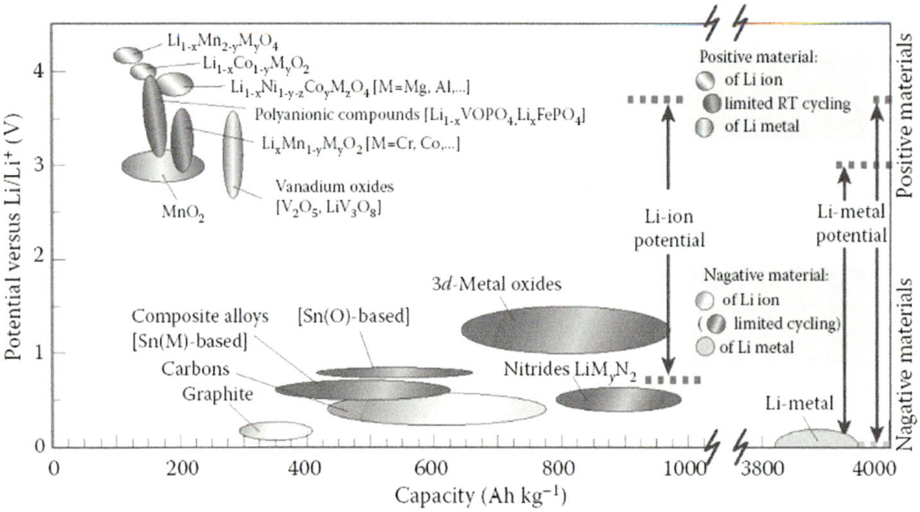

FIGURE 3.22 Voltage versus capacity for positive- and negative-electrode materials of rechargeable Li-based cells. The output voltage values for Li-ion cells or Li-metal cells are represented. (Reprinted with permission from Tarascon, J. M.; Armand, M. *Nature* 2001, *414*, 359–367.)

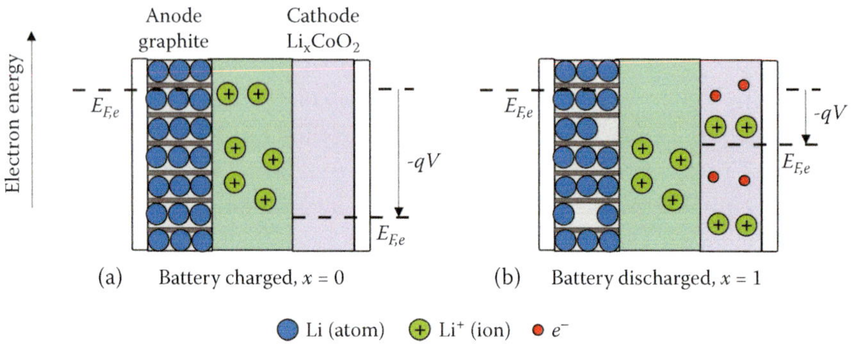

FIGURE 3.23 Basic structure of an LIB. The voltage in the battery is the difference of electrochemical potential of electrons at the $LiCoO_2$ cathode with respect to that in the graphite anode. The battery discharges by transference of ions through the cathode to the anode, while electrons are transferred via the external circuit. It is assumed that the electrochemical potential of electrons in the anode, made of graphite (or Li), is stable. The electrochemical potential of electrons in the cathode changes depending on number of Li^+ ions in the structure, and provides the change of voltage. When there are few Li^+ ions intercalated in the cathode framework, the voltage is very positive and the cell is charged. When Li^+ ions flow across the electrolyte and into the $LiCoO_2$ cathode, the voltage decreases (still positive) and the battery is discharged.

at each electrode is determined by the chemical potential of Li atoms at the respective electrode, as we show later in Section 5.10. The anode potential corresponds to electrochemical potential of pure lithium metal or graphite, which is stable. In contrast, the chemical potential of Li at the Li_xMO_y, determining the Fermi level of electrons at the cathode depends on Li^+ content. In the charged state, the graphite anode is full of Li ions that would rather occupy the vacant states of the metal oxide cathode. However, the favored chemical reaction involving the transference of Li atoms from anode to cathode is prevented because the electrolyte blocks the passage of Li atoms and only allows the transference of Li^+ ions. This occurs only when electrons go around the circuit from anode to cathode, and thus provide external work. Discharge curves, showing the voltage with respect to the concentration (often labeled by an index x that indicates the relative amount of Li in the lattice), are shown in Figure 3.24 for several cathode materials and for graphite. Figure 3.25 shows the changes of voltage in the LiAl electrode with respect to the lithium content.

In the study of electrode materials for the lithium battery, both anodes and cathodes, metallic lithium is typically used as reference electrode (Verbrugge and Koch, 1996). Lithium is extremely reactive, and the surface is commonly covered with reaction products, but nonetheless elemental lithium is a highly reliable potential reference in lithium-based electrochemical systems.

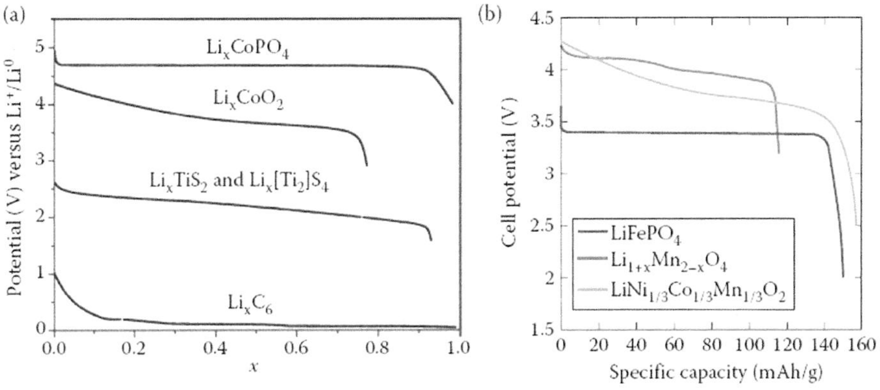

FIGURE 3.24 (a) Equilibrium discharge curves for several cathode and anode materials. (Reproduced with permission from Goodenough, J. B.; Kim, Y. *Chemistry of Materials* 2010, *22*, 587–603.) (b) Equilibrium discharge curves for several cathode materials. The voltage is given with respect to Li^+/Li^0 potential. (Reproduced with permission from Doeff, M. M. *Batteries: Overview of Battery Cathodes*; Lawrence Berkeley National Laboratory LBNL Paper LBNL-3652E. : http://escholarship.ucop.edu/uc/item/1n55870s, 2011.)

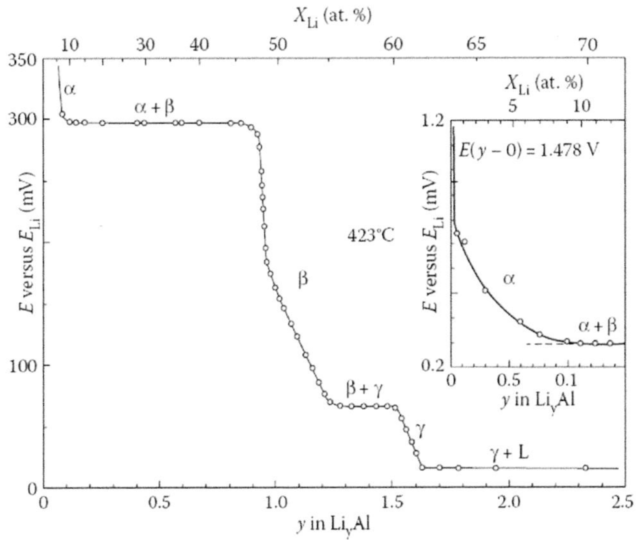

FIGURE 3.25 Voltage-composition curve for lithium-aluminum system showing three single phases, that is, α, β, and γ, at 423°C. (Reproduced by permission from Wen, C. J. et al. *Journal of the Electrochemical Society* 1979, *126*, 2258–2266.)

The potentials and theoretical capacity of a variety of anode and cathode materials are shown in Table 3.2. Figure 3.26 shows the potential of several cathode materials and the stability window of electrolyte.

The cathode of the LIB usually consists of a compound formed by electrochemical reaction between a transition metal oxide MO_x or phosphate and lithium. The oxides of titanium, vanadium, manganese, iron, cobalt, and manganese have been widely investigated and applied in commercial batteries. There are two main classes of Li-ion storage in the material: insertion and phase transformation of the host material.

Insertion has been discussed earlier. It occurs without chemical modification of the host but with a change of occupation of vacant or interlayer sites in the structure. The change of voltage between charge and discharge is associated with the concentration of positive Li^+ ions and electrons entering the material. Although the potential associated with the reductive insertion of lithium ions is high in the intercalation oxides, the number of charges that can be accommodated is very limited, since it is restricted to 1 Li or less per 3d metal atom in $LiCoO_2$. Consequently, the energy storage per mass in cell, measured by the capacity, is usually low in intercalation materials. Intercalation can produce an order–disorder phase transformation. For example, the change of Li content in Li_xMO_y often involves first-order phase transformations in which two phases coexist. During coexistence the chemical potential of the Li_xMO_y and Li_x is constant, so that the voltage-state of charge curve forms a large plateau which keeps the battery voltage constant. These processes are further discussed in Section 5.11.

The high reactivity of the lithium anode toward the electrolyte was an impediment in early attempts to its use. Graphite has a layered structure that offers an intercalation anode to replace Li. It is possible to reversibly store only one atom of lithium per six atoms of carbon (i.e., LiC_6), into the graphene planes as shown in Figure 3.27, corresponding to a theoretical capacity of about 370 mAh g^{-1}. A major breakthrough that allowed the commercialization of LIBs (in 1990) was the development of electrolytes that caused the formation of a solid electrolyte interface (SEI) on surfaces of the carbon particles. The SEI acts as a Li-ion conductor but reduces the reactivity of the active material, as indicated in Figure 3.28. Standard electrolytes for Li-ion batteries

TABLE 3.2

Characteristics of Li-Ion and Li-Air Battery Materials

Material	Potential versus Li/Li$^+$, V	Specific Capacity, Ah kg^{-1}	Specific Energy, Wh kg^{-1}
Li	0	3860	
LiC_6	0.2	370	
LiSi		3579	
LiGe		1600	
Li_2O	2.91	1794	5222 (11,140[a])
Li_2O_2	2.96	1168	3460
$LiCoO_2$	3.9	140	546
$LiMn_2O_4$	4.1	100–120	410–490
$LiFePO_4$	3.45	150–170	520–588

[a] Excluding O_2.

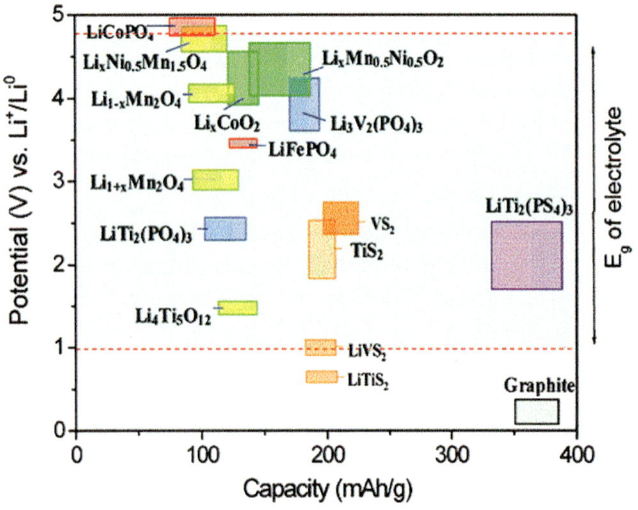

FIGURE 3.26 Voltage relative to Li$^+$/Li^0 potential versus capacity of several electrode materials. The window of stability of the electrolyte 1M $LiPF_6$ in EC/DEC (1:1) is indicated. (Reproduced with permission from Goodenough, J. B.; Kim, Y. *Chemistry of Materials* 2010, *22*, 587–603.)

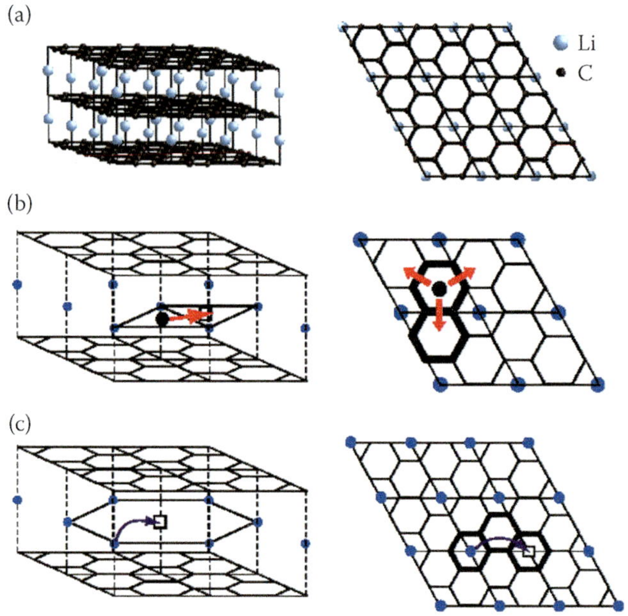

FIGURE 3.27 (a) Crystal structure of LiC_6. Schematic drawings of lithium migrations by (b) the interstitial, and (c) the vacancy mechanisms. (Reproduced with permission from Toyoura, K. et al. *Physical Review B* 2008, *78*, 214303.)

are mixtures of the Li salt lithium hexafluorophosphate ($LiPF_6$) with solvent ethylene carbonate (EC) combined with dimethyl carbonate (DMC), ethyl methyl carbonate (EMC), or diethyl carbonate (DEC). As is shown in Figure 3.26, the potential of the carbon electrode is nearly 1 V negative to that of the LUMO of a carbonate electrolyte, and hence readily reduces the electrolyte. However, incorporation of reduction products into the carbonate electrolyte promotes formation of an SEI layer on the carbon that provides the kinetic stability.

The second kind of Li storage applied in batteries is a *phase transformation* that changes the chemical composition of the host. This transformation may occur by the total phase change, or by decomposition reaction in which the reaction product separates into different phases, metal, and Li_2O. This last mechanism is known as *conversion reaction*. In the conversion reaction, the lithium can react with a range of transition metal oxides storing up to 3 Li per metal atom and providing capacities up to 1000 mAh g^{-1}. However, it is often found that upon first charging only about 70% of Li$^+$ can be extracted from Li_2O. The high irreversibility limits the application of conversion reaction materials.

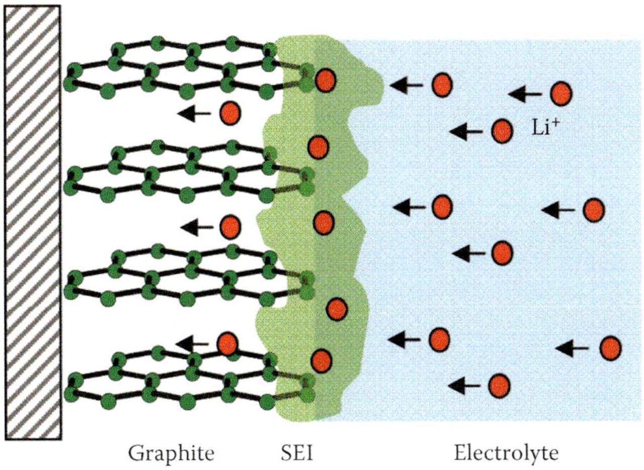

FIGURE 3.28 Schematic presentation of the formation of the SEI layer by decomposition of ethylene carbonate-based electrolytes. (Reproduced with permission from Goodenough, J. B.; Kim, Y. *Chemistry of Materials* 2010, *22*, 587–603.)

Elements that alloy with Li and have much larger capacities than graphitic materials and are very promising for their application as anodes, viz., Sn: 990 mAh g^{-1}, SnO_2: 780 mAh g^{-1}, Sb: 660 mAh g^{-1}, Si: 3600 mAh g^{-1}, and Ge: 1600 mAh g^{-1}. These materials undergo reversible alloying with Li but these processes produce a large volume expansion of about 300%, which leads to pulverization of bulk crystals and battery failure.

GENERAL REFERENCES

Definition of voltage: Rickert (1982), Riess (1997).

Electrochemistry: Vetter (1967), Gerischer et al. (1978), Schmickler (1996), Sato (1998), Bard and Faulkner (2001).

The absolute electrode potential: Trasatti (1986), Trasatti (1990).

Redox potential, electrode potential and work functions: Trasatti (1976), Gerischer and Ekardt (1983), Reiss (1985), Reiss and Heller (1985), Trasatti (1986), Hansen and Hansen (1987a, b), Trasatti (1995), Jaegermann (1996), Sato (1998), Tsiplakides and Vayenas (2001).

Reference electrodes: Huggins (2000), Cardona et al. (2011).

Batteries: Vincent and Scrosati (1997), Huggins (2009).

Zinc-silver battery: Karpinski et al. (1999), Crompton (2000), Salkind et al. (2009), Braam et al. (2012).

Li batteries: Thackeray (1997), Besenhard and Winter (1998), Poizot et al. (2000), Tarascon and Armand (2001), Ralph et al. (2004), Matthieu et al. (2006), Ellis et al. (2010), Goodenough and Kim (2010), Etacheri et al. (2011), Maier (2013), van der Ven et al. (2013).

REFERENCES

Bard, A. J.; Faulkner, L. R. *Electrochemical Methods, Fundamentals and Applications*, 2nd ed.; John Wiley and Sons: Weinheim, 2001.

Besenhard, J. O.; Winter, M. Insertion reactions in advanced electrochemical energy storage. *Pure and Applied Chemistry* 1998, *70*, 603–608.

Bito, A. Overview of the sodium-sulfur battery for the IEEE Stationary Battery Committee. *Power Engineering Society General Meeting* 2005, *2*, 1232–1235. IEEE.

Braam, K. T.; Volkman, S. K.; Subramanian, V. Characterization and optimization of a printed, primary silver/zinc battery. *Journal of Power Sources* 2012, *199*, 367–372.

Cardona, C. M.; Li, W.; Kaifer, A. E.; Stockdale, D.; Bazan, G. C. Electrochemical considerations for determining absolute frontier orbital energy levels of conjugated polymers for solar cell applications. *Advanced Materials* 2011, *23*, 2367–2371.

Colin, W.; Riccardo, R.; Robert, A. H.; Yi, C. Investigations of the electrochemical stability of aqueous electrolytes for lithium battery applications. *Electrochemical and Solid-State Letters* 2010, *13*, A59–A61.

Crompton, T. R. *Battery Reference Book*, 3rd ed.; Newness: Oxford, 2000.

Doeff, M. M. *Batteries: Overview of Battery Cathodes*; Lawrence Berkeley National Laboratory LBNL Paper LBNL-3652E. http://escholarship.ucop.edu/uc/item/1n55870s, 2011.

Ellis, B. L.; Lee, K. T.; Nazar, L. F. Positive electrode materials for Li-ion and Li-batteries. *Chemistry of Materials* 2010, *22*, 691–714.

Etacheri, V.; Marom, R.; Elazari, R.; Salitra, G.; Aurbach, D. Challenges in the development of advanced Li-ion batteries: A review. *Energy and Environmental Science* 2011, *4*, 3243–3262.

Farrell, T. W.; Please, C. P.; McElwain, D. L. S.; Swinkels, D. A. J. Primary alkaline battery cathodes: A three-scale model. *Journal of the Electrochemical Society* 2000, *147*, 4034–4044.

Fawcett, W. R. The ionic work function and its role in estimating absolute electrode potentials. *Langmuir* 2008, *24*, 9868–9875.

Fongy, C.; Gaillot, A. C.; Jouanneau, S.; Guyomard, D.; Lestriez, B. Ionic vs electronic power limitations and analysis of the fraction of wired grains in $LiFePO_4$ composite electrodes. *Journal of the Electrochemical Society* 2010, *157*, A885–A891.

Fröhlich, H. *Theory of Dielectrics*, 2nd ed.; Oxford University Press: Oxford, 1958.

Gerischer, H.; Ekardt, W. Fermi levels in electrolytes and the absolute scale of redox potentials. *Applied Physics Letters* 1983, *43*, 393–395.

Gerischer, H.; Kolb, D. M.; Sass, J. K. The study of solid surfaces by electrochemical methods. *Advances in Physics* 1978, *27*, 437–498.

Girishkumar, G.; McCloskey, B.; Luntz, A. C.; Swanson, S.; Wilcke, W. Lithium–air battery: Promise and challenges. *Journal of Physical Chemistry Letters* 2010, *1*, 2193–2203.

Goodenough, J. B.; Kim, Y. Challenges for rechargeable Li batteries. *Chemistry of Materials* 2010, *22*, 587–603.

Grahame, D. C. Fiftieth anniversary: Mathematical theory of the faradaic admittance. *Journal of the Electrochemical Society* 1952, *99*, 370C–385C.

Hansen, W. N.; Hansen, G. J. Absolute half-cell potential: A simple direct measurement. *Physical Review A* 1987a, *36*, 1396–1402.

Hansen, W. N.; Hansen, G. J. Reference states for absolute half-cell potentials. *Physical Review Letters* 1987b, *59*, 1049–1052.

Huggins, R. A. Reference electrodes and the Gibbs phase rule. *Solid State Ionics* 2000, *136/137*, 1321–1328.

Huggins, R. A. *Advanced Batteries*; Springer: New York, 2009.

Jaegermann, W. The semiconductor/electrolyte interface: A surface science approach. *Modern Aspects of Electrochemistry* 1996, *30*, 1–186.

Karpinski, A. P.; Makovetski, B.; Russell, S. J.; Serenyi, J. R.; Williams, D. C. Silver-zinc: Status of technology and applications. *Journal of Power Sources* 1999, *80*, 53–60.

Kätz, E. R.; Neff, H.; Müller, K. A UPS, XPS and work function study of emersed silver, platinum and gold electrodes. *Journal of Electroanalytical Chemistry and Interfacial Electrochemistry* 1986, *215*, 331–344.

Liu, R.; Duay, J.; Lee, S. B. Heterogeneous nanostructured electrode materials for electrochemical energy storage. *Chemical Communications* 2010, *47*, 1384–1404.

Maier, J. Thermodynamics of electrochemical Lithium storage. *Angewandte Chemie International Edition* 2013, *52*, 4998–5026.

Matthieu, D.; Vojtech, S.; Ruey, H.; Bor Yann, L. Incremental capacity analysis and close-to-equilibrium OCV measurements to quantify capacity fade in commercial rechargeable lithium batteries. *Electrochemical and Solid-State Letters* 2006, *9*, A454–A457.

Oldham, K. B.; Myland, J. C. *Fundamentals of Electrochemical Science*; Academic Press: San Diego, 1994.

Poizot, P.; Lamelle, S.; Grugeon, S.; Dupont, L.; Tarascon, J.-M. Nano-sized transition-metal oxides as negative-electrode materials for lithium-ion batteries. *Nature* 2000, *407*, 496–499.

Ralph, J. B.; Kathryn, R. B.; Randolph, A. L.; Richard, L. M.; John, R. M.; Esther, T. Batteries, 1977–2002. *Journal of the Electrochemical Society* 2004, *151*, K1–K11.

Reiss, H. The Fermi level and the redox potential. *The Journal of Physical Chemistry* 1985, *89*, 3783–3791.

Reiss, H.; Heller, A. The absolute potential of the standard hydrogen electrode: A new estimate. *The Journal of Physical Chemistry* 1985, *89*, 4207–4213.

Rickert, H. *Electrochemistry of Solids*; Springer Verlag: Berlin, 1982.

Riess, I. What does a voltmeter measure? *Solid State Ionics* 1997, *95*, 327–328.

Salkind, A. J.; Karpinski, A. P.; Serenyi, J. R. *Encyclopedia of Electrochemical Power Sources*; Jargen, G., Ed.; Elsevier: Amsterdam, 2009; pp. 513–523.

Saslow, W. E. Voltaic cells for physicists: Two surface pumps and an internal resistance. *American Journal of Physics* 1999, *67*, 574–583.

Sato, N. *Electrochemistry at Metal and Semiconductor Electrodes*; Elsevier: Amsterdam, 1998.

Schmickler, W. *Interfacial Electrochemistry*; Oxford University Press: New York, 1996.

Tarascon, J. M.; Armand, M. Issues and challenges facing rechargeable lithium batteries. *Nature* 2001, *414*, 359–367.

Thackeray, M. M. Manganese oxides for lithium batteries. *Progress in Solid State Chemistry* 1997, *25*, 1–71.

Toyoura, K.; Koyama, Y.; Kuwabara, A.; Oba, F.; Tanaka, I. First-principles approach to chemical diffusion of lithium atoms in a graphite intercalation compound. *Physical Review B* 2008, *78*, 214303.

Trasatti, S. *Advances in Electrochemistry and Electrochemical Engineering*; Gerischer, H., Tobias, C. W., Eds.; Wiley Interscience: New York, 1976; pp. 213–321.

Trasatti, S. The absolute electrode potential: An explanatory note. *Pure and Applied Chemistry* 1986, *58*, 955–966.

Trasatti, S. The absolute electrode potential: The end of the story. *Electrochimica Acta* 1990, *35*, 269–271.

Trasatti, S. Surface science and electrochemistry: Concepts and problems. *Surface Science* 1995, *335*, 1–9.

Tripkovic, V.; Bjarketun, M. E.; Skalason, E.; Rossmeisl, J. Standard hydrogen electrode and potential of zero charge in density functional calculations. *Physical Review B* 2011, *84*, 115452.

Tsiplakides, D.; Vayenas, C. G. Electrode work function and absolute potential scale in solid-state electrochemistry. *Journal of the Electrochemical Society* 2001, *148*, E189–E202.

van der Ven, A.; Bhattacharya, J.; Belak, A. A. Understanding Li diffusion in Li-intercalation compounds. *Accounts of Chemical Research* 2013, *46*, 1216–1225.

Varney, R. N.; Fischer, L. H. Electromotive force: Volta's forgotten concept. *American Journal of Physics* 1980, *48*, 405–408.

Verbrugge, M. W.; Koch, B. J. Modeling modeling lithium intercalation of single-fiber carbon microelectrodes. *Journal of The Electrochemical Society* 1996, *143*, 600–608.

Vetter, K. J. *Electrochemical Kinetics*; Academic Press: New York, 1967.

Vincent, C. A.; Scrosati, B. *Modern Batteries: An Introduction to Electrochemical Power Sources*, 2nd ed.; Arnold: London, 1997.

Wagemaker, M.; Mulder, F. M. Properties and promises of nanosized insertion materials for Li-ion batteries. *Accounts of Chemical Research* 2011, *46*, 1206–1215.

Wen, C. J.; Boukamp, B. A.; Huggins, A. J. H.; Weppner, W. Thermodynamic and mass transport properties of LiAl. *Journal of the Electrochemical Society* 1979, *126*, 2258–2266.

4 Work Functions and Injection Barriers

The knowledge of the work function of metals and organic and inorganic semiconductor layers is a fundamental step in determining the functional properties of devices. The alignment of Fermi level of the phases in contact forms energetic differences between the materials that compose the junction and establish the structure of energy barriers. We discuss here the main experimental methods to measure the work function: thermionic emission, the Kelvin Probe, and photoelectron emission spectroscopy, which provide information on electron energies such as conduction band and valence band edges. The description of the experimental tools provides deeper insight into the properties of metal/semiconductor junctions, and enables detailed investigation of injection barriers at the semiconductor contacts. We examine the properties of the junctions, starting from the simple assumption of the vacuum level alignment, and then take into account the effect of dipole layers and the charge-neutrality level that governs occupancy of interface states.

4.1 INJECTION TO VACUUM IN THERMIONIC EMISSION

In modern electronic, optoelectronic, and energy devices, the contacts inject electrons into organic and inorganic semiconductors, and in electrochemical devices such as batteries, charge is shuttled by redox or ionic species in solution. The notion of ejection of electrons from the electrode to vacuum, or to have an "electron at rest in vacuum" as the reference for the absolute energy scale, may sound academic, since modern devices realize conduction of electrons and ions in solid and liquid media. However, devices based on injection to vacuum motivated many of the scientific discoveries of quantum phenomena and revealed the properties of atoms and radiation, at the transition between the nineteenth and twentieth century. Thermionic cathodes dominated electronics for more than 50 years. They are still widely used in highly efficient fluorescent tubes, though semiconductor-based devices (LEDs) are increasingly occupying the lighting market.

Thermionic emission is a flow of electrons from a metal to vacuum induced by the thermal energy of the carriers (see Figure 4.1). The emission of electrons from a cathode to vacuum is controlled by the probability of release of electrons that in turn depends on the work function of the metal. The released electrons can be withdrawn from the metal surface by an electric field and moved to an anode that is positively biased. Thermionic emission also occurs by overcoming a potential barrier at metal/semiconductor contact, with additional charge transfer factors that will be discussed in Section 6.7.

In a metal cathode at finite temperature, Figure 4.1, the electrons tail upward from the Fermi level following a Boltzmann distribution as discussed in Figure 5.1. Remember from Chapter 2 that the work function of a metal Φ_m is given by the energy difference between the Fermi level and the electron at rest in vacuum, just outside the surface (see Figure 4.1). The thermal probability for electrons in the metal to overcome the barrier at the metal surface is given by the Boltzmann factor that provides the number of electrons of energy E_{vac} inside the metal. It follows that the current shows the dependence

$$j \propto e^{-q\Phi_m/k_BT} \tag{4.1}$$

More concretely, it is shown below that the current in thermionic emission has the expression of the *Richardson–Dushmann equation*

$$j = AT^2 e^{-q\Phi_m/k_BT} \tag{4.2}$$

The Richardson constant

$$A = \frac{4\pi m_0 q k_B^2}{h^3} \tag{4.3}$$

being m_0 the mass of the electron, is a universal constant, that takes the value

$$A = 1.20 \times 10^6 \ \mathrm{A/m^2K^2} \tag{4.4}$$

The first systematic studies of thermionic emission were done by Thomas Edison in 1880 in the context of his attempts to develop a long-lasting incandescent light bulb. In 1883, he made a lamp with an additional electrode to collect the charges emitted from the negative pole of the filament. He found that current flowed to the extra electrode when this electrode was connected to the *positive* side of the filament. This is because electrons start from the heated cathode and current flows only when the anode is biased positively with respect to the cathode. Thus was the *vacuum diode* invented, although

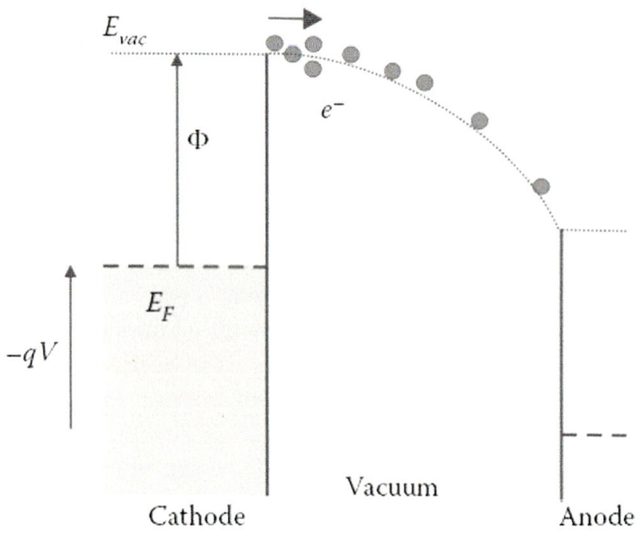

FIGURE 4.1 Thermionic emission from a cathode of work function Φ, and space-charge limited transport across the vacuum toward the anode. The electron concentration is highest at the cathode surface. The electric potential increases away from the cathode, thus electron energy decreases, as indicated by the bending of the vacuum level.

it was not recognized at that time. In 1897, J. J. Thomson proved that the ray emission from a cathode was composed of negatively charged particles that we now call electrons. In 1890, A. Fleming reported the first observation of rectification. By about 1900, the diode began to be used as a detector of radio signals. Another subject of intensive investigation was the spectrum of the light emitted by the heated filament. This led Planck to the discovery of the blackbody spectrum using the hypothesis of quanta. Soon after this, Einstein proved the reality of light quanta by his explanation of the photoelectric effect, again related to cathodic emission.

Around the year 1920, there was still ambiguity about whether thermionic currents could exist in a perfect vacuum from a pure metal, or whether the observed current was due to ionization of residual impurities. The controversy was clarified by contributions of Richardson, Child, and Langmuir. As mentioned earlier, the electrons emitted from the heated cathode will be transported to the anode by the electric field that is produced by the applied voltage. If there are few electrons in the vacuum, then the field is linear $\mathbf{E} = V/L$ and the current is determined by the emission rate indicated in Equation 4.2. But as electrons enter the region between the electrodes, they modify the field in the vacuum region, as implied by Poisson Equation 2.3. Since the electric current is spatially continuous by conservation of charge, the velocity of the electrons (and the field) is larger in the regions with less charge. This coupling of field and charge defines the space-charge limited current ((SCLC) that will be studied in Chapter 15. In this transport regime, the electron density is high at the cathode surface and decreases toward the anode, while the field is close to zero at the cathode surface and increases toward the anode, as indicated in Figure 4.1. (Note that the local field is proportional to the gradient of the VL.) Given a thickness and a voltage, these transport properties impose a fixed current that can be drawn from the cathode. In summary, the current can be limited either by the emission rate or by the transport rate. The problem of distinguishing limitations to current by either injection from the cathode or transport between the electrodes was solved in measurements by Langmuir (1913). In general, it is a difficult problem that persists to this day in all kinds of organic diodes (Shen et al., 2004).

4.2 RICHARDSON–DUSHMAN EQUATION

The thermionic current in Equation 4.2 is usually derived by integration of the product of electron occupation function and the density of states in the metal (Equation 2.12). However, another approach (Herring and Nichols, 1949; Reiss, 1985) is based on thermodynamic arguments and is very useful to illustrate the principle of detailed balance that will be further discussed in Chapter 6. First, with no voltage applied between anode and cathode, a gas of emitted electrons may form in the vacuum, that will attain equilibrium with the electrons in the metal. No net current flows then from the cathode, since the same number of electrons flows in both directions, outside and inside the metal. This derivation makes the assumption that the current flowing *in equilibrium* in both directions is the same. The equilibrium current is calculated using the properties of the electron gas in the vacuum, then, it is postulated that the same current will flow when the electrons in the gas are removed by an external field.

The equilibrium of electrons across the metal surface requires that the electrochemical potentials of electrons be the same in the two phases. The electrons in the metal have the energy $E_F = -\Phi_m$, with respect to the SVL, while the chemical potential of the electron gas, given in Equation 2.33, is

$$\mu_n = k_B T \ln\left(\frac{n}{N_g}\right) \tag{4.5}$$

Here, n is the electron density, and N_g is an effective density of states given by the expression

$$N_g = 2\lambda_{th}^{-3} \tag{4.6}$$

where the factor 2 is due to a multiplicity of spin states, and

$$\lambda_{th} = \frac{h}{\sqrt{2\pi m_0 k_B T}} \quad (4.7)$$

is the thermal de Broglie wavelength (Baierlein, 1999). The equilibrium condition $E_F = \mu_n$ imposes the following concentration of electrons

$$n = \frac{2}{\lambda_{th}^3} e^{-\Phi_m / k_B T} \quad (4.8)$$

The current incident on the metal plane from the gas side can be readily calculated by elementary kinetic theory. The number of collisions made by particles at one side of a surface placed in a gas is $nv_{th}/4$ per unit area per unit time, where the average thermal velocity for a Maxwell distribution is

$$v_{th} = \left(\frac{8k_B T}{\pi m_0} \right)^{1/2} \quad (4.9)$$

Therefore, the current results

$$j = \frac{1}{4} qnv_{th} = qn \left(\frac{k_B T}{2\pi m_0} \right)^{1/2} \quad (4.10)$$

Including Equation 4.8 in 4.10 provides the Richardson–Dushman Equation 4.3.

Measurement of thermionic current at a known temperature allows determining an "effective" work function of a metal that is defined as the value derived from the Richardson–Dushman equation by assuming the value of Equation 4.4 for the Richardson constant.

In the simple model of the vacuum thermionic device described previously, the rate of emission from the cathode is not affected by the applied potential at all. The metal emits electrons at a given rate fixed by the thermal excitation of electrons at the Fermi level above the SVL. When we consider electrons injected to a solid material, a number of additional effects must be considered. The electric field that exists between the two contacts in the device active layer does affect the extraction of electrons from the metal by increasing the rate of electron tunneling and also facilitating electron transfer at the interface. These features are discussed in Section 6.7.

The transport of injected electrons depends on the overall distribution of charges in the bulk material. Mott and Gurney (1940) first proposed by theoretical prediction that SCLC can be made to flow in insulators. To make the solid equivalent of the vacuum thermionic diode, it was required to recognize the influence of electron traps in the insulator, and the observation of SCLC took several years (Smith and Rose, 1955). It was also required to define an *ohmic contact* to an insulator. This is also a very important problem today in many classes of organic devices, especially OLEDs. When using an insulator to induce SCLC, a contact that is a source of electrons or holes is required. Smith (1955) defined a metal–semiconductor contact as *ohmic* when "the metal serves as a reservoir of carriers with free access to the conduction band of the semiconductor." In these early studies, it was observed that the critical property to form such an ohmic contact is that the metal has a low work function. This feature will be further discussed in the analysis of injection barriers below.

4.3 KELVIN PROBE METHOD

We have discussed in Chapter 2 that two materials of different work function have a different position of the Fermi level with respect to SVL. If such materials are brought to electronic equilibrium, then a Volta potential difference is established between the negatively and positively charged surfaces of the two conductors, that is the contact potential difference (CPD), which equals the original difference of work functions. Measuring the CPD of a sample with respect to a reference material of known work function allows determining the work function of the sample. It has been emphasized in Section 3.3 that the CPD cannot be simply measured with a voltmeter.

The Kelvin Probe (KP) is an experimental method to determine the CPD. As shown in Figures 3.6 and 3.7, in a capacitor arrangement we can modulate the surface charges and the Volta potential difference by changing the applied voltage. A parallel plate capacitor with a conductive substrate, coated with the material of interest, is shown schematically in Figure 4.2. A reference electrode is placed in front of the sample surface separated by a gap, and connected with the sample by an external circuit. There is a special situation in which there are no charges in the plates, that happens when the applied voltage exactly cancels the built-in potential or initial CPD.

The KP method allows detecting the situation in which the built-in field, due to the difference of Volta potentials, is nulled. When the system is biased at a value V_{app}, by Equation 3.11 the charge in the capacitor is

$$Q = C\left(V_{app} + V_{bi}\right) \quad (4.11)$$

Now the reference electrode vibrates as indicated in Figure 4.2 while keeping the voltage fixed. Since the

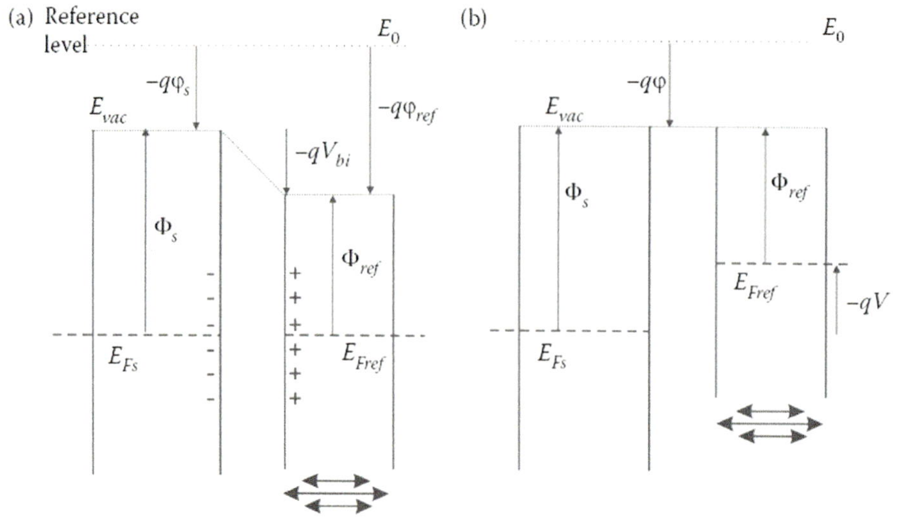

FIGURE 4.2 Measurement of the work function of a sample (*s*) with respect to a reference (*ref*) metal. (a) The two metals are connected and reach equilibrium. The voltage between the surfaces is the CPD, $\Delta\varphi = V_{bi}$. If the reference electrode vibrates, a capacitive current flows in the external circuit. (b) An applied voltage $V_{ex} = \Delta\varphi$ equilibrates the CPD. In this situation, no current is measured in the circuit.

capacitance is changing, a displacement current flows between the plates given by

$$I = \frac{dC}{dt}\left(V_{app} + V_{bi}\right) \tag{4.12}$$

The current vanishes when the external voltage compensates the contact potential. The contact potential is thus given by $V_{app} = -V_{bi}$. With respect to the general scheme of Figure 2.16, the KP measures the voltage V that nullifies the initial Volta potential difference existing in equilibrium at $V = 0$, namely, $\Delta\psi_{CPD} = \psi_1^{eq} - \psi_2^{eq}$.

The KP is a simple measurement operationally, but it has the disadvantage of poor spatial resolution. The plate electrodes have sizes of the order of mm, and therefore they provide an average of the surface properties. To obtain a high resolution of the CPD at the nanometer scale, one may attempt to reduce the size of the capacitor plate using a tip of atomic force microscopy (AFM). However, this procedure usually results in poor resolution due to small currents. The technique of Kelvin Probe Force Microscopy (KPFM) successfully determines the CPD between a conducting tip of AFM tip and a sample by measuring the electrostatic force. Note that in KPFM the voltage measured is the difference between the VL at the sample surface and the VL at the surface of the tip, $\Delta\psi_{CPD} = \psi_{sample}^{eq} - \psi_{ref}^{eq}$.

The use of a small perturbation modulated voltage allows one to resolve the local capacitance in addition to the local CPD. As a result, the KPFM-based techniques permit to monitor the spatial variations of the CPD and therefore provide laterally resolved images of a sample's surface potential or charge distribution. As example, Figure 4.3 shows the local voltage drop obtained by a measurement of the CPD by KPFM in a functionalized

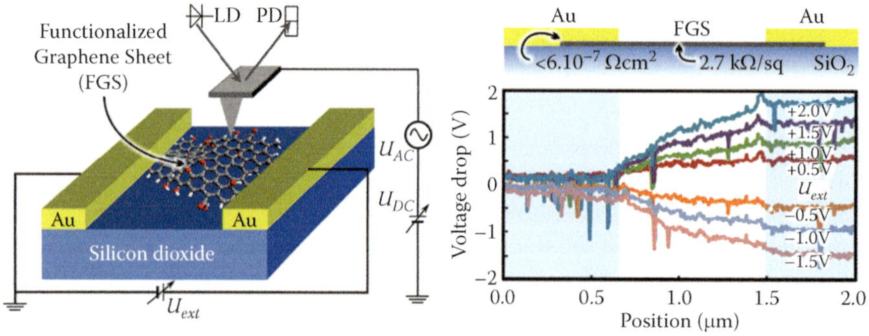

FIGURE 4.3 Scheme of a functionalized graphene oxide on a silicon dioxide substrate contacted with two gold pads, indicating the KPFM setup. Measured voltage drop with external bias ranging from +1.5 to +2.0 V. (Reproduced with permission from Yan, L. et al. *Nano Letters* 2011, *11*, 3543–3549).

graphene sheet of sub-µm size under external bias conditions.

4.4 PHOTOELECTRON EMISSION SPECTROSCOPY

Photoelectron spectroscopies (PES) consist of a range of methods in which electrons in a solid are excited by high-energy photons. Carriers excited above the SVL abandon the material with certain kinetic energy, and the resulting spectrum of photoelectrons as a function of the kinetic energy is measured in a detector. These methods provide a detailed picture of electronic states in the material from the Fermi level downward. In Inverse Photoemission Spectroscopy (IPES) incident, electrons with controlled kinetic energy (5–15 eV) penetrate the solid above the VL and decay in an empty state. The decay of the electron produces a photon that is recorded by the detector. The IPES provides a view of the unoccupied electronic states of the conduction band.

The PES methods are widely used to determine the structure of energy alignment at metal/organic interface or in general between any two semiconductors, either organic or inorganic. VL shifts can be determined upon formation of the junction and correspond to the formation of either interface dipole or band bending. To address the energetics of a particular junction, PES measures the electron levels of a substrate, first, and then the progressive development of electron levels when the organic material is deposited, by successive molecular layering. The evolution of HOMO, LUMO, and Fermi level of the materials with respect to vacuum that occurs, when covering the substrate with a required material is monitored. The drawback is that measurements are done on ultrahigh vacuum conditions, to avoid surface modification by adsorbates, and also to allow the photoemitted electrons to reach the detector without appreciable scattering. The technique allows probing each of the relevant interfaces that one is interested in, but they must be prepared separately and not integrated in devices.

In ultraviolet photoelectron spectroscopy (UPS), a sample in vacuum is irradiated with high-energy monochromatic light, and the energy distribution of emitted electrons is measured by a detector that has a common Fermi level with the sample. The source of photons often used is HeI ($h\nu$ = 21.2 V) or HeII radiation ($h\nu$ = 40.8 eV). These high-energy photons excite electrons from the valence band of the irradiated material. The sources of photons in x-ray photoelectron spectroscopy (XPS) are x-ray tubes that produce more energetic photons between 1.2 and 1.5 keV. XPS can access, in addition to the valence band region, deeply bound core-level electronic states that are characteristic of the specific chemical element.

Figure 4.4 gives an outline of the UPS method and the main energy levels and energy differences involved in the measurement. In Figure 4.4a we show the extraction of electrons from the valence band of a metal by photons of energy $h\nu$. The density of states below the Fermi level that are occupied with electrons is indicated, as well as the SVL of the sample and the detector. Provided that the photon energy is larger than the work function of the sample, electrons can be extracted and arrive to the detector with a kinetic energy that has the following value:

$$E_{kin} = h\nu - E_{binding} - \Phi_{det} \tag{4.13}$$

Here, $E_{binding}$ is the binding energy (BE) of the electron with respect to the Fermi energy of the sample, and Φ_{det} is the work function of the detector, that is normally adjusted setting the Fermi edge emission of a clean metal sample to zero-binding energy. The distribution of kinetic energies is shown at the top right of Figure 4.4a and as a function of the binding energy in Figure 4.4c. The spectra have the following characteristic features (Cahen and Kahn, 2003):

a. The highest energy at which electrons are extracted is E_k^{max} and corresponds to electrons at the Fermi level for metals, or of the valence band maximum/HOMO for semiconductors and insulators.
b. At lower kinetic energies, and higher BE, the spectra of photoemitted electrons follow the shape of the occupied DOS of the valence band. At large binding energies, peaks of core levels can be recognized.
c. The energy of electrons with *lowest* kinetic energy E_k^{min} at the detector, corresponding to the electrons with largest binding energy, is called the secondary electrons cutoff. These electrons leave the surface with zero kinetic energy and correspond to the electrons emitted with a final energy equal to the vacuum level at the sample surface. In the low E_k region there are two contributions to the emitted electrons, a small number of primary electrons that have been excited by the photon of energy $h\nu$, and secondary electrons which result from collisions of the primary electrons. These secondary electrons produce a peak in the emitted distribution, just above the cutoff. The energy of the secondary cutoff indicates changes in the VL, as will be discussed later in this chapter.

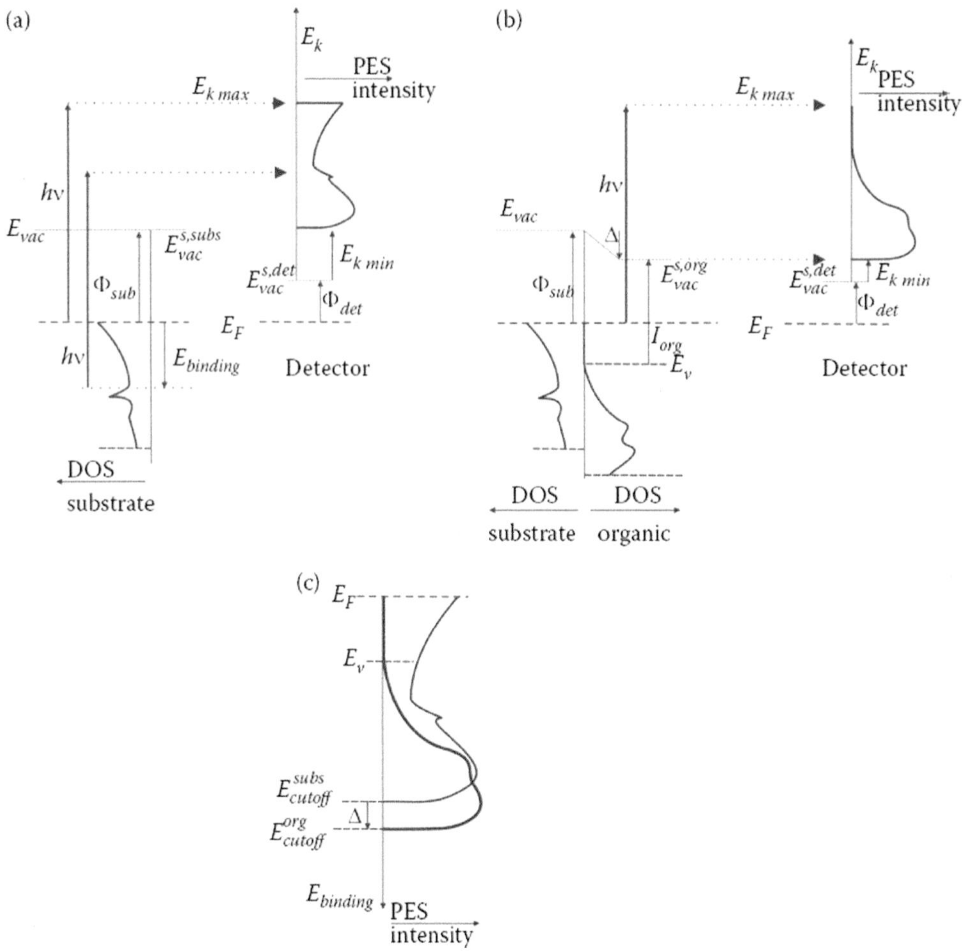

FIGURE 4.4 Principle of the UPS study of a metal (a) and metal/organic interface (b). The metal DOS (bottom left) and UPS spectrum (top right) is shown in (a). The following energy levels are indicated: energy of incident photon, vacuum level, the Fermi level, the work function for the metal substrate and the detector, and the range of kinetic energies of the emitted electrons. (b) An organic layer is deposited on top of the metal substrate. A dipole shifts down the vacuum level with respect to the VL in the substrate metal. At the detector, electrons arrive with smaller kinetic energy. (c) UPS spectrum in the scale of binding energy with respect to the Fermi level. The secondary cutoff corresponds to electrons emitted with zero kinetic energy from the deepest states in the valence band. The shift of the secondary cutoff corresponds to the displacement of the VL, Δ.

As shown in Chapter 2, the metal Fermi level lies at distance Φ_m deep with respect to the SVL. Since the electrons at the Fermi level appear with an energy E_k^{max}, the work function of the metal is given by the difference between the photon energy, $h\nu$, and the total extent of binding energies, that is the kinetic energy of electrons at the Fermi level minus that at the secondary cutoff (Hirooka et al., 1973), as given by the relation (Figure 4.4a)

$$\Phi_m = h\nu - \left(E_k^{max} - E_k^{min}\right) \tag{4.14}$$

For semiconductors, since there are no electrons at the Fermi level, the width of BEs is given by the difference: BE(E_F) − BE (secondary cutoff). The UPS provides an absolute determination of the work function of the solid, in contrast to the KP method. For semiconductors, E_k^{max} is not at E_F since there are no electrons at the Fermi level. The parenthesis in Equation (4.14) must be replaced by: BE(E_F) − BE (secondary cutoff). As all samples are electrically connected, E_F is still at zero-binding energy when the spectrometer is properly calibrated (Fermi levels are aligned).

In Chapter 2 we saw that the VL outside the surface—that is, the reference for the work function—and the VL far from the surface differ by a finite amount, the Volta potential difference, as follows:

$$E_{vac}^{\infty} - E_{vac}^{s} = -q\Delta\psi \tag{4.15}$$

In UPS the energy of those electrons that have sufficient energy to escape from the solid is measured, because they arrive at the surface with an energy $E_{vac}^{s,sample}$ or larger. The minimal kinetic energy that is detected is

$$E_k^{min} = E_{vac}^{s,sample} - E_{vac}^{s,det} \quad (4.16)$$

where $E_{vac}^{s,det}$ is the SVL at the detector, as illustrated in Figure 4.4a. Equation 4.16 is the energy that corresponds in Figure 2.16 to the Volta potential difference, $q\Delta\psi_{CPD} = q\left(\psi_{sample}^{eq} - \psi_{ref}^{eq}\right)$, that is measured by KP method. The energy of an electron at rest far from the crystal, E_{vac}^{∞}, has been introduced as a useful conceptual reference in Chapter 2, but it does not intervene in the measurement of UPS.

A typical experiment outlined in Figure 4.4 consists in measuring a clean metallic sample as a reference, in which the required material such as organic or inorganic semiconductor is deposited in a stepwise manner, for example, spin-coating a polymer film, or *in situ* condensation of a molecular monolayer. The UPS spectrum is measured for increasing thickness (Braun et al., 2009). Examples of the formation of a dipole at the metal/organic interface are shown in Figures 4.5 and 4.6. Any change of the SVL of the sample produces a change of E_k^{min}, measured in the detector, as shown in Figure 4.4b. Therefore, the low energy cutoff represents the vacuum level of the organic layer. The observation that E_{cutoff} has a different value from that obtained in the reference metal substrate indicates a dipole that shifts the VL, as shown in Figure 4.4c, or alternatively that band bending has taken place.

For an organic semiconductor deposited on the metal substrate, the Fermi level lies in the gap, as shown in Figure 4.4b; hence, the electron emitted at lowest energy indicates the edge of the HOMO or valence band. The position of the HOMO is taken as the intercept between the tangent of the onset feature and the base-line of the spectrum (Hwang et al., 2009). Combined analysis of UPS and IPES allows forming the full density of states as shown in Figure 4.7.

A modification of the Fermi level in the semiconductor layer corresponds to a change of E_k^{max}. But since the Fermi level of the sample is adjusted to the Fermi level of the detector, in terms of binding energy of measured photoelectrons the change of Fermi level will appear as a displacement of the energy levels as a whole. This is illustrated for the case of a semiconductor in Figure 4.8. According to Equation 2.45, we have

$$\Delta\Phi_{sample} = \Delta\chi_n - \Delta\zeta_n \quad (4.17)$$

The changes of the effective work function of a semiconductor (the Fermi level with respect to SVL) is due

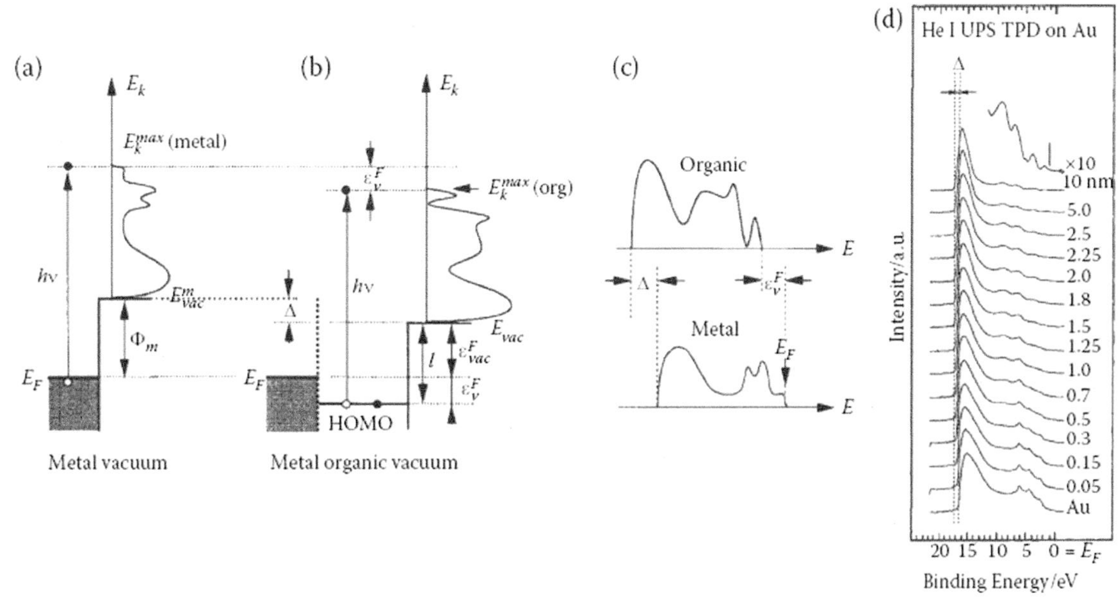

FIGURE 4.5 UPS measurement of an organic/metal interface. (a) Photoemission from the metal. (b) Photoemission from the organic layer deposited on the metal substrate. E_k: kinetic energy of photoelectron, E_k^{max} (metal) maximum kinetic energy of photoelectron from the metal, E_k^{max} (org): maximum kinetic energy of photoelectron form the organic layer. (c) Presentation of the UPS spectra of metal and organic material with the energy of an emitted electron with an arbitrary origin as the abscissa. (d) UPS spectra of TPD incrementally deposited on Au substrate as a function of film thickness. The shift of the left-hand cutoff corresponds to the VL shift Δ in (b). (Reproduced with permission from Ishii, H. et al. *Advanced Materials* 1999, *11*, 605).

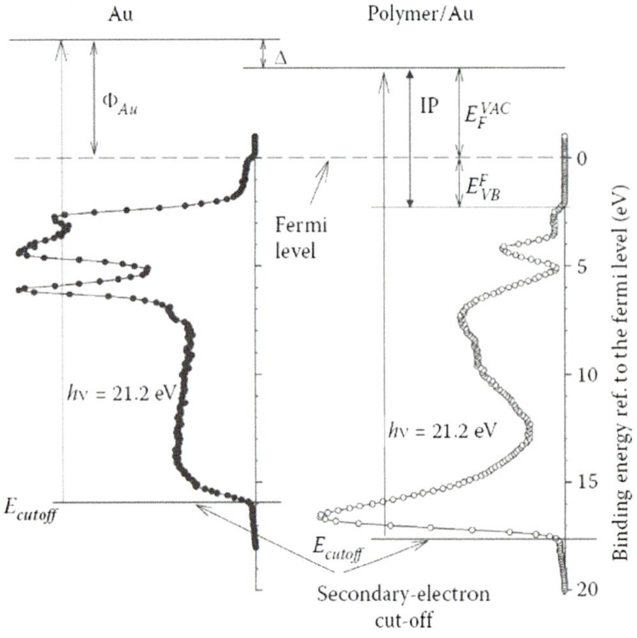

FIGURE 4.6 Schematic illustration of some of the important parameters derived from UPS characterization of surfaces and interfaces. (Reproduced with permission from Braun, S. et al. *Advanced Materials* 2009, *21*, 1450–1472).

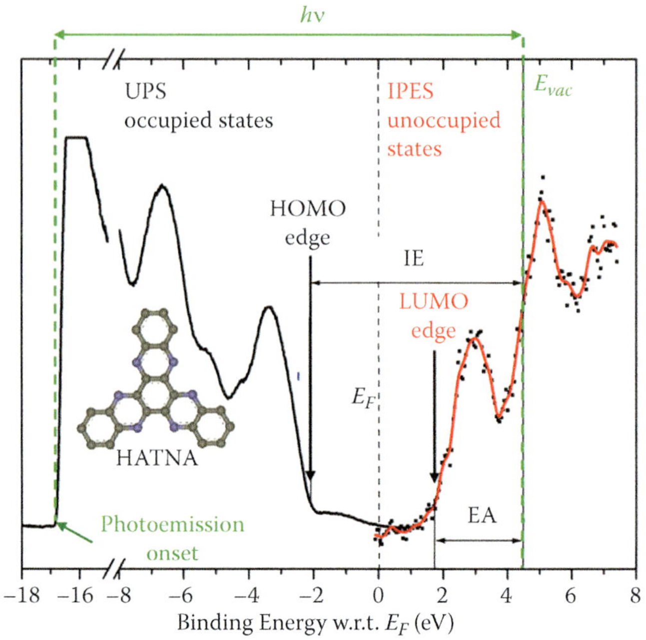

FIGURE 4.7 Combined UPS and IPES spectra from a 6 nm film of HATNA deposited on Au. The chemical structure of the molecule is shown in inset. The photoemission onset, the vacuum level (E_{vac}), the HOMO and LUMO edges, and the ionization energy (IE) and electron affinity (EA) are indicated. (Reproduced with permission from Hwang, J.; Wan, A.; Kahn, A. *Materials Science and Engineering: R: Reports* 2009, *64*, 1–31).

to several possible causes, indicated in Figure 4.8: (a) the creation of a dipole at the substrate/semiconductor interface, that changes the affinity $\Delta\chi$; (b) the formation of band bending; or (c) when the semiconductor doping is modified, there is a change of the Fermi level with respect to the conduction band edge, that is the chemical potential is modified as $\Delta\zeta_n$. These changes present different signatures in UPS, as shown in the lower panels of Figure 4.8.

The formation of interfacial dipoles for a combination of organic layers CBP/m-MTDATA on different substrates can also be followed by the displacement of E_{cutoff} as shown in Figure 4.9.

4.5 INJECTION BARRIERS

The rate of transference of carriers at the contact between a metal and an organic or inorganic semiconductor is often governed by the injection barrier that is defined as the energy difference between the metal Fermi level and the semiconductor level where carriers are injected. Figure 4.10 shows the formation of a metal–semiconductor junction that determines the height of the barrier to electron injection, $\Phi_{B,n}$, while at the semiconductor side we obtain a Schottky barrier (SB), as explained in detail later in Chapter 9. As discussed in Chapter 2, in equilibrium, the Fermi levels of the two materials come to equilibrium. This equilibration requires a total shift of the vacuum level by an amount $\Phi_m - \Phi_{sc}$, the difference between the metal and semiconductor work functions. The equilibration may take place by two basic mechanisms that produce different results for the size of the injection barrier (Cowley and Sze, 1965). The change of the VL can be achieved by the formation of extended space charge (band bending) in the semiconductor, by the formation of a dipole layer of atomic size at the interface between the metal and the semiconductor, or by a combination of the two.

We first treat the situation in which there is no interfacial dipole formation. The energy levels (conduction band, valence band) are not relatively shifted at both sides of the interface, and the vacuum level at the interface is common. This is called the *vacuum level alignment rule* or the *Mott–Schottky (MS) rule*. In contrast to this assumption, the formation of a dipole produces an abrupt shift of the vacuum level at the interface, as discussed in Chapter 2. This situation is very common at the metal/ organic interface and will be discussed in Section 4.6.

As shown in Figure 4.10b, under VL alignment rule the barrier height for injection of electrons from the

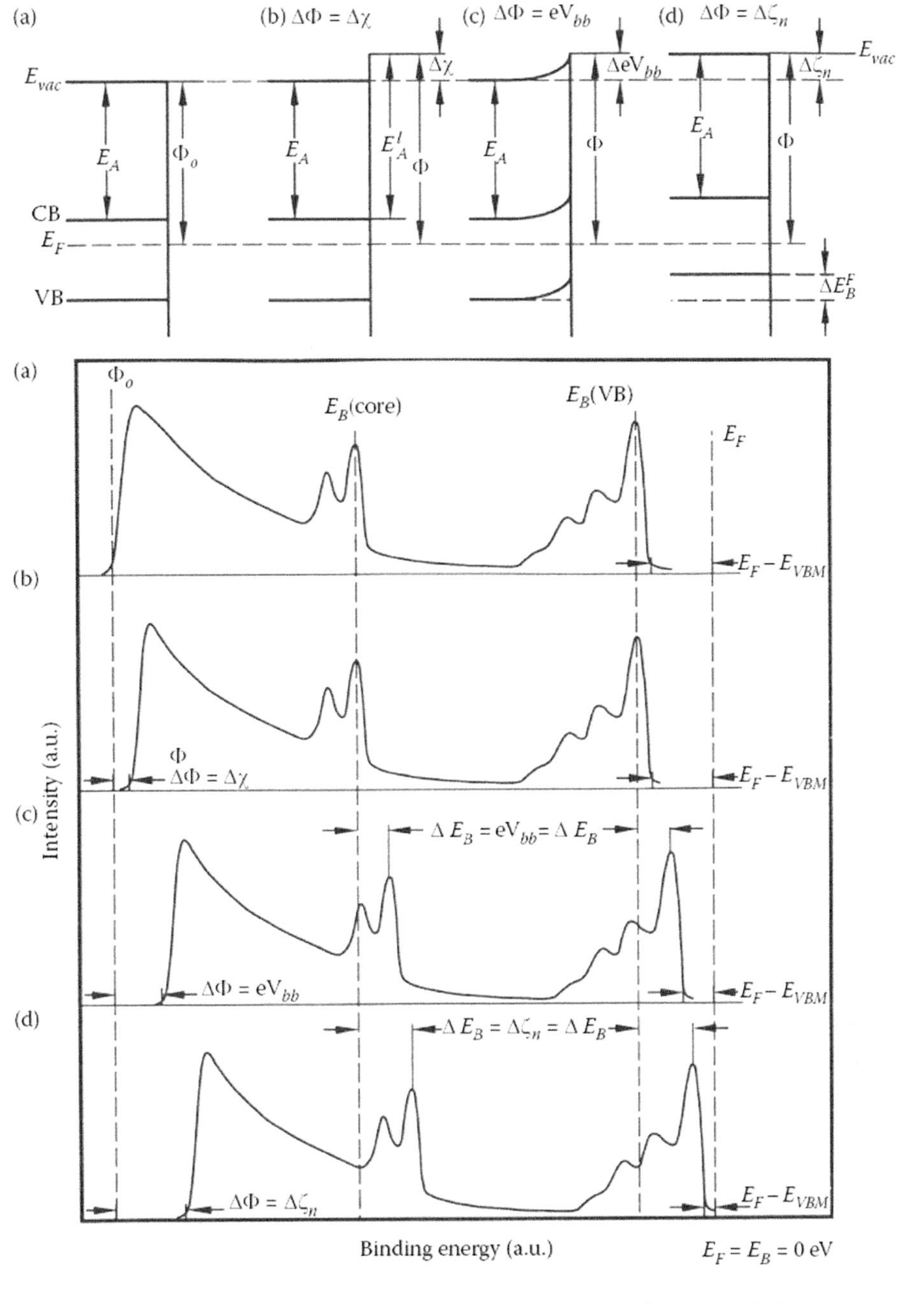

FIGURE 4.8 Schematic changes of energetic conditions at semiconductor interfaces and their consequences for the valence-band spectra. The change of the surface dipole only shifts $\chi_{sc}(E_A)$ and therefore the work function Φ. Band bending, qV_{bb}, and changes of the doping, $\Delta\zeta_n$, shift both Φ and the Fermi level with respect to the valence band edge. (Reproduced with permission from Jaegermann, W. *Modern Aspects of Electrochemistry* 1996, *30*, 1–186.)

metal is given by the difference between the metal work function and the semiconductor electron affinity:

$$\Phi_{B,n} = \Phi_m - \chi_n \tag{4.18}$$

The barrier for hole injection is given by the difference of hole affinity and the substrate work function:

$$\Phi_{B,p} = \chi_p - \Phi_m \tag{4.19}$$

In Chapter 3, we introduced the idea of the built-in potential, V_{bi}, as the CPD between two metals with a different work function. At the metal/semiconductor contact, V_{bi} is associated with the voltage change across the semiconductor space-charge layer (Ishii et al., 2004), and it is otherwise denoted as the *diffusion voltage* V_D (Sze, 1981). Here, we adopt the first denomination (built-in potential). In general, this V_{bi} does not coincide with the CPD, due to the presence of

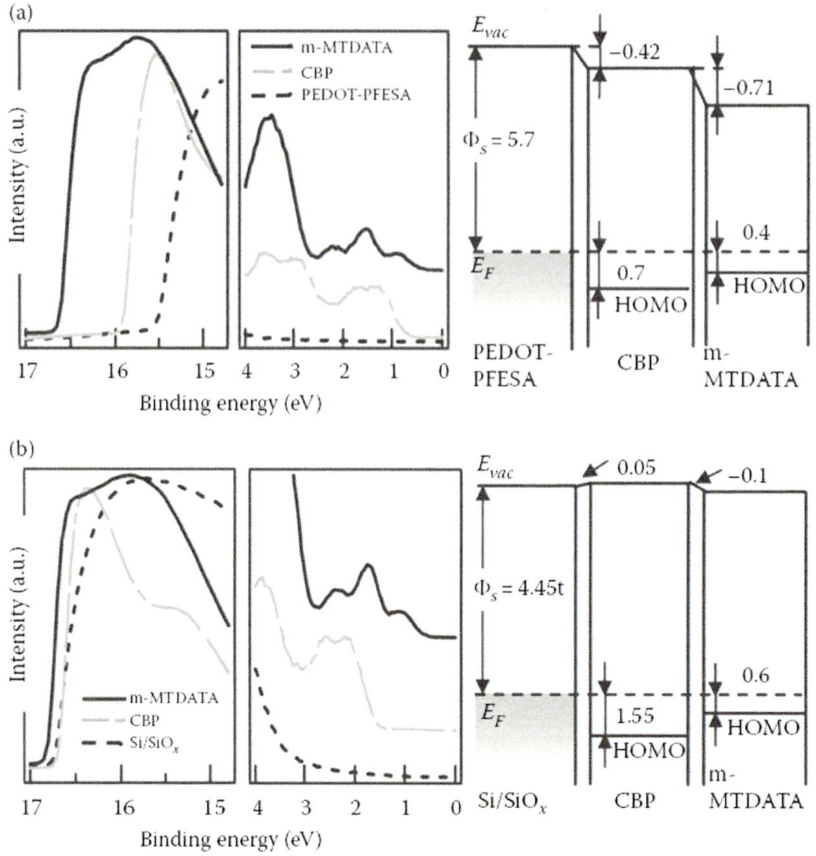

FIGURE 4.9 UPS spectra and the corresponding diagrams of energy level alignment of interfaces (a) PEDOT-PFESA/CBP/m-MTDATA (b) Si/SiO$_x$/CBP/*m*-MTDATA. (Reproduced with permission from Braun, S. et al. *Physics Letters* 2007, *91*, 202108).

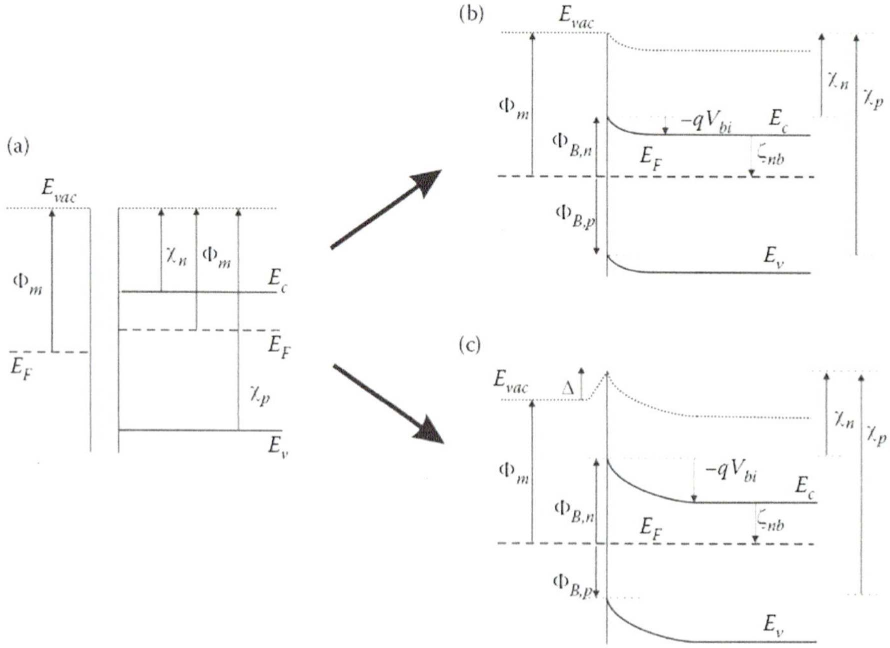

FIGURE 4.10 Equilibration of a metal with a semiconductor showing different types of contacts and the corresponding energy barriers for injection of electrons ($\Phi_{B,n}$) and holes ($\Phi_{B,P}$) at the metal–semiconductor interface and the built-in potential (V_{bi}). (a) The separate materials. (b) Common VL at the interface. (c) An interfacial dipole raises the levels of the semiconductor and increases the barrier for electron injection and the built-in potential.

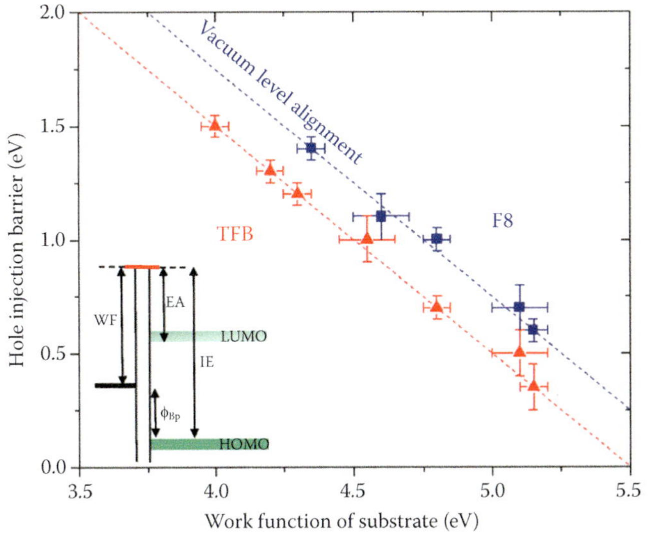

FIGURE 4.11 Hole barrier measured by UPS at polymer-on-substrate interfaces plotted as a function of substrate work function for F8 (square) and TFB (triangle). The data points fit on the vacuum level alignment, or Schottky–Mott, line. Inset: schematic energy diagram. (Reproduced with permission from Hwang, J.; Wan, A.; Kahn, A. *Materials Science and Engineering: R: Reports* 2009, *64*, 1–31).

the surface dipole. However, in vacuum level alignment situation, we have

$$V_{bi} = \frac{1}{q}\left(\Phi_m - \Phi_{sc}\right) \tag{4.20}$$

Usually, V_{bi} is reported as a positive quantity, $V_{bi} = |\Phi_m - \Phi_{sc}|/q$.

According to Equation 4.18, the MS rule predicts the following change of the barrier height with the metal work function:

$$\frac{\partial \Phi_{B,n}}{\partial \Phi_m} = 1 \tag{4.21}$$

An example of perfect VL alignment is shown in Figure 4.11. Here, the hole barrier for organic materials F8 and TFB (see Appendix 3 for the nomenclature) determined by UPS varies linearly with a slope 1 with respect to the work function of the substrate as indicated in Equation 4.21. A similar relationship for the contact of a semiconductor and electrolyte is discussed in Section 9.7.

In Section 4.1 we introduced the idea of an ohmic contact, as that contact which may supply unlimited charge for the required transport process. The ohmic contact has the property of small impedance, relative to other impedances in the device, and therefore the Fermi level is nearly flat across the ohmic contact. This is opposed to the idea of rectification, as in an SB, or polarization, in which a contact takes the step of Fermi levels corresponding to the applied voltage.

One way to form an ohmic contact is to obtain a small injection barrier (Shen et al., 2004) so that the current will flow easily for any slight difference of Fermi levels across the contact. The facile kinetics at the interface is quantified by the exchange current density j_0 (this will be discussed in Section 6.6). A metal with a low work function can easily align to the electron transport level in organic or inorganic semiconductors, and is therefore a preferred candidate to make an ohmic contact to electrons, while a large work function material matches better the edge of the valence band for hole transport. However, in some cases, the contact mechanisms among metal, metal oxide, and organic materials are more complex.

The size of the semiconductor band bending in equilibrium depends on the position of the Fermi level with respect to the semiconductor conduction band, determined by the amount of doping as commented later in Chapter 9. The chemical potential of electrons in the flat region is

$$\zeta_{nb} = \left(E_{Fn} - E_c\right)_{bulk} \tag{4.22}$$

Therefore, the built-in potential depends on doping as

$$V_{bi} = -\frac{1}{q}\left(\Phi_m - \chi_n + \zeta_{nb}\right) \tag{4.23}$$

when the doping level increases the band bending increases accordingly.

4.6 PINNING OF THE FERMI LEVEL AND CHARGE-NEUTRALITY LEVEL

In general, even in very clean interfaces a dipole layer is formed at the contact between the metal and the semiconductor. Such dipole layer plays a key role in the process of charge injection through metal/organic interfaces. The effect of the dipole at the semiconductor/liquid junction will be discussed in Sections 9.7 and 9.11.

Figure 4.10c shows the effect of the dipole layer at the metal/semiconductor contact for $\Delta > 0$. The barrier for electron injection is modified to

$$\Phi_{B,n} = \Phi_m - \chi_n + \Delta \tag{4.24}$$

and the barrier for hole injection is

$$\Phi_{B,p} = \chi_p - \Phi_m - \Delta \tag{4.25}$$

As explained earlier in UPS experiments, a very thin organic layer is deposited on a metal. A dipole is formed by the first molecular layers on the metal which provokes the change of the metal work function. As shown in Equation 2.56, the effective work function of the metal is

$$\Phi_{eff} = \Phi_m + \Delta \tag{4.26}$$

In the case shown in Figure 4.10c, the effective work function of the metal has been increased with $\Delta > 0$. The barrier for electron injection in Equation 4.24 can be expressed as

$$\Phi_{B,n} = \Phi_{eff} - \chi_n \tag{4.27}$$

If, in addition, the semiconductor layer is large, it forms a space-charge region as indicated in Figure 4.10c and the built-in voltage of Equation 4.20 is increased as

$$V_{bi} = \frac{1}{q}\left|\Phi_m + \Delta - \Phi_{sc}\right| \tag{4.28}$$

It should be remarked, however, that some texts define V_{bi} as the whole CPD at the metal/semiconductor contact.

A central question about injection barriers at metal/semiconductor interface is to predict the dipole formation, if any, when metals of progressively larger work function are deposited on the semiconductor. For example, Figure 4.12 shows the formation of a positive dipole at the 8-hydroxyquinoline aluminum (Alq) contacted with Al, and a negative dipole at Mg/Alq interface. It is required to obtain the dependence of $\Phi_{B,n}$ on Φ_m, according to material and surface properties.

In many cases it is found a linear dependence of the injection barrier, on a given semiconductor, with respect to the metal work function. The change of the barrier is regulated by *the index of interface behavior* S_Φ

$$S_\Phi = \frac{\partial \Phi_{B,n}}{\partial \Phi_m} \tag{4.29}$$

The value of this parameter is 1 under MS rule. When S_Φ is very small the height of the barrier $\Phi_{B,n}$ is insensitive to the metal work function, meaning that variations of the difference $\Phi_m - \Phi_{sc}$ are accommodated as an interfacial dipole. This behavior is known as *Fermi level pinning*. It is associated with the presence of interface states that are able to produce a strong variation of the charge at the interface with a small variation of the Fermi level at the semiconductor side of the interface. In Section 7.2, we will describe the properties of a distribution of interface states in the bandgap of a semiconductor. This property is used in the model

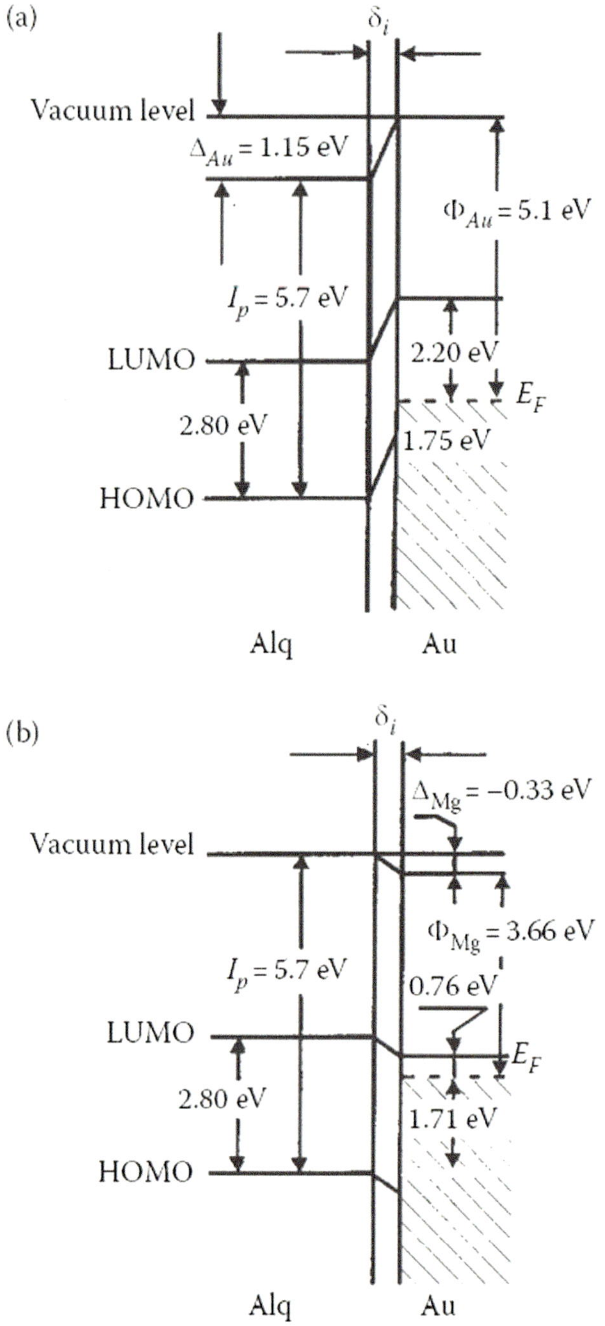

FIGURE 4.12 The energy level alignment for (a) Au/alq and (b) Mg/Alq interfaces. (Reproduced with permission from Lee, S. T. et al. *Applied Physics Letters* 1998, *72*, 1593).

(Cowley and Sze, 1965) to account for Equation 4.29. The model describes the surface dipole in terms of an interfacial density of states D_{is} (E) that can accommodate charge at the interface depending on the position of the Fermi level. The pinning parameter is

$$S_\Phi = \frac{1}{1 + q^2 \delta\, D_{is}\left(F_F\right) / \varepsilon_i \varepsilon_0} \tag{4.30}$$

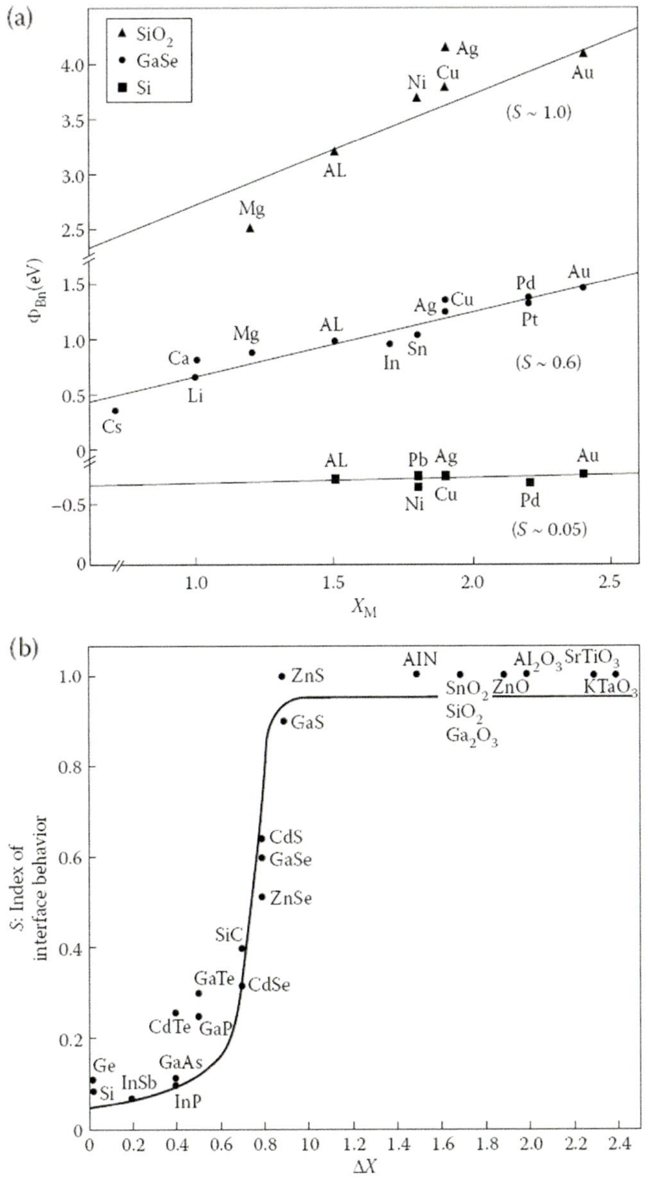

FIGURE 4.13 (a) Barrier height versus electronegativity of metals deposited on Si, GaSe, and SiO_2. (b) Behavior of index of interface S as a function of electronegativity difference of the semiconductors. (Reprinted with permission from Kurtin, S. et al. *Physical Review Letters* 1969, *22*, 1433–1436).

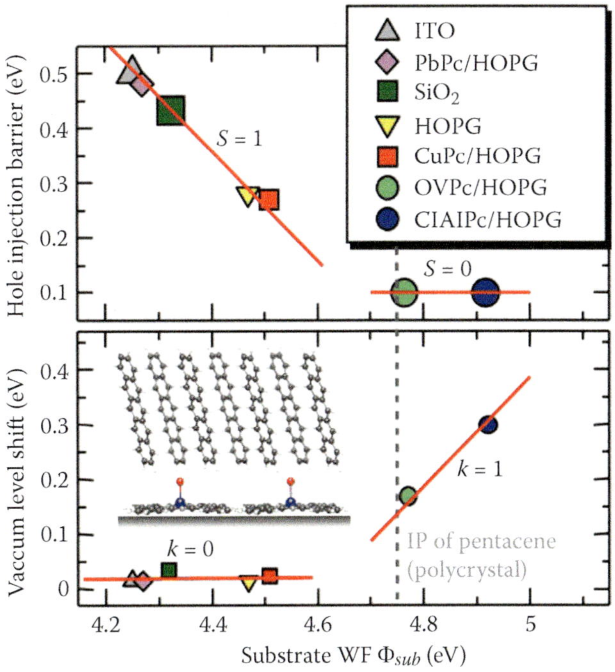

FIGURE 4.14 HOMO leading edge (or hole injection barrier, upper panel) and vacuum level shift (lower panel) for pentacene deposited on various substrates as a function of substrate work function Φ_{sub}. (Reproduced with permission from Fukagawa, H. et al. *Advanced Materials* 2007, *19*, 665–668).

where δ is the width of the dipole, and ε_i is the relative dielectric constant at the interface. It is observed that for a large DOS the interfacial charge can absorb the variation of the work function and $S_\Phi \approx 0$. The measurements of the interface Alq with Mg and Al shown in Figure 4.12 indicate that the Fermi level is effectively "pinned." If, however, D_{is} is small, then the VL alignment is satisfied, and the changes of $\Phi_m - \chi_n$ produce a variation of the injection barrier and a change of the band bending in the semiconductor.

Values of S_Φ in inorganic semiconductors range from quite small (≈0.1) for covalent semiconductors such as Ge and Si, to nearly 1 for ionic compound semiconductors, as shown in Figure 4.13. Si/metal contacts normally result in the equilibration of the Fermi level position, at an energy 0.4–0.9 eV from the bottom of the conduction band edge (Kurtin et al., 1969; Sze, 1981). An example of equilibration of organic layers is shown in Figure 4.14 that depicts hole injection barrier and VL shift (size of interfacial dipole) of a pentacene monolayer deposited on different phthalocyanine monolayer modified substrates. The results show two different behaviors. When the substrate work function is smaller than the ionization energy of the pentacene, the vacuum level alignment can be observed. However, the HOMO level is pinned close to the Fermi level when the substrate work function exceeds the ionization energy of the pentacene.

A model for the buildup of the interfacial dipole that is satisfied in many cases is given in Figure 4.15 and introduces a key parameter accounting for the interface equilibration: the charge-neutrality level (CNL). In this model, first suggested by Bardeen (1947), the direction of the dipole Δ is that which makes the Fermi level of the substrate to approach the CNL.

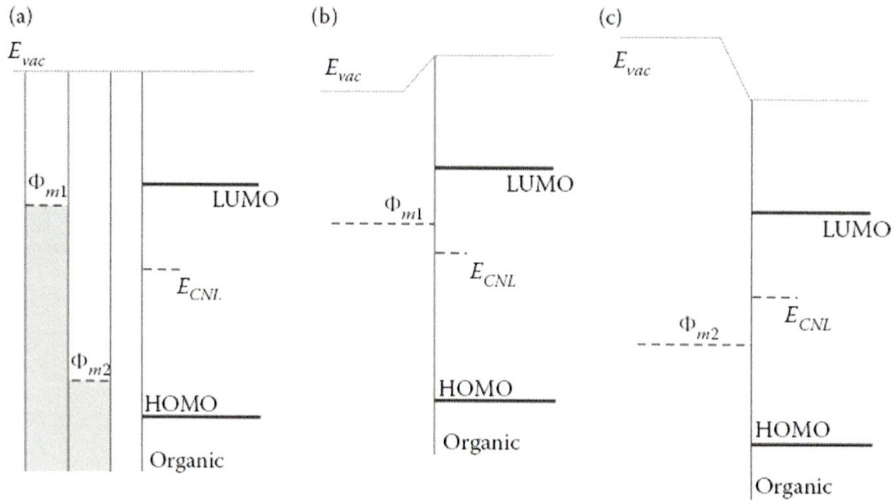

FIGURE 4.15 Contact of an organic layer with either of two metals with low and high work function. (a) Scheme of the energy levels of the three materials. (b) and (c) The interface dipole tends to align the metal Fermi level with the CNL.

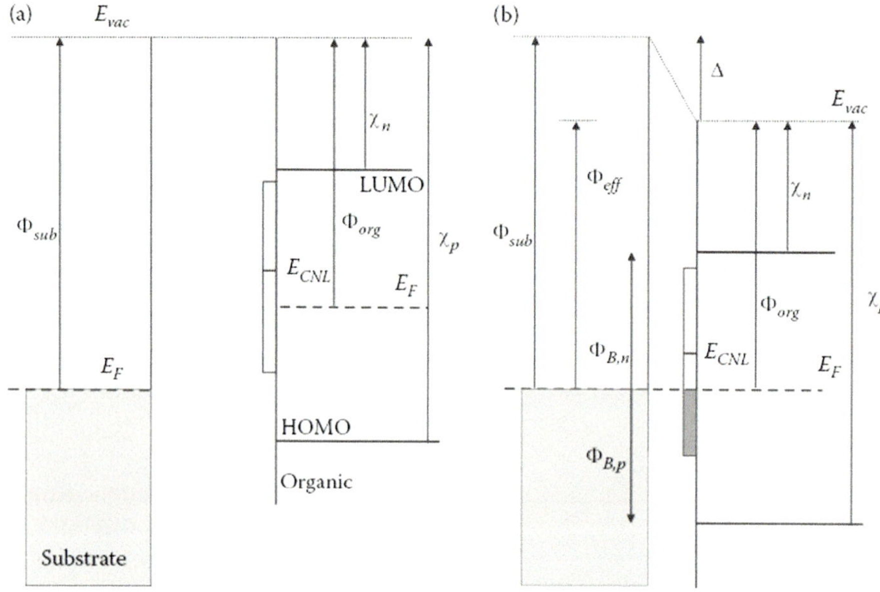

FIGURE 4.16 (a) Energy scheme of a metal substrate and organic layer. (b) When the materials come into contact, the interfacial density of gap states favors the creation of a dipole layer that makes the charge-neutrality level with energy, E_{CNL}, approach the Fermi level of the substrate.

It has been observed that the CNL plays a determining role in the interface equilibration at metal–organic interfaces. The scheme in Figure 4.16 shows in more detail that the contact of a metal substrate and an organic layer promotes the charge transfer between an interfacial density of gap states and the substrate, what leads to the formation of an interfacial dipole (Hwang et al., 2009). In this model, interfacial bandgap states are assumed at the surface in contact with the metal. The character of the surface states changes from mostly acceptor type approaching the bottom of the conduction band to mostly donor type close to the top of the valence band. The CNL is located at a given position within the interfacial DOS. E_{CNL} represents the energy that marks the separation between the two classes of states within the interfacial DOS.

The equilibration of the metal and the organic occurs by a charge transfer process that pulls the metal Fermi level Φ_m toward CNL. The sign of the dipole layer Δ, therefore, depends on the energy difference $E_{CNL} - \Phi_m$. If the Fermi level of the metal is initially above E_{CNL}, then electrons are transferred to the interfacial bandgap states, while the opposite is true if the Fermi level of the metal lies below CNL. The net consequence of the transference is that the metal E_F will tend to align with the charge-neutrality level. The injection barriers are reduced, with respect to simple vacuum alignment, or increased, according to the following expressions (Braun et al., 2009):

$$\Phi_{org} - E_{CNL} = S_{\Phi}\left(\Phi_{sub} - E_{CNL}\right) \tag{4.31}$$

$$\Delta = \left(1 - S_{\Phi}\right)\left(\Phi_{sub} - E_{CNL}\right) \tag{4.32}$$

An important assumption of this model is that the interface states are in intimate connection (equilibrium) with the metal Fermi level through a very thin separating layer (<10 Å). The dipole layer then is transparent to electrons while it can withstand a potential difference across it. The fact that interfaces of the organic with different metals can be compared is due to the relatively weak sensitivity of the organic CNL to the metal/organic interaction (Vázquez et al., 2004). If the density of interface states is very high, all metal Fermi levels will be pulled to the same place, E_{CNL}, and the interface undergoes Fermi level pinning. In this case, the interfacial states are able to accommodate all of the charges required to equilibrate a large portion of offset in work functions $\Phi_m - \Phi_{sc}$.

GENERAL REFERENCES

Charge-neutrality level: Bardeen (1947), Vázquez et al. (2004), Braun et al. (2009), Flores et al. (2009), Cakir et al. (2012).

Interfacial dipole: Tersoff (1984a, b), Baldo and Forrest (2001), Tung (2001), Ishii et al. (2004), Helander et al. (2008), Hwang et al. (2009).

Kelvin Probe and KPFM: Binnig et al. (1986), Nonnenmacher et al. (1991), Pfeiffer et al. (1996), Kronik and Shapira (1999), Hayashi et al. (2002), Bocquet et al. (2008), Melitz et al. (2011).

Photoelectron spectroscopies: Apker et al. (1948), Jaegermann (1996), Ishii et al. (1999), Cahen and Kahn (2003), Scott (2003), Braun et al. (2009), Hwang et al. (2009), Tonti et al. (2009), Klein et al. (2012).

Thermionic emission: Hensley (1961), Briere and Sommer (1977), Bonzel and Kleint (1995), Redhead (1998), Angiolillo (2009).

REFERENCES

Angiolillo, P. J. On thermionic emission and the use of vacuum tubes in the advanced physics laboratory. *American Journal of Physics* 2009, *77*, 1102–1106.

Apker, L.; Taft, E.; Dickey, J. Photoelectric emission and contact potentials of semiconductors. *Physical Review* 1948, *74*, 1462.

Baierlein, R. *Thermal Physics*; Cambridge University Press: Cambridge, 1999.

Baldo, M. A.; Forrest, S. R. Interface-limited injection in amorphous organic semiconductors. *Physical Review B* 2001, *64*, 085201.

Bardeen, J. Surface states and rectification at a metal semiconductor contact. *Physical Review* 1947, *71*, 717–727.

Binnig, G.; Quate, C. F.; Gerber, C. Atomic force microscope. *Physical Review Letters* 1986, *56*, 930–933.

Bocquet, F.; Nony, L.; Loppacher, C.; Glatzel, T. Analytical approach to the local contact potential difference on (001) ionic surfaces: Implications for Kelvin probe force microscopy. *Physical Review B* 2008, *78*, 035410.

Bonzel, H. P.; Kleint, C. On the history of photoemission. *Progress in Surface Science* 1995, *49*, 107–153.

Braun, S.; Salaneck, W. R.; Fahlman, M. Energy-level alignment at organic/metal and organic/organic interfaces. *Advanced Materials* 2009, *21*, 1450–1472.

Briere, T. R.; Sommer, A. H. Low-work-function surfaces produced by cesium carbonate decomposition. *Journal of Applied Physics* 1977, *48*, 3547–3451.

Cahen, D.; Kahn, A. Electron energetics at surfaces and interfaces: Concepts and experiments. *Advanced Materials* 2003, *15*, 271–277.

Cakir, D.; Bokdam, M.; Jong, M. P. D.; Fahlman, M.; Brocks, G. Modeling charge transfer at organic donor-acceptor semiconductor interfaces. *Applied Physics Letters* 2012, *100*, 203302.

Cowley, A. M.; Sze, S. M. Surface states and barrier height of metal-semiconductor systems. *Journal of Applied Physics* 1965, *36*, 3212–3220.

Flores, F.; Ortega, J.; Vazquez, H. Modelling energy level alignment at organic interfaces and density functional theory. *Physical Chemistry Chemical Physics* 2009, *11*, 8658–8675.

Fukagawa, H.; Kera, S.; Kataoka, T.; Hosoumi, S.; Watanabe, Y.; Kudo, K.; Ueno, N. The role of the ionization potential in vacuum-level alignment at organic semiconductor interfaces. *Advanced Materials* 2007, *19*, 665–668.

Hayashi, N.; Ishii, H.; Ouchi, Y.; Seki, K. Examination of band bending at buckminsterfullerene C60/metal interfaces by the Kelvin probe method. *Journal of Applied Physics* 2002, *92*, 3784–3793.

Helander, M. G.; Wang, Z. B.; Qiu, J.; Lu, Z. H. Band alignment at metal/organic and metal/oxide/organic interfaces. *Applied Physics Letters* 2008, *93*, 193310–193313.

Hensley, E. B. Thermionic emission constants and their interpretation. *Journal of Applied Physics* 1961, *32*, 301.

Herring, C.; Nichols, M. H. Thermionic emission. *Reviews of Modern Physics* 1949, *21*, 185–270.

Hirooka, T.; Tanaka, K.; Kuchitsu, K.; Fujihira, M.; Inokuchi, H.; Harada, Y. Photoelectron spectroscopy of naphthacene crystal in the vacuum ultraviolet region. *Chemical Physics Letters* 1973, *18*, 390–393.

Hwang, J.; Wan, A.; Kahn, A. Energetics of metal/organic interfaces: New experiments and assessment of the field. *Materials Science and Engineering: R: Reports* 2009, *64*, 1–31.

Ishii, H.; Hayashi, N.; Ito, E.; Washizu, Y.; Sugi, K.; Kimura, Y.; Niwano, M.; et al. Kelvin probe study of band bending at organic semiconductor/metal interfaces: Examination of Fermi level alignment. *Physica Status Solidi (a)* 2004, *201*, 1075–1094.

Ishii, H.; Sugiyama, K.; Ito, E.; Seki, K. Energy level alignment and interfacial electronic structures at organic/metal and organic/organic interfaces. *Advanced Materials* 1999, *11*, 605.

Jaegermann, W. The semiconductor/electrolyte interface: A surface science approach. *Modern Aspects of Electrochemistry* 1996, *30*, 1–186.

Klein, A.; Mayer, T.; Thissen, A.; Jaegermann, W. *Methods in Physical Chemistry*; Wiley-VCH Verlag GmbH and Co. KGaA, Berlin, 2012; 477–512.

Kronik, L.; Shapira, Y. Surface photovoltage phenomena: Theory, experiment, and applications. *Surface Science Reports* 1999, *37*, 1–206.

Kurtin, S.; McGill, T. C.; Mead, C. A. Fundamental transition in the electronic nature of solids. *Physical Review Letters* 1969, *22*, 1433–1436.

Langmuir, I. The effect of space charge and residual gases on thermionic currents in high vacuum. *Physical Review* 1913, *2*, 450–486.

Melitz, W.; Shen, J.; Kummel, A. C.; Lee, S. Kelvin probe force microscopy and its application. *Surface Science Reports* 2011, *66*, 1–27.

Mott, N. F.; Gurney, R. W. *Electronic Processes in Ionic Crystals*, 2nd ed.; Oxford University Press: London, 1940.

Nonnenmacher, M.; O'Boyle, M. P.; Wickramasinghe, H. K. Kelvin probe force microscopy. *Applied Physics Letters* 1991, *58*, 2921–2923.

Pfeiffer, M.; Leo, K.; Karl, N. Fermi level determination in organic thin films by the Kelvin probe method. *Journal of Applied Physics* 1996, *80*, 6880–6883.

Redhead, P. A. The birth of electronics: Thermionic emission and vacuum. *Journal of Vacuum Science and Technology* 1998, *16*, 1394–1401.

Reiss, H. The Fermi level and the redox potential. *The Journal of Physical Chemistry* 1985, *89*, 3783–3791.

Scott, J. C. Metal—organic interface and charge injection in organic electronic devices. *Journal of Vacuum Science & Technology A: Vacuum, Surfaces, and Films* 2003, *21*, 521–531.

Shen, Y.; Hosseini, A. R.; Wong, M. H.; Malliaras, G. G. How to make ohmic contacts to organic semiconductors. *Chem.Phys.Chem.* 2004, *5*, 16–25.

Smith, R. W. Properties of ohmic contacts to Cadmium Sulfide single crystals. *Physical Review* 1955, *97*, 1525–1530.

Smith, R. W.; Rose, A. Space-charge-limited currents in single crystals of Cadmium Sulfide. *Physical Review* 1955, *97*, 1531–1535.

Sze, S. M. *Physics of Semiconductor Devizes*, 2nd ed.; John Wiley and Sons: New York, 1981.

Tersoff, J. Schottky barrier heights and the continuum of gap states. *Physical Review Letters* 1984a, *52*, 465–468.

Tersoff, J. Theory of semiconductor heterojunctions: The role of quantum dipoles. *Physical Review B* 1984b, *30*, 4874–4877.

Tonti, D.; Zanoni, R.; Garche, J. R. *Encyclopedia of Electrochemical Power Sources*; Elsevier: Amsterdam, 2009; pp. 673–695.

Tung, R. T. Recent advances in Schottky barrier concepts. *Materials Science and Engineering: R: Reports* 2001, *35*, 1–138.

Vázquez, H.; Flores, F.; Oszwaldowski, R.; Ortega, J.; Pérez, R.; Kahn, A. Barrier formation at metal-organic interfaces: Dipole formation and the charge neutrality level. *Applied Surface Science* 2004, *234*, 107–112.

5 Thermal Distribution of Electrons, Holes, and Ions in Solids

The electrochemical potential (Fermi level) is the physical quantity that sets the essential transduction between electron density in a semiconductor and the voltage at the contacts. In this chapter, we examine the relationship among the density of states, the occupation of such states, and the voltage, and also several important cases for energy conversion and storage devices. Based on statistical concepts, we provide the interpretation of the variation of the Fermi level in terms of the change of entropy of an ensemble of electrons. We analyze various practical instances such as the occupation of band edges in doped semiconductors, the degenerate semiconductors used in highly conducting transparent oxides, and the properties of hot electrons. A device of fundamental importance in many applications is the diode that we describe in a simple model in terms of selective contacts and homogeneous Fermi levels. The recombination diode model develops a detailed analysis of an electronic device as a thermal machine that uses the available energy to produce a quantity of work, conserving or decreasing the total entropy. In the final section of the chapter, we analyze in more depth the origin of voltage in the Li battery cell. To this end, we discuss the energy diagram containing a mixed ionic/electronic conductor that incorporates variations of chemical potential of both types of carrier. The model for the chemical potential of weakly interacting particles is applied to ions in the lattice gas model. Finally, to explain the voltage in more realistic situations, configurational phase transitions and the consequences of the changes of oxidation state of the intercalation material are discussed.

5.1 EQUILIBRATION OF THE ELECTROCHEMICAL POTENTIAL OF ELECTRONS

In crystalline semiconductors used in electronic and optoelectronic devices, the conduction band and valence band states play a very important role, as these states govern the transport of electrons. The quantum mechanical states are Bloch functions of infinite extension reflecting the periodicity of the lattice. The extended states over long distance in the semiconductor lattice allow for fast electronic transport, only interrupted by scattering with the lattice imperfections and with phonons.

In Chapter 2, we discussed a semiconductor with electrons in a rigid band of lowest energy level E_c. We introduced the notion of the electron Fermi level E_{Fn}, which is equivalent to the electrochemical potential of the electrons η_n. In the analysis of semiconductor devices, the Fermi level is most useful in two specific aspects: it determines the work and voltage that a collection of carriers in the semiconductor can deliver, and it provides a determination of the *occupancy* of the available electronic states, depending on the energy of the states.

Let us establish an important equilibrium condition that states that the electrochemical potential of electrons in two regions in contact that achieve thermal equilibrium must be equal (see Equation 2.60). To calculate the Fermi level of an ensemble of electrons, we start from the general expression of the Helmholtz free energy F in terms of the total energy of the electrons E_{tot}, the total configurational entropy S_n, and the absolute temperature T:

$$F = E_{tot} - TS_n \tag{5.1}$$

The entropy of electrons is calculated from the number of available configurations for the electrons ensemble and a specific model is given in Section 5.2. In general, the electrochemical potential is defined as

$$\eta_n = \frac{\partial F}{\partial n} = \frac{\partial E_{tot}}{\partial n} - T\frac{\partial S_n}{\partial n} \tag{5.2}$$

The energy of one electron can be identified with the edge of the conduction band,

$$E_c = \frac{\partial E_{tot}}{\partial n} \tag{5.3}$$

and the configurational part of the chemical potential has the expression

$$\zeta_n = -T\frac{\partial S_n}{\partial n} \tag{5.4}$$

Thus, we obtain Equation 2.35.

Consider two electron reservoirs that come into contact. The free energy of the whole system must be a minimum with respect to all other configurations that would be obtained by isothermally transferring electrons from one system to the other, that is

$$\Delta F = \frac{\partial F_1}{\partial n_1}\Delta n_1 + \frac{\partial F_2}{\partial n_2}\Delta n_2 = 0 \tag{5.5}$$

By the conservation of the number of electrons $\Delta n_1 = -\Delta n_2$, we obtain the equilibrium condition $\eta_{n1} = \eta_{n2}$ (Equation 2.60).

5.2 CONFIGURATIONAL ENTROPY OF WEAKLY INTERACTING PARTICLES

We calculate the thermodynamic function of the ensemble of electrons that gives their electrochemical potential as a function of the concentration. The entropy and free energy of a set of electrons is calculated by the methods of statistical mechanics and depends on the multiplicity of configurations and on the interactions of carriers that determine the energy of each configuration. We neglect any interaction of electrons like the Coulomb repulsion, which is a good approximation in a shielded or charge-compensated system at moderate concentrations (Section 5.4). We discuss the most basic model that consists of a total number of N_c electronic states, all with the same energy E_c, in which electrons occupying the states are considered as noninteracting entities, and each of the N_c states can be occupied by one electron, at most. For Bloch functions of infinite extension, the number of states in a solid has been given in Equation 2.14, and for electrons in vacuum, it is given in Equation 4.6. Both expressions have the same form only differing in the effective mass of the electron.

If n is the number of electrons, then n sites are occupied and $(N_c - n)$ are empty. The number of configurations is

$$P_n = \frac{N_c!}{n!(N_c - n)!} \tag{5.6}$$

Using the Stirling approximation valid for large numbers, the entropy of such distribution can be calculated from the general definition

$$S_n = -k_B \ln P_n \tag{5.7}$$

and is given by

$$S_n = -k_B N_v \left\{ \frac{n}{N_c} \ln\left(\frac{n}{N_c}\right) + \left(\frac{N_c - n}{N_c}\right) \ln\left(\frac{N_c - n}{N_c}\right) \right\} \tag{5.8}$$

The entropy in Equation 5.8 is configurational entropy of the system that arises from the random distribution of nonidentical particles. In physical chemistry, it is known as the *entropy of mixing*. From the expression of the entropy, we obtain the configurational part of the chemical potential

$$\zeta_n = -T\frac{\partial S_n}{\partial n} = k_B T \ln\left(\frac{n}{N_c - n}\right) \tag{5.9}$$

Thus, Equation 2.31 can be written as

$$\mu_n = \mu_n^0 + k_B T \ln\left(\frac{n}{N_c - n}\right) \tag{5.10}$$

This result completes the derivation of the Fermi–Dirac distribution function (see Equation 5.13). This model is valid for any system based on randomly distributed weakly interacting particles, either ionic or electronic. In Equation 2.32, we provided expressions that are restricted to the case of a diluted system, $n \ll N_c$, but now Equation 5.10 is effective at high occupation numbers as well. However, when the average distance of separation between particles are small, the energy levels of the system are affected by the interaction. The chemical potential then requires correction by an "activity coefficient." In this case the reference state, which is taken μ_n^0 when the Fermi level is at the band edge in Chapter 2 has to be more suitably defined (Harvey, 1962).

5.3 EQUILIBRIUM OCCUPANCY OF CONDUCTION BAND AND VALENCE BAND STATES

The density of states (DOS) in the conduction and valence bands are distributed over a wide range of energies (see Figure 2.2), but the carriers thermalize to the lowest (higher) energy levels in the conduction (valence) band. According to Equation 2.13, the width of the occupancy in the energy axis is of the order $2k_BT$. Therefore, often we can consider that the electrons are located at the lowest level with energy E_c, with the effective density of states N_c, which for crystalline semiconductors is $N_c \approx 10^{19}$–10^{21} cm^{-3}. From Equations 2.35 and 5.10, we can write the expression of the electrochemical potential as

$$\eta_n = E_{Fn} = E_c + k_B T \ln\left(\frac{n}{N_c - n}\right) \tag{5.11}$$

and for holes occupying a total of each of N_v states

$$-\eta_p = E_{Fp} = E_v + k_B T \ln\left(\frac{p}{N_v - p}\right) \tag{5.12}$$

The electron density at the energy level E is determined by the product of the mean occupation number, $0 \leq f \leq 1$, and number density of such state, N_v

$$n = N_v f \tag{5.13}$$

Equation 5.11 can be written in the form (5.13), being $f = F(E - E_{Fn})$, the Fermi–Dirac distribution function that has the form

$$F(E - E_F) = \frac{1}{1 + e^{(E - E_F)/k_B T}} \tag{5.14}$$

This result has been obtained from the model of weakly interacting particles but the form of the Fermi–Dirac distribution function $F(E - E_F)$ is quite general. The ratio of occupied to empty sites is

$$\frac{F(E - E_F)}{1 - F(E - E_F)} = e^{(E - E_F)/k_B T} \tag{5.15}$$

For the occupation of level at the energy E_c, we obtain

$$n = N_c \frac{1}{1 + e^{(E_c - E_F)/k_B T}} \tag{5.16}$$

Figure 5.1a plots the Fermi–Dirac distribution of Equation 5.16 at the temperature of 300 K. The plot gives the occupation of any energy level (including that of localized states in the bandgap, further discussed in Chapter 7), when the Fermi level has a certain value. As the Fermi energy *increases* in the energy scale (at a fixed temperature), the occupancy of an electronic state *increases*. An electronic state at energy E_0 will be occupied with 0.5 probability when $E_{Fn} = E_0$. The occupation function passes rapidly from 0, for $E_{Fn} \ll E_0$, to 1 at $E_{Fn} \gg E_0$.

If the Fermi level of the electrons stays below the conduction band edge, so that $E_c - E_{Fn} \gg k_B T$, then Equation 5.16 can be described by the *Boltzmann distribution*

$$n = N_c e^{(E_{Fn} - E_c)/k_B T} \tag{5.17}$$

(see Equation 2.36). In Equations 5.16 and 5.17, we may observe that the electron density in the conduction band is determined by the distance of E_{Fn} to E_c, that is, the chemical potential of the electrons $\zeta_n = E_{Fn} - E_c$, and we have

$$n = N_c e^{\zeta_n / k_B T} \tag{5.18}$$

When $E_{Fn} \ll E_c$ the occupancy of the conduction band determined by Equation 5.17 is a small number, so that only a very small fraction of the effective density of states, N_c, is occupied with electrons (see Figure 5.1a). However, such small occupancy is highly significant, as this population makes the semiconductor material an electronic conductor. Figure 5.1b shows how the Boltzmann tail creates free carriers in the conduction band in the situations of Fermi level position as indicated in Figure 5.1a.

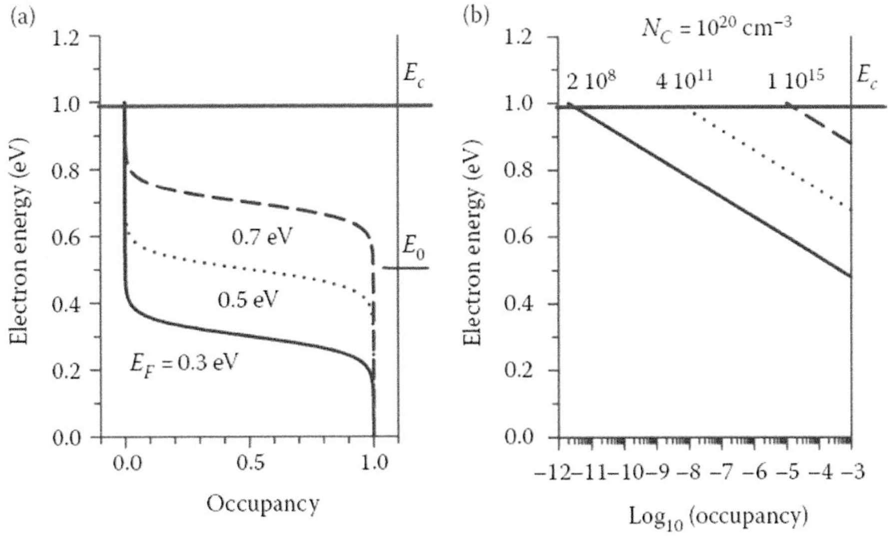

FIGURE 5.1 (a) The Fermi–Dirac distribution at different values of the Fermi level at $T = 300$ K. The vertical axis is the energy level and the lower axis is the occupancy. At the right is shown a particular level $E_0 = 0.5$ eV representing a localized level in the bandgap. At $E_F = E_0$, the trap level is exactly half-occupied. (b) The same distributions in log scale, showing the occupancy of the conduction band level. The numbers on the top axis is the electron density for a total density of states $N_c = 10^{20}$ cm^{-3}.

When the semiconductor contains states in the vicinity of the Fermi level, it is said that the Fermi level separates occupied from unoccupied states. However, this statement must be made relative to the temperature of the electron distribution. Figure 5.2a shows the large impact of the temperature on the Fermi distribution. At low temperature, it is quite accurate to assume that the Fermi level separates states occupied (below E_{Fn}) and empty (above E_{Fn}) of electrons. But at room temperature and higher the thermal energy of the carriers broadens the transition from 1 to 0, the Boltzmann tail reaches much farther so that the conduction band becomes rapidly populated with free carriers.

The process of displacing the Fermi level in a semiconductor is illustrated in Figure 5.3. We denote the Fermi level in the absence of applied external voltage E_{F0}, and this is called the "equilibrium Fermi level." Correspondingly, the electron density at equilibrium is

$$n_0 = N_c e^{(E_{F0}-E_c)/k_BT} \tag{5.19}$$

and from Equations 5.17 and 5.19, we may write

$$n = n_0 e^{(E_{Fn}-E_{F0})/k_BT} \tag{5.20}$$

This last relationship shows explicitly that the electron density increases exponentially when the Fermi level rises, as outlined in Figures 2.8 and 5.3b. Note that the electrons reside in the level E_c; there are no electrons at the Fermi level since this is the bandgap of the semiconductor and those energy states are forbidden. The situation is different when there are trap states in the bandgap. In this case there is a density of electrons right at the Fermi level, as discussed in Chapter 7.

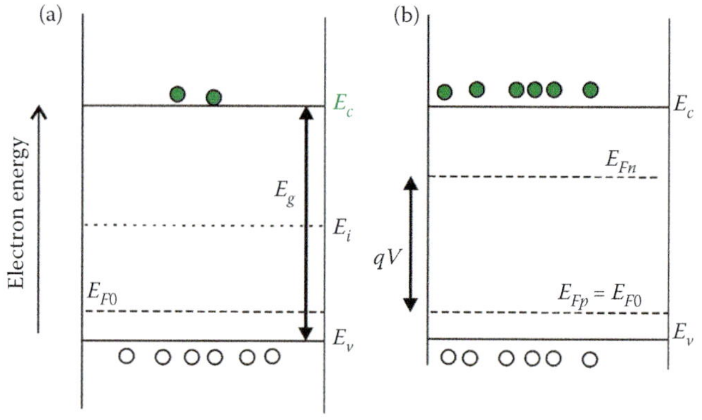

FIGURE 5.3 Energy diagram of p-type semiconductor layer. E_g is the bandgap energy, E_c is the energy of the conduction band, E_v is the energy of the valence band, E_i is the intrinsic energy level, E_{F0} is the equilibrium Fermi level, E_{Fn} is the electron Fermi level, E_{Fp} is the hole Fermi level, q is the elementary charge, and V is the voltage. (a) Zero bias. (b) Forward bias increases the population of electrons in the conduction band.

The density of holes in the valence band level (at energy E_v) in relation to the holes Fermi level E_{Fp} is given by the expression

$$p = N_v\left[1 - f\left(E_v - E_{Fp}\right)\right] = N_v \frac{1}{1 + e^{-(E_v-E_{Fp})/k_BT}} \tag{5.21}$$

Note that the concentration of holes increases when the hole Fermi level gets deeper into the band gap, moving

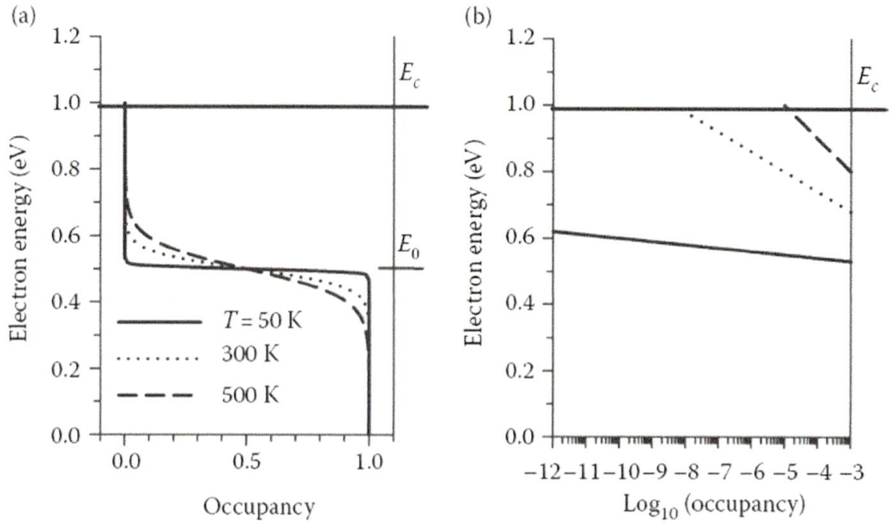

FIGURE 5.2 (a) The Fermi–Dirac distribution at different temperatures for a Fermi level at $E_F = 0.5$ eV. The vertical axis is the energy level and the lower axis is the occupancy. At the right is shown a particular level $E_0 = 0.5$ eV representing a localized level in the bandgap. At $E_F = E_0$, the trap level is exactly half-occupied. (b) The same distributions in log scale, showing the increasing occupancy of the conduction band level as $1/k_BT$ becomes smaller.

away from the VL and approaching the energy of the valence band. In the Boltzmann approximation, and using the chemical potential of holes $\zeta_p = E_v - E_{FP}$

$$p = N_v e^{(E_v - E_{Fp})/k_B T} = N_v e^{\zeta_p / k_B T} \tag{5.22}$$

and we also have

$$p = p_0 e^{-(E_{Fp} - E_{F0})/k_B T} \tag{5.23}$$

The number of electrons and holes that a semiconductor contains in equilibrium depends on the dopants concentration, as discussed in Section 2.1. It is useful to introduce the carrier density of the intrinsic, nondoped semiconductor that is produced just by thermal excitation of carriers. The intrinsic carrier density n_i and intrinsic Fermi energy E_i are determined by the two equations:

$$n_i = n_0 \Big|_{E_{Fn} = E_i} = N_c e^{(E_i - E_c)/k_B T} \tag{5.24}$$

$$n_i = p_0 \Big|_{E_{Fp} = E_i} = N_v e^{(E_v - E_i)/k_B T} \tag{5.25}$$

The intrinsic Fermi energy is typically situated close to the midgap energy (Figure 5.3), and has the value

$$E_i = \frac{E_c + E_v}{2} + \frac{1}{2} k_B T \ln\left(\frac{N_v}{N_c}\right) \tag{5.26}$$

The intrinsic carrier density can be expressed as

$$n_i = \sqrt{N_c N_v} e^{-E_g / 2 k_B T} \tag{5.27}$$

It has a strong temperature dependence as shown in Figure 5.4 and described by the expression

$$n_i(T) = 2\left(\frac{2\pi k_B T}{h^2}\right)^{3/2} \left(m_e^* m_h^*\right)^{3/4} e^{-E_g / 2 k_B T} \tag{5.28}$$

The rise of occupation with temperature is due to the increase of thermal occupancy as shown in Figure 5.2b.

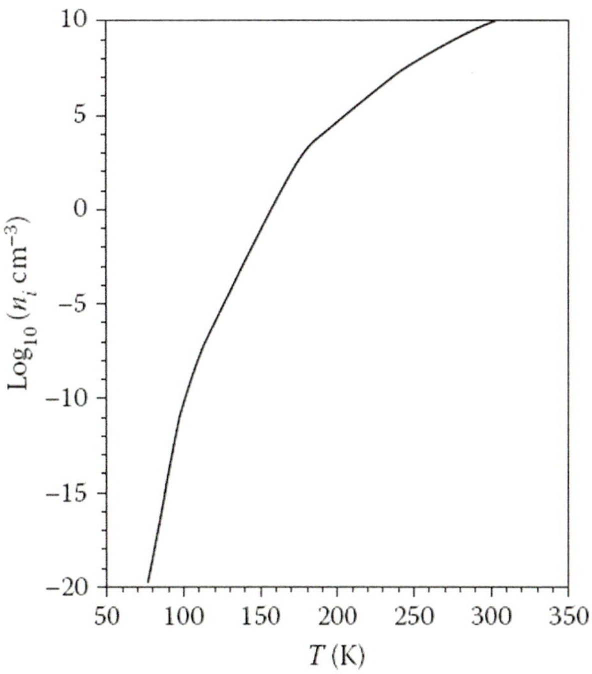

FIGURE 5.4 The intrinsic carrier concentration of silicon as function of temperature.

5.4 EQUILIBRIUM FERMI LEVEL AND THE CARRIER NUMBER IN SEMICONDUCTORS

In semiconductors, the carrier density is highly variable, and it can be controlled by the voltage in devices such as diodes and transistors that lie at the core of semiconductor electronics or photogeneration. If the Fermi level of a carrier is close to the respective band, the density is large and the conductivity associated with that carrier is also large. A concentration of electrons of the order of $n \approx 10^{16}$ cm^{-3} usually provides fairly good conductivity, but the conductivity depends on the carrier mobility as well. Conversely, when the Fermi level moves away from the associated band the carrier density decreases, and the conductivity becomes poor. In insulators the Fermi level remains far from the band edges, although it is possible to inject a large quantity of carriers that drift in SCL regime.

The condition of electrochemical equilibrium already shown in Equation 2.60 states that different materials in contact must reach a state where the Fermi level has the unique equilibrium value. If there are several species in the same material that interact by chemical reaction or recombination processes, they must reach equilibrium of the electrochemical potential. In particular, the chemical equilibrium of electrons and holes in the same material requires that the Fermi level of all the species is the same, E_{F0} (see Figure 5.3a):

$$E_{Fn} = E_{Fp} = E_{F0} \tag{5.29}$$

Traditional semiconductor devices are based on semiconductors that are doped. This means that the semiconductor possesses a homogeneous distribution of charged impurities that produce a correspondent distribution of carriers in the band, provided that Fermi level lies below the electron donor level (Blackmore, 1962). For a density of charged donors N_D, the number of free electrons is

$$n_0 \approx N_D \tag{5.30}$$

and for a density of charged acceptors N_A the number of free holes that exist in the valence band is

$$p_0 \approx N_A \tag{5.31}$$

In the bulk of a doped semiconductor, the free carrier density is fixed by the donor density. The region where the band is flat (in opposition to regions of band bending close to the surface) is called the *neutral* or *quasineutral* region (Figure 2.1). The distance between the Fermi level and the conduction band corresponds to the chemical potential of this carrier ζ_{nb}, as indicated in Equation 4.22:

$$\zeta_{nb} = \left(E_{Fn} - E_c\right)_{bulk} \tag{5.32}$$

In quasineutral region, free carriers are found with the concentration of Equations 5.30 and 5.31, hence

$$\zeta_{nb} = k_B T \ln\left(N_D/N_c\right) \tag{5.33}$$

The most abundant carrier in the semiconductor is called the *majority carrier*, for example, holes in the *p*-type doped semiconductor as shown in Figure 5.3a, in which the equilibrium Fermi level E_{F0} lies close to the valence band edge. By photogeneration or voltage injection it is possible to introduce a significant quantity of the opposite carrier, electrons in this case, which is called the *minority carrier*, while the majority carrier Fermi level is not modified (Figure 5.3b).

Equation 5.29 indicates a situation in which the assembly of electrons in the conduction band is in equilibrium with the assembly of *electrons* in the valence band (determining the *hole* distribution). When $E_c - E_{F0} \gg k_B T$ and $E_{F0} - E_v \gg k_B T$, the following relationship is satisfied:

$$n_0 p_0 = n_i^2 \tag{5.34}$$

that is known as the *mass action law*.

Carriers in their respective bands may obey the thermal distribution of the ambient temperature, as we discuss here, or not, in the case of hot electrons treated in Section 5.6. Consider the DOS of the conduction band $D_c(E)$ that has been introduced in Equation 2.10, and the correspondent DOS for the valence band of a semiconductor, $D_v(E)$. The electron density in the conduction band at energy level E above E_c is given by the product of the DOS and the occupation function (see Equation 2.12)

$$n(E)dE = D_c(E) f(E - E_{Fn}) dE \tag{5.35}$$

and the density of holes at a level below E_v is

$$p(E)dE = D_v(E)\left[1 - f(E - E_{Fp})\right]dE \tag{5.36}$$

Figure 5.5 shows different examples of the distribution of thermalized electrons and holes in the conduction and valence band, respectively, in a DOS described in Figure 2.2 and shown in Figure 5.5a. The carriers extend into the conduction (or valence) band over an energy range of approximately $2k_B T = 0.05$ eV. Figure 5.5b shows a case in which the two ensembles of carrier maintain electrochemical equilibrium at the same Fermi level E_{F0}.

In the analysis of devices we typically consider situations out of equilibrium, in which a finite difference of Fermi levels, or voltage $V \neq 0$, occurs. There are *separate equilibria* to the assemblies of electrons in the conduction band, on the one hand, and those in the valence band, in the other. That is, the relative populations of the states *within each* band will be of the same form as the equilibrium distribution at temperature T, dictated by Fermi–Dirac distributions functions in the available DOS.

Such situation is described by *separate* electrochemical potentials for both types of carriers, introduced by Shockley (1950). When the two groups of carriers are not in equilibrium with each other, the Fermi levels are often denoted with a prefix "quasi." In the applications discussed in this book, we do not use such prefix and speak normally of the separate *Fermi levels* of electrons and holes, E_{Fn} and E_{Fp} (see Figure 5.3b). Our special denomination for the Fermi level in the case of total equilibrium of all carriers, electrons, and holes is the *equilibrium Fermi level*, E_{F0}, of Equation 5.29. The equilibrium of Fermi levels more generally includes any other carrier states in the device, such as the redox carriers in the semiconductor/electrolyte system, implying $E_{F0} = E_{redox}$.

In Figure 5.5c the separate populations of carriers are thermalized to the semiconductor temperature at separate Fermi levels. Since both Fermi levels are closer to the band edges than in Figure 5.5b, the concentrations are much larger in Figure 5.5c (note the different scales of N). The stability of such distribution that provides an internal separation of electrochemical potentials, and therefore must tend to equilibrium at E_{F0}, is determined by the relationship of the thermalization time (discussed in Section 5.6), with respect to the carrier lifetime for

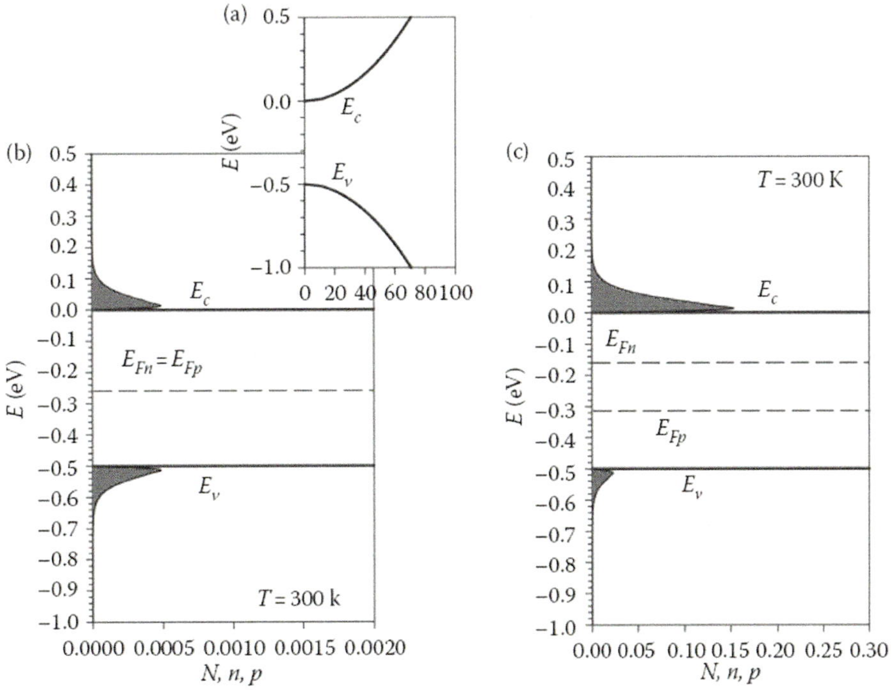

FIGURE 5.5 (a) DOS in the conduction and valence bands. (b) Energy distribution of the populations of electrons and holes in electrochemical equilibrium at room temperature. (c) Electrons and holes thermalized in the respective bands at room temperature at separate Fermi levels.

recombination of electrons and holes. Usually the former is in the picosecond, and the latter is in the nanosecond domain, so that the thermalized population at high voltage can be maintained over a significant timeframe for the required applications. The internal voltage can also be maintained by an external stimulus as light irradiation. The voltage is established by the excess carriers resulting from the balance of generation and recombination.

In summary, equilibrium of an ensemble of carriers means that they have been distributed according to the thermal distribution corresponding to the temperature of the surrounding material, and a quasi-Fermi level can be attributed to the ensemble. A quasiequilibrium distribution is then wholly specified by either parameter, the carrier density or the Fermi level, these being related by Equations 5.16 and 5.21.

The product of the carrier densities, np, is of interest for several aspects of device analysis such as recombination rates. We have out of equilibrium

$$np = N_c N_v e^{-E_g/k_BT} e^{(E_{Fn}-E_{Fp})/k_BT} = N_c N_v e^{(\zeta_n-\zeta_p)/k_BT} \quad (5.37)$$

In terms of the intrinsic carrier density of Equation 5.27, we may write

$$np = n_i^2 e^{(E_{Fn}-E_{Fp})/k_BT} \quad (5.38)$$

5.5 TRANSPARENT CONDUCTING OXIDES

In metals, Fermi level is fixed at a certain energy level inside the semiconductor band and the density of free electrons is very high. Since the whole set of electrons can gain kinetic energy in available states right above the Fermi level, the conductivity is large. If the Fermi level of a semiconductor is brought into the conduction band, we have $n \approx N_c$ and $E_{Fn} \geq E_c$. The semiconductor contains a large carrier density that requires a more general type of statistics than Equation 5.15 and is called a *degenerate semiconductor.*

Development of applications in the area of transparent thin film electronics and optoelectronics require highly conducting transparent conductors. This is usually achieved with a thin layer (about 200 nm) of a highly doped transparent conducting oxide (TCO) such as indium-doped tin oxide (ITO) or fluor-doped tin oxide (FTO) on highly transparent glass. Band diagrams of ZnO, In_2O_3, and SnO_2 determined from in situ photoelectron spectroscopy measurements of films prepared by magnetron sputtering are shown in Figure 5.6. These materials can be so highly doped that the Fermi level is placed deeply into the conduction band. The large variation in the Fermi level position of around 1 eV is obtained by preparation of films using different amounts of doping and by varying the oxygen content in the sputter gas.

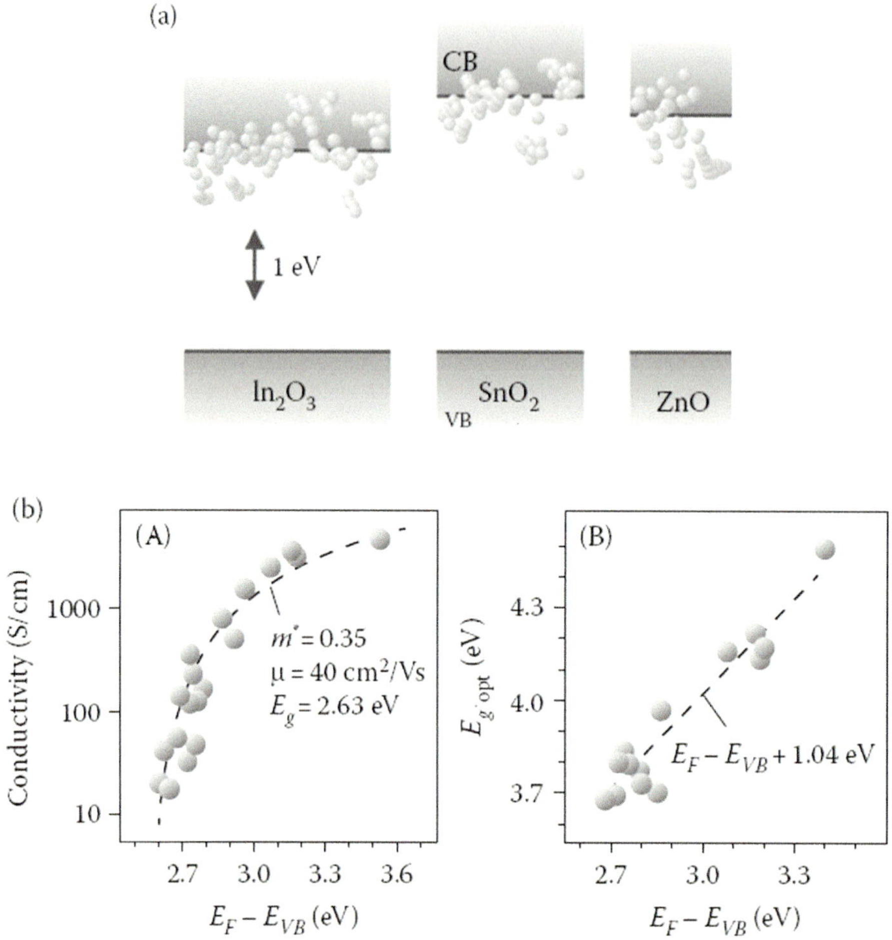

FIGURE 5.6 (a) Fermi levels of ZnO, In_2O_3, and SnO_2 determined from in situ photoelectron spectroscopy measurements of films prepared by magnetron sputtering. Each data point has been recorded from a separate film. (b) Correlation between electrical and optical properties and the Fermi level position obtained from in situ XPS measurements of the same films performed right after deposition: (A) electrical conductivity, (B) optical bandgap. The dashed line in (A) corresponds to a calculated conductivity using a parabolic conduction band. (Reproduced with permission from Klein, A. *Journal of the American Ceramic Society* 2013, *96*, 331–345.)

The sheet resistance R_{sh} of the TCO layer of thickness d is obtained from the resistivity ρ_n as

$$R_{sh} = \frac{\rho_n}{d} \tag{5.39}$$

It is often given in units of Ω/square (Ω/sq), and it is required to be below 10 Ω in many applications. With a typical carrier mobility of n-type TCO of $u_n = 30\ cm^2\ s^{-1}\ V^{-1}$ a carrier density of $2.1 \times 10^{21}\ cm^{-3}$ is required, to provide a resistivity of $\rho_n = 10^{-4}\ \Omega$ cm, for a film thickness of 100 nm, in order to obtain a sheet resistance of 10 Ω. Such resistivity is at the limit of what is possible with state-of-the-art TCO materials (Klein, 2013).

5.6 HOT ELECTRONS

Photoexcitation of electrons to the conduction band produces carriers that are not thermalized to the lattice temperature and have large kinetic energies, and are therefore called *hot carriers* (see Figure 5.7). The different steps involved in hot carrier relaxation are shown in Figure 5.8. The carriers may equilibrate among themselves, at a higher temperature than the ambient temperature. Then, the velocities follow a Maxwell–Boltzmann distribution, with average velocity indicated in Equation 4.9, so that the equilibrated case is called *Maxwellian*. Figure 5.9 shows the distribution of the carriers thermalized at the temperature of 3400 K. In this case $k_BT = 0.29$ eV, so that the carriers extend over a range of 1 eV into the respective bands (compare Figure 5.5).

However, if the hot carriers are not equilibrated among themselves then neither a temperature nor a Fermi level can be attributed to such carriers. Some studies show that the injected carrier distribution significantly departs from Maxwellian form (Neges et al., 2006). Experimental methods to determine hot electron

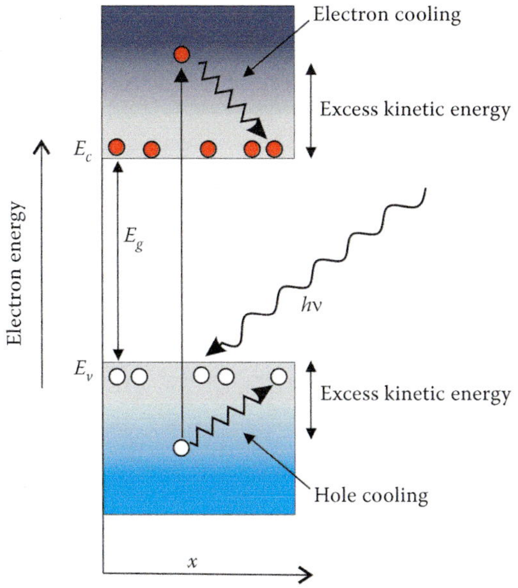

FIGURE 5.7 The absorption of a photon of energy $h\nu$ promotes the photoexcitation of an electron to the conduction band that lies in a high-energy state for a short time. Thereafter, the hot electron cools to the bottom of the conduction band in states of energy E_c.

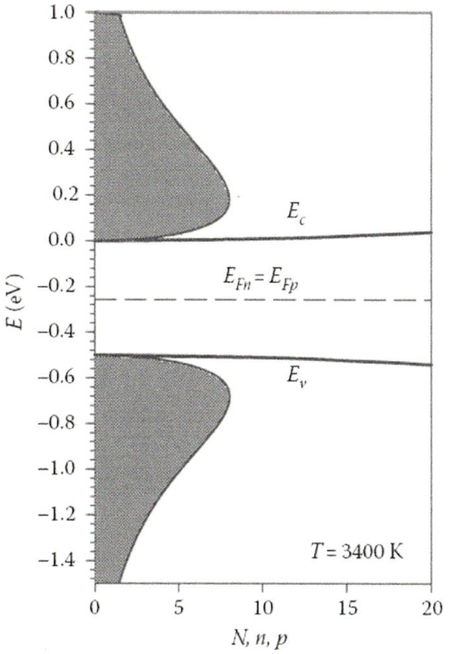

FIGURE 5.9 Electrons and holes in electrochemical equilibrium at high temperature.

effects at molecular processes in the metal surface are discussed in Gadzuk (2002) and Toben et al. (2005).

When the hot carrier distribution is generated in an environment at normal room temperature, the high-temperature thermalized Maxwellian distribution shown in Figure 5.9 only exists for a very short time interval. After a time in the scale of about 10^{-11} s the electrons relax in the lattice and since that instant on, it is well justified to use the distributions thermalized to the ambient temperature.

The short-lived high-energy carriers or excitons as shown in Figure 5.9 provide an opportunity to convert a large fraction of the initial photon energy into voltage by stabilizing a large Fermi level difference, and therefore to design high-efficiency solar cells. However, the utilization of the nonthermalized carriers at high energy requires ultrafast injection to a material with a stable Fermi level, via energy-selective contacts. In molecular or nanoscale absorbers, it is feasible that the time for carrier thermalization is longer than that for carrier extraction or recombination, for example, by suppressing the phonons that are emitted in the relaxation of the hot carriers (Conibeer et al., 2008). In practice, overcoming ultrafast relaxation of photogenerated carriers to the semiconductor band edges is an enormous challenge and such devices remain at an early stage of research. The use of plasmon-induced hot electrons in metals, injected to a semiconductor, is another resource that has been explored to make photovoltaic devices (Clavero, 2014).

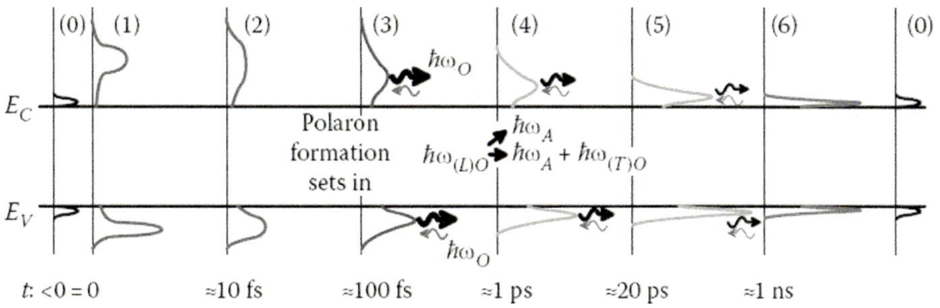

FIGURE 5.8 Carrier cooling kinetics in bulk semiconductor: Thermal equilibrium (0); immediately after optical generation (1); carrier–carrier scattering, impact ionization, re-normalization of carrier energies, Fermi–Dirac statistics (2); optical phonon emission (re-absorption) (3); decay of optical into acoustic phonons (4); further phonon emission (5), to thermal equilibrium, onset of carrier recombination (6). (Reproduced with permission from Konig, D. et al. *Physica E: Low-Dimensional Systems and Nanostructures* 2010, *42*, 2862–2866.)

5.7 SCREENING

Coulomb interaction has a long range, with a dependence $1/r$ on the distance of the interacting particles. Therefore, a collection of carriers in a material is likely to strongly interact if the density is high. In many situations however, the electrical attraction or repulsion between charges is modified by properties of the medium, and the range of the interaction is considerably shortened. The reduction of carrier–carrier interaction, and the weakening of internal electric field, is termed *screening*. The orientational polarizability produced by the orientation of dipolar molecules in a polar medium introduces the screening of local fields. The potential around a point charge Q inversely decreases with the increase of the static dielectric constant ε_r,

$$\varphi(r) = \frac{Q}{4\pi\varepsilon_r\varepsilon_0 r} \tag{5.40}$$

Another type of screening is associated with the presence of mobile charge in a conducting medium, both electrons and holes in a semiconductor, or ions in a liquid or solid electrolyte. In the presence of mobile charges in the medium with density n_0, the local potential increases the density of charge around a point charge, and the potential of the central charge becomes screened. Using Boltzmann statistics and Poisson equation, it can be shown that the potential is modified to the expression

$$\varphi(r) = \frac{Q}{4\pi\varepsilon_r\varepsilon_0 r}\exp\left(-\frac{r}{\lambda_D}\right) \tag{5.41}$$

where

$$\lambda_D = \left[\frac{\varepsilon_r\varepsilon_0 k_B T}{q^2 n_0}\right]^{1/2} \tag{5.42}$$

is the Debye screening length. Figure 5.10 gives the size of the Debye screening length as function of the background-free carrier density in the medium and of the dielectric constant of the medium.

The physical meaning of Debye screening length is directly related to the balance between electrostatic force among the carriers in the medium and the random thermal force that gives rise to diffusion. The competition between the electrostatic attraction and thermal repulsion (through diffusion force) establishes a minimum distance over which it is possible to set local electric fields by space charge distribution. Electric fields cannot be supported over distances *smaller* than λ_D, because the thermal motion destroys the ordering of charges that sustains such field by Poisson equation. Hence, space charge formation of size less than λ_D is rapidly shattered by the thermal force. The time scale in which local space charge is annihilated is the dielectric relaxation time,

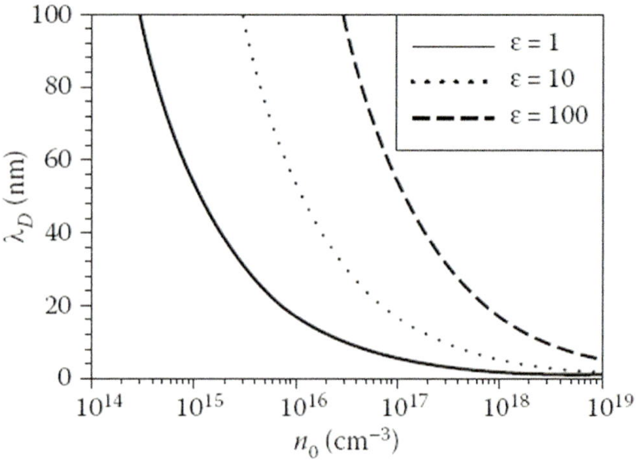

FIGURE 5.10 The Debye screening length as function of background density of carriers for different values of dielectric constant of the medium.

$$\tau_d = \frac{\varepsilon_r\varepsilon_0}{\sigma_n} \tag{5.43}$$

Here, σ_n is the conductivity.

For example, in the depletion layer of a Schottky barrier, it is required that the layer thickness be $w \gg \lambda_D$ as discussed in Chapter 9. In equilibrium, the width of the SB can be written in terms of the built-in potential V_{bi}, as follows:

$$w_0 = \lambda_D\left[\frac{2qV_{bi}}{k_B T}\right]^{1/2} \tag{5.44}$$

We observe that for $qV_{bi} \gg k_B T$ the depletion layer is much larger than the Debye length.

It is instructive to normalize the local potential to the thermal energy value

$$\Phi = q\varphi/k_B T \tag{5.45}$$

Then, Poisson Equation 2.7 takes the form

$$\frac{\partial^2\Phi}{\partial x^2} = -\frac{1}{\lambda_D^2} \tag{5.46}$$

In electrolytes, the concentration of ions c is typically very high. For a 1:1 salt of 0.1 M concentration, we have $c_0 = 10^{20}$ cm^{-3}. The electrolyte of such concentration has very small Debye length. Any space charge and spatial charge separation is rapidly destroyed, except at the surfaces of the electrodes. The Helmholtz layer at the electrode surface accommodates the potential drops, as commented in Section 3.7.

The expression in Equation 5.42 is restricted to systems in which the free carriers display Boltzmann statistics. The Debye screening length can be expressed for arbitrary statistics of carriers as follows (Banyai and Koch, 1986):

$$\lambda_2 = \left[\frac{\varepsilon_r \varepsilon_0}{q^2}\left(\frac{\partial n}{\partial E_{Fn}}\right)^{-1}\right]^{1/2} \tag{5.47}$$

The screening length correlates with the chemical capacitance, c_μ, which will be introduced in Chapter 7

$$\lambda_D = \left[\frac{\varepsilon_r \varepsilon_0}{c_\mu}\right]^{1/2} \tag{5.48}$$

5.8 THE RECTIFIER AT FORWARD AND REVERSE VOLTAGE

In this section we present a preliminary overview of some concepts that are central to understanding the operation of solar energy conversion in photovoltaic cells. This topic lies at the heart of this book and it will be explained in more systematic terms in Chapter 20.

We have previously analyzed the modification of carrier density and Fermi levels in the example of a *p*-type semiconductor of Figure 5.3. As discussed in Section 3.1, to convert such slab into a voltage-operated device, we need to supply contacts to the slab. We consider now an archetype device denoted a *diode* that is formed by adding *carrier-selective* contacts to a semiconductor. The meaning of such type of contact in a semiconductor is that the Fermi level of the current collector equilibrates to the Fermi level of either kind of carrier, electrons, or holes. The operation of such contacts is shown schematically in Figure 5.11 and with further details in Figure 5.12.

When we apply a bias voltage departing from the zero bias situations, different and asymmetric current responses of the device occur depending on the sign of voltage. The reason for the asymmetric behavior of contacts is discussed later, and the main consequence is the characteristic current–voltage curve shown in Figure 5.13. In the direction denoted "forward bias," the diode produces a large flow of carriers and current

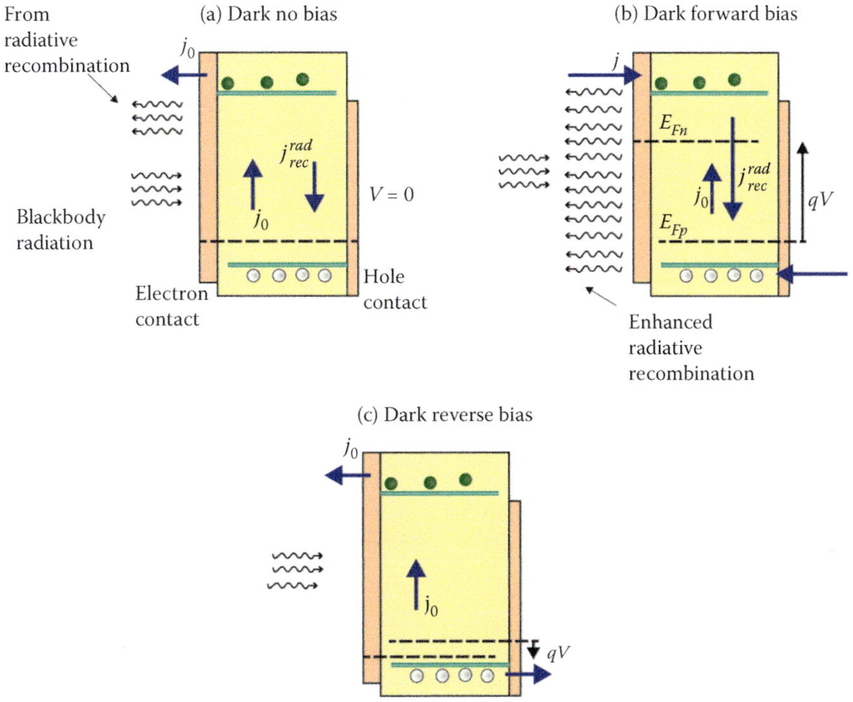

FIGURE 5.11 Model diode, of *p*-type semiconductor material, with an electron selective contact at the left side and a hole selective contact at the right side, indicating the voltage, Fermi levels, internal and external currents, and the balance of incoming and outgoing photons absorbed or produced in the material. (a) The diode in dark equilibrium. The generation by incoming thermal radiation is balanced by recombination that corresponds to reverse saturation current. All recombination is radiative. (b) Under applied bias voltage in the dark carriers are injected (a positive voltage increases minority carrier density) and the recombination of carriers is greatly increased. The diode emits light corresponding to the energy of the bang gap. (c) At reverse voltage, the internal current generated in equilibrium is extracted (reverse saturation current).

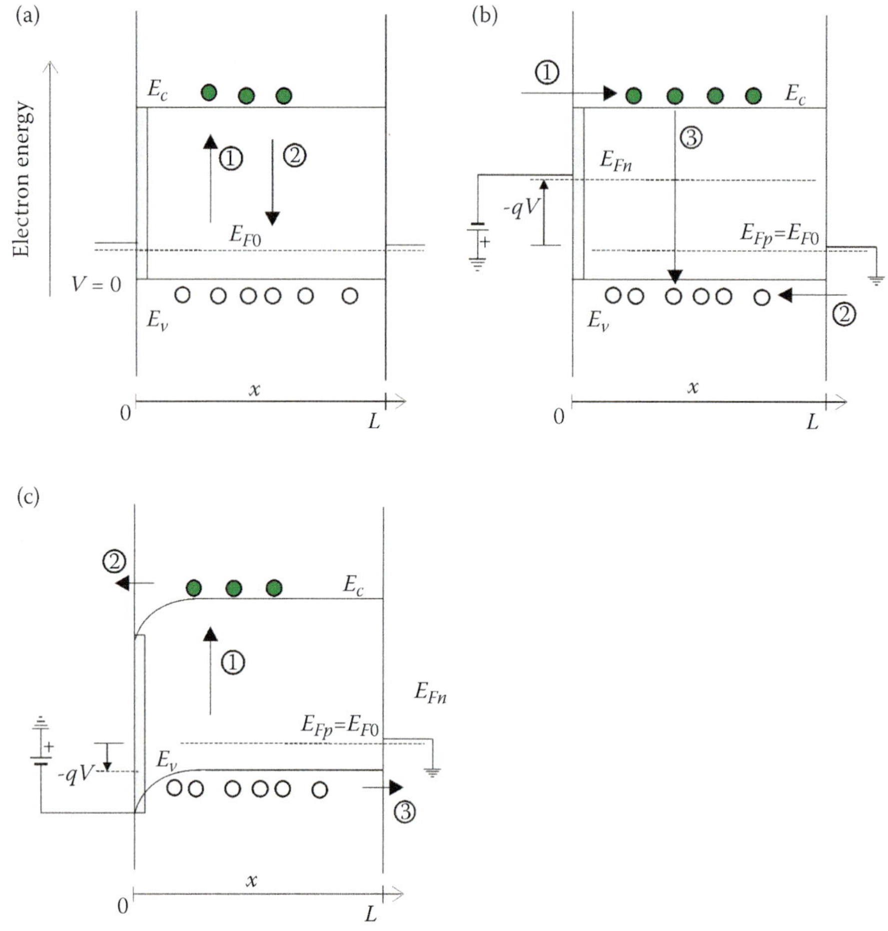

FIGURE 5.12 Energy diagram of a p-type semiconductor layer with electron selective contact at the left side and hole selective contact at the right side. q is the elementary charge, V is the voltage at the left terminal, E_c is the energy of the conduction band, E_v is the energy of the valence band, E_{F0} is the equilibrium Fermi level, E_{Fn} is the electron Fermi level, E_{Fp} is the hole Fermi level. (a) Zero bias in dark. Blackbody radiation produces generation of electron–hole pairs (1). The dark generation achieves equilibrium with recombination (2) of electrons and holes. (b) Forward bias injects electrons (1) and holes (2) and causes recombination (3). (c) Reverse bias causes modification of the interfaces, the only current is due to thermal generation of carriers that are extracted. The Fermi level of electrons is not shown in the semiconductor material.

increases exponentially, while in "reverse" bias the diode limits the flow into a constant, relatively small current of a value j_0, the *reverse saturation current*. This current density–voltage characteristic can be expressed as

$$j = j_0\left(e^{qV/k_BT} - 1\right) \tag{5.49}$$

This asymmetric response of the jV characteristic allows us to convert alternating current (ac) into direct current (dc) by suppressing either positive or negative half wave, in a process known as *rectification*, which is a central element of modern electronics. For the purpose of this book, the significance of rectification is that it allows us to manipulate the interplay between current, carriers, and voltage in a targeted way. Electronic devices for energy delivery such as solar cells require rectification, because the production of *carriers* in an absorber material that is arranged with rectifying properties can be converted into useful voltage, current, or to perform desired chemical reactions.

In order to obtain insight into the reason of the jV curve of the diode operation, we discuss the actual variations of the electron and hole Fermi levels across the active layer, induced by the applied voltage. As noted earlier in this chapter, in the p-type material the equilibrium Fermi level E_{F0} is closer to the valence band than to the conduction band so that the density of majority carrier holes is much larger than the density of electrons. Therefore, in this model we only consider variation of minority carrier density. Furthermore, we assume that the Fermi levels can be manipulated from the contacts but remain flat throughout the active layer.

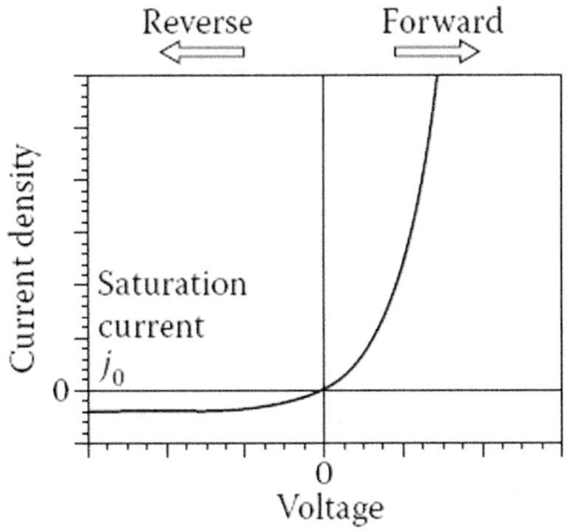

FIGURE 5.13 Current density–voltage characteristics of a diode with reverse saturation current j_0.

The voltage applied in the device corresponds to a difference of Fermi levels at the contacts, the Fermi level of electrons at the electron selective contact, and the Fermi level of holes at the whole selective contact

$$-qV = E_{Fn}(0) - E_{Fp}(L) \tag{5.50}$$

The application of the negative potential at the left contact raises the Fermi level of electrons and therefore produces changes of electron density as shown in Figures 5.11b and 5.12b. To produce this effect, special provisions are required in the regions close to the contact, that is, the contacts have to be *carrier-selective* as mentioned earlier. The left contact of Figure 5.12b has the property that electrons may cross the contact without their Fermi level being altered, and in addition, the holes "bounce" at this contact. Therefore, it behaves as an ohmic contact to electrons. This is called a *perfectly selective contact* for electrons, and is a key component of the device structure. A similar situation, for the opposite carriers, is found at $x = L$. The contact easily transmits the Fermi level of holes but does not equilibrate to the Fermi level of electrons.

At *low injection condition*, the majority carrier Fermi level remains stationary, $E_{Fp} = E_{F0}$, as shown in Figures 5.11b and 5.12b. Then, we obtain from Equations 5.20 and 5.50 the dependence of the minority carrier concentration with the voltage

$$n = n_0 e^{qV/k_BT} \tag{5.51}$$

All the applied voltage appears at the junction that forms the electron (minority) selective contact (see Figure 7.1b). This type of diode is called a *minority carrier diode*. The forward current is due to recombination of the minorities injected from the contact; hence, this is sometimes called a *recombination diode*.

Equation 5.50 describes the fact that the negative voltage on the left contact of Figure 5.12b increases the electron Fermi level. This is called a forward voltage. In general, a *forward voltage* is

- a negative potential applied to the electron selective contact or
- a positive potential applied to the hole selective contact

The *reverse voltage* is defined as

- a positive potential applied to the electron selective contact or
- a negative potential applied to the hole selective contact

The forward voltage (negative in the case of Figure 5.12) increases the number of carriers in the bulk semiconductor, while the reverse bias voltage usually causes a situation of unfavorable energetics at the contacts and tends to remove the carriers from the semiconductor.

If the metal/semiconductor contact consists of a Schottky barrier, the voltage makes the size of the depletion region at the surface to increase (reverse bias) or decrease (forward bias) (Figure 5.12c). At reverse bias, the Fermi level is displaced away from the band in the depletion region, with respect to the quasineutral region where $E_{Fn} - E_c = \zeta_{nb}$ is constant.

It is often convenient to use a *positive* voltage V for the forward voltage, whatever the type of minorities. In many types of nanostructured energy devices, minority carriers are electrons and the forward bias is a negative voltage at the electron contact. In this case, we invert the *sign of the voltage* of Equation 5.50 using the following definitions:

$$qV = E_{Fn}(0) - E_{Fp}(L) \tag{5.52}$$

$$n = n_0 e^{qV/k_BT} \tag{5.53}$$

This convention is shown in Figure 5.11.

Under *high injection conditions*, the concentration of both majority and minority carrier can be modified. We consider the product np, given in Equation 5.37. At a voltage V, defined as Equation 5.52, we have

$$qV = E_g + k_BT \ln\left(\frac{np}{N_c N_v}\right) \tag{5.54}$$

By Equation 5.37

$$E_g = qV - \zeta_n - \zeta_p \quad (5.55)$$

Equation 5.38 writes

$$np = n_i^2 e^{qV/k_BT} \quad (5.56)$$

5.9 SEMICONDUCTOR DEVICES AS THERMAL MACHINES THAT REALIZE USEFUL WORK

A device like a battery or solar cell performs electric work on its surrounding by electrons that travel from one contact to the other through an external circuit. The work these electrons can do (in reversible conditions) is proportional to the voltage in the device. The connection of the voltage to the Fermi level has been already indicated in Chapter 3 (see Figures 3.2 and 3.3), with respect to the Fermi levels at the contacts. In the following we inquire into the fundamental relation among energy, entropy, and work that the ensemble of thermalized electrons and holes *inside a semiconductor* can perform.

Consider a thermal machine that operates in a cycle that is a sequence of processes that returns to the initial equilibrium state. By the second law, the difference between energy lost and work performed in the cycle is associated with the entropy that must be released to compensate for the extraction of heat from a reservoir that acts as the energy source. Suppose that the machine takes a quantity of heat Q_1 from a hot reservoir at the temperature T_h, that it uses to provide a quantity of work W. The heat extraction generates a quantity of entropy $\Delta S = Q_1/T_h$. The machine must transfer heat $Q_2 = T\Delta S = Q_1 T/T_h$ to a colder reservoir in order to conserve the total entropy of the universe. Therefore, maximal work $W = Q_1 - Q_2$ provides the Carnot efficiency factor $(1 - T/T_h)$.

In relation to these concepts, we discuss different energy and entropy transference processes in the model diode presented in Figure 5.12. In this model, we assume that energy levels of electrons and holes in the semiconductor (E_c, E_v) are not modified during operation. The applied voltage to the diode can be perfectly translated into a separation of Fermi levels inside the device so that all changes induced by the voltage have effect in the chemical potentials ζ_η, ζ_p.

In Figure 5.14 we observe a device that delivers electrons from the conduction band at electrochemical potential (Fermi level) E_{FA} and receives them at the valence band at electrochemical potential E_{FB}. When an electron goes round the external circuit, a net transference is realized, of an electron from the conduction band to the valence band. This process can be viewed as an effective recombination event that takes place outside of the semiconductor. The energy lost from the electrons and holes ensemble is $E_g = E_c - E_v$. From the amount of energy E_g, a part can be used to do useful work on an external load, and a part is wasted as heat, in order to obey the second law of thermodynamics. By transfer of Δn electrons from A to B at voltage V, a quantity of work $W = q\ \Delta nV$ is provided; thus the work obtained is $W = (E_{FB} - E_{FA})\Delta n$. When the electrons pass from a stable energy in the semiconductor E_c, to a stable energy level $E \approx E_{FA}$ in the metal, a dissipation of heat occurs with amount $Q_{lat} \approx E_c - E_{FA} = \zeta_\eta \Delta_n$, that is the wasted energy is determined by the chemical potential of the electrons.

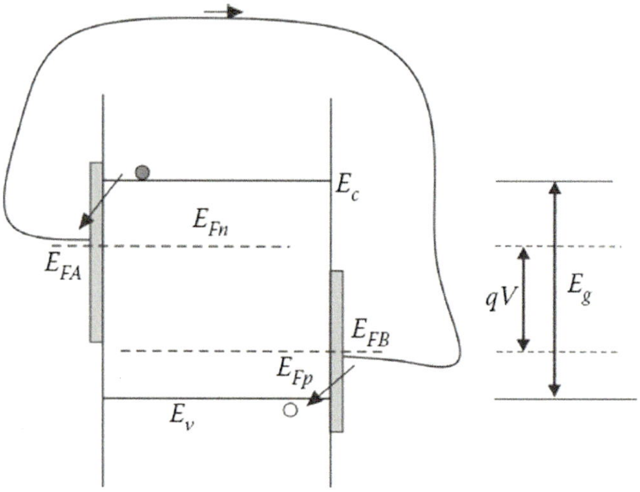

FIGURE 5.14 Operation of a semiconductor device by transference of electrons through the external circuit.

We now describe various specific processes involving the three subsystems that intervene in electron and energy exchange in electronic devices: an external load or battery; the electron and hole system; and the semiconductor lattice that acts as a thermal reservoir at temperature T (Badescu and Landsberg, 1995a). We also assume that the electron and hole distributions are thermalized at lattice temperature T.

Starting from equilibrium situation of electrons and holes, application of a bias voltage V achieves a separation of electron and hole Fermi levels in the carrier system, with the creation of Δn carriers of each type (Figure 5.15a). The internal energy gained in the electron system is $\Delta U_{el} = \Delta n E_g$, while the work received from the battery is $W_{ext} = -\Delta nqV$. The remaining energy required to satisfy conservation of energy is taken from the lattice

$$Q_{lat} = \Delta n\left(E_g - qV\right) \quad (5.57)$$

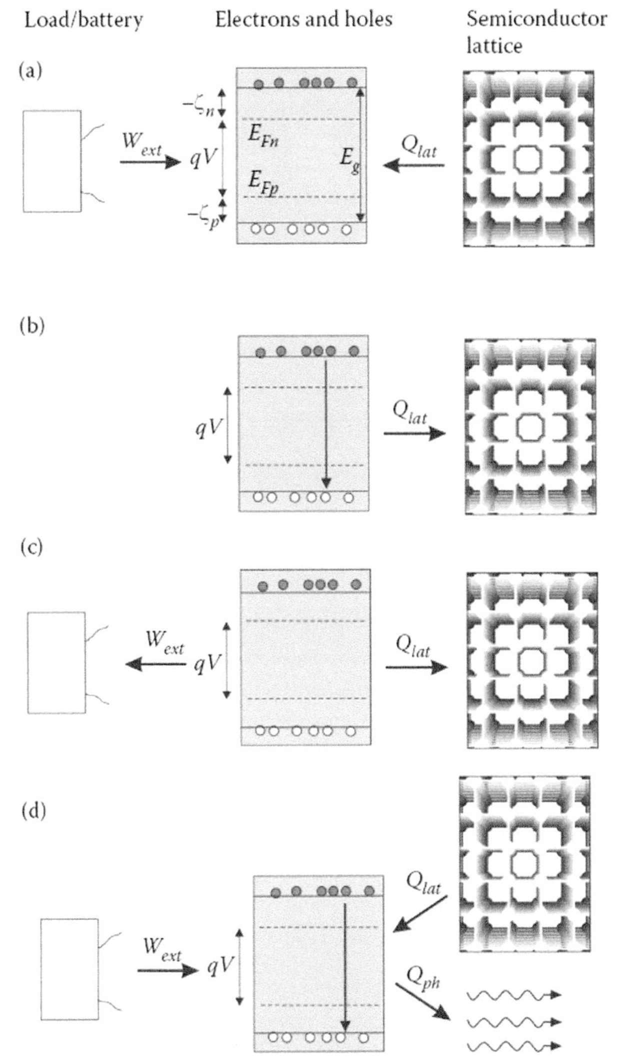

FIGURE 5.15 Energy exchanges in difference processes involving an external load of battery, the set of electrons and holes, and the semiconductor lattice. (a) Application of the voltage creates carriers to split the Fermi levels. (b) Nonradiative recombination, converts carriers to heat. (c) The electron system does work on the load. (d) Radiative recombination while the number of electrons and holes is kept constant by the applied voltage.

The entropy change of the electron subsystem is obtained from Equation 5.9

$$\Delta S_{el} = \Delta\left(S_n + S_p\right) = -\frac{\zeta_n + \zeta_p}{T}\Delta n \tag{5.58}$$

Using Equation 5.55

$$\Delta S_{el} = \frac{E_g - qV}{T}\Delta n \tag{5.59}$$

we obtain that the increase of entropy of the electron subsystem is the same as the loss of entropy in the heat bath, $\Delta S_{lat} = -Q_{lat}/T$. The process can be operated in reversible conditions, by matching the external voltage from the potentiostat to the value of the internal voltage V of the semiconductor, and passing an infinitesimal current, so that no entropy-producing Joule effect heat losses occur in the load or in the semiconductor. In the reversible operation of the device in Figure 5.15a, the lattice is actually cooled.

The second process that we discuss, Figure 5.15b, is a nonradiative recombination event, occurring at Fermi level separation qV that produces the loss of Δn carriers of each type. Internal energy is lost in the electron subsystem $\Delta U_{el} = -\Delta n\ E_g$, while the phonon-mediated recombination process transfers the same amount of energy to the lattice as heat, $\Delta U_{lat} = \Delta n\ E_g$. Since the number of electrons is reduced, there is a reduction of the entropy of the electron system, which is given by Equation 5.59 with a changed sign

$$\Delta S_{el} = -\frac{E_g - qV}{T}\Delta n \tag{5.60}$$

However, the entropy increase in the heat bath is $\Delta S_{lat} = (E_g\ \Delta n)/T$. Therefore, the process is spontaneous but irreversible, as the net quantity of entropy is created

$$\Delta S = \Delta S_{el} + \Delta S_{lat} = \frac{qV}{T}\Delta n \tag{5.61}$$

This entropy is related to the free energy of the Δn electrons, $qV\Delta n$, that *could have been used* to do work, but instead it is wasted in heat by the nonradiative recombination process.

So let us assume that the electrons instead of being lost by recombination are extracted round the external circuit as in Figure 5.13. This process is shown in Figure 5.15c and it is just the reversal of Figure 5.15a. The electrons do a quantity of work $W_{ext} = \Delta nqV$ on the external load. The energy $Q_{lat} = \Delta n(E_g - qV)$ is transferred to the lattice which is effectively heated. The process conserves entropy and can be operated reversibly.

Finally, we view in Figure 5.15d a process in which Δn electron and hole pairs are converted into photons of energy $hv = E_g$ by radiative recombination, while the voltage source supplies Δn carriers again, so that the process is maintained in a steady state (this is an LED). As there is no change of the electron subsystem, $\Delta S_{el} = 0$, and the conservation of energy implies

$$\Delta U_{el} = W_{ext} + \Delta U_{ph} + Q_{lat} = 0 \tag{5.62}$$

The work received from the battery is $W_{et} = -qV\ \Delta n$, and the work delivered as outgoing photon energy is

$\Delta U_{ph} = qE_g\,\Delta n$. Thus, a quantity of heat is taken from the lattice

$$Q_{lat} = \left(E_g - qV\right)\Delta n \tag{5.63}$$

This arrangement can be operated as an electroluminescent refrigerator (Dousmanis et al., 1964). The entropy decrease in the thermal bath

$$\Delta S_{lat} = -\frac{Q_{lat}}{T} = -\frac{E_g - qV}{T}\Delta n \tag{5.64}$$

can be compensated in excess by the entropy that the photons carry away (Weinstein, 1960),

$$\Delta S_{ph} = \frac{\Delta U_{ph}}{T^*} = \frac{E_g}{T^*}\Delta n \tag{5.65}$$

where T^* is the effective temperature of the radiation (typically such radiation is not thermalized). For the cyclic operation of the process, it is required that

$$\Delta S_{lat} + \Delta S_{ph} = \left(-\frac{E_g - qV}{T} + \frac{E_g}{T^*}\right)\Delta n > 0 \tag{5.66}$$

The process of Figure 5.15d can be reversed if a light source is available that contains photons that can be absorbed by the semiconductor generating electron–hole pairs. Then, work is delivered to the external load and a residual heat goes to the lattice, as indicated in Figure 5.15c. This is the photovoltaic effect in which the device operates as a battery, called a solar cell that produces work from the energy of the incoming light, which is the heat reservoir of the solar cell operating as a thermal machine.

5.10 CELL POTENTIAL IN THE LITHIUM ION BATTERY

The changes of composition of a solid material can be controlled electrochemically allowing to form Li ion batteries (LIB) as discussed in Section 3.15. Intercalation typically consists of an insertion of Li^+ ions in interstitial energy minima that can be reached by low-activation energy routes from the surface in contact with a Li-rich electrolyte. Electrons enter from the metal contact to compensate for cationic charge. Figure 5.16 shows the equilibrium sites for $A[B_2]X_4$ spinels that contain B cations in octahedral sites and A cations in tetrahedral sites of a close-packed-cubic X-atom array, for example, as in $Li_xTi_2S_4$ and $Li_xMn_2O_4$. The intercalation host structure typically contains a connected space that is devoid of

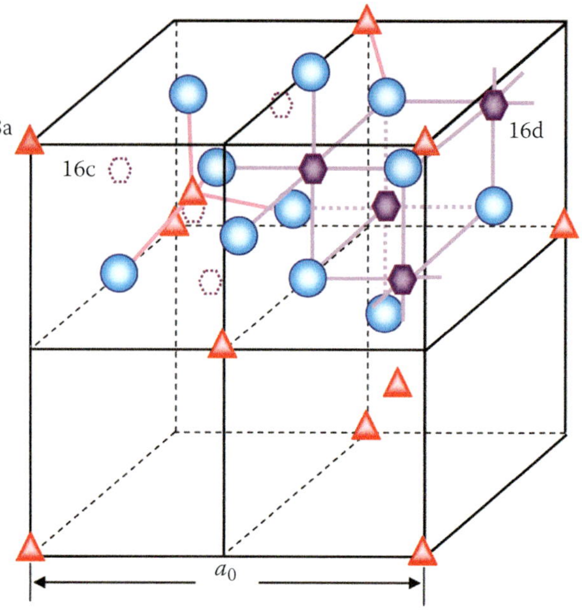

FIGURE 5.16 Two quadrants of the cubic spinel $A[B_2]X_4$ showing the occupied tetrahedral sites (8a), occupied octahedral sites (16d), and unoccupied octahedral sites (16c). The Li species of $Li_{1-x}[B_2]O_4$ occupy 8a tetrahedral sites, and those of $Li_{1+x}[B_2]O_4$ occupy only unoccupied octahedral sites (16c). The Li species of $Li_x[Ti_2]S_4$ occupy only unoccupied octahedral sites (16c) for all x of $0 \le x \le 2$. (Reproduced with permission from Goodenough, J. B.; Kim, Y. *Chemistry of Materials* 2010, 22, 587–603.)

any atom or charge density. The insertion path can be of several dimensionalities: Li ions can move in twodimensional planes in a layered structure as in V_2O_5 or graphite shown in Figure 3.27, or in one-dimensional channels of the structure, as in Li_xFePO_4, that shows a preference for lithium ion transport along the *b*-axis as indicated in Figure 5.17.

The principle of the LIB has been discussed in Section 3.15 (see Figure 3.23). We discuss in more detail the physical origin of the voltage in the cell shown in Figure 5.18, with structure M | el | MB composed of an alkali metal (M) anode (a), an electrolyte (el) containing metal cations (M^+), and a cathode (c) that is a mixed ion/electron conductor metal compound MB such as a layered metal oxide. As remarked in Equation 3.67, the voltage is related to the Gibbs free energy of reaction of the neutral compounds. For the analysis of electrochemical equilibrium (at open circuit), it is convenient to treat separately electrons (e) and alkali ions. We will assume that electrochemical potentials of the different species in the cell can be separately stated as Equation 2.41. This is a simplifying assumption that is nonetheless useful if the host material is not substantially modified by the intercalation process. We have the following electrochemical

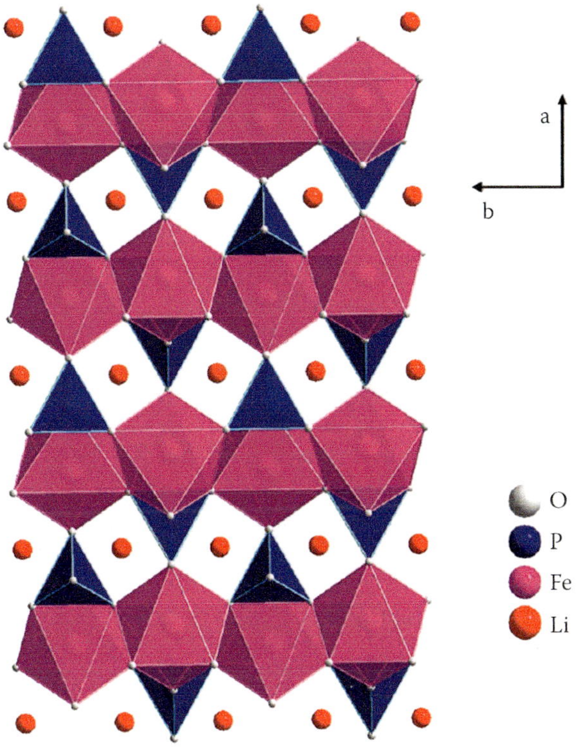

FIGURE 5.17 Polyhedral representation of the structure of LiFePO4. (Reproduced with permission from Goodenough, J. B.; Kim, Y. *Chemistry of Materials* 2010, 22, 587–603.)

potentials, in terms of the chemical potentials of the species and the Galvani potential of each phase (see Figure 5.18):

$$\eta_e^a = \mu_e^a - q\varphi_a \tag{5.67}$$

$$\eta_e^c = \mu_e^c - q\varphi_c \tag{5.68}$$

$$\eta_{M^+}^a = \mu_{M^+}^a - q\varphi_a \tag{5.69}$$

$$\eta_{M^+}^{el} = \mu_{M^+}^{el} - q\varphi_{el} \tag{5.70}$$

$$\eta_{M^+}^c = \mu_{M^+}^c - q\varphi_c \tag{5.71}$$

The cell voltage is given by the difference of Fermi levels of electrons,

$$-qV = \eta_e^c - \eta_e^a \tag{5.72}$$

therefore

$$-qV = \mu_e^c - \mu_e^a - (\varphi_c - \varphi_a) \tag{5.73}$$

Since the electrolyte is an insulator to electrons, the species that maintains the equilibrium across in the three phases of the cell is the metal ions, so that we have a constant electrochemical potential:

$$\eta_{M^+}^a = \eta_{M^+}^{el} = \eta_{M^+}^c \tag{5.74}$$

We therefore obtain that the Galvani potential difference between anode and cathode equals the difference of chemical potential of metal ions

$$-(\varphi_c - \varphi_a) = \mu_{M^+}^c - \mu_{M^+}^a \tag{5.75}$$

Thus, the voltage in the cell can be expressed in terms of two contributions that are the difference of chemical potentials of electrons, and ions, in the two electrodes (Gerischer et al., 1994).

$$-qV = \Delta\mu_{M^+} + \Delta\mu_e \tag{5.76}$$

The difference of chemical potentials in the Equations 5.77 and 5.78 are

$$\Delta\mu_{M^+} = \mu_{M^+}^c - \mu_{M^+}^a \tag{5.77}$$

$$\Delta\mu_e = \mu_e^c - \mu_e^a \tag{5.78}$$

Normally, the anode is made of pure lithium or another material such as graphite that maintains a chemical potential close to Li under cycling. Furthermore, as remarked in Section 3.14, in electrochemical characterization metallic lithium is used as reference electrode. As it is a phase of pure composition $\mu_{M^+}^a$ and μ_e^a are constants correspondent to the standard value $\mu_{M^+}^{a,0}$ and $\mu_e^{a,e}$. Therefore, we discuss the cell voltage based on the properties of the cathode of Figure 5.18 that may also represent the working electrode in the case of a three electrode cell.

The voltage in the cell in the state of charge can be understood in terms of electrochemical potential of electrons in the active redox couple in the cathode, for example, Co^{3+}/Co^{2+}, with respect to the Li^+/Li^0, as shown in Figure 5.19 (Goodenough and Kim, 2010). Upon discharge of the cell, there occurs simultaneous incorporation of Li^+ cations and electrons to the cathode. This process can be viewed as an ionization of Li atoms in which the incoming electrons fill the available DOS in the host material, as shown in Figure 5.20, in which the host electronic levels are not modified by lithiation of the cathode. This assumption is termed a rigid band model.

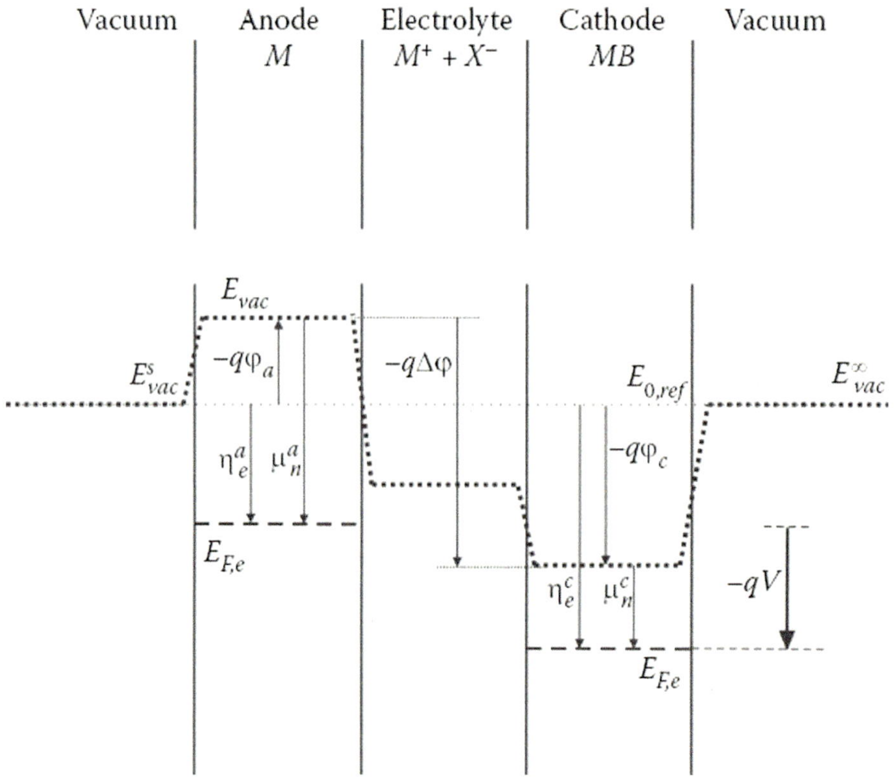

FIGURE 5.18 Energy diagram of a LIB composed of lithium metal anode and metal oxide cathode as shown in the top of the scheme. $E_{F,e}$ is the Fermi level of electrons at each electrode.

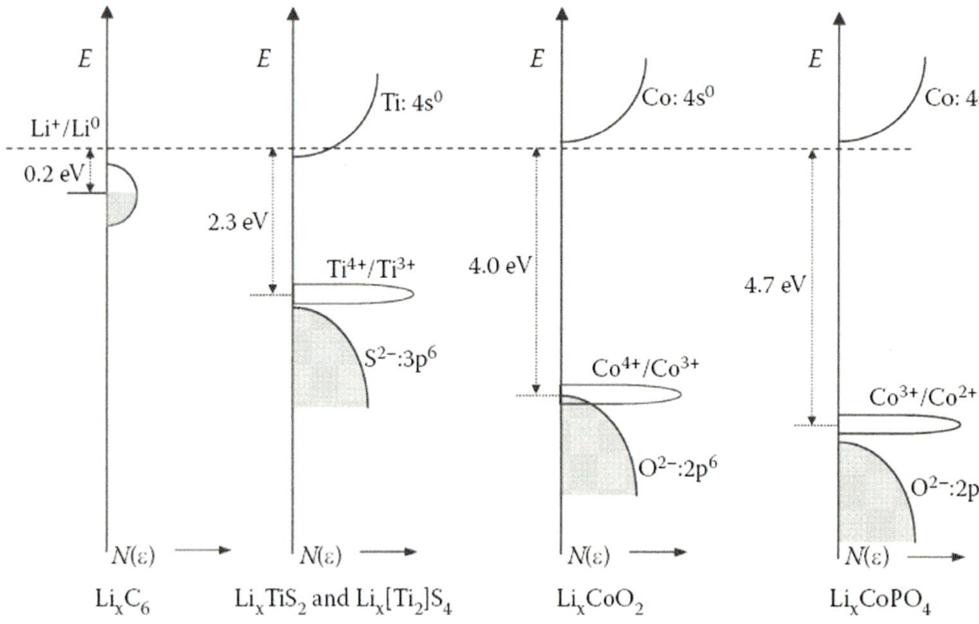

FIGURE 5.19 Electron energy diagram versus density of states for several anode and cathode aterials for which discharge curves are shown in Figure 3.24a, showing the positions of the Fermi energy with respect to Li^+/Li, in an itinerant electron band for Li_xC_6, the Ti^{4+}/ Ti^{3+}redox couple for Li_xTiS_2 and $Li_x[Ti_2]S_4$, the Co^{4+}/Co^{3+}redox couple for Li_xCoO_2, and the Co^{3+}/Co^{2+} redox couple for Li_xCoPO_4. (Reproduced with permission from Goodenough, J. B.; Kim, Y. *Chemistry of Materials* 2010, 22, 587–603.)

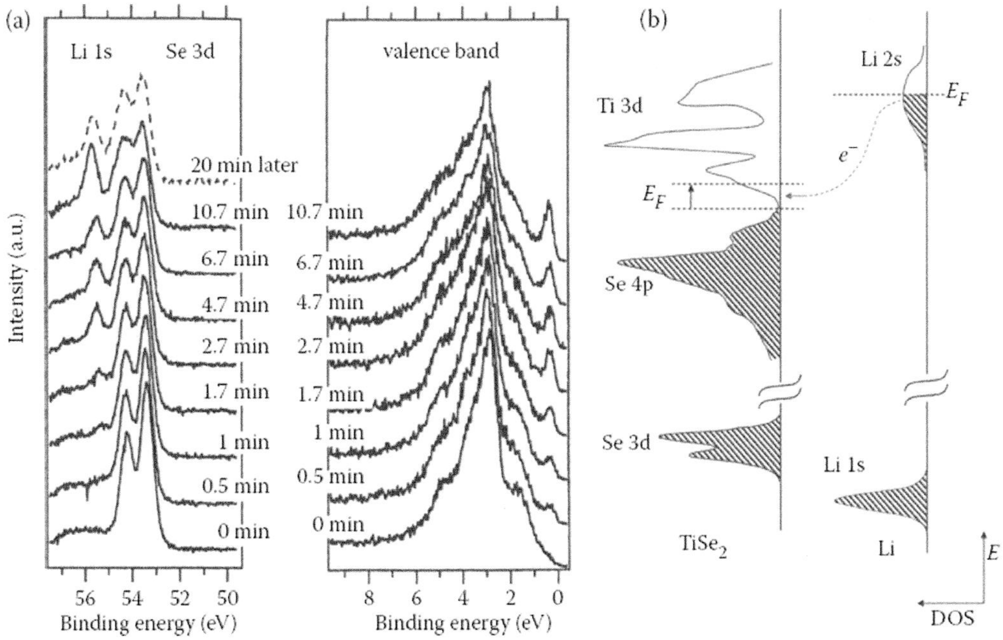

FIGURE 5.20 Soft x-ray photoelectron spectroscopy (SXPS) spectra from (a) Li 1s, Se 3d, and (middle) the valence band (VB) of $TiSe_2(0001)$ intercalated at room temperature with Li by sublimation in ultrahigh vacuum for the reported times. The adsorbed Li is readily intercalated into the substrate without interface decomposition, as shown by the sharp, single component of Li 1 s peak and its binding energy (BE) value. The VB spectra are dominated by the occupied Se 4p band, which remains mostly unchanged in shape, showing only a small shift to larger BEs. The evident increased emission at the Fermi level (BE = 0 eV) indicates filling of previously empty Ti 3d states, due to charge transfer from Li, as sketched in the scheme of energy versus density of states (b). (Reproduced with permission from Tonti, D.; Zanoni, R. *Encyclopedia of Electrochemical Power Sources;* Elsevier: Amsterdam, 2009; 673–695.)

In the context of rigid electronic bands, there are two main causes of the variation of the cell voltage, as indicated in Equation 5.76. The first is a change of the chemical potential of electrons, in which the Fermi level of electrons in the cathode moves toward the LVL and the cell voltage decreases. This feature is identical to the variation of the voltage in the semiconductor diode of Figure 5.12. However, the chemical potential of Li^+ ions in the cathode depends on the composition in terms of the activity as follows:

$$\mu^c_{M^+} = \mu^{c,0}_{M^+} + k_B T \ln a_{M^+} \qquad (5.79)$$

A number of specific models for Equation 5.79 that allow to describe variations of the cell potential with the extent of intercalation, x, are discussed in Section 5.11. The variation of composition of the cathode, that changes M^+ chemical potential, produces a correspondent change of the Galvani potential of the cathode, to keep the electrochemical potential of alkali ions constant across the cell, hence Equation 5.75. The change of φ_c brought about by cation insertion, moves the electron energy levels as a whole, which produces the same change of $E_{F,e}$ at the cathode, without modification of the chemical potential of electrons.

If the electron DOS is large around the Fermi level, and if local electrostatic interactions between ions and electrons do not have significant effects in the electrochemical potentials, then electronic contributions to the variations of the cell potential are neglected and variations of cell voltage are attributed exclusively to the chemical potential of intercalated Li^+ ions, as discussed in Section 5.11.

However, if there is a blank space in the DOS of electrons, there will be a sudden step of the cell voltage due to $\Delta\mu_e$ (McKinnon and Selwyn, 1987). A change of $\Delta\mu_e$ occurs as well if it the DOS is modified during the Li intercalation process, by alteration of the structure that changes the electronic bands. An example shown in Figure 5.21 is the step of voltage between the 4.1 V and 4.8 V plateaus in $Li/Li_yCr_xMn_{2-x}O_4$. This step corresponds to a sudden change of the chemical potential of electrons that occurs when the number of Li^+ removed from the lattice is $(1 - x)$ that corresponds to the oxidation of all the $(1 - x)Mn^{3+}$ions to Mn^{4+} (Obrovac et al., 1998). The step of the voltage due to a gap in the electronic DOS is further discussed in Figure 7.6.

As we have mentioned often, the treatment of charged ionic and electronic species in solids as independent entities, although illustrative for a starting model, must be

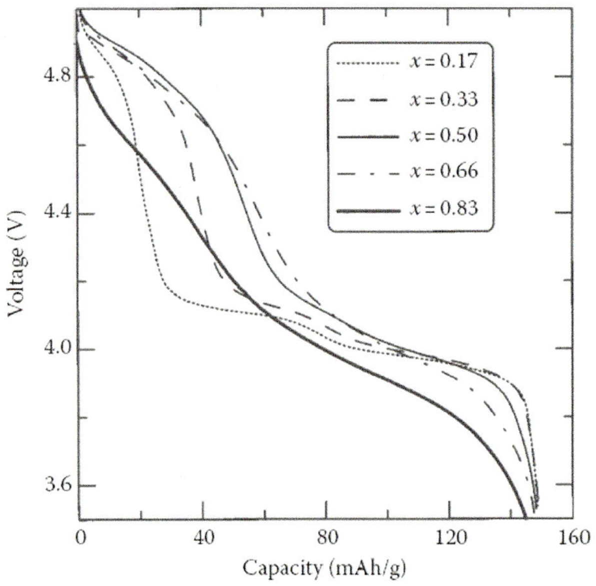

FIGURE 5.21 Discharge curves of $Li_yCr_xMn_{2-x}O_4$ ($0.17 \leq x \leq 0.83$). (Reproduced with permission from Obrovac, M. N.; Gao, Y.; Dahn, J. R. *Physical Review B* 1998, 57, 5728–5733.)

taken with caution for the interpretation of experimental results. In the case of intercalation materials, in the rigid band model the inserted ions may be treated as ionized atoms that provide electrons to the host band (Figure 5.20), and produce an increase of the Fermi level of electrons. However, this simple reasoning neglects all electrostatic short range and even longer range interactions that depend on how the positive ionic charge is screened, according to Debye length. Furthermore, the electrons may remain bound to an Li^+ atom, and in this case they may not be computed in the statistical average of the chemical potential. The distribution of electrons also moderates the strength of guest Li^+–Li^+ interactions and affects the chemical potential of the ions (Mackinnon and Haering, 1983).

5.11 INSERTION OF IONS: THE LATTICE GAS MODEL

Intercalation compounds include spaces in the structure where small alkali ions can diffuse inside the solid. We can consider that a guest atom stays most of the time at an equilibrium position of the host material effecting small vibrations, as shown in Figure 5.22a. Eventually, the atom gets enough thermal energy from the surrounding to perform a hop to a neighboring equilibrium site, as shown in Figure 5.22b. The energy cost of hopping may differ substantially according to the occupancy of the neighbor sites, as indicated in Figures 5.22c and 5.22d. Since the ions occupy fixed sites on the lattice, the system on Li^+ ions can be treated as particles that are randomly distributed in the equilibrium sites (Berlinsky et al., 1979; McKinnon and Haering, 1983). This type of description is called a lattice gas model. In this model all the properties are related to the ordering of the ions in the different sites of a rigid guest lattice. One can thus describe *topotactic* compounds in which the different phases that appear during intercalation are characterized

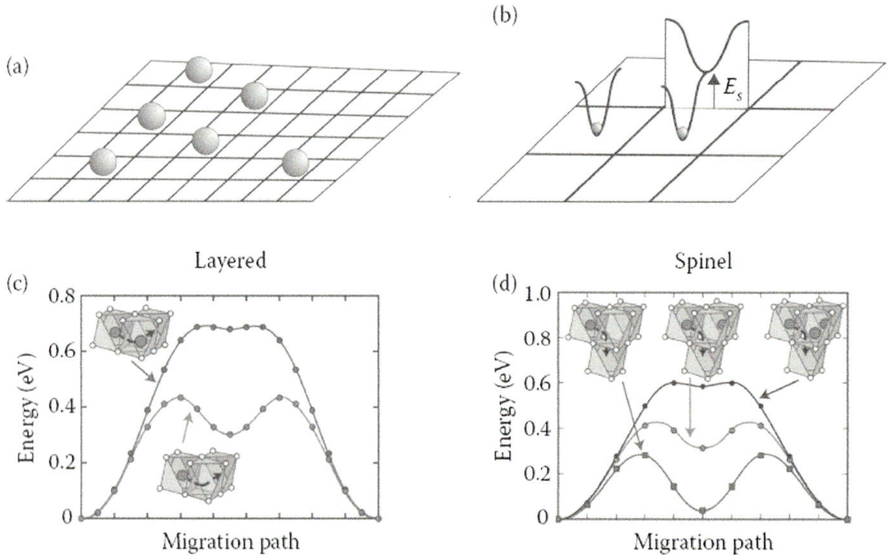

FIGURE 5.22 (a) Ions in equilibrium sites in a square lattice. (b) Detail of energetic landscape for ion hopping to a neighbor site, showing the minimum of free energy at two lattice sites, and the saddle point for hopping to a neighbor site with activation energy E_S. (c) and (d) Li migration barriers for hops between neighboring octahedral sites in layered and spinel $LiTiS_2$ are very sensitive to the occupancy of sites adjacent to the intermediate tetrahedral site of the hop. The barrier for hops into isolated vacancies is significantly larger than for divacancy hops and triple vacancy hops. (Reproduced with permission from van der Ven, A.; Bhattacharya, J.; Belak, A. A. *Accounts of Chemical Research* 2013, *46*, 1216–1225.)

by a different concentration, such that the local Li concentration can serve as a field variable (Han et al., 2004).

For ions, Coulomb repulsion plays the same role as the Pauli principle for electrons making possible the occupation of each site by one ion. Therefore, in a first approximation we can treat the ions in the lattice using the same model that has been derived in Section 5.2 for weakly interacting electrons, namely, considering all sites at the same energy and the chemical potential given by the entropy of mixing of the ions. Using Equation 5.10 in the general expression (5.79), the following expression for the chemical potential of ions in the cathode is obtained:

$$\mu^{c}_{M^+} = \mu^{c,0}_{M^+} + k_B T \ln \frac{c_{M^+}}{c^0_{M^+} - c_{M^+}} \tag{5.80}$$

The fractional concentration of ions with respect to total sites $c^0 = N_{Li}$ is $x = c/N_{Li}$, that is termed the *extent of intercalation*. The cell voltage in Equation 5.76 can be expressed as

$$V = V^0 - \frac{k_B T}{q} \ln \frac{x}{1-x} \tag{5.81}$$

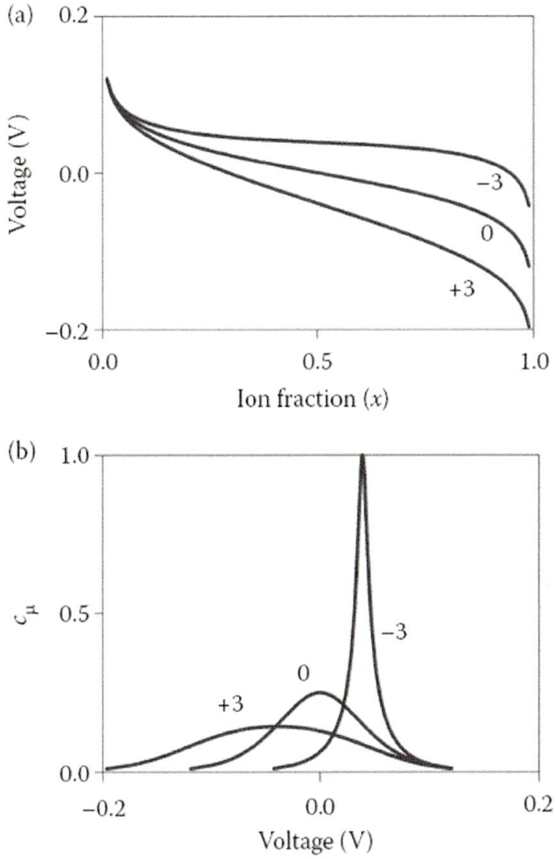

FIGURE 5.23 Plots of voltage–composition curve (a) and equilibrium chemical capacitance c_μ (b) versus voltage of the cathode for the lattice gas model with interaction (mean field), using different values of the interaction parameter, *g*, as indicated.

where the standard voltage is

$$-qV^0 = \left(\mu^{c,0}_{M^+} - \mu^{a,0}_{M^+}\right) + \left(\mu^c_e - \mu^a_e\right) \tag{5.82}$$

The concentration depends on voltage as

$$x = \frac{1}{1 + e^{\left(qV - V^0\right)/k_B T}} \tag{5.83}$$

In the diluted limit, close to full charge of the cathode ($x \approx 0$), we obtain

$$x = e^{\left(qV - V^0\right)/k_B T} \tag{5.84}$$

The dependence of voltage with concentration is shown in the middle curve of Figure 5.23a (labeled 0), assuming that $\Delta\mu_e = \mu^c_e - \mu^a_e$ does not change with composition. One example in which the configurational entropy dominates the changes of cell voltage is shown in Figure 5.24.

In general, the chemical potential of ions is strongly influenced by the interactions between intercalated ions and by strain fields caused by expansion or contraction of the lattice. The lattice gas model allows to treat pairwise interactions between the Li^+ ions. Usually interactions between neighbor and next-near neighbor particles are considered, with either attractive or repulsive interactions that lead to a number of *configurational* transitions. Such features can be modeled in some cases with analytical approximations of the statistical functions such as Bethe approximation or else using Monte Carlo simulations in which arbitrary specific site energy distributions and interparticle interactions can be postulated. The decomposition of the system into domains of different

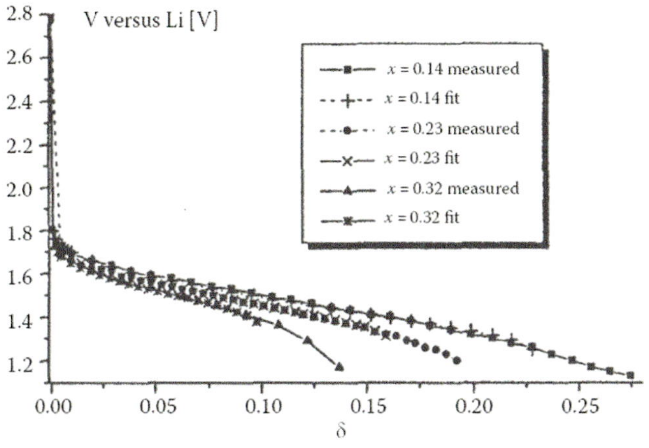

FIGURE 5.24 Discharge curves of $Li_{x+\delta}La_{(2-x)/3}TiO_3$ compared to the fitting to lattice gas model theoretical expression of voltage, including a term for the chemical potential of electrons. (Reproduced with permission from Klingler, M.; Chu, W. F.; Weppner, W. *Ionics* 1997, *3*, 289–291.)

concentrations leads to first-order phase transitions. Below the critical temperature, the system separates into a rarefied phase of low concentration and a condensed phase of high concentration. Phase-field models such as Cahn–Hilliard theory allow to deal with sharp composition gradients. In addition, due to Li^+ intercalation the host may undergo *structural* transformations, like a volume expansion, or staging, as in graphite, in which a periodic array of unoccupied layer gaps at a low concentration of guest is formed, or a host lattice reconstruction, which will severely affect the energy of electrons as well.

Here we examine only a number of simple models that modify the noninteracting lattice gas toward a more realistic description. One standard approach to include interactions uses an effective potential that depends on the concentration. The interaction strength is set by a parameter g

$$V = V^0 - k_B T \ln\left(\frac{x}{1-x}\right) + g k_B T x \tag{5.85}$$

Equation 5.85 belongs to the class of statistical models of mean field interactions, known as the Bragg–Williams approximation and also as the molecular field approximation. The parameter gk_BT represents the average interaction of an ion with its neighbors, either for attractive ($g < 0$) or repulsive interactions ($g > 0$). The critical value of this parameter is $g = -4$. The model provides a qualitative description of the thermodynamic properties of intercalation materials based on a simple average description of the concentration, as shown in Figure 5.23a that plots the voltage–composition curves for different values of g. Note that the case $g = 0$ is equivalent to the Langmuir absorption isotherm, while $g \neq 0$ is the Frumkin isotherm.

Let us consider that the Li^+ ions enter the lattice in two different types of structural states, 1 and 2. We have first N_1 sites at standard voltage V_1^0 that are occupied with concentration c_1 and a level at more negative standard voltage V_2^0 sites and concentration c_2. The difference in both V_k^0 values may be attributed to a step change of the chemical potential of electrons as discussed above, or by a phase transition. The intercalation of atoms first produces the occupation of the energy favored state 1 and then another plateau at the state 2 as shown in Figure 5.25a. The concentration of intercalated ions is $c = c_1 + c_2$, and the partial occupancy of each type of site is $y_1 = c_1/N_1$ and $y_2 = c_2/N_2$. Hence, the extent of intercalation is defined with respect to sublattice 1 as

$$y = \frac{c}{N_1} = y_1 + \frac{N_2}{N_1} y_2 \tag{5.86}$$

The equilibrium voltage is defined by the condition that ions in both sublattices are in equilibrium, therefore

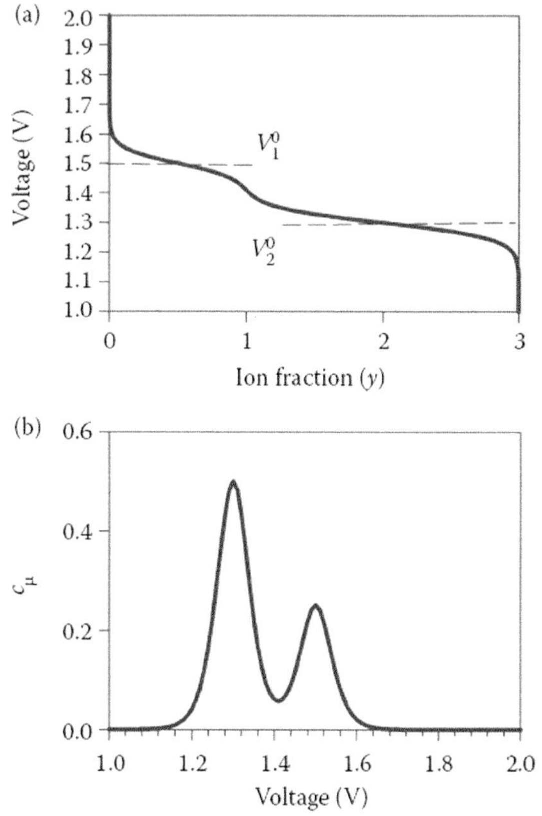

FIGURE 5.25 (a) Plot of voltage–composition curve for a two-level system with N_1 sites of energy $V_1^0 = 1.5\,\text{eV}$ and $2N_1$ sites of energy $V_2^0 = 1.3\,\text{eV}$. (b) Equilibrium chemical capacitance c_μ.

$$V = V_1^0 + k_B T \ln\left(\frac{y_1}{1-y_1}\right) = V_2^0 + k_B T \ln\left(\frac{y_2}{1-y_2}\right) \tag{5.87}$$

We obtain the expression of the voltage–composition curve shown in Figure 5.25a as follows:

$$y = \frac{1}{1+e^{\left(qV-V_1^0\right)/k_BT}} + \frac{N_2}{N_1}\frac{1}{1+e^{\left(qV-V_2^0\right)/k_BT}} \tag{5.88}$$

Let us analyze the causes of voltage change in one specific material. Intercalation of lithium in manganese oxides has been reviewed by Thackeray (1997). The discharge of $Li_xMn_2O_4$ proceeds in two main features, one around 4 V and another at 3 V, as shown in Figure 5.26. Only the 4 V plateau is used in operation of this cathode material. Such plateau corresponds to accommodating one lithium per formula unit from the spinel $Li_xMn_2O_4$ until the rock salt composition $LiMnO_2$ is reached. As Li^+ and electrons enter Mn_2O_4, the manganese ions are progressively reduced from Mn^{4+} to Mn^{3+}. The complete reduction involves an electrode capacity of 308 mA h g^{-1}. The insertion of ions produces a moderate distortion of the lattice due to Jahn–Teller effect that occurs

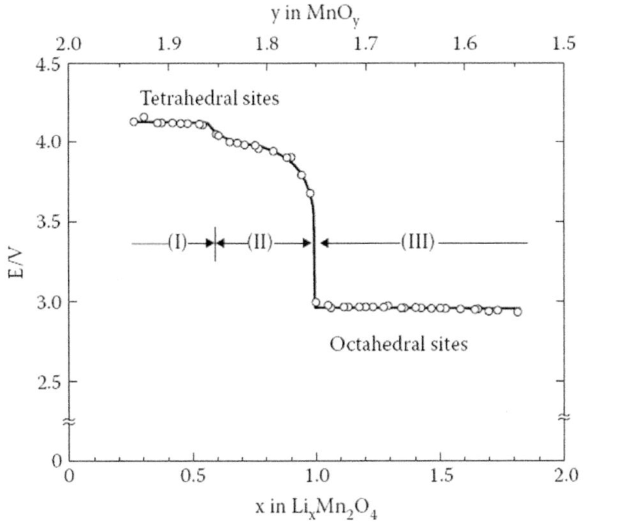

FIGURE 5.26 Equilibrium discharge curve for $Li_xMn_2O_4$. The discharge of the tetrahedral sites produces a capacity of 120 mA h/g, as shown in Figure 3.24c. (Reproduced with permission from Ohzuku, T.; Kitagawa, M.; Hirai, T. *Journal of The Electrochemical Society* 1990, *137*, 769–775.)

when the manganese ion oxidation state is about 3.5. However, the structure can be stabilized for the cycling of the cathode at 4 V with respect to lithium up to a rechargeable capacity of approximately 120 mA h g^{-1} (see Figure 3.23b). As indicated in Figures 5.16 and 5.26, the first filling of $Li_xMn_2O_4$ with Li^+ for $0 \leq x \leq 1$ occurs in tetrahedral (8a) sites of the spinel structure. Such filling process occurs in two stages separated about 0.150 V that correspond to ordering of the lithium ions on one-half of the tetrahedral sites 8a. This small step therefore corresponds to a change of the chemical potential of ions, as further discussed in the next paragraph. When the rock salt stoichiometry $LiMn_2O_4$ is reached, a cooperative displacement of Li^+ ions from 8a tetrahedral to 16c octahedral sites by repulsive Coulomb interaction generates a two-phase electrode. This reaction is a first-order phase transition that takes place by a reaction front that moves from the surface toward the particle center at constant voltage of 3 V as shown in Figure 5.26. This reaction is accompanied by strong Jahn–Teller distortion that changes the crystal symmetry from cubic to tetragonal. Due to the severe structural change, the electrochemical cycling destroys the integrity of the cathode particles resulting in capacity loss. This is why the practical capacity for the battery operation is restricted to 120 mA h g^{-1} with the potential at 4 V.

The two-stage feature at 4 V of $Li_xMn_2O_4$ ($0 \leq x \leq 1$) has been described with a two-state lattice gas model with mean field interactions. The interaction between the lithium ions can be modeled by considering two sublattices. Each lithium ion in one sublattice has 4 nearest neighbors in the other sublattice and 12 second-nearest neighbors within the same sublattice. The lattice gas model can be solved considering the free energy of each sublattice using the Brag–Williams approximation (Gao et al., 1996; Pyun and Kim, 2000). Such interactions in the lattice gas model can be well represented using a Monte Carlo simulation. The results of simulation considering interaction energies $\mathbf{J}_1$ = 37.5 meV (the repulsive nearest neighbor interaction) and $\mathbf{J}_2$ = 4.0 meV (the attractive next-nearest neighbor interaction) are shown in Figures 5.27 and 5.28. Figure 5.27a indicates the evolution of the occupation of two sublattices, and

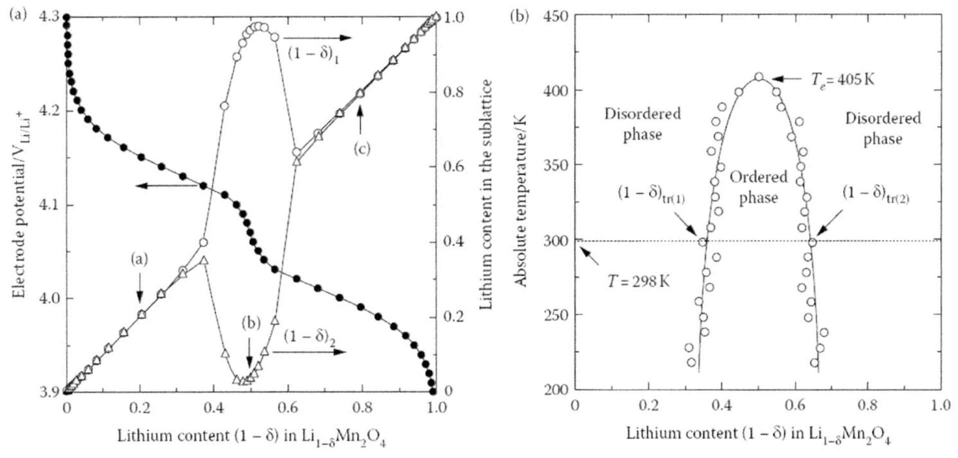

FIGURE 5.27 (a) Electrode potential versus lithium content curve and the plots of the occupation of two separate sublattices obtained theoretically from the Monte Carlo simulation, (b) The theoretical phase diagram representing the order to disorder transition in the $LiMn_2O_4$ electrode. The dotted line is the isothermal line at T = 298 K. $(1-\delta)_1$ and $(1-\delta)_2$ indicate the disorder to order and the order to disorder transition points, respectively. The disordered phase is stable over the entire range of lithium content above the critical temperature T_c. (Reproduced with permission from Kim, S. W.; Pyun, S. I. *Electrochimica Acta* 2001, *46*, 987–997.)

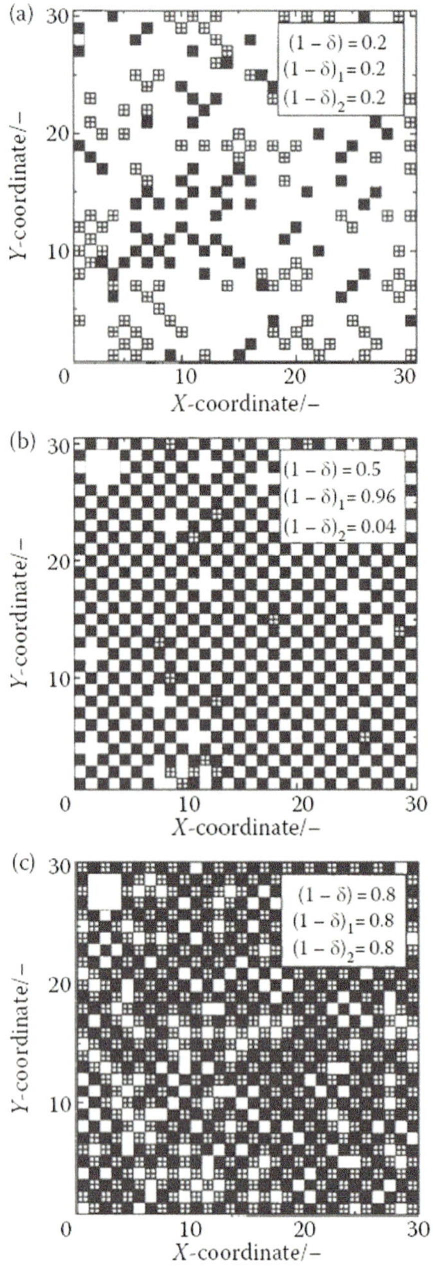

FIGURE 5.28 Equilibrium configuration of the cubic lattice obtained from the Monte Carlo simulation at (a) lithium content $(1 - \delta) = 0.2$, (b) 0.5, and (c) 0.8. The closed square and the cross-centered square symbols represent lithium ions at the sites of sublattices 1 and 2, respectively. (Reproduced with permission from Kim, S. W.; Pyun, S. I. *Electrochimica Acta* 2001, *46*, 987–997.)

the corresponding evolution of the electrode potential. In Figure 5.28 the ions are randomly distributed at low and high concentrations, but at intermediate concentration the Li^+ intercalated ions occupy preferentially one sublattice. The ordering causes a second-order phase transition, accompanied by a steep potential drop at $x = 1 - \delta = 0.5$.

GENERAL REFERENCES

Diffusion of ions in a solid: (Morgan et al., 2004; Ellis et al., 2010; Persson et al., 2010).

Highly doped semiconductors and TCO: (Blackmore, 1962; Mönch, 1993; Klein, 2013).

Hot electrons: (Shah and Leite, 1969; Ross and Nozik, 1982; Lyon, 1986; Moon et al., 1994; Nozik, 2002; Konig et al., 2010; Le Bris et al., 2012).

Lattice gas model and two-state model: (Strässler and Kittel, 1965; Berlinsky et al., 1979; McKinnon and Haering, 1983; Gao et al., 1996; Strömme, 1998; Levi and Aurbach, 1999; Yang et al., 1999; Kim et al., 2001; Bisquert and Vikhrenko, 2002; Kobayashi et al., 2008; Lai and Ciucci, 2010).

Origin of the cell potential: (McKinnon and Selwyn, 1987; Gerischer et al., 1994; Jaegermann et al., 1994; Tonti et al., 2004; Wu, 2006).

Statistics of electrons: (Shockley, 1950; Rosenberg, 1960; Blackmore, 1962; Harvey, 1962; Kubo, 1965; Parrott, 1996).

Screening: (Chazalviel, 1999; Raja and Eccleston, 2011).

Semiconductors as a thermal machine: (Rose, 1960; Weinstein, 1960; Landsberg, 1983; Berdahl, 1985; Craig, 1992; Badescu and Landsberg, 1995; Markvart and Landsberg, 2002; Bent, 2013).

Thermodynamic functions of ions in the lattice: (McKinnon and Selwyn, 1987; Zheng and Dahn, 1997; Besenhard and Winter, 1998; Kudo and Hibino, 1998; Darling and Newman, 1999; Kim and Pyun, 2001; Wong and Newman, 2002; Han et al., 2004; Groda et al., 2005; Burch and Bazant, 2009; Lasovsky et al., 2011; Vakarin and Badiali, 2011; Ciucci and Lai, 2012).

REFERENCES

Badescu, V.; Landsberg, P. T. Statistical thermodynamic foundation for photovoltaic and photothermal conversion. I. Theory. *Journal of Applied Physics* 1995a, *78*, 2782–2792.

Badescu, V.; Landsberg, P. T. Statistical thermodynamics foundation for photovoltaic and photothermal conversion. II. Application to photovoltaic conversion. *Journal of Applied Physics* 1995b, *78*, 2793–02802.

Banyai, L.; Koch, S. W. Absorption blue shift in laser-excited semiconductor microspheres. *Physical Review Letters* 1986, *57*, 2722–2724.

Bent, H. A. *The Second Law: An Introduction to Classical and Statistical Thermodynamics*; Dover Publications: New York, 2013.

Berdahl, P. Radiant refrigeration by semiconductor diodes. *Journal of Applied Physics* 1985, *58*, 1369–1374.

Berlinsky, A. J.; Unruh, W. G.; McKinnon, W. R.; Haering, R. R. Theory of Lithium ordering in $LixTiS_2$. *Solid State Communications* 1979, *31*, 135–138.

Besenhard, J. O.; Winter, M. Insertion reactions in advanced electrochemical energy storage. *Pure and Applied Chemistry* 1998, *70*, 603–608.

Bisquert, J.; Vikhrenko, V. S. Analysis of the kinetics of ion intercalation. Two state model describing the coupling of solid state ion diffusion and ion binding processes. *Electrochimica Acta* 2002, *47*, 3977–3988.

Blackmore, J. S. *Semiconductor Statistics*; Dover Publications: New York, 1962.

Burch, D.; Bazant, M. Z. Size-dependent spinodal and miscibility gaps for intercalation in nanoparticles. *Nano Letters* 2009, *9*, 3795–3800.

Ciucci, F.; Lai, W. Electrochemical impedance spectroscopy of phase transition materials. *Electrochimica Acta* 2012, *81*, 205–216.

Clavero, C. Plasmon-induced hot-electron generation at nanoparticle/metal-oxide interfaces for photovoltaic and photocatalytic devices. *Nature Photonics* 2014, doi: 10.1038/nphoton.2013.1238.

Conibeer, G. J.; Konig, D.; Green, M. A.; Guillemoles, J. F. Slowing of carrier cooling in hot carrier solar cells. *Thin Solid Films* 2008, *516*, 6948–6953.

Craig, N. C. *Entropy Analysis*; VCH Publishers: New York, 1992.

Chazalviel, J. N. *Coulomb Screening by Mobile Charges*; BirkHäuser: Boston, 1999.

Darling, R.; Newman, J. Dynamic Monte Carlo simulations of diffusion in $Li_yMn_2O_4$. *Journal of the Electrochemical Society* 1999, *146*, 3765–3772.

Dousmanis, G. C.; Mueller, C. W.; Nelson, H.; Petzinger, K. G. Evidence of refrigerating action by means of photon emission in semiconductor diodes. *Physical Review* 1964, *133*, A316.

Ellis, B. L.; Lee, K. T.; Nazar, L. F. Positive electrode materials for Li-ion and Li-batteries. *Chemistry of Materials* 2010, *22*, 691–714.

Gadzuk, J. W. On the detection of chemically-induced hot electrons in surface processes: From X-ray edges to Schottky barriers. *The Journal of Physical Chemistry B* 2002, *106*, 8265–8270.

Gao, Y.; Reimers, J. N.; Dahn, J. R. Changes in the voltage profile of $Li/Li_{1+x}Mn_{2-x}O_4$ cells as a function of x. *Physical Review B* 1996, *54*, 3878–3883.

Gerischer, H.; Decker, F.; Scrosati, B. The electronic and ionic contribution to the free energy of Alkali Metals in intercalation compounds. *Journal of the Electrochemical Society* 1994, *141*, 2297–2300.

Goodenough, J. B.; Kim, Y. Challenges for rechargeable Li batteries. *Chemistry of Materials* 2010, *22*, 587–603.

Groda, Y. G.; Lasovsky, R. N.; Vikhrenko, V. S. Equilibrium and diffusion properties of two-level lattice systems: Quasi-chemical and diagrammatic approximation versus Monte Carlo simulation results. *Solid State Ionics* 2005, *176*, 1675–1680.

Han, B. C.; Van der Ven, A.; Morgan, D.; Ceder, G. Electrochemical modeling of intercalation processes with phase field models. *Electrochimica Acta* 2004, *49*, 4691–4699.

Harvey, W. W. The relation between the chemical potential of electrons and energy parameters of the band theory as applied to semiconductors. *Journal of Physics and Chemistry of Solids* 1962, *23*, 1545–1548.

Jaegermann, W.; Pettenkofer, C.; Schellenberger, A.; Papageorgopoulos, C. A.; Kamaratos, M.; Vlachos, D.; Tomm, Y. Photoelectron spectroscopy of UHV *in situ* intercalated $Li/TiSe_2$. Experimental proof of the rigid band model. *Chemical Physics Letters* 1994, *221*, 441–446.

Kim, J.-S.; Prakash, J.; Selman, J. R. Thermal Characteristics of $LixMn_2O_4$ Spinel. *Electrochemical and Solid-State Letters* 2001, *4*, A141–A144.

Kim, S. W.; Pyun, S. I. Thermodynamic and kinetic approaches to lithium intercalation into a $Li_{1-x}Mn_2O_4$ electrode using Monte Carlo simulation. *Electrochimica Acta* 2001, *46*, 987–997.

Klein, A. Transparent conducting oxides: Electronic structure–property relationship from Photoelectron Spectroscopy with *in situ* sample preparation. *Journal of the American Ceramic Society* 2013, *96*, 331–345.

Klingler, M.; Chu, W. F.; Weppner, W. Coulometric titration of substituted $Li_xLa_{(2-x)/3}$ TiO_3. *Ionics* 1997, *3*, 289–291.

Kobayashi, Y.; Mita, Y.; Seki, S.; Ohno, Y.; Miyashiro, H.; Nakayama, M.; Wakihara, M. Configurational entropy of lithium manganese oxide and related materials. *Journal of the Electrochemical Society* 2008, *155*, A14–A19.

Konig, D.; Casalenuovo, K.; Takeda, Y.; Conibeer, G.; Guillemoles, J. F.; Patterson, R.; Huang, L. M.; et al. Hot carrier solar cells: Principles, materials and design. *Physica E: Low-dimensional Systems and Nanostructures* 2010, *42*, 2862–2866.

Kubo, R. *Statistical Mechanics*; North-Holland: Amsterdam, 1965.

Kudo, T.; Hibino, M. Theoretical dependences of the free energy and chemical potential upon composition in intercalation systems with repulsive interaction between guest ions. *Electrochimica Acta* 1998, *43*, 781–789.

Lai, W.; Ciucci, F. Thermodynamics and kinetics of phase transformation in intercalation battery electrodes phenomenological modeling. *Electrochimica Acta* 2010, *56*, 531–542.

Landsberg, P. T. Some maximal thermodynamic efficiencies for the conversion of blackbody radiation. *Journal of Applied Physics* 1983, *54*, 2841–2843.

Lasovsky, R. N.; Bokun, G. S.; Vikhrenko, V. S. Phase transition kinetics in lattice models of intercalation compounds. *Solid State Ionics* 2011, *188*, 15–20.

Le Bris, A.; Lombez, L.; Laribi, S.; Boissier, G.; Christol, P.; Guillemoles, J. F. Thermalisation rate study of GaSb-based heterostructures by continuous wave photoluminescence and their potential as hot carrier solar cell absorbers. *Energy and Environmental Science* 2012, *5*, 6225–6232.

Levi, M. D.; Aurbach, D. Frumkin intercalation isotherm—A tool for the elucidation of Li ion solid state diffusion into host materials and charge transfer kinetics. A review. *Electrochimica Acta* 1999, *45*, 167–185.

Lyon, S. A. Spectroscopy of hot carriers in semiconductors. *Journal of Luminescence* 1986, *35*, 121–154.

Mackinnon, W. R.; Haering, R. R. *Modern Aspects of Electrochemistry*; White, R. E., Bockris, J. O. M., Conway, B. E., Eds.; Plenum Press: New York, 1983; Vol. 15; pp. 235–304.

Markvart, T.; Landsberg, P. T. Thermodynamics and reciprocity of solar energy conversion. *Physica E* 2002, *14*, 71–77.

McKinnon, W. R.; Haering, R. R. *Modern Aspects of Electrochemistry*; White, R. E., Bockris, J. O. M., Conway, B. E., Eds.; Plenum Press: New York, 1983; Vol. 15; pp. 235–304.

McKinnon, W. R.; Selwyn, L. S. Ionic and electronic contributions to the Li chemical potential. *Physical Review B* 1987, *35*, 7275–7278.

Mönch, W. *Semiconductor Surfaces and Interfaces*; Springer: Berlin, 1993.

Moon, J. A.; Tauc, J.; Lee, J. K.; Schiff, E. A.; Wickboldt, P.; Paul, W. Femtosecond photomodulation spectroscopy of a-Si:H and a-Si:Ge:H alloys in the midinfrared. *Physical Review B* 1994, *50*, 10608–10618.

Morgan, D.; van der Ven, A.; Ceder, G. Li conductivity in Li_xMPO_4 (M = Mn, Fe, Co, Ni) Olivine materials. *Electrochemical and Solid State Letters* 2004, *7*, A30–A32.

Neges, M.; Schwarzburg, K.; Willig, F. Monte Carlo simulation of energy loss and collection of hot charge carriers, first step towards a more realistic hot-carrier solar energy converter. *Solar Energy Materials and Solar Cells* 2006, *90*, 2107–2128.

Nozik, A. J. Quantum dot solar cells. *Physica E* 2002, *14*, 115–200.

Obrovac, M. N.; Gao, Y.; Dahn, J. R. Explanation for the 4.8-V plateau in $LiCr_xMn_{2-x}O_4$. *Physical Review B* 1998, *57*, 5728–5733.

Ohzuku, T.; Kitagawa, M.; Hirai, T. Electrochemistry of manganese dioxide in lithium nonaqueous cell. *Journal of The Electrochemical Society* 1990, *137*, 769–775.

Parrott, J. E. Thermodynamic theory of transport processes in semiconductors. *IEEE Transactions on Electron Devices* 1996, *43*, 809–826.

Persson, K.; Sethuraman, V. A.; Hardwick, L. J.; Hinuma, Y.; Meng, Y. S.; van der Ven, A.; Srinivasan, V.; et al. Lithium diffusion in graphitic carbon. *The Journal of Physical Chemistry Letters* 2010, *1*, 1176–1180.

Pyun, S.-I.; Kim, S.-W. Thermodynamic approach to electrochemical lithium intercalation into $Li(1\text{-}y)Mn_2O_4$ electrode prepared by sol-gel method. *Molecular Crystals and Liquid Crystals Science and Technology, Section A: Molecular Crystals and Liquid Crystals* 2000, *341*, 155–162.

Raja, M.; Eccleston, B. The significance of Debye length in disordered doped organic devices. *Journal of Applied Physics* 2011, *110*, 114524.

Rose, A. L. Photovoltaic effect derived from the Carnot cycle. *Journal of Applied Physics* 1960, *31*, 1640–1641.

Rosenberg, A. J. Activity coefficients of electrons and holes at high concentrations. *The Journal of Chemical Physics* 1960, *33*, 665–667.

Ross, R. T.; Nozik, A. J. Efficiency of hot carrier solar energy converters. *Journal of Applied Physics* 1982, *53*, 3813–3818.

Shah, J.; Leite, R. C. C. Radiative recombination from photoexcited hot carriers in GaAs. *Physical Review Letters* 1969, *22*, 1304–1307.

Shockley, W. *Electrons and Holes in Semiconductors*; Van Nostrand: Princeton, 1950.

Strässler, S.; Kittel, C. Degeneracy and the order of phase transformation in the molecular field approximation. *Physical Review A* 1965, *139*, 758–760.

Strömme, M. Cation intercalation in sputter-deposited W oxide films. *Physical Review B* 1998, *58*, 11015–11022.

Thackeray, M. M. Manganese oxides for lithium batteries. *Progress in Solid State Chemistry* 1997, *25*, 1–71.

Toben, L.; Gundlach, L.; Ernstorfer, R.; Eichberger, R.; Hannappel, T.; Willig, F.; Zeiser, A.; et al. Femtosecond transfer dynamics of photogenerated electrons at a surface resonance of reconstructed InP(100). *Physical Review Letters* 2005, *94*, 067601.

Tonti, D.; Pettenkofer, C.; Jaegermann, W. Origin of the electrochemical potential in intercalation electrodes: Experimental estimation of the electronic and ionic contributions for Na intercalated into TiS_2. *The Journal of Physical Chemistry B* 2004, *108*, 16093–16099.

Vakarin, E. V.; Badiali, J. P. Towards a unified description of the host–guest coupling in the course of insertion processes. *Journal of Solid State Electrochemistry* 2011, *15*, 917–929.

van der Ven, A.; Bhattacharya, J.; Belak, A. A. Understanding Li diffusion in Li-intercalation compounds. *Accounts of Chemical Research* 2013, *46*, 1216–1225.

Weinstein, M. A. Thermodynamics of radiative emission processes. *Physical Review* 1960, *119*, 499.

Wong, W. C.; Newman, J. Monte. Carlo simulation of the open-circuit potential and the entropy of reaction in lithium manganese oxide. *Journal of the Electrochemical Society* 2002, *149*, 493–498.

Wu, Q.-H. The utility of photoemission spectroscopy in the study of intercalation reactions. *Surface and Interface Analysis* 2006, *38*, 1179–1185.

Yang, X. Q.; Sun, X.; Lee, S. J.; McBreen, J.; Mukerjee, S.; Daroux, M. L.; Xing, X. K. In situ synchrotron X-ray diffraction studies of the phase transitions in $Li_xMn_2O_4$ cathode materials. *Electrochemical and Solid-State Letters* 1999, *2*, 157–160.

Zheng, T.; Dahn, J. R. Lattice-gas model understand voltage profiles of $LiNi_xMn_{2-x}O_4$/Li electrochemical cells. *Physical Review B* 1997, *56*, 3800–3805.

6 Interfacial Kinetics and Hopping Transitions

So far in this book we have paid great attention to the distribution of carriers in the energy axis and to the equilibration of a carrier distribution across a contact between different materials. For the production of useful action in energy devices, it is necessary to consider the displacement of carriers in a material and their flux across the interfaces. These features are generally termed *kinetics* and they lie at the heart of the observed electric current and the rate of production of a given compound by electrochemical reaction. There are many important aspects to the electric current in practical devices, but here we will focus on the physical origin of the current and its suitable description based on phenomenological models for electron transference mechanisms between molecules, inside a transport medium, or at interfaces between different materials. Electron transfer in a particular circumstance could be determined from first principle calculations by solving Schrödinger's equation for the electron's wave function, provided that sufficient information is available about the interactions in the electron environment. However, often we cannot afford a useful device description based on first principle alone. In addition, following the Franck–Condon principle, the transition of a carrier is so fast that the environmental degrees of freedom can be considered as frozen. We will focus here on jumping or hopping models in which the transition is viewed as an instantaneous event between states that obey a thermal distribution. As a fundamental basis to the analysis of physical characteristics of electron transfer between different media, or between different states in a single material, a very general detailed balance principle will be explained. This principle helps to establish specific forms of carrier transfer rates. The properties of trapping and detrapping of electrons in semiconductors are discussed specifically, and we derive the rate of recombination of electrons and holes mediated by a trap in the center of the bandgap. Then we provide an overview of the Marcus model that starts from the important fact that transference of charge provokes a large change of equilibrium state of a molecule, both for the internal state corresponding to the change in bond-length, as well as for the surrounding polarization of the molecule if it lies in a polar medium. The polarization of the medium also has a great influence in the displacement of a carrier in the phenomenon of polaron hopping. In the final section of the chapter, we present models of macroscopic current flow at interfaces in various instances. We first determine the rate of electron transfer at the metal/solution interface that allows us to derive the central expression for current-voltage characteristics of electrodes in electrochemical cells. Then, we analyze the different types of electron transfer at the metal-semiconductor contact that is ubiquitous in electronic devices. Finally, we describe the specific characteristics of electron transfer at the semiconductor-electrolyte contact that explain primary characteristics of photoelectrochemical cells.

6.1 PRINCIPLE OF DETAILED BALANCE

The frequency of electron transfer determines important properties such as electron transport rate along such type of states or the rate of electrochemical reaction, for example, in a biased electrode or in an organic conductor where the electron switches between different localized electronic states (see also Section 7.2). We are therefore interested in determining the rate of the transition of the electron from an initial to a final state. The transition rate, compounded with the local concentration, will serve to formulate specific phenomenological assumptions for electron flux in diffusion or in reaction across interfaces. Here, we aim to identify a set of fundamental physical requirements on phenomenological transition rates that are shared by a wide class of models. In particular, we establish a basic and general connection between statistics of occupancy in equilibrium and jump rate kinetics connecting two equilibrium states, set by the *principle of detailed balance*.

Detailed balance uses the equilibrium situation to calculate fundamental rates and these are then employed in the more interesting, nonequilibrium situation in which current flows. Examples of application of detailed balance to derive specific rates in physical and chemical phenomena are Langmuir's original derivation of the absorption isotherm by detailed balance between the film and its vapor; the demonstration of transfer rates in thermionic emission by Herring and Nichols (1949), reviewed in Section 4.2, and kinetic

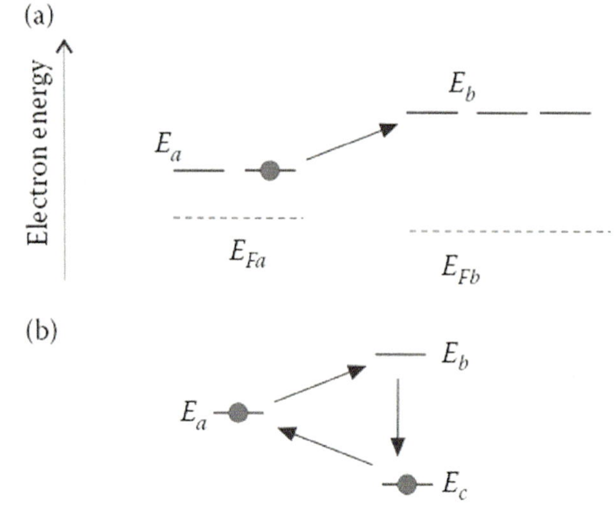

FIGURE 6.1 (a) Carrier transfer between two types of electronic states, at energies E_a and E_b, with local Fermi levels E_{Fa} and E_{Fb}, respectively. (b) Carrier transfer between three different types of states.

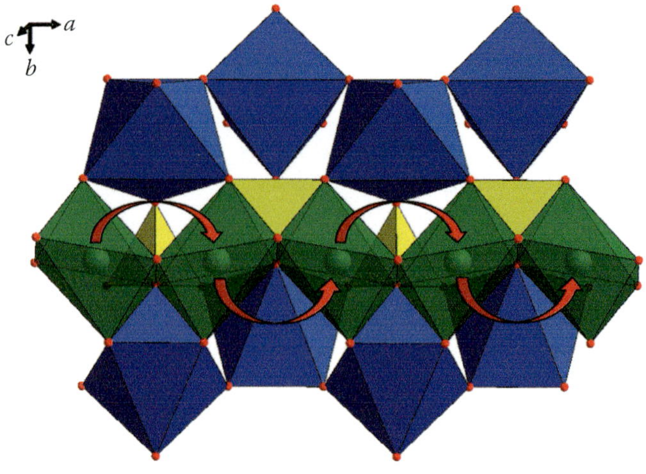

FIGURE 6.2 Structure of $LiFePO_4$ (Figure 5.17) depicting the curved trajectory of Li-ion transport along the *b*-axis, shown with arrows. (Reproduced with permission from Ellis, B. L.; Lee, K. T.; Nazar, L. F. *Chemistry of Materials* 2010, *22*, 691–714.)

factors of recombination in defects in semiconductors (Shockley and Read, 1952), which we will discuss in Section 6.3. More recently, Scott and Malliaras (1999) used the principle of detailed balance to connect the injection rate to the carrier mobility at the metal/organic contact. The formulation of this general principle for macroscopic fluxes was adopted by Onsager (1931) in the context of his reciprocity relations used in irreversible thermodynamics.

The principle of detailed balance is an extension of the fundamental microscopic time reversibility, according to which for each quantum transition the exact reversal is also possible. Therefore, in equilibrium the forward rate must equal the reverse rate of that process. However, the application of this principle to phenomenological rates of charge transfer and chemical reaction, that actually involve macroscopic fluxes instead of individual transition events, requires a considerable leap of generalization. When we consider a reaction rate between reactants and products, a thermal average of both the starting and final states is required. If the reaction is described by classical trajectories going from reactants to products, there are the time-reversed trajectories going from products to reactants. Averaging over an ensemble of initial states, the rate constant for the forward reaction is thereby related to that for the backward reaction and the ratio between the two rate constants is the equilibrium constant. In general, detailed balance condition imposes the constraint that in the equilibrium steady state, the flux of forward and backward transitions between a pair of states *balance in detail*.

In the displacement of a carrier in a solid medium, the interactions in the environment may lead to a *decoherence* of the quantum state (Hinrichsen, 2006). Then, on a fundamental level, an electron or an ion in a solid behaves as a classical object that hops from one localized site to another. Furthermore, we will assume that the average occupation of each type of state follows a thermal distribution. These types of models are called single jump models or *hopping models*. One example of carrier jump between two types of states, *a* and *b*, is shown in Figure 6.1a. The models assume the absence of coherent quantum transport, as well as hot carriers effects in which the carrier can realize additional transitions before thermalization occurs.

In the diffusion of ions inside a solid, the ion stays most of the time in an equilibrium position and realizes hops between such positions, assisted by thermal energy from the surrounding, as shown in Figure 5.22b. One case is the diffusion process of Li in one-dimensional channels of Li_xFePO_4 as shown in Figure 6.2. The activation energy for such hop E_s is given by the difference between the barrier minimum and the saddle point between the two equilibrium positions. This hopping model leads to the jump rate ν of the *Arrhenius form* (Vineyard, 1957)

$$\nu = \nu_0 \exp\left[-E_s/k_B T\right] \tag{6.1}$$

Here, ν_0 is called an *attempt frequency* and corresponds to a typical vibrational frequency of the atom in the equilibrium site, times the number of equivalent saddle points in the equilibrium position ($n_s = 4$ in the two-dimensional model of Figure 5.22b). The form in Equation 6.1 is widely used for chemical reaction, in terms of a *reaction coordinate*, that is the most probable path between the reactants and the products in the general profile of Gibbs energy. Equation 6.1 is also used widely in the study of ion hopping in solids, and can be justified from transition state theory, that asserts that the rate of transition is proportional to the concentration of the "activated complex" at the saddle point of the potential surface. Equation 6.1 is thus associated to the Boltzmann population of the transition state.

Figure 6.1a represents a set of localized electronic states. An electron transition in this system may occur by quantum mechanical tunneling, or by a thermal-assisted hopping induced by phonons, or by a combination of both. The transition probabilities between two localized states are given by the following upward and downward jump rates (Miller and Abrahams, 1960; Ambegaokar et al., 1971), which contain a tunneling factor dependent on the distance and a thermally activated term for upward transitions:

$$\nu_{\uparrow} = \nu_0 \exp\left[-2\frac{r_{ij}}{\alpha_l} - \frac{E_j - E_i}{k_B T}\right]\left(E_j > E_i\right),$$
$$\nu_{\downarrow} = \nu_0 \exp\left[-2\frac{r_{ij}}{\alpha_l}\right]\left(E_j \le E_i\right) \tag{6.2}$$

where ν_0 is again an attempt frequency, r_{ij} is the distance between sites, α_l is the localization radius, and E_j, E_i are the energies of the target and starting sites. After some manipulations of Equation 6.2, the same expression can be written in the form

$$\nu_{ij} = \bar{\nu}_0 \exp\left[-2\frac{r_{ij}}{\alpha_l} - \frac{E_{ij}}{k_B T}\right] \tag{6.3}$$

where

$$E_{ij} = \left(\left|E_i - E_{Fn}\right| + \left|E_j - E_{Fn}\right| + \left|E_j - E_i\right|\right)/2 \tag{6.4}$$

Here, E_{Fn} is the electron Fermi level.

The energy change of the electron in an upward or downward jump is shown in Figure 6.1a. The same quantity of energy that an electron loses by the transition from the conduction band to a trap at midgap, or by hopping between two traps, is released to the vibrational phonon modes. The Miller-Abrahams jump rate in Equation 6.2 strictly applies to a hop in which *only one* acoustic phonon is emitted or absorbed (Emin, 1974). In contrast to this mechanism, another type of electron transference mechanism requires the combination of many phonons to produce a matched state of donor and acceptor, involving reorganization effects of the surrounding of the electrons, and these correspond to the Marcus model for charge transfer and polaron hopping that will be reviewed in Sections 6.4 and 6.5.

We analyze a simple electron-transfer rate based on the example of Figure 6.1a, where we denote f_i as the occupancy of each site, $0 \le f_i \le 1$, so that $(1 - f_i)$ represents the fractional concentration of vacant electronic states. We formulate the rate of change of occupancy in terms of forward and backward fluxes J_{ij} between a and b. We assume that each of the states in the system can be described by a thermal distribution, so that the flux depends on the local Fermi levels of the states shown in Figure 6.1, and we obtain

$$\frac{df_b}{dt} = J_{ab}\left(E_{Fa}, E_{Fb}\right) - J_{ba}\left(E_{Fa}, E_{Fb}\right) \tag{6.5}$$

The carrier fluxes are given by the expressions

$$J_{ab} = \nu_{a,b} f_a \left(1 - f_b\right) \tag{6.6}$$

$$J_{ba} = \nu_{b,a} f_b \left(1 - f_a\right) \tag{6.7}$$

The transition rate $\nu_{a,b}$ is a probability of carrier transfer per second. Equation 6.5 is usually written in the form

$$\frac{df_b}{dt} = +\nu_{a,b} f_a \left(1 - f_b\right) - \nu_{b,a} f_b \left(1 - f_a\right) \tag{6.8}$$

which is denoted a *master equation*. It states the system kinetics by loss–gain for the probabilities of the separate states. More generally, the master equation is established for transitions between different configurational states of the whole system.

In the above example, the equilibrium occupancy terms f_i are specified by the Fermi level of each state, while the kinetics, stating the velocity or rates of transfer, is contained in the transition rate $\nu_{a,b}$. The principle of detailed balance connects both types of quantities. In Equation 6.5, the fluxes should be equal in equilibrium

$$J_{ab}\left(E_{F0}\right) = J_{ba}\left(E_{F0}\right) \tag{6.9}$$

This condition requires that

$$\frac{\nu_{b,a}}{\nu_{a,b}} = \frac{f_a^{eq}\left(1 - f_b^{eq}\right)}{f_b^{eq}\left(1 - f_a^{eq}\right)} \tag{6.10}$$

Each f_i^{eq} is given by the Fermi–Dirac distribution at equilibrium Fermi level (the system is unbiased). By Equation 5.15, we obtain

$$\frac{\nu_{b,a}}{\nu_{a,b}} = e^{(E_b - E_a)/k_B T} \tag{6.11}$$

Therefore, detailed balance implies that the ratio of rates for a transition and its time reverse is given by the Boltzmann factor of the energy cost of the transition. Note that the constraint (6.11) does not entirely determine the transition rates, and a very wide variety of types of transition rates are possible, as seen in the examples of Equations 6.1 and 6.2 above. The rates of a physical situation could in principle be obtained from a detailed knowledge of the microscopic properties of the system, but often such calculation is not feasible, and phenomenological assumptions are adopted. Some models will be reviewed in the next section.

It is important to remark that Equation 6.11 is a statement of the unbiased equilibrium situation in which the transitions between any pair of microstates are exactly balanced. A dynamical system is said to be *out of equilibrium* when the fluxes are not equilibrated; thus violating detailed balance (Hinrichsen, 2006). Typically, in an energy device the current flow under applied voltage is a nonequilibrium situation, since a net flux results from out-of-balance of microstates. However, once the rates consistent with Equation 6.11 have been stated, it is assumed that the *rates* hold for the transitions in any required condition in Equation 6.8, also far beyond the equilibrium situation.

Consider the more general situation of Figure 6.1b in which state a may transfer electrons to states of different energy b and c. Electrons may also effect upward and downward transitions between b and c. Now the master equation has the form

$$\frac{df_b}{dt} = J_{ab}\left(E_{Fa}, E_{Fb}\right) - J_{bc}\left(E_{Fa}, E_{Fb}\right) + J_{cb}\left(E_{Fc}, E_{Fb}\right) - J_{bc}\left(E_{Fc}, E_{Fb}\right) \tag{6.12}$$

and the solution of the equilibrium situation provides a link between the six rate constants. However, the principle of detailed balance is stronger than just a statement of the conservation of particle number, as the equilibrium condition applies to *each pair of states separately*. Thus, for each of the three transitions, the rate of the forward process must be equal to the rate of the backward transition at equilibrium so that constraint (6.11) is supplemented of two similar conditions for ac and bc transitions. This formulation has been established by Onsager (1931) in order to avoid that a net circular flux $a \to b \to c \to a$ can be maintained in equilibrium. This statement of detailed balance also ensures that the unbiased system will reach an equilibrium state in the long time (Miller et al., 2009).

6.2 FORM OF THE TRANSITION RATES

In order to discuss the implications of detailed balance (Equation 6.11), and obtain the acceptable forms of transition rates in master equation models, we split the transition rates into two factors as follows:

$$\nu_{a,b} = \Gamma_0 \Lambda\left(E_a, E_b\right) \tag{6.13}$$

$$\nu_{b,a} = \Gamma_0 \Lambda\left(E_b, E_a\right) \tag{6.14}$$

where Γ_0 is a constant frequency that indicates the timescale of the transitions, and Λ is a function of the states energies that determines the weight of the probability of transitions in each direction. According to Equation 6.11, the adimensional energy function satisfies

$$\frac{\Lambda\left(E_b, E_a\right)}{\Lambda\left(E_a, E_b\right)} = e^{(E_b - E_a)/k_B T} \tag{6.15}$$

Note that these terms Λ in Equation 6.15 are equilibrium quantities, while all the dynamic information in ν is included in the prefactor Γ_0.

Different prescriptions can be applied to obtain a physically plausible model for the transition rate that realizes Equation 6.15. Let us discuss specific examples of Λ that are broadly used in phenomenological models. We can distinguish various situations of increasing complexity.

First of all in the Arrhenius form of the rates, the transition probability is determined by the height of the barrier. As mentioned above, it is assumed that thermal equilibrium exists between the bottom of the well and the top of the barrier (the activated state of energy E_s), hence

$$\Lambda\left(E_a, E_b\right) = e^{-(E_s - E_a)/k_B T} \tag{6.16}$$

This model is useful in diffusion problems in which all sites have equivalent energy, which is the case if there is only a very weak interaction between the diffusing particles.

While Equation 6.16 is independent of the energy of the target site, in another, broadly used model, the energy barrier to jumping is raised or lowered by half the energy change on hopping, as follows:

$$\Lambda\left(E_a - E_b\right)e^{-(E_b-E_a)/k_BT} \tag{6.17}$$

Equation 6.17 is characteristic of hopping events that depend on thermal fluctuation of the environment of the particle, as in the Marcus transfer model that is discussed in detail in Section 6.4, and in polaronic diffusion as described in Section 6.5.

In electrochemistry it is usual to consider the electron transfer between the Fermi level of the electronically conducting part of the electrode and oxidized species in the electrolyte (see Figure 6.3). The factor $\Lambda(E_a, E_b)$ is obtained by dividing the energy difference across the interface in two parts, $\beta\Delta E$ and $(1 - \beta)\Delta E$, introducing the "transfer coefficient" or "symmetry factor" $\beta < 1$

$$\begin{aligned}\Lambda\left(E_a,E_b\right) &= e^{-\beta(E_b-E_a)/k_BT}\\ \Lambda\left(E_b,E_a\right) &= e^{(1-\beta)(E_b-E_a)/k_BT}\end{aligned} \tag{6.18}$$

The energy difference $E_b - E_a$ in Equation 6.18 has the meaning of the overpotential $-qV_\eta = E_F^{WE} - E_{redox}$ as shown in Figure 3.9b. The fundamental reason for the charge transfer model in Equation 6.18 is that a variation of the electric field in the Helmholtz layer influences the position of the saddle point of the electron transition as well as its height, as shown in more detail in Figure 6.18.

Another form of the transition rate between two localized energy levels is

$$\begin{aligned}\Lambda\left(E_a,E_b\right) &= 1 \text{ if } E_b < E_a\\ \Lambda\left(E_b,E_a\right) &= e^{-(E_b-E_a)/k_BT} \text{ if } E_b > E_a\end{aligned} \tag{6.19}$$

so that a downward transition occurs with probability one, and is limited only by the vibration frequency Γ_0, while the upward transition is thermally activated. This approach includes the Miller-Abraham hopping rate of Equation 6.2, the Shockley-Read-Hall model for electron traps (Section 6.3), and the standard Metropolis rate often used in the Monte Carlo simulation (Metropolis et al., 1953; Fichthorn and Weinberg, 1991).

Some simulation of dynamics in the lattice gas model, in which the displacement of ions is stochastic in character, has been presented in Section 5.11. In this type of simulation, the evolution of the system is described in terms of a probability function P of any configuration of the system. From an initial condition A_0 the transition rates determine the likelihood of successive transitions, say between A and B, with the probabilities $w(A, B)$. The dynamics of the system is calculated by selecting one particle at each time and comparing the probability of transition to an empty state with a randomly generated number, which determines whether the hop occurs or not.

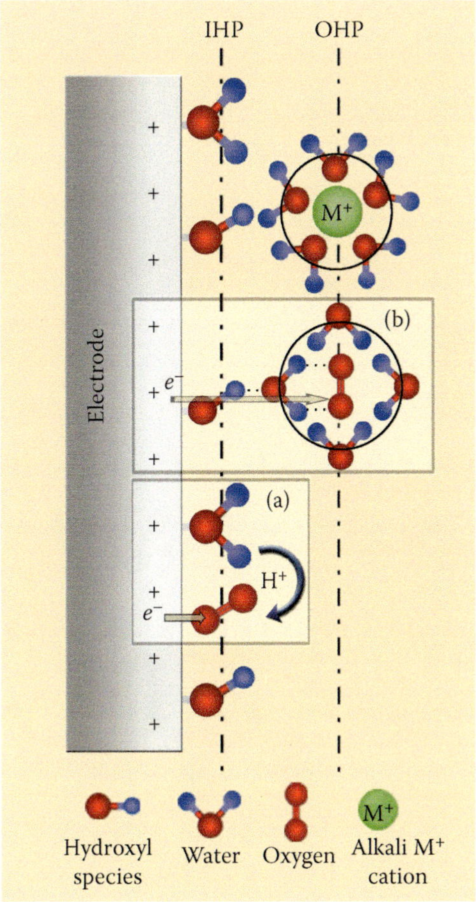

FIGURE 6.3 Schematic illustration of the double-layer structure during oxygen reduction reaction in alkaline media, showing different species in solution. Inset (a) illustrates the inner-sphere electron-transfer processes for the reduction of an absorbed molecule at the electrode surface (inner Helmholtz plane). Inset (b) shows an outer-sphere electron transfer for a solvated molecule close to the surface (outer Helmholtz plane). (Reproduced with permission from Ramaswamy, N.; Mukerjee, S. *Advances in Physical Chemistry* 2012, *2012*, 491604.)

It was discussed in Section 5.11 that the energy barrier for hopping from a given site depends on the occupation of neighbor and often near-neighbor sites (see Figure 5.22). For the Monte Carlo simulation of systems, as those of Figure 5.28, it is necessary to formulate a dependence of the transition rate on initial and final state of the system consistent with the detailed balance. One example is the Metropolis rate mentioned earlier. The *transition dynamics algorithm* is more complete in that the transition proceeds via the intermediate state I between A and B, of energy E_I. The transition probability for each jump is written as the product

$$w(A,B)=w(A,I)w(I,B) \tag{6.20}$$

The energy E_I has to be chosen to represent the attempt jump in the presence of interactions, and one example is

$$E_I = E_0 + \frac{E_A + E_B}{2} \tag{6.21}$$

where E_0 is the activation energy in the low occupation limit, and E_A, E_B the energies of the initial and final configuration.

A specific example of the transition rate under interactions in the lattice gas is formulated starting from the mean-field model of Section 5.11. Suppose that we consider the charge transfer between a phase a described by noninteracting lattice gas with an electrochemical potential and a phase b that is an interacting lattice gas in mean-field approximation, as in Equation 5.85

$$\eta_a = E_a + k_B T \ln \frac{x_a}{1-x_a} \tag{6.22}$$

$$\eta_b = E_b + g x_b + k_B T \ln \frac{x_b}{1-x_b} \tag{6.23}$$

where g is the interaction parameter. As a specific choice of the transition rates in Equation 6.13, we consider that the jump rate is $\Gamma_0 = \nu_0 \exp[-E_s/k_B T]$, E_s being the saddle point energy, as indicated in Equation 6.1. Following Equation 6.17, we assume that the energy barrier to jumping is raised or lowered by half the energy change on hopping, so that the rate of jumping becomes

$$\nu_{a,b} = \Gamma_0 \,\text{expt}\left[-\Delta E_{ab}/2k_B T\right] \tag{6.24}$$

In order to determine ΔE_{ab} consistently with detailed balance, we take the prescription that the energy level in phase B is determined by grouping the two first terms in Equation 6.23. Therefore, we obtain that the transition rate depends on concentration as (Levi and Aurbach, 1999)

$$\nu_{a,b} = \nu_0 \exp\left[-E_s/k_B T - \left(E_b + gx - E_a\right)/2k_B T\right] \tag{6.25}$$

6.3 KINETICS OF LOCALIZED STATES: SHOCKLEY-READ-HALL RECOMBINATION MODEL

A localized state in the bandgap can exchange carriers with the bands, provided that the energy is conserved by phonon collisions. In terms of the kinetic properties, localized states in the bandgap are also called *traps*, if the carriers in these states mainly make transitions to one of the bands. A localized state that captures carriers from both bands is called a *recombination center* (Figure 6.4).

Let us describe the change of mean occupancy of a trap f by trapping and release from the conduction band. The conservation equation has the form

$$\frac{\partial f}{\partial t} = \beta_n n_c \left[1-f\right] - \varepsilon_n f \tag{6.26}$$

Equation 6.26 is one realization of Equation 6.8. The capture rate is proportional to the number of electrons present in the conduction band, n_c, and to $(1-f)$, which is the fractional number of unoccupied traps, or "holes." β_n is the time constant for electron capture, usually given in terms of the thermal velocity v_{th} of the electrons and capture cross section:

$$\beta_n = v_{th}\,\sigma_{nc} \tag{6.27}$$

The release of an electron corresponds to a spontaneous emission to the conduction band, with the rate constant ε_n.

In equilibrium without applied bias we obtain from Equation 6.26

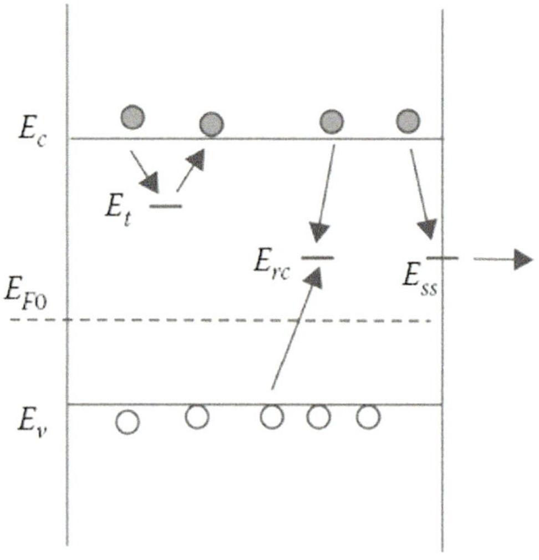

FIGURE 6.4 Electronic processes in localized states: a trap, a recombination center, and a surface state.

$$f_0 = \frac{1}{1+\varepsilon_n/(\beta_n n_0)} \tag{6.28}$$

This last equation can be equated with Fermi–Dirac occupation function in equilibrium and provides the following constraint:

$$\varepsilon_n(E_t) = \beta_n n_0 e^{(E_t - E_{F0})/k_B T} = \beta_n N_c e^{(E_t - E_c)/k_B T} \tag{6.29}$$

Alternatively, Equation 6.10 directly gives Equation 6.29. The detailed balance condition, therefore, provides a relationship between β_n and ε_n that holds for arbitrary occupations n_c and f (Shockley and Read, 1952).

A Shockley-Read-Hall (SRH) recombination is a mechanism that restores thermal equilibrium by de-excitation of electrons in the conduction band via an intermediate impurity level between the conduction band and the valence band. The energy of an excited carrier is dissipated by the emission of several phonons (Lang and Henry, 1975). The interaction of conduction band levels and the impurity level is strong if the latter is a highly localized state. Several schemes of utilization of intermediate bandgap levels, for visible sensitization or intermediate bandgap solar cells, need to reduce the extent to which the impurity level enhances recombination. SRH is reduced if the intermediate levels have a large density so that an extended state is formed at midgap (Luque et al., 2012).

To obtain the rate of SRH recombination, consider the equations for capture and release of electrons in the conduction band and holes in the valence band, to a recombination center of concentration N_t (Figure 6.4),

$$\frac{\partial n}{\partial t} = -\beta_n N_t (1-f) n + \varepsilon_n N_t f \tag{6.30}$$

$$\frac{\partial p}{\partial t} + \varepsilon_p N_t (1-f) - \beta_p N_t f p \tag{6.31}$$

$$N_t \frac{\partial f}{\partial t} = -\frac{\partial n}{\partial t} + \frac{\partial p}{\partial t} \tag{6.32}$$

Taking into account the intrinsic carrier density n_i of Equation 5.24, the detailed balance constraints are expressed as

$$\frac{\varepsilon_n}{n_i \beta_n} = \frac{n_i \beta_p}{\varepsilon_p} = e^{(E_t - E_i)/k_B T} \tag{6.33}$$

In steady state, the rate of change of f must be zero. Setting $\partial n/\partial t = \partial p/\partial t$, we find that the occupation of the trap is given by

$$f = \frac{\beta_n n + \varepsilon_p}{\beta_n n + \beta_p p + \varepsilon_n + \varepsilon_p} \tag{6.34}$$

The SRH recombination rate of electrons and holes $U_{SRH} = -\partial n/\partial t = -\partial p/\partial t$ can be stated as

$$\begin{aligned} U_{\mathrm{SRH}} &= \beta_n N_t (1-f) n - \varepsilon_n N_t f \\ &= N_t \beta_n \beta_p \frac{np - n_i^2}{\beta_n n + \beta_p p + \varepsilon_n + \varepsilon_p} \end{aligned} \tag{6.35}$$

6.4 REORGANIZATION EFFECTS IN CHARGE TRANSFER: THE MARCUS MODEL

When an electron is transferred between two molecules in a polar medium, the change of state of charge of the molecule that occurs on the electron-transfer event strongly affects the internal state of the molecule as well as the polarization state in the surrounding medium. Such change of the internal and surrounding configurational properties involves a modification of the energy of the carrier that imposes a significant barrier to electron transfer, which is usually explained following the ideas of the seminal work of Rudolph A. Marcus. Related effects occur in an electron transfer between a metal or semiconductor electrode and a molecule in solution, or in general for electron transfer events in "soft media," as in transport of hole carriers in an organic material. In the jump process of an electron between localized states, the conservation of energy in a downward or upward transition may be achieved by the collision of a single phonon, as in the Miller-Abrahams jump. But due to the sluggish dynamics of vibrational and polarization modes, the energy loss of the electron in a solvent or polar medium cannot be readily dissipated.

The Franck–Condon principle states that the electronic quantum transition is so fast that the nuclear degrees of freedom of the molecule and its surrounding can be considered immobile during the transition of the electron. If the energies of the donor and the acceptor states are very different, the jump of the electron may occur assisted by a photon, as shown in Figure 5.7. In a molecular or polar system, the jump corresponds to a *vertical transition* that occurs in the configuration of the ground state of the system, as will be discussed further later. Then, the molecular system has to rearrange to achieve the equilibrium configuration of the excited state.

In the absence of energy input by photons the electron-transfer event may occur only if another mechanism establishes the conservation of energy. Therefore, as discussed in the theory of electron-transfer reactions by Marcus (1956), before an electron can jump from one molecule to the other, there first has to be a "reorganization" of the solvent molecules around the molecule. The electron-transfer rate is thus limited by the probability that the thermal fluctuations of all the vibrational and polarization modes achieve a nonequilibrium configuration in which the energy is conserved between the starting and final state of the electron transfer, satisfying the Franck–Condon principle.

We adopt the chemical nomenclature in order to distinguish an initial situation of "reactants" (R) and a final state of "products" (P), both of which consist of all the molecules that comprise the charge transfer system, as well as their surroundings, and they differ in the location of one electron charge, which implies different equilibrium configurations for each R and P state. The electron-transfer theory described here applies in cases in which the electron-transfer tunnels from one complex to another without bond making or bond breaking, usually across a solvation layer, in a type of oxidation-reduction reaction known as an *outer-sphere* electron transfer (see Figure 6.3) (Bard, 2010). In contrast, in *inner-sphere* reactions, the electron transfer occurs in an activated complex where a ligand is shared between the donor and acceptor molecules. For outer-sphere reactions from an electrode, the reactants, products, and intermediates do not interact strongly with the electrode material, while in heterogeneous inner-sphere reactions a specific interaction with the electrode surface occurs and these are often called *electrocatalytic* reactions. In addition, if there is significant participation of high-frequency vibration modes of the molecules, the quantum mechanical behavior of such modes should also be taken into account.

Marcus's electron-transfer theory, originally developed for reduction-oxidation reactions, is also used in many situations in which there is no chemical change but just the funneling of electrons, as in the donor-acceptor system in organic solar cells (Liu and Troisi, 2011, 2012), or in charge and energy transport in organic conductors (Köhler and Bässler, 2011; Ruhle et al., 2011; Schrader et al., 2012) and in electron transfer in dye-sensitized solar cells (Bisquert and Marcus, 2013).

Here, we discuss the reorganization characteristics of electron transfer following the example of a donor and acceptor system, shown in Figure 6.5. This model system represents well the charge separation step in an organic solar cell, as indicated in Figure 1.7e. Prior to electron transfer, the D unit is photoexcited to D* to form an electron-hole pair. The electron relaxes to the first excited state of D* as described later in Figure 6.9, while a hole remains in the ground state. The D-A energy levels are suitably arranged so that the electron is transferred from the excited state D* to the LUMO of A. We are concerned here only with the dynamics of electron transfer from the photoexcited neutral pair to the charge separated state. Therefore, R is the state of the system D^*A^0, and P is the stabilized charge separated state D^+A^-. As an example, the transference from a light-absorbing polymer molecule to an electron acceptor modified fullerene is shown in Figure 6.6.

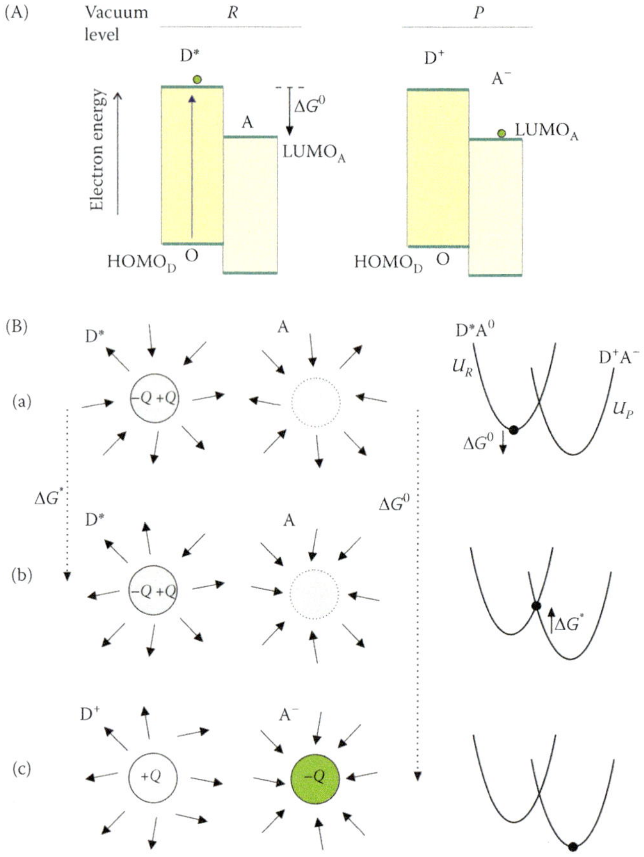

FIGURE 6.5 (A): Energy diagram of a donor-acceptor system, showing, at the left, the "reactant" state which is the relaxed excited state D^*A^0, and at the right, the charge separated ("product") state D^+A^-. (B): polarization in different stages of the DA system. In the center of the sphere is indicated the state of charge Q of each element. At the right is shown the potential energy of the reactants U_R and the products state U_P, as a function of the nuclear coordinates. (a) The solvent in equilibrium with the reactant. (b) A nonequilibrium transition state in which the energy of the reactant state equals that of the product. (c) Solvent in equilibrium with the product.

It is observed in Figure 6.5 that the step of energy levels provides a driving force for the transfer of a photoexcited electron by a (negative) free energy difference ΔG^0.

However, an instantaneous electron transfer from D^* to A^0 does not conserve energy and cannot occur

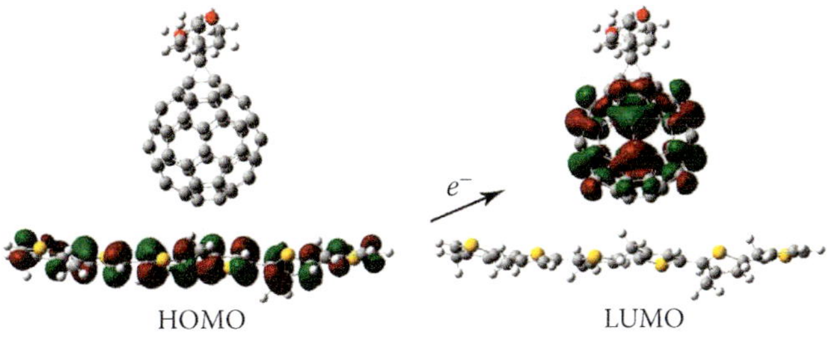

FIGURE 6.6 Model of the electron transfer between the P3HT and PCBM interface. One PCBM molecule and one oligomer (regioregular) containing six thiophene rings (hexyl groups on P3HT are replaced by methyl groups) are displayed, showing the frontier orbitals involved in the excitation processes obtained by the DFT method: an electron in the LUMO of P3HT and electron in the LUMO of PCBM. (Reproduced with permission Liu, T.; Troisi, A. *The Journal of Physical Chemistry C* 2011, *115*, 2406–2415.)

unless a suitable change of D^*A^0 to D^+A^- configuration by an intermediate nonequilibrium state, facilitated by the modification of the surrounding, takes place. Let us consider in more detail the configuration that allows the transference of the electron. In Figure 6.5B, we can observe a sketch of the DA system in which each D and A molecule is represented by a charged or neutral sphere. The molecule is surrounded by solvent dipoles that can adjust their orientation to the prevalent local electric field, while at the same time those dipoles have a random component that represents the thermal fluctuations. The arrows suggest the different polarization states. The D^*A^0 (Figure 6.5Ba) complex is neutral, thus a random distribution of the surrounding dipoles occurs in the equilibrium configuration, whereas in D^+A^- (Figure 6.5Bc), dipoles point in correlation toward A^- and away from D^+.

The crucial observation of the Marcus theory is that for the electron tunneling to occur an intermediate situation has to be obtained, by the thermal fluctuation, in which the energy of the nonequilibrium system D^*A^0 is the same as that of a nonequilibrium state of D^+A^-. Such intermediate state is suggested in Figure 6.5Bb. It can be regarded as the transition state for electron transfer involving an energy barrier ΔG^*, which is the free energy cost of bringing the system from Figure 6.5Ba to 6.5Bb as indicated in Figure 6.5. According to Equation 6.1 the rate constant for electron transfer has the form

$$k_{et} = A_{et}\exp\left[-\Delta G^*/k_B T\right] \tag{6.36}$$

where A_{et} is a prefactor that depends on the probability of electron tunneling, once the transition state has been reached.

The goal of the electron-transfer theory is to provide a suitable formalism to describe the reorganization changes in order to calculate the free energy barrier ΔG^*. Such finality requires evaluating the nonequilibrium free energy of the solute–solvent interactions of the electron from being localized in one site to being localized in the other. The fluctuation of potential energy of the reactant and products are described with respect to the relevant nuclear and solvent degrees of freedom that affect the energy of the electron. Since the solvent polarization modes are treated classically but the electron tunneling is established quantum mechanically, this approach is denoted *semiclassical electron-transfer theory*. The semiclassical treatment evaluates quantum mechanically the crossing probability, as a function of the energy gap that is found along the classical trajectories of the solute–solvent system (Warshel, 1982).

Intramolecular normal vibrational coordinates would be denoted $\mathbf{q}_i$ and the orientation-vibrational dielectric polarization of the solvent at each point $\mathbf{r}$ of the system may be denoted by a vectorial quantity $\mathbf{P}_u(\mathbf{r})$ (Marcus, 1956). We describe all these coordinates by a single generic coordinate $\mathbf{q}$, in which $\mathbf{q}_R$ denotes the equilibrium configuration of the reactants and $\mathbf{q}_P$ that of the products. For example, in Figure 6.5B, $\mathbf{q}$ stands for a specific distribution of the polarization around both *D* and A. Note that the equilibrium distribution of Figure 6.5Ba is very different to that of Figure 6.5Bc, as discussed above, and such difference is labeled in terms of the representative nuclear coordinate by the quantity

$$\Delta_0\mathbf{q} = \mathbf{q}_P - \mathbf{q}_R \tag{6.37}$$

In general, it is a good approximation to represent the increase of energy, as $\mathbf{q}$ departs from the specific equilibrium value of either *R* or *P*, using the parabolic potential energy, as shown in Figure 6.7. Therefore, we can write the expressions

$$U_R(\mathbf{q}) = U_R(\mathbf{q}_R) + \frac{1}{2}k_F(\mathbf{q}-\mathbf{q}_R)^2 \tag{6.38}$$

$$U_P(\mathbf{q}) = U_P(\mathbf{q}_P) + \frac{1}{2}k_F(\mathbf{q}-\mathbf{q}_P)^2 \tag{6.39}$$

where k_F is a force constant. Detailed molecular calculations show that the curvature of the paraboli for reactions and products are often nearly the same (Gregory and Arieh, 1990), but this is not always the case (see Figure 6.18).

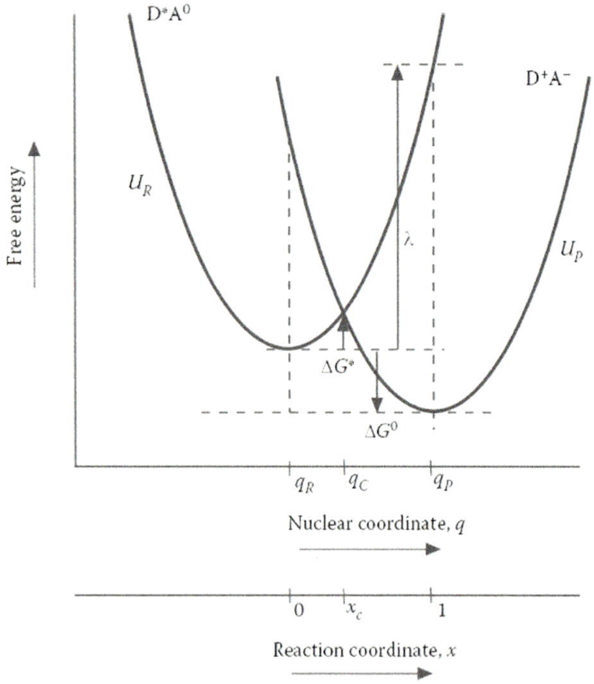

FIGURE 6.7 Free energy curves for an electron-transfer reaction showing the potential energy of the reactant state D^*A^0, U_R, and the products state D^+A^-, U_P, as a function of the nuclear coordinates and reaction coordinate. ΔG^0 indicates the "standard" change of free energy from reactants (R) to products (P) and is negative in this case. The transition state activation free energy is ΔG^*, which is actually the free energy barrier for the electron transfer. The reorganization energy is λ.

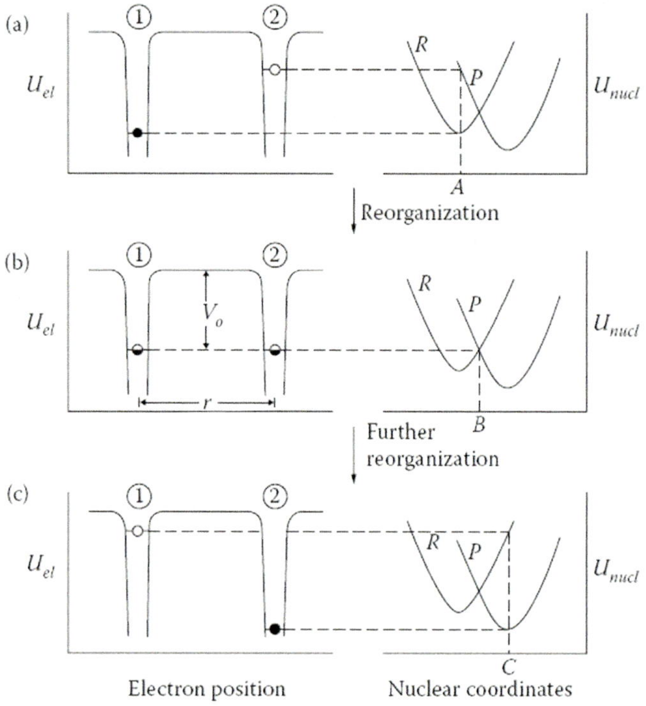

FIGURE 6.8 Electronic potential-energy curves (on left) and the corresponding (nuclear) potential-energy curves on the right. The electronic energies in the two wells are indicated on the left for three nuclear configurations A, B, and C for the reaction $red_1 + ox_2 \rightarrow ox_1 + red_2$. The levels in the wells are vertical ionization energies: the filled and open circles denote, respectively, ionization of the reduced state at its equilibrium nuclear configuration and at the equilibrium configuration appropriate to its oxidized state; the half-filled circles refer to ionization of the reduced state at the nuclear configuration appropriate to the intersection region. The level in well 1 is initially occupied in (a) and the level in well 2 in (c). (Reproduced with permission from Marcus, R. A.; Sutin, N. *Biochimica et Biophysica Acta (BBA)—Reviews on Bioenergetics* 1985, *811*, 265–322.)

The difference between the energy minima, ΔG^0, is the "standard" change of free energy from R to P,

$$\Delta G^0 = U_P(\mathbf{q}_P) - U_R(\mathbf{q}_R) \tag{6.40}$$

ΔG^0 is often called the "driving force" for the reaction or, more generally, for the charge transfer event. It is also useful to introduce a normalized nuclear coordinate, also called "reaction coordinate," x, that represents a change of the system from the nuclear and solvent configuration in equilibrium state of the reactants at $x = 0$ to that of the products at $x = 1$, as defined by the expression

$$q = q_R + x\Delta_0 q \tag{6.41}$$

This coordinate is also shown in Figure 6.7. It is also important to introduce the following quantity

$$\lambda = \frac{1}{2}k_F(\Delta_0\mathbf{q})^2 \tag{6.42}$$

It is termed the *reorganization energy* and it plays a key role in the theory of electron transfer. In Equation 6.42 we observe that λ is the energy associated to stretch the R state from the equilibrium value $\mathbf{q}_R$ to that in which the products are in equilibrium, $\mathbf{q}_P$, as shown in Figure 6.7.

Figure 6.8 illustrates the equilibrium states R and P and the transition state for a redox system, indicating the electron potential well, in the left column, and the nuclear coordinate diagram, in the right column. It is observed that in each of the R and P situations the oxidized state is far from the equilibrium value. The evolution along free energy curves leading to the electron transfer is also indicated in the right column of Figure 6.8.

We may express the potential-energy curves of Equation 6.38 and 6.39 as

$$U_R(x) = U_R(0) + \lambda x^2 \tag{6.43}$$

$$U_P(x) = U_P(1) + \lambda(1-x)^2 \tag{6.44}$$

The charge transfer event is regulated by the occurrence of fluctuations of all the coordinates, starting from $x = 0$ in the R curve, transitioning to the P curve at the crossing point, and proceeding on the P curve to the region at $x = 1$. The free energy barrier for the electron transfer is the difference between the potential energy at the crossing point, x_C, and the bottom of the reactant free energy parabola

$$\Delta G^* = U_R(x_C) - U_R(0) \tag{6.45}$$

The crossing point occurs at a position determined by the expression

$$U_R(x_C) = U_P(x_C) \tag{6.46}$$

The value of x at the transition state is

$$x_C = \frac{1}{2}\left(1 + \frac{\Delta G^0}{\lambda}\right) \tag{6.47}$$

Finally, from Equation 6.45 the free energy barrier is given by the expression

$$\Delta G^* = \frac{1}{4\lambda}\left(\Delta G^0 + \lambda\right)^2 \tag{6.48}$$

By Equation 6.36, the rate constant can be written in the form

$$k_{et} = A_{et} \exp\left[-\frac{\left(\Delta G^0 + \lambda\right)^2}{4\lambda k_B T}\right] \tag{6.49}$$

that is, the famous Marcus formula for electron-transfer rate.

For a photon-induced transition, commented above, the molecule changes suddenly in the vertical transition, as indicated in Figure 6.9. Just after absorption, the "hot" electron-molecule system is in the nuclear configuration of the ground state, and then relaxes to the bottom of the potential well of the excited state by an amount corresponding to the reorganization energy. Subsequently, the molecule may decay to the ground state by emitting a photon that is red shifted with respect to the maximal absorbance, and subsequently followed by a relaxation again by the energy λ. This is the process of *fluorescence*, and the relative displacement of the absorption and emission spectra is termed a *Stokes shift*.

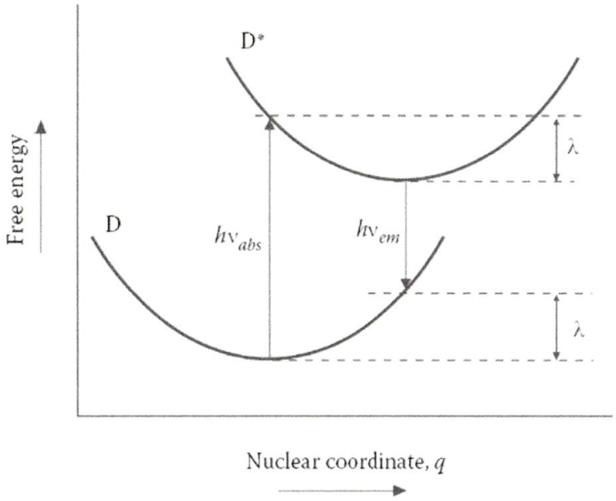

FIGURE 6.9 Free energy curves for the absorption and emission of photons between the ground (D) and excited state (D*) of a molecule. The photon energy for the maximal absorbance and emission are indicated, the reorganization energy λ for the relaxation of the first excited state after absorption, and the ground state after the emission of a photon by fluorescence.

Another important aspect of the outer-sphere electron-transfer model is expressed in the tunneling factor that is included in the prefactor A_{et} of Equation 6.49. For an assessment of the probability of the quantum transition, it is important to distinguish the *nonadiabatic* and *adiabatic* electron-transfer cases, which depend on the extent of coupling of the reactants and products electrons' wave functions. In the nonadiabatic transfer, the electronic coupling is small and so the probability of electron transfer when the system reaches the crossing point of the potential-energy curves is also small. For a nonadiabatic reaction, it can be written (Marcus and Sutin, 1985)

$$A_{et} = \frac{2\pi}{\hbar} \frac{|H|^2}{\sqrt{4\pi\lambda k_B T}} \tag{6.50}$$

where H is the electronic matrix element coupling the two reactants, and A_{et} has units s^{-1}. One approximation for a coupling bridge of length R is (Marcus and Sutin, 1985)

$$|H|^2 \approx 10^{13} e^{-\beta_b R} \tag{6.51}$$

where the coupling attenuation factor β_b depends on the nature of the bridge between the reactants (Chidsey, 1991).

In contrast to the former case, in an adiabatic electron transfer, the electronic coupling between the reactants' and the products' wave functions near or at the intersection is substantial. In such a case, nearly every system reaching the crossing point undergoes an electron transfer. An approximate expression for the transition probability is given by the Landau-Zener expression (Landau, 1932; Zener, 1933). Reviews of adiabatic–nonadiabatic effects have been given by Feldberg and Sutin (2006) and Emin (2013a).

Let us analyze in more detail the interpretation of the reorganization energy, which can be separated in two contributions as

$$\lambda = \lambda_0 + \lambda_i \tag{6.52}$$

Here, λ_0 describes the outer reorganization energy of the dielectric medium while the inner reorganization energy λ_i arises from the change in equilibrium values of vibrational coordinates of the reactants that involves bond lengths and bond angles. For example, if the reactants undergo a change $\Delta \mathbf{q}_i$ in the equilibrium value of some collective coordinate, a "normal coordinate" of a reactant, and if k_j^R and k_j^P are the "force constants" of that normal mode for the reactant and for the product, respectively, then classically (Marcus, 1965)

$$\lambda_i = \sum_j \left[\frac{k_j^R k_j^P}{k_j^R + k_j^P} \right] (\Delta q_j)^2 \tag{6.53}$$

However, the dielectric continuum treatment for the reorganization energy λ_0 gives the result (Marcus, 1956)

$$\lambda_0 = \frac{(\Delta z)^2 q^2}{4\pi\varepsilon_0} \left(\frac{1}{2a_1} + \frac{1}{2a_2} - \frac{1}{R} \right) \left(\frac{1}{\varepsilon_{op}} - \frac{1}{\varepsilon_s} \right) \tag{6.54}$$

where Δz is the number of charges transferred, the two a_i's are the radii of the two reactants at distance R, ε_{op} is the relative optical dielectric constant (square of the refractive index) of the solvent, and ε_s is its static dielectric constant. Equation 6.54 is often simplified to the case of two identical reactants in solution as (Hamann et al., 2005)

$$\lambda_0 = \frac{(\Delta z)^2 q^2}{4\pi\varepsilon_0 a} \left(\frac{1}{\varepsilon_{op}} - \frac{1}{\varepsilon_s} \right) \tag{6.55}$$

There are some important features of the reorganization energy depending on the size of the reacting ions, the distance that separates their centers, and the dielectric constant. For example, the smaller the radii a_i of the reactants, the larger is λ_0, due to the fact that the orientating force of the ionic charges on the nearby solvent molecules is larger. When the distance R is small, the reorganization energy λ_0 is smaller, as it turns out that distant solvent molecules hardly see a change of electric field accompanying the electron transfer. When the solvent is nonpolar, that is when $\varepsilon_s = \varepsilon_{op}$, the solvent contribution to λ_0 arising from the orientational and vibrational part of its dielectric polarization disappears. Therefore, the larger λ_0, the more polar the medium.

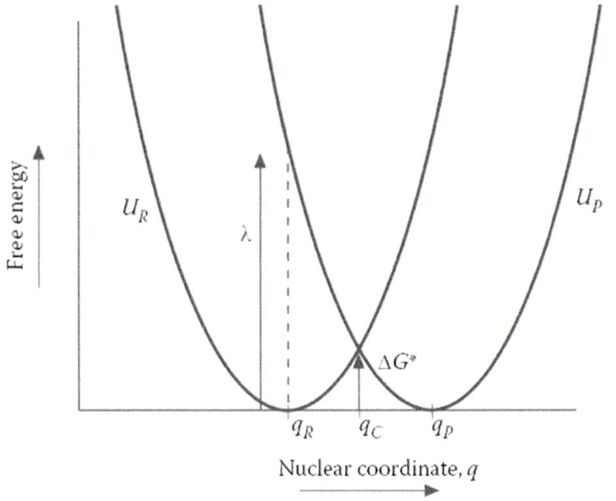

FIGURE 6.10 Free energy curves for an electron-transfer reaction as a function of the nuclear coordinates and reaction coordinate for a self-exchange reaction in which the reactants and products are the same.

The simplest application of the theory of Equation 6.49 is that in which $\Delta G^0 = 0$, which occurs in a class of reactions known as "self-exchange," in which the free energy barrier to reaction ΔG^* is $\lambda/4$, as shown in Figure 6.10. This case corresponds to polaron hopping and will be more extensively discussed in the next section.

The most significant success of the Marcus electron-transfer theory is the prediction of an *inverted region*, in which the rate of electron transfer *decreases* for increasing driving force $|\Delta G^0|$, as shown in Figure 6.11. The observation of the inverted region in metal electrodes was reported by Closs and Miller (1988) and Chidsey (1991) and for semiconductor electrodes by Hamann et al. (2005).

6.5 POLARON HOPPING

An electronic charge in a solid environment can become stabilized by interaction with the surrounding medium. The charge of the electron distorts the lattice which

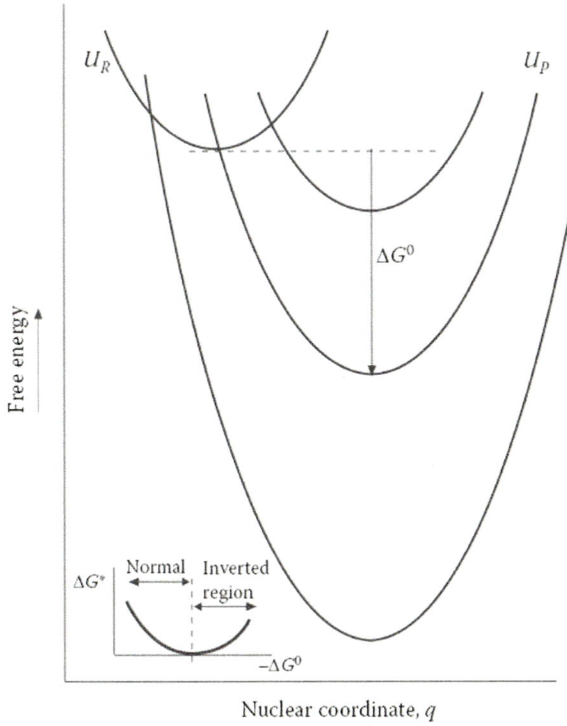

FIGURE 6.11 Free energy curves for an electron-transfer reaction as a function of the nuclear coordinates. The curves show a change of the product for an increasingly large driving force, ΔG^0, until the transition state activation free energy, ΔG^*, passes through value 0 and starts to increase in the Marcus-inverted region, as shown in the inset.

in turn acts on the electron and lowers its energy. The remark that an electron can "dig its own potential well" by interaction with the surrounding lattice was first made by Landau (1933). The self-stabilized electronic charge in a deformable polar medium, by strong interaction with phonons, is termed a *polaron*, and its slow motion, that drags the lattice distortion, is defined as polaron hopping.

The main effect of polaron hopping is that the local potential well is stabilized after each carrier jump to a new site. Let $\mathbf{q}_1$ be a suitable coordinate representing the distortion of the near surroundings to site 1, see Figure 6.12a. $\mathbf{q}_1$ collectively represents the normal vibrational modes of the solid. The potential energy without the added carrier is given by (Austin and Mott, 1969)

$$U^0\left(\mathbf{q}_1 = \frac{1}{2}k_F\mathbf{q}_1^2\right) \tag{6.56}$$

The interaction energy of the carrier with its surroundings can be suitably described by a linear dependence of the carrier energy on the lattice displacements

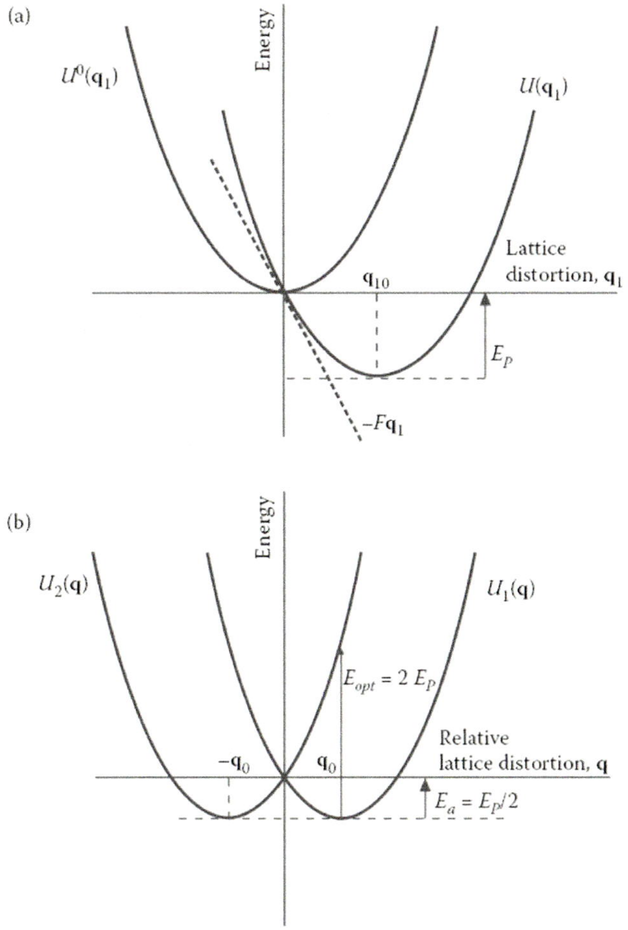

FIGURE 6.12 (a) Stabilization of a carrier in a solid medium by interaction with phonons, to form a polaron. The graph shows the modification of the elastic energy for a single carrier for the interaction of the carrier with the lattice. (b) Effective potential wells for two sites, showing in each case the situation when the carrier is within the site. Also indicated is a vertical optical transition.

$E_{\text{int}}(q_1) = -F\mathbf{q}_1$. The total energy of the carrier and the distorted surrounding is

$$\begin{aligned} U(\mathbf{q}_1) &= \frac{1}{2}k_F\mathbf{q}_1^2 - F\mathbf{q}_1 \\ &= -E_P + \frac{1}{2}k_F(\mathbf{q}_1 - \mathbf{q}_{10})^2 \end{aligned} \tag{6.57}$$

where $\mathbf{q}_{10} = F/k_F$. As shown in Figure 6.12a the energy of the carrier-lattice system is stabilized by the quantity

$$E_p = \frac{1}{2}k_F\mathbf{q}_{10}^2 = \frac{1}{2}\frac{F^2}{k_F} \tag{6.58}$$

This is the polaron binding energy, that relates to reorganization energy as $E_P = \lambda/2$.

We now consider the hopping of a carrier between neighbor sites 1 and 2. The total energy of the two sites is (Faust et al., 1994)

$$U(\mathbf{q}) = \frac{1}{2}k_F\left(\mathbf{q}_1^2 + \mathbf{q}_2^2\right) - F\left(n_1\mathbf{q}_1 + n_2\mathbf{q}_2\right) \quad (6.59)$$

where the indexes $n_1 = 1$, $n_2 = 0$ if the carrier is in site 1, and $n_1 = 0$ and $n_2 = 1$ if it is in site 2. Introducing the coordinates $\mathbf{q} = \mathbf{q}_1 - \mathbf{q}_2$ and $\mathbf{Q} = \mathbf{q}_1 + \mathbf{q}_2$, the total energy can be expressed as

$$U(\mathbf{q}) = \frac{1}{4}k_F\left(\mathbf{q}^2 + \mathbf{Q}^2\right) - \frac{F}{2}(n_1 - n_2)\mathbf{q} - \frac{F}{2}\mathbf{Q} \quad (6.60)$$

The terms in $\mathbf{Q}$ do not change in a carrier jump. Hence, we may distinguish the two situations

$$U_1(\mathbf{q}) = \frac{1}{4}k_F\mathbf{q}^2 - \frac{F}{2}\mathbf{q}$$
$$U_2(\mathbf{q}) = \frac{1}{4}k_F\mathbf{q}^2 + \frac{F}{2}\mathbf{q} \quad (6.61)$$

for $n_1 = 1$ and $n_2 = 1$. These energy wells are shown in Figure 6.12b for the case in which the carrier is within the respective well. Using the same arguments as in the Marcus theory, it is observed that the activation energy for thermally activated hopping is

$$E_a = \frac{1}{2}E_p = \frac{F^2}{4k_F} \quad (6.62)$$

The Arrhenius temperature dependence dominates polaron hopping when the temperature is high enough so that the vibration of lattice atoms can be treated as classical. This occurs when the thermal energy is much larger than the energy of the phonons.

Figure 6.12b also shows an optical vertical transition, that requires the photon energy $E_{opt} = 2E_p$. The self-trapping polaronic effect can be revealed by the Stokes shift between light absorption and emission, indicated in Figure 6.9. This is shown in Figure 6.13 for anatase TiO_2 crystal at a very low temperature. The excitation by a photon is a vertical transition that takes place above 3.2 eV, the bandgap of anatase.

But the light emission occurs from the self-trapped state and has a maximum at 2.3 eV. The difference between onset absorption and emission is associated with the stabilization of the electron-lattice system in the excited state. There are different possible causes for the broad luminescence peak observed at an energy of 2.3 eV in anatase TiO_2 nanoparticles, but in recent work it is usually attributed to self-trapped excitons in combination with a surface recombination (Cavigli et al., 2009, Pallotti et al., 2013). The stabilization process of the electron and hole carriers in anatase TiO_2 is illustrated in Figure 6.14. First principle calculations by Di Valentin and Selloni (2011) indicate that the electron becomes self-trapped at a Ti lattice site (Ti^{4+}), which then becomes formally Ti^{3+}, with an energy gain (trapping energy) associated with the polaronic distortion $\Delta E_{trap} = 0.23$ eV, while the self-trapping energy of a hole is 0.74 eV.

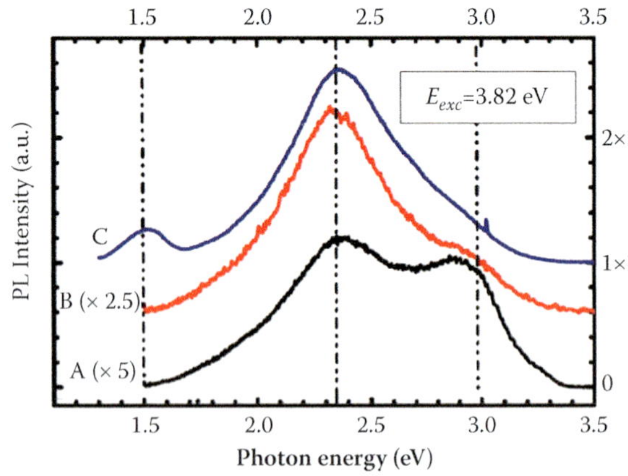

FIGURE 6.13 Luminescence (dotted line) and excitation (solid line) spectra of anatase TiO_2 at low temperature (a) and luminescence at room temperature of several samples (b). (Reproduced with permission from Watanabe, M.; Hayashi, T. *Journal of Luminescence* 2005, *112*, 88–91; Pallotti, D. K. et al. *Journal of Applied Physics* 2013, *114*, 043503.)

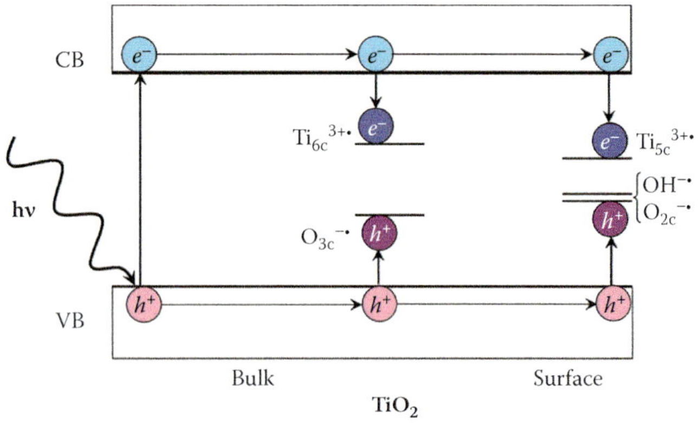

FIGURE 6.14 Schematic representation of the initial stages of a photoinduced process in anatase TiO_2: creation of a bulk exciton upon light irradiation, formation of self-trapped polarons at bulk lattice sites, and polarons trapped at surface sites. (Reproduced with permission from Di Valentin, C.; Selloni, A. *The Journal of Physical Chemistry Letters* 2011, *2*, 2223–2228.)

In most organic materials the electronic carrier is strongly stabilized by the vibronic coupling so that the reorganization effects may be dominated by the inner reorganization energy λ_i rather than polarization effects. The motion of the carrier is determined by a combination of two effects, namely, the polaronic stabilization of energy and the energy difference between sites caused by the disorder. The latter is not a consequence of the interaction with phonons, but it is due to a disordered arrangement of the solid and spatial fluctuation of the dielectric polarization of the surrounding (Austin and Mott, 1969, Köhler and Bässler, 2011) see Section 8.6. Owing to disorder, the charge transfer between neighbor sites occurs across the energy difference ($E_j - E_i$). The charge transfer rate for hopping is then written in terms of the semiclassical electron-transfer theory stated above in Equation 6.49 as follows:

$$\nu_{ij} = \frac{\mathbf{J}_{ij}^2}{\hbar}\left(\frac{\pi}{\lambda k_B T}\right)^{1/2} \exp\left[-\frac{\left(E_j - E_i + \lambda\right)^2}{4\lambda k_B T}\right] \tag{6.63}$$

where $\mathbf{J}_{ij}$ is the electronic coupling element, a transfer integral. Equation 6.63 has been used extensively in simulation of carrier transport in soft disordered materials.

We should recall that we have two different frameworks for the displacement of electrons in a disordered material. The expression of Miller and Abrahams (1960) in Equation 6.2 is essentially a single phonon process, that is thermally activated. On the contrary, the polaron or Marcus conduction in Equation 6.63 is a multiphonon process involving strong reorganization effects. A generalized expression that bridges the gap between the two regimes was given by Emin (1974), see further discussion in Köhler and Bässler (2011).

6.6 RATE OF ELECTRODE REACTION: BUTLER-VOLMER EQUATION

The process of electron transfer at the electrode interface is represented in Figure 6.15, which gives more details on Figure 6.3 discussed earlier. Figure 6.15 shows the molecular species that can be reduced and oxidized by electron transfer from or to the metal part of the electrode. The scheme also shows inert ions of both signs of charge that form the Helmholtz layer as discussed earlier in Figure 3.8. Figure 6.15b indicates that a modification of the electrode potential, associated with a difference of the electrochemical potential of electrons in the metal and in redox ions in the electrolyte, causes an increase of the voltage across the Helmholtz layer. These changes

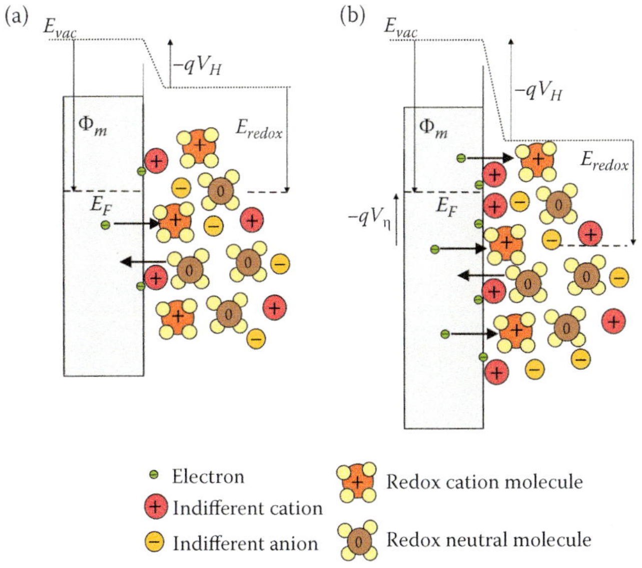

FIGURE 6.15 Structure of the Helmholtz layer at the contact between the metal electrode and an electrolyte, containing solvated redox species and salt anions and cations, and the events of interfacial charge transfer that constitute electrochemical reactions. (a) At zero applied potential, a contact potential difference is built, consisting of a voltage drop V_H, across the interfacial dipole formed by electrons at the metal surface and ions at the solution side. The reduction and oxidation reactions flow at the same rates. (b) With a negative applied potential, the amount of positive and negative charge at the surface increases, and so does the interfacial potential drop V_H. The increase of the voltage across the interface increases the rate of the cathodic (reduction) reaction.

occur in a very short distance of the surface, normally less than 1 nm. The change of the Helmholtz potential modifies position and height of the barrier to electron transition and induces a net interfacial current flow, as will be discussed later.

In Figure 6.15a we show the situation of electronic equilibrium between the metal and electrolyte. We denote the electrode potential $V = V_r$, and the redox potential of the electrolyte, V_{redox}, both measured with respect to an RE as shown in Figure 6.16. The deviation of the electrode potential from the equilibrium potential is termed the *overvoltage*, V_η (Figure 3.9), as follows:

$$V_\eta = V_r - V_{redox} \tag{6.64}$$

More generally, the overvoltage is the excess voltage with respect to thermodynamic equilibrium value, required for current flow, due to different kinetic limitations.

If the electrode potential is displaced to a value that is more negative than V_{redox}, then a negative overvoltage results and the cathodic current is promoted, in which the electrons of the metal (or, the electrode) flow into

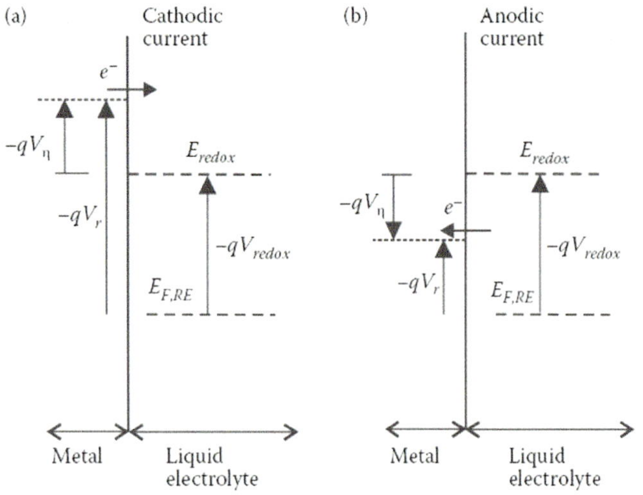

FIGURE 6.16 Scheme of electrode reactions in electron energy scale. The potential V_r is the voltage applied to the electrode with respect to a reference electrode in solution. The overpotential V_η is the difference between the electrode potential and the equilibrium redox potential V_{redox}. (a) The overpotential V_η is negative and promotes the cathodic reaction, since the electron energy level is above the redox level. (b) The overpotential V_η is positive and promotes the anodic reaction at the electrode surface, as the electron energy at the metal is below the redox level.

the empty orbitals (LUMO) of the electroactive species present in solution, as shown in Figure 6.16a, while the anodic current is largely suppressed. Conversely, if the overvoltage is positive, then the anodic reaction and anodic current are favored as indicated in Figure 6.16b. In general, a given potential is not "cathodic" or "anodic" per se, as this quality depends on the relative value of V_r with respect V_{redox}.

The complete electrochemical cell is shown in Figure 6.17a. In the electrolyte, the current is carried by the electroactive species and a steady-state electric current is maintained in the cell. The current is determined by electron-transfer rate at the metal surface, since ions cannot enter the metal electrode. The electron flux across the metal/electrolyte interface has the form

$$J_{ct} = k'_{ct} n_s C^s_{ox} \tag{6.65}$$

The rate is determined by two types of factors: (1) the availability of the reacting species, namely, electrons in the metal side (concentration n_s) and the active ions at the electrolyte side c_{ox} (for the reduction flux), and (2) the kinetic constant k'_{ct} for the probability of charge transfer of an electron to an *ox* ion. A similar expression may be written for the rate of transfer of electrons from the reduced half of the redox couple to the metal:

$$J_{ct} = k_{ct} P_s C^s_{red} \tag{6.66}$$

where P_s is the density of electron vacancies (holes) at the surface, and c_{red} is the concentration of the species that is oxidized in the surface reaction. Taking the anodic current positive, the charge transfer current is given by

$$j_{ct} = q\left(K_{ct} P_s C^s_{red} - k'_{ct} n_s C^s_{ox}\right) \tag{6.67}$$

We will now discuss the different elements of Equation 6.67.

6.6.1 Availability of Electronic Species

Since the metal Fermi level occurs at a high density of states, there are many electrons (below E_F) available to jump to the solution acceptor under cathodic overvoltage, and vice versa, there are also many empty states (holes, above the Fermi level) that can accept an electron from the reduced ionic species. Therefore, in metal electrodes the electron/hole density at the surface is not a matter of concern, and the terms n_s and p_s are simply incorporated to the rate constants as follows:

$$j_{ct} = q\left(K_{ct} C^s_{red} - k'_{ct} C^s_{ox}\right) \tag{6.68}$$

The situation is different when considering the charge transfer between a semiconductor electrode and a redox couple, since the electron density at the surface is highly variable and is a main factor in the rate of the overall interfacial reaction (see Section 6.8).

6.6.2 Availability of Redox Species

If a reacting species is exhausted at the electrode surface, the concentrations at the surface, C^s_{red} and C^s_{ox}, may differ substantially from the bulk concentrations C^b_{red} and C^b_{ox}. In a controlled experiment in a liquid electrolyte, the natural convection tends to stir the electrolyte in an uncontrolled fashion, and concentrations above a certain distance from the electrode are maintained uniform at the bulk electrolyte value. The inert salt ions, shown in Figure 6.15, have an important role in facilitating screening of electric field in the bulk solution, so that the transport to the electrode surface for the supply of reactants is carried by diffusion. A diffusion boundary layer, known as the *Nernst diffusion layer*, is established, as shown in Figure 6.17b. In this layer, the concentration gradients are essentially linear, and the thickness δ is defined as the distance through which the linear portion of the concentration profile of the reacting species reaches the bulk value, c^b. A process is electrochemically *reversible*

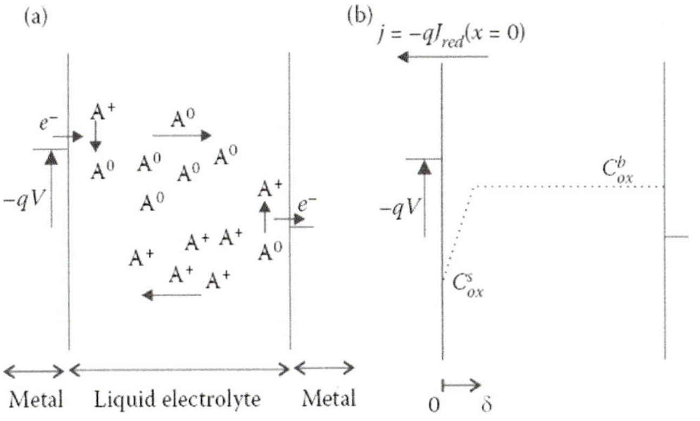

FIGURE 6.17 Scheme of an electrochemical cell, consisting of an electrolyte between two metal plates of the same material. (a) At the surface of the negative biased (left) electrode, the oxidized A^+ ionic species in solution is reduced to A^0. At the positive biased (right) electrode, the reduced species A^0 is oxidized to A^+. The flow of A^0 and A^+ in the electrolyte is indicated. (b) The electric current is equal to the discharge of electrons at the negative electrode, which is proportional to the surface concentration of A^+ at the surface of the electrode. Mass transport in the electrolyte is controlled by diffusion across a boundary layer of thickness δ.

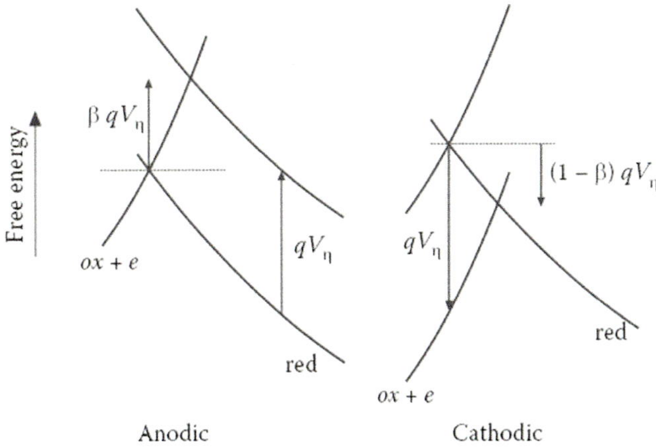

FIGURE 6.18 Free energy diagram for a one electron transfer showing the change of free energy curves of *ox* and red species as a function of the overpotential V_η. β is the symmetry factor.

if the rate of the electron transfer is higher than the rate of the mass transport. The process is defined as electrochemically irreversible if the rate of the electron transfer is lower than the rate of the mass transport.

6.6.3 The Kinetic Constant for Charge Transfer

This factor is determined by the quantum mechanical probability of electron tunneling from the metal to the frontier orbitals of the ion, and several additional factors concerning, among others, the polarization of the *red* and *ox* states by the surrounding solvent. In order to apply the Marcus transition rate of Equation 6.49, we note that for a cathodic reaction at an electrode surface, the "standard" change of free energy can be stated as

$$\Delta G^0 = E_{redox} - E_{Fn} = qV_\eta \tag{6.69}$$

Therefore,

$$\frac{\left(\Delta G^0 + \lambda\right)^2}{4\lambda k_B T} = \frac{\lambda}{4k_B T} + \frac{1 qV_\eta}{2k_B T} + \frac{q^2 V_\eta^2}{4\lambda k_B T} \tag{6.70}$$

Assuming $|qV_\eta| \ll \lambda$, we have from Equation 6.49

$$k_{et} = k_r \exp\left(-\frac{1}{2}\frac{qV_\eta}{k_B T}\right) \tag{6.71}$$

The effect of the voltage is to relatively displace the free energy curves and for a moderate overvoltage, the interfacial rate constant depends exponentially on V_η. More generally, if the free energy paraboli have different curvatures, the change of the height of the transition state point, as well as its position, depends asymmetrically on voltage as indicated in Figure 6.18. The coefficient 1/2 in the exponent of Equation 6.71 can be more generally replaced by a number $(1 - \beta)$, where $\beta < 1$ is called the *transfer coefficient*. Then, we obtain the rate constant for the anodic direction

$$k_{ct} = k_r \exp\left[\beta q\left(V - V^0\right)/k_B T\right] \tag{6.72}$$

where V^0 is a reference voltage. For the cathodic direction, we have

$$k'_{ct} = k_r \exp\left[-(1-\beta) q\left(V - V^0\right)/k_B T\right] \tag{6.73}$$

These time constants for the forward and reverse direction obey detailed balance as shown in Equation 6.18, provided that a suitable reference of potential V^0 is chosen.

In order to determine the current dependence on overvoltage, we check that Equation 6.68 obeys detailed balance. In equilibrium forward and reverse flux is the same, $V = V_{redox}$, and the surface concentration of ions equals their bulk concentration. Using Equations 6.72 and 6.73, we obtain the condition

$$\frac{C^b_{ox}}{C^b_{red}} = e^q\left(V_{redox} - V^0\right)/k_B T \tag{6.74}$$

Note that the kinetic constants are removed from the condition of equilibrium, since they were already taken care of in Equation 6.18. We observe that Equation 6.74 matches the Nernst equation of the electrolyte (see Equation 3.49), provided that $V^0 = V^0_{redox}$, the standard potential of the redox couple.

Having determined the equilibrium constants of the expression of the current, we apply simple transformations that allow us to write Equation 6.68 as

$$j_{ct}(V_\eta) = qk_r\left(c^b_{ox}\right)^\beta\left(c^b_{red}\right)^{1-\beta}\left[\frac{c^s_{red}}{c^b_{red}}e^{q\beta V_\eta/k_BT} - \frac{c^s_{ox}}{c^b_{ox}}e^{-q(1-\beta)V_\eta/k_BT}\right] \quad (6.75)$$

The prefactor of Equation 6.75 is termed the *exchange current density*

$$j_0 = qk_r\left(c^b_{ox}\right)^\beta\left(c^b_{red}\right)^{1-\beta} \quad (6.76)$$

This is the amount of current flowing in both cathodic and anodic directions under equilibrium at $V_\eta = 0$. Equation 6.75 can be expressed in the form

$$j_{ct}(V_\eta) = j_0\left[\frac{c^s_{red}}{c^b_{red}}e^{q\beta V_\eta/k_BT} - \frac{c^s_{ox}}{c^b_{ox}}e^{-q(1-\beta)V_\eta/k_BT}\right] \quad (6.77)$$

that is known as the *current-overpotential equation*. If there is no limitation of supply of reactants toward the surface, then the ionic concentrations take the bulk values and Equation 6.77 is simplified as

$$j_{ct}(V_\eta) = j_0\left[e^{q\beta V_\eta/k_BT} - e^{-q(1-\beta)V_\eta/k_BT}\right] \quad (6.78)$$

This is the Butler-Volmer equation. If the elementary transference consists of z electrons, then

$$j_{ct}(V) = j_0\left[e^{qz\beta V_\eta/k_BT} - e^{-qz(1-\beta)V_\eta/k_BT}\right] \quad (6.79)$$

and in equilibrium the Nernst Equation 3.49 will be recovered.

The shape of the current-overpotential curve according to Equation 6.78 is shown in Figure 6.19a. The influence of the exchange current density parameter is shown in Figure 6.20a. For a large j_0, the current increases at a lower overpotential. Therefore, j_0 contains the essential features of the electrode kinetics. As j_0 is reduced, increasing overvoltage is required to obtain a given current, say 10 mA cm^{-2}.

The kinetics of charge transfer is accelerated by using a catalyst, whose role is precisely to enhance the charge transfer constant of the reaction. Several functions of the catalyst layer can be distinguished that may actuate separately or in combination: providing reaction sites, decreasing activation energies, increasing the binding strength of the reactant, and facilitating the dissociation of molecules; thus favoring electron discharge to the adsorbed species. It should be taken into account that important reactions such as water splitting usually contain a multistep mechanism and it is difficult to precisely determine the role of the catalyst. Especially for energy conversion electrochemical cells, the presence of a considerable overvoltage in the desired reactions is a major drawback, as this voltage is subtracted from the available driving force and does not lead to useful energy. This point is further commented in Section 6.8. On the contrary, it is sometimes required to slow down the reaction rate to avoid undesired reactions. The use of protective layers increases the overvoltage for solvent decomposition at electrodes in battery cells, and provides a larger cell voltage, as commented in Section 3.12.

As shown in Figure 6.20c, the curves at different values of j_0 have the same shape. The current increases

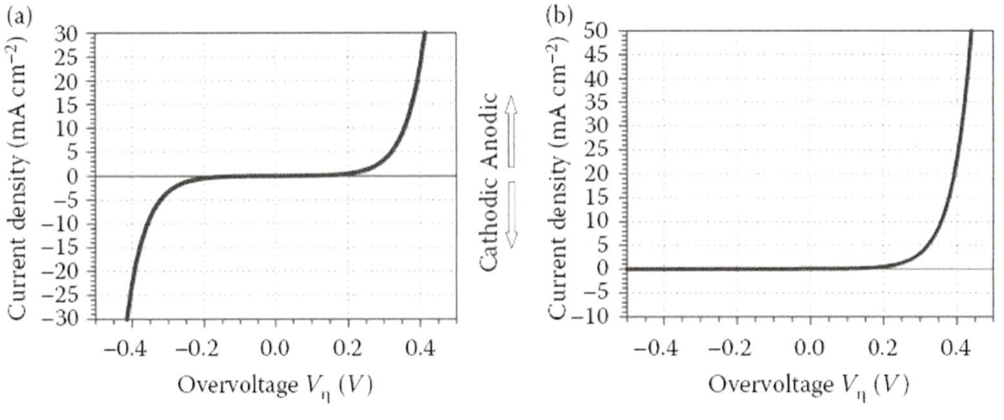

FIGURE 6.19 (a) Current-overvoltage curve for the cathodic and anodic reactions at the metal/electrolyte interface. The exchange current density is $j_0 = 10^{-2}$ mA cm^{-2}, $\beta = 0.5$, and $T = 300$ K. (b) Current-overvoltage curve for the cathodic reaction.

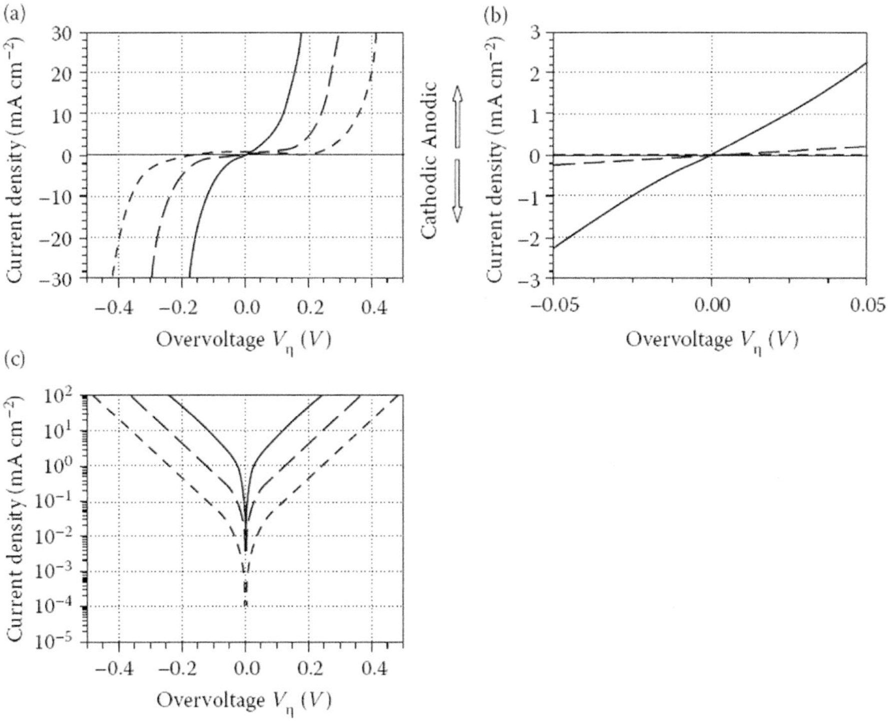

FIGURE 6.20 (a) Current-overvoltage curve for the cathodic and anodic reactions at the metal/electrolyte interface. The exchange current density is indicated, $\beta = 0.5$ and $T = 300$ K (- - - - $j_0 = 0.01$ mA cm^{-2}, — — — — $j_0 = 0.1$ mA cm^{-2} ———— $j_0 = 1$ mA cm^{-2}). (b) The current at low overvoltage. (c) The current in log scale.

exponentially for both large positive and negative potential, and at low overvoltage there is a region that is practically linear as shown in Figure 6.20b. To check the linearity of the curve, we calculate its slope. This is a differential resistance that can be measured and it is called the *charge transfer resistance* R_{ct}.

$$R_{ct} = \left(\frac{\partial j_{ct}}{\partial V_\eta} \right)^{-1} \tag{6.80}$$

and we obtain

$$\frac{1}{R_{ct}} = \frac{q j_0}{k_B T} \left[\beta e^{q\beta V_\eta / k_B T} + (1-\beta) e^{-q(1-\beta)V_\eta / k_B T} \right] \tag{6.81}$$

In general, the charge transfer resistance increases exponentially for both positive and negative overpotentials. But for small overpotential $V_\eta \le k_B T/q$ (corresponding to the linear region in Figure 6.19b), the resistance is a constant

$$R_{ct} = \frac{k_B T}{q j_0} \tag{6.82}$$

This is termed the *ohmic region*. At equilibrium, while the dc current is zero, there are still currents of magnitude j_0 that cross the interface in both directions and cancel out. In general, if the demand of current is $\approx j_0$, then a small overvoltage is required, that remains in the ohmic region, and therefore the contact is ohmic. However, at larger voltage rectification or nonlinear properties of the charge transfer at the particular interface may appear.

If we write Equation 6.82 in terms of the fundamental kinetic constants, we obtain

$$R_{ct} = \frac{k_B T}{q^2 k_r \left(c_{ox}^b\right)^{\beta} \left(c_{red}^b\right)^{1-\beta}} \tag{6.83}$$

When the combination of the specific electrode material and reactant species in the solution effectively allows the reaction to reversibly occur in both directions, cathodic and anodic, then the shape of Figure 6.19a is obtained. If only the anodic reaction is permitted, then the current-voltage relation reduces to the expression

$$j_{ct}(V) = j_0^{q\beta V_\eta / k_B T} \tag{6.84}$$

This dependence is shown in Figure 6.19b. Here, we must notice the similarity of the current-overvoltage Equation 6.84 with the diode Equation 5.49 that was obtained for a minority carrier recombination diode. In both cases the current increases exponentially at "forward" bias and decreases at reverse bias. However, we must also remark some differences. In the recombination diode, the current increases because the electron density increases in the material. In contrast to this, in the metal electrode there is no increase of electron density and what is enhanced by the voltage is the rate of charge transfer as represented by k_{ct} in Equation 6.72. In reverse, there is no saturation current in Equation 6.84 as the back current is totally suppressed. The rectification is a purely interfacial effect.

6.7 ELECTRON TRANSFER AT METAL-SEMICONDUCTOR CONTACT

The emission of electrons from a metal contact (the current collector) to a semiconductor and vice versa is a process of great significance for the operation of electronic and optoelectronic devices. The transference of electrons between the metal and a crystalline semiconductor with well-defined energy levels can be described by a modification of the thermionic emission model, discussed in Section 4.1. If the electric field in the space charge region is sufficiently large, then the current is limited only by the thermal velocity of carriers in the semiconductor. According to Bethe (1942), the criterion for the predominance of thermionic emission is that the potential changes at least k_BT/q along a mean free path λ_{fp} of the electron. Since the injection current to the metal corresponds to thermal motion of carriers in the semiconductor, this model uses the effective electron mass $m_e^* = m_e m_0$. The following expression applies for the equilibrium current density:

$$j_0 = Am_eT^2e^{-q\Phi_b/k_BT} \tag{6.85}$$

Here, A is the effective Richardson constant (Equation 4.3), and Φ_b is the effective barrier height. By taking into account the effective density of states of the conduction band (Equation 2.14), and introducing the following expression for the carrier charge transfer velocity (Equation 4.9)

$$v_n^{th} = \left(\frac{8k_BT}{\pi m_e^*}\right)^{1/2} \tag{6.86}$$

the Equation 4.10 can be generalized to

$$j_0 = \frac{1}{4}qN_cv_n^{th}e^{-q\Phi_b/k_BT} \tag{6.87}$$

Then, we obtain the standard expression

$$j_0 = A^*T^2e^{-q\Phi_b/k_BT} \tag{6.88}$$

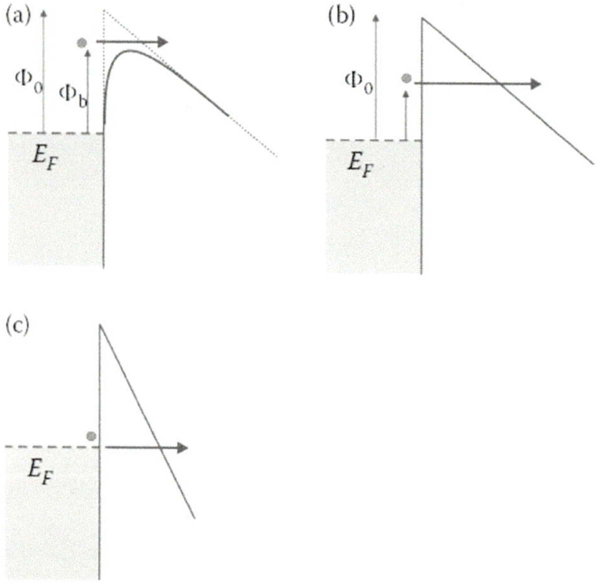

FIGURE 6.21 (a) The lowering of interfacial barrier by image force effect. (b) Thermally assisted Fowler-Nordheim tunneling. (c) Tunneling at the Fermi level.

where A^* is the same as Richardson constant of Equation 4.3, except for the substitution of the semiconductor effective mass m_e^* for the free electron mass m_0

$$A^* = \frac{4\pi m_e^* qk_B^2}{h^3} \tag{6.89}$$

To account for the barrier and energy levels close to the interface, it is necessary to consider the effects of image charge, as stated by Schottky. The effective barrier height, Φ_b, corresponds to the total barrier decreased by the image force potential, as shown in Figure 6.21a

$$\Phi_b = \Phi_0 = \sqrt{\frac{q\mathbf{E}}{4\pi\varepsilon_r\varepsilon_0}} \tag{6.90}$$

Here, $\mathbf{E}$ is the maximum field at the junction (a detailed description of the calculation of $\mathbf{E}$ is given in Chapter 9). The current at the contact can be further affected by field emission and by tunneling assisted by interface states existing at the junction (Figure 6.21b). The probability of tunneling outside of the metal increases when the barrier is thinner, and it is facilitated if the electric field increases. This mode of carrier injection is termed *field emission*. Fowler and Nordheim (1928) introduced the model of wave mechanical tunneling of electrons through a rounded, field induced, and triangular barrier. The current is given by

$$j = \frac{q(q\mathbf{E})^2}{8\pi h\Phi_0}\exp\left[\frac{-4\gamma(\Phi_0)^{3/2}}{3q\mathbf{E}}\right] \tag{6.91}$$

where $\gamma^2 = 2m_0/\hbar^2$. The model is usually presented in plots $\left(j/\mathbf{E}^2\right)$ versus $1/\mathbf{E}$ (Forbes, 2009). At low temperatures, there is no thermal energy and the carriers tunnel from the Fermi level (cold field emission); this is also the case at strong fields (see Figure 6.21c). At intermediate temperatures and moderate fields, the carriers tunnel from an energy that is higher than the Fermi energy (Figure 6.21b), and the mechanism is a combination of thermionic emission and field emission. At very high temperature, the thermal activation over the barrier becomes the dominant emission mechanism.

A generalization of Equation 6.87 can be stated in terms of a transfer velocity v_n introduced by Crowell and Sze (1966), that is also denoted as "recombination velocity," and the concentration of electrons at the surface, n_s,

$$j_0 = qn_s v_n \tag{6.92}$$

Equation 6.92 shows that the current across the barrier has the standard form of a charge transfer flux in terms of carrier density and transfer rate (see Equation 6.67). The majority carrier transfer at metal/semiconductor interface may also be limited by other effects, so that v_n may depend on the product of the thermal velocity in Equation 6.86 and the probability that carriers will cross the interface when they reach it, given by the transmission coefficient κ. The model of Equation 6.87 can be formulated more generally (Mönch, 1993)

$$j_0 = \kappa A T^2 e^{-q\Phi_b/k_BT} \tag{6.93}$$

For metal/semiconductor interfaces it is usually taken $\kappa = 1$, since the heterogeneous charge transfer event is usually not a limiting factor. However, κ can be much less than unit for localized systems. In addition in localized systems such as an organic conductor or a redox couple in an electrolyte, the charge transfer rate may be completely limited by the product of the rate constant for heterogeneous charge transfer k_{et}, involving reorganization effects as noted in Section 6.6. The rate of transference may also be limited by the match of the density of states at both sides of the interface, as in the contact of an inorganic semiconductor with a conjugated polymer (Lonergan, 2004). In conclusion, the carrier charge transfer velocity, v_n, in Equation 6.92 can be regarded as a purely thermal velocity, v_n^{th}, only in certain situations in which the interface allows a fast transference of carriers that pass into a crystalline material. In general, this velocity is a function of specific transfer mechanism at the interface, including disorder, reorganization, and so on.

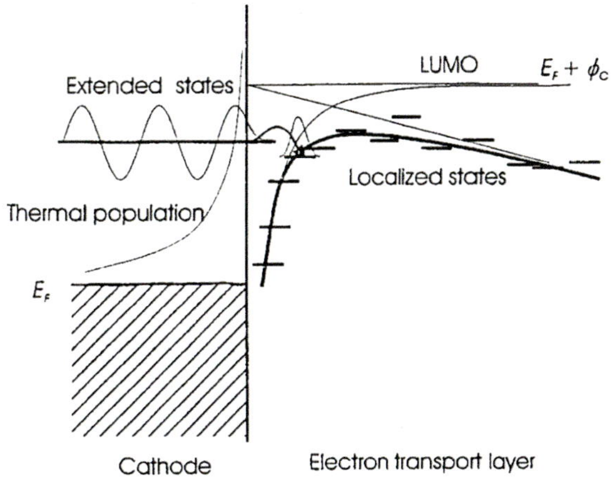

FIGURE 6.22 Energetics of the charge injection process. Electrons in extended states in the metal, including those thermally excited above the Fermi level, tunnel into localized states in the semiconductor, where there are four contributions to the energy: the barrier height, the image potential, the potential due to the interfacial electric field, and a random component due to disorder. (Reproduced with permission from Scott, J. C. *Journal of Vacuum Science & Technology A: Vacuum, Surfaces, and Films* 2003, *21*, 521–531.)

At the metal/organic interface, the electronic states in the organic conductor are highly localized and show a wide distribution of energies. A series of studies considered the effect of disorder by dividing the injection process into two steps: the first is the tunneling from the metal into one of the localized states of the organic semiconductor (Figure 6.22), and the second is the escape from that state into the bulk of the organic film (Bässler et al., 1998). Baldo and Forrest (2001) consider a different degree of disorder right at the interface, using a model shown in Figure 6.23 in which the interfacial region includes a random contribution of interface dipoles that significantly broadens the energy distribution of organic transport states in the interfacial layer.

6.8 ELECTRON TRANSFER AT THE SEMICONDUCTOR/ELECTROLYTE INTERFACE

The exchange of electronic carriers between a semiconductor surface and redox species in an electrolyte is an electrochemical reaction of great significance for energy-related applications such as photoelectrochemical photovoltaic cells and fuel production cells. Some of the electron-transfer processes are shown in Figure 6.24.

It is interesting to compare two different paradigms of electrochemical electron transfer. In a metal electrode shown in Figure 6.15, the electrode potential V_r

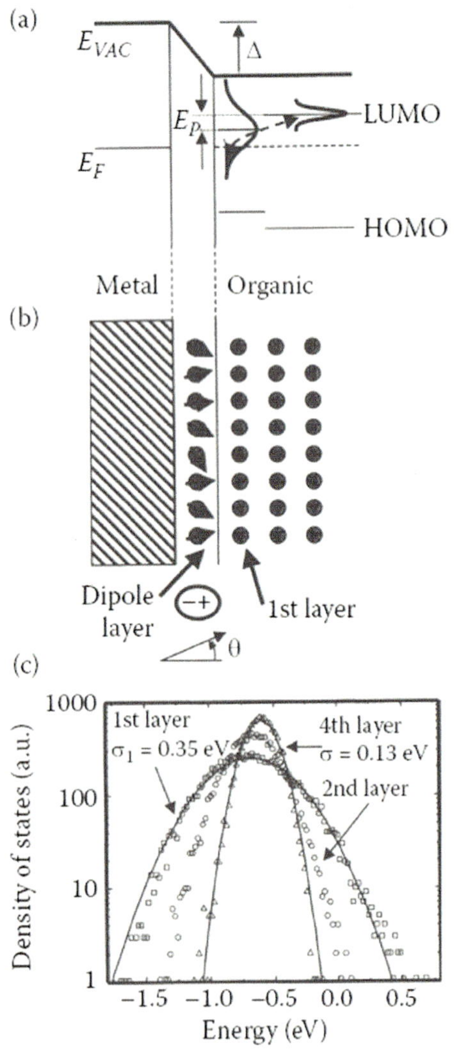

FIGURE 6.23 (a) Two-step process injection model. Charges are initially injected into an interface region where the distribution of energy sites is broadened by interface dipoles. The limiting step is the hop from the interfacial region into the bulk distribution. (b) The interface is modeled by a collection of point dipoles. In the bulk, each site is randomly oriented but at the interface the dipoles are oriented in a Gaussian distribution about the perpendicular ($\theta = 0$). (c) The effect of the interface dipoles on the electron-energy distribution in the bulk is calculated. The distribution in the first few organic monolayers is considerably broadened. With increasing distance, the energetic distribution approaches the bulk distribution. (Reproduced with permission from Baldo, M. A.; Forrest, S. R. *Physical Review B* 2001, *64*, 085201.)

changes the Helmholtz layer potential V_H, as viewed in the change of the VL at the metal/electrolyte interface, and displaces the charge transfer rates k_{ct} as discussed in Section 6.6. But in a semiconductor electrode, the voltage moves the Fermi level of electrons at the semiconductor surface, and this changes the concentration of electrons that are able to participate in charge transfer

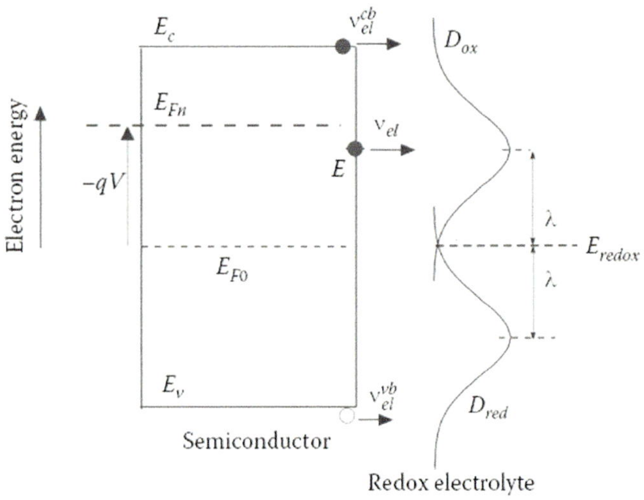

FIGURE 6.24 Charge transfer from a semiconductor, to the fluctuating energy levels of the *ox* and *red* species in electrolyte with a reorganization energy λ.

(see Figure 6.24). In this case, the change of the VL is either at the contact with the current collector, or at the depletion layer at the semiconductor electrode surface, as discussed in Chapter 9. However, the Helmholtz potential may remain static; hence the rate constant for electron transfer, denoted ν_{el} for the semiconductor surface, is fixed under a variation of the electrode potential.

The electron flux across the interface, which is the rate of the electrochemical reaction, is written in the same form as Equation 6.65:

$$J = \nu_{el} n_s c_{ox} \tag{6.94}$$

The starting energy level in the semiconductor may either be the conduction band for the transference of electrons, or the valence band for the transference of holes. Surface states in the energy bandgap may also transfer electrons to an oxidized redox species, or holes to a reduced species of a redox couple, as shown in Figure 6.24.

The flux has three components: the concentration of electrons at the semiconductor surface, n_s, the concentration of oxidized ions in the electrolyte side of the interface, c_{ox}, and the rate constant ν_{el} which can also be viewed as the probability of charge transfer. Since both n_s and c_{ox} are given in cm^{-3}, and the flux is in units cm^{-2} s^{-1}, the rate constant ν_{el} has units cm^4 s^{-1}. An equivalent expression to Equation 6.94 can be stated for the reverse flux from reduced species to the semiconductor surface.

The variations $n_s(V_r)$ are treated in Chapter 9. Here, we discuss the treatment of reorganization effects that influence the electron-transfer rate constant. The probability for an elementary electron-transfer event, ν_{el}, has been described by Gosavi and Marcus (2000). For a planar interface

$$v_{el}(E) = \frac{2\pi}{\hbar}|H|^2 \frac{l_{sc}}{\beta_{sc} d_{sc}^{2/3} (6/\pi)^{1/3}} FC \quad (6.95)$$

Equation 6.95 contains a coupling factor H introduced in Equation 6.50. The geometric terms l_{sc}, β_{sc}, and d_{sc}, characterizing the interaction of electrons and redox species, have the following meaning: l_{sc} (cm) is the effective coupling length between the oxidized redox ion and the electrode, β_{sc} (cm^{-1}) is the coupling attenuation factor as in Equation 6.51, and d_{sc} (cm^{-3}) is the density of the atoms that contribute to the density of states of either the surface states or the band of concern (Chidsey, 1991; Royea et al., 1997). The Franck–Condon factor includes the reorganization effects and is given by

$$FC = \frac{1}{\sqrt{4\pi k_B T \lambda}} \exp\left[-\frac{(\lambda + \Delta G)^2}{4\lambda k_B T}\right] \quad (6.96)$$

The starting state of the electron in the semiconductor surface is generally denoted E, while the free energy of electrons in the electrolyte is E_{redox}.

$$\Delta G = E_{redox} - E \quad (6.97)$$

ΔG is negative for the surface state shown in Figure 6.24.

When we consider the charge transfer between a semiconductor electrode and a redox electrolyte, the expression for the outer-sphere reorganization energy λ_0 (Equation 6.52) takes into account a reorganization in the semiconductor electrode itself (Hamann et al., 2005):

$$\lambda_0 = \frac{(\Delta z)^2 q^2}{8\pi\varepsilon_0 a}\left(\frac{1}{\varepsilon_1^{op}} - \frac{1}{\varepsilon_1^s}\right) - \frac{(\Delta z)^2 q^2}{16\pi\varepsilon_0 R}\left(\frac{\varepsilon_2^{op} - \varepsilon_1^{op}}{\varepsilon_2^{op} + \varepsilon_1^{op}}\frac{1}{\varepsilon_1^{op}} - \frac{\varepsilon_2^s - \varepsilon_1^s}{\varepsilon_2^s + \varepsilon_1^s}\frac{1}{\varepsilon_1^s}\right) \quad (6.98)$$

Here, ε_1^i refers to the relative optical and static dielectric constants of the electrolyte, and ε_2^i those of the semiconductor.

The electron-transfer rate constant is, therefore, given by the following formula:

$$v_{el}(E) = k_0 \exp\left[-\frac{(E_{redox} - E + \lambda)^2}{4\lambda k_B T}\right] \quad (6.99)$$

where

$$k_0 = \frac{2\pi}{\hbar}\frac{l_{sc}}{\beta_{sc} d_{sc}^{2/3} (6/\pi)^{1/3}}\frac{|H|^2}{\sqrt{4\pi\lambda k_B T}} \quad (6.100)$$

so that k_0 is measured in units $cm^4\ s^{-1}$.

As mentioned earlier, the carrier transference from a semiconductor surface can occur from a variety of energy levels, and in general the semiconductor may show a broad distribution of surface states as discussed in Chapters 7 and 8. According to Equation 6.99, the optimal rate of charge transfer occurs when the electron donor level E in the semiconductor and the redox energy in the electrolyte E_{redox} differs by the energy λ. Around this maximum the probability of charge transfer decreases as a Gaussian shape as a function of the semiconductor energy level E.

A suggestive image of the rate constant for charge transfer can be drawn in the energy diagram as in Figure 6.24, by introducing effective density of states of the electron acceptor (*ox*) and donor (*red*) levels (Gerischer, 1990). The normalized distributions correspond to the FC factor in Equation 6.96, and they have the shape of Gaussian distributions as follows:

$$D_{ox} = \frac{1}{\sqrt{4\pi\lambda k_B T}} \exp\left[-\frac{(E - E_{ox})^2}{4\lambda k_B T}\right] \quad (6.101)$$

$$D_{red} = \frac{1}{\sqrt{4\pi\lambda k_B T}} \exp\left[-\frac{(E - E_{red})^2}{4\lambda k_B T}\right] \quad (6.102)$$

Here, $E_{ox} = E_{redox} + \lambda$ and $E_{red} = E_{redox} - \lambda$ are the most probable energy levels for the oxidized and reduced electrolyte species. The interfacial flux for cathodic current in Equation 6.94 is

$$J = k_r n_s c_{ox} D_{ox} \quad (6.103)$$

where

$$k_r = k_0 \sqrt{4\pi\lambda k_B T} \quad (6.104)$$

The determination of the effective DOS of ferrocyanic ions in solution is shown in Figure 6.25. Figure 6.26 shows the dependence of electron-transfer rate constants on the driving force for interfacial charge transfer from *n*-type ZnO electrodes in aqueous solutions to a series of nonadsorbing, one electron, outer-sphere redox couples with formal reduction potentials that span approximately 900 mV. Differential capacitance versus potential and current density versus potential measurements were used to determine the energetics and kinetics, respectively, of the interfacial electron-transfer processes. The results show the Gaussian shape of Equation 6.96 displaying both the normal and Marcus-inverted regions of interfacial electron-transfer process.

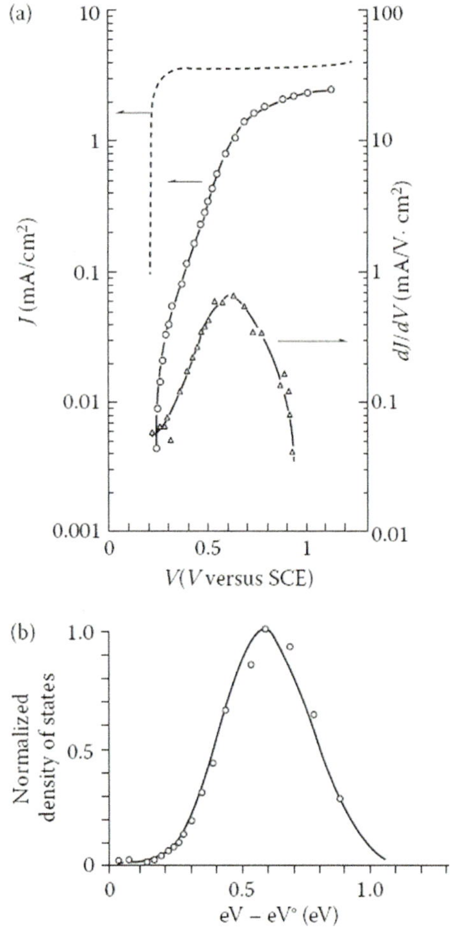

FIGURE 6.25 (a) Current density versus surface potential characteristics, as well as the potential derivatives (dj/dV) of an oxide-covered Pt-silicide electrode measured in a $0.1M-Fe(CN)^{3-/4-}$ and 0.5M-Na_2SO_4 electrolyte. The broken line is an *I-V* characteristic for a Pt electrode measured in the same electrolyte. (b) Effective density of electronic states for ferrocyanic ions deduced from a method that uses (dj/dV). The continuous line is the theoretical expression of the Marcus–Gerischer model, using a reorganization energy of 0.6 eV. (Reproduced with permission from Morisaki, H.; Ono, H.; Yazawa, K. *Journal of The Electrochemical Society* 1988, *135*, 381–383.)

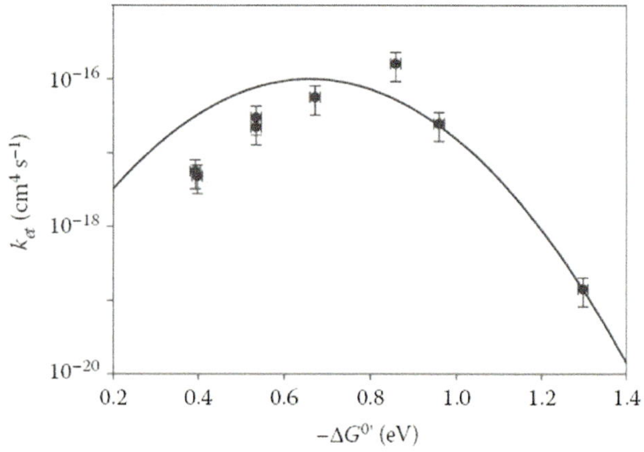

FIGURE 6.26 The dependence of electron-transfer rate constants on the driving force for interfacial charge transfer from *n*-type ZnO electrodes in aqueous solutions to a series of nonadsorbing, one electron, outer-sphere redox couples with formal reduction potentials that span approximately 900 mV. (Reproduced with permission from Hamann, T. W. et al. *Journal of the American Chemical Society* 2005, *127*, 7815–7824.)

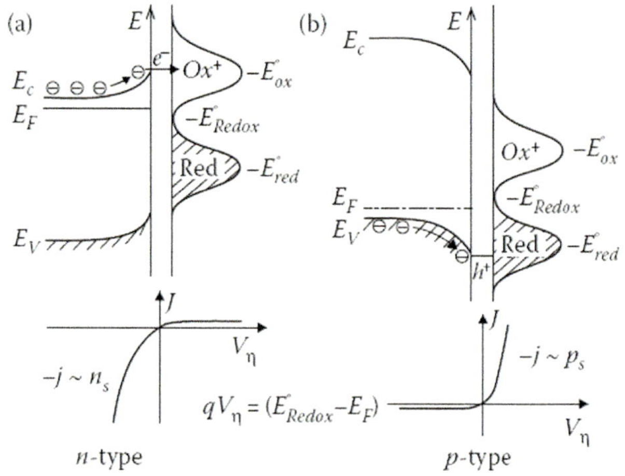

FIGURE 6.27 Redox reactions at *n*-type (a) and *p*-type (b) semiconductor electrodes. Above: typical energy correlations at a particular voltage V_η. Below: current-voltage curves. (Reproduced with permission from Gerischer, H. *Electrochimica Acta* 1990, *35*, 1677–1699.)

It should be emphasized that the model for electron transfer considers a *single* electronic state in the electrolyte. D_{ox} and D_{red} represent fluctuations out of equilibrium but are not real densities of states (Morrison, 1980). The fluctuating energy levels describe the match of the redox levels with the semiconductor energy levels. It is a useful picture for drawing charge transfer energy diagrams, and it is widely used, but not universally adopted.

The current-voltage characteristics of the semiconductor-electrolyte junction are shown in Figure 6.27. Depending on the match of electronic levels with the effective DOS of the electrolyte, and the energy values of band edges for electrons and holes, either cathodic or anodic current can be favored when the density of surface carriers is modified by the applied voltage. This system shows rectification properties governed by the majority carrier interfacial current with no role of the minorities (in the dark).

The enhancement of either electron or hole transfer at the semiconductor-liquid contact, by preferential kinetic match of one of the species, allows to form a selective contact. An arrangement for selectivity to electrons in the conduction band is shown in Figure 6.28. Even if

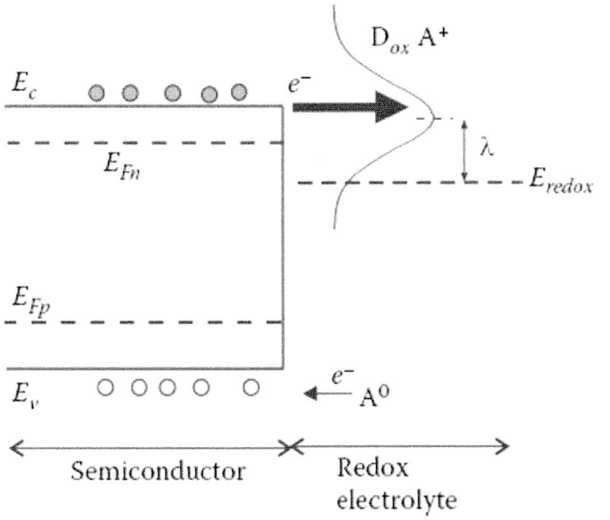

FIGURE 6.28 The semiconductor/electrolyte interface forms a selective contact to electrons in the semiconductor when the energy level of the semiconductor conduction band minimum matches well those of oxidized species in the electrolyte, while the transference of holes across the interface is hindered by slow kinetics.

both electrons and holes are equally abundant at the interface, the transference to solution of either type of carrier dominates, while the other is blocked by the surface.

When a semiconductor is illuminated, a substantial number of minority carriers are created and they can be used for surface reaction, while the majorities are extracted by the current collector toward a complementary electrode. A semiconductor anode with adequate electron-hole charge separation properties, and the suitable energy levels for electrochemical reactions, can be used to produce a photoelectrochemical cell that converts solar photons into a chemical fuel as shown in Figure 6.29. This is a water-splitting cell that serves to produce hydrogen from water, combining oxygen evolution reaction at the anode and proton reduction at the cathode.

The thermodynamic values of the hydrogen reduction and oxygen evolution potentials are shown in Figure 3.11. In the photoelectrolysis cell shown in Figure 6.29, the semiconductor photoanode must have the maximum of

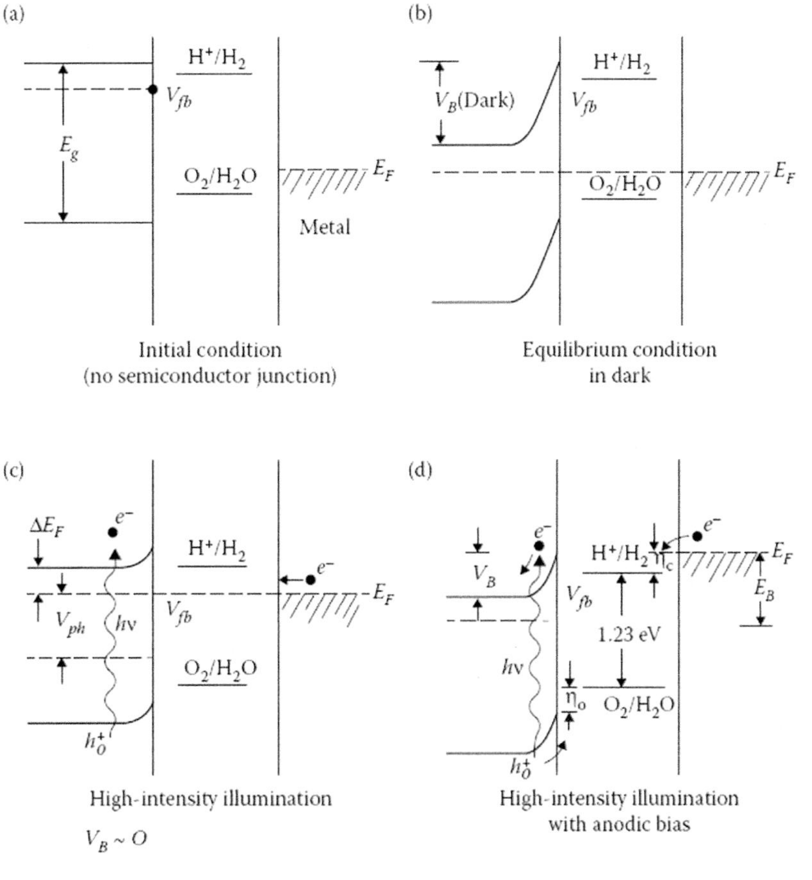

FIGURE 6.29 Sequence of energy level diagrams for a semiconductor-metal photoelectrolysis cell. (a) No contact and no electrochemical potential equilibration, (b) contact in dark equilibrium, (c) photoelectrolysis of water without any external bias, and (d) photoelectrolysis of water with an external bias. (Reproduced with permission from Nozik, A. J.; Memming, R. *The Journal of Physical Chemistry* 1996, *100*, 13061–13078.)

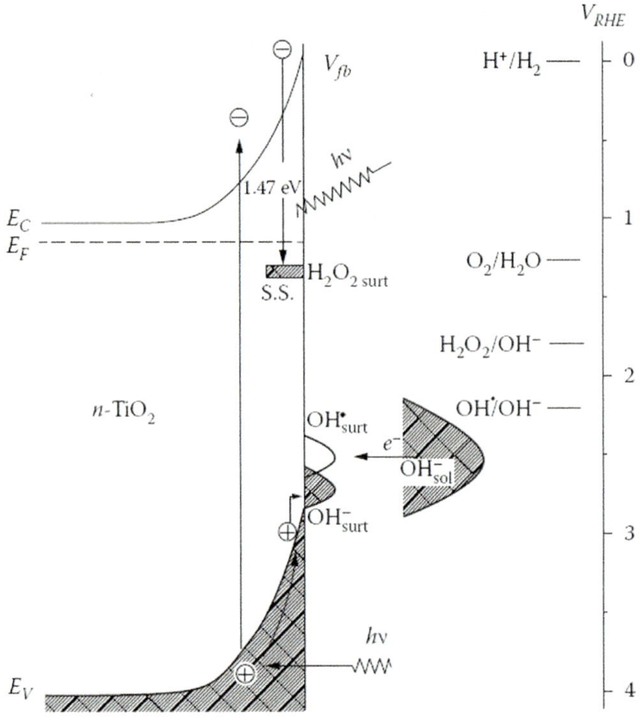

FIGURE 6.30 Energy level diagram of the n-TiO_2/electrolyte interface including redox potentials of electrolyte couples. (Reproduced with permission from Salvador, P.; Gutierrez, C. *The Journal of Physical Chemistry* 1984, *88*, 3696–3698.)

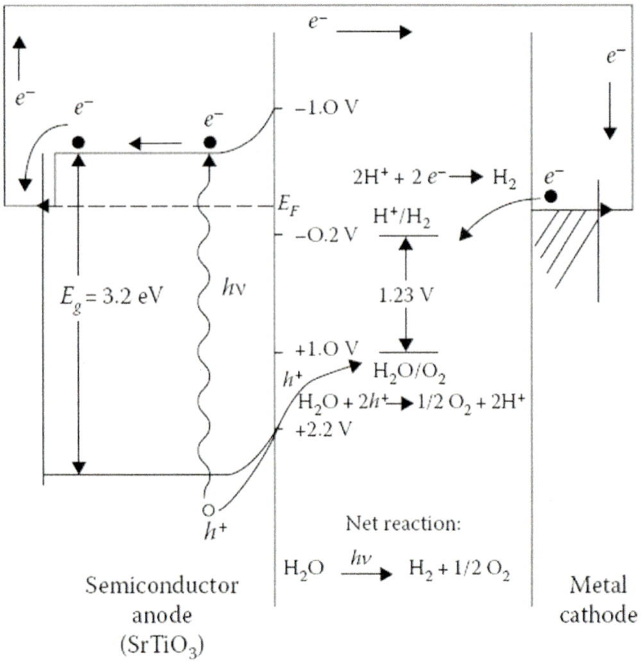

FIGURE 6.31 Energy level diagram for the photoelectrolysis of water using the single, wide bandgap semiconductor strontium titanate ($SrTiO_3$). The photoelectrolysis reaction can proceed without an external bias because the conduction and valence band edges straddle the redox potentials for the hydrogen and oxygen evolution reactions. (Reproduced with permission from Nozik, A. J.; Memming, R. *The Journal of Physical Chemistry* 1996, *100*, 13061–13078.)

the valence band lower than water oxidation potential, which is the case in Figure 6.29c. However, the Fermi level of the cathode metal is at the value of the Fermi level of majority electrons in the bulk semiconductor, and in Figure 6.29c this value is not sufficient to permit the hydrogen reduction at the metal cathode. This is a property of TiO_2, which has the flatband potential (conduction band potential) very close to H^+/H_2 potential, as shown in Figure 6.30 (see Figure 9.12) for the band edge values of different semiconductors.

In addition, an overpotential of 400 mV is required for breaking H_2O molecules to produce O_2, and 50 mV for the cathodic reaction producing H_2 from protons (Weber and Dignam, 1984) as indicated in Figure 6.29d. It should also be emphasized that water splitting reaction is a four electron inner-sphere reaction on the electrode surface, so it is not just thermodynamics of products that should matter but also the intermediate states of the reaction (Liao et al., 2012).

When a negative voltage is applied to the cathode, the metal energy levels raise up and the cell may proceed to produce the required reactions. However, the value of gained fuel energy is diminished by the cost of energy consumption by the applied voltage. It is clearly desirable that the photoelectrolysis occurs unassisted by voltage, just by conversion of solar photon energy to fuel. This is possible provided that the semiconductor valence band edge and Fermi level straddle the water potentials, with an excess to account for the required overvoltages. Figure 6.31 shows a cell with a semiconductor anode ($SrTiO_3$) that possesses adequate energetics for photoelectrochemical production of the water splitting and hydrogen production reactions without external bias. However, the wide bandgap semiconductor harvests a small proportion of the solar spectrum. Semiconductors with better characteristics for water photoelectrolysis are the object of research (Prevot and Sivula, 2013).

GENERAL REFERENCES

Diffusion/hopping transitions: Glasstone et al. (1941), Fröhlich (1958), Ambegaokar et al. (1971), Talkner et al. (1990), Philibert (1991), Uebing and Gomer (1991), Ala-Nissila et al. (2002), Hinrichsen (2006).

Electron transfer at semiconductor/electrolyte interface: Gerischer (1961), Morrison (1980), Gerischer (1990), Morisaki et al. (1990), Sato (1998), Gao et al. (2000), Memming (2001), Bisquert et al. (2002).

Marcus model: Marcus (1956), Marcus (1959), Marcus (1960), Hush (1961), Marcus (1965), Brunschwig et al. (1980), Marcus and Sutin (1985), Liu and Newton (1995), Barbara et al. (1996), Bard (2010), Feldberg (2010).

Metal-semiconductor contact: Padovani and Stratton (1966), Rhoderick and Williams (1988), Mönch (1993), Bässler et al. (1998), Bässler et al. (1999), Daniels-Hafer et al. (2002).

Polaron hopping: Pekar (1954), Holstein (1959), Appel et al. (1968), Austin and Mott (1969), Sildos et al. (2000), Watanabe and Hayashi (2005), Henderson (2011), Emin (2013).

Polarons in organic materials: Cornil et al. (2001), Brédas et al. (2002), Fishchuk et al. (2004), Nagata and Lennartz (2008), Ruhle et al. (2011), Schrader et al. (2012).

Principle of detailed balance: Bridgman (1928), Onsager (1931), Boyd (1977), Alberty (2004), Pekar (2007), Astumian (2009).

Reorganization energy at semiconductor electrodes: Marcus (1990), Gao et al. (2000), Hamann et al. (2005), Royea et al. (2006).

REFERENCES

Ala-Nissila, T.; Ferrando, R.; Ying, S. C. Collective and single particle diffusion on surfaces. *Advances in Physics* 2002, *51*, 949–1078.

Alberty, R. A. Principle of detailed balance in kinetics. *Journal of Chemical Education* 2004, *81*, 1206–1208.

Ambegaokar, V.; Halperin, B. I.; Langer, J. S. Hopping conductivity in disorded systems. *Physical Review B* 1971, *4*, 2612–2620.

Appel, J.; Frederick Seitz, D. T.; Henry, E. *Solid State Physics*; Academic Press: New York, 1968; Vol. 21; pp. 193–391.

Astumian, R. D. Comment: Detailed balance revisited. *Physical Chemistry Chemical Physics* 2009, *11*, 9592–9594.

Austin, I. G.; Mott, N. F. Polarons in crystalline and non-crystalline materials. *Advances in Physics* 1969, *18*, 41–102.

Baldo, M. A.; Forrest, S. R. Interface-limited injection in amorphous organic semiconductors. *Physical Review B* 2001, *64*, 085201.

Barbara, P. F.; Meyer, T. J.; Ratner, M. A. Contemporary issues in electron transfer research. *The Journal of Physical Chemistry* 1996, *100*, 13148–13168.

Bard, A. J. Inner-sphere heterogeneous electrode reactions, electrocatalysis and photocatalysis: The challenge. *Journal of the American Chemical Society* 2010, *132*, 7559–7567.

Bässler, H.; Arkhipov, V. I.; Emelianova, E. V.; Tak, Y. H. Charge injection into light-emitting diodes: Theory and experiment. *Journal of Applied Physics* 1998, *84*, 848–856.

Bässler, H.; Wolf, K.; Arkhipov, V. I. Current injection from a metal to a disordered hopping system. II. Comparison between analytic theory and simulation. *Physical Review B* 1999, *59*, 7514–7520.

Bethe, H. A. Theory of the boundary layer of crystal rectifiers. *MIT Radiation Laboratory Report* 1942, *43*, 12.

Bisquert, J.; Marcus, R. A. Device modeling of dye-sensitized solar cells. *Topics in Current Chemistry* 2013, doi: 10.1007/1128_2013_1471.

Bisquert, J.; Zaban, A.; Salvador, P. Analysis of the mechanism of electron recombination in nanoporous TiO_2 dye-sensitized solar cells. Nonequilibrium steady state statistics and transfer rate of electrons in surface states. *The Journal of Physical Chemistry B* 2002, *106*, 8774–8782.

Boyd, R. K. Macroscopic and microscopic restrictions on chemical kinetics. *Chemical Reviews* 1977, *77*, 93–119.

Brédas, J. L.; Calbert, J. P.; da Silva Filho, D. A.; Cornil, J. Organic semiconductors: A theoretical characterization of the basic parameters governing charge transport. *Proceedings of the National Academy of Sciences* 2002, *99*, 5804–5809.

Bridgman, P. W. Note on the principle of detailed balancing. *Physical Review* 1928, *31*, 101–102.

Brunschwig, B. S.; Logan, J.; Newton, M. D.; Sutin, N. A semiclassical treatment of electron-exchange reactions. Application to the hexaaquoiron(II)-hexaaquoiron(III) system. *Journal of the American Chemical Society* 1980, *102*, 5798–5809.

Cavigli, L.; Bogani, F.; Vinattieri, A.; Faso, V.; Baldi, G. Volume versus surface-mediated recombination in anatase TiO_2 nanoparticles. *Journal of Applied Physics* 2009, *106*, 053516.

Closs, G. L.; Miller, J. R. Intramolecular long-distance electron transfer in organic molecules. *Science* 1988, *240*, 440–447.

Cornil, J.; Beljonne, D.; Calbert, J. P.; Brédas, J. L. Interchain interactions in organic π-conjugated materials: Impact on electronic structure, optical response, and charge transport. *Advanced Materials* 2001, *13*, 1053–1067.

Crowell, C. R.; Sze, S. M. Current transport in metal semiconductor barriers. *Solid-State Electronics* 1966, *9*, 1035–1048.

Chidsey, C. E. D. Free energy and temperature dependence of electron transfer at the metal-electrolyte interface. *Science* 1991, *251*, 918–922.

Daniels-Hafer, C.; Jang, M.; Boettcher, S. W.; Danner, R. G.; Lonergan, M. C. Tuning charge transport at the interface between indium phosphide and a polypyrrole-phosphomolyb-date hybrid through manipulation of electrochemical potential. *The Journal of Physical Chemistry B* 2002, *106*, 1622–1636.

Di Valentin, C.; Selloni, A. Bulk and surface polarons in photoexcited anatase TiO_2. *The Journal of Physical Chemistry Letters* 2011, *2*, 2223–2228.

Ellis, B. L.; Lee, K. T.; Nazar, L. F. Positive electrode materials for Li-ion and Li-batteries. *Chemistry of Materials* 2010, *22*, 691–714.

Emin, D. Phonon-assisted jump rate in noncrystalline solids. *Physical Review Letters* 1974, *32*, 303–307.

Emin, D. *Polarons*; Cambridge University Press: Cambridge, 2013a.

Emin, D. Theory of Meyer-Neldel compensation for adiabatic charge transfer. *Monatshefte fur Chemie—Chemical Monthly* 2013b, *144*, 3–10.

Faust, B.; Muller, H.; Schirmer, O. F. Free small polarons in $LiNbO_3$. *Ferroelectrics* 1994, *153*, 297–302.

Feldberg, S. W. The theory of electron transfer. *Journal of Solid State Electrochemistry* 2010, *14*, 705–739.

Feldberg, S. W.; Sutin, N. Distance dependence of heterogeneous electron transfer through the nonadiabatic and adiabatic regimes. *Chemical Physics* 2006, *324*, 216–225.

Fichthorn, K. A.; Weinberg, W. H. Theoretical foundations of dynamical Monte Carlo simulations. *Journal of Chemical Physics* 1991, *95*, 1090–1097.

Fishchuk, I. I.; Kadashchuk, A.; Bässler, H.; Abkowitz, M. Low-field charge-carrier hopping transport in energetically and positionally disordered organic materials. *Physical Review B* 2004, *70*, 245212.

Forbes, R. G. Use of Millikan-Lauritsen plots, rather than Fowler-Nordheim plots, to analyze field emission current-voltage data. *Journal of Applied Physics* 2009, *105*, 114313.

Fowler, R. H.; Nordheim, L. Electron emission in intense electric fields. *Proceedings of the Royal Society of London: Series A* 1928, *119*, 173–181.

Fröhlich, H. *Theory of Dielectrics*, 2nd ed.; Oxford University Press: Oxford, 1958.

Gao, Y. Q.; Georgievskii, Y.; Marcus, R. A. On the theory of electron transfer reactions at semiconductor electrode/liquid interfaces. *The Journal of Chemical Physics* 2000, *112*, 3358–3369.

Gerischer, H. Semiconductor electrode reactions. *Advances in Electrochemistry and Electrochemical Engineering* 1961, *1*, 139–232.

Gerischer, H. The impact of semiconductors on the concepts of electrochemistry. *Electrochimica Acta* 1990, *35*, 1677–1699.

Glasstone, S.; Laidler, K. J.; Eyring, H. *The Theory of Rate Processes*; McGraw-Hill: New York, 1941.

Gosavi, S.; Marcus, R. A. Nonadiabatic electron transfer at metal surfaces. *The Journal of Physical Chemistry B* 2000, *104*, 2067–2072.

Gregory, K.; Arieh, W. Investigation of the free energy functions for electron transfer reactions. *The Journal of Chemical Physics* 1990, *93*, 8682–8692.

Hamann, T. W.; Gstrein, F.; Brunschwig, B. S.; Lewis, N. S. Measurement of the dependence of interfacial charge-transfer rate constants on the reorganization energy of redox species at n-ZnO/H_2O interfaces. *Journal of the American Chemical Society* 2005a, *127*, 13949–13954.

Hamann, T. W.; Gstrein, F.; Brunschwig, B. S.; Lewis, N. S. Measurement of the free-energy dependence of interfacial charge-transfer rate constants using ZnO/H_2O semiconductor/liquid contacts. *Journal of the American Chemical Society* 2005b, *127*, 7815–7824.

Henderson, M. A. A surface science perspective on photocatalysis. *Surface Science Reports* 2011, *66*, 185–297.

Herring, C.; Nichols, M. H. Thermionic emission. *Reviews of Modern Physics* 1949, *21*, 185–270.

Hinrichsen, H. Non-equilibrium phase transitions. *Physica A: Statistical Mechanics and Its Applications* 2006, *369*, 1–28.

Holstein, T. Studies of polaron motion, Part I: The molecular-crystal model. *Annals of Physics (NY)* 1959, *8*, 325–342.

Hush, N. S. Adiabatic theory of outer sphere electron-transfer reactions in solution. *Transactions of the Faraday Society* 1961, *57*, 557–580.

Köhler, A.; Bässler, H. What controls triplet exciton transfer in organic semiconductors? *Journal of Materials Chemistry* 2011, *21*, 4003–4011.

Landau, L. D. Assotsiatsiya dvukhatomnykh molekul. *Sovetskii Fizicheskii Zhurnal* 1932, *2*, 46–52.

Landau, L. D. *Physikalische Zeitschrift der Sowjetunion* 1933, *3*, 644.

Lang, D. V.; Henry, C. H. Nonradiative recombination at deep levels in GaAs and GaP by lattice-relaxation multiphonon emission. *Physical Review Letters* 1975, *35*, 1525–1528.

Levi, M. D.; Aurbach, D. Frumkin intercalation isotherm—A tool for the elucidation of Li ion solid state diffusion into host materials and charge transfer kinetics. A review. *Electrochimica Acta* 1999, *45*, 167–185.

Liao, P.; Keith, J. A.; Carter, E. A. Water oxidation on pure and doped hematite (0001) surfaces: Prediction of Co and Ni as effective dopants for electrocatalysis. *Journal of the American Chemical Society* 2012, *134*, 13296–13309.

Liu, T.; Troisi, A. Absolute rate of charge separation and recombination in a molecular model of the P3HT/PCBM interface. *The Journal of Physical Chemistry C* 2011, *115*, 2406–2415.

Liu, T.; Troisi, A. What makes fullerene acceptors special as electron acceptors in organic solar cells and how to replace them. *Advanced Materials* 2012, *25*, 1038–1041.

Liu, Y.-P.; Newton, M. D. Solvent reorganization and donor/acceptor coupling in electron-transfer processes: Self-consistent reaction field theory and ab initio applications. *The Journal of Physical Chemistry* 1995, *99*, 12382–12386.

Lonergan, M. Charge transport at conjugated polymer-inorganic semiconductor and conjugated polymer-metal interfaces. *Annual Review of Physical Chemistry* 2004, *55*, 257–298.

Luque, A.; Marti, A.; Stanley, C. Understanding intermediate-band solar cells. *Nature Photonics* 2012, *6*, 146–152.

Marcus, R. A. On the theory of oxidation-reduction reactions involving electron transfer. I. *The Journal of Chemical Physics* 1956, *24*, 966–979.

Marcus, R. A. On the theory of electrochemical and chemical electron transfer processes. *Canadian Journal of Chemistry* 1959, *37*, 155–163.

Marcus, R. A. Exchange reactions and electron transfer reactions including isotopic exchange. Theory of oxidation-reduction reactions involving electron transfer. Part 4. *Faraday Discussion Chemical Society* 1960, *29*, 21–31.

Marcus, R. A. On the theory of electron-transfer reactions. VI. Unified treatment for homogeneous and electrode reactions. *The Journal of Chemical Physics* 1965, *43*, 679–702.

Marcus, R. A. Reorganization free energy for electron transfers at liquid-liquid and dielectric semiconductor-liquid interfaces. *The Journal of Physical Chemistry* 1990, *94*, 1050–1055.

Marcus, R. A.; Sutin, N. Electron transfers in chemistry and biology. *Biochimica et Biophysica Acta (BBA)—Reviews on Bioenergetics* 1985, *811*, 265–322.

Memming, R. *Semiconductor Electrochemistry*; Wiley-VCH: Weinheim, 2001.

Metropolis, N.; Rosenbluth, A. W.; Rosenbluth, M. N.; Teller, A. H.; Teller, E. *The Journal of Chemical Physics* 1953, *21*, 1087.

Miller, A.; Abrahams, S. Impurity conduction at low concentrations. *Physical Review* 1960, *120*, 745.

Miller, J. A.; Klippenstein, S. J.; Robertson, S. H.; Pillingc, M. J.; Greend, N. J. B. Detailed balance in multiple-well chemical reactions. *Physical Chemistry Chemical Physics* 2009, *11*, 1128–1137.

Mönch, W. *Semiconductor Surfaces and Interfaces*; Springer: Berlin, 1993.

Morisaki, H.; Ono, H.; Yazawa, K. Measurement of the density of electronic states in aqueous electrolytes using oxide-covered Pt-silicide electrodes. *Journal of The Electrochemical Society* 1988, *135*, 381–383.

Morisaki, H.; Nishikawa, A.; Ono, H.; Yazawa, K. Electronic state densities of ferrocene and anthracene in nonaqueous solvents determined by electrochemical tunneling spectroscopy. *Journal of The Electrochemical Society* 1990, *137*, 2759–2763.

Morrison, S. R. *Electrochemistry at Semiconductor and Oxidized Metal Electrodes*; Plenum Press: New York, 1980.

Nagata, Y.; Lennartz, C. Atomistic simulation on charge mobility of amorphous tris(8-hydroxyquinoline) aluminum (Alq 3): Origin of Poole—Frenkel—type behavior. *The Journal of Chemical Physics* 2008, *129*, 034709.

Nozik, A. J.; Memming, R. Physical chemistry of semiconductor-liquid interfaces. *The Journal of Physical Chemistry* 1996, *100*, 13061–13078.

Onsager, L. Reciprocal relation in irreversible processes. I. *Physical Review* 1931, *37*, 405.

Padovani, F. A.; Stratton, R. Field and thermionic-field emission in Schottky barriers. *Solid-State Electronics* 1966, *9*, 695–707.

Pallotti, D. K.; Orabona, E.; Amoruso, S.; Aruta, C.; Bruzzese, R.; Chiarella, F.; Tuzi, S.; et al. Multi-band photoluminescence in TiO_2 nanoparticles-assem-bled films produced by femtosecond pulsed laser deposition. *Journal of Applied Physics* 2013, *114*, 043503.

Pekar, M. Detailed balance in reaction kinetics—Consequence of mass conservation? *Reaction Kinetics and Catalysis Letters* 2007, *90*, 323–329.

Pekar, S. I. *Untersuchungen über die Electronentheorie der Kristalle*; Akademie Verlag: Berlin, 1954.

Philibert, J. *Atom movements: Diffusion and Mass Transport in Solids*; Les Editions de Physique: Les Ulis, 1991.

Prévot, M. S.; Sivula, K. Photoelectrochemical tandem cells for solar water splitting. *The Journal of Physical Chemistry C* 2013, *117*, 17879–17893.

Ramaswamy, N.; Mukerjee, S. Fundamental mechanistic understanding of electrocatalysis of oxygen reduction on Pt and Non-Pt Surfaces: Acid versus alkaline media. *Advances in Physical Chemistry* 2012, *2012*, 491604.

Rhoderick, E. H.; Williams, R. H. *Metal-Semiconductor Contacts*, 2nd ed.; Clarendon Press: Oxford, 1988.

Royea, W. J.; Fajardo, A. M.; Lewis, N. S. Fermi golden rule approach to evaluating outer-sphere electron-transfer rate constants at semiconductor/liquid interfaces. *The Journal of Physical Chemistry B* 1997, *101*, 11152–11159.

Royea, W. J.; Hamann, T. W.; Brunschwig, B. S.; Lewis, N. S. A comparison between interfacial electron-transfer rate constants at metallic and graphite electrodes. *The Journal of Physical Chemistry B* 2006, *110*, 19433–19442.

Ruhle, V.; Lukyanov, A.; May, F.; Schrader, M.; Vehoff, T.; Kirkpatrick, J.; Baumeier, B.; et al. Microscopic simulations of charge transport in disordered organic semiconductors. *Journal of Chemical Theory and Computation* 2011, *7*, 3335–3345.

Salvador, P.; Gutierrez, C. The nature of surface states involved in the photo- and electroluminescence spectra of n-titanium dioxide electrodes. *The Journal of Physical Chemistry* 1984, *88*, 3696–3698.

Sato, N. *Electrochemistry at Metal and Semiconductor Electrodes*; Elsevier: Amsterdam, 1998.

Scott, J. C. Metal-Organic interface and charge injection in organic electronic devices. *Journal of Vacuum Science & Technology A: Vacuum, Surfaces, and Films* 2003, *21*, 521–531.

Scott, J. C.; Malliaras, G. G. Charge injection and recombination at the metal-organic interface. *Chemical Physics Letters* 1999, *299*, 115.

Schrader, M.; Fitzner, R.; Hein, M.; Elschner, C.; Baumeier, B.; Leo, K.; Riede, M.; et al. Comparative study of microscopic charge dynamics in crystalline acceptor-substituted oligothiophenes. *Journal of the American Chemical Society* 2012, *134*, 6052–6056.

Shockley, W.; Read, W. T. Statistics of the recombinations of holes and electrons. *Physical Review* 1952, *87*, 835–842.

Sildos, I.; Suisalu, A.; Aarik, J.; Sekiya, T.; Kurita, S. Self-trapped exciton emission in crystalline anatase. *Journal of Luminescence* 2000, *87–89*, 290–292.

Talkner, P.; Borkovec, M.; Hänggi, P. Reaction-rate theory: Fifty years after Kramers. *Reviews of Modern Physics* 1990, *62*, 251.

Uebing, C.; Gomer, R. A Monte Carlo study of surface diffusion coefficients in the presence of adsorbate-adsorbate interactions, I: Repulsive interactions. *Journal of Chemical Physics* 1991, *95*, 7626–7652.

Vineyard, G. H. Frequency factors and isotope effects in solid state rate processes. *Journal of Physical Chemistry of Solids* 1957, *3*, 121.

Warshel, A. Dynamics of reactions in polar solvents. Semiclassical trajectory studies of electron-transfer and proton-transfer reactions. *The Journal of Physical Chemistry* 1982, *86*, 2218–2224.

Watanabe, M.; Hayashi, T. Time-resolved study of self-trapped exciton luminescence in anatase TiO_2 under two-photon excitation. *Journal of Luminescence* 2005, *112*, 88–91.

Weber, M. F.; Dignam, M. J. Efficiency of splitting water with semiconducting electrodes. *Journal of the Electrochemical Society* 1984, *131*, 1258–1265.

Zener, C. Dissociation of excited diatomic molecules by external perturbations. *Proceedings of the Royal Society of London: Series A* 1933, *140*,660–668.

7 The Chemical Capacitance

The chemical capacitance is a basic thermodynamic quantity related to charge and energy storage mechanisms. This quantity appears in the experimental response of a great variety of systems, whenever a small perturbation measurement of a semiconductor, in which a Fermi level displacement changes the occupation of the density of states, is performed. In this chapter, we derive the basic theory of the chemical capacitance. The central result is that the chemical capacitance provides a direct measurement of the density of states, although the relationship is not satisfied in some situations. This feature finds many applications in the analysis of materials that show a broad density of states, especially in amorphous or disordered materials that contain localized states in the bandgap. We also review the features of voltage-charge curves in intercalation materials, and the quantum capacitance of graphene.

7.1 CARRIER ACCUMULATION AND ENERGY STORAGE IN THE CHEMICAL CAPACITANCE

In Chapter 3, we presented the properties of a capacitor formed by two parallel metal plates separated a distance d and filled with a dielectric (see Section 3.4). This device is called a *dielectric capacitor*, and the capacitance, per unit area (F cm^{-2}), is given by

$$C = \frac{\varepsilon_r \varepsilon_0}{d} \tag{7.1}$$

It was shown that the dielectric capacitance relates to a displacement current that flows while charging the spatially separated plates of the capacitor in ac or transient current mode. This type of capacitance is ubiquitous at interfaces with space charge such as the surface depletion layer in a semiconductor and the Helmholtz layer at the solid/solution interface.

The mechanism of charge and energy storage in a dielectric capacitor is illustrated in Figure 7.1a. The applied voltage is associated with a difference of electrostatic potential between the two plates, indicated by the tilt of the vacuum level (VL). The energy storage is related to the buildup of an electrical field, as function of the spatial separation of positive and negative charge. The actual value of the electrical field **E** depends on the dielectric constant of the material between the plates ε_r. We showed in Section 3.6 that the free energy F stored in the dielectric capacitor, when the amount of charge dQ is brought at the voltage V, is given by the internal energy of Equation 3.32. In the dielectric capacitor of Figure 7.1a, free energy is associated to the buildup of electrostatic potential difference and energy is stored in the form of energy density in the electrical field as given in Equation 3.34.

Since the capacitance may become a function of the voltage, in general we are interested in the *differential capacitance* that is measured by small perturbation experimental methods, and can be written as

$$C = \frac{dQ}{dV} \tag{7.2}$$

Equation 7.1 is one instance of the differential capacitance in the case of a constant dielectric capacitor, but there are other cases as we see in the following.

If the space between the contacts is filled with a medium that can conduct electronic carriers, the applied voltage can produce a different type of effect that gives rise to a *chemical capacitance*. An example is the device structure shown in Figure 5.12. The capacitive characteristics are shown in Figure 7.1b. Here, a piece of semiconductor is supplied with a contact that is reversible to electrons, in the left side, and a selective contact to holes in the right side. If the Fermi level of the majority carrier (hole) contact is stationary, the negative voltage applied at the electron contact has the effect to approach the electron Fermi level E_{Fn} to the bottom of the conduction band at E_c. In this configuration, the effect of the voltage is to change the *chemical potential* of electrons in the semiconductor, ζ_n. The number of electrons increases, and the charge in the semiconductor (number of carriers × q) consequently increases, so that this effect allows to obtain a change of charge with respect to the voltage. This device is therefore a capacitor, according to Equation 7.2. The capacitive mechanism is entirely related to the change of the chemical potential of the carriers and thus the denomination of *chemical capacitance*.

An increase of negative charge without modification of the resulting space charge (leaving the VL constant) requires that electroneutrality is satisfied. This can be

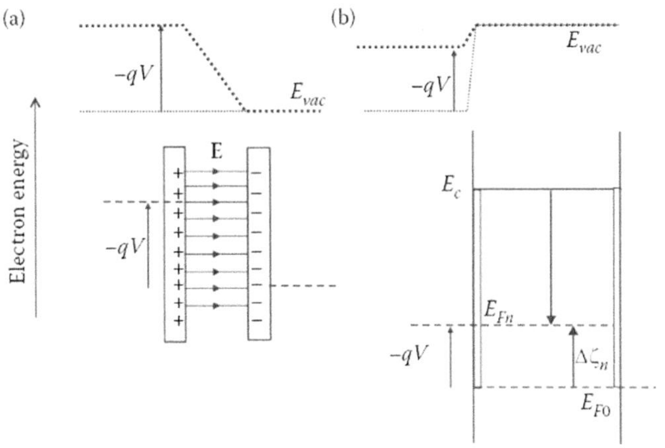

FIGURE 7.1 (a) Representation of a dielectric capacitor, showing the VL, the electrical charge at the plates, and the electrical field, E, in the space between the plates. (b) A semiconductor with selective contacts. Under applied bias the position of the Fermi level of minority carrier changes, which produces variation of the chemical potential of the electrons, ζ_n.

achieved by several basic mechanisms, and we mention two significant modes:

1. In a compact doped semiconductor, the compensation of charge is established by dopants and majority carriers as indicated in Equations 5.30 and 5.31. When the voltage increases the minority carrier density, the previous balance is only slightly disturbed, so that compensation remains effective.
2. The voltage increases carrier density in nanostructured porous semiconductors, where the excess charge is compensated by charge of opposite sign in the medium filling the pores. Reference examples of such configuration are the nanostructured semiconductor electrodes shown in Figures 1.5 and 3.1b.

Since the operation of the chemical capacitance occurs in a situation of screened charge, very often we will find diffusion current associated with the transient operation of a chemical capacitor, though this is not strictly necessary.

Let us consider the effect of variation of voltage in the situation of Figures 5.12b and 7.1b. The voltage is given by a difference of Fermi levels, or electrochemical potentials, of the electrons and holes at the respective selective contacts. From Equation 5.50, a change of the voltage changes the Fermi level as

$$dV = -dE_{Fn}/q \tag{7.3}$$

However, in a weakly interacting system we can adopt the separation of the electrochemical potential E_{Fn} of the electrons, into two components, as shown in Equation 2.35:

$$E_{Fn} = -\chi_n - q\varphi + \zeta_n \tag{7.4}$$

where φ is the local electrostatic potential. Thus, a change of voltage can be associated with either a modification of electrostatic energy or to a change of the chemical potential. If we assume that shielding of internal fields is effective, the local electrostatic level, that is, the conduction band position (Equation 2.23), is not modified by variation of Fermi level. Then a displacement of Fermi level can be identified with a variation of the chemical potential, $dE_{Fn} = d\zeta_n$. Thus,

$$dV = -d\zeta_n/q \tag{7.5}$$

We should remark the difference between the dielectric and chemical capacitor. The first type is based on spatially separated charges building an electrical field between them. Therefore, geometric "plates" can be recognized. In contrast, the chemical capacitor may function in conditions of charge screening. The "plates" of the capacitor do not need spatial separation, and in fact the chemical capacitance may occur in a single bulk material. If we operate this capacitor in ac mode, there is no electrical field variation in the bulk region and hence no displacement current. Indeed one may associate the "plates" of the capacitor to the energy levels at which the chemical potential is modified, and such electronic levels are charged by conduction current.

We may also distinguish different modes of storing free energy, either by the buildup of an electrical field, as indicated before for the dielectric capacitor, or by a reduction of entropy. As E_{Fn} increases as in Figure 7.1b, ζ_n increases, by reduction of the configurational entropy of the electrons. Such influx of negative entropy is produced by an external agent such as the voltage, as discussed in Section 5.9. The energy persists stored so far as some internal barrier prevents the irreversible process of entropy increase from occurring. When the barrier is removed, the device will tend to return to the higher entropy situation. It can then do work on an external system.

Recombination in an electronic device is a process that restores thermal equilibrium internally. It should be pointed out that a system with internal recombination as in Figure 5.12 is not suitable to form a practical capacitor for energy storage applications. As in a battery, the capacitor requires an electronic insulator in the internal pathway between the two contacts. This can be achieved in a porous structure as in Figure 3.1b provided that interfacial charge transfer at the solid/electrolyte interface is suppressed.

Let us define the chemical capacitance in conditions in which Equation 7.5 is satisfied. If L is the thickness of the layer, the charge accumulated in the semiconductor, per unit area, is $Q = -Lnq$. Hence, we obtain the *chemical capacitance* per unit area (F cm^{-2}), that is defined as follows (Bisquert, 2003):

$$C_\mu = Lq^2 \frac{dn}{dE_{Fn}} \tag{7.6}$$

It is noted in Equation 7.6 that for a film of homogeneous material, the chemical capacitance is simply proportional to the volume (or quantity of material). Therefore, a specific chemical capacitance per unit volume (F cm^{-3}) can be defined as

$$c_\mu = \frac{C_\mu}{L} \tag{7.7}$$

$$c_\mu = q^2 \frac{dn}{dE_{Fn}} \tag{7.8}$$

Sometimes, however, the neat distinction between electrostatic and chemical potential is not obtained, normally because interactions of carriers are significant. Then one retains Equation 7.3 but not Equation 7.4. In such a case, Equation 7.6 still serves as a good definition but in more precise terminology the quantity C_μ should be called the *electrochemical capacitance* (Gopar et al., 1996).

Since the chemical capacitance is such a fundamental quantity, it has been independently found in different fields and it obtains many varied denominations. One early example is the precise definition of capacitance for conduction band electrons, valence band holes, and electrons in traps by Shockley (1958). These correspond to the chemical capacitance of a single electronic state in a semiconductor, as discussed in Section 7.2. The same type of capacitance, which was conventionally denominated "pseudocapacitance," will be found if a redox molecule is attached to the surface of an electrode. When redox molecules form a film that can be charged from the current collector, we obtain the "redox capacitance" (Chidsey and Murray, 1986), which is another instance of the chemical capacitance as first recognized by Hong and Mauzerall (1974). A closely related idea, developed for single atoms or molecules, is the *chemical hardness* (Parr and Pearson, 1983). Further examples are presented in the next sections of this chapter and also in Chapter 8.

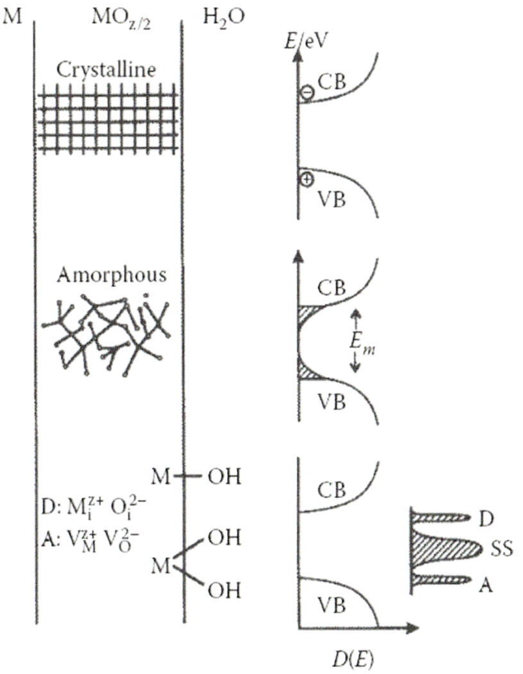

FIGURE 7.2 Schematic diagram of crystal structures and electronic states in a semiconductor film. (Reproduced with permission from Schultze, J. W.; Lohrengel, M. M. *Electrochimica Acta* 2000, *45*, 2499–2513.)

7.2 LOCALIZED ELECTRONIC STATES IN DISORDERED MATERIALS AND SURFACE STATES

In a crystalline semiconductor, the DOS is well described by delocalized states in the conduction and valence bands, whose occupation by charge carriers has been described in Chapter 5. In crystalline semiconductors, the bandgap is a region devoid of electronic states, and electrons and holes accumulate at the edges of the respective bands.

In amorphous or disordered semiconductors such as amorphous silicon and most organic materials, the extended states may or may not exist. Localized states, in which the wave function is strongly localized, occur in disordered solids, at defects or impurities of the crystal lattice in the bulk crystalline semiconductor as interstitials or vacancies and at the surface of semiconductors (interface or surface states). These different types of distributions of states in the energy axis are indicated in Figure 7.2. The energy levels of localized states can be discretely or continuously distributed. In the latter case, the localized states are attributed a characteristic DOS, $g(E)$ (see Figure 7.3). In amorphous oxides, the band tails enter the gap and contribute to processes at lower energies than E_g. In hydrogenated amorphous silicon

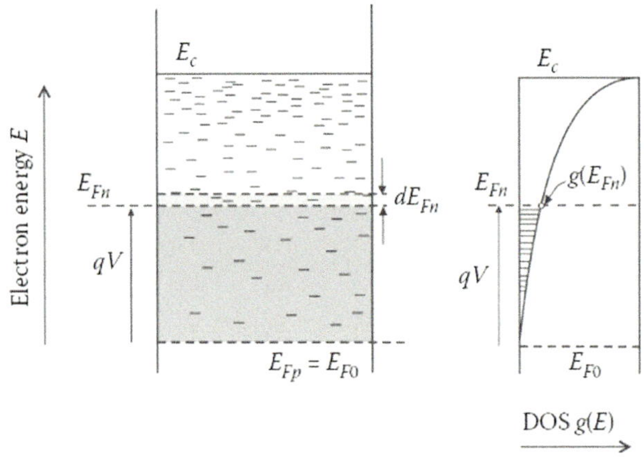

FIGURE 7.3 Schematic energy diagram of the occupancy of a distribution of localized states in the bandgap in an amorphous semiconductor. The shaded region indicates the bandgap states that are occupied with electrons below the Fermi level.

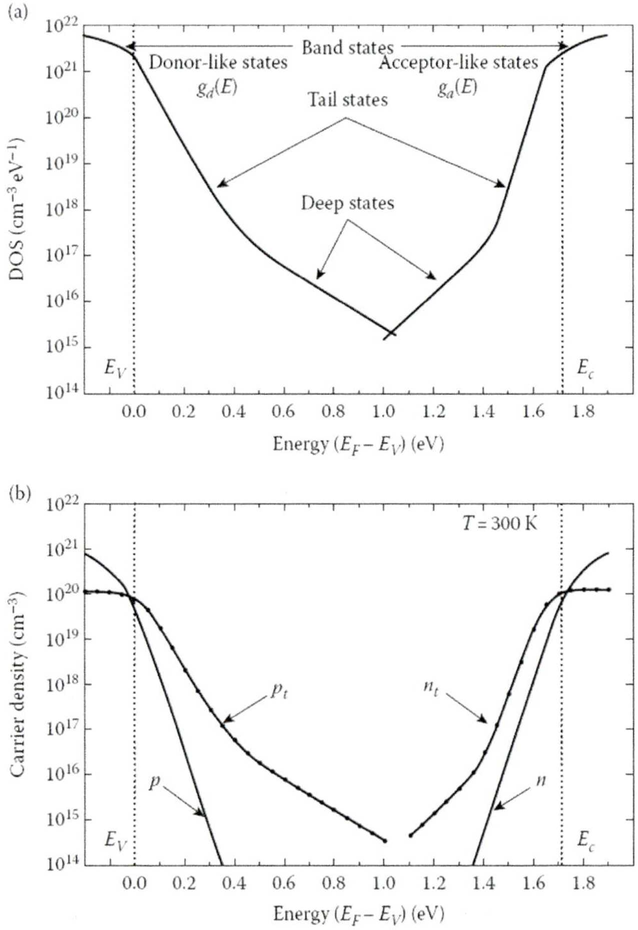

FIGURE 7.4 (a) Density of states in the bandgap of a-Si:H. edge. (b) Free and trapped charge in the bandgap for the DOS distribution. (Reproduced with permission from Shaw, J. G.; Hack, M. *Journal of Applied Physics* 1988, *64*, 4562–4566.)

(Figure 7.4), there is not a sharp division between a forbidden gap and an energy continuum of allowed states, as in crystalline semiconductors. Instead, we have in the energy axis regions with a large density of states, similar to bands, and regions with a very low density of available states for electrons, similar to the gap of a crystalline semiconductor.

The presence of bandgap states has a great influence in the electronic properties occurring in a semiconductor, both at the bulk and at the surface. The main effects caused by localized electronic states are to modify the transport rate by trapping processes; induce bandgap mediated recombination as shown in Section 6.3; to modify the chemical capacitance and displacement of the Fermi level when charge is injected; to modify space charge distribution by the influence of trapped charge.

When electronic states are highly localized, the transport of carriers occurs by a hopping mechanism. In the regions of a low DOS, since the distance between localized states is large, the carriers trapped in localized states cannot hop to close neighbors. An important concept for the physics of disordered semiconductors is the *mobility edge* that separates the transport states, from localized states, or traps. The mobility edge may correspond to a true distinction between states where the electron wave function is in effect extended in space, and localized states (Mott, 1987). Alternatively, the mobility edge may be derived from the hopping conduction model. For a DOS that decreases rapidly toward the center of the gap, carriers can tunnel at energies at which the density of states is large, and long-range conduction occurs, but at some point the average distance between traps is too large and carriers become localized. Since the transition from the band to the deep traps occurs gradually, the region close to the band is called a *band tail* (see Figure 7.4). The tail feature is called the Urbach tail, and it was first discovered in the absorption spectrum of crystalline semiconductors, where it is a consequence of thermal fluctuations.

At the surface of the semiconductor the periodic crystal symmetry is broken, as a result of which electronic states in the bandgap exist which are called *surface states*. There are different classes of surface states depending on the ionic or covalent character of the semiconductor. Ionic or Tamm states have energetic positions close to the band edges, while in covalent semiconductors it is frequent to find Shockley states corresponding to dangling bonds that bond in pairs forming states of bonding and antibonding, which tend to occur at the center of the gap, see Figure 7.2. Extrinsic surface states result from interactions of the electronic states at the semiconductor surface with external phases such as absorbates, metals, or electrolytes. For instance, at the metal/semiconductor interface, the wave function of the electrons in the metal mixes with the solution of the wave equation in the semiconductor giving rise to states in the forbidden gap. These interface states have a large impact on the properties of the junction as remarked in Section 4.6. On the other hand, in photoelectrochemical systems reaction products as OH- and H_2O_2 become attached to the surface as shown in Figure 6.30 and behave as surface states as they are able to accept carriers from the valence and conduction bands of the semiconductor (Salvador and Gutierrez, 1984).

7.3 CHEMICAL CAPACITANCE OF A SINGLE STATE

The simplest case of the chemical capacitance is to consider one specific electronic state uniformly distributed over the material volume, characterized by the energy E_0. Later on we deal with the chemical capacitance of a *distribution* of states, by the summation of the capacitances of each energy level.

Let us consider the occupation of a localized state with total density N_t. For the state to reach equilibrium occupancy, it is necessary that it exchanges electrons with an electron reservoir. We discuss this process in more detail in the next section, and here we simply assume the trap establishes equilibrium by exchange of electrons with the conduction band, as shown in Figure 6.4. Then, both systems of electrons, free and trapped, are characterized by the same Fermi level.

The occupancy function of the localized state is given by $f = F(E_0 - E_{Fn})$ the Fermi–Dirac distribution function in Equation 5.14, which has the following properties as discussed in Section 5.3:

$$f(E - E_F) = \begin{cases} e^{-(E-E_F)/k_BT} & \text{When } E \text{ is above } E_F, E - E_F \gg k_BT \\ 0.5 & \text{When } E = E_F \\ 1 & \text{When } E \text{ is below } E_F, E_F - E \gg k_BT \end{cases} \tag{7.9}$$

Figure 5.1a shows that a localized state in the bandgap at energy $E_0 = 0.5$ eV, will be first nearly empty (at $E_{Fn} = 0.3$ eV), then half occupied ($E_{Fn} = 0.5$ eV), and then almost completely fully occupied ($E_{Fn} = 0.7$ eV).

A displacement of the Fermi level causes a variation of the state occupancy in the following way:

$$\frac{df}{dE_{Fn}} = \frac{1}{k_BT} f(1 - f) \tag{7.10}$$

The number of carriers in this state will be

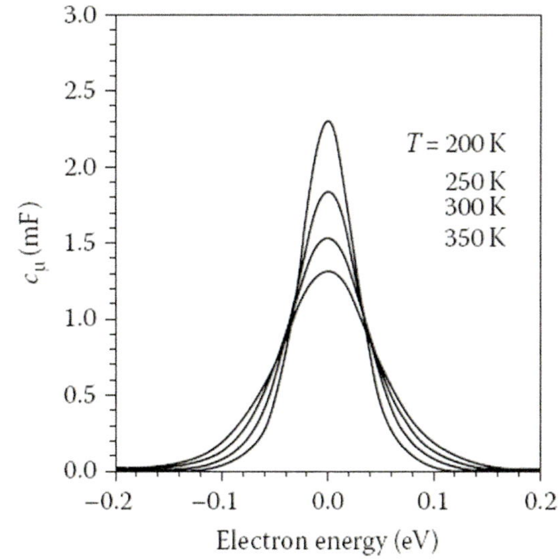

FIGURE 7.5 Representation of the chemical capacitance of a monoenergetic energy level at $E = 0$, as a function of the temperature as indicated.

$$n_t = N_t f(E_0 - E_{Fn}) \tag{7.11}$$

Therefore, the chemical capacitance, per unit volume, of this system is

$$c_\mu = q^2 N_t \frac{\partial f}{\partial E_{Fn}} = \frac{N_t q^2}{k_BT} f(1 - f) \tag{7.12}$$

The capacitance of a single state is shown in Figure 7.5. The capacitance traces a peak at $E_{Fn} = E_0$. The peak is broadened by the thermal distribution of carriers around the Fermi level, shown in Figure 5.2, so that the peak becomes sharper as the temperature decreases. Since the Nernst formula (3.53) is identical with Fermi–Dirac statistics, Equation 7.12 and Figure 7.5 describe also the chemical capacitance of a redox species as a function of the Fermi level, either in solution or attached to the semiconductor surface.

Let us suppose that the Fermi level lies far below the energy of the state E_0. In this case, we have $f \ll 1$ and Equation 7.12 leads to

$$c_\mu = \frac{N_t q^2}{k_BT} f \tag{7.13}$$

This situation is equivalent to using the Boltzmann statistics. In terms of the electron density in the conduction band, n_c, we have

$$c_\mu = \frac{n_c q^2}{k_BT} \tag{7.14}$$

Equation 7.14 is sometimes called the "conduction band capacitance." It was first calculated by Shockley (1958),

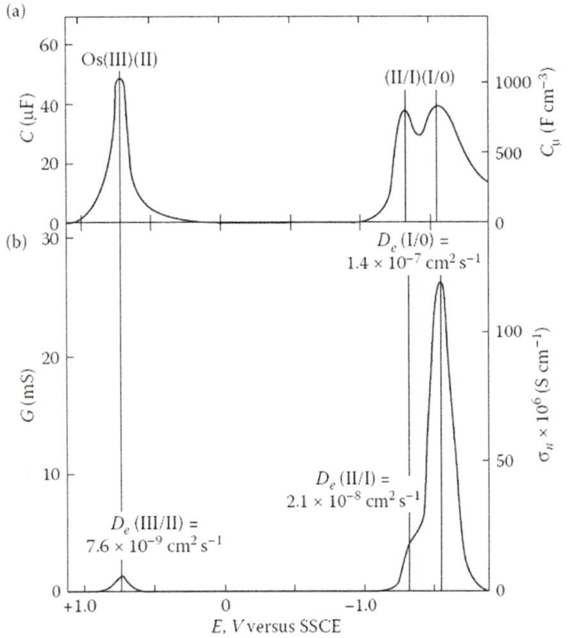

FIGURE 7.6 (a) Film capacitance, C, and redox capacity (chemical capacitance), C_μ, and (b) film conductance, G, and dc electron conductivity, σ_n, of a polymer poly[Os(bpy)$_2$(vpy)$_2$] $(ClO_4)_x$ sandwiched between Pt and porous Au electrodes and immersed in 0.1 M Et_4NClO_4/CH_3CN. (Reproduced with permission from Chidsey, C. E. D.; Murray, R. W. *The Journal of Physical Chemistry* 1986, *90*, 1479.)

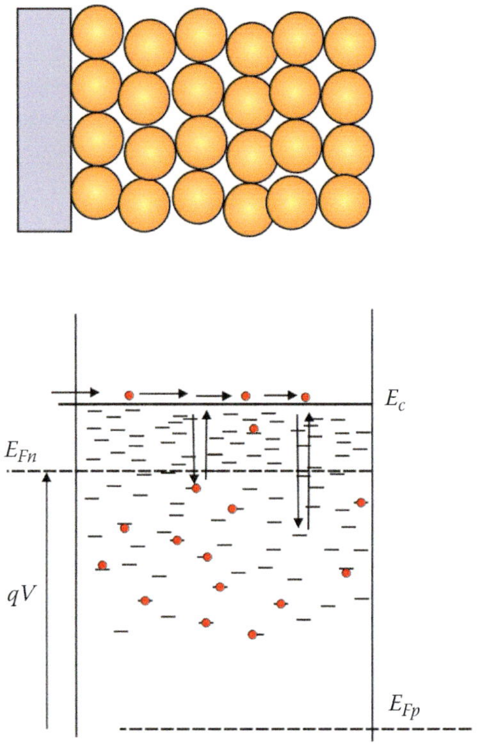

FIGURE 7.7 Model of a nanostructured semiconductor film with a transport state at energy E_c and a broad distribution of localized states in the bandgap. The voltage determines the Fermi level of electrons E_{Fn}.

and it came to be known as the "diffusion capacitance" in semiconductor device physics (Sze, 1981). Using Equation 5.51, Equation 7.14 can be stated as a function of voltage as follows:

$$c_\mu = \frac{n_0 q^2}{k_B T} e^{-qV/k_B T} \tag{7.15}$$

The chemical capacitance depends exponentially on the voltage.

An example of the chemical capacitance of an organic film is shown in Figure 7.6. Figure 7.6a shows the charging of a redox polymer film, containing osmium. It is indicated the metal-centered reduction of Os(III) to Os(II) at +0.73 V, and further reductions labeled Os(I) and Os(O), respectively. This is a characteristic behavior of conducting polymer films (Heinze et al., 2010). It should also be emphasized that an increase of the conductivity is strongly correlated with the peaks of the chemical capacitance, as shown in Figure 7.6b.

7.4 CHEMICAL CAPACITANCE OF A BROAD DOS

There are many instances in which organic and inorganic semiconductors show a broad distribution of states in the bandgap. Examples of the band tails in amorphous semiconductor are shown in Figures 7.3 and 7.4, and the general features of disordered solids will be discussed in detail in Chapter 8. We expect that the chemical capacitance is obtained by convolution of the different states in the distribution. We will now derive general expressions that apply in different classes of disordered solids.

Let us consider a nanostructured system, as shown in the scheme of Figure 7.7, composed of two classes of electronic states: (1) an extended state E_c that allows fast carrier transport, and (2) a broad distribution of localized states, characterized by the DOS, in which the density of states per cm^3 per eV is $g(E)$. Only the extended states can exchange electrons with the metal contact. Therefore, the voltage, V, states the position of the Fermi level and the occupancy of the extended state. Furthermore, the extended states allow redistribution of carriers across the film. The precise transport mechanism is not an essential requirement for the next derivation, provided that all the different states can reach a condition of equilibrium (in which detailed balance is satisfied by all microstates). The kinetics of electrons in a localized state has been already described in Section 6.3. The time change of the fractional occupancy f_L of one particular state is due to trapping and release from a transport state, as given in Equation 6.26.

We arrive at the conclusion that when the voltage is stationary, after an equilibration time, according to Equation 6.26, all the states reach the situation of equilibrium occupancy by a *single common Fermi level E_{Fn}*. The number density of carriers in localized states, n_L, in the energy interval dE at energy level E can be described as

$$n_L(E)dE = g(E)f(E - E_{Fn})dE \qquad (7.16)$$

The total density of carriers in localized states is found as

$$n_L = \int_{-\infty}^{+\infty} g(E)f(E - E_{Fn})dE \qquad (7.17)$$

Therefore, when a wide DOS exists in the material, the Fermi level determines the occupancy of the states as indicated in Equation 7.9. The occupation of an exponential density of states that decreases toward the center of the bandgap is indicated in Figure 7.3. Figure 7.8 shows the full picture of the exponential DOS and the occupancy in thermal equilibrium.

There are different methods to measure the chemical capacitance, but they all involve a small displacement of the Fermi level. Therefore, there is a direct connection between c_μ and $g(E)$. It was shown in Figure 5.2 that when the absolute temperature of the system decreases toward $T = 0$, the Fermi function becomes a step function. In this case, we will obtain a simpler distribution, in which the states below the Fermi level are fully occupied, and those above are empty. In the zero-temperature limit, the chemical capacitance coincides with the definition of the DOS $g(E)$, as stated in Equation 2.10. The zero-temperature approximation is often a useful tool to compute the occupation of a DOS at room temperature, though not in all cases.

The chemical capacitance for a general DOS, calculated from Equations 7.8 and 7.17, takes the form of an integral over all the states in the distribution:

$$c_\mu = q^2 \int_{-\infty}^{+\infty} g(E) \frac{df}{dE_{Fn}}(E - E_{Fn})dE \qquad (7.18)$$

Taking the derivative in Equation 7.18, we obtain that the total chemical capacitance is obtained as an integral of the individual peaks of Equation 7.10:

$$c_\mu = \frac{q^2}{k_B T} \int_{-\infty}^{+\infty} g(E)f(E - E_{Fn})\big[1 - f(E - E_{Fn})\big]dE \qquad (7.19)$$

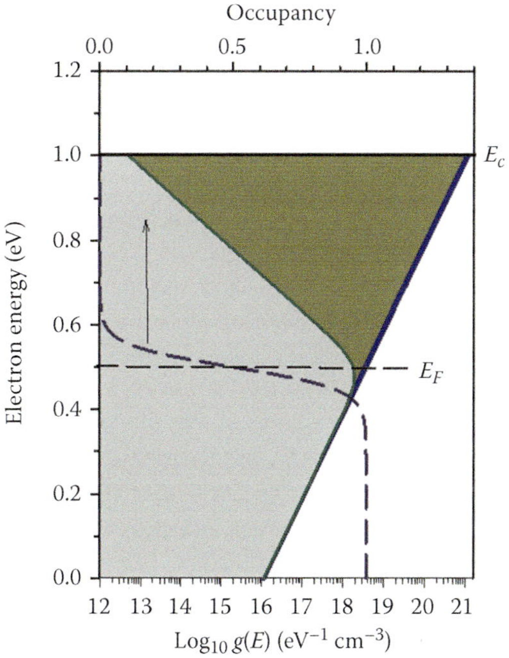

FIGURE 7.8 Energy diagram showing the thermal occupation at temperature $T = 300$ K of an exponential DOS in the bandgap. E_F is the Fermi level of electrons and E_c is the conduction band energy. The clear shaded region indicates the bandgap states that are occupied with electrons, and the dark shaded region indicates empty levels, as determined by the Fermi–Dirac distribution function (dashed line). Parameters used in the calculation: $T = 300$ K, $T_0 = 800$ K, $N_L = 10^{20}$ cm^{-3}.

If the DOS is sufficiently broad, the individual peaks disappear and c_μ has the shape of the DOS itself as we show in the following. If we use the identity

$$\frac{df(E - E_{Fn})}{dF_{Fn}} = -\frac{df(E - E_{Fn})}{dE} \qquad (7.20)$$

in Equation 7.18 and integrate by parts we arrive at

$$c_\mu = q^2 \int_{-\infty}^{+\infty} \frac{dg}{dE} f(E - E_{Fn})dE \qquad (7.21)$$

Furthermore, we use the approximation of the zero-temperature limit of the Fermi function, then

$$c_\mu = q^2 \int_{-\infty}^{E_{Fn}} \frac{dg}{dE} dE \qquad (7.22)$$

Therefore,

$$c_\mu = q^2 g(E_{Fn}) \qquad (7.23)$$

This result already anticipated earlier indicates that the chemical capacitance is proportional to the DOS at the Fermi level. Equation 7.23 can also be obtained from Equation 7.18 observing that the derivative of the step function is a δ-function. In this approximation, displacing the Fermi level by dE_F simply fills with carriers a slice of the DOS: $dn = g(E_F)dE_F$. Such slice is marked in Figure 7.3. Figure 7.9 shows that the combination of the individual peaks associated with chemical capacitances of the states in the distribution allows us to reconstruct the DOS. On the contrary, the exact thermal expression (Equation 7.19), c_μ, is denoted a *thermodynamic density of states* (Lee, 1982), as opposed to $g(E)$, the true DOS. For small conductors the electrochemical capacitance contains in addition to the DOS a term associated to the interactions of the charges induced by the change of electrochemical potential (Büttiker, 1993).

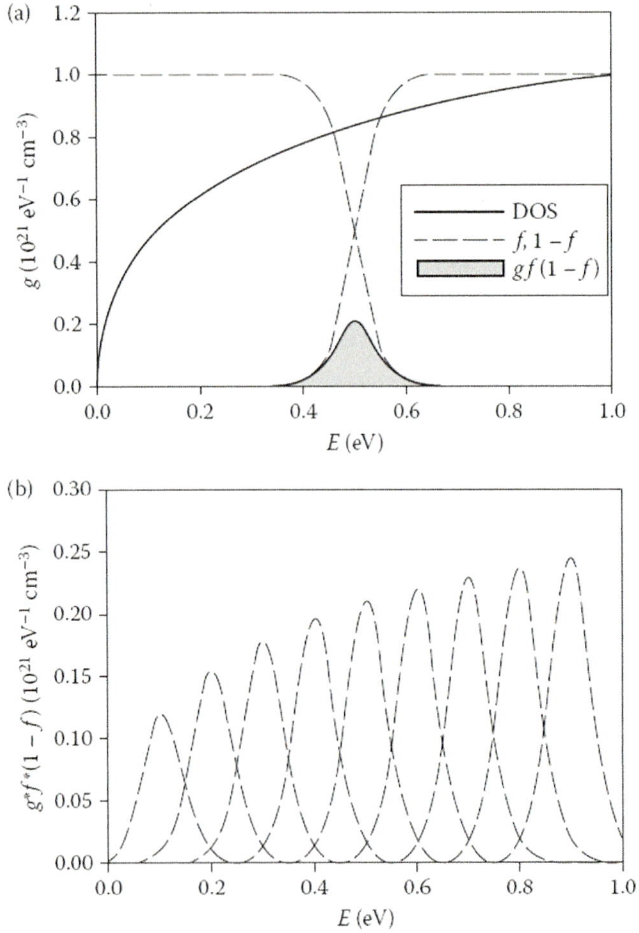

FIGURE 7.9 (a) Representation of an arbitrary DOS ($g(E) = AE^{0.5}(1 - E^{0.3}/B)$, $A = B = 2$), the Fermi–Dirac function for occupation of electrons and holes, and the product $g(E)f(E - E_F)[1 - f(E - E_F)]$ with $E_F = 0.5$ eV. (b) Representation of $g(E)f(E - E_F)[1 - f(E - E_F)]$ for E_F varying at steps of 0.1 eV from 0.1 eV to 0.9 eV.

The method suggested by Equation 7.23 has been applied in many works for investigation of the DOS in organic conductors (Bisquert et al., 2004; Hulea et al., 2004), quantum dot solids (Roest et al., 2003), and porous semiconductors (Bisquert et al., 2008). In the remaining parts of this chapter, we will present some instances of the applications of the chemical capacitance. Main properties of disordered inorganic semiconductors are presented in Chapter 8.

7.5 FILLING A DOS WITH CARRIERS: THE VOLTAGE AND THE CONDUCTIVITY

In previous chapters, we emphasized the connection of the occupancy of the DOS to the position of the Fermi level. This physical property becomes even more interesting when a DOS has a finite extent in the energy axis. An example that consists of a double Gaussian is shown in Figure 7.10. The properties of the Gaussian will be more generally analyzed in Section 8.6, but a simple application of Equation 7.17 reveals some important properties. We consider that the DOS is progressively filled with electrons in this example, so that the Fermi level increases in the energy axis and the occupation of carriers is shown in shaded areas. When the lower Gaussian is filled, the forthcoming electrons have to occupy the top Gaussian, and there will be a sudden step of the Fermi level. In terms of a device, an increase of voltage will be observed as indicated in Figure 7.10a, just with a moderate addition of carriers. This feature accounts for the voltage step associated with the electrons chemical potential in some lithium intercalation materials, as remarked in Section 5.13.

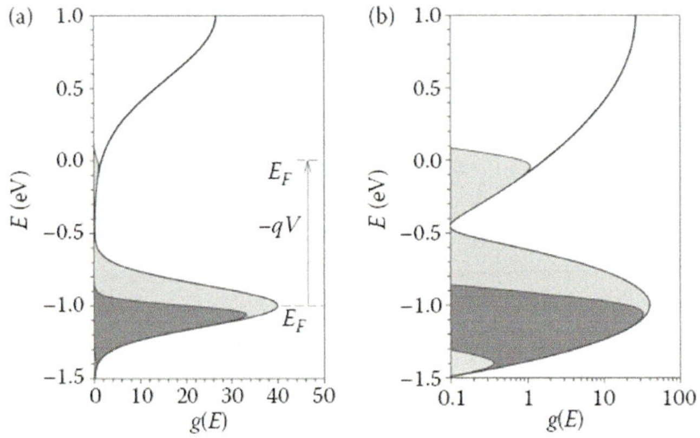

FIGURE 7.10 Energy diagram of a double Gaussian DOS in linear (a) and logarithmic (b) scale of the density of states. The bottom Gaussian has center $E_1 = -1$ eV, total density $N_1 = 10$, and dispersion parameter $\sigma_1 = 0.1$ eV. For the top Gaussian $E_2 = 1$ eV, $N_2 = 30$, and $\sigma_2 = 0.2$ eV. The filled areas correspond to the occupied states at $T = 300$ K at the Fermi level positions −1.4, −1.0, and 0.0 eV, from bottom to top.

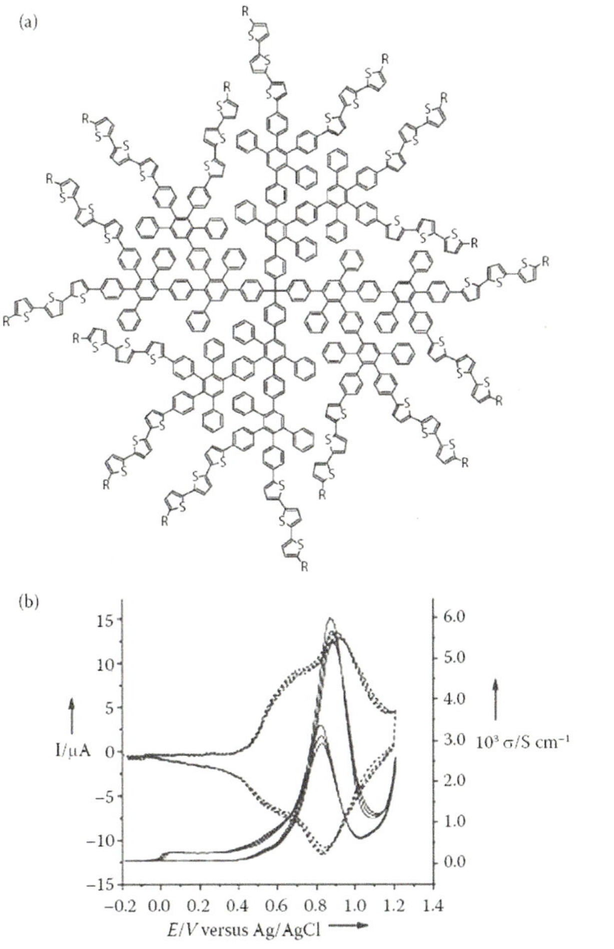

FIGURE 7.11 (a) Terthiophene functionalized polyphenylene core dendrimer. (b) The charging (chemical capacitance, dashed line) and in situ conductivity measurement (thin line) of a thin film at T = 300 K. (Reproduced with permission from John, H. et al. *Angewandte Chemie International Edition* 2005, *44*, 2447–2451.)

Another important feature of the finite DOS is the behavior of the conductivity of the carriers. The connection between chemical capacitance and conductivity of organic films has been often remarked (John et al., 2005; Heinze et al., 2010). In the conduction process, the carriers must gain energy in an electrical field. So the conductivity is not simply controlled by the number of carriers and the mobility. It is necessary that the carriers have free available states in order to gain kinetic energy. Therefore, the conductivity is mostly realized by carriers around the Fermi level, and it is given by the expression (Shapiro and Adler, 1985; Bisquert, 2008; Palenskis, 2013)

$$\sigma_n(E_F) = D_n c_\mu(E_{Fn}) = \frac{q^2 D_n}{k_B T} \int_{-\infty}^{+\infty} g(E) f(E - E_{Fn}) \left[1 - f(E - E_{Fn})\right] dF \tag{7.24}$$

If the diffusion coefficient D_n is constant, when the Fermi level is modified the conductivity follows the shape of the DOS. If the DOS decreases above the Fermi level, the conductivity decreases even when the carrier number increases, due to the lack of free quantum levels represented by the factor $[1 - f(E - E_{Fn})]$ in Equation 7.24. These features are shown in Figure 7.6 and in Figure 7.11 for another thin film of conducting polymer.

The significance of the capacitance peak for electron transport in nanostructures has been emphasized by Datta (2012).

7.6 CHEMICAL CAPACITANCE OF LI INTERCALATION MATERIALS

In previous sections, we have discussed the chemical capacitance as a thermodynamic feature applied to the process of electrons that fill a certain DOS with an increase of their electrochemical potential. Since the chemical capacitance has been generally defined, it may apply also to other types of carriers. In Section 5.11, we observed that the variation of voltage in Li intercalation cathodes may be associated with the concentration of ions in the solid material, via the variation of the chemical potential of ions, when the contribution of electrons chemical potential can be neglected. In this case, the capacitance of the electrode provides the chemical capacitance of ions in the solid lattice.

In the field of batteries, the derivative dQ/dV obtained from voltage–composition curves has been widely used for assessing the thermodynamic properties of materials (McKinnon and Haering, 1983; Li et al., 1992; Gao et al., 1996; Strömme, 1998). Such derivative corresponds to the general concept of a chemical capacitance (although this name was not generally applied), provided that kinetic limitations can be neglected. If c is the concentration of ions

$$c_\mu = q^2 \frac{dc}{d\eta_{Li^+}} = \frac{dQ}{dV} \tag{7.25}$$

In terms of the extent of intercalation x, we have

$$c_\mu = qc^0 \frac{dx}{dV} \tag{7.26}$$

The derivative of voltage–composition curve of a lattice-gas model in Equation 5.85 is calculated as follows:

$$\frac{dV}{dx} = \frac{k_B T}{q} \left[\frac{1}{x} + \frac{1}{1-x} - 8 \right] \tag{7.27}$$

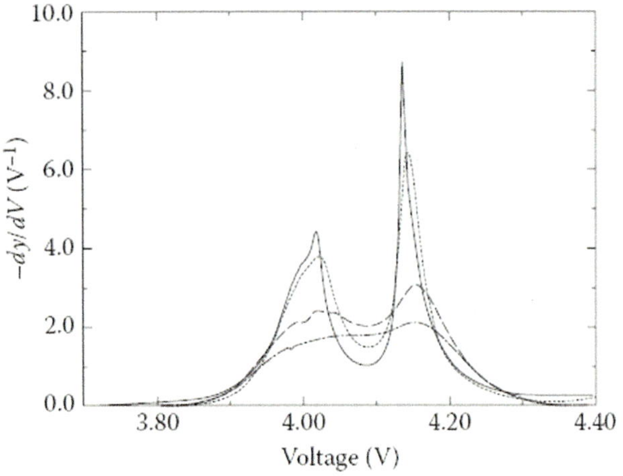

FIGURE 7.12 Experimental derivative curves $-dy/dV$ for $Li_xMn_2O_4$. (Reproduced with permission from Gao, Y.; Reimers, J. N.; Dahn, J. R. *Physical Review B* 1996, *54*, 3878–3883.)

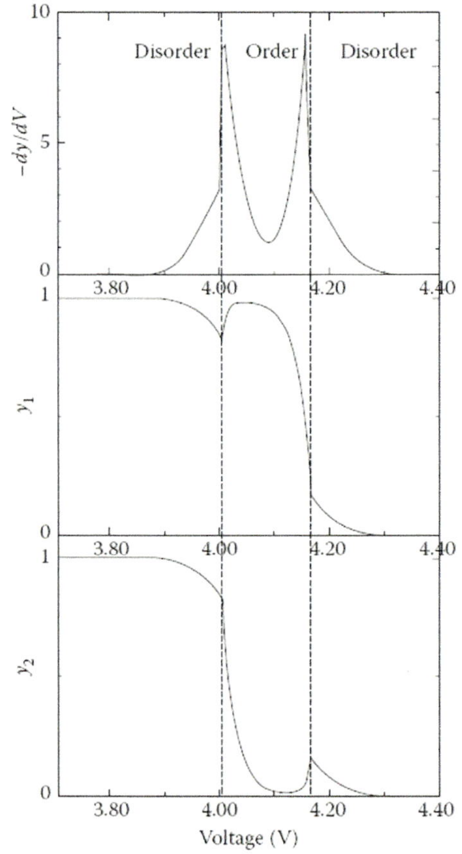

FIGURE 7.13 The top panel shows the calculated derivative curve $-dy/dV$ for $Li_xMn_2O_4$, and the bottom two panels show the occupations of the two sublattices y_1 and y_2. They are all plotted versus the voltage. The two vertical dashed lines represent the boundaries between ordered and disordered phases. (Reproduced with permission from Gao, Y.; Reimers, J. N.; Dahn, J. R. *Physical Review B* 1996, *54*, 3878–3883.)

and we obtain the chemical capacitance

$$c_\mu = \frac{q^2 c^0}{k_B T}\left[\frac{1}{x} + \frac{1}{1-x} - g\right]^{-1} \tag{7.28}$$

The shape of the capacitance and the modifications introduced by the mean field interactions are shown in Figure 5.23b. The capacitance of a single-state lattice gas is equivalent to that shown earlier in Figure 7.5. For a two-state model the chemical capacitance shows two peaks at the voltages that match the energies of the individual states (see Figure 5.25b).

An example of measurement of dQ/dV of $Li_xMn_2O_4$, showing the double peak feature of the two-state model is shown in Figure 7.12. Figure 7.13 shows that the capacitance is associated with the features of order–disorder transition that have been described in Section 5.11. Further examples for intercalation of several cations in crystalline WO_3 are shown in Figure 7.14, again in comparison with a two-state lattice-gas model theory.

7.7 CHEMICAL CAPACITANCE OF GRAPHENE

In semiconductor devices that include very narrow layers, showing strong quantization effects, the (chemical) capacitance associated with adding carriers to the band structure of the semiconductor, is called the *quantum capacitance*. This denomination is, of course, appropriate because the electrons are put into quantum states, when the electrochemical potential is modified.

The Fermi energy in graphene depends on the carrier concentration n^s (in units of cm^{-2}) as

$$E_{Fn} = \hbar |v_F| \sqrt{\pi n^s} \tag{7.29}$$

where the Fermi velocity of a Dirac electron relates to the speed of light as $v_F \approx c/300$. Therefore, the chemical capacitance is (Das et al., 2008)

$$c_\mu = q^2 \frac{dn^s}{dE_{Fn}} = \frac{2q^2 E_{Fn}}{\pi(\hbar v_F)^2} = \frac{2q^2}{\hbar v_F \sqrt{\pi}} \sqrt{n^s} \tag{7.30}$$

The quantum capacitance c_μ of 2D materials such as graphene is usually evaluated in a transistor structure, shown in Figure 7.15a. Figure 7.15b shows the

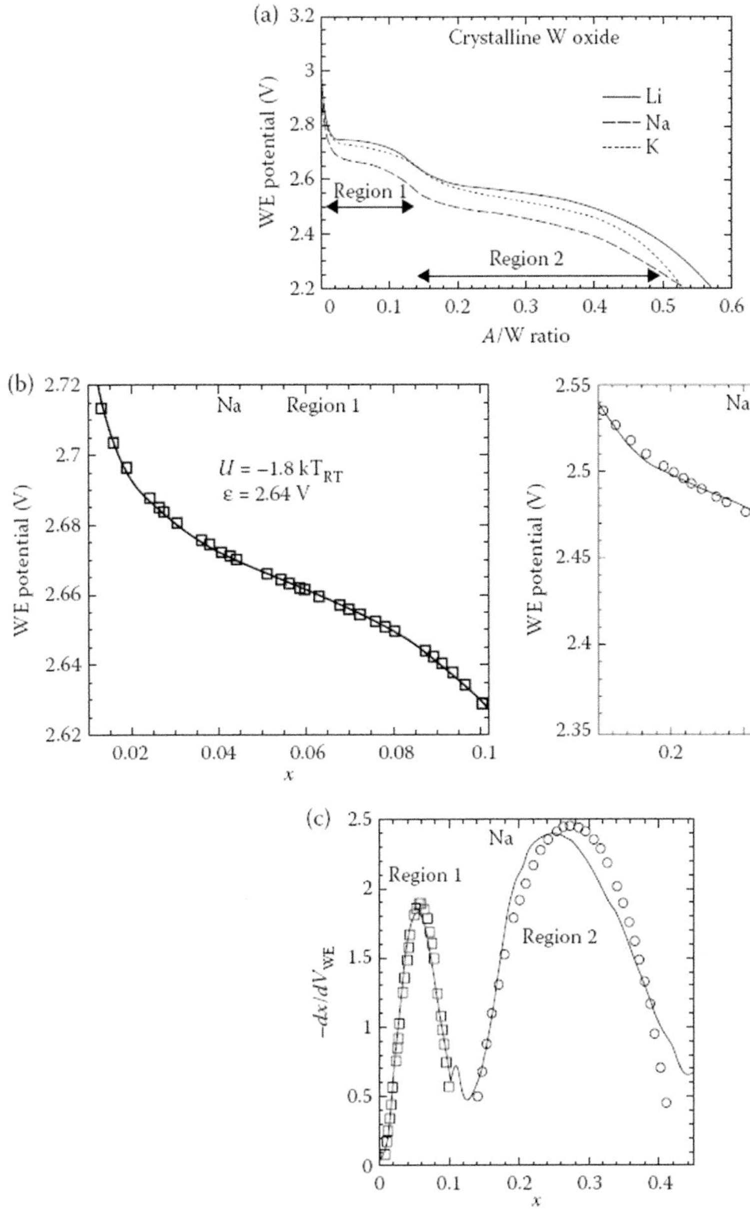

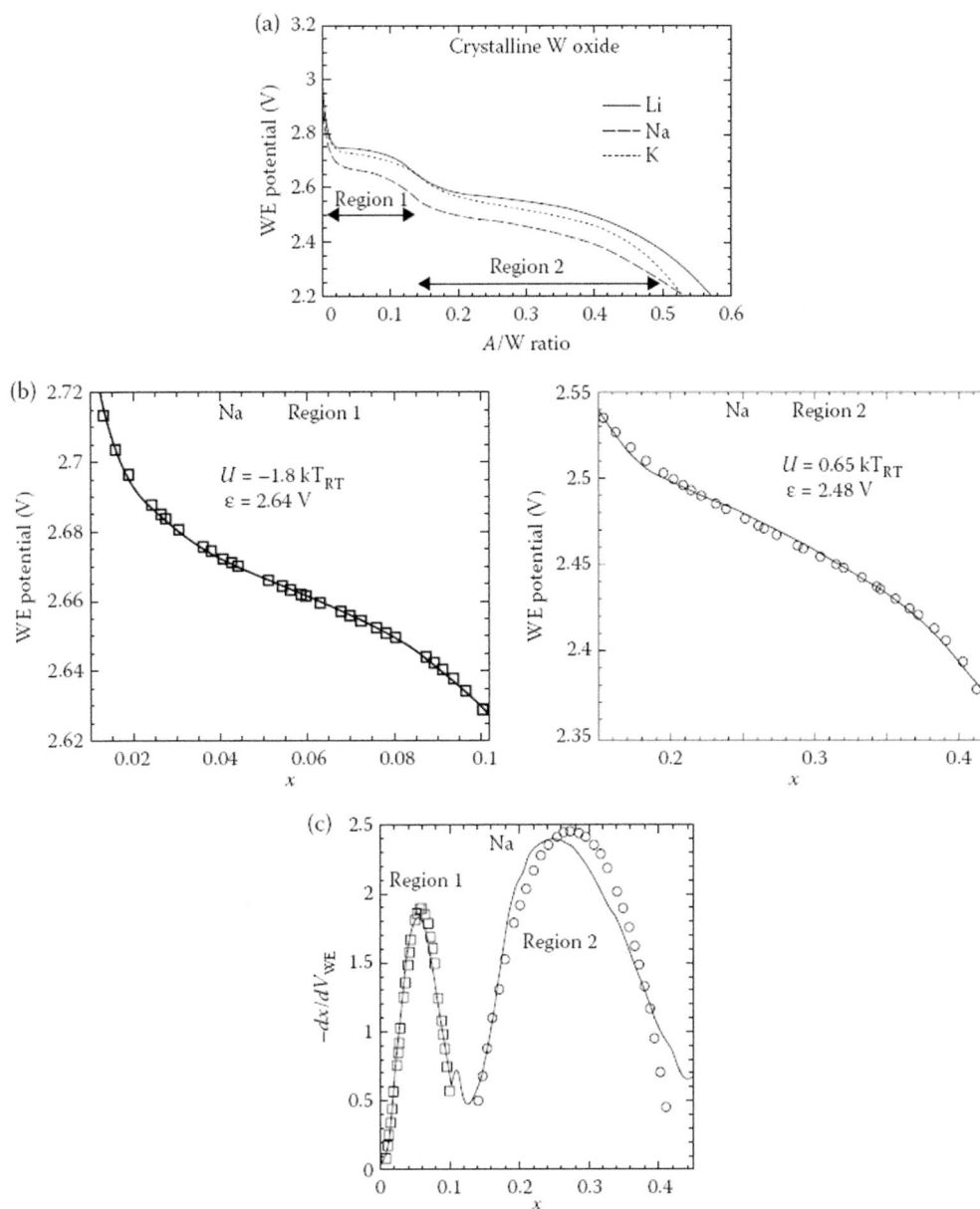

FIGURE 7.14 (a) Electrode potential as a function of the A/W ratio x (A is Li, Na, or K) as obtained under constant current intercalation into crystalline W oxide. The two regions marked out are analyzed in (b), where full lines represent measured chronopotentiometry and squares (composition region) and circles (composition region 2) are data obtained by fitting to the lattice-gas model. The interaction parameter U and the site energy ε are displayed. (c) Derivative of the composition x with respect to working electrode potential. (Reproduced with permission from Strömme, M. *Physical Review B.* 1998, *58*, 11015–11022.)

measurement of the gate capacitance and Figure 7.15d shows the chemical capacitance of graphene derived via Equation 7.30. Figure 7.16 shows the chemical (quantum) capacitance of graphene that is directly proportional to the electronic DOS. It has a zero minimum at the Dirac point and a linear increase on both sides, with symmetric electron and holes branches around the Dirac point. Comparing the theory and the measurements shown in Figure 7.15d, it is observed that this theory of quantum capacitance only agrees well at large channel potential while significant deviation occurs near the Dirac point. The deviation is mainly due to the fact that the local potential fluctuates spatially due to charged impurities above and below the graphene layer (Xia et al., 2009; Xu et al., 2011b). The quantum capacitance of graphene has been also measured using the electrochemical transistor configuration (see Figure 7.17).

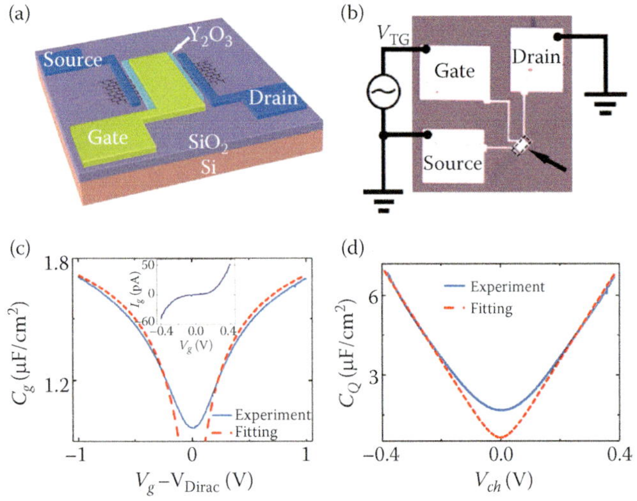

FIGURE 7.15 (a) Diagram showing the geometry of a graphene metal-oxide-semiconductor structure, where a graphene-Y_2O_3–Ti/Au capacitor lies on an SiO_2/Si substrate. (b) Optical image showing a device and experimental set up for measurements. The graphene channel is in the dashed frame indicated by the black arrow. The channel area covered by top gate electrode is W/L = 4.5 μm/7 μm. The separation between source/drain electrodes to the top gate electrode is 12 μm. (c) Total gate capacitance of the fabricated graphene FET. Inset: gate leakage current density as a function of gate voltage. (d) Quantum capacitance of graphene extracted from (c). The dashed line is the theoretical value of quantum capacitance of graphene at 300 K with a Fermi velocity 1.15×10^6 m s^{-1}. (Reproduced with permission from Xu, H.; Zhang, Z.; Peng, L.-M. *Applied Physics Letters* 2011, *98*, 133122.)

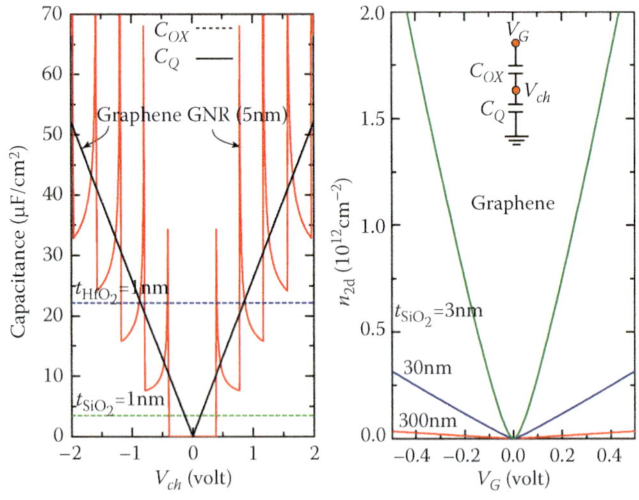

FIGURE 7.16 (a) Quantum capacitance of 2D graphene and a 5 nm graphene nanoribbons (GNR) compared with the parallel-plate capacitance of 1 nm SiO_2 and HfO_2. (b) 2D carrier density in a graphene sheet as a function of gate voltage for different oxide thicknesses. (Reproduced with permission from Fang, T. et al. *Applied Physics Letters* 2007, *91*, 092109.)

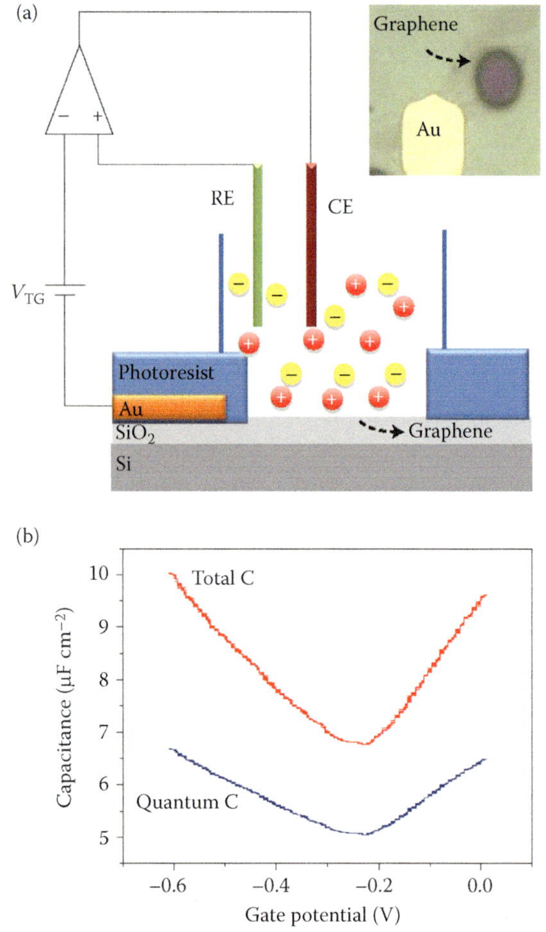

FIGURE 7.17 (a) Schematic of the quantum capacitance measurement setup in which a graphene sheet on a silicon/SiO_2 substrate is connected to a gold electrode. The edge of the graphene sheet is covered with an insulation layer so that only the top surface is exposed to an ionic liquid electrolyte. The potential of the graphene is controlled and varied with respect to a platinum RE using a three-electrode electrochemical configuration, electrochemical gate voltage. The inset is an optical micrograph of the graphene device. (b) Total capacitance and quantum capacitance of graphene measured in ionic liquid. (Reproduced with permission from Xia, J. et al. *Nature Nanotechnology* 2009, *4*, 505–509.)

GENERAL REFERENCES

Chemical capacitance of graphene: Giannazzo et al. (2008), Xia et al. (2009), Xu et al. (2011).

Chemical capacitance: Lindholm–Sethson et al. (1995), Jamnik and Maier (2001).

Li intercalation capacitance: Li et al. (1992), Gao et al. (1996), Strömme (1998), Matthieu et al. (2006), Dubarry and Liaw (2009), Smith et al. (2012).

Surface and interface states: Bardeen (1947), Brattain and Boddy (1962), Mönch (1993), Jaegermann (1996).

The quantum capacitance: Lee (1982), Luryi (1988), Datta (1995), John et al. (2004), Xia et al. (2009), Arash et al. (2011).

REFERENCES

Arash, H.; Joseph, A. S.; Georgi, D.; David, G.-G.; Wong, H. S. P. An integrated capacitance bridge for high-resolution, wide temperature range quantum capacitance measurements. *Review of Scientific Instruments* 2011, *82*, 053904.

Bardeen, J. Surface states and rectification at a metal semiconductor contact. *Physical Review* 1947, *71*, 717–727.

Bisquert, J. Chemical capacitance of nanostructured semiconductors: Its origin and significance for heterogeneous solar cells. *Physical Chemistry Chemical Physics* 2003, *5*, 5360–5364.

Bisquert, J. Interpretation of electron diffusion coefficient in organic and inorganic semiconductors with broad distributions of states. *Physical Chemistry Chemical Physics* 2008, *10*, 3175–3194.

Bisquert, J.; Fabregat-Santiago, F.; Mora-Seró, I.; Garcia-Belmonte, G.; Barea, E. M.; Palomares, E. A review of recent results on electrochemical determination of the density of electronic states of nanostructured metal-oxide semiconductors and organic hole conductors. *Inorganica Chimica Acta* 2008, *361*, 684–698.

Bisquert, J.; Garcia-Belmonte, G.; García-Cañadas, J. Effects of the Gaussian energy dispersion on the statistics of polarons and bipolarons in conducting polymers. *The Journal of Chemical Physics* 2004, *120*, 6726.

Brattain, W. H.; Boddy, P. J. The interface between germanium and a purified neutral electrolyte. *Journal of the Electrochemical Society* 1962, *109*, 574–582.

Büttiker, M. Capacitance, admittance and rectification properties of small conductors. *Journal of Physics: Condensed Matter* 1993, *5*, 9361.

Chidsey, C. E. D.; Murray, R. W. Redox capacity and direct current electron conductivity in electroactive materials. *The Journal of Physical Chemistry* 1986, *90*, 1479.

Das, A.; Pisana, S.; Chakraborty, B.; Piscanec, S.; Saha, S. K.; Waghmare, U. V.; Novoselov, K. S.; et al. Monitoring dopants by Raman scattering in an electrochemically top-gated graphene transistor. *Nature Nanotechnology* 2008, *3*, 210–215.

Datta, S. *Electronic Transport in Mesoscopic Systems*; Cambridge University Press: London, 1995.

Datta, S. *Lessons from Nanoelectronics: A New Perspective on Transport*; World Scientific: Singapore, 2012.

Dubarry, M.; Liaw, B. Y. Identify capacity fading mechanism in a commercial $LiFePO_4$ cell. *Journal of Power Sources* 2009, *194*, 541–549.

Fang, T.; Konar, A.; Xing, H.; Jena, D. Carrier statistics and quantum capacitance of graphene sheets and ribbons. *Applied Physics Letters* 2007, *91*, 092109.

Gao, Y.; Reimers, J. N.; Dahn, J. R. Changes in the voltage profile of $Li/Li_{1+x}Mn_{2-x}O_4$ cells as a function of x. *Physical Review B* 1996, *54*, 3878–3883.

Giannazzo, F.; Sonde, S.; Raineri, V.; Rimini, E. Screening length and quantum capacitance in graphene by Scanning Probe Microscopy. *Nano Letters* 2008, *9*, 23–29.

Gopar, V. A.; Mello, P. A.; Büttiker, M. Mesoscopic capacitors: A statistical analysis. *Physical Review Letters* 1996, *77*, 3005.

Heinze, J.; Frontana-Uribe, B.; Ludwigs, S. Electrochemistry of conducting polymers: Persistent models and new concepts. *Chemical Reviews* 2010, *110*, 4724–4771.

Hong, F. T.; Mauzerall, D. Interfacial photoreactions and chemical capacitance in lipid bilayers. *Proceedings of the National Academy of Sciences* 1974, *71*, 1564–1568.

Hulea, I. N.; Brom, H. B.; Houtepen, A. J.; Vanmaekelbergh, D.; Kelly, J. J.; Meulenkamp, E. A. Wide energy-window view on the density of states and hole mobility in poly(p-phenylene vinylene). *Physical Review Letters* 2004, *93*, 166601.

Jaegermann, W. The semiconductor/electrolyte interface: A surface science approach. *Modern Aspects of Electrochemistry* 1996, *30*, 1–186.

Jamnik, J.; Maier, J. Generalised equivalent circuits for mass and charge transport: Chemical capacitance and its implications. *Physical Chemistry Chemical Physics* 2001, *3*, 1668–1678.

John, D. L.; Castro, L. C.; Pulfrey, D. L. Quantum capacitance in nanoscale device modeling. *Journal of Applied Physics* 2004, *96*, 5180–5184.

John, H.; Bauer, R.; Espindola, P.; Sonar, P.; Heinze, J.; Müllen, K. 3D-Hybrid networks with controllable electrical conductivity from the electrochemical deposition of terthiophenefunctionalized polyphenylene dendrimers. *Angewandte Chemie International Edition* 2005, *44*, 2447–2451.

Lee, P. A. Density of states and screening near the mobility edge. *Physical Review B* 1982, *26*, 5882–5885.

Li, W.; Reimers, J. N.; Dahn, J. R. Crystal structure of $Li_xNi_{2-x}O_2$ and a lattice-gas model for the order-disorder transition. *Physical Review B* 1992, *46*, 3236–3246.

Lindholm-Sethson, B.; Tjärnhage, T.; Sharp, M. On the measurements of pseudocapacitances in thin polymer films on electrode surfaces. *Electrochimica Acta* 1995, *40*, 1675–1679.

Luryi, S. Quantum capacitance devices. *Applied Physics Letters* 1988, *52*, 501–503.

Matthieu, D.; Vojtech, S.; Ruey, H.; Bor Yann, L. Incremental capacity analysis and close-to-equilibrium OCV measurements to quantify capacity fade in commercial rechargeable lithium batteries. *Electrochemical and Solid-State Letters* 2006, *9*, A454–A457.

McKinnon, W. R.; Haering, R. R. *Modern Aspects of Electrochemistry*; White, R. E., Bockris, J. O. M., Conway, B. E., Eds.; Plenum Press: New York, 1983; Vol. 15; pp. 235–304.

Mönch, W. *Semiconductor Surfaces and Interfaces*; Springer: Berlin, 1993.

Mott, N. F. *Conduction in Non-Crystalline Solids*, 2nd ed.; Oxford University Press: Oxford, 1987.

Palenskis, V. Einstein relation and other related parameters of randomly moving charge carriers in materials with degenerated electron gas. *American Journal of Modern Physics* 2013, *2*, 155–167.

Parr, R. G.; Pearson, R. G. Absolute hardness: Companion parameter to absolute electronegativity. *Journal of the American Chemical Society* 1983, *105*, 7512–7516.

Roest, A. L.; Kelly, J. J.; Vanmaekelbergh, D. Coulomb blockade of electron transport in a ZnO quantum-dot solid. *Applied Physics Letters* 2003, *83*, 5530.

Salvador, P.; Gutierrez, C. The nature of surface states involved in the photo- and electrolumi-nescence spectra of n-titanium dioxide electrodes. *The Journal of Physical Chemistry* 1984, *88*, 3696–3698.

Schultze, J. W.; Lohrengel, M. M. Stability, reactivity and breakdown of passive films. Problems of recent and future research. *Electrochimica Acta* 2000, *45*, 2499–2513.

Shapiro, F. R.; Adler, D. Equilibrium transport in amorphous semiconductors. *Journal of NonCrystalline Solids* 1985, *74*, 189–194.

Shaw, J. G.; Hack, M. An analytic model for calculating trapped charge in amorphous silicon. *Journal of Applied Physics* 1988, *64*, 4562–4566.

Shockley, W. Electrons, holes, and traps. *Proceedings of the IRE* 1958, *46*, 973–990.

Smith, A. J.; Dahn, H. M.; Burns, J. C.; Dahn, J. R. Long-term low-rate cycling of $LiCoO_2$/Graphite Li-ion cells at 55 C. *Journal of the Electrochemical Society* 2012, *159*, A705–A710.

Strömme, M. Cation intercalation in sputter-deposited W oxide films. *Physical Review B* 1998, *58*, 11015–11022.

Sze, S. M. *Physics of Semiconductor Devices*, 2nd ed.; John Wiley and Sons: New York, 1981.

Xia, J.; Chen, F.; Li, J.; Tao, N. Measurement of the quantum capacitance of graphene. *Nature Nanotechnology* 2009, *4*, 505–509.

Xu, H.; Zhang, Z.; Peng, L.-M. Measurements and microscopic model of quantum capacitance in graphene. *Applied Physics Letters* 2011a, *98*, 133122.

Xu, H.; Zhang, Z.; Wang, Z.; Wang, S.; Liang, X.; Peng, L.-M. Quantum capacitance limited vertical scaling of graphene field-effect transistor. *ACS Nano* 2011b, *5*, 2340–2347.

8 The Density of States in Disordered Inorganic and Organic Conductors

In the previous chapters we have emphasized the equilibrium properties of an electronic or ionic system, represented by the density of states in a single phase, and by built-in potential at interfaces. We have also discussed the general kinetic properties associated with electron transport across interfaces. We begin this chapter addressing the actual measurement of both aspects of the response of a material film deposited on a current collector, employing as a reference experimental method the technique of cyclic voltammetry. We show how both kinetic and energetic aspects can be separately characterized, and we address how to distinguish them in the experimental response. Thereafter, we will treat some important classes of materials that present the most characteristic distributions of states, namely, the exponential and Gaussian. We discuss the observation of band tailing in energetically disordered semiconductors, as amorphous silicon, and then we review the capacitive properties of nanocrystalline TiO_2 electrodes, used in mesoporous solar cells and photocatalysis. We follow up with a brief introduction to organic conducting materials and a description of the main properties of the Gaussian distribution as well as some instances of the observation in hole-conducting organic films.

8.1 CAPACITIVE AND REACTIVE CURRENT IN CYCLIC VOLTAMMETRY

Cyclic voltammetry (CV) is a technique that explores the electrochemical response of materials and molecules as a function of the electrode potential in a very wide voltage range. CV is a rapid method that signals specific effects that occur at each value of the potential. It is therefore an excellent technique for an exploratory approach of the processes occurring in a system and it can also give quantitative mechanistic information (Greef et al., 1985).

In linear sweep voltammetry, the electric current is measured while the electrode potential is swept between two values V_{in} and V_1 at a rate $\nu = dV/dt$. Then the direction of variation of electrode potential is reversed $dV/dt = -\nu$ until the voltage returns to V_{in}. The variation of the potential with time is

$$0 < t < t_1 \quad V = V_{in} + vt \tag{8.1}$$

$$t_1 < t < t_2 \quad V = V_{in} + 2vt_1 - vt \tag{8.2}$$

where

$$t_1 = \frac{V_1 - V_{in}}{v} \tag{8.3}$$

Often, a number of complete cycles are performed to check the stability of the response, see Figure 7.11.

Let us consider a metal electrode in contact with a solution that contains the molecules that have to be investigated, as in Figures 6.3 and 6.15. The electrochemical system is further represented in Figure 8.1a. We perform the CV and obtain the current due to electrochemical reaction when the molecules are reduced and oxidized at the metal surface. A cathodic reaction, associated with reduction of the species situated close to the electrode as indicated in Figures 6.15b and 6.16a, is identified by the occurrence of a negative current. Conversely, the anodic reaction is indicated by the positive voltammetric current. When the reaction occurs at very negative potential with respect to the RE, it means that the compound has a low electron affinity and it can easily reduce other materials. Therefore, CV is a method useful to investigate the electron affinities and ionization potential, in conjunction with other methods like UPS, which was discussed in Chapter 4.

To approach the interpretation of CV it is useful to start with a distinction of *two types* of processes, the electrochemical *charge transfer* reaction, and the *capacitive charging*. We consider them in turn and later we describe a combination of both.

Let us suppose that the current is associated with an electrochemical reaction at the electrode surface as described in Section 6.6, see Figure 6.15. In addition, we assume that charge transfer rate at the interface is just a function of the electrode potential, with no additional kinetic limitations, such as diffusion of the reactive species to the surface. Then the current density can be stated as

$$j = j_{ct}(V) \tag{8.4}$$

which does not depend on the time scale of the measurement, as it is the voltage at any instant, V, the only

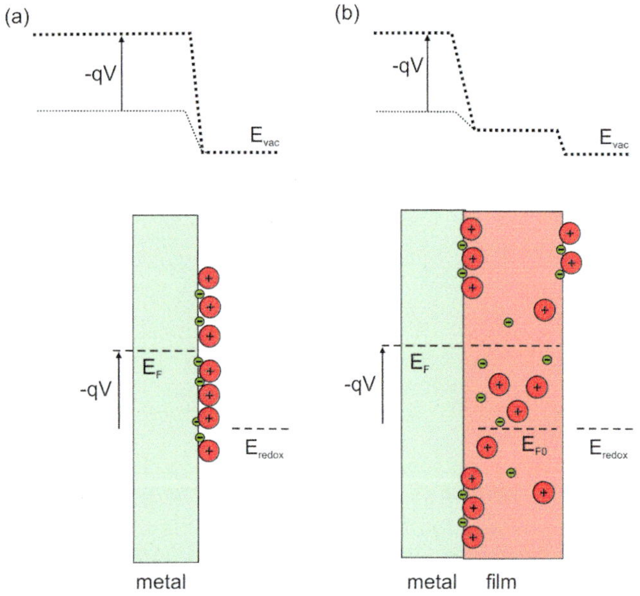

FIGURE 8.1 Model systems for capacitive charging, under applied bias, indicating the shift of the VL by the applied voltage. (a) Metal/electrolyte contact. (b) Metal/active film/solution system.

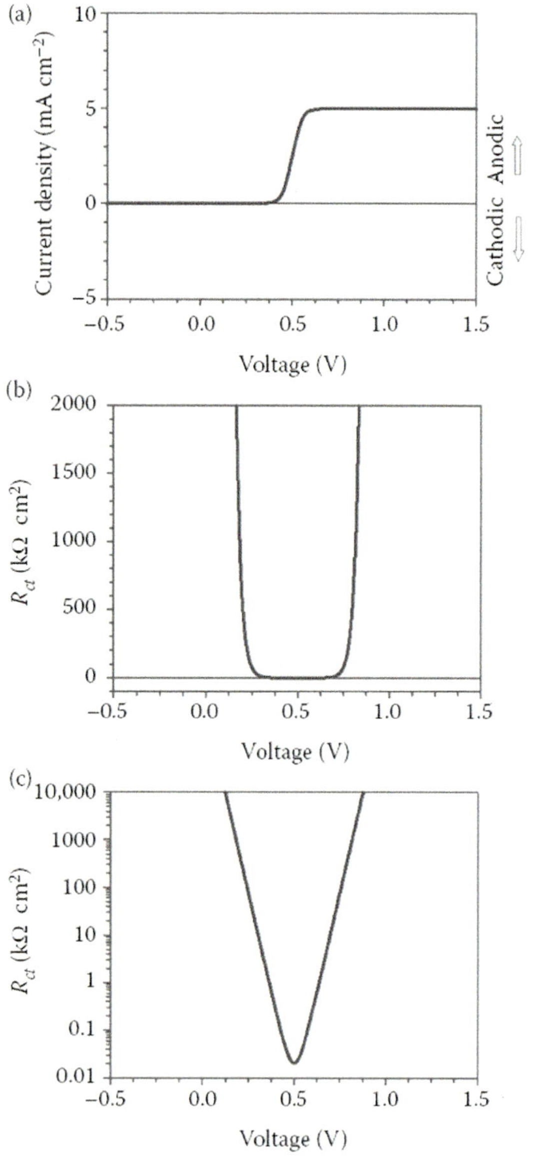

FIGURE 8.2 Model system electrochemical current (a) Steady-state current density. (b) Charge transfer resistance. (c) Same as (b) in log vertical scale. Parameters: j_m = 5 mAcm^{-2}, V^0 = 0.5 V.

quantity that determines the current. Consequently, the CV is *independent on the scan rate*. It follows that the go and return scans will give the same result. If the current is totally reactive, both scans will overlap perfectly and the different methods should give equivalent results. We remark that the capacitance itself can be time-dependent as in dielectric relaxation. Such property adds a term VdC/dt to the capacitive current, as in KP measurement in Equation 4.12. Here we ignore such variations for simplicity.

Instead of the standard Butler-Volmer expression given in Section 6.6, we now use an arbitrary example of $j(V)$ characteristic at the electrode surface, to illustrate the behavior of CV, described by the following model:

$$j_{ct} = j_m \frac{\theta}{1+\theta} \tag{8.5}$$

where

$$\theta(V) = \exp\left[q\left(V - V^0\right)/k_B T\right] \tag{8.6}$$

Here, the potential V is measured with respect to an RE in solution. The model of Equation 8.5 shows a saturation at a current density j_m at positive potentials, with a characteristic charge transfer potential V^0, as indicated in Figure 8.2a. When we cycle the voltage as in Equations 8.1 and 8.2, we always obtain the same result as Figure 8.2a, indicating that the current is totally reactive. We also note that the current is antisymmetric with respect to V^0, hence the charge transfer resistance of Equation 6.80 is symmetric about this point, and it increases when the potential departs from this value, see Figure 8.2b and c.

The second component of the basic classification of CV responses is a capacitive current. In this case there is no electron transfer to solution, as the electrochemical reaction at the film/solution interface is inhibited by the properties or the electrolyte or by the presence of blocking layers. The measured current is directed to accumulate charge, in the surface of the electrode, as in double-layer charging or absorption, modifying the voltage at the Helmholtz layer as shown in Figure 8.1a, or else to charge a film that contacts the current collector, by filling the electronic DOS, or by ion intercalation, as indicated in Figure 8.1b. In this latter case, there is obviously a charge transfer from the metal surface to the interior of the film. But the electrons remain in the film, charge compensated by ionic species from solution, as in a battery material described in Section 5.11. The process of charging a nanostructured electrode in contact with solution, in which charge compensating ions do not need to enter the semiconductor structure, is discussed in Section 8.3.

Let Q be the charge per unit area that can be accumulated at the electrode. The capacitive current in CV is given by

$$j_{cap} = \frac{dQ}{dt} = \frac{\partial Q}{\partial V}\frac{dV}{dt} \quad (8.7)$$

Therefore, current can be expressed in terms of the differential capacitance defined in Equation 7.2 as follows:

$$j(V) = C(V)v \quad (8.8)$$

In Section 7.1, we discussed the different types of capacitances that can be associated with electrodes and electroactive films. Basically, we may have the dielectric capacitance as in Helmholtz layer, Figure 8.1a or the chemical capacitance as in absorption or film charging, Figure 8.1b. In either case, we may assume that the system possesses a capacitance that is a unique function of the electrode potential $C(V)$. We call this function the *equilibrium capacitance* of the electrode or film. It is usually derived by knowledge of the metal–solution interface, or else from thermodynamic properties of the system such as the chemical potential function, or more generally combining the variation of ionic and electronic properties of the active film as illustrated in Figure 5.18. Somewhat paradoxically, although the capacitance is determined by thermodynamic properties, a capacitance cannot be measured in a steady-state method. Some type of transient measurement, such as CV or ac impedance spectroscopy (IS) as discussed in Section 3.5, is necessary, and the different methods should give equivalent results. We remark that the capacitance itself can be time-dependent as in dielectric relaxation, or in KP measurement in Equation 4.12. Here we ignore such variations for simplicity.

Equation 8.8 indicates that the current in the capacitive CV regime is proportional to the scan rate. Since the equilibrium capacitance of the film is a unique function of the voltage, the current changes sign between anodic and cathodic wave (as dV/dt is inverted) and should show completely symmetrical and mirror-image waves with identical peak potentials and current levels. When the film or system is purely capacitive, then the CV current directly reflects the shape of the capacitance dependence on voltage. Thus, an exponentially increasing capacitance will provide an exponential raising current, while a constant capacitance displays a square voltammogram, see Figure 8.3.

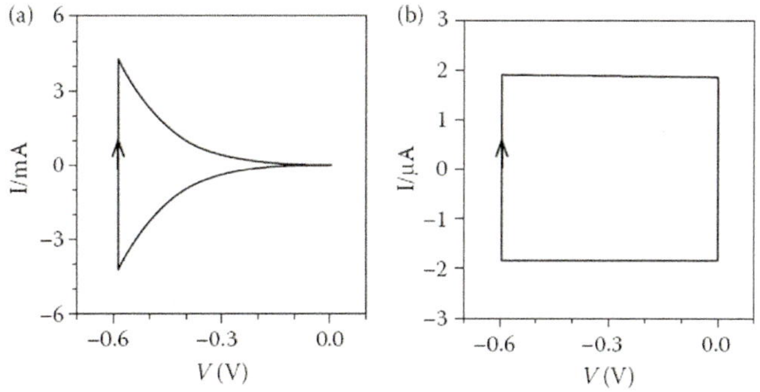

FIGURE 8.3 Simulation of cyclic voltammograms, between $V_1 = 0$ V and $V_2 = -0.6$ V, for an exponential capacitance (a) and a constant capacitance $C = 0.01$ mF (b) at scan rate $v = 200$ mV · s^{-1}.

Let us examine a few specific examples. A symmetric, totally capacitive CV for deposition of Pb adatoms on polycrystalline Au is shown in Figure 8.4. The bottom of Figure 8.4 shows the progressive covering of the gold surface. The symmetric pattern of CV of a conducting polymer film is shown in Figure 8.5 (see also Figure 7.11), where the inset shows that the peak height is proportional to scan rate, which confirms fully capacitive behavior as stated in Equation 8.8. However, there is some distortion of the peaks, which was attributed to additional kinetic steps associated with the increment of the obstruction of ions and electrons transfer in the thick film. These effects are discussed in the next section. CVs of electrocatalytic metal oxides RuO_2 and IrO_2 are shown in Figure 8.6. These materials have been investigated for electroreduction and supercapacitor applications (Trasatti and Buzzanca, 1971; Burke and Healy, 1981; Burke, 2000). The potential window of investigation is limited by hydrogen evolution at negative potential and oxygen evolution (water oxidation) at positive potential. Between these limits, RuO_2 exhibits three quasi-reversible peaks corresponding to the surface Ru(3+/2+), Ru(4+/3+), and Ru(5+/4+) couples, while IrO_2 shows only the surface Ir(3+/4+) couple.

Let us discuss in more detail the voltammetric response during charging and discharging of redox-active thin films, as shown in Figure 7.6. We consider a film in which electron charge $Q = -Lnq$ per unit film

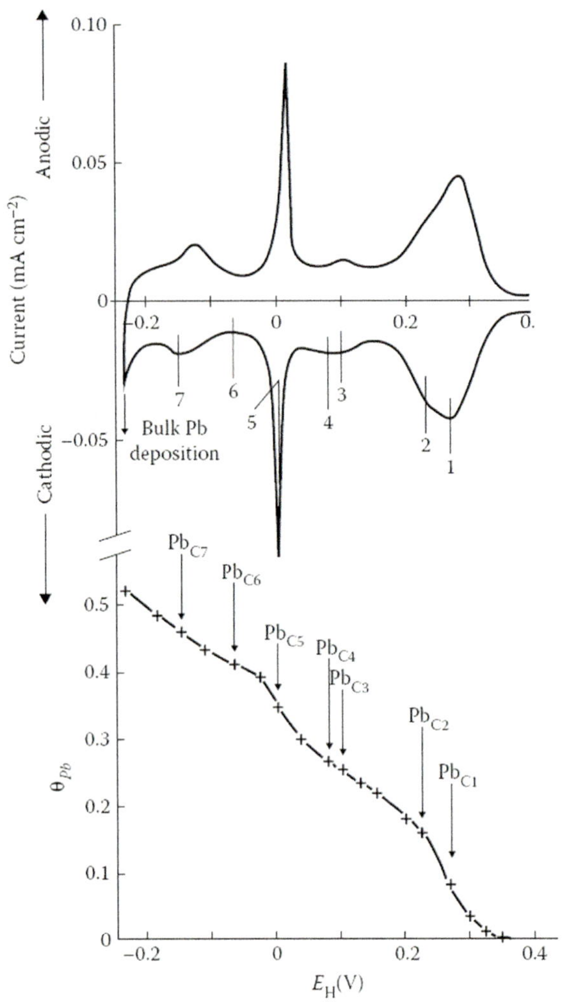

FIGURE 8.4 CV profile for deposition of Pb adatoms on polycrystalline Au from aqueous $HClO_4$ solution, exhibiting seven distinguishable states below a monolayer. Lower curve is integral of the cathodic sweep profile. (Reproduced with permission from Conway, B. E.; Pell, W. G. *Journal of Solid State Electrochemistry* 2003, *7*, 637–644.)

area can be stored, where n is the volume density of the reduced species,

$$n = N_t f \tag{8.9}$$

with f the fractional occupancy described by the Fermi–Dirac distribution function of Equation 5.13,

$$f(V - V_0) = \frac{1}{1 + \exp\left[q\left(V - V^0\right)/k_B T\right]} \tag{8.10}$$

in terms of the potential corresponding to the redox energy of the state, V^0. We may write

$$f(V - V_0) = \frac{1}{1 + \theta} \tag{8.11}$$

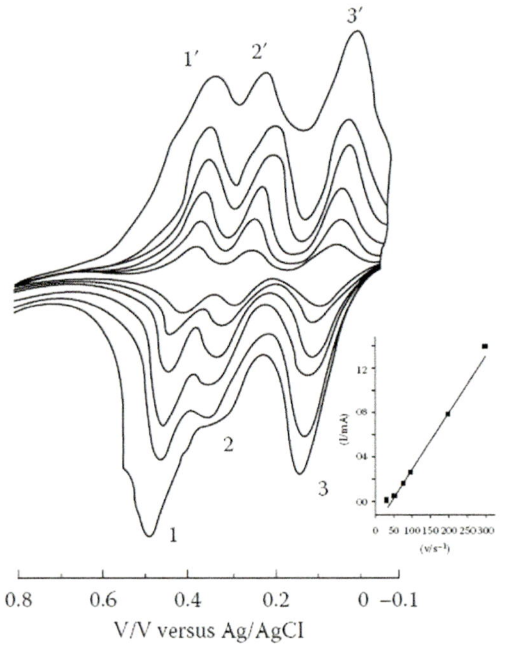

FIGURE 8.5 Cyclic voltammograms for a 1:12 phosphomolybdic acid $H_7PMo_{12}O_{42} \cdot xH_2O$-PPy/glassy carbon electrode in 0.50 M H_2SO_4 solution with different scan rates (from inner to outer: 30, 50, 80, 100, 200, and 300 mV s^{-1}, respectively). The dependence of peak currents with scan rate is shown in the inset. (Reproduced with permission from Guo, M. D.; Guo, H. X. *Journal of Electroanalytical Chemistry* 2005, *585*, 28–34.)

where $\theta(V)$ is defined in Equation 8.6. The capacitive current is

$$j = \frac{dQ}{dt} = qLvc_\mu \tag{8.12}$$

where c_μ is the chemical capacitance of a single state given in Equation 7.12. Equation 8.12 can also be expressed as

$$c_\mu = \frac{q^2 N_t}{k_B T} = \frac{1}{(1+\theta)^2} \tag{8.13}$$

This last equation is a standard expression of the reversible redox peak (Hubbard, 1969; Heinze et al., 2010). If the film contains a distribution of redox states, then we observe a collection of redox peaks (Figure 8.5). If the individual redox element peaks are closely spaced, then a continuous distribution will be measured, as shown in the example of Figure 7.11 and further discussed in Section 8.5. Using Equations 7.23 and 8.8, we obtain a direct relation between CV current and the DOS, $g(E_{Fn})$, at the Fermi level established by the applied voltage V in the film

$$g(E_{Fn}) = \frac{j(V)}{q^2 v} \tag{8.14}$$

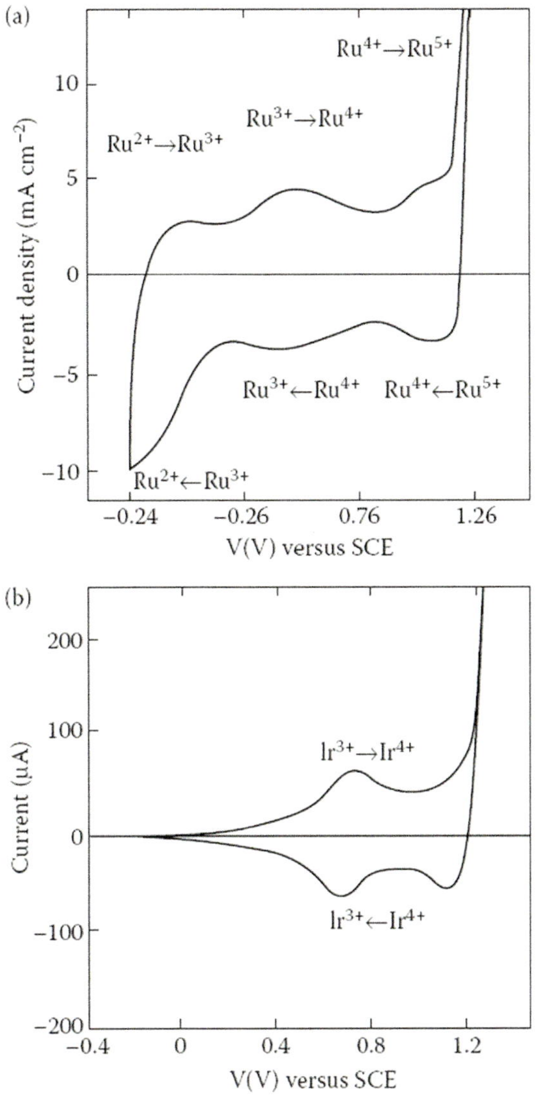

FIGURE 8.6 Room temperature cyclic voltammogram for (a) RuO_2-coated Ti substrate in 1 M H_2SO_4 and (b) IrO_2 in 0.5 M H_2SO_4. (Reproduced with permission from Goodenough, J. B.; Manoharan, R.; *Journal of the American Chemical Society* 1990, *112*, 2076–2082.)

However, this last expression must be used with caution for the Gaussian distribution at low charging level as discussed in Section 8.5.

8.2 KINETIC EFFECTS IN CV RESPONSE

Let us follow up the example of Equation 8.5 and Figure 8.2a in order to include the capacitive effect. We add to this model the capacitance of the film, given by a Gaussian function

$$C(V) = \frac{C_0}{\sqrt{2\pi}\sigma_d} \exp\left[-\frac{(V - V_{ca})^2}{2\sigma_d^2}\right] \quad (8.15)$$

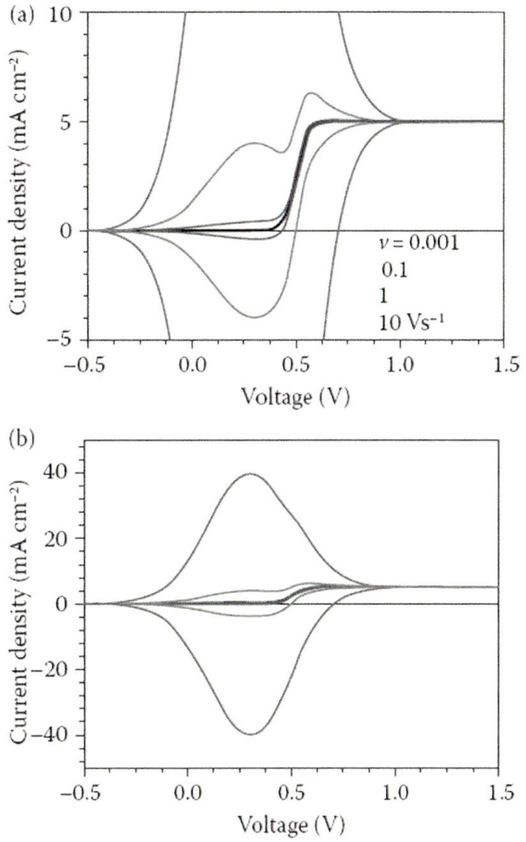

FIGURE 8.7 Voltammetry of a model system including anodic current, in parallel with capacitive charging. (a) CV at different scan rates as indicated. (b) Same as (a) shown at different scale. Parameters: j_m = 5mAcm^{-2}, C_0 = 2mF, V_{ca} = 0.3V, $V_{ct}^0 = 0.5$V, V_{in} = −1V, V_1 = 1.5V.

that is described by parameters C_0, V_{ca} (the voltage at which the capacitance peaks), and σ_d (the spread of the capacitance in the voltage axis). Now the CV current is composed of both charge transfer and capacitive currents. This distinction was already elaborated in Section 3.7 in connection with the faradaic and nonfaradaic currents at the metal–electrolyte interface. Since both currents are produced by the same voltage, they are added as

$$j_{tot}(V) = j_{ct} + j_{cap} \quad (8.16)$$

where j_{cap} is given by Equation 8.8. Figure 8.7a shows the effect of the capacitive component on the CV of Figure 8.2a. At very low scan rate, the capacitive component is small but it already produces an opening of the resistive line into two branches. As the sweep rate increases, the capacitive current becomes huge, and finally the CV is totally dominated by the symmetric shape of capacitive charging, as shown in Figure 8.7b.

In general, the CV is influenced by a variety of kinetic factors that must be suitably controlled in order to extract the required information. One important factor limiting the reaction at an electrode is the mass transport rate. If the electrode current is caused by reduction or oxidation of a redox species, that is governed by diffusion to the electrode surface, the CV has asymmetrical shape with current j proportional to $\nu^{1/2}$. The peak potential separation ΔV_p is $0.059/z$ V at all scan rates at 25°C (Greef et al., 1985). When we analyze capacitive charging of a film and the film thickness increases, eventually we must reach transport effects within the film. As a consequence, the voltammetric response gradually shifts from mirror symmetrical diagrams to the diffusion pattern in which the peaks are separated in the voltage axis. In general, the kinetic effects may distort the voltammogram from the symmetric shape, but if electron transfer of electrons to solution is forbidden, then the number of electrons, injected to the film in the cathodic scan, is recovered in the anodic one. So even if the shape is distorted, the areas of forward and backward potential sweeps will be the same.

If the transport of electronic or ionic species is a significant effect, then the required gradients of the respective electrochemical potentials take a fraction of the potential difference in the external circuit. One can write the electrode potential as

$$V = V_F + V_T \tag{8.17}$$

where V_F is the "Fermi level voltage" that controls the capacitor as shown in Figure 8.1, and V_T is associated with the transport components (Pell and Conway, 2001). In order to treat such features in a simple way, one may consider a model that contains a constant series resistance R_S related to transport, in addition to the film capacitance, so that when the CV current is I, we have $V_T = IR_s$, thus

$$V = V_F + IR_s \tag{8.18}$$

This is a suitable description of transport in the bulk electrolyte outside the film, and is usually termed the *ohmic drop*. Figure 8.8 shows that correction of IR drops in the measurement of the differential capacitance of commercial $LiFePO_4$ battery cells (Section 7.6). This correction allows to plot the capacitive peaks with respect to the internal potential of the battery electrodes, so that the peaks at different scan rates nearly overlap and show the internal characteristic charging of the materials in the anode and cathode of the battery.

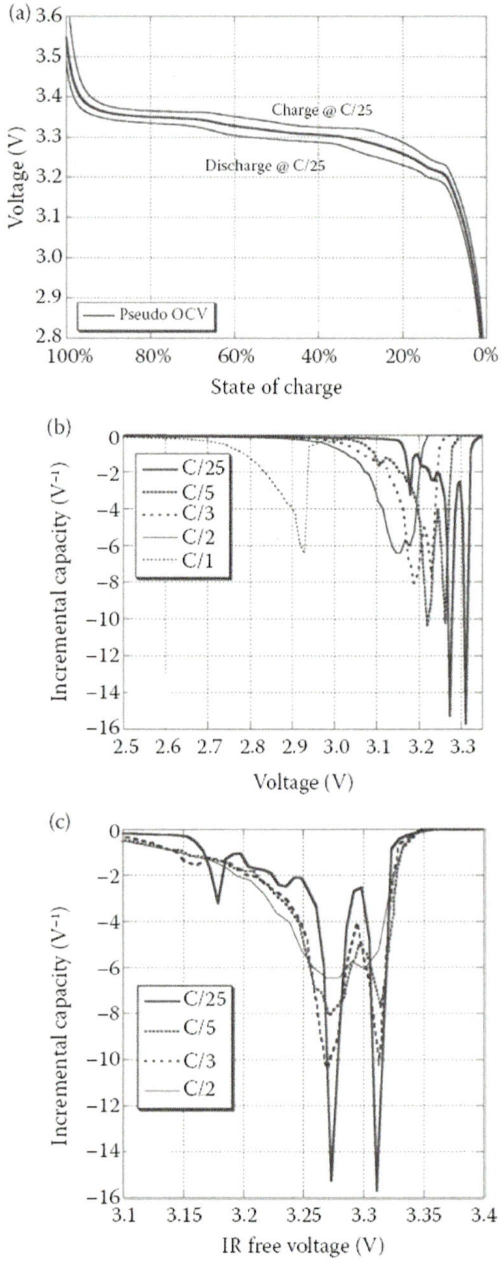

FIGURE 8.8 (a) C/25 charge and discharge curve of a commercial $LiFePO_4$-based lithium ion cell from the test cell and the reconstructed stationary voltage versus state of charge curve. (b) Differential capacity peaks calculated from the discharge curves of various rates and (c) the same peaks with a polarization correction. (Reproduced with permission from Dubarry, M.; Liaw, B. Y. *Journal of Power Sources* 2009, *194*, 541–549.)

8.3 THE EXPONENTIAL DOS IN AMORPHOUS SEMICONDUCTORS

General properties of the DOS of amorphous solids were discussed in Section 7.2. We will now review the features of bulk semiconductors showing an exponential DOS.

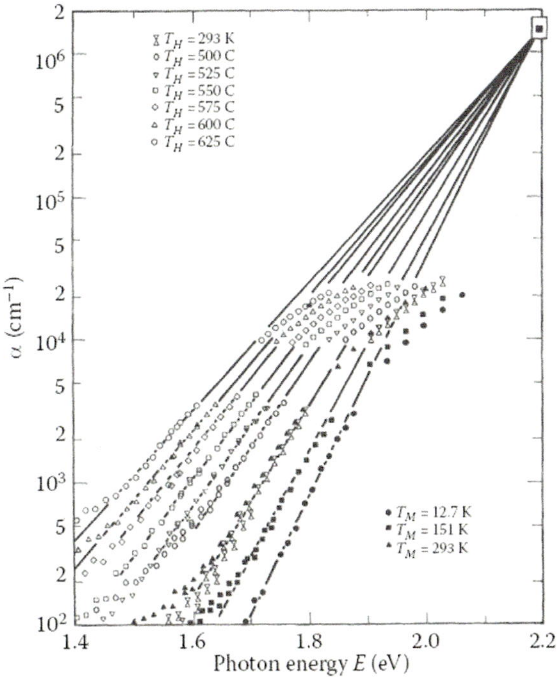

FIGURE 8.9 Optical absorption coefficient of a-Si:H, as a function of photon energy, at different temperatures as indicated. (Reproduced with permission from Cody, G. D. *Physical Review Letters* 1981, *47*, 1480–1483.)

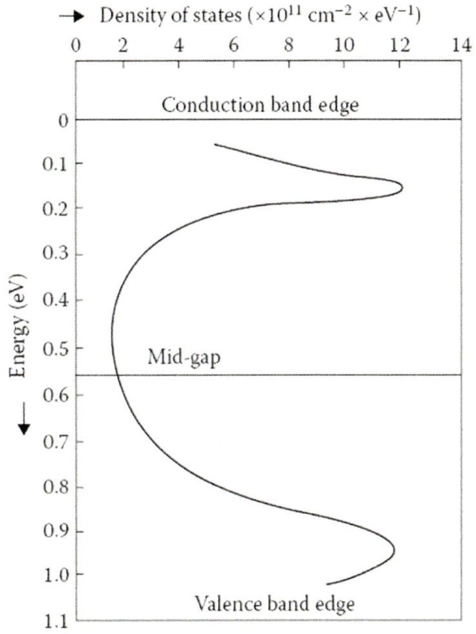

FIGURE 8.10 Typical curve of surface state density at the Si-SiO_2 interface versus energy in Si bandgap. (Reproduced with permission from Goetzberger, A.; Heine, V.; Nicollian, E. H. *Applied Physics Letters* 1968, *12*, 95–97.)

In the standard behavior of crystalline semiconductors, the absorption coefficient often shows a band tailing featured termed Urbach tail, Figure 8.9. The Urbach tail is either impurity-induced or a consequence of dynamic disorder caused by local coulombic interactions, thermal fluctuations and polaronic effects (Pankove, 1971). This topic will be more amply discussed in Chapter 19, Equation 19.9, and in Section 25.5 for the implications on photovoltaic efficiency of semiconductors.

Distinct to crystalline solids, in the amorphous semiconductors the transition of available electronic states to the gap in the energy axis is not sharp, but the DOS decreases gradually. Still, many amorphous semiconductors show a bandgap, and this can be explained for example in terms of certain models of amorphous covalent solids (Weaire and Thorpe, 1971). However in amorphous semiconductors the tail has a dominant component due to the structural disorder. The tail of localized states enters the gap of the semiconductor, usually in the form of an exponential distribution of the type

$$g(E)=\frac{N_L}{k_BT_0}\exp\left[\left(E-E_C\right)/k_BT_0\right] \qquad (8.19)$$

Here, E_c is the conduction band minimum, or mobility edge (Section 7.2), N_L is the total density of localized states and T_0 is a parameter with temperature units that determines the depth of the distribution, which can be alternatively expressed as a coefficient $\alpha = T/T_0$ or as an energy $E = k_BT_0$. In some materials, Equation 8.19 is obeyed almost exactly over many decades of the carrier concentration (Monroe and Kastner, 1986).

The exponential DOS has been amply described for the states in the gap of hydrogenated amorphous silicon, a material that is widely used in thin film solar cells. The characteristic value of the bandtail parameter is $T_0 \approx 250 - 300$ K. Figure 7.4 shows the typical structure of the DOS in a-Si:H. It is composed of a high density of states at the conduction and valence band (defined by the respective mobility edges) and an exponential tail of both electron and hole states toward the center of the gap. Figure 8.10 shows the characteristic distribution of surface states at silicon-silicon oxide interface.

The exponential distribution of traps in amorphous silicon was established by evidence on the multiple trapping transport model. The exponential distribution is found as well in other materials based on different experimental techniques such as photoinduced absorption and transient photoconductivity. The DOS can also be obtained by capacitance–voltage characteristics, measured at different frequencies and temperatures, in deep reverse conditions (Balberg and Gal, 1985) and by the space-charge limited current at different temperatures (Krellner et al., 2007).

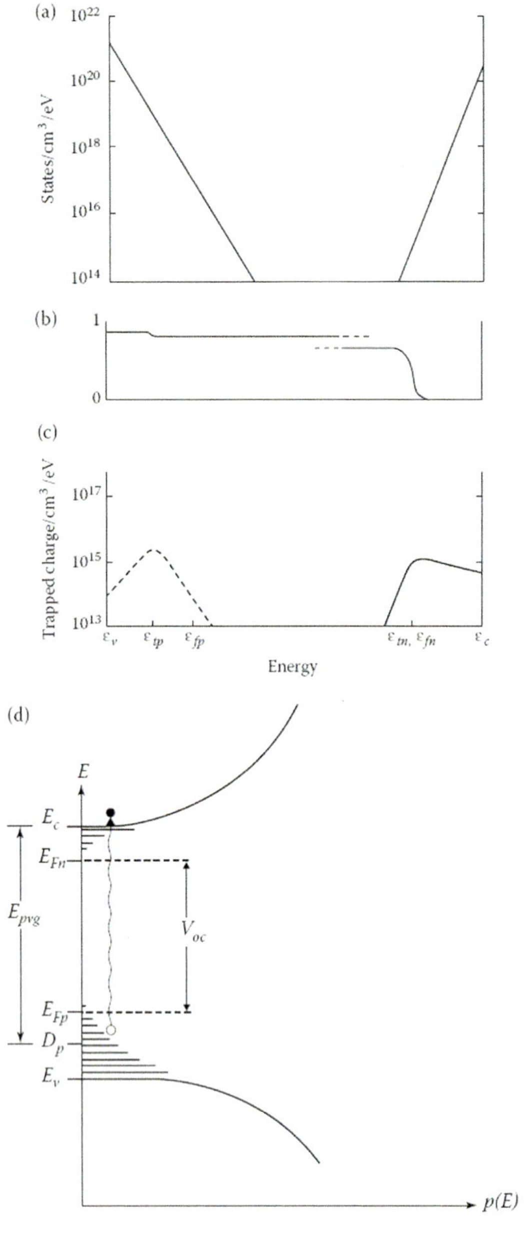

FIGURE 8.11 (a) Density of states in the gap (b) occupation function for the gap states and (c) distribution of photogenerated trapped electron and holes. (d) Energy-level diagram of a-Si:H under illumination. (Reproduced with permission from Tiedje, T. *Applied Physics Letters* 1982, *40*, 627–629; Yablonovitch, E.; Tiedje, T.; Witzke, H. *Applied Physics Letters* 1982, *41*, 953–955.)

Since the voltage that can be extracted from a given material, in a solar cell configuration, is determined by the Fermi level separation, in these materials the tail states become the relevant ones for the production of photovoltage, rather than mobility edge energies. This is a general characteristic for photovoltaic cells that use disordered materials: a larger DOS in the bandgap implies a lower Fermi level (other things being equal), and reduces the photovoltage (Garcia-Belmonte et al., 2010). In a-Si:H cells the amorphous structure of the absorber and the resulting tail states imposes a price of several 100 mV, as compared to perfect crystalline cells, see Figure 8.11.

8.4 THE EXPONENTIAL DOS IN NANOCRYSTALLINE METAL OXIDES

Random nanoparticulate networks, formed by sintering undoped metal-oxide nanoparticles, as in Figure 1.1, are widely used in photovoltaic devices such as dye-sensitized solar cells (DSC) (O'Regan and Grätzel, 1991; Wang et al., 2014) and solar photocatalysis applications (Kubacka et al., 2012). The main method to prepare TiO_2 nanocrystalline paste for DSC is by hydrolysis of titanium tetraisopropoxide with the addition of a binder such as ethyl cellulose in α-terpineol (O'Regan et al., 1990). The TiO_2 layers are deposited using the doctor blading technique on transparent conducting oxide (TCO) glass. The resulting photoelectrodes of about 10 µm thickness are sintered at 450°C and then immersed in $TiCl_4$ solution at 70°C followed by a new calcination at 450°C. TiO_2 anatase nanoparticles consist of a truncated bipyramid as shown in Figures 1.2 and 1.3.

Here, we comment on the characteristic behavior of nanocrystalline metal-oxide semiconductor films immersed in the electrolyte solution. The application of bias voltage at the TCO substrate raises the Fermi level of electrons in the semiconductor film, as shown in Figure 7.7. The increasing electron charge is compensated by excess ionic charge at the particle's surface as indicated in Figure 8.12. Electrons can be injected either from the conducting substrate or by dye sensitization of the surface as shown in Figure 8.12a, which is the core phenomenon exploited in a DSC. TiO_2 and ZnO nanoparticles form an electrical insulator matrix, but injected electrons raise the conductivity (Hoyer and Weller, 1995; Abayev et al., 2003) and the electrons Fermi level becomes nearly homogeneous in the film, Figure 7.7. The use of techniques such as voltammetry or impedance spectroscopy allows to measure the chemical capacitance of the metal-oxide nanostructure.

The DOS of nanostructured TiO_2 films shows several characteristic features indicated in Figure 8.13. The first is the representation of extended states, in terms of the conduction band (CB) minimum level E_c. The position of the CB in the energy axis is a crucial factor for the injection of electrons from the adsorbed chromophores in DSC. The dominant feature of the DOS is an exponential distribution of localized states that starts from the edge of the CB and decreases toward the center of the gap. Another frequent feature in this system is a much less intense peak that appears at about 0.4 eV below E_c,

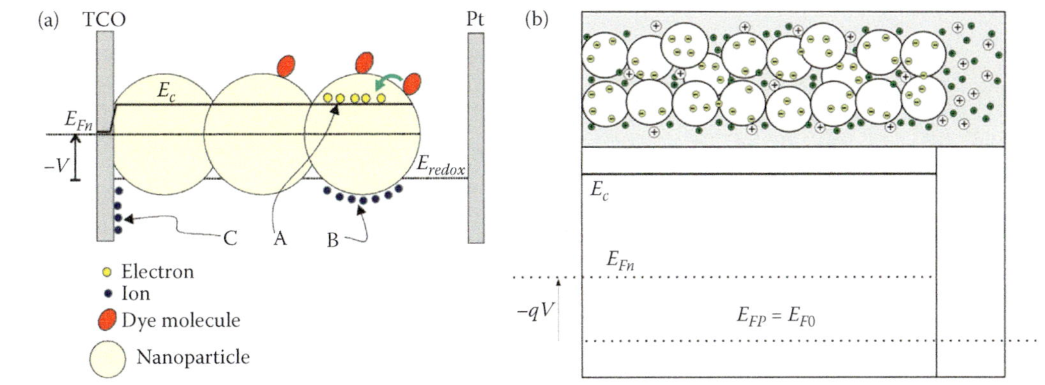

FIGURE 8.12 (a) Scheme of a dye-sensitized nanostructured semiconductor, consisting on dye molecules adsorbed on nanoparticulate framework that is deposited over a transparent conducting oxide (TCO). Photoinjection, or application of voltage, increases the chemical potential (concentration) of electrons in the TiO_2 phase, A. The electron Fermi level, E_{Fn}, is displaced with respect to the lower edge of the conduction band, E_c. The electrode potential, V, is given by the difference between E_{Fn} and the redox level E_{redox}. The increasing negative charge in the semiconductor nanoparticles is compensated by positive ionic charge at the surface, B. With the change of E_{Fn} also changes the electrostatic potential of the Helmholtz layer and semiconductor band bending at the interface between the exposed surface of the transparent conducting oxide substrate and the electrolyte, C. (b) Energy diagram showing electron accumulation in a nanocrystalline semiconductor electrode, associated to the increase of the Fermi level with respect to equilibrium level E_{F0}, and the compensating positive charge in the electrolyte as well as freely diffusing cations.

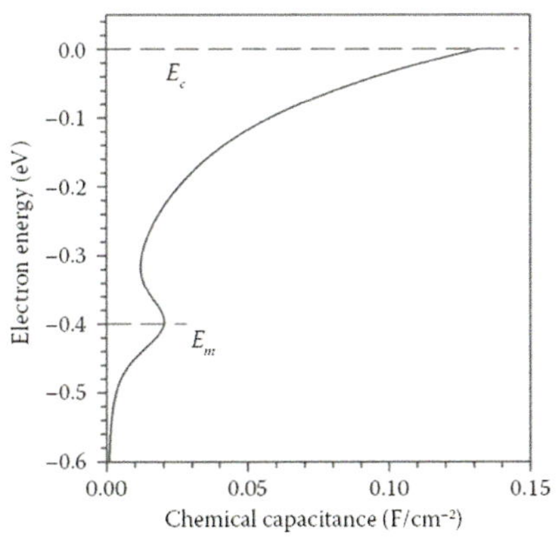

FIGURE 8.13 Representation of the chemical capacitance of an exponential distribution of states with density $N_{exp} = 10^{20}$ cm^{-3} below the lower edge of the conduction band ($E_C = 0$ eV) and a monoenergetic energy level with density $N_m = 10^{18}$ cm^{-3} ($E_m = -0.4$ eV) at temperature $T = 300$ K.

corresponding to a nearly monoenergetic localized state situated inside the gap.

The exponential DOS can directly be measured by determination of the chemical capacitance of traps in the bandgap, for example, applying Equation 8.14. Following the Equations 7.23 and 8.19, the chemical capacitance displays an exponential dependence on the Fermi level:

$$c_\mu = \frac{N_L q^2}{k_B T_0} \exp\left[\left(E_{Fn} - E_c\right)/k_B T_0\right] \tag{8.20}$$

Taking into account the relationship of Equation 5.52 of voltage to the Fermi level, we obtain

$$c_\mu = \frac{n_0 q^2}{k_B T_0} \exp\left[qV/k_B T_0\right] \tag{8.21}$$

where n_0 is given in Equation 5.19 (replacing N_c by N_L). Equation 8.21 shows that the chemical capacitance of the exponential DOS displays a slope q/k_BT_0 in log-linear representation with respect to voltage, independent of the temperature (Wang et al., 2006). Examples of the chemical capacitance, showing the exponential DOS, measured by IS in a collection of DSC, in a voltage scale with respect to redox potential I_3^-/I^-, are shown in Figure 8.14. This capacitance is routinely observed and reported for nanostructured TiO_2 in electrolyte solution, using several techniques such as IS, CV, and electron extraction as a function of open-circuit potential. The bandtail parameter has a characteristic value $T_0 \approx 1000$ K.

As mentioned before in a slow CV (to avoid kinetic distortions), the capacitance is obtained from the voltammetry current by Equation 8.8. Thus, the CV current is directly proportional to the DOS shape, Equation 8.14, provided that the capacitive behavior, consisting of a symmetric cyclic voltammogram with respect to voltage axis, is observed. In nanostructured TiO_2 in aqueous solution, the symmetric shape of CV with the exponential increase is indeed obtained in good quality electrodes, and appears in a large number of reports, since the first one reported by Kavan et al. (1995). Some examples are shown in Figure 8.15. The symmetric shape is

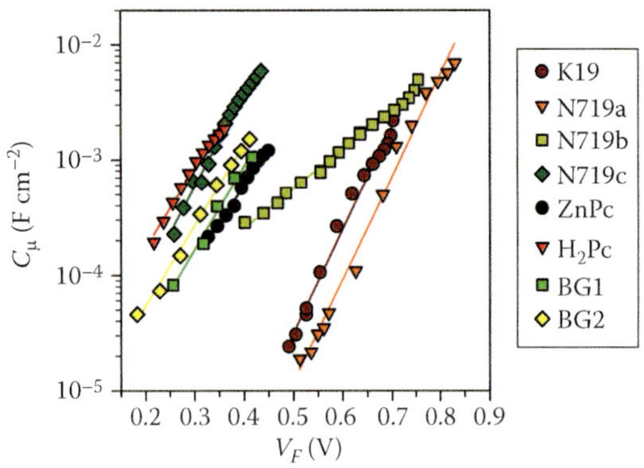

FIGURE 8.14 The chemical capacitance of TiO_2 nanostructured electrodes used in DSC in different conditions of the electrolyte and the absorbed dye. The reference of voltage is I_3^-/I^-. (Reproduced with permission from Fabregat-Santiago, F. et al. *Physical Chemistry Chemical Physics* 2011, *13*, 9083–9118.)

sometimes distorted due to the modification of the voltage scale by a series resistance *IR* term.

Another component of the chemical capacitance is the "conduction band capacitance" that was defined in Equation 7.14. This capacitance is determined by the Boltzmann tail above the Fermi level, Figure 7.8, and it should show a slope q/k_BT in log-linear representation with respect to voltage, strongly dependent on the temperature. A transition from the traps to the conduction band capacitance should be expected as the Fermi level approaches the conduction band, but this feature is not usually observed, as the negative potential induces strong charging which causes a shift of the conduction band (see Section 9.11). It is however possible to observe the conduction band capacitance using crystalline silicon. An example in which the slope of the capacitance is very close to k_BT at forward bias (when the depletion layer of the Si zone is overcome) is shown in Figure 8.16a.

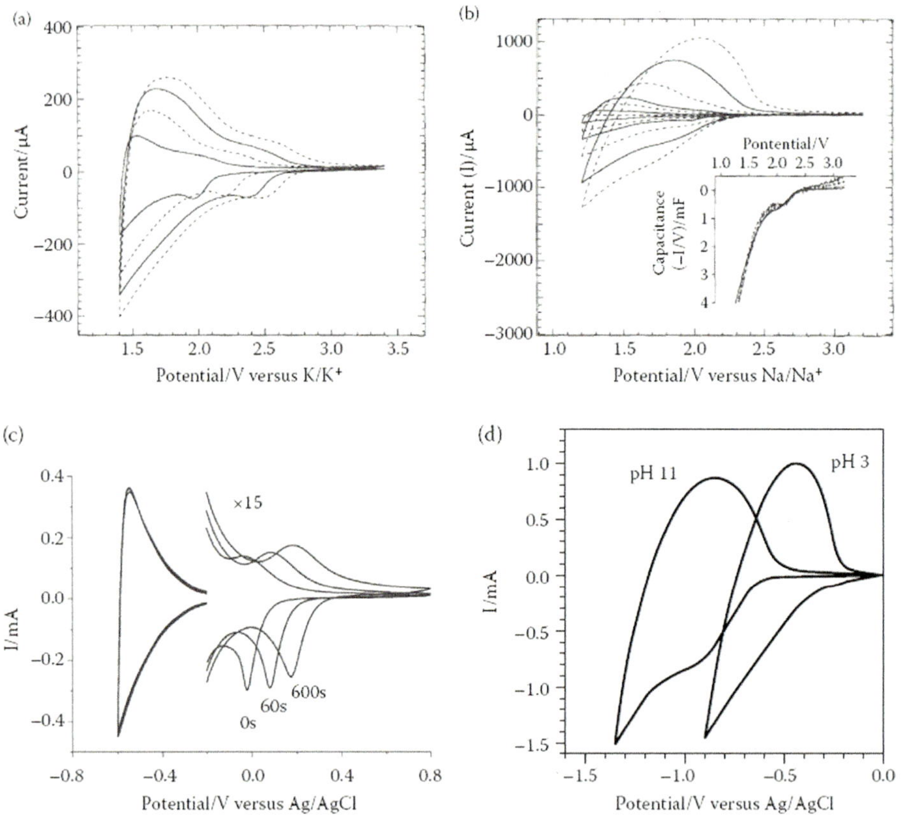

FIGURE 8.15 Cyclic voltammograms of nanostructured TiO_2 electrodes in different conditions. (a) In 0.5 M KCF_3SO_3 + propylene carbonate at a scan rate of 0.1 V s^{-1}. The curves (from right to left) correspond to application of potential scans between 1.4 and 3.5 V for 0, 2, 5 and 24 h. (Reproduced with permission from Kavan, L.; Kratochvilova, K.; Grätzel, M. *Journal of Electroanalytical Chemistry* 1995, *394*, 93–102.) (b) In 0.5 M NaCF3SO3$^+$ propylene carbonate at scan rates ν of 1000, 500, 200, 100, 50, 20, and 0.5 mV s^{-1}. The inset shows *I/V* versus potential for the negative sweep at 50, 20, 10, and 5 mV s^{-1}. (Reproduced with permission from Kavan, L.; Kratochvilova, K.; Grätzel, M. *Journal of Electroanalytical Chemistry* 1995, *394*, 93–102.) (c) (50% anatase + 50% rutile) in a solution 0.1 M HCl_4 purged with N_2 as a function of electroreduction time at −0.6 V. (Reproduced with permission from Berger, T.; *The Journal of Physical Chemistry C* 2007, *111*, 9936–9942.) (d) CVs of nanocrystalline TiO_2 in aqueous solution at different pHs.

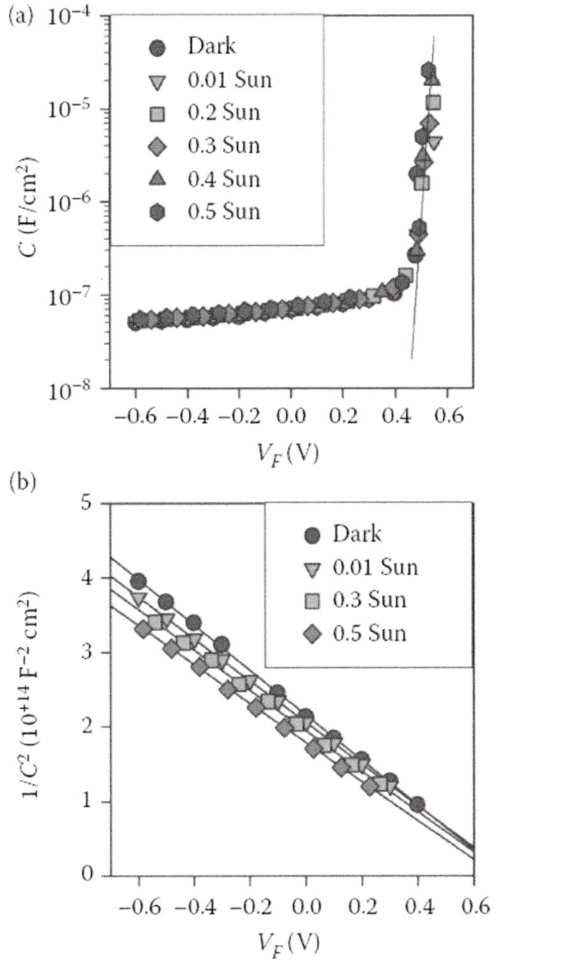

FIGURE 8.16 (a) Capacitance and (b) Mott–Schottky plot of a crystalline silicon solar cell (p-silicon wafer, N_A = 4 × 10^{16} cm^{-3}). The line in (a) indicates the region where the chemical capacitance of minority carriers (with slope $(k_BT)^{-1}$) is measured. (Reprinted with permission from Mora-Seró, I. et al. *Energy and Environmental Science* 2009, *2*, 678–686.)

The third feature remarked in Figure 8.13 is the localized peak measured in the nanocrystalline TiO_2 as shown in several examples of Figure 8.15, associated with the filling of the surface states below the conduction band edge. These states are interpreted as under-coordinated surface Ti^{4+} atoms mainly lying at the (100) edges found at the intersections between (101) surfaces of anatase nanocrystals (Nunzi et al., 2013). Experimentally it has been observed that the surface state changes by electrochemical manipulation, as shown in Figures 8.15c and 8.17. It is likely that the observed peak is influenced by kinetic conditions of the measurement (Bertoluzzi et al., 2014).

The characteristic shape of the capacitance of a nanostructured TiO_2 electrode used in DSC is shown in Figure 8.18. The capacitance as measured is shown in Figure 8.18a and it displays an exponentially raising region at intermediate voltage, associated with the chemical capacitance, and in addition, two regions of nearly constant capacitance at extreme voltages. As mentioned before, the potential drop at a series resistance is an important effect that distorts the voltage axis at large current density. When the voltage is corrected to V_F, Equation 8.18, it is possible to observe the exponential shape of the chemical capacitance, as indicated in Figures 8.14 and 8.18b.

The capacitance in Figure 8.18a shows additional features that are not related to the chemical capacitance of the semiconductor, and they will be more fully treated in Chapter 9. At low voltage applied to the TiO_2 electrode, a nearly constant capacitance is found, $c_{substrate}$, corresponding to the exposed surface of the conducting substrate. This feature is observed also in the voltammetries reported in previous figures. At these voltages the TiO_2 nanostructure shows a negligible electron density, and it is not electronically active to participate in the measured

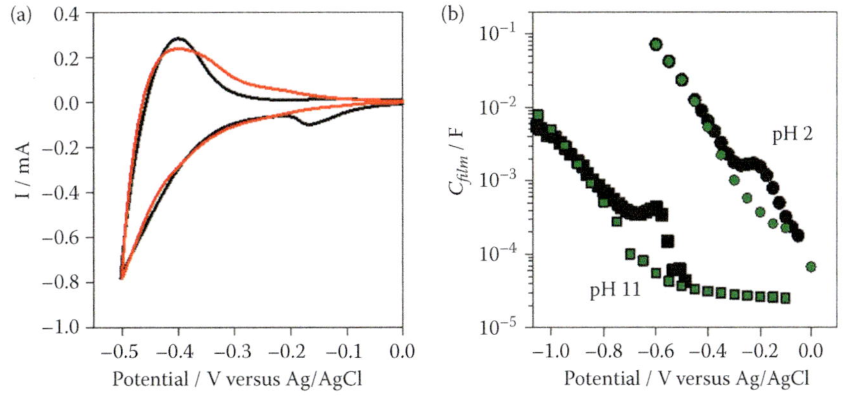

FIGURE 8.17 (a) Cyclic voltammograms in aqueous solution at pH 2 and (b) capacitance from IS at pH 2 and pH 11 of nanostructured TiO_2 electrode. The peak at low potentials disappears after aging the sample.

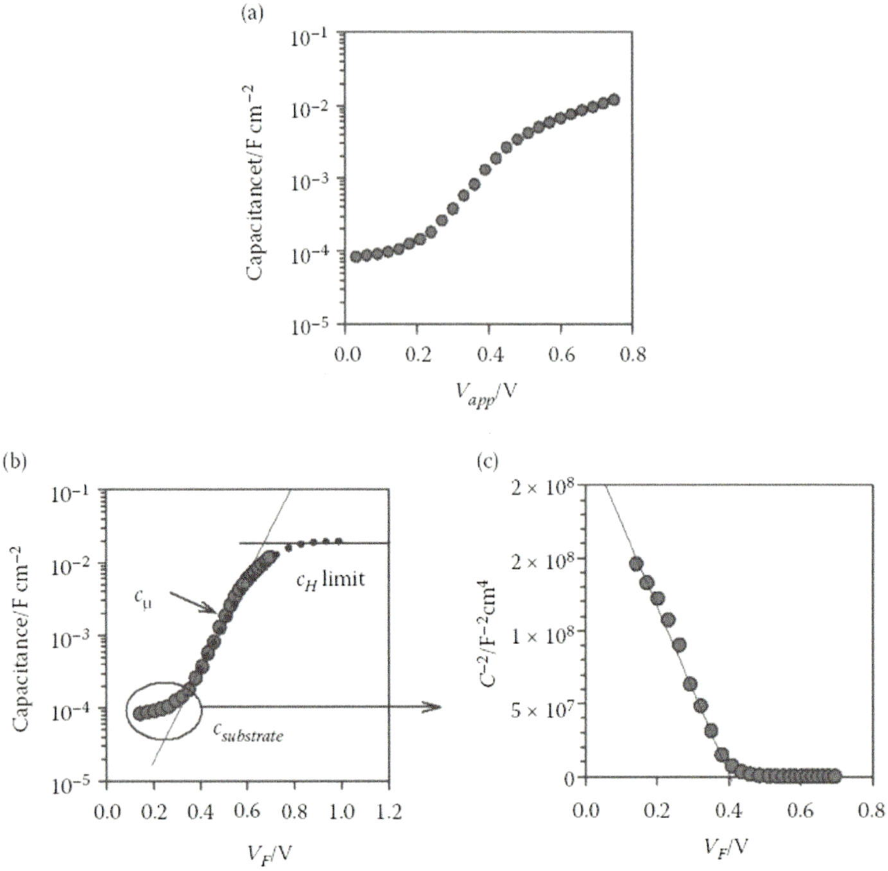

FIGURE 8.18 The capacitance of a DSC as a function of voltage. (a) Measured capacitance as a function of cell potential. (b) The capacitance versus the potential corrected for series resistance drop. (c) Mott-Shottky plot of the low voltage region. Characteristics: Thickness of TiO_2 layer 10 μm; porosity 0.5, area 0.5 cm^2; dye N719; electrolyte 0.5 M methylpropyl-imidazolium iodide, 0.5 M tert-butylpyridine, and 0.05 M I2 in 3-methoxypropionitrile. The capacitance used to describe the saturation region at high voltage is $c_H = 20$ μF cm^{-2} × *rf* where roughness factor is 1000. (Data courtesy of Francisco Fabregat-Santiago.)

capacitance. The capacitance therefore corresponds to the interface between the conducting substrate and the electrolyte, as indicated by C in Figure 8.12. It is a depletion capacitance and exhibits Mott–Schottky characteristics as shown in Figure 8.18c (Fabregat-Santiago et al., 2003a). Finally, at very high potentials, close to the conduction band edge, the capacitance is partially controlled by Helmholtz capacitance at the inner surface, as indicated by B in Figure 8.12, and becomes nearly constant with voltage. Here, the film enters into band unpinning, Section 9.11.

8.5 BASIC PROPERTIES OF ORGANIC LAYERS

Organic compounds are materials based on carbon that have traditionally been regarded as being insulating. With the development of materials highly conjugated, this picture has changed dramatically and, presently, organic-based materials provide a variety of excellent characteristics to realize optoelectronic and electronic devices. They combine electronic properties of semiconductors with the advantages of plastics such as the versatility of chemical synthesis, flexibility, and ease of processing. These features make possible their production in large areas at low processing costs. The scope of their applications is vast as they can be used as organic light-emitting diodes (OLEDs), organic field effect transistors (OFETs), solar cells, sensors, e-paper, and radio frequency identification tags.

Organic semiconducting materials include both small molecular units and *polymers*. Polymers are large molecules made up of simple repeating units called monomers and if only a few repeating units are present these are denoted as *oligomers*. Solid organic layers used in electrooptical devices can be crystalline or amorphous. Three types of organic solids can be distinguished according to Köhler and Bässler (2009) as shown in Figure 8.19:

a. *Molecular crystals* consist of a crystal formed by molecules. They are useful to study the basic physics of the organic materials, but the crystals are brittle and require a certain minimum thickness for sampling handling.
b. *Molecular assemblies* can be deposited as an amorphous film by evaporation or spin coating. Thin amorphous films formed by low-molecular weight molecules deposited by thermal evaporation under high vacuum are employed in LEDs and solar cells, and doped molecular layers are used as selective contacts.
c. *Polymer films* are formed by conjugated polymers that can be deposited by solution-based methods such as spin coating or, preferably for inexpensive material deposition on large area substrate, printing.

To achieve electronic conductivity, we use conjugated organic molecules that contain localized σ bonds that can be regarded as the "frame" of the polymer and delocalized electrons in a series of connected *p*-orbitals, the π-bonds, formed by alternating double and single bonds. By using this, design conjugated polymers have delocalized π molecular orbitals which form quasi-continuous energy bands that can support mobile charge carriers. The three main parameters that define an organic semiconductor are the highest occupied molecular orbital (HOMO), the lowest unoccupied molecular orbital (LUMO) and the band gap (E_g). The HOMO is the highest bonding state filled by electrons that will be a π orbital. This molecular orbital is equivalent to the valence band edge in traditional inorganic semiconductors. The LUMO is the first molecular orbital empty available, represents an antibonding state π* and it is able to carry free charges behaving as the conduction band edge in inorganic semiconductors. Finally, the band gap is the difference between the HOMO and LUMO level. Figure 8.20a shows an oligomer of polythiophene with the delocalized molecular orbitals of the HOMO level. Due to the π conjugation, in the perfect isolated oligomeric chain, the delocalized π electron cloud extends along the whole length of the chain. Molecular systems show strong reorganization and polaronic effects so that the affinity and ionization energy may depart substantially from the one-electron LUMO and HOMO levels (Savoie et al., 2013).

From the synthetic point of view, organic molecules offer a great versatility in their design to render the desired properties. For example, in organic photovoltaics (OPV) both small molecular entities and polymer can be used as electron transport materials ($PCB_{60}M$, TCNQ, or PTCDA), light absorber and hole transporter (ZnTPP, DCV4T, P3HT, PCDTBT, or PTB7) as shown in Figure 8.21. Similarly, the same is true for OLEDs where the molecules can be chosen to be hole transporter (TPD), emitter (Alq3, MEH-PPV), electron transporter (Biphen), or host material (poly-9, 9′-dioctylfluorene). In general depending on the position of their HOMO and LUMO level, they can be used with a specific role and these values can greatly vary depending on the chemical structure. For example, different materials are shown in Figure 8.22 where the LUMO level values can vary over a range of 4 eV.

Extensive experience has been accumulated both in the preparation of solution-processed organic electronic devices and in the characterization of their functional properties. It has been established that the morphology of the material or blend impacts very strongly on the properties (electrical, optical, or chemical) and function of the device. Solution-processed organic materials contain diverse levels of organization, at the molecular level, the supramolecular level, and the mesoscopic blend. The

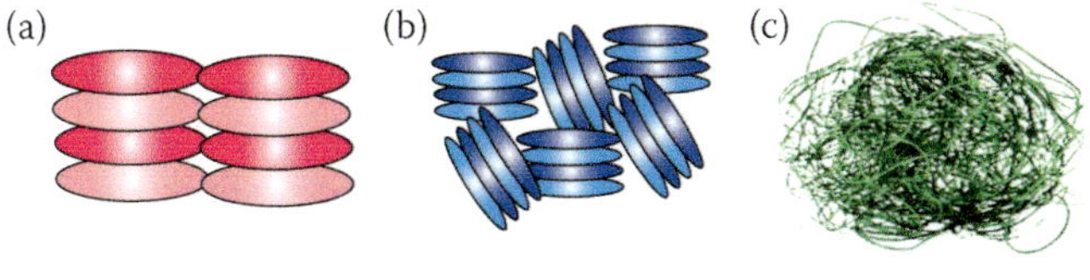

FIGURE 8.19 Different types of organic solids: (a) Molecular crystal, (b) amorphous molecular assembly, and (c) polymer films.

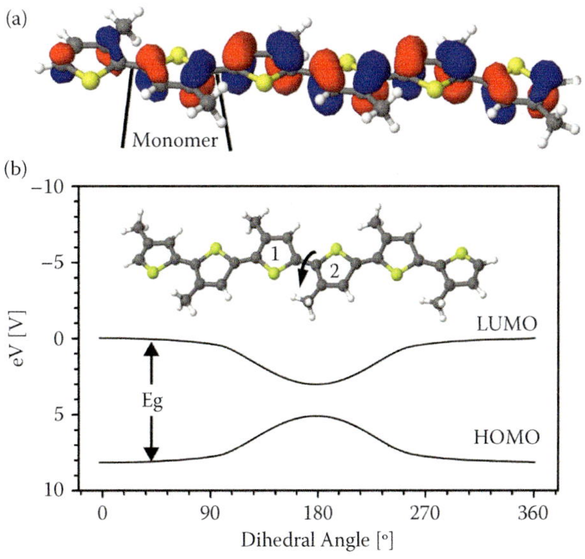

FIGURE 8.20 (a) Oligomer of thiophene showing the molecular orbitals of the HOMO level. (b) HOMO and LUMO level variation as a function of the torsion angle of the dihedral angle defined by the rings 1 and 2.

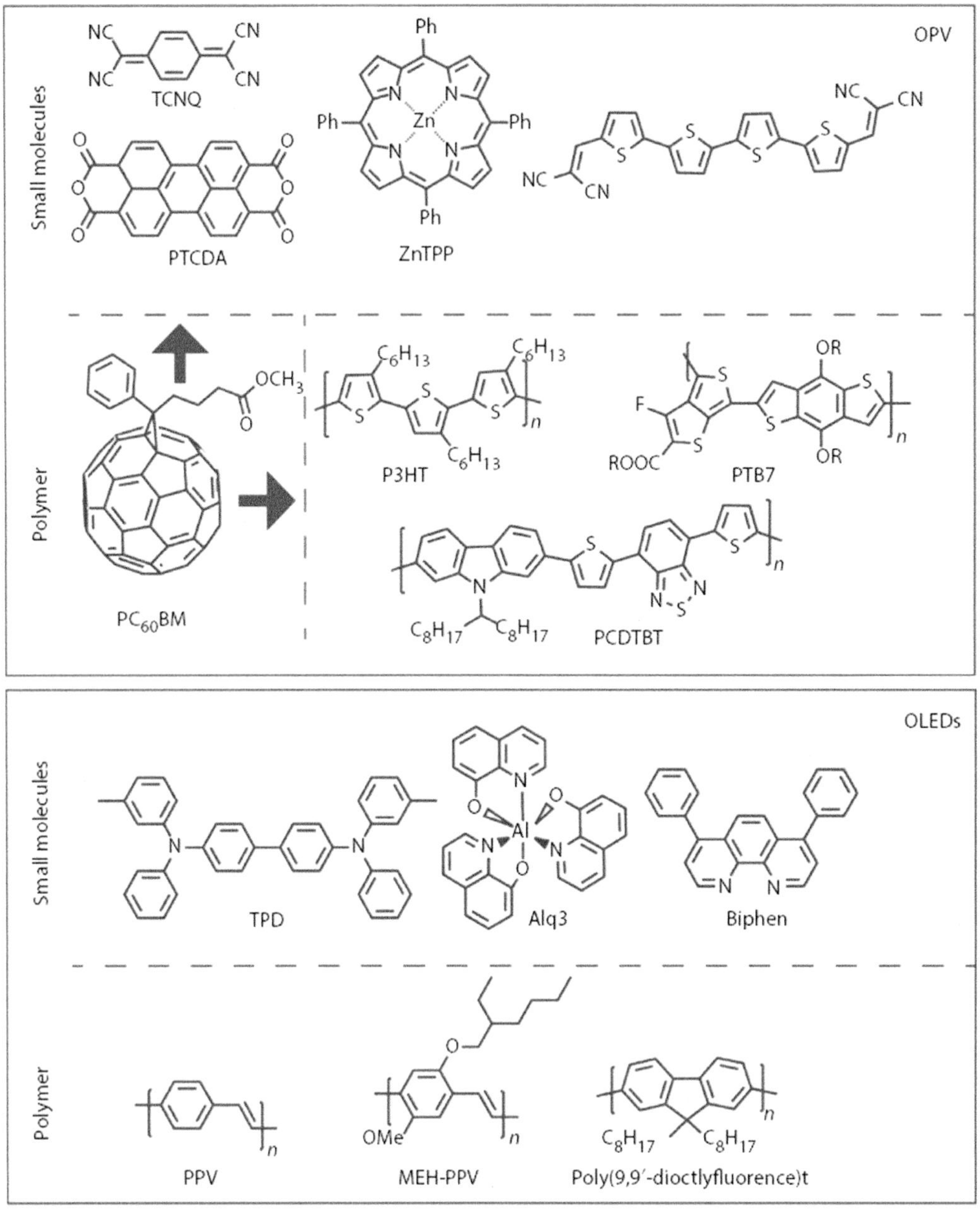

FIGURE 8.21 Representative small molecular units and polymers used in organic photovoltaics and organic LEDs.

benefits for preparation of large area organic electronic devices in terms of processability are accompanied by the difficulty to maintain a control of the morphology at all levels as required for full optimization of physical properties at bulk or interfaces (Groves et al., 2009). Properties like the mobility, or interface characteristics as the average electrical dipole, depend on the atomistic details of the structural organization. Small variations of the synthesis or process conditions can modify dramatically the electronic and optical properties of organic layers (Braun et al., 2009). For example, if we look at polymeric materials, although the chains may be quite long, the number of monomers that are actually conjugated is usually restricted to a small number (two to seven units) defining the effective conjugation length of the polymer; (see Figure 8.23a). If we then take into account that transport of carriers takes place by hopping between these conjugated units (see Figure 8.23b), then it is clear that the transport will highly depend on the morphology of the polymer. Controlling the growth and the extent of mixing, the connectivity and the crystallinity of nanosized domains in multicomponent systems (blends), is a very difficult challenge, and often requires post-processing annealing methods to equilibrate the film morphology toward the required configuration. For example, in spin-casting method, the morphology

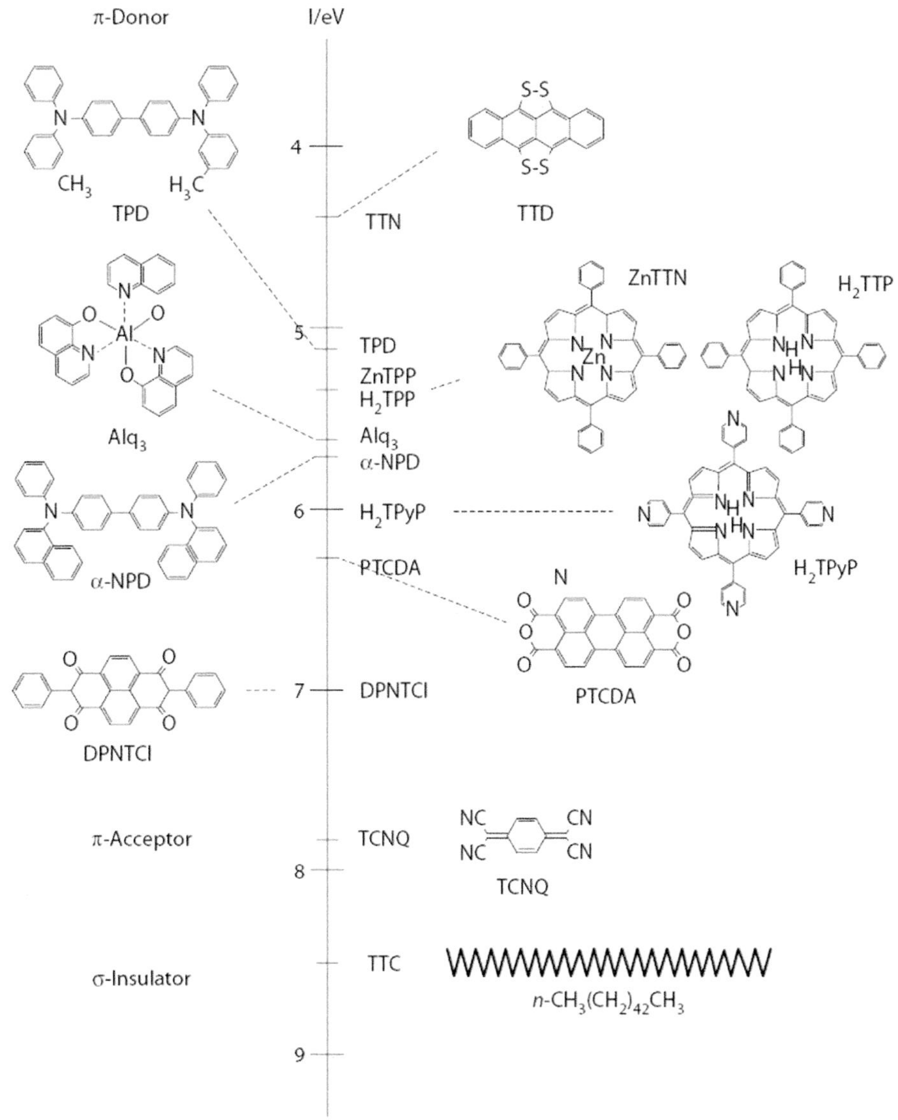

FIGURE 8.22 The structures and ionization threshold energies I (in eV) of organic materials. (Reproduced with permission from Ishii, H. et al. *Advanced Materials* 1999, *11*, 605.)

of films is strongly dependent on the solvent used. The polymer chains can become entangled before the evaporation of the solvent is complete and the degree of ordering domains is affected (Sheridan et al., 2000).

The classical picture of a polymer chain is that of spaghetti where torsion of the chains or kinks partially break the conjugation at different parts of the polymer chain (see Figure 8.19c). HOMO and LUMO are highly structure-dependent, as shown in Figure 8.20b. In this graph the variation of the HOMO and LUMO level is represented as a function of the dihedral angle defined by the rings 1 and 2 of this oligomer of polythiophene. It is observed that the interaction between neighboring units is the strongest when both rings are in the same plane (dihedral angle = 0°) and the system is more delocalized being smaller than the energy gap between the HOMO and LUMO. Additionally, in real materials other sources of defects/impurities may be present, that is, atoms eliminating the double bonds among others. Therefore, it is clear why solution-processed organic materials are usually disordered or amorphous. They exhibit energy disorder to a large extent, and some properties of the energy distribution of states, the DOS, are discussed in the next section. The length of the polymer segment may vary randomly and that is a major reason for energetic disorder implying inhomogeneous properties and a relatively broad DOS.

In organic semiconductors, the Fermi level can be shifted toward the LUMO conduction band by *n*-doping, or toward the HOMO valence band by *p*-doping. These modifications allow to obtain higher conductivity or to form the desired type on junction at the interface with

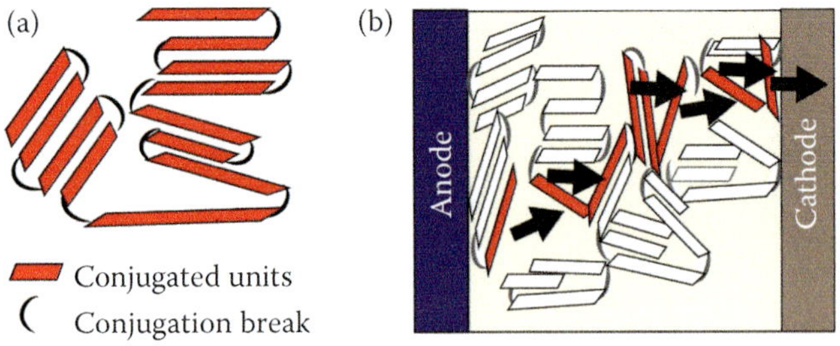

FIGURE 8.23 (a) A sketch of a polymer that shows the units that define the conjugation length and structural defects that break this conjugation. (b) A diagram that shows the typical hopping mechanism of transport in organic molecules.

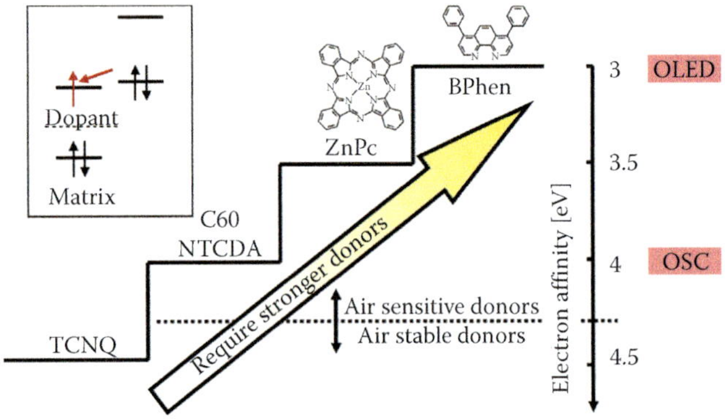

FIGURE 8.24 n-Type doping of molecular materials relevant for OLED and organic solar cells requires different electron affinities. While organic solar cells may be doped with donors having a HOMO around 4.0 eV, OLEDs require stronger donors with a HOMO at about 3.0 eV. Such materials are increasingly unstable in air, which requires handling under protective atmosphere only. The "staircase" shows typical host materials for OLEDs and organic solar cells. (Reproduced with permission from Walzer, K. et al. *Chemical Reviews* 2007, *107*, 1233–1271.)

another material (Mayer et al., 2012). Doping is especially useful for utilization as injection and transport layers that form selective contacts in OLEDS and organic solar cells (Walzer et al., 2007). In general, the organic materials have a high HOMO and it is not difficult to find a suitable electron acceptor that will remove electrons from the valence band leaving a hole behind. In this way, the conductivity can be increased enormously. In contrast, for efficient n-doping of an organic material, the HOMO level of the dopant must be energetically above the LUMO level of the material, which makes such dopant material unstable against oxygen. For solar cells, the electron transport material does not have a very high lying LUMO, and the formation of electron transport layers is feasible, but OLED materials have a rather low electron affinity of around 3.0 eV, and the dopants become oxidized by air. The range of electron affinities

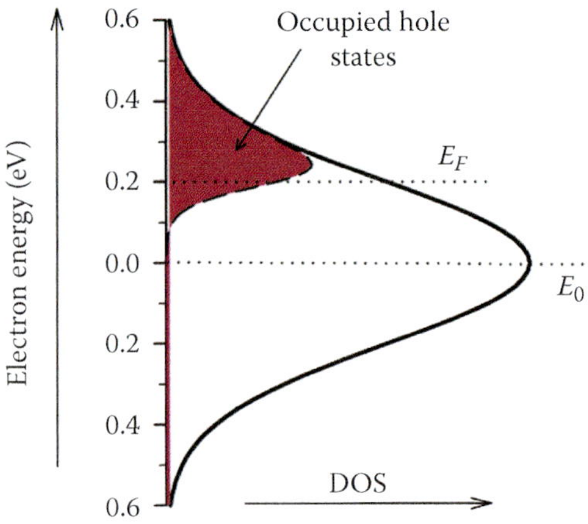

FIGURE 8.25 Representation of a Gaussian DOS (center of the DOS $E_0 = 0$ eV, and width parameter $\sigma_1 = 0.2$ eV). The shaded area shows the occupancy of the DOS with holes when the Fermi level is at 0.2 eV at $T = 300$ K.

in the two cases as well as the domain of stability is shown in Figure 8.24.

8.6 THE GAUSSIAN DOS

A frequent shape of the DOS in disordered organic semiconductors is a Gaussian distribution (Bässler, 1993)

$$g(E,E_0) = \frac{N_L}{\sqrt{2\pi}\sigma_1}\exp\left[-\frac{(E_0-E)^2}{2\sigma_1^2}\right] \tag{8.22}$$

where N_L is the volume density of localized states, E_0 is the center of the distribution and σ_1 is the disorder parameter that gives the width. This type of distribution is usually attributed to disorder or structural correlations with correlation lengths of a few intermolecular distances that lead to a dispersion of energies. Correlations can be caused by the fluctuation of the lattice polarization energies, dipole interactions, and molecular geometry fluctuations.

Figure 8.25 shows the full occupation (by holes) of the upper tail of the DOS. Figure 8.26 shows the properties of charging the Gaussian DOS with electrons. The carrier density is found as

$$n = \int_{-\infty}^{+\infty} g(E) f(E - E_{Fn})\,dE \tag{8.23}$$

where f is the Fermi–Dirac function. The chemical capacitance is given by Equation 7.19. It is convenient to distinguish two regimes, a high (>10%) charging level, and a low charging level. At high charging the capacitance

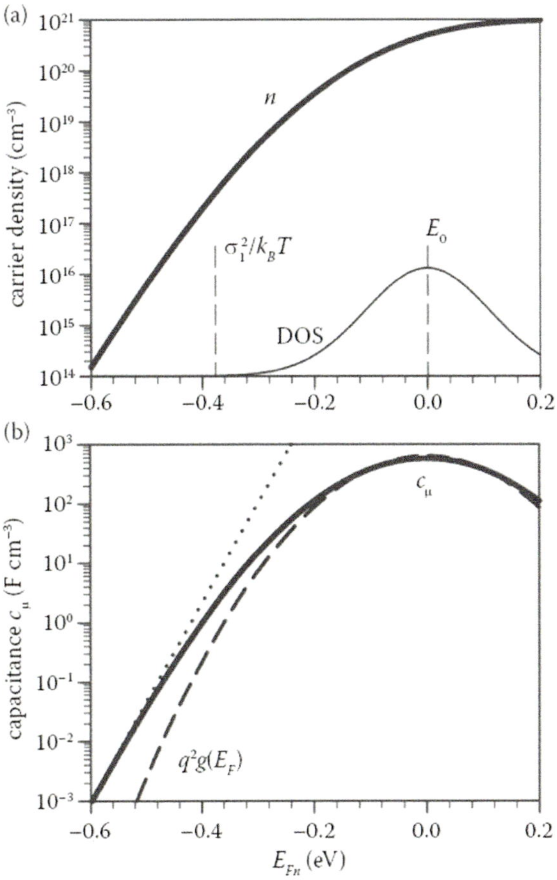

FIGURE 8.26 Carrier accumulation in a Gaussian DOS as a function of the Fermi level of electrons. (a) Electron density. The Gaussian DOS is also shown in linear scale. (b) Capacitance. The thin dashed line is the capacitance according to the approximation $c_\mu = q^2g(E_F)$. The thin line is the low-density approximation (exponential function). The following set of parameters was used in the calculation: $N_{vol} = 1.0 \times 10^{21}$ cm^{-3}, $E_0 = 0$eV, $\sigma_1 = 0.1$ eV, $T = 300$ K.

reflects the shape of the DOS, Figure 8.26b, and displays a parabolic form in semi-logarithmic representation. This is expected in the standard approximation (7.23) used in charging experiments to determine the DOS, as stated in Equation 8.15.

We also compute the carrier distribution when the Fermi level is low enough below E_0 that the occupancy is well described by Boltzmann distribution. The carrier distribution is given by

$$n(E,E_{Fn}) = g(E,E_0)\exp\left[-(E-E_{Fn})/k_BT\right] \quad (8.24)$$

By algebraic manipulation of Equation 8.24, we obtain

$$n(E,E_{Fn}) = g(E,E_m)\exp\left[-(E_\sigma-E_{Fn})/k_BT\right] \quad (8.25)$$

where we have defined

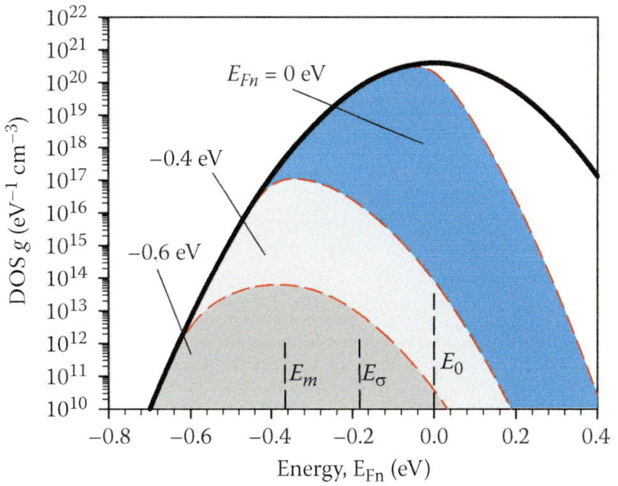

FIGURE 8.27 Charge accumulation in a Gaussian DOS centered at $E_0 = 0$ eV with dispersion $\sigma_1 = 0.1$ eV at $T = 300$ K. E_{Fn} is the Fermi level of electrons, $E_m = \sigma_1^2/k_BT$. The dashed lines indicate the occupied states at different values of Fermi level.

$$E_m = E_0 - \frac{\sigma_1^2}{k_BT} \quad (8.26)$$

and

$$E_\sigma = E_0 - \frac{\sigma_1^2}{2k_BT} \quad (8.27)$$

In Equation 8.25, valid when $E_{Fn} \ll E_m$, the exponential factor does not depend on E. Therefore, the carrier occupancy forms a Gaussian distribution of width σ_1 centered at the energy level E_m, independently of the Fermi level value (Bässler, 1993; Arkhipov et al., 2001; Coehoorn et al., 2005), as shown in Figure 8.27; see also the occupation of the lower Gaussian in Figure 7.10b at E_{Fn} = −1.4 eV. At $E_{Fn} \ll E_m$ the shape of the occupied states distribution is invariant but the total number is larger when E_{Fn} increases. In addition, using Equation 8.25 we obtain that the total carrier density in localized states is given by

$$n(E_{Fn}) = \int_{-\infty}^{+\infty} n(E,E_{Fn})dE = N_L\exp\left[-(E_\sigma-E_{Fn})/k_BT\right] \quad (8.28)$$

According to Equation 8.28, the number of carriers in a Gaussian DOS when $E_{Fn} \ll E_m$ is the same as in a monoenergetic level at E_σ with total density N_L. From Equation 7.14, the chemical capacitance has the value

$$c_\mu = \frac{q^2N_L}{k_BT}\exp\left[-(E_\sigma-E_{Fn})/k_BT\right] \quad (8.29)$$

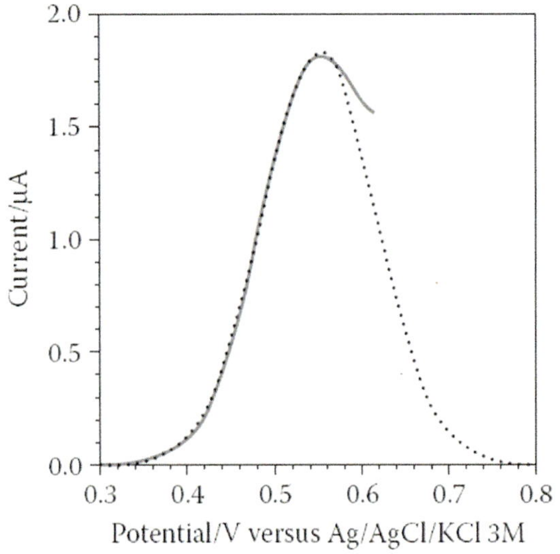

FIGURE 8.28 Comparison between CV for the first oxidation peak of OMeTAD film onto ITO substrate (solid) and fit (dotted) using a Gaussian distribution of states (σ = 65 meV, E_0 = 552 meV and $N = 2.9 \times 10^{19}$ cm^{-3}). (Reproduced with permission from García-Cañadas, J. et al. *Synthetic Metalls* 2006, *156*, 944.)

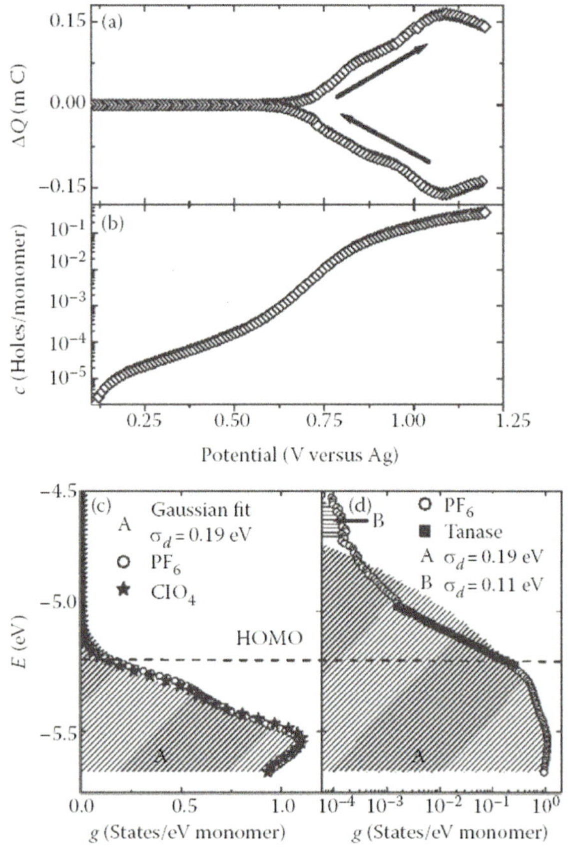

FIGURE 8.29 Top panel: Electrochemical injection of holes into a PPV film. (a) The differential charge stored in PPV when successive 10 mV steps are applied. The potential is defined with respect to the Ag reference electrode. The arrows show the two scanning directions: doping and dedoping. (b) Doping per monomer calculated from the data in (a). The reversible doping range extends over four orders of magnitude. Bottom panel: The energy dependence of the experimentally determined DOS $g(E)$ where E is with respect to the vacuum level. The horizontal dashed line marks the HOMO position found from CV (a) E versus g(E) using PF_6 and ClO_4 as anions. A Gaussian function with 0.19 eV (area A) fits the data well. (b) E versus $g(E)$ for PF_6 on a linear-log scale. The filled squares are the PPV FET—data by Tanase, C.; Meijer, E. J.; Blom, P. W. M.; de Leeuw, D. M. Unification of the hole transport in polymeric field effect-transistors and LED. *Physical Review Letters* 2003, *91*, 216601, which are well described by an exponential function. (Reproduced with permission from Hulea, I. N. et al. *Physical Review Letters* 2004, *93*, 166601.)

Therefore, when only the deep tail of the DOS is occupied, the capacitance is exponential, as indicated in Figure 8.26b. Note that the zero-temperature approximation of Equation 7.23 (which requires that only states below the Fermi level are occupied) is invalid in this region in which the majority of carriers do not lie below the Fermi level, but instead, are mainly located above the Fermi level (Arkhipov et al., 2001), symmetrically distributed around E_m, as shown in Figure 8.27.

The Gaussian shape in the distribution of electronic states in disordered organic conductors has been observed by capacitance measurements. Results of determination of the DOS by CV for a film of spiro-OMeTAD are shown in Figure 8.28. Figure 8.29 shows the Gaussian distribution measured in PPV with different counterions, in electrochemical transistor configuration, over a very wide range of doping density covering four orders of magnitude (Hulea et al., 2004). Figure 8.29d indicates that the tail of the Gaussian is well described as an exponential distribution (Tanase et al., 2004).

GENERAL REFERENCES

DOS of amorphous semiconductors: Cody et al. (1981), Orenstein and Kastner (1981), Tiedje and Rose (1981), Marshall et al. (1986), Shaw and Hack (1988), Street (1991), Tang et al. (1995), Schiff (2004).

DOS of nanocrystalline TiO_2: Peter et al. (2002), Bisquert (2003), Fabregat-Santiago et al. (2003, 2005, 2011), Bailes et al. (2005), Boschloo and Hagfeldt (2005), Niinobe et al. (2005), Zhang et al. (2005), Barea et al. (2010), Wang et al. (2006), Berger et al. (2007, 2012), Bisquert et al. (2008), Nunzi et al. (2013).

DOS of organic materials: Richert et al. (1989), Bässler (1993), Dieckmann et al. (1993), Arkhipov et al. (2001), Yu et al. (2001), Bisquert et al. (2004), Hulea et al. (2004), Coehoorn et al. (2005), García-Cañadas et al. (2006).

TiO_2 surface states: Kavan et al. (1995), de Jongh and Vanmaekelbergh (1997), Boschloo and Fitzmaurice (1999), Wang et al. (2001), Wang and Lewis (2009).

REFERENCES

Abayev, I.; Zaban, A.; Fabregat-Santiago, F.; Bisquert, J. Electronic conductivity in nanostructured TiO_2 films permeated with electrolyte. *Physics Status Solidii (a)* 2003, *196*, R4–R6.

Arkhipov, V. I.; Heremans, P.; Emelianova, E. V.; Adriaenssens, G. J. SCLC in materials with Gaussian energy distribution of localized states. *Applied Physics Letters* 2001, *79*, 4154.

Bailes, M.; Cameron, P. J.; Lobato, K.; Peter, L. M. Determination density and energetic distribution of electron traps in dye-sensitized nanocrystalline solar cells. *The Journal of Physical Chemistry B* 2005, *109*, 15429–15435.

Balberg, I.; Gal, E. Determination of distribution of states in hydrogenated amorphous silicon from capacitance–voltage characteristics. *Journal of Applied Physics* 1985, *58*, 2617–2627.

Barea, E. M.; Ortiz, J.; Payá, F. J.; Fernández-Lázaro, F.; Fabregat-Santiago, F.; Sastre-Santos, A.; Bisquert, J. Energetic factors governing injection, regeneration and recombination in dye solar cells with phthalocyanine sensitizers. *Energy and Environmental Science* 2010, *3*, 1985–1994.

Bässler, H. Charge transport in disordered organic photoconductors. *Physics Status Solidii (b)* 1993, *175*, 15–56.

Berger, T.; Lana-Villarreal, T.; Monllor-Satoca, D.; Gomez, R. An electrochemical study on the nature of trap states in nanocrystalline rutile thin films. *The Journal of Physical Chemistry C* 2007, *111*, 9936–9942.

Berger, T.; Monllor-Satoca, D.; Jankulovska, M.; Lana-Villarreal, T.; Gomez, R. The electrochemistry of nanostructured titanium dioxide electrodes. *ChemPhysChem* 2012, *13*, 2824–2875.

Bertoluzzi, L.; Herraiz-Cardona, I.; Gottesman, R.; Zaban, A.; Bisquert, J. Relaxation of electron carriers in the density of states of nanocrystalline TiO_2. *The Journal of Physical Chemistry Letters* 2014, *5*, 689–694.

Bisquert, J. Chemical capacitance of nanostructured semiconductors: Its origin and significance for heterogeneous solar cells. *Physical Chemistry Chemical Physics* 2003, *5*, 5360–5364.

Bisquert, J.; Fabregat-Santiago, F.; Mora-Seró, I.; Garcia-Belmonte, G.; Barea, E. M.; Palomares, E. A review of recent results on electrochemical determination of the density of electronic states of nanostructured metal-oxide semiconductors and organic hole conductors. *Inorganica Chimica Acta* 2008, *361*, 684–698.

Bisquert, J.; Garcia-Belmonte, G.; García-Cañadas, J. Effect of the Gaussian energy dispersion on the statistics of polarons and bipolarons in conducting polymers. *The Journal of Chemical Physics* 2004, *120*, 6726–6733.

Boschloo, G.; Fitzmaurice, D. Spectroelectrochemical investigation of surface states in nanostructured TiO_2 electrodes. *The Journal of Physical Chemistry B* 1999, *103*, 2228–2231.

Boschloo, G.; Hagfeldt, A. Activation energy of electron transport in dye-sensitized TiO_2 solar cells. *The Journal of Physical Chemistry B* 2005, *109*, 12093.

Braun, S.; Salaneck, W. R.; Fahlman, M. Energy-level alignment at organic/metal and organic/organic interfaces. *Advanced Materials* 2009, *21*, 1450–1472.

Burke, A. Ultracapacitors: Why, how, and where is the technology. *Journal of Power Sources* 2000, *91*, 37–50.

Burke, L. D.; Healy, J. F. The importance of reactive surface groups with regard to the electrocatalytic behaviour of oxide (especially RuO_2) anodes. *Journal of Electroanalytical Chemistry* 1981, *124*, 327.

Cody, G. D.; Tiedje, T.; Abeles, B.; Brooks, B.; Goldstein, Y. Disorder and the optical-absorption edge of hydrogenated amorphous silicon. *Physical Review Letters* 1981, *47*, 1480–1483.

Coehoorn, R.; Pasveer, W. F.; Bobbert, P. A.; Michels, C. J. Charge-carrier concentration dependence of the hopping mobility in organic materials with Gaussian disorder. *Physical Review B* 2005, *72*, 155206.

Conway, B. E.; Pell, W. G. Double-layer and pseudocapacitance types of electrochemical capacitors and their applications to the development of hybrid devices. *Journal of Solid State Electrochemistry* 2003, *7*, 637–644.

de Jongh, P. E.; Vanmaekelbergh, D. Investigation of the electronic properties of nanocrystalline particulate TiO_2 electrodes by intensity-modulated photocurrent spectroscopy. *The Journal of Physical Chemistry B* 1997, *101*, 2716–2722.

Dieckmann, A.; Bässler, H.; Borsenberg, P. M. An assessment of the role of dipoles on the density-of-states function of disordered molecular solids. *The Journal of Chemical Physics* 1993, *99*, 8136.

Dubarry, M.; Liaw, B. Y. Identifying capacity fading mechanism in a commercial LiFePO4 cell. *Journal of Power Sources* 2009, *194*, 541–549.

Fabregat-Santiago, F.; Bisquert, J.; Garcia-Belmonte, G.; Boschloo, G.; Hagfeldt, A. Impedance spectroscopy study of the influence of electrolyte conditions in parameters of transport and recombination in dye-sensitized solar cells. *Solar Energy Materials and Solar Cells* 2005, *87*, 117–131.

Fabregat-Santiago, F.; Garcia-Belmonte, G.; Bisquert, J.; Bogdanoff, P.; Zaban, A. Mott-Schottky analysis of nanoporous semiconductor electrodes in the dielectric state deposited on $SnO_2(F)$ conducting substrates. *Journal of the Electrochemical Society* 2003a, *150*, E293–E298.

Fabregat-Santiago, F.; Garcia-Belmonte, G.; Mora-Seró, I.; Bisquert, J. Characterization of nanostructured hybrid and organic solar cells by impedance spectroscopy. *Physical Chemistry Chemical Physics* 2011, *13*, 9083–9118.

Fabregat-Santiago, F.; Mora-Seró, I.; Garcia-Belmonte, G.; Bisquert, J. Cyclic voltammetry studies of nanoporous semiconductor electrodes. Models and application to nanocrystalline TiO_2 in aqueous electrolyte. *The Journal of Physical Chemistry B* 2003b, *107*, 758–769.

Garcia-Belmonte, G.; Boix, P. P.; Bisquert, J.; Lenes, M.; Bolink, H. J.; La Rosa, A.; Filippone, S.; et al. Influence of the intermediate density-of-states occupancy on open-circuit voltage of bulk heterojunction solar cells with different fullerene acceptors. *Journal of Physical Chemistry Letters* 2010, *1*, 2566–2571.

García-Cañadas, J.; Fabregat-Santiago, F.; Bolink, H.; Palomares, E.; Garcia-Belmonte, G.; Bisquert, J. Determination of electron and hole energy levels in mesoporous nanocrystalline TiO_2 solid-state dye solar cell. *Synthetic Metalls* 2006, *156*, 944.

Goetzberger, A.; Heine, V.; Nicollian, E. H. Surface states in silicon from charges in the oxide coating. *Applied Physics Letters* 1968, *12*, 95–97.

Goodenough, J. B.; Manoharan, R.; Paranthaman, M. Surface protonation and electrochemical activity of oxides in aqueous solution. *Journal of the American Chemical Society* 1990, *112*, 2076–2082.

Greef, R.; Peat, R.; Peter, L. M.; Pletcher, D.; Robinson, J. *Instrumental Methods of Electrochemistry*; Ellis Horwood Limited: Chichester, 1985.

Groves, C.; Reid, O. G.; Ginger, D. S. Heterogeneity in polymer solar cells: Local morphology and performance in organic photovoltaics studied with scanning probe microscopy. *Accounts of Chemical Research* 2009, *43*, 612–620.

Guo, M. D.; Guo, H. X. Voltammetric behaviour study of creatinine at phosphomolybdic-polypyrrole film modified electrode. *Journal of Electroanalytical Chemistry* 2005, *585*, 28–34.

Heinze, J.; Frontana-Uribe, B.; Ludwigs, S. Electrochemistry of conducting polymers: Persistent models and new concepts. *Chemical Reviews* 2010, *110*, 4724–4771.

Hoyer, P.; Weller, H. Potential dependent electron injection in nanoporous colloidal ZnO films. *The Journal of Physical Chemistry* 1995, *99*, 14096–14100.

Hubbard, A. T. Study of the kinetics of electrochemical reactions by thin-layer voltammetry: I. Theory. *Journal of Electroanalytical Chemistry and Interfacial Electrochemistry* 1969, *22*, 165–174.

Hulea, I. N.; Brom, H. B.; Houtepen, A. J.; Vanmaekelbergh, D.; Kelly, J. J.; Meulenkamp, E. A. Wide energy-window view on the density of states and hole mobility in poly(p-phenylene vinylene). *Physical Review Letters* 2004, *93*, 166601.

Ishii, H.; Sugiyama, K.; Ito, E.; Seki, K. Energy level alignment and interfacial electronic structures at organic/metal and organic/organic interfaces. *Advanced Materials* 1999, *11*, 605.

Kavan, L.; Kratochvilova, K.; Grätzel, M. Study of nanocrystalline TiO_2 (anatase) electrode in the accumulation regime. *Journal of Electroanalytical Chemistry* 1995, *394*, 93–102.

Köhler, A.; Bässler, H. Triplet states in organic semiconductors. *Materials Science and Engineering, R: Reports* 2009, *66*, 71–109.

Krellner, C.; Haas, S.; Goldmann, C.; Pernstich, K. P.; Gundlach, D. J.; Batlogg, B. Density of bulk trap states in organic semiconductor crystals: Discrete levels induced by oxygen in rubrene. *Physical Review B* 2007, *75*, 245115.

Kubacka, A.; Fernandez-Garcia, M.; Colon, G. Advanced nanoarchitectures for solar photocatalytic applications. *Chemical Reviews* 2012, *112*, 1555–1614.

Marshall, J. M.; Street, R. A.; Thompson, M. J. Electron drift mobility in amorphous Si:H. *Philosophical Magazine B* 1986, *54*, 51–60.

Mayer, T.; Hein, C.; Mankel, E.; Jaegermann, W.; Muller, M. M.; Kleebe, H.-J. Fermi level positioning in organic semiconductor phase mixed composites: The internal interface charge transfer doping model. *Organic Electronics* 2012, *13*, 1356–1364.

Monroe, D.; Kastner, M. A. Exactly exponential band tail in a glassy semiconductor. *Physical Review B* 1986, *33*, 8881–8884.

Mora-Seró, I.; Garcia-Belmonte, G.; Boix, P. P.; Vázquez, M. A.; Bisquert, J. Impedance characterisation of highly efficient silicon solar cell under different light illumination intensities. *Energy and Environmental Science* 2009, *2*, 678–686.

Niinobe, D.; Makari, Y.; Kitamura, T.; Wada, Y.; Yanagida, S. Origin of enhancement in open-circuit voltage by adding ZnO to nanocrystalline SnO_2 in dye-sensitized solar cells. *The Journal of Physical Chemistry B* 2005, *109*, 17892–17900.

Nunzi, F.; Mosconi, E.; Storchi, L.; Ronca, E.; Selloni, A.; Grätzel, M.; De Angelis, F. Inherent electronic trap states in TiO_2 nanocrystals: Effect of saturation and sintering. *Energy and Environmental Science* 2013, *6*, 1221–1229.

O'Regan, B.; Grätzel, M. A low-cost high-efficiency solar cell based on dye-sensitized colloidal TiO_2 films. *Nature* 1991, *353*, 737–740.

O'Regan, B.; Moser, J.; Anderson, M.; Grätzel, M. Vectorial electron injection into transparent semiconductor membranes and electric field effects on the dynamics of light-induced charge separation. *The Journal of Physical Chemistry* 1990, *94*, 8720–8726.

Orenstein, J.; Kastner, M. Photocurrent transient spectroscopy: Measurement of the density of localized states in a-As_2Se_3. *Physical Review Letters* 1981, *46*, 1421–1424.

Pankove, J. I. *Optical Processes in Semiconductors*; Prentice-Hall: Englewood Cliffs, 1971.

Pell, W. G.; Conway, B. E. Analysis of power limitations at porous supercapacitor electrodes under cyclic voltammetry modulation and dc charge. *Journal of Power Sources* 2001, *96*, 57–67.

Peter, L. M.; Duffy, N. W.; Wang, R. L.; Wijayantha, K. G. U. Transport and interfacial charge transfer of electrons in dye-sensitized nanocrystalline solar cells. *Journal of Electroanalytical Chemistry* 2002, *524–525*, 127–136.

Richert, R.; Pautmeier, L.; Bässler, H. Diffusion and drift of charge carriers in a random potential: Deviation from Einstein's law. *Physical Review Letters* 1989, *63*, 547.

Savoie, B. M.; Jackson, N. E.; Marks, T. J.; Ratner, M. A. Reassessing the use of one-electron energetics in the design and characterization of organic photovoltaics. *Physical Chemistry Chemical Physics* 2013, *15*, 4538–4547.

Schiff, E. A. Drift-mobility measurements and mobility edges in disordered silicons. *Journal of Physics: Condensed Matter* 2004, *16*, S5265.

Shaw, J. G.; Hack, M. An analytic model for calculating trapped charge in amorphous silicon. *Journal of Applied Physics* 1988, *64*, 4562–4566.

Sheridan, A. K.; Lupton, J. M.; Samuel, I. D. W.; Bradley, D. D. C. Effect of temperature on the spectral line-narrowing in MEH-PPV. *Chemical Physics Letters* 2000, *322*, 51–56.

Street, R. A. *Hydrogenated Amorphous Silicon*; Cambridge University Press: Cambridge, 1991.

Tanase, C.; Blom, P. W. M.; de Leeuw, D. M.; Meijer, E. J. Charge carrier density dependence of the hole mobility in poly(p-phenylene vinylene). *Physics Status Solidii (a)* 2004, *201*, 1236.

Tanase, C.; Meijer, E. J.; Blom, P. W. M.; de Leeuw, D. M. Unification of the hole transport in polymeric field effect-transistors and LED. *Physical Review Letters* 2003, *91*, 216601.

Tang, H.; Levy, F.; Berger, H.; Schmid, P. E. Urbach tail of anatase TiO_2. *Physical Review B* 1995, *52*, 7771–7774.

Tiedje, T.; Rose, A. A physical interpretation of dispersive transport in disordered semiconductors. *Solid State Communications* 1981, *37*, 49.

Tiedje, T. Band tail recombination limit to the output voltage of amorphous silicon solar cells. *Applied Physics Letters* 1982, *40*, 627–629.

Trasatti, S.; Buzzanca, G. Ruthenium dioxide: A new interesting electrode material. Solid state structure and electrochemical behaviour. *Journal of Electroanalytical Chemistry and Interfacial Electrochemistry* 1971, *29*, A1–A5.

Walzer, K.; Maennig, B.; Pfeiffer, M.; Leo, K. Highly efficient organic devices based on electrically doped transport layers. *Chemical Reviews* 2007, *107*, 1233–1271.

Wang, H.; He, J.; Boschloo, G.; Lindström, H.; Hagfeldt, A.; Lindquist, S. Electrochemical investigation of traps in a nanostructured TiO_2 film. *The Journal of Physical Chemistry B* 2001, *105*, 2529–2533.

Wang, H.; Lewis, J. P. Localization of frontier orbitals on anatase nanoparticles impacts water adsorption. *The Journal of Physical Chemistry C* 2009, *113*, 16631–16637.

Wang, P.; Bai, Y.; Mora-Seró, I.; De Angelis, F.; Bisquert, J. Titanium dioxide nanomaterials for photovoltaic applications. *Chemical Reviews* 2014, doi:10.1021/cr400606n.

Wang, Q.; Ito, S.; Grätzel, M.; Fabregat-Santiago, F.; Mora-Seró, I.; Bisquert, J.; Bessho, T.; et al. Characteristics of high efficiency dye-sensitized solar cells. *The Journal of Physical Chemistry B* 2006, *110*, 19406–19411.

Weaire, D.; Thorpe, M. F. Electronic properties of an amorphous solid. I. A simple tight-binding theory. *Physical Review B* 1971, *4*, 2508–2520.

Yu, Z. G.; Smith, D. L.; Saxena, A.; Martin, R. L.; Bishop, A. R. Molecular geometry fluctuations and field-dependent mobility in conjugated polymers. *Physical Review B* 2001, *63*, 085202.

Yablonovitch, E.; Tiedje, T.; Witzke, H. Meaning of the photovoltaic band gap for amorphous semiconductors. *Applied Physics Letters* 1982, *41*, 953–955.

Zhang, Z.; Zakeeruddin, S. M.; O'Regan, B. C.; Humphry-Baker, R.; Grätzel, M. Influence of 4-guanidinobutyric acid as coadsorbent in reducing recombination in dye-sensitized solar cells. *The Journal of Physical Chemistry B* 2005, *109*, 21818–21824.

9 Planar and Nanostructured Semiconductor Junctions

The potential barrier at the semiconductor surface in contact with a metal or an electrolyte is a determining factor in solid-state device physics and in photoelectrochemistry. Since the height of the barrier can be modified by the applied voltage, the flow of electrons is enhanced in forward mode and impeded in reverse. Therefore, this provides a useful selective contact that is a key component of many electronic and optoelectronic devices. In the first section of this chapter, we treat quantitatively the shape of the barrier at the semiconductor surface in contact with metal (or any solid contact). We stress the importance of the capacitance measurements that reveal different properties of the semiconductor and the barrier. The next section of the chapter provides a detailed view of the properties of semiconductor/electrolyte junctions that are used in photoelectrochemical cells. A general energy diagram is presented to combine the interpretation of the electrochemical measurement with the changes of the local vacuum level. Then, we analyze generic properties of various specific semiconductor junctions, as the p–n junction and the heterojunction. Finally, we discuss small-size nanostructures, such as nanoparticles and nanowires, first considering the depletion layer at the internal surface, and then the nanostructures that operate by homogeneous displacement of the Fermi level.

9.1 STRUCTURE OF THE SCHOTTKY BARRIER AT A METAL/SEMICONDUCTOR CONTACTS

The discussion of the properties of a metal/semiconductor junction starts from the energy diagram of the separate materials shown in Figure 9.1a. When the materials are brought into contact (Figure 9.1b), electronic equilibrium is achieved by the formation of an interfacial barrier, provided that the interface permits the flow of electrons. The initial driving force for diffusive flow of electrons is the difference in electrochemical potentials in both phases. When the equilibrium is achieved, both electrochemical potentials equalize. In thermal equilibrium, there exists an electric potential energy barrier that neutralizes the internal chemical potential gradient.

In Figure 2.1, we have described the characteristic energy barrier that can be formed at the surface of a doped semiconductor. In Figure 9.1a, the junction is formed by a metal of work function Φ_m and a semiconductor with electron affinity χ_n and work function Φ_{sc}. The doped n-type semiconductor is assumed to have a uniform distribution of cations N_D, which are compensated by the same quantity of conduction band electrons $n_0 = N_D$ in the quasi-neutral region, where the band edge is flat. ζ_{nb}, the chemical potential of electrons, is the distance between the conduction band and the Fermi level in the quasi-neutral region (Equation 5.32). Owing to the difference of work functions, a contact potential difference or Volta potential difference is established between the separate materials, as indicated in Figure 2.15. In Equation 4.20, we have termed such difference of potentials the *built-in potential* V_{bi}. In semiconductor photoelectrochemistry, this quantity is termed the *flatband potential*, V_{fb}, because application of such voltage causes the bands to be flat.

When the materials are brought into contact, they reach equilibrium of the Fermi levels as shown in Figure 9.1b. It is assumed that the position of the semiconductor conduction band at the surface is not modified, that is, there are no charging effects of dipole or surface state. This is the vacuum level alignment rule or the Mott–Schottky rule discussed in Chapter 4. The height of the barrier from the metal side is given in Equation 4.18. In Figure 9.1b, electrons diffuse from the semiconductor to the metal, which has a larger work function. The potential barrier in the semiconductor is caused by the removal of carriers that have been transferred to the metal. In the semiconductor there remains the ionic background charge, forming a space-charge region (SCR) termed the *depletion region*. The same amount of charge but with opposite (negative) sign is located at the interface, at the metal contact, as indicated in Figure 9.1c. The charge transferred across the metal/semiconductor interface produces a dipolar structure. This structure forms a dielectric capacitor.

In order to determine the structure of the barrier as a function of the applied voltage, we consider the local electrostatic potential φ in Figure 9.1b. As indicated in Chapter 4, V_{bi} is defined as the voltage change across the

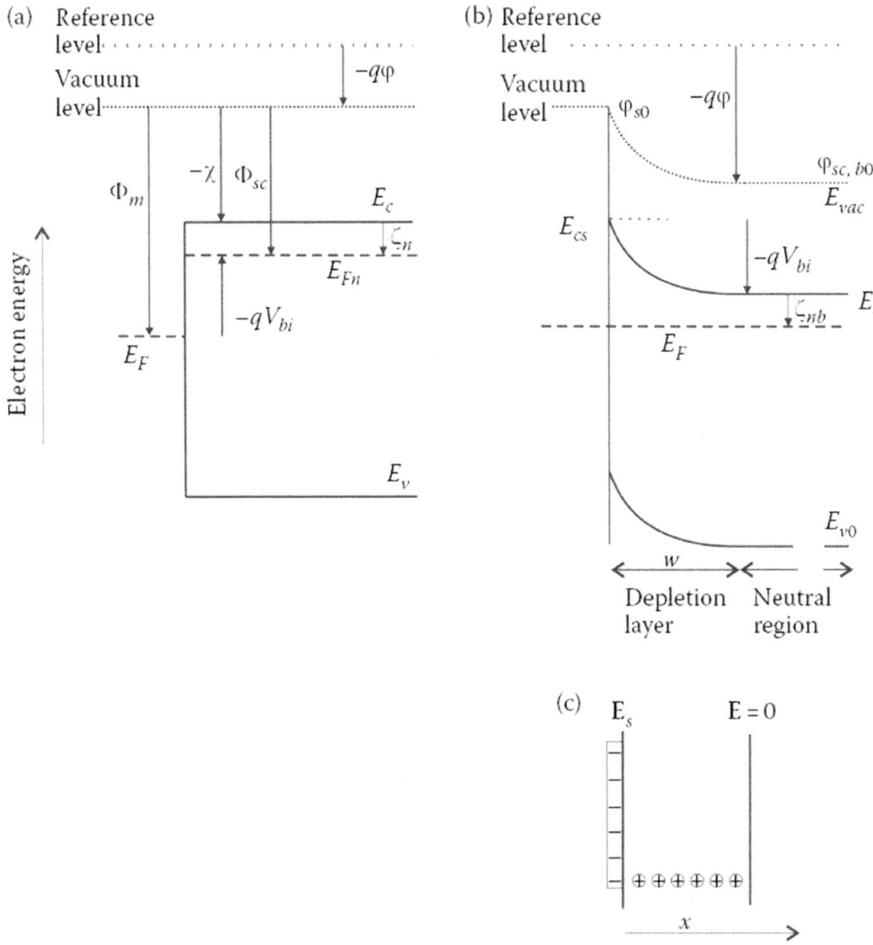

FIGURE 9.1 Scheme of a Schottky barrier at the semiconductor surface. (a) Separate energy levels of metal and *n*-type semiconductor. E_c is the energy of the conduction band, E_v is the energy of the valence band, E_{Fn} is the electron Fermi level in the semiconductor. (b) Structure of the barrier at equilibrium. w is the width of the depletion zone. (c) Spatial distribution of charge.

semiconductor space-charge layer in equilibrium (Ishii et al., 2004); therefore,

$$V_{bi} = \varphi_{sc,b0} - \varphi_{s0} \tag{9.1}$$

Here, $\varphi_{sc,\,b}$ is the electrostatic potential in the quasi-neutral region (denoted by subscript *b*), and φ_s is the potential at the semiconductor surface (subscript *s*). The subscript 0 denotes the equilibrium (zero-bias) case. In the absence of surface dipoles, the potential at the semiconductor surface, φ_s is the same as that in the metal, $\varphi_s = \varphi_m$, and Equation 4.20 is obtained. The contact potential difference (CPD) is accommodated as band bending in the semiconductor, as shown in Figure 9.1b. More generally, the CPD is formed by the combination of a built-in potential and interfacial dipole (Equation 4.28).

The displacement of the Fermi levels of the separate materials, toward the Fermi level equilibration, occurs in both, semiconductor and metal. It implies that, due to the initial Volta potential difference, the LVL in both materials has to move with respect to the vacuum level (VL) at infinite. However, the change of the LVL is not the same in the two materials. It will change less in the most electronically populated material, or in the material that has a larger DOS at the Fermi level, so that a small displacement of the Fermi level produces a large change of surface charge. Due to the very high density of states in the metal, the displacement of the metal Fermi level at the semiconductor/metal contact is very small.

9.2 CHANGES OF THE SCHOTTKY BARRIER BY THE APPLIED VOLTAGE

We now discuss the modification of the band bending under a bias potential. The voltage V is applied to the metal/semiconductor contact as indicated in Figure 9.2. The polarization (Fermi level step) occurs at the rectifying contact where the barrier is formed. It is assumed that electrons remain in thermal equilibrium throughout the whole semiconductor, including the SCR, so that the

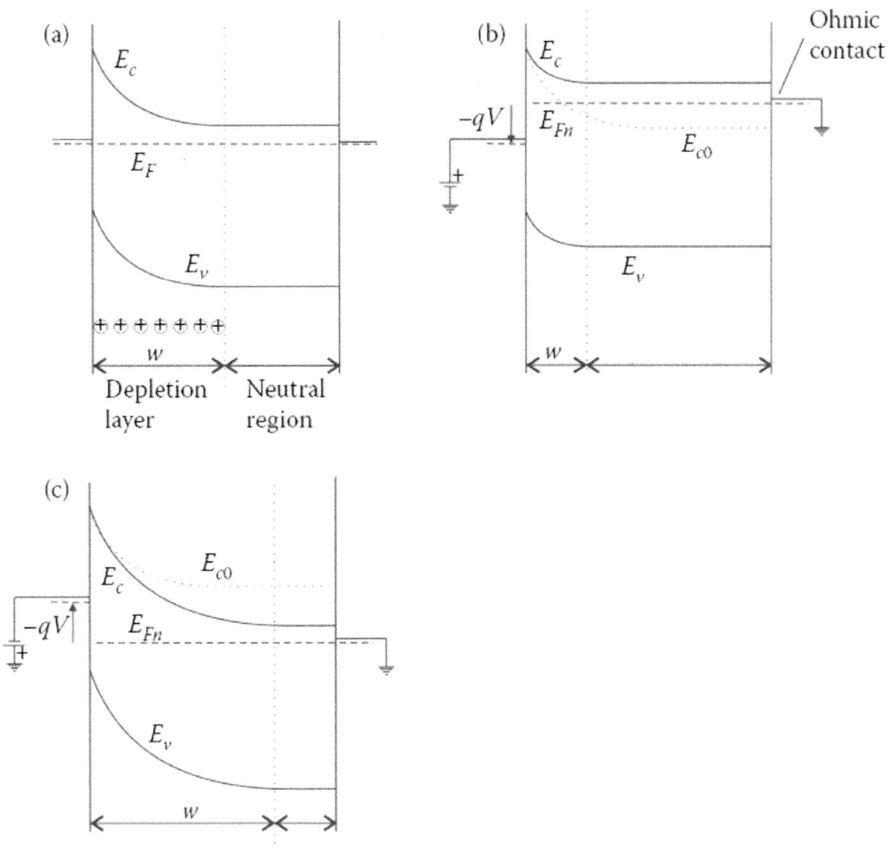

FIGURE 9.2 Energy diagram of an n-type semiconductor forming a Schottky barrier at the left contact, and an ohmic contact at the right for majority carriers. V is the voltage applied at the left contact, E_c is the energy of the conduction band, E_{c0} is the energy of the conduction band at zero bias, E_v is the energy of the valence band, E_F is the Fermi level, E_{Fn} is the electron Fermi level, w is the width of the depletion zone. (a) Zero bias, (b) forward bias, and (c) reverse bias.

Fermi level is homogeneous in this region. The forward bias occurs when the voltage is made negative at the metal/semiconductor contact with respect to the ohmic contact of the semiconductor, leading to a reduction of the size of the SCR (Figure 9.2b). At reverse bias the size of the depletion region increases along with the height of the barrier, as shown in Figure 9.2c. However, when the Fermi level at the semiconductor surface surpasses the midgap energy, there starts the creation of abundant concentration of minority carriers. Here ends the regime of depletion and the semiconductor enters inversion as discussed in Section 9.8.

The voltage applied across the junction is related to the difference of Fermi levels between metal and semiconductor as

$$-qV = E_{Fm} - E_{Fn} \tag{9.2}$$

The applied voltage can also be related to modification of the difference of potentials between the contacts, as shown in Figure 9.3:

$$V = \varphi_{sc,b} - \varphi_s - \left(\varphi_{sc,b0} - \varphi_{s0}\right) \tag{9.3}$$

Note that in the situation shown in Figure 9.3a, we assume that the electron affinity of the semiconductor at the surface, E_{cs}, is not modified by the applied voltage.

This assumption does not hold if there is Fermi level pinning, as described in Chapter 4 and later in Section 9.7.

The potential difference across the barrier in the semiconductor surface (band bending) can be written as

$$V_{sc} = \varphi_{sc,b} - \varphi_s \tag{9.4}$$

In equilibrium, as in Figure 9.1b

$$V_{sc0} = V_{bi} \tag{9.5}$$

Under applied voltage, the band bending is

$$E_{cs} - E_{cb} = -q\varphi_s + q\varphi_{sc,b} = qV_{sc} \tag{9.6}$$

and in equilibrium

$$E_{cs0} - E_{c0} = -q\varphi_{s0} + q\varphi_{sc,b0} = qV_{bi} \tag{9.7}$$

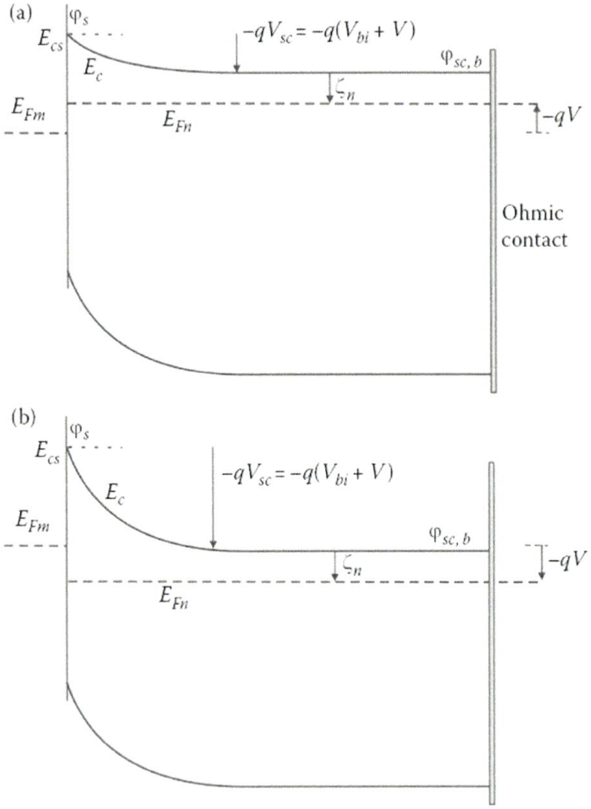

FIGURE 9.3 Energy scheme of a Schottky barrier at the semiconductor surface: (a) forward bias, (b) reverse bias voltage.

We can write

$$V_{sc} = V + V_{bi} \tag{9.8}$$

In Equation 9.8 we observe that if we apply a voltage $-V_{bi}$, we cancel the barrier and recover the situation of Figure 9.1a in which the materials were not in contact, that is, the flatband condition.

9.3 PROPERTIES OF THE PLANAR DEPLETION LAYER

We have remarked that the formation of the Schottky barrier (SB) involves the removal of free carriers from the barrier region, leaving behind only the fixed (donor or acceptor) ionic charges (see Figures 9.1 and 9.2). In order to determine the shape of the barrier quantitatively, we now adopt the *depletion approximation*, in which the density of charge is assumed to be a step function. The charge density is abruptly separated between the SCR, of thickness w, where the electron carriers are removed, and the neutral region. Hence,

$$\rho(x) = \begin{cases} qN_D & 0 < x < w \\ 0 & x > w \end{cases} \tag{9.9}$$

Poisson Equation 2.4 give

$$\frac{\partial^2 \varphi}{\partial x^2} = -\frac{qN_D}{\varepsilon} \tag{9.10}$$

where $\varepsilon = \varepsilon_r \varepsilon_0$. This equation has the solution

$$\varphi = \varphi_s - \mathbf{E}_s x - \frac{qN_D}{2\varepsilon} x^2 \tag{9.11}$$

that is defined by two initial constants, namely, the potential φ_s, and the field $\mathbf{E}_s$, taken at the semiconductor surface at $x = 0$. The band bending follows the shape of the potential, hence

$$E_c = E_{cs} + q\mathbf{E}_s x + \frac{q^2 N_D}{2\varepsilon} x^2 \tag{9.12}$$

The form of the barrier is parabolic. When arriving at the edge of the depletion region, at $x = w$, we enter the quasi-neutral region, and the band is flat. Therefore, no electric field exists at this point. In other words, we can write

$$\frac{\partial E_c}{\partial x}(w) = \mathbf{E}_s + \frac{qN_D}{\varepsilon} w = 0 \tag{9.13}$$

It means that the electric field at the contact with the metal is

$$\mathbf{E}_s = \frac{qN_D}{\varepsilon} w \tag{9.14}$$

Therefore,

$$E_c = E_{cs} - \frac{q^2 N_D}{\varepsilon} x \left(w - \frac{x}{2} \right) \tag{9.15}$$

Thus, the field changes from $\mathbf{E}_s = \mathbf{E}(0)$ to 0, and the height of the energy barrier is

$$E_{sc} - E_{cb} = \frac{q^2 N_D w^2}{2\varepsilon} \tag{9.16}$$

The total space charge enclosed in thickness w is

$$Q_{sc} = qN_D w \tag{9.17}$$

Hence, Equation 9.14 is the result that can be obtained directly by Gauss law (see Figure 9.1c).

Using Equations 9.6 and 9.16, we obtain

$$V_{sc} = V + V_{bi} = \frac{qN_D}{2\varepsilon} w^2 \tag{9.18}$$

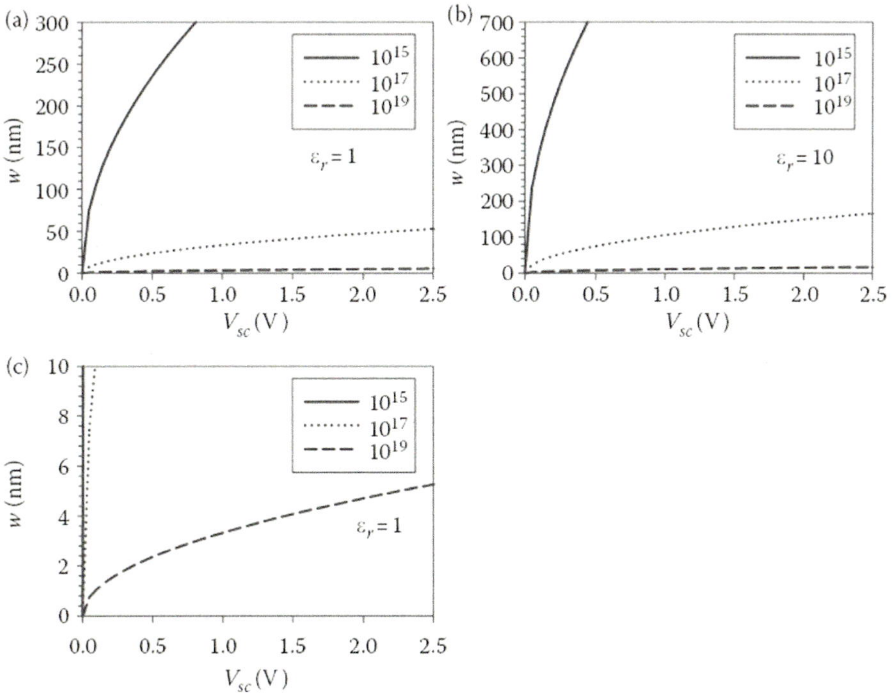

FIGURE 9.4 Size of the depletion region, as a function of the voltage across the barrier, for different values of doping density and dielectric constant in (a) and (b). Panel (c) shows the short *w* region in (a).

The size of the depletion region is

$$w = \left[\frac{2\varepsilon V_{sc}}{qN_D}\right]^{1/2} \tag{9.19}$$

In equilibrium, we have the value

$$w_0 = \left[\frac{2\varepsilon V_{bi}}{qN_D}\right]^{1/2} \tag{9.20}$$

Under applied bias, the size of the depletion layer is

$$w = \left[\frac{2\varepsilon}{qN_D}\left(V + V_{bi}\right)\right]^{1/2} \tag{9.21}$$

Note that the SCR vanishes when the applied forward potential (negative bias) equals the built-in potential, as mentioned earlier.

Practical doping of semiconductors ranges from low doping of 10^{15} cm^{-3} to very high doping density of 10^{19} cm^{-3}. In Figure 9.4, we observe the corresponding size of the depletion region as a function of the size of the barrier (in volts). For a low-doped organic semiconductor, with a low dielectric constant, and the equilibrium potential of 0.5 V, the size of the SCR is about 200 nm, which is the characteristic size of organic layers in some optoelectronic devices. It means that practical organic layers can contain a Schottky barrier or be totally depleted.

9.4 MOTT–SCHOTTKY PLOTS

The spatial separation of positive and negative charge forms a dielectric capacitor at the interface of the metal/semiconductor contact, as indicated in Figure 9.1c. We calculate the differential capacitance as

$$C_{sc} = \frac{\partial Q}{\partial V} = qN_D\frac{\partial w}{\partial V} \tag{9.22}$$

From Equation 9.19, the result is

$$C_{sc} = \frac{\varepsilon}{w} = \left(\frac{qN_D\varepsilon}{2}\right)^{1/2} V_{sc}^{-1/2} \tag{9.23}$$

It turns out that

$$C_{sc}^{-2} = \frac{2}{q\varepsilon N_D}\left(V + V_{bi}\right) \tag{9.24}$$

Therefore, a representation of the reciprocal square capacitance, as a function of applied voltage, contains important information about the semiconductor. The slope gives the doping density (provided that the dielectric constant is known). The intercept to the *x* axis provides the built-in potential and allows establishing the semiconductor conduction band level with respect to the reference of potential. The representation of Equation 9.24 is called a *Mott–Schottky* (MS) plot (see

Figures 8.16b and 8.18c). In solid junctions, we can take as a reference the metal Fermi level, if the work function is known, which provides a full energy diagram in the physical scale.

Let us analyze some examples of the application of the MS analysis to the characterization of thin semiconductor layers. In Figure 9.5a we show two examples of MS plots, of semiconductor layers contacted with Al (A) and Au (B) electrodes. The characteristics of the layers are indicated in Table 9.1.

Equation 9.24 can be applied to extract information about the energetic structure of the layer and the interface with the metal contact. We first observe that the capacitances of both layers show a flat region in large reverse polarization; this feature must correspond to full depletion of the thin layer. In fact, in Equation 9.23, we can insert the value of the depletion layer $w = d$ that corresponds to the size of the layer and we obtain

$$C_{sc} = \frac{\varepsilon}{d} \tag{9.25}$$

that is, simply, the dielectric capacitance of the layer, Equation 3.6. As we can measure experimentally the thickness of the layer, the constant value at reverse provides the dielectric constant of the semiconductor. In both layers the result is $\varepsilon_r = 3.5$ in this example.

From the slope of the MS plot we can now obtain the doping density, which is different in the two cases (see Table 9.1). We also obtain the flatband value from the intercept to the horizontal axis. We can determine the parameter ζ_{nb} assuming that the effective density of states in the conduction band is $N_c = 10^{20}$ cm^{-3}. It is then possible to establish the position of the conduction band at the contact with the metal, E_{cs}, and then the full structure of the Schottky barrier. The results are shown in Figures 9.5b and c. If the semiconductor is the same, we conclude that the doping has experienced a significant variation. In addition, the two depositions differ strongly on the value E_{cs}, meaning that a surface dipole has been formed at least in one of the cases.

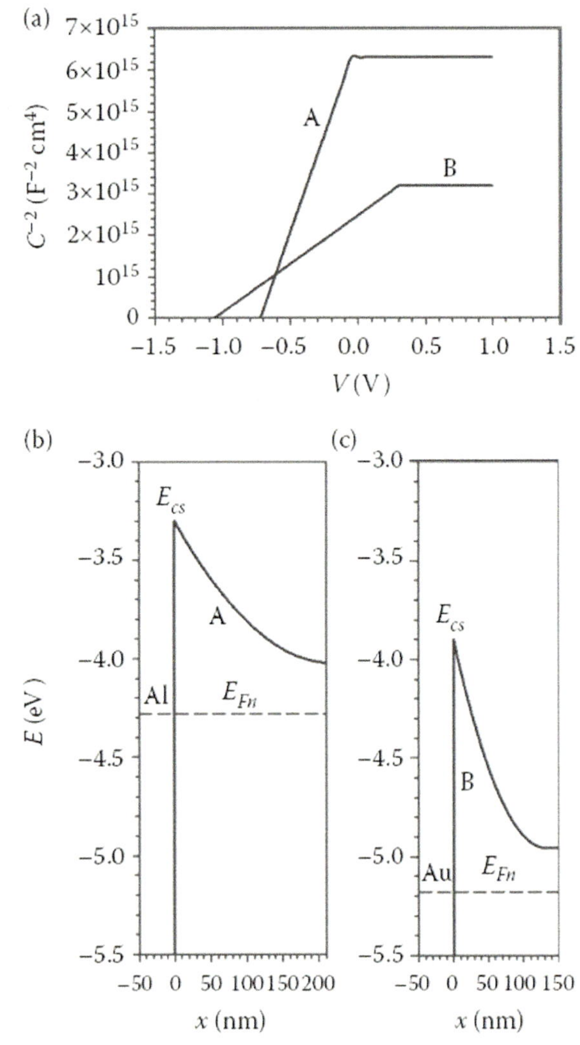

FIGURE 9.5 (a) Mott–Schottky plots of two different simulated semiconductor thin layers, each one in contact with a different metal: Al (A) and Au (B). (b) and (c) are the energy diagrams recovered from (a) for each contact.

9.5 CAPACITANCE RESPONSE OF DEFECT LEVELS AND SURFACE STATES

The MS plot obtained in Equation 9.24 has been derived on the assumption in Equation 9.9 that the SCR terminates abruptly. A more accurate analysis considering the

TABLE 9.1
Characteristics of Semiconductor Layers

Layer	Thickness (nm)	Contact	Work Function Φ_m (eV)	Flatband Voltage (V)	Doping (cm^{-3})	ζ_{nb} (eV)	E_{cs} (eV)
A	210	Al	4.28	0.722	5 10^{15}	0.257	3.30
B	150	Au	5.18	1.059	2 10^{16}	0.221	3.90

statistics of electrons (Sze, 1981) provides the following result for the size of the depletion region

$$w = \left[\frac{2\varepsilon}{qN_D} \left(V_{bi} + V - \frac{k_B T}{q} \right) \right]^{1/2} \tag{9.26}$$

Therefore,

$$C_{sc}^{-2} = \frac{2}{q\varepsilon N_D} \left(V_{bi} + V - \frac{k_B T}{q} \right) \tag{9.27}$$

When the interfacial barrier is of the order k_BT, special care has to be taken to interpret the capacitance measurement. In fact, at these small voltages the capacitance makes a peak that can be used for the determination of the built-in voltage (Germs et al., 2012).

The MS analysis can more generally resolve a variable doping profile in the semiconductor as follows:

$$\frac{d\left(C_{sc}^{-2}\right)}{dV} = \frac{2}{q\varepsilon N_D(w)} \tag{9.28}$$

The derivative gives the doping at the edge of the depletion region, $N_D(w)$. However, this method only provides a spatial resolution of the order of a Debye length λ_D (Johnson and Panousis, 1971).

If the material contains a defect level in the gap (see Figure 9.6), the bias voltage changes the occupancy of the deep level and produces changes of capacitance and conductance (Vincent et al., 1975). The determination of capacitance-voltage characteristics, by measuring at different ac frequencies and temperatures, in deep reverse conditions, allows one to obtain the density of states in part of the gap of a material. Another widely used method to scan deep levels in Schottky barriers is "admittance spectroscopy" and consists in measuring the capacitance at a fixed frequency with varying temperatures (Losee, 1975).

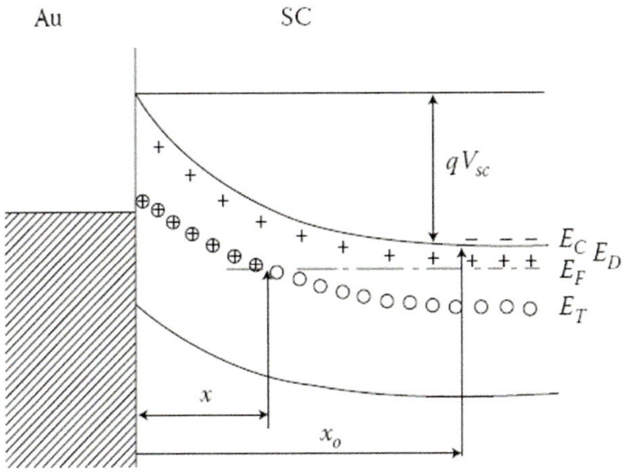

FIGURE 9.6 Schottky barrier for an *n*-type semiconductor containing a level of defect states. The ionic charge in the depletion layer becomes active when the defect level emerges above the Fermi level because of the band bending. (Reproduced with permission from Vincent, G.; Bois, D.; Pinard, P. *Journal of Applied Physics* 1975, *46*, 5173.)

The changes of electron affinity of the semiconductor are a frequent occurrence and can even be controlled by purposeful production of dipolar layers or by partial Fermi level pinning as explained in Chapter 4. Several characteristic situations in which the Schottky barrier is modified by contact effects are shown in Figure 9.7. In particular, the molecular modification of silicon has attained a great degree of control with reproducible samples that allows investigating semiconductor-molecule electronics (Vilan et al., 2010). In the surface of a semiconductor, without any contact, band bending is affected by surface states (Yablonovitch et al., 1989). Surface photovoltage technique is used to determine the position of the band edges (Kronik and Shapira, 2001; Reshchikov et al., 2010).

9.6 SEMICONDUCTOR ELECTRODES AND THE FLATBAND POTENTIAL

The kinetics of transference of electronic carriers at the semiconductor/electrolyte junction has been discussed in Section 6.8. Here, we take a detailed look at the energetic properties of the junction. The photoelectrochemical cell (PEC) is formed by the semiconductor electrode, the electrolyte, and a counterelectrode (CE), as shown in Figure 9.8. The PEC can be used to convert sunlight either to electricity with a redox carrier in solution, or to chemical fuel such as hydrogen, when photogenerated electron–hole pairs drive interfacial reactions as photoelectrolysis, as shown in Figure 6.30.

The semiconductor/electrolyte junction is the core element of the PEC as it takes the polarization and controls the rate of the electrochemical reactions. In Figure 9.8, it is observed that the semiconductor electrode can be biased against the redox electrolyte, and the changes of applied potential affect the band bending at the semiconductor/electrolyte interface. This modification has a large impact on electron density at the semiconductor/liquid interface and enhances or reduces the electrochemical reaction rate. A detailed description of the metal/semiconductor contact has been provided in the previous sections of this chapter. When the semiconductor is contacted by a redox electrolyte, similar concepts apply in the semiconductor side of the system, but new aspects appear by the presence of the electrolyte.

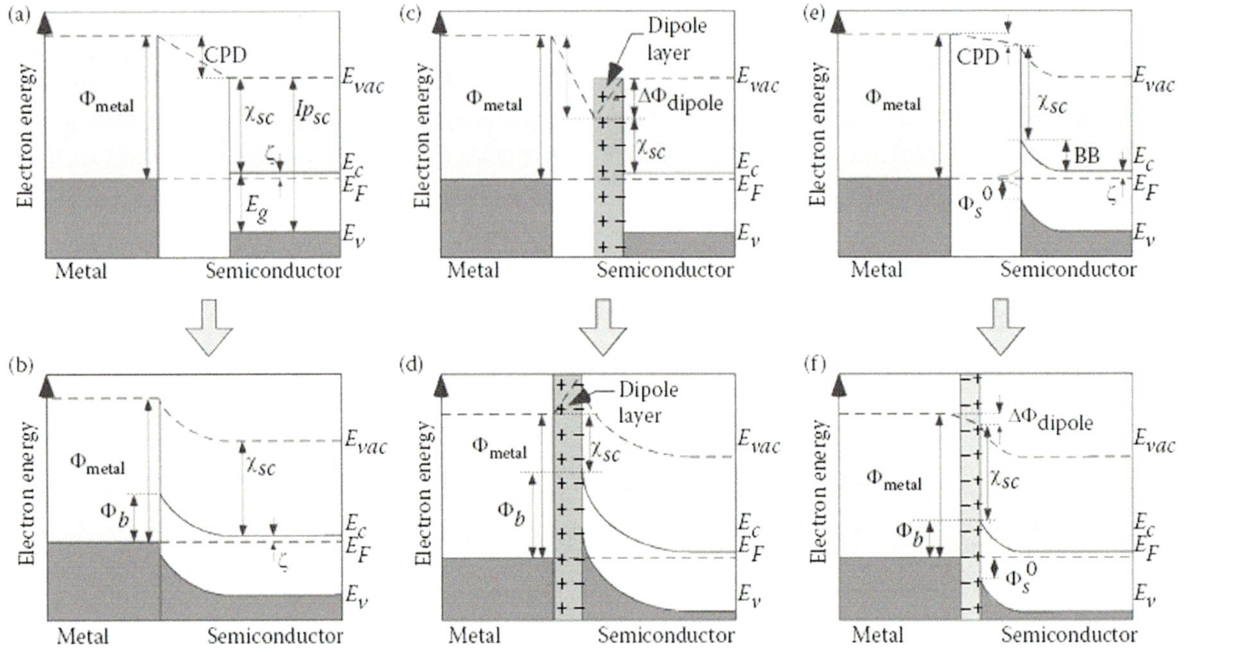

FIGURE 9.7 Schematic description of possible sources of a molecular effect on the junction's barrier height. The potential energy for one electron is plotted as a function of distance into the metal (left) or n-type semiconductor (right). The panels present situations before (a, c, e) and after contact (b, d, f), with (c, d) and without (a, b) an added dipole layer at the interface, and for partial Fermi level pinning (e, f). In panel (a), characteristic energy parameters of the semiconductor surface are presented as well. (Reprinted with permission from Vilan, A.; Ghabboun, J.; Cahen, D. *The Journal of Physical Chemistry B* 2003, *107*, 6360–6376.)

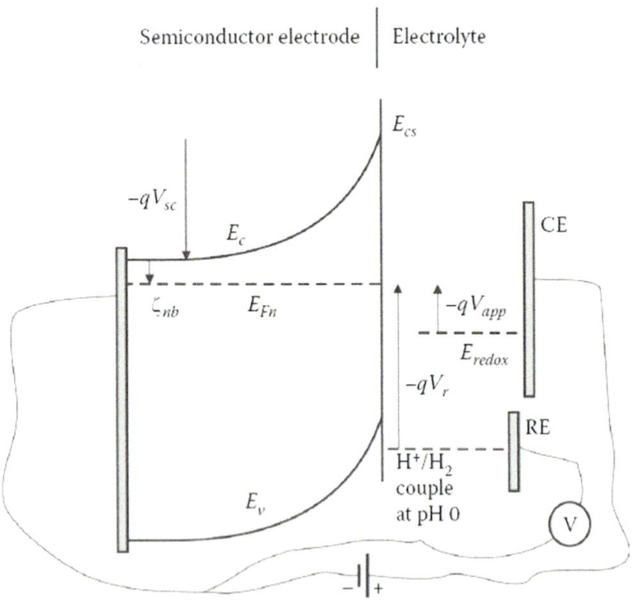

FIGURE 9.8 Schematic representation of a photoelectrochemical cell, in which an n-type semiconductor is biased against a redox electrolyte. The semiconductor electrode potential is measured with respect to a reference electrode in solution. (Reproduced with permission from Bisquert, J. et al. *The Journal of Physical Chemistry Letters* 2014, *5*, 205–207.)

As discussed in Chapter 3, for electrochemical measurements it is usually convenient to measure the voltage applied to the electrode with respect to a reference electrode (RE) in solution that is indicated in Figure 9.8. The standard reference of the potential in the solution is the SHE defined in Section 3.9 in terms of the H^+/H_2 couple at pH = 0, that is shifted from the physical scale, the electron at rest in vacuum, by the energy $E_0^{SHE} = -4.44\,\text{eV}$ (see Equation 3.54). The *measured voltage* with respect to the RE is termed V_r. However, we reserve the denomination "applied voltage" to the external voltage between the semiconductor and CE (see Figure 9.8). The energy level of ions in solution is E_{redox}. If we neglect any potential drop at the surface of the CE, we have

$$V_{app} = -\frac{1}{q}\left(E_{Fn} - E_{redox}\right) \tag{9.29}$$

We normally assume that the vacuum level of the electrolyte is immutable. The changes of voltage V_r produce a displacement of the semiconductor Fermi level that causes modification of the semiconductor barrier height at the depletion layer, V_{sc}. One particularly important aspect of the study of the semiconductor/electrolyte system is to determine the position of the edge of the conduction band at the semiconductor surface, E_{cs}. We have earlier introduced the idea of the flatband potential in connection with the metal/semiconductor junction. For the present case, the measured flatband voltage is connected to the energy of the conduction band in a particular surface condition. In addition, the flatband potential situation marks a point of the electrode potential variation where the transference of photogenerated electrons, either to the surface or toward the bulk, is inverted. Therefore, V_{fb} is a limit to the open-circuit voltage when the PEC is operated as a solar cell or solar fuel device.

The measurement with respect to RE only takes the potential at the surface of the solid, but not in the interior. On the contrary, in a semiconductor film that is composed of several junctions, as in semiconductor nano-heterostructures including catalytic layers, it is convenient to use the energy scale with reference to the VL, as discussed in Section 2.4. This type of reference in the energy diagram is useful to track the local variations of energy, electrostatic potential, and Fermi energy inside the solid. In the following we provide a consistent set of relationships that allows one to combine both types of conventions (Bisquert et al., 2014).

The energy diagram of an n-type semiconductor electrode in contact with electrolyte is shown in Figure 9.9. As mentioned previously, subindex s is for a surface quantity and subindex b is for a bulk semiconductor quantity (where electrons and holes remain in dark equilibrium value).

Following Equation 3.38, the measured voltage of the semiconductor electrode with respect to an RE is

$$V_r = -\frac{E_{Fn,b} - E_0^{\mathrm{SHE}}}{q} \tag{9.30}$$

The measured voltage has the following expression in terms of the semiconductor parameters, the barrier height, and the Helmholtz potential defined in Equation 3.35.

$$V_r = \frac{E_0^{\mathrm{SHE}} + \chi + \zeta_{nb}}{q} + V_{sc} + V_H \tag{9.31}$$

The conduction band edge at position x in the semiconductor is modified by the local electrostatic potential

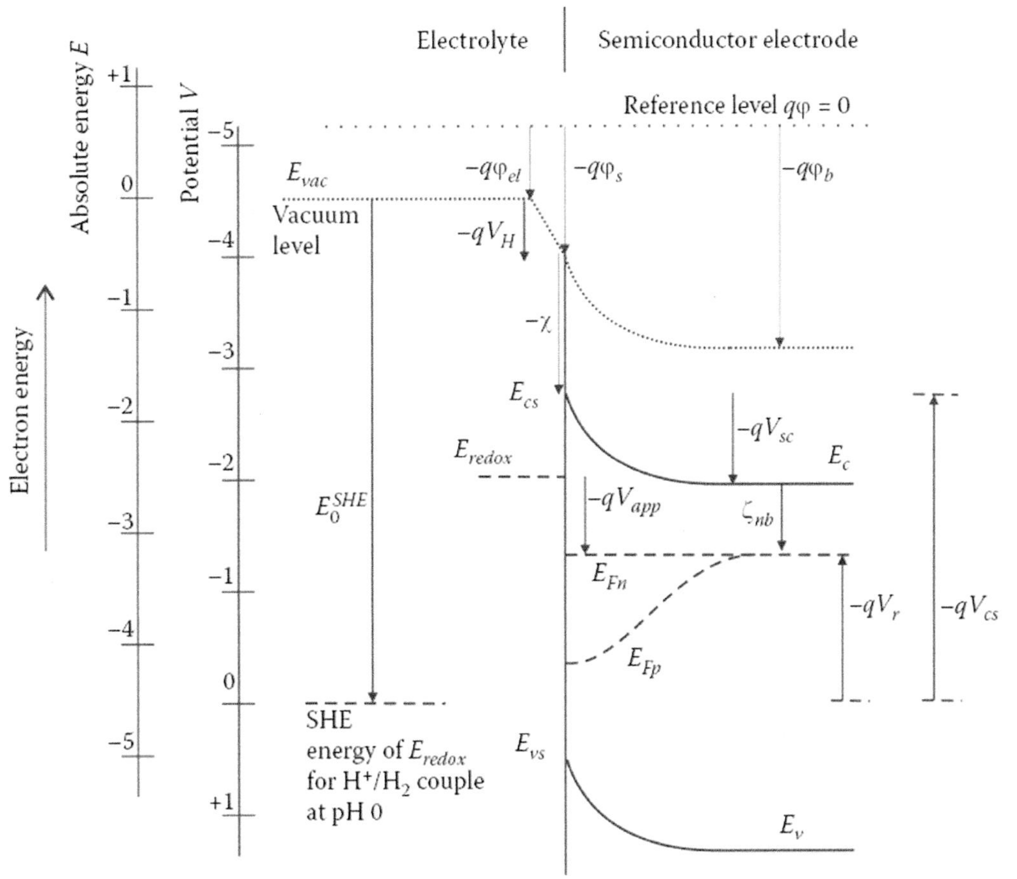

FIGURE 9.9 Scheme of an n-type semiconductor electrode, indicating energies in the absolute energy scale (with respect to vacuum level), and potentials in the electrochemical scale, with respect to the Standard Hydrogen Electrode (SHE). (Reproduced with permission from Bisquert, J. et al. *The Journal of Physical Chemistry Letters* 2014, 5, 205–207.)

$$E_c(x) = -q\left[\varphi(x) - \varphi_{el}\right] - \chi \tag{9.32}$$

Absolute energies are given with respect to VL in the electrolyte. The *energy* of the edge of the conduction band at the surface is

$$E_{cs} = -\chi - qV_H \tag{9.33}$$

The *potential* of the conduction band edge (with respect to RE), V_{cs}, is determined by the expression

$$E_{cs} = E_0^{\mathrm{SHE}} - qV_{cs} \tag{9.34}$$

Therefore,

$$V_{cs} = \frac{E_0^{\mathrm{SHE}} + \chi}{q} + V_H \tag{9.35}$$

The flatband potential is the voltage at which $V_{sc} = 0$,

$$V_{fb} = V_r\big|_{V_{sc}=0} \tag{9.36}$$

Hence, from Equation 9.31 (Butler and Ginley, 1978),

$$V_{fb} = \frac{E_0^{\mathrm{SHE}} + \chi + \zeta_{nb}}{q} + V_H^{fb} \tag{9.37}$$

In the presence of surface states or specifically adsorbed ions at the semiconductor surface, V_H^{fb} may not be zero when $V_{sc} = 0$, depending on the potential of zero charge (pzc). The following expression, obtained from Equations 9.31 and 9.37, is a useful way of writing the voltage V_r:

$$V_r = V_{fb} + V_{sc} + \left(V_H - V_H^{fb}\right) \tag{9.38}$$

Using Equations 9.35 and 9.37, we obtain

$$V_{fb} = V_{sc} - \left(V_H - V_H^{fb}\right) + \frac{\zeta_{nb}}{q} \tag{9.39}$$

This last expression shows that the flatband potential directly correlates with the position of the edge of the conduction band, with respect to the given reference electrode.

The applied voltage of Equation 9.29 can be stated as follows:

$$V_{app} = V_r - V_{r0} \tag{9.40}$$

where the equilibrium value V_{r0} coincides with the potential of the redox couple with respect to the RE, V_{redox}, therefore,

$$V_{app} = V_r - V_{redox} \tag{9.41}$$

We also have

$$V_{app} = \varphi_b - \varphi_{b0} \tag{9.42}$$

and in terms of built-in potential, using Equations 9.38 and 9.40

$$V_{app} = V_{sc} - V_{bi} + \left(V_H - V_{H0}\right) \tag{9.43}$$

Alternatively, in terms of the flatband potential, we may write from Equations 9.40 and 9.43

$$V_{app} = V_{sc} + V_{fb} - V_{redox} + \left(V_H - V_H^{fb}\right) \tag{9.44}$$

Note that V_{fb} has been defined above with respect to RE. If we define the applied potential in the special situation in which the band is flat, we have

$$V_{app}^{fb} = V_{fb} - V_{redox} \tag{9.45}$$

Therefore, we can write

$$V_{app} = V_{sc} + V_{app}^{fb} + \left(V_H - V_H^{fb}\right) \tag{9.46}$$

The applied voltage V_{app} is negative in Figure 9.8 (forward bias) and positive in Figure 9.9 (reverse polarization). Using Equation 9.39, we obtain

$$V_{app}^{fb} = V_{cs} - V_{redox} - \left(V_H - V_H^{fb}\right) + \frac{\zeta_{nb}}{q} \tag{9.47}$$

The relationship of the flatband potential to the built-in potential directly comes from Equation 9.43:

$$V_{app}^{fb} = -V_{bi} + V_H^{fb} - V_{H0} \tag{9.48}$$

In Figure 9.9, a negative potential V_{app}^{fb} must be applied to the semiconductor in equilibrium with the electrolyte, in order to flatten the band.

9.7 CHANGES OF REDOX LEVEL AND BAND UNPINNING

When a semiconductor achieves equilibrium of Fermi level with a redox electrolyte, the size of the surface barrier V_{sc0} is established. In the condition of *band pinning*, we assume that changes in the Helmholtz layer can be neglected. From Equation 9.48, we have

$$V_{sc0} = V_{redox} - V_{fb} \tag{9.49}$$

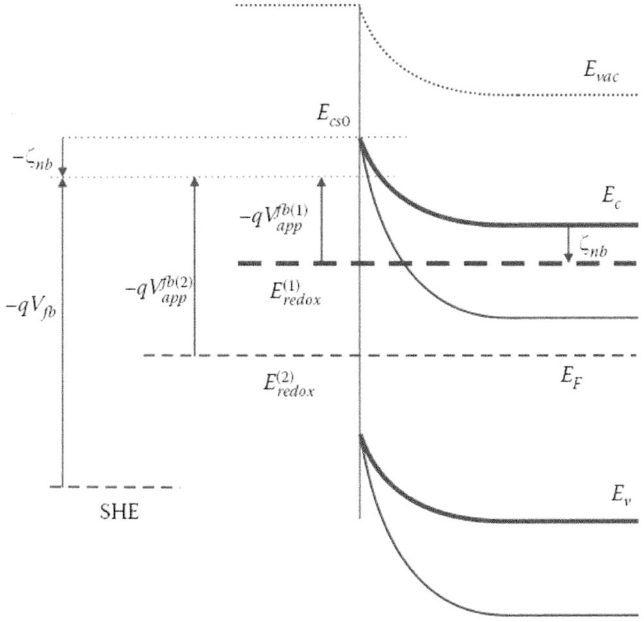

FIGURE 9.10 The Schottky barrier at the n-type semiconductor surface increases linearly with the redox potential in solution (in the absence of surface dipole).

If we change the redox level in the electrolyte, the barrier at the semiconductor surface is accordingly adjusted, while V_{fb} is not modified, as it depends only on the potential of the conduction band edge at the surface (see Figure 9.10). Therefore, a linear relation with slope 1 is expected for the barrier size of the semiconductor in different redox electrolytes

$$\frac{d(V_{sc0})}{dV_{redox}} = 1 \quad (9.50)$$

The same relationship is obtained for the flatband potential measured in a two electrode configuration, V_{app}^{fb}, with respect to changes in V_{redox} (see Equation 9.47 and Figure 9.10).

Equation 9.50 is equivalent to Equation 4.21, which expresses the vacuum alignment rule at the metal-semiconductor contact. Equation 9.50 is precisely illustrated in Figure 9.11a for the contact of n-type silicon electrode with the redox couple viologen$^{2+/+}$ for varying viologen$^{2+/+}$ redox potential. Another example of the same effect is shown in Figure 9.11b but in this case the quantity measured is the open-circuit photovoltage, (V_{oc}), that corresponds to the surface barrier, V_{sc0}, in this type of PEC cell, and correlates with the redox potential in perfect agreement with Equation 9.50.

The position of the conduction and valence bands of a range of semiconductors is shown in Figure 9.12. In many ionic semiconductors in aqueous solution, the surface potential is determined by the absorption of [H$^+$] at the surface. In general, in systems in which H$^+$/OH$^-$ is the potential determining redox couple, the point of zero charge is given in terms of the pH. Changes in pH of the solution displace the conduction band according to the expression

$$E_{cb} = E_{cb0} - 0.059\,\text{pH} \quad (\text{at } T = 300\,\text{K}) \quad (9.51)$$

The displacement of the flatband potential follows the trend of Figure 3.11. This is a characteristic feature of metal-oxide semiconductors (Bolts and Wrighton, 1976; Morrison, 1980). Figure 9.13a shows that the flatband potential determined by Mott–Schottky plot is shifted

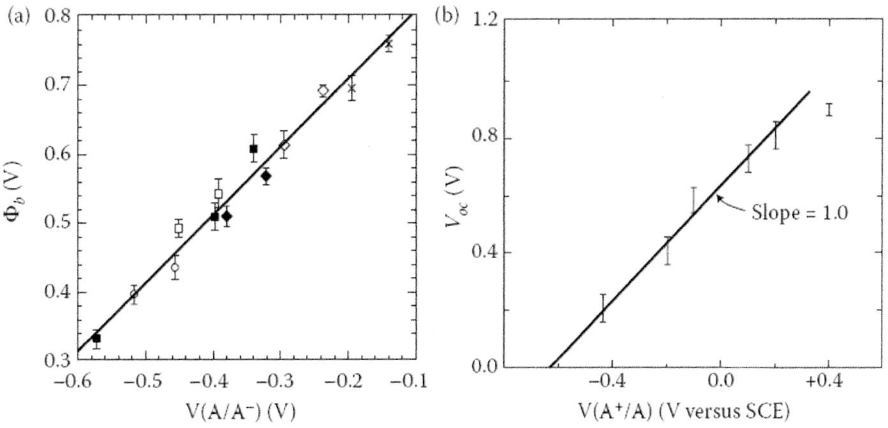

FIGURE 9.11 (a) Plot of the barrier height at the semiconductor depletion layer versus the Nernstian solution potential for n-Si/CH_3OH-viologen$^{2+,+}$ contacts for varying viologen$^{2+,+}$ redox potential. (Reproduced with permission from Fajardo, A. M.; Lewis, N. S. *The Journal of Physical Chemistry B* 1997, *101*, 11136–11151.) (b) Open-circuit photovoltage for the a-Si:H/CH_3OH interface as a function of solution redox potential. (Reproduced with permission from Gronet, C. M. et al. *Journal of the Electrochemical Society* 1984, *131*, 2873–2880.)

to negative potentials at increasing pH. The figure also shows the determination of band positions of Ta_2O_5, TaON, and Ta_2N_5 by UPS, which provides excellent agreement with the electrochemical measurement, see Figure 9.13c.

Fermi level pinning has been described in Section 4.6. At the semiconductor/ liquid contact, it is the effect in which V_{sc0} is independent of the redox potential (Bard et al., 1980). This is because all the CPD can be absorbed at the Helmholtz layer, or at least a part of it, as indicated in Equation 4.29. The change of the Helmholtz potential causes band unpinning. This feature is usually related to the presence of surface states so that the interfacial voltage can be located in the Helmholtz capacitor C_H, similar to the metal-semiconductor contact, where the voltage that shifts the VL occurs in the thin capacitance at the semiconductor surface.

If the influence of minority carriers trapping at the surface and majority carrier transfer to the solution can be neglected, we can assume that semiconductor surface states remain in equilibrium with the majority carriers in the corresponding band. A change of applied voltage modifies the occupation of surface states and produces band unpinning (Lincot and Vedel, 1987). To describe the modification of the Helmholtz potential by the charging of a surface state with electrons, we need to formulate the electrostatic conditions of the interface. The charge (per unit area) in the surface states with total density N_{ss} and occupation function f is

$$Q_{ss} = qN_{ss}f \tag{9.52}$$

The charge in the depletion region is obtained from Equations 9.17 and 9.19

$$Q_{sc} = qN_D w_0 \left(\frac{V_{sc}}{V_{sc0}} \right)^{1/2} \tag{9.53}$$

And the charge in the outer Helmholtz plane is

$$Q_H = C_H V_H \tag{9.54}$$

The compensation of charge can be stated as

$$Q_H = Q_{ss} = Q_{sc} \tag{9.55}$$

Therefore, the potential V_H is determined from the equation

$$C_H V_H + qN_{ss}f = +qN_D w_0 \left(\frac{V_{sc}}{V_{sc0}} \right)^{1/2} \tag{9.56}$$

In Figure 9.14a, the semiconductor electrode is in deep reverse conditions and the surface state is discharged. When a more negative potential is applied (Figure 9.14b), the charge of the surface state turns the Helmholtz potential more negative, to the value V_H^{fb}, and the conduction band shifts up. The value of the flatband potential as measured by MS analysis will change. The value close to flatband condition V'_{fb} is more negative than the value at large reverse V_{fb}, as shown in Figure 9.14c. The difference is given by the total charging of the Helmholtz layer

$$\Delta V_H = \frac{qN_{ss}}{C_H} \tag{9.57}$$

During Fermi level pinning, the capacitance remains constant, as shown in Figure 9.14c.

It should also be pointed out that in the analysis of this type of system by impedance spectroscopy, the capacitance in the MS plot is the high frequency capacitance, as the low-frequency capacitance will contain mainly the contribution of the chemical capacitance of the surface states (Bertoluzzi and Bisquert, 2012; Klahr et al., 2012).

According to these considerations, Equation 9.50 becomes strongly modified in the presence of surface states. As an example, Figure 9.15 shows that the V_{oc} of pyrite (FeS_2)/electrolyte junction remains almost constant at a value between 0.2 and 0.3 V over the redox potential range of 0.22–1.1 V (SCE). The fact that redox couples with different thermodynamic potentials give rise to nearly the same output photovoltage is attributed to Fermi level pinning by the surface states.

Figure 9.16a shows the photovoltage of *p*-type $CuInS_2$ electrode with respect to $V^{2+/3+}$ redox potential for different surface conditions. For the untreated sample the Fermi level is nearly completely pinned, but when the sample is etched for successively longer times, the voltage becomes finally nearly proportional to redox potential as indicated in Equation 9.50. The change of slope of the successive measurements in Figure 9.16a indicates a continuous and constant density of surface states that is progressively passivated by the etch treatment.

Another important effect of the surface states is the band unpinning under illumination that also causes a modification of the flatband potential (Kelly and Memming, 1982; Allongue and Cachet, 1985). In contrast to Figure 9.14, band unpinning under illumination is caused by the trapping of *minority carriers* at the surface states, as shown in Figure 9.17.

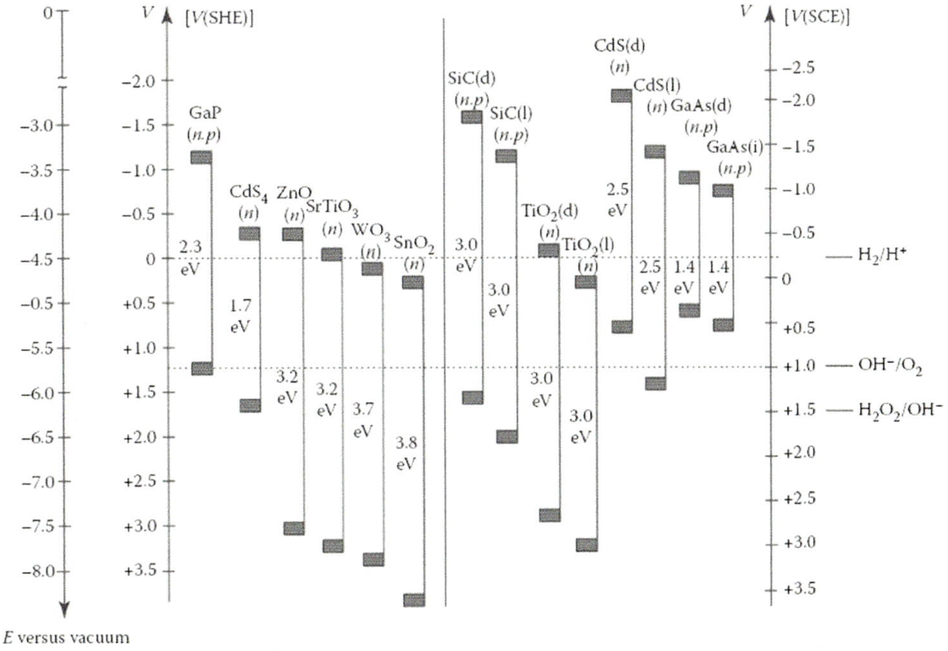

FIGURE 9.12 Band position of different semiconductors in dark (d) and under illumination (l) with respect to the SHE and SCE electrochemical scales (V) and the vacuum reference energy level (E). (Reproduced with permission from Dwayne Miller, R. J.; Memming, R. In *Nanostructured and Photoelectrochemical Systems for Solar Photon Conversion*; Archer, M. D., Nozik, A. J., Eds.; Imperial College Press: London, 2008.)

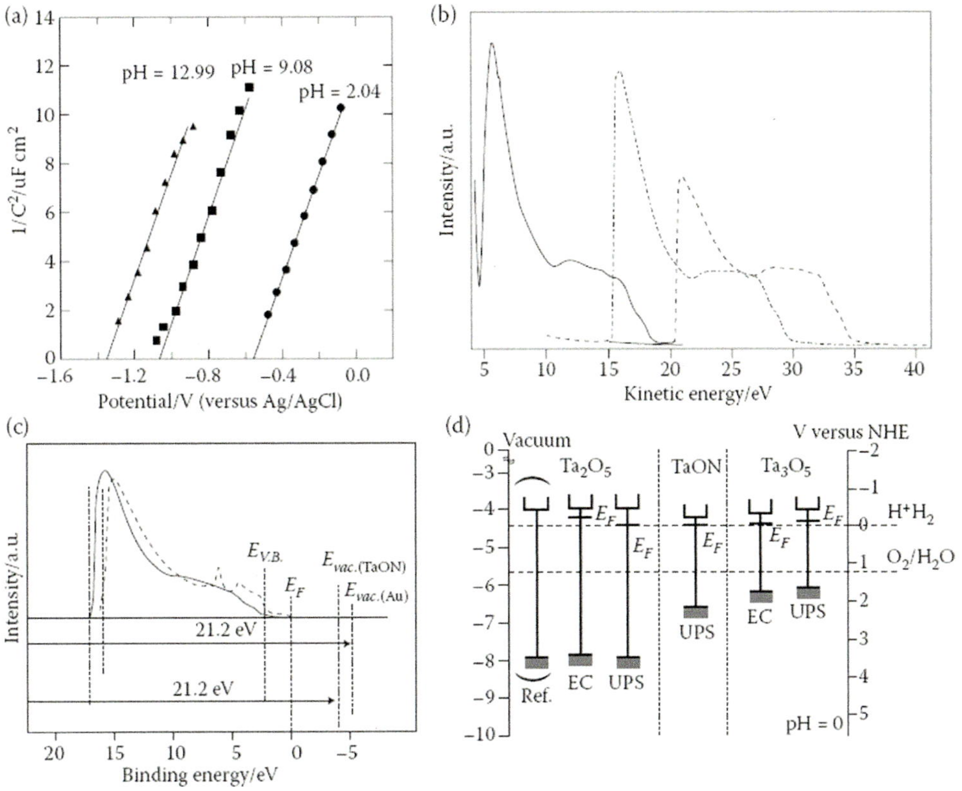

FIGURE 9.13 (a) Mott–Schottky plots of Ta_2O_5. (b) UPS spectra of TaON/Pt measured at various sample biases. Solid line: 5 V. Dash-dotted line: 10 V. Dashed line: 15 V. (c) Comparison and relationship of E_{VB}, E_F, and E_{vac} of UPS spectra. Solid line: TaON/Pt. Dashed line: Au on TaON/Pt. (d) Band positions of Ta_2O_5, TaON, and Ta_3N_5 determined by electrochemical analysis and UPS measurements. (Reproduced with permission from Chun, W.-J. et al. *The Journal of Physical Chemistry B* 2003, *107*, 1798–1803.)

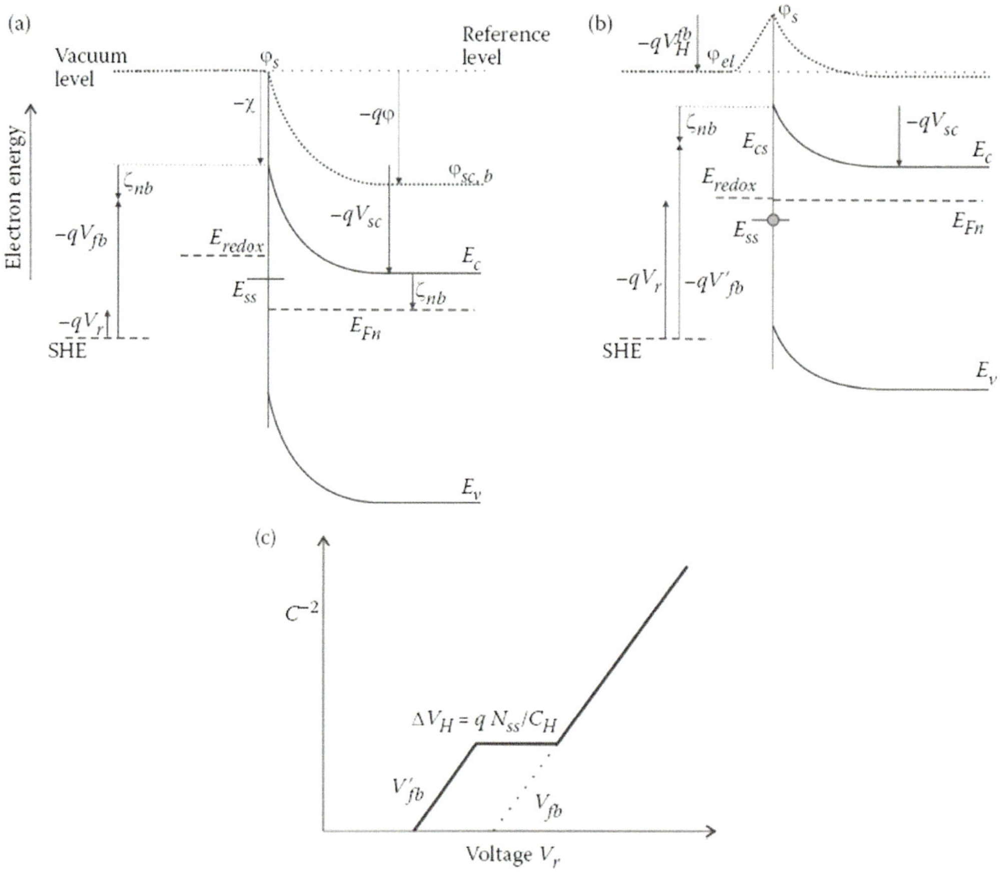

FIGURE 9.14 (a) Energy diagram of a semiconductor/electrolyte junction, in which the semiconductor has a surface state that remains in equilibrium with the majority carrier band. The semiconductor Fermi level is below the E_{SS} level, so that the surface state is empty of electrons. (b) When the Fermi level raises, at more negative potential, the band shifts up (Fermi level pinning) until the surface state is occupied and the band bending *V*sc changes again. (c) Mott–Schottky plot with different regions of capacitance.

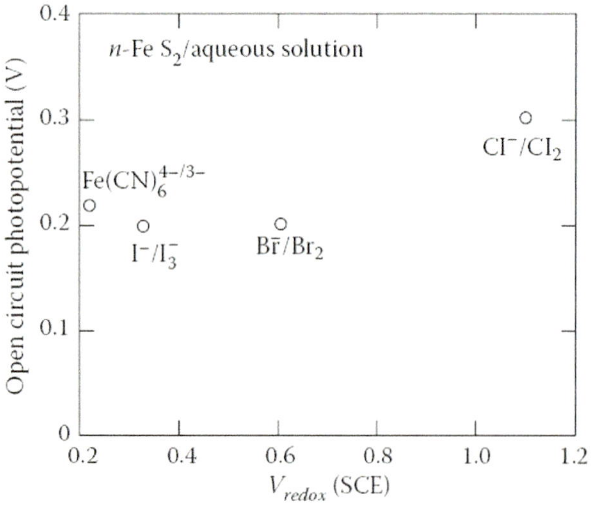

FIGURE 9.15 Open circuit photovoltage for the pyrite (FeS_2)/electrolyte junction as a function of solution redox potential. (Reproduced with permission from Mishra, K. K.; Osseo-Asare, K. *Journal of the Electrochemical Society* 1992, *139*, 749-752.)

9.8 INVERSION AND ACCUMULATION LAYER

The application of bias voltage across a Schottky barrier produces different charge distributions in the SCR:

a. Accumulation mode, in which the Fermi level of the majority carrier approaches the respective band edge.
b. Depletion mode, fully discussed in Section 9.3.
c. Inversion mode, in which the Fermi level of the minority carrier approaches the respective band edge.

These three domains are shown in Figure 9.18 for a *p*-type material in a metal-insulator-semiconductor (MIS) field-effect transistor. Under strong forward bias, the Fermi level enters the band edge and the semiconductor is in accumulation regime. In solar cells and in photoelectrochemical cells, the domain of main interest is the depletion region that assists in charge separation

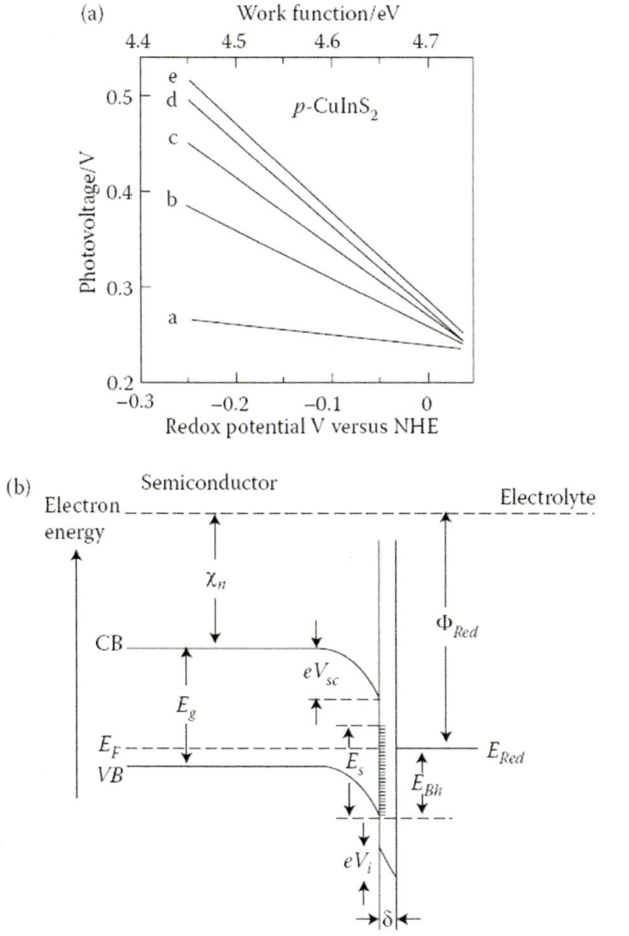

FIGURE 9.16 (a) Dependence of the photovoltage of p-$CuInS_2$ on $V^{2+/3+}$ redox potential for different surface conditions, *a* is the untreated sample and for *b-e*, samples are etched for successively longer times. (b) The energy picture to explain the partial Fermi level pinning in terms of a constant distribution of surface states. (Reproduced with permission from Lewerenz, H. J.; Goslowsky, H. *Journal of Applied Physics* 1988, *63*, 2420424.)

and contact selectivity. However, the depletion voltage is finally limited by the formation of an inversion region. When the reverse bias makes the minority carrier Fermi level approach the band edge of this carrier, a large density of the minorities is formed at the semiconductor surface (Figure 9.18c). The practical application of the inversion layer is often closely related to the transistor configuration. Let us discuss some specific examples.

In Figure 9.19 is shown the metal/electrolyte interface in a photoelectrochemical cell. The inversion layer creates minority carriers and constitutes a p–n junction at the semiconductor surface that prevents charge transfer of electrons to the electrolyte. Figure 9.20 displays an MIS device in which the semiconductor is p-type. When the metal is polarized positively, there is a constant electric field across the insulator as in a thin film transistor. The band is strongly bent at the contact with the insulator, and the Fermi level in the semiconductor approaches the conduction band forming an inversion layer populated of minority carriers. In the device of Figure 9.20, that is an early example of electroluminescent device, the inversion layer is removed by sudden switch of the voltage, causing luminescence between the voltage-generated minority and the majority carriers that refill the depletion layer.

9.9 HETEROJUNCTIONS

A semiconductor homojunction is termed p–n junction when made of n- and p-type doped pieces of the same material. The formation of the junction and the energetic structure across the contact is shown in Figure 2.18. The possibility to form such junctions depends on the facility to dope a material obtaining both types, which is possible for Si and most III–V semiconductors, though difficult with metal oxides, for example, due to the very large ionization potential.

The central assumption of device operation in a biased p–n junction is that the Fermi level of each carrier is conserved across the contact between the n- and p-type materials. The voltage applied to the junction, V, is given by

$$-qV = E_{Fn} - E_{Fp} \tag{9.58}$$

All the applied voltage between the outer contacts appears at the junction. The bands are not tilted elsewhere. In Figure 9.21a, the junction is drawn intentionally small, to emphasize the distribution of carriers in the neutral regions, which is most relevant for solar cell applications. In Figure 9.21b, it is observed that electrons at the left side in the n-doped region are majority carriers, and have a stable Fermi level which is independent of bias voltage. But the Fermi level of electrons at the p side of the junction increases when applying a negative bias since $E_{Fn} = E_{Fp} - qV$, causing the electron density to increase in the p region. Here, the electrons are minority carriers. The conversion of minorities to majorities plays a central role in the operation of a p–n junction as selective contact in inorganic solar cells.

A heterojunction is a junction between crystalline semiconductors that have different bandgaps. Band alignment rules are necessary to predict the properties of heterojunctions. An example is shown in Figure 9.22. This figure shows the significance of the different energy levels that were discussed in Chapter 2: the VL,

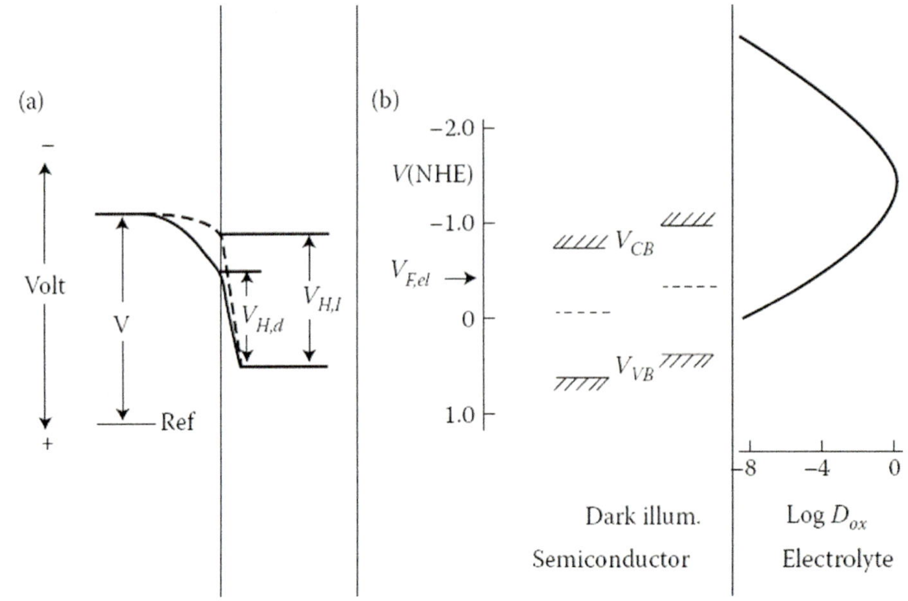

FIGURE 9.17 Behavior of a p-type semiconductor under illumination. Photogenerated minority carrier electrons are trapped at surface states increasing the potential of the Helmholtz layer (a) so that the band shifts up (b). (Reproduced with permission from Kelly, J. J.; Memming, R. *Journal of the Electrochemical Society* 1982, *192*, 730–738.)

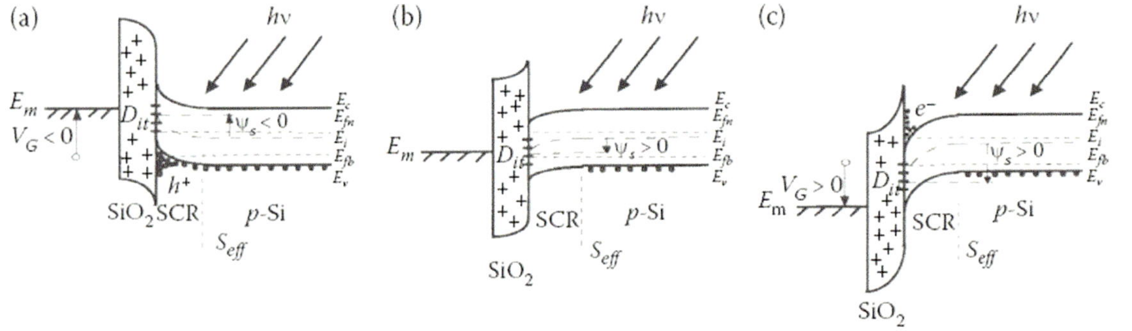

FIGURE 9.18 Band diagram illustration of a metal-oxide semiconductor system consisting on SiO_2 on p-type Si, under a nonequilibrium condition. Three different voltage biasing conditions are shown: (a) accumulation mode, (b) depletion mode, and (c) inversion mode. Positive charges are shown uniformly distributed in the gate oxide. The separation of electron and hole Fermi levels is due to illumination. (Reproduced with permission from Bai, Y.; Phillips, J. E.; Barnett, A. M. *Electron Devices, IEEE Transactions on* 1998, *45*, 1784–1790.)

the conduction band (and valence band) level, and the Fermi level. The difference of the Fermi levels is a built-in potential that has to be transferred at the interface by an equivalent displacement of the VL so that the Fermi level across the junction in equilibrium is flat. In addition to the difference of Fermi levels, the separate materials present an offset of the conduction band levels, obtained by taking the difference between their electron affinities, $\Delta\chi$. The simplest rule to establish the energetic structure of the heterojunction is the *electron affinity rule*, in which the initial conduction band offset is preserved (Capasso and Margaritondo, 1987). Neglecting any surface dipoles and charges, the offset of the VL that is established corresponds to band bending in the semiconductors, as indicated in Figure 9.22b.

In general, a rearrangement of atoms occurs at the semiconductor surface, with respect to the bulk, in order to reduce the total energy of the surface, and this modification is termed surface reconstruction. The reconstruction involves the outward or inward displacement of surface atoms, which leads to the change of the surface dipole layer accompanied by a change of the effective affinity, as discussed in Chapter 2. These effects may produce a considerable variation of the electron affinity up to ≥1 eV. For example, for metal oxides, a metal termination leads to a low and an oxygen termination to

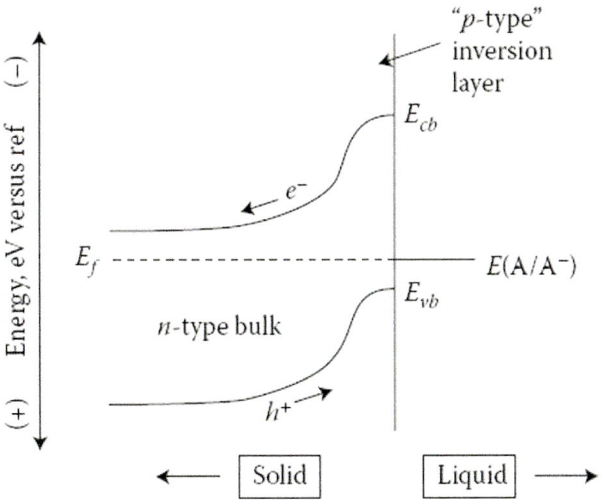

FIGURE 9.19 Band diagram of an n-type semiconductor/liquid junction in which the nearsurface region of the semiconductor contains an inversion layer. In this situation, the semiconductor/liquid junction device acts as an "in situ" n–p homojunction device. (Reproduced with permission from Laibinis, P. E.; Stanton, C. E.; Lewis, N. S. *The Journal of Physical Chemistry* 1994, *98*, 8765–8774.)

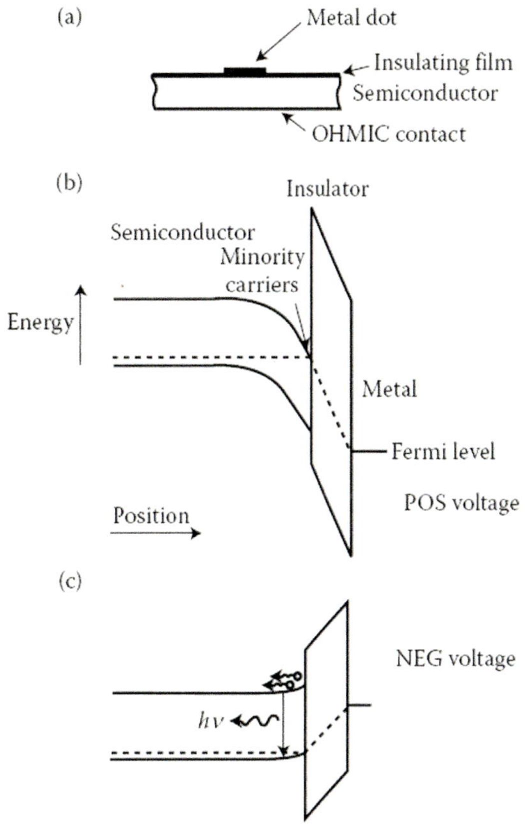

FIGURE 9.20 (a) Metal-insulator-semiconductor (MIS) structure. (b) Band diagram at positive metal voltage inducing an inversion layer by creation of minority carriers at the semiconductor surface. (c) Band diagram at negative metal voltage approaching flatband conditions. This is a sudden inversion of the voltage that sweeps the minority carriers away from the surface and promotes recombination and luminescence. (Reproduced with permission from Berglund, C. N. *Applied Physics Letters* 1966, *9*, 441–444.)

a high electron affinity (Klein, 2012). At semiconductor–semiconductor interfaces, the interface reconstruction will be different than the surface reconstruction. In these cases a dipole occurs at the interface and the electron affinity rule does not hold true. Measurements of XPS and UPS are necessary to obtain an accurate energy diagram of the heterojunction.

In Section 4.4, we showed how the UPS method is used to determine the energy values and surface dipole of a thin layer that is progressively grown on a metal substrate that serves as a reference. This was done by measuring the energy of electrons ejected from the Fermi level and deeper levels in the conduction band. This method is also applied to determine the band diagram of heterojunctions and an example is shown in Figure 9.23 for CdS layer on Cu_2S substrate. First, the emission of electrons from the Fermi level and below for Cu_2S layer allows to determine that the valence band maximum is 0.1 eV below the Fermi level, indicating high p-type doping of the Cu_2S film (Figure 9.23a). However, when the layer of CdS is deposited on top of Cu_2S, the emission of electrons from the conduction band of the two materials will be mixed.

To obtain more reliable information, the emission of core levels from separate atoms is monitored (Figure 9.23b). The XPS core level and valence band spectra are measured after each growth step. The measurements indicate the variation of energy levels from the Cu_2S surface to a complete coverage by the CdS layer. It must be recalled that the energy reference is the Fermi level that is flat for the equilibrium situation. In Figure 9.23b, there is a pronounced shift of all core level lines to larger binding energies, which indicates the formation of a space-charge layer by band bending in the substrate. The Cu 2p emission was gradually decreased in intensity and the Cd 3d intensity increases during the deposition process. The final line up contains band bending in both Cu_2S substrate and CdS layer as well as the interfacial dipole of 0.3 eV (Figure 9.23c).

9.10 EFFECT OF VOLTAGE ON HIGHLY DOPED NANOCRYSTALLINE SEMICONDUCTORS

A nanoporous thin film is one contiguous material that contains porosity at the nanoscale. When filled with an electrolyte the liquid medium floods the pores and

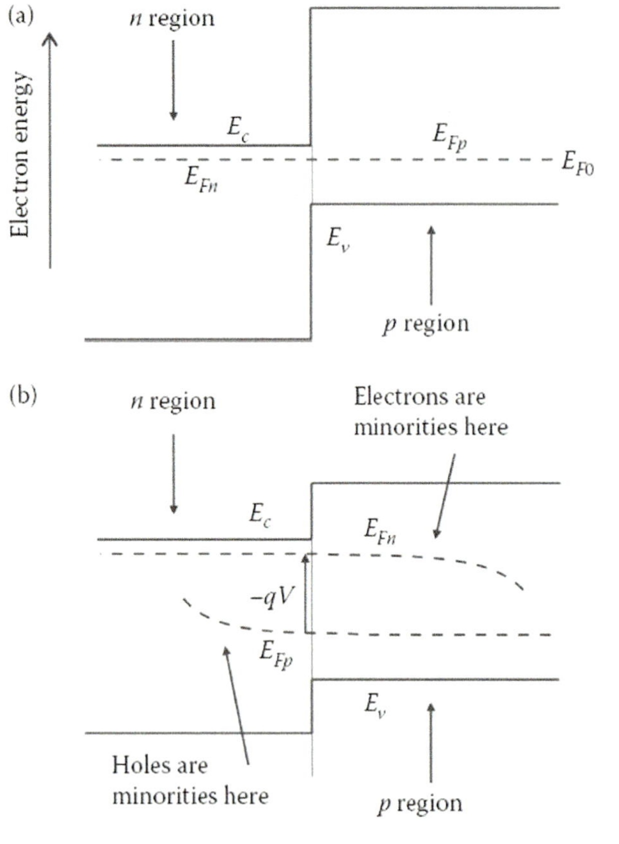

FIGURE 9.21 Energy diagram of p–n junction. Depletion layers existing in the contact region (where the band bending produces the steps of E_c and E_v) are not shown, these layers are assumed to be extremely thin. (a) No bias voltage. (b) The n-type side is negatively biased (forward bias). The voltage causes the separation of electron and hole Fermi levels at the contact between the n- and p-type materials. The population of minorities increases at both sides of the junction. The diagram shows that electrons entering the p-type side become minority carriers, since $E_c - E_{Fn} \gg E_{Fp} - E_v$ and electrons are much less abundant than holes, and vice versa for holes entering the n-type region.

immediately forms a perfect electrical junction, as indicated in Figure 8.12. Nanoporous semiconductor films allowing the transport of electronic carriers from a conducting substrate to a large internal area are the basis for a wide variety of energy, optoelectronic, and functional devices such as solar cells, sensors, supercapacitors, and biodevices. For instance, the semiconductor nanowires and nanotubes have been amply investigated due to their potential applications in several optoelectronic devices and sensors.

In the nanocrystalline film electrode, we can find different types of behavior associated with the size of the nanoparticles. Figure 9.1 shows that for a planar, semi-infinite semiconductor, the Fermi level difference between the contacting media is accommodated as an interfacial depletion barrier. But in the case of mesoporous semiconductors, a critical parameter that comes into play is the size of structural units as nanoparticles or nanowires. The standard Schottky barrier requires that the size of the depletion layer be smaller than the nanoparticle size. The physical yardstick whereby the voltage can be accommodated in a semiconductor element is the Debye length in the semiconductor, as discussed in Section 5.7 (see Equation 5.44). In Figure 5.10, we observe that the parameters $n_0 = 10^{16}$ cm^{-3} and $\varepsilon_r = 10$ lead to $\lambda_D \approx 50$ nm. At this doping density, it is not possible to sustain a 0.2 V barrier in a nanocrystalline semiconductor in which the radius of nanoparticles is $a = 10$ nm. If however the semiconductor is highly doped to $n_0 = 10^{19}$ cm^{-3}, then band bending into the nanoparticles is possible, since $\lambda_D \approx 2$ nm. The Debye length in the semiconductor nanoparticles, therefore, sets how the nanocrystalline film responds to an applied voltage. In the following, we analyze the case of a high doping density implying $\lambda_D \ll a$, and the opposite case is discussed in the next section.

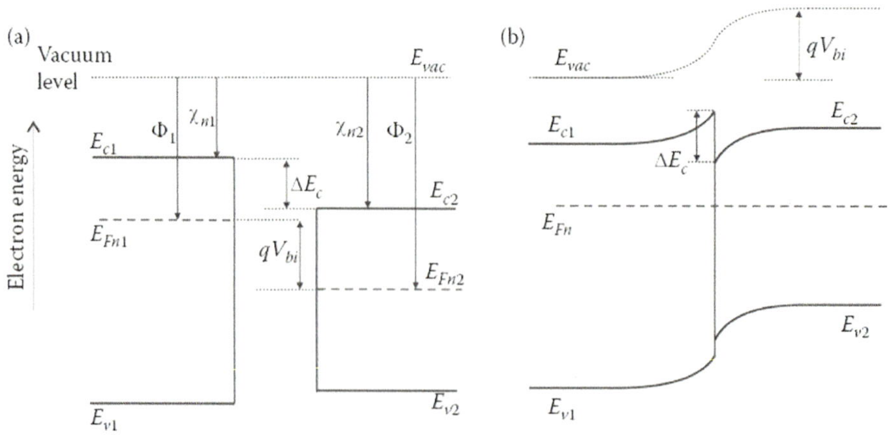

FIGURE 9.22 Energy levels of a heterojunction before (a) and after (b) contacting both semiconductors following the electron affinity rule, in which the starting difference of electron affinities is preserved after contact.

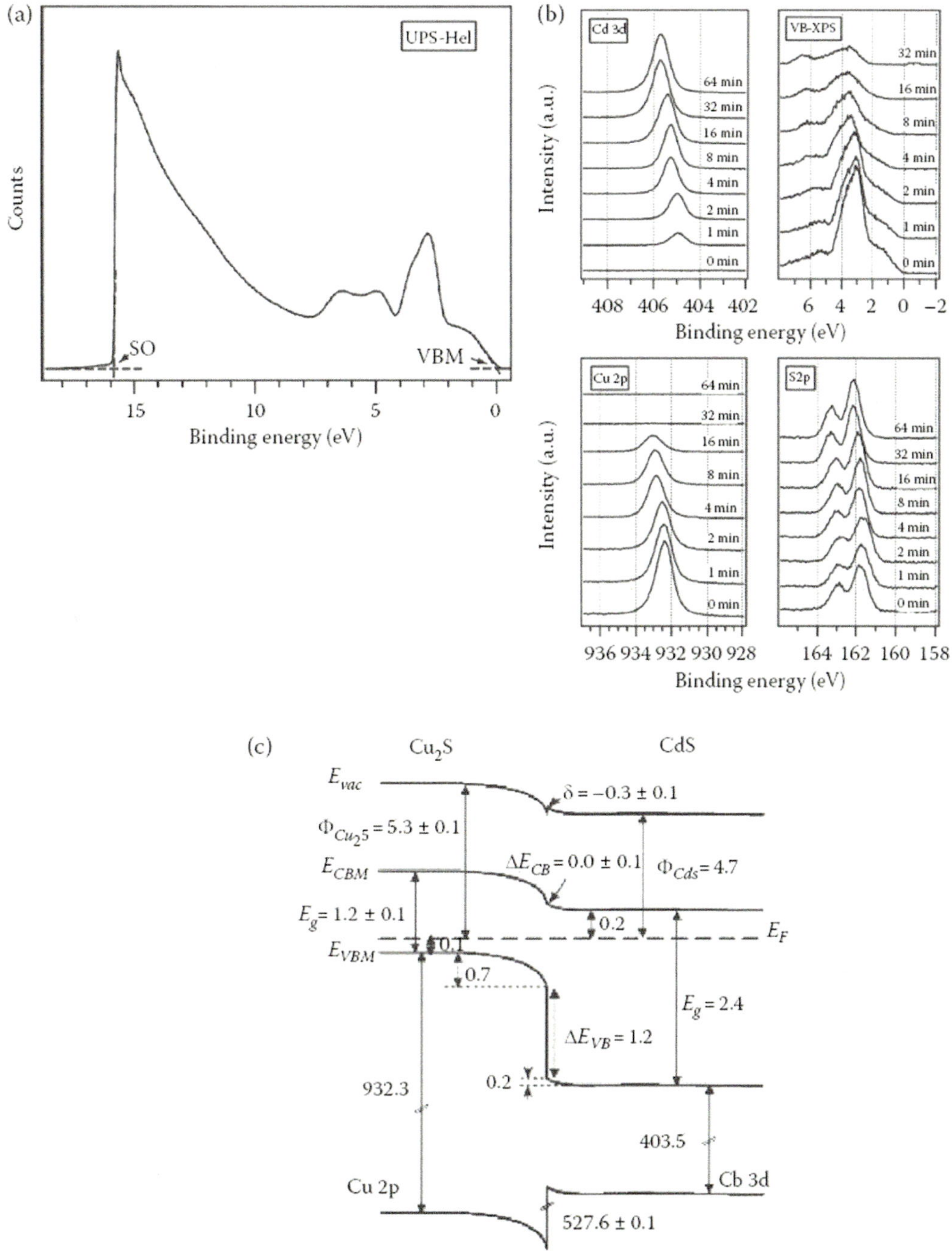

FIGURE 9.23 (a) Secondary emission UPS (He I) of a 100 nm Cu_xS film deposited on SnO_2. (b) Photoemission spectra of core levels and valence band measured after each CdS growth step on Cu_2S substrate. (c) Band diagram in the Cu_2S/CdS system fabricated by vacuum evaporation of CdS on Cu_2S under UHV condition. (Reproduced with permission from Liu, G. et al. *Thin Solid Films* 2003, *431–432*, 477–482.)

When $\lambda_D \ll a$ is satisfied, the nanoparticle can sustain the band bending associated with a substantial potential drop V_{sc}. The *n*-type nanoparticulate semiconductor situation is shown in Figure 9.24a. The control of the voltage is obtained by interconnection of the nanoparticles that are deposited on top of a conducting substrate, as indicated in Figure 9.25. By applying positive voltage to the substrate, we increase the surface depletion region. An interesting feature of the system is that the maximum amount of band bending is limited to the situation in which the central quasi-neutral region shrinks to one point, as indicated in Figure 9.24b.

To determine the amount of band bending, we use the depletion approximation in a spherical particle, so that full depletion occurs in the layer $r_1 \le r \le a$ (Figure 9.24a)

(Albery and Bartlett, 1984; Bisquert et al., 1999). The potential is distributed according to the Poisson equation

$$\frac{1}{r^2}\frac{\partial}{\partial r}\left(r^2\frac{\partial\varphi}{\partial r}\right)=-\frac{qN_D}{\varepsilon} \tag{9.59}$$

It has the solution

$$\varphi=-\frac{qN_D}{6\varepsilon}\left[r^2-3r_1^2+2\frac{r_1^3}{r}\right] \quad (r_1\le r\le a) \tag{9.60}$$

By putting $r = a$ and $r_1 = 0$, the maximum band bending is therefore

$$V_{sc,max}=\frac{k_BT}{6q}\left(\frac{a}{\lambda_D}\right)^2 \tag{9.61}$$

This formula also shows that the condition $\lambda_D \ll a$ is required to obtain $V_{sc} \gg k_BT/q$.

If a voltage $|v| > V_{sc,\ max}$ is applied, as shown in Figure 9.24c, then there is a missing voltage drop, since the Fermi level in the semiconductor is brought down but this is not accompanied by a potential drop in the nanoparticle. To view the situation it is convenient to use a simpler columnar geometry that represents the nanocrystalline electrode, as indicated in Figure 9.25b. The result of applying a voltage difference between the base of the column and the other faces (that have the potential of the wetting electrolyte) is shown in Figure 9.26, for a column in which $\lambda_D = a$ (Bisquert et al., 1999). It is observed that any voltage in excess of $V_{sc,\ max}$ falls in a short region close to the substrate. Due to the constraint of the size of the semiconductor, the electric field cannot penetrate along the axis of the cylinder.

When an array of semiconductor nanorods is immersed in solution, the outer surface of the rods is depleted of carriers, forming surface band bending in the radial direction, while the central region of the rods is a conducting (quasi-neutral) tubular region connected to the substrate, as indicated in Figure 9.27. This structure may facilitate the channeling of electrons toward the current collector, avoiding recombination at the surface. Since the whole surface of the rods is an equipotential, the surface barrier can be manipulated by modifying the voltage of the substrate with respect to solution, as noted earlier. Furthermore, the depletion layer has a circular shape, and this introduces a strong modification to the standard linear MS relationship of Equation 9.24 (Mora-Seró et al., 2006; Tena-Zaera et al., 2008). We discuss the calculation of

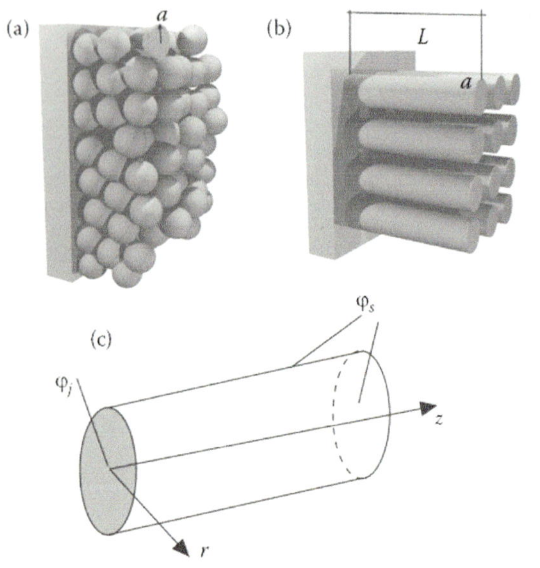

FIGURE 9.25 Schematic representations of nanocrystalline electrodes. (a) An array of connected spherical particles. (b) A bundle of columnar particles. (c) View of a columnar particle showing a coordinate system. The shaded circle represents the junction with the substrate.

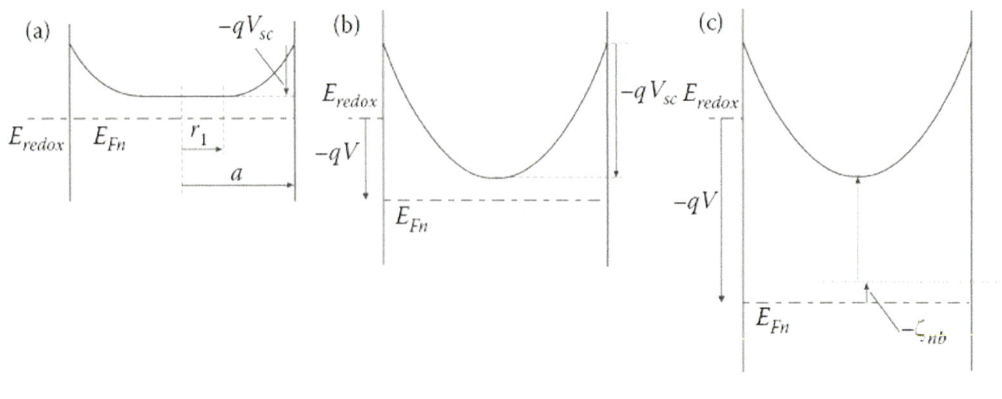

FIGURE 9.24 Energy band edges in a spherical n-type semiconductor particle of radius a. Three cases are shown: (a) Partially depleted particle. (b) Fully depleted particle by application of a positive voltage to the nanocrystalline film. (c) Fully depleted particle with the Fermi level well below depletion level.

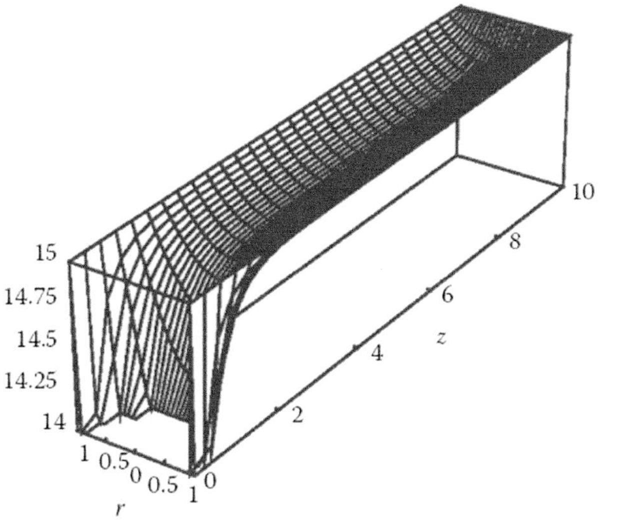

FIGURE 9.26 The local potential (conduction band energy) in a longitudinal section of a fully depleted cylindrical semiconductor particle for $a = \lambda_D$, $L = 10a$, $\varphi_j = 5k_BT/q$, and $\varphi_s = 15k_BT/q$.

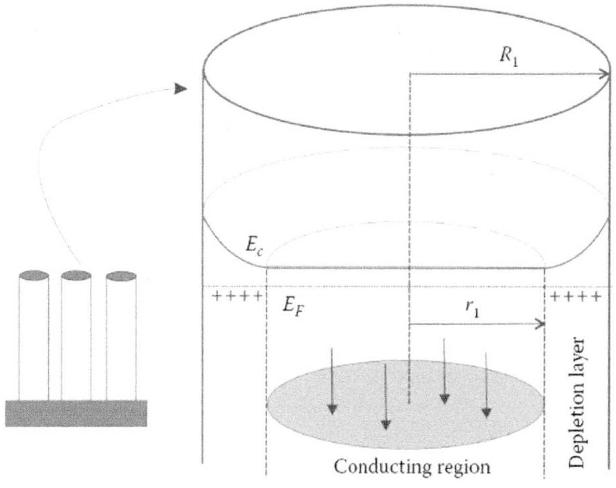

FIGURE 9.27 Band bending and carrier distribution in a semiconductor rod of radius R_1 in contact with electrolyte. The surface depletion layer and the central conducting region (radius r_1) are indicated.

band bending structure in semiconductor nanotubes as shown in Figure 9.27. This calculation works for any geometry whose cross section can be approximated by ring geometry. The main restriction of this model is that the axial dimension must be much larger than the radial one.

We count the charge inside a cylinder of radius r with $r_1 \le r \le R_1$, being R_1 the radius of the external surface of the tube, and r_1 the edge of the external depletion region

$$Q_1 = qN_D\pi L\left(r^2 - r_1^2\right) \tag{9.62}$$

To apply Gauss law, we remember that the electric field at the edge of the depletion region ($r = r_1$) is 0 and the result is

$$\mathbf{E}(r) = \frac{Q_1}{2\pi rL\varepsilon} = \frac{qN_D}{2\varepsilon_r\varepsilon_0}\left(r - \frac{r_1^2}{r}\right) \tag{9.63}$$

The potential across the external depletion layer is

$$V_{sc} = V_{s1} - V(r_1) = -\int_{r_1}^{R_1}\mathbf{E}dr = \frac{qN_D}{2\varepsilon}\left[\frac{1}{2}\left(r_1^2 - R_1^2\right) + r_1^2\ln\frac{R_1}{r_1}\right] \tag{9.64}$$

Now we take the derivatives

$$\frac{dQ_1}{dr_1} = -2qN_D\pi Lr_1 \tag{9.65}$$

$$\frac{dV_{sc}}{dr_1} = \frac{qN_Dr_1}{\varepsilon}\ln\frac{R_1}{r_1} \tag{9.66}$$

and, therefore, the capacitance of the rod wall is given by

$$C_1 = -\frac{dQ_1}{dV_s} = \frac{2\pi L\varepsilon}{\ln\left(R_1/r_1\right)} \tag{9.67}$$

and the specific capacitance per surface unit is

$$c_1 = \frac{\varepsilon}{R_1\ln\left(R_1/r_1\right)} \tag{9.68}$$

Combination of Equations 9.64 and 9.68 provides the voltage-dependence of the capacitance, which is illustrated in Figure 9.28 for different values of the tube radius. For the thinnest tubes, the effect of the confined size is very significant and the MS plot shows a large curvature due to the restriction of the depletion layer in the tubular geometry. As the radius increases, the depletion layer w is confined in a short region close to the surface and the straight MS line (Equation 9.24) is recovered. For the case in which the depletion layer is much thinner than $R_1(r_1 \approx R_1)$, an expansion of the denominator of Equation 9.67 gives the following result:

$$C_1 \approx \frac{2\pi Lr_1}{R_1 - r_1} \tag{9.69}$$

Experimentally, it has been found that electrochemically grown ZnO rod arrays show a very high doping of the order 10^{20} cm^{-3}, which decreases considerably by thermal annealing (Mora-Seró et al., 2006). When the

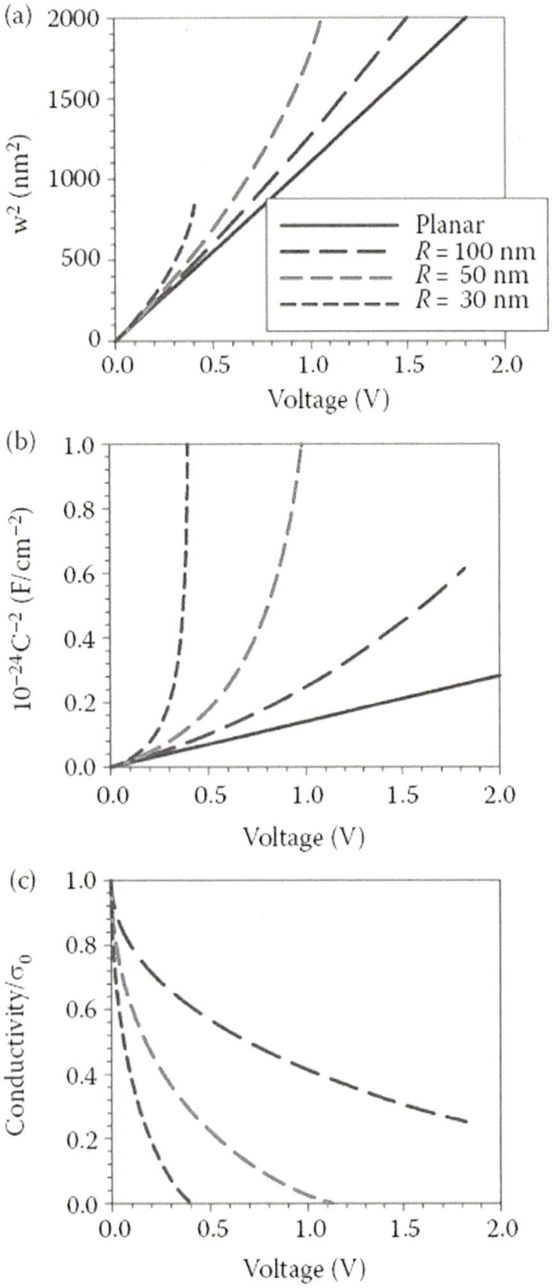

FIGURE 9.28 Simulation of (a) square of the depletion layer thickness, (b) inverse square (specific) capacitance and (c) relative axial conductivity, as a function of voltage across the barrier, for semiconductor rods of different radii as indicated. Parameters used in the simulation: $\varepsilon_r = 10$, $N_D = 10^{18}$ cm^{-3}.

depletion layer grows toward the center of the wire, the central conducting region shrinks, and the resistance of the wire, measured from end to end, increases accordingly. The resistance of a single wire can be measured provided that adequate contacts are formed as shown in Figure 9.29. The resistance changes drastically and depends on the oxygen content of the surrounding

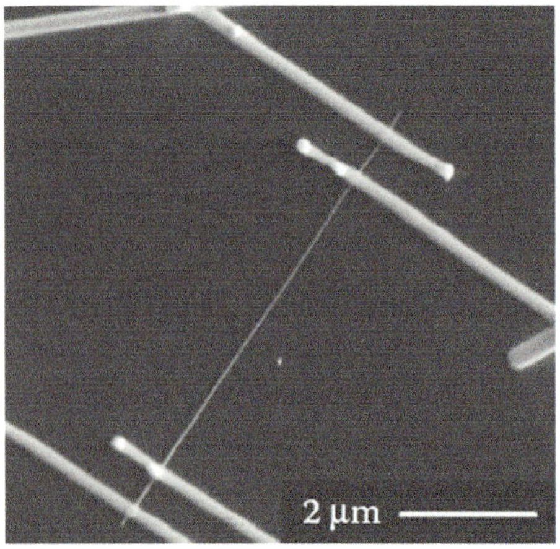

FIGURE 9.29 SnO_2 nanowire of radius $r = 27 \pm 3$ nm and length $L = 11$ μm electrically contacted using dual-beam focused ion beam nanolithography techniques. (Reproduced with permission from Hernandez-Ramirez, F. et al. *Physical Review B* 2007, *76*, 085429.)

atmosphere, which provides opportunities for sensing. The variations of the conductivity depend mainly on two factors: surface defects that are modified by ion adsorption, and ion diffusion into the metal-oxide structure (Jin et al., 2006; Ajay and Aslam, 2012).

In nanotubes or cylindrical nanowires, the center is hollow and the depletion layer can grow from both, the external and internal facets. For the internal depletion layer in the nanotube of Figure 9.30, a similar calculation as above gives the following results:

$$V_{sc} = \frac{qN_D}{2\varepsilon}\left[\frac{1}{2}\left(r_2^2 - R_2^2\right) + r_2^2 \ln\frac{R_2}{r_2}\right] \tag{9.70}$$

$$C_2 = \frac{2\pi L\varepsilon}{\ln\left(R_2/r_2\right)} \tag{9.71}$$

9.11 HOMOGENEOUS CARRIER ACCUMULATION IN LOW-DOPED NANOCRYSTALLINE SEMICONDUCTORS

In the first section of this chapter, we showed that the modification of the majority carrier Fermi level by application of a voltage bias causes a change in the semiconductor depletion layer. In Section 9.10, we analyzed the highly doped nanocrystalline films in which an internal field distribution, associated with band bending, can be recognized. Now we treat the opposite situation of low-doped nanoparticles that is widely applied in devices such as dye-sensitized solar cells. In Section 8.3, we

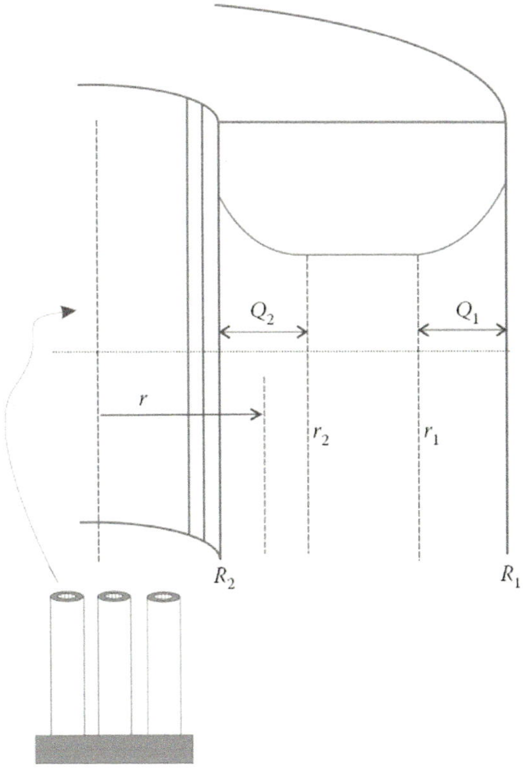

FIGURE 9.30 Electric potential, band bending, and carrier distribution in a semiconductor tube in contact with electrolyte.

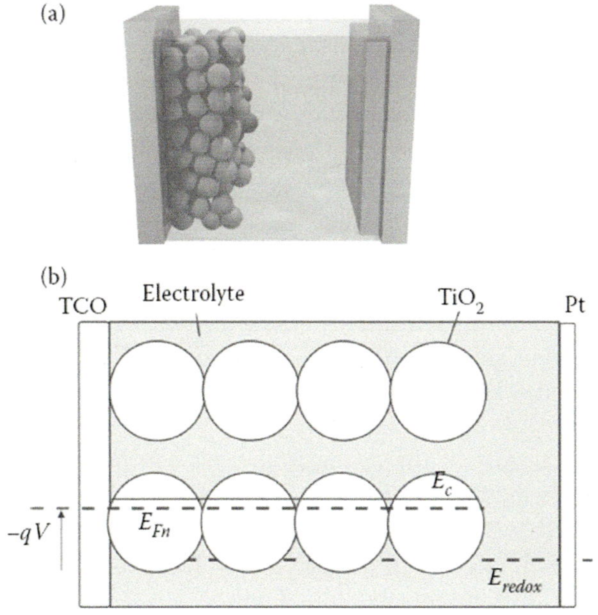

FIGURE 9.31 (a) Scheme of a nanocrystalline metal-oxide electrode, immersed in electrolyte, and a counterelectrode forming an electrochemical cell. (b) The Fermi level of electrons in the nanoparticles is controlled from the transparent conducting oxide substrate. The nanoparticles are smaller than Debye length, hence conduction band level is homogeneous.

discussed the charging of nanocrystalline TiO_2 with special emphasis of the measurement of the chemical capacitance, related to the DOS of the nanoparticulate semiconductor. We will now revisit the system shown in Figure 8.12 to obtain more insight into the operation mechanisms.

Figure 9.31 represents a nanocrystalline semiconductor that has a low doping density, which means that the Debye length is very long in comparison with the size of the semiconductor elements $\lambda_D^{sc} \gg a$. If the liquid medium filling the nanopores is a highly concentrated electrolyte, the Debye length of the electrolyte is rather small, of the order of nanometers. This means that the combined effective homogenous medium consisting of nanocrystalline semiconductor and electrolyte cannot sustain long-range electric fields, because the electrolyte rapidly produces a redistribution of ions that will shatter space charge (Paasch et al., 1993). The macrohomogeneous system must obey electroneutrality at scales larger than several times the structural unit size, a (Zaban et al., 1997). Since the electrolyte contains an ionic salt in high concentration, an increased concentration of electron density in the semiconductor, caused by voltage-injected or photogenerated carriers, is compensated by an increase of local positive charge, as indicated in Figure 8.12b. In fact, the nanocrystalline semiconductor in contact with electrolyte behaves like a doped semiconductor. In the nanostructure, there is only one kind of carrier, for example, electrons, that can be considered minority carriers, and the positive ions in solution are the majority carriers. If the salt concentration in the electrolyte is too low to balance the injected electron charge, a failure of charge compensation occurs and the device will not work properly (Zheng and Jow, 1997; Robinson et al., 2010).

The long Debye length property implies that the displacement of the Fermi level in Figure 9.31, by modification of the voltage applied at the substrate, cannot be accommodated as band bending, in contrast to the case discussed above. Details of band bending in individual nanoparticle elements may not be resolved simply by a voltage measurement. Thus, we assume that the band of the semiconductor is homogeneous, and the lowest energy state is the conduction band minimum, E_c. Hence, the displacement of the Fermi level causes accumulation of electron carriers in the nanocrystalline semiconductor. Expressions for electron density as a function of voltage have been given in Section 5.4.

We observe an important effect, shown in Figures 8.14, 8.15d, and 8.17b. The capacitance line that represents the exponential DOS of anatase TiO_2 can be displaced left and right in the voltage scale, which means that the DOS is shifted down or up in the energy scale. This effect is known as a *shift of the conduction band* and it has already been described in Section 9.7, where we showed that it produces a displacement of the flatband potential in an MS plot. For a low-doped nanocrystalline semiconductor, all the energy levels of the semiconductor nanoparticles are effectively displaced with respect to the redox level or reference electrode that is taken as reference in solution.

The shift of the conduction band can be intentionally formed in several ways. In Figure 9.32, the shift is caused by a surface adsorbed dipole. If the dipole points outward, the band is displaced upward, then the capacitive CV is negatively shifted in the voltage scale (see Figure 9.33Ba and 9.33Bc). Another way to modify the position of the conduction band is to change the surface pH in aqueous solution, according to Equation 9.51. In Figure 8.15d, the shift is approximately 450 mV, in good agreement with 59 mV/pH unit predicted. A similar shift is observed also in Figure 8.17b, however in this case the shift is accompanied by a change in the slope, which indicates a modification of the DOS. A shift of the conduction band occurs also if the size or concentration of cations added to a polar solution is modified, due to the modification of the Helmholtz layer (Boschloo et al., 2006). In particular, the presence of Li^+ ions produces a downward shift of the conduction band of TiO_2.

In the previous mechanism, the shift of the conduction band is permanent because the modification of surface conditions remains constant at all potentials. A different effect is shown in Figure 9.33b. Here, the shift is caused by charging of the Helmholtz layer due to ionic

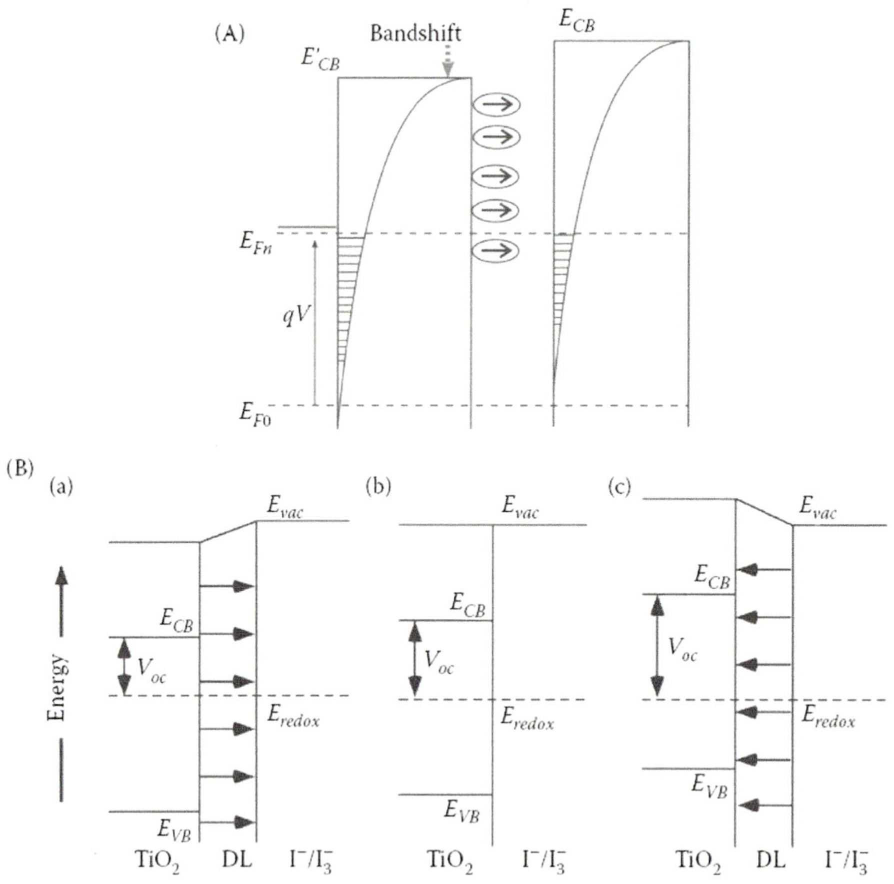

FIGURE 9.32 (A) Scheme showing a downward shift of the conduction band of TiO_2 that displaces the exponential DOS. (B) Energy band diagrams of TiO_2 nanoparticle/electrolyte interface with (a) a dipole layer composed of molecules with positive dipole moments at the interface (arrow points toward negative pole), (b) without dipole layer at the interface, and (c) with a dipole layer of negative dipole moment molecules at the interface: E_{vac}, local vacuum level; E_{CB}, bottom of the conduction band; E_{vb}, top of the valence band; E_{redox}, redox potential of the electrolyte; V_{oc}, open-circuit voltage. (Reproduced with permission from Ruhle, S.; Greenshtein, M. et al. *The Journal of Physical Chemistry B* 2005, *109*, 18907–18913.)

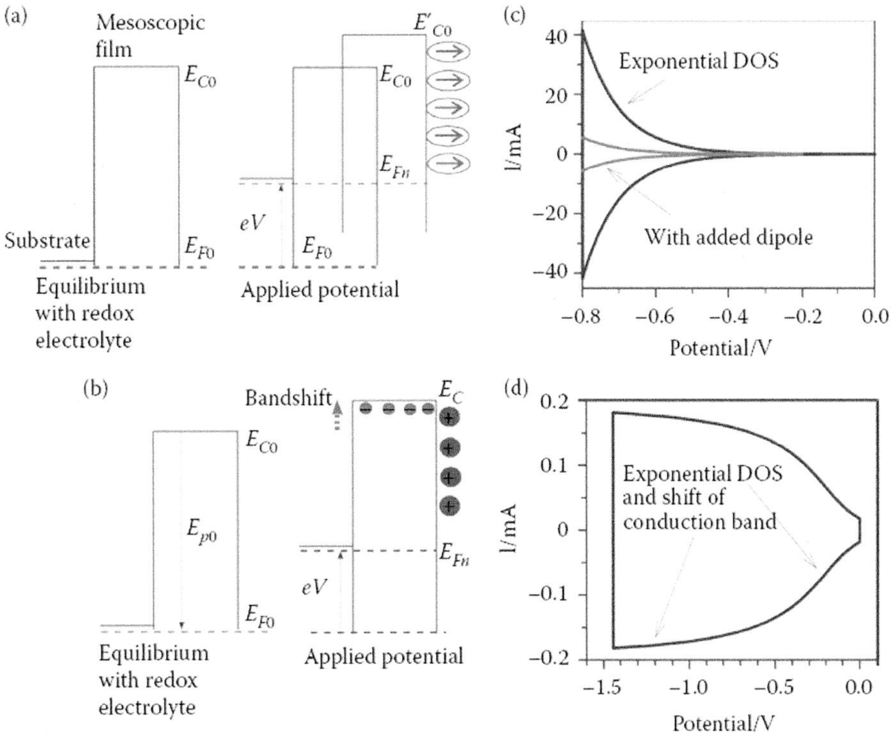

FIGURE 9.33 The left column, (a) and (b), shows the schematic energy diagram of a nanostructured semiconductor under application of a potential that rises homogeneously the Fermi level in the film. (a) Stationary energy levels, and another case is shown in which modification of the semiconductor surface with dipolar molecules induces a permanent shift of the energy levels. (b) Shift of the energy levels along with an increasing electron density by modification of the Helmholtz potential at the nanoparticles surface by ionic compensating charge. (c) and (d) show the corresponding CVs for an exponential distribution of bandgap states in the nanoparticles.

charge that compensates electronic charge injected to the semiconductor, and it is called *band edge movement* (Schlichthörl et al., 1997). The extent of the band displacement is not constant but proportional to the charge at the surface. In addition to the accumulation of electron charge in the chemical capacitance, C_μ, ionic accumulation at the particle's surface charges the Helmholtz capacitance, C_H, which is usually a constant, and is connected in series. The total capacitance becomes

$$C = \left(C_\mu^{-1} + C_H^{-1}\right)^{-1} \tag{9.72}$$

Equation 9.72 is useful to describe the distribution of the applied voltage, between a Fermi level in the semiconductor, V_F, and interfacial potential drop at the Helmholtz layer, V_H. At strong forward bias the chemical capacitance becomes large and it is usual to find that the capacitance saturates toward C_H, implying that the increasing potential modifies the voltage in the Helmholtz layer, so that the band shifts, as indicated in Figure 9.33d, and observed experimentally in Figure 8.18b. This is the reason why it is difficult to probe the potential of the conduction band by electrochemical techniques: at high electron accumulation the band shifts and the Fermi level cannot reach E_c.

Monitoring the position of the CB is a crucial tool for the analysis of the causes of performance of dye-sensitized solar cells (Barea et al., 2010). An upward shift of the conduction band usually improves the photovoltage of the solar cell, but reduces the photocurrent due to a lowering of injection from the dye, and increases recombination. As an example, Figure 9.34 shows the CVs for a layer of 20-nm-sized TiO_2 particles, first sensitized with an amphiphilic ruthenium sensitizer coded as K-19, and then cografted 4-guanidinobutyric acid (GBA) as coadsorbent. The CVs reveal that the addition of GBA shifts negatively the conduction band of TiO_2. This shift produces an increase of photovoltage. Grätzel and coworkers (Zhang et al., 2005) showed that the bandshift by coadsorption of GBA is small enough not to affect the electron injection from the excited sensitizer into the conduction band of TiO_2. Furthermore, GBA also inhibits recombination. These modifications cause an increase of the solar cell efficiency.

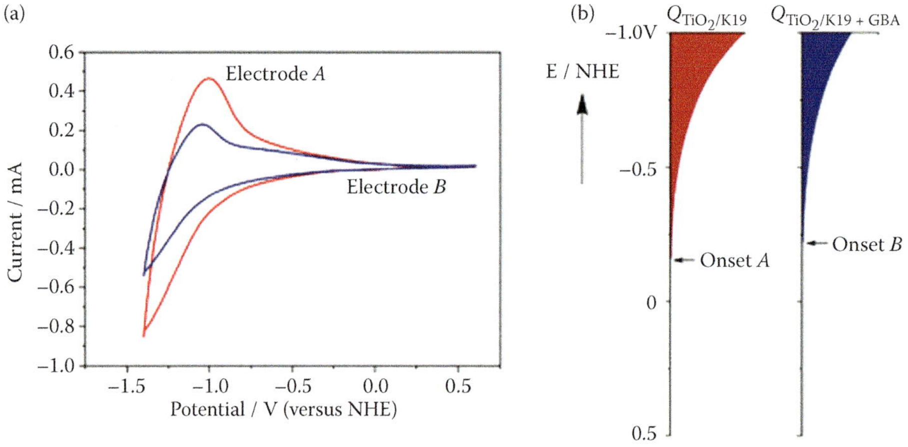

FIGURE 9.34 (a) Cyclic voltammetries of TiO_2 electrode A (with absorbed sensitizer K-19) and electrode B (with sensitizer and GBA cografted). (b) Energy levels at the mesoscopic TiO_2/electrolyte interface. (Reprinted with permission from Zhang, Z. et al. *The Journal of Physical Chemistry B* 2005, *109*, 21818–21824.)

GENERAL REFERENCES

Capacitance-voltage characteristics: Kimerling (1974), Otterloo and Gerritsen (1978), Balberg (1985), Balberg and Gal (1985), Ishida and Ikoma (1993), Walter et al. (1996), Afanasev (2008), Boix et al. (2009).

Photoelectrochemistry: Fujishima and Honda (1972), Morrison (1980), Finklea (1988), Morrison (1980), Gerischer (1990), Bard et al. (1991), Sato (1998), Memming (2001).

Schottky barrier and flatband potential: Garrett and Brattain (1955), Cowley and Sze (1965), Cowley (1966), Butler and Ginley (1978), Sze (1981), Trasatti (1986), Mönch (1993), Gerischer et al. (1994), Tung (2001).

Semiconductor nanowires and nanotubes: Cui and Lieber (2001), Cui et al. (2001), Baxter and Aydil (2006), Mor et al. (2006), Hernandez-Ramirez et al. (2007), Mor et al. (2007), Hernandez-Ramirez et al. (2008), Garnett and Yang (2010), Foley et al. (2012), Fitch et al. (2013).

Surface states in semiconductors: Frese and Morrison (1979), Allongue and Cachet (1984), Lincot and Vedel (1987), Hoffmann et al. (1998), Scurtu et al. (2009).

REFERENCES

Afanasev, V. V. *Internal Photoemission Spectroscopy*; Elsevier: Amsterdam, 2008.

Ajay, K.; Aslam, M. Defect induced high photocurrent in solution grown vertically aligned ZnO nanowire array films. *Journal of Applied Physics* 2012, *112*, 054316.

Albery, W. J.; Bartlett, P. N. The transport and kinetics of photogenerated carriers in colloidal semiconductor electrode particles. *Journal of the Electrochemical Society* 1984, *131*, 315–325.

Allongue, P.; Cachet, H. I–V curve and surface state capacitance at illuminated semiconductor/liquid contacts. *Journal of Electroanalytical Chemistry* 1984, *176*, 369–375.

Allongue, P.; Cachet, H. Band-edge shift and surface charges illuminated n-GaAs/aqueous electrolyte junctions. *Journal of the Electrochemical Society* 1985, *132*, 45–52.

Balberg, I. Relation between distribution of states and the space-charge-region capacitance in semiconductors. *Journal of Applied Physics* 1985, *58*, 2603–2616.

Balberg, I.; Gal, E. Determination of distribution of states in hydrogenated amorphous silicon from capacitance-voltage characteristics. *Journal of Applied Physics* 1985, *58*, 2617–2627.

Bard, A. J.; Bocarsly, A. B.; Fan, F.-R. F.; Walton, E. G.; Wrighton, M. S. The concept of Fermi level pinning at semiconductor/liquid juctions. Consequences for energy conversion efficiency and selection of useful solution redox couples in solar devices. *Journal of the American Chemical Society* 1980, *102*, 3671–3677.

Bard, A. J.; Memming, R.; Miller, B. Terminology in semiconductor electrochemistry and photoelectrochemical energy conversion. *International Union of Pure and Applied Chemistry* 1991, *63*, 569–596.

Barea, E. M.; Ortiz, J.; Payá, F. J.; Fernández-Lázaro, F.; Fabregat-Santiago, F.; Sastre-Santos, A.; Bisquert, J. Energetic factors governing injection, regeneration and recombination in dye solar cells with phthalocyanine sensitizers. *Energy and Environmental Science* 2010, *3*, 1985–1994.

Baxter, J. B.; Aydil, E. S. Nanowire-based dye-sensitized solar cells. *Solar Energy Materials and Solar Cells* 2006, *90*, 607–622.

Bertoluzzi, L.; Bisquert, J. Equivalent circuit of electrons and holes in thin semiconductor films for photoelectrochemical water splitting applications. *The Journal of Physical Chemistry Letters* 2012, *3*, 2517–2522.

Bisquert, J.; Cendula, P.; Bertoluzzi, L.; Gimenez, S. Energy diagram of semiconductor/electrolyte junctions. *The Journal of Physical Chemistry Letters* 2014, *5*, 205–207.

Bisquert, J.; Garcia-Belmonte, G.; Fabregat Santiago, F. Modeling the electric potential distribution in the dark in nanoporous semiconductor electrodes. *Journal of Solid State Electrochemistry* 1999, *3*, 337–347.

Boix, P. P.; Garcia-Belmonte, G.; Munecas, U.; Neophytou, M.; Waldauf, C.; Pacios, R. Determination of gap defect states in organic bulk heterojunction solar cells from capacitance measurements. *Applied Physics Letters* 2009, *95*, 233302.

Bolts, J.; Wrighton, M. Correlation of photocurrent-voltage curves with flat-band potential stable photoelectrodes for the photoelectrolysis of water. *The Journal of Physical Chemistry* 1976, *80*, 2641–2645.

Boschloo, G.; Haggman, L.; Hagfeldt, A. Quantification of the effect of 4-tert-butylpyridine addition to I_3^-/I^- redox electrolytes in dye-sensitized nanostructured TiO_2 solar cells. *The Journal of Physical Chemistry B* 2006, *110*, 13144–13150.

Butler, M. A.; Ginley, D. S. Prediction of flatband potentials at semiconductor/electrolyte interfaces from atomic electronegativities. *Journal of The Electrochemical Society* 1978, *125*, 228–232.

Capasso, F.; Margaritondo, G. *Heterojunction Band Discontonuities*; Elsevier: Amsterdam, 1987.

Cowley, A. M. Depletion capacitance and diffusion potential of gallium phosphide Schottkybarries diode. *Journal of Applied Physics* 1966, *37*, 3024–3032.

Cowley, A. M.; Sze, S. M. Surface states and barrier height of metal-semiconductor systems. *Journal of Applied Physics* 1965, *36*, 3212–3220.

Cui, Y.; Lieber, C. M. Functional nanoscale electronic devices assembled using silicon nanowire building blocks. *Science* 2001, *291*, 851–853.

Cui, Y.; Wei, Q.; Park, H.; Lieber, C. M. Nanowire nanosensors for highly sensitive and selective detection of biological and chemical species. *Science* 2001, *293*, 1289–1292.

Finklea, H. O. *Semiconductor Electrodes*; Elsevier: Amsterdam, 1988.

Fitch, A. G.; Strandwitz, N.; Brunschwig, B. S.; Lewis, N. S. A comparison of the behavior of single crystalline and nanowire array ZnO photoanodes. *The Journal of Physical Chemistry C* 2013, *117*, 2008–2015.

Foley, J. M.; Price, M. J.; Feldblyum, J. I.; Maldonado, S. Analysis of the operation of thin nanowire photoelectrodes for solar energy conversion. *Energy and Environmental Science* 2012, *5*, 5203–5220.

Frese, K. W.; Morrison, S. R. Electrochemical measurements of interface states at the GaAs/ Oxide interface. *Journal of the Electrochemical Society* 1979, *126*, 1235–1241.

Fujishima, A.; Honda, K. Electrochemical photolysis of water at a semiconductor electrode. *Nature* 1972, *238*, 37–38.

Garnett, E.; Yang, P. Light trapping in silicon nanowire solar cells. *Nano Letters* 2010, *10*, 1082–1087.

Garrett, C. G. B.; Brattain, W. H. Physical theory of semiconductor surfaces. *Physical Review* 1955, *99*, 376–387.

Gerischer, H. The impact of semiconductors on the concepts of electrochemistry. *Electrochimica Acta* 1990, *35*, 1677–1699.

Gerischer, H.; Decker, F.; Scrosati, B. The electronic and ionic contribution to the free energy of Alkali metals in intercalation compounds. *Journal of the Electrochemical Society* 1994, *141*, 2297–2300.

Germs, W. C.; Mensfoort, S. L. M. V.; Vries, R. J. D.; Coehoorn, R. Effects of energetic disorder on the low-frequency differential capacitance of organic light emitting diodes. *Journal of Applied Physics* 2012, *111*, 074506.

Hernandez-Ramirez, F.; Prades, J. D.; Tarancon, A.; Barth, S.; Casals, O.; Jimenez-Diaz, R.; Pellicer, E.; et al. Insight into the role of oxygen diffusion in the sensing mechanisms of SnO_2 nanowires. *Advanced Functional Materials* 2008, *18*, 2990–2994.

Hernandez-Ramirez, F.; Tarancon, A.; Casals, O.; Pellicer, E.; Rodriguez, J.; Romano-Rodriguez, A.; Morante, J. R.; et al. Electrical properties of individual tin oxide nanowires contacted to platinum electrodes. *Physical Review B* 2007, *76*, 085429.

Hoffmann, P. M.; Oskam, G.; Searson, P. C. Analysis of the impedance response due to surface states at the semiconductor/solution interface. *Journal of Applied Physics* 1998, *83*, 4309.

Ishida, T.; Ikoma, H. Bias dependence of Schottky barrier height in GaAs from internal photoemission and current-voltage characteristics. *Journal of Applied Physics* 1993, *74*, 3977–3982.

Ishii, H.; Hayashi, N.; Ito, E.; Washizu, Y.; Sugi, K.; Kimura, Y.; Niwano, M.; et al. Kelvin probe study of band bending at organic semiconductor/metal interfaces: Examination of Fermi level alignment. *Physica Status Solidi (a)* 2004, *201*, 1075–1094.

Jin, L.; Puxian, G.; Wenjie, M.; Changshi, L.; Zhong, L. W.; Rao, T. Quantifying oxygen diffusion in ZnO nanobelt. *Applied Physics Letters* 2006, *89*, 063125.

Johnson, W. C.; Panousis, P. T. The influence of Debye length on the C-V measurement of doping profiles. *Electron Devices, IEEE Transactions on* 1971, *18*, 965–973.

Kelly, J. J.; Memming, R. The influence of surface recombination and trapping on the cathodic photocurrent at p-type III-V electrodes. *Journal of the Electrochemical Society* 1982, *192*, 730–738.

Kimerling, L. C. Influence of deep traps on the measurement of free-carrier distributions in semiconductors by junction capacitance techniques. *Journal of Applied Physics* 1974, *45*, 1839–1845.

Klahr, B.; Gimenez, S.; Fabregat-Santiago, F.; Hamann, T.; Bisquert, J. Water oxidation at hematite photoelectrodes: The role of surface states. *Journal of the American Chemical Society* 2012, *134*, 4294–4302.

Klein, A. Energy band alignment at interfaces of semiconductor oxides. *Thin Solid Films* 2012, *520*, 3721–3728.

Kronik, L.; Shapira, Y. Surface photovoltage spectroscopy of semiconductor structures. *Surface and Interface Analysis* 2001, *31*, 954.

Lincot, D.; Vedel, J. Recombination and charge transfer at the illuminated n-CdTe/electrolyte interface: Simplified kinetic model. *Journal of Electroanalytical Chemistry and Interfacial Electrochemistry* 1987, *220*, 179–200.

Losee, D. L. Admittance spectroscopy of impurity levels in Schottky barriers. *Journal of Applied Physics* 1975, *46*, 2204.

Memming, R. *Semiconductor Electrochemistry*; Wiley-VCH: Weinheim, 2001.

Mönch, W. *Semiconductor Surfaces and Interfaces*; Springer: Berlin, 1993.

Mor, G. K.; Shankar, K.; Paulose, M.; Varghese, O. K.; Grimes, C. A. High efficiency double heterojunction polymer photovoltaic cells using highly ordered TiO_2 nanotube arrays. *Applied Physics Letters* 2007, *91*, 152111.

Mor, G. K.; Varghese, O. K.; Paulose, M.; Shankar, K.; Grimes, C. A. A review on highly ordered, vertically oriented TiO_2 nanotube arrays: Fabrication, material properties, and solar energy applications. *Solar Energy Materials and Solar Cells* 2006, *90*, 2011–2075.

Mora-Seró, I.; Fabregat-Santiago, F.; Denier, B.; Bisquert, J.; Tena-Zaera, R.; Elias, J.; Lévy-Clement, C. Determination of carrier density of ZnO nanowires by electrochemical techniques. *Applied Physics Letters* 2006, *89*, 203117.

Morrison, S. R. *Electrochemistry at Semiconductor and Oxidized Metal Electrodes*; Plenum Press: New York, 1980.

Otterloo, J. D. V.; Gerritsen, L. J. The accuracy of Schottky-barrier-height measurements on clean-cleaved silicon. *Journal of Applied Physics* 1978, *49*, 723–729.

Paasch, G.; Micka, K.; Gersdorf, P. Theory of the electrochemical impedance of macrohomogeneous porous electrodes. *Electrochimica Acta* 1993, *38*, 2653–2662.

Reshchikov, M. A.; Foussekis, M.; Baski, A. A. Surface photovoltage in undoped n-type GaN. *Journal of Applied Physics* 2010, *107*, 113535.

Robinson, D. B.; Wu, C.-A. M.; Jacobs, B. W. Effect of salt depletion on charging dynamics in nanoporous electrodes. *Journal of The Electrochemical Society* 2010, *157*, A912–A918.

Ruhle, S.; Greenshtein, M.; Chen, S.-G.; Merson, A.; Pizem, H.; Sukenik, C. S.; Cahen, D.; et al. Molecular adjustment of the electronic properties of nanoporous electrodes in dye-sensitized solar cells. *The Journal of Physical Chemistry B* 2005, *109*, 18907–18913.

Sato, N. *Electrochemistry at Metal and Semiconductor Electrodes*; Elsevier: Amsterdam, 1998.

Scurtu, R.; Ionescu, N. I.; Lazarescu, M.; Lazarescu, V. Surface states- and field-effects at p- and n-doped GaAs(111)A/solution interface. *Physical Chemistry Chemical Physics* 2009, *11*, 1765–1770.

Schlichthörl, G.; Huang, S. Y.; Sprague, J.; Frank, A. J. Band edge movement and recombination kinetics in dye-sensitized nanocrystalline TiO_2 solar cells: A study by intensity modulated photovoltage spectroscopy. *The Journal of Physical Chemistry B* 1997, *101*, 8141–8155.

Sze, S. M. *Physics of Semiconductor Devices*, 2nd ed.; John Wiley and Sons: New York, 1981.

Tena-Zaera, R.; Elias, J.; Lévy-Clement, C.; Bekeny, C.; Voss, T.; Mora-Seró, I.; Bisquert, J. Influence of the potassium chloride concentration on the physical properties of electrodeposited ZnO nanowire arrays. *The Journal of Physical Chemistry C* 2008, *112*, 16318–16323.

Trasatti, S. The absolute electrode potential: An explanatory note. *Pure and Applied Chemistry* 1986, *58*, 955–966.

Tung, R. T. Recent advances in Schottky barrier concepts. *Materials Science and Engineering: R: Reports* 2001, *35*, 1–138.

Vilan, A.; Yaffe, O.; Biller, A.; Salomon, A.; Kahn, A.; Cahen, D. Molecules on Si: Electronics with chemistry. *Advanced Materials* 2010, *22*, 140–159.

Vincent, G.; Bois, D.; Pinard, P. Conductance and capacitance studies in GaP Schottky barriers. *Journal of Applied Physics* 1975, *46*, 5173.

Walter, T.; Herberholz, R.; Muller, C.; Schock, H. W. Determination of defect distributions from admittance measurements and application to Cu(In,Ga)Se_2 based heterojunctions. *Journal of Applied Physics* 1996, *80*, 4411–4420.

Yablonovitch, E.; Skromme, B. J.; Bhat, R.; Harbison, J. P.; Gmitter, T. J. Band bending, Fermi level pinning, and surface fixed charge on chemically prepared GaAs surfaces. *Applied Physics Letters* 1989, *54*, 555–557.

Zaban, A.; Meier, A.; Gregg, B. A. Electric potential distribution and short range screening in nanoporous TiO_2 electrodes. *The Journal of Physical Chemistry B* 1997, *101*, 7985–7990.

Zhang, Z.; Zakeeruddin, S. M.; O'Regan, B. C.; Humphry-Baker, R.; Grätzel, M. Influence of 4-guanidinobutyric acid as coadsorbent in reducing recombination in dye-sensitized solar cells. *The Journal of Physical Chemistry B* 2005, *109*, 21818–21824.

Zheng, J. P.; Jow, T. R. The effect of salt concentration in electrolytes on the maximum energy storage for double layer capacitors. *Journal of The Electrochemical Society* 1997, *144*, 2417–2420.

Part II

Foundations of Carrier Transport

10 Carrier Injection and Drift Transport

We begin the analysis of transport of carriers in a material layer, focusing on the drift of carriers in a local electrical field. We start with the formulation of the main relationships that provide the current density as a function of conductivity and electrical field. Another significant factor that determines the total electrical current is the injection of carriers at the contact, which leads us to discuss effects related to injection in organic conductors. We then analyze in detail the metal-insulator-metal model, in which the transport is driven by a constant field governed by the applied voltage. It is a simple device model, but nonetheless it is illustrative of many device features including contact formation properties.

10.1 TRANSPORT BY DRIFT IN THE ELECTRICAL FIELD

The electrical current density has two basic components, the conduction current and the displacement current:

$$j_{tot} = j_{cond} + j_{disp}. \tag{10.1}$$

The displacement current was discussed in Section 3.10 and it is associated with the time variation of the electrical field **E**:

$$j_{disp} = \varepsilon \frac{\partial \mathbf{E}}{\partial t}. \tag{10.2}$$

Normally, the displacement current is a transient effect associated with the charging of dielectric capacitances in the system.

The conduction current density (hereafter denoted j) is related to the transport of carriers by the application of some driving force. The current density at a cross-section of the layer is a combination of the current carried by the different charge carriers in the system:

$$j = \sum_i j_i. \tag{10.3}$$

The electrical current is a continuous magnitude that has the same value at all points of the x-axis. In the absence of recombination, the current is conserved:

$$\frac{\partial j}{\partial x} = 0. \tag{10.4}$$

The total electrical current across the device is

$$I = Aj = A\sum_i j_i, \tag{10.5}$$

where A is the area of the electrode.

The average velocity of a species is v_i and is determined by the product of the particle mechanical mobility B_i and the force F_i. The velocity can be given as

$$v_i = B_i F_i. \tag{10.6}$$

We consider the transport of a charge carrier with concentration c_i (cm^{-3}) and a charge number z_i, so that its electrical charge is $Q_i = z_i q$ in terms of the elementary charge q. For ionic species, we can have $z_i = \pm 1, \pm 2, \ldots$. The electrical force on a particle is

$$F_i = q z_i \mathbf{E}. \tag{10.7}$$

For electronic carriers, the holes drift in the direction of the electrical field and the electrons against the field. The direction of the flux of each electronic carrier is indicated in Figure 10.1. Equation 10.6 leads to

$$v_i = B_i q z_i \mathbf{E}. \tag{10.8}$$

The carrier mobility u_i ($cm^2\ V^{-1}\ s^{-1}$) is defined as the velocity in a unit electrical field

$$v_i = \frac{z_i}{|z_i|} u_i \mathbf{E}. \tag{10.9}$$

Comparing Equation 10.8, we observe that

$$u_i = |z_i| q B_i. \tag{10.10}$$

In general, mobility is a function of the electrical field or charge density and the dependence becomes strongly nonlinear in large fields, as will be discussed in Chapters 13 and 14. We may state the definition of mobility in terms of the average carrier velocity $\langle v(\mathbf{E})\rangle$ at low field values as

$$u_i = \frac{z_i \langle dv_i(\mathbf{E})\rangle}{|z_i| d\mathbf{E}}\Big|_{\mathbf{E}=0} \tag{10.11}$$

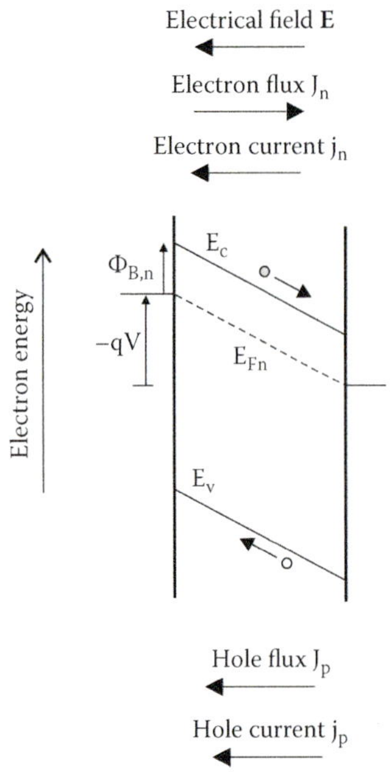

FIGURE 10.1 Energy diagram of an electron-conducting layer material with applied bias potential. q is the elementary charge, V is the voltage applied at the left terminal, E_c is the energy of the conduction band, E_v is the energy of the valence band, and E_{Fn} is the Fermi level. All labeled quantities are represented with respect to the electron energy scale shown on the left side. The arrows at the top and bottom show the vector direction of the indicated magnitudes.

In the case of hopping transport, mobility can be defined on the basis of the difference of effective charge carrier jump probability in the direction along and against the electric field (Fishchuk et al., 2003).

The electronic carrier mobility in semiconductors is strongly dependent on scattering processes. Mobility is almost constant for low doping concentrations and is basically affected by phonon scattering. At higher doping concentrations, mobility decreases due to scattering by ionized impurities. Values of electron mobility in crystalline semiconductors can be as large as 10^4 cm^2 V^{-1} s^{-1}. The mobility in organic layers is often found to depend on the value of the electric field and it is well described by an expression of the type

$$u(\mathbf{E}) = u_0 \exp\left(\gamma\sqrt{\mathbf{E}}\right). \tag{10.12}$$

Here u_0 is a constant, the mobility of electrons or holes at zero field, and γ is the parameter describing the field dependence (Pai, 1970; Davids et al., 1997).

The *carrier flux density* J is defined as the number of carriers crossing a unit area per second

$$J_i = c_i v_i. \tag{10.13}$$

The flux can originate from drift, diffusion, or other mechanisms such as convection. The motion of the carrier due to the direct action of the electrical field is normally denoted as drift transport. The flux density associated with drift transport is given by

$$J_i = \frac{z_i}{|z_i|} c_i u_i \mathbf{E}. \tag{10.14}$$

The electrical current density j carried by species i is

$$J_i = z_i q J_i = q|z_i| c_i u_i \mathbf{E}. \tag{10.15}$$

In a situation of transport by drift in an electrical field, the electrical conductivity

σ_i of one specific carrier is defined by the relationship

$$j_i = \sigma_i \mathbf{E}. \tag{10.16}$$

From Equation 10.11, we find that the conductivity of carrier i has the expression

$$\sigma_i = |z_i| q c_i u_i. \tag{10.17}$$

The total electrical conductivity is henceforth given by

$$\sigma = \sum_i \sigma_i = \sum_i |z_i| q c_i u_i. \tag{10.18}$$

10.2 INJECTION AT CONTACTS

We consider a simple device formed by a single material and two contacts as shown in Figure 10.1. We examine the factors that determine whether there will be a substantial current flow in such a device when a voltage is applied between the contacts and the magnitude of such current. To obtain steady current flow, two effects must occur efficiently: (1) carrier injection and extraction at the contacts and (2) conductivity at all points of the internal material. In this section, we discuss the first of the two aspects with a particular emphasis on the contact with organic materials.

Carrier injection, discussed in Chapters 4 and 6, concerns the interfaces of the device and particularly the contact between the metal and the active material. Figure 10.2 shows the injection of either electrons or holes at electrochemical contacts, previously described in Figure 6.27. In general, the current flow may be due to carriers that flow across the device, are injected at one contact and extracted at the opposite one, or the current may be established by injection of electrons and holes at opposite contacts that recombine inside the active layer.

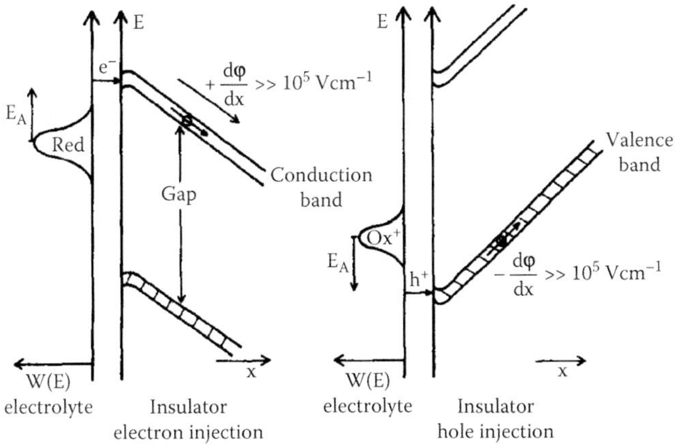

FIGURE 10.2 Energy scheme for electron or hole injection into insulators from redox species and for the transport of the injected charge carriers through the insulator by strong electric fields. (Reproduced with permission from Gerischer, 1990.)

This last case forms the basis of the recombination diode that will be discussed in Chapter 20. In this section, we focus on a discussion of transport features neglecting recombination. If only one contact injects and the other one blocks the carriers, the carriers can be accumulated in the material and there is a transient current until the active material is fully charged.

As the applied voltage in a device increases, the driving force for transport increases, and the current across the device normally increases. However, the amount of current that flows through the sample can be limited by the carrier flow provided by injection at the contacts. This is depicted with the dashed line in Figure 10.3. The current achieves a lower magnitude than that of the bulk-limited current and is called an *injection-limited current*, j_{inj}. Its functional dependence is determined by the mechanism of injection as discussed in Section 6.7.

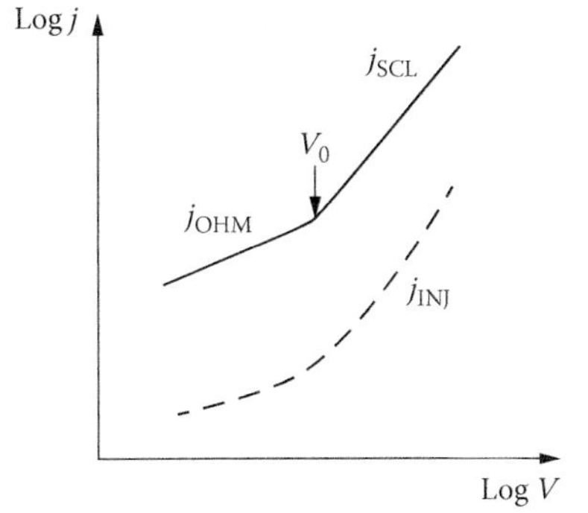

FIGURE 10.3 Bulk-limited (solid line) and injection-limited (dashed line) current density versus voltage characteristics for a trap-free semiconductor. The threshold voltage V_0 when the bulk-limited current turns from ohmic to space-charge-limited current is indicated. (Reproduced with permission from Shen et al., 2004.)

The main rule for a facile injection of carriers is to have a small injection barrier at the contact, as was discussed in Section 4.6. A contact is *ohmic* if the resistance across the interface is independent of the voltage. However, the low resistance value is to be taken relative to the impediments to current flow in other parts of the device, that is, the impedances due to bulk-limited current. Normally, the actual current-potential characteristics of the ohmic contact are not measured or analyzed. If the resistance is negligible, it is generally referred to as ohmic. An ohmic contact also has the property that it discharges to the metal, the current of the specific carrier that it receives from the semiconductor side, that is, it transmits the current from the internal surface of the device to the metal contact, without impediment. The degree of ohmicity of a hole-injection contact is often well correlated with the injection barrier as shown in Figure 10.4. One useful definition of the ohmic contact is with respect to the exchange current density, j_0, which is the current flowing in equilibrium in both senses, across the interface; see Section 6.6. If the demand of current is of the order $\approx j_0$, then little voltage is required to drive the current and the contact is ohmic. It is usually suggested that if the barrier at a metal–organic interface is greater than 0.3 eV at zero electric field, then the current is limited by injection. If the barrier is less than 0.3 eV, then charges can be injected efficiently into the device and the current will be controlled by transport (Lopez Varo et al., 2012).

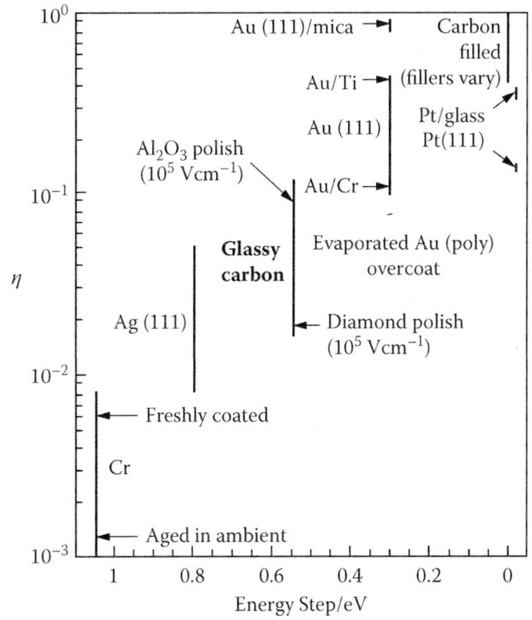

FIGURE 10.4 Hole-injection efficiency versus energy barrier at the contact for substrate contacts of varying work function. (Reproduced with permission from Abkowitz et al., 1998.)

In devices using organic semiconductors, such as organic light-emitting diodes (OLEDs) and solar cells, a low barrier for electron injection/extraction should result whenever the Fermi energy of the cathode closely matches the lowest unoccupied molecular orbital (LUMO) energy of the organic electron conductor. Metals with a low work function also have a small electron affinity and will tend to release electrons to the contacting material. These materials form low injection barriers to the conduction band of a semiconductor and are therefore appropriate for injecting or extracting electrons, especially in the case of organic semiconductors, which usually show low electron affinities as well. In OLEDs, low work function metals such as magnesium and calcium or their alloys with silver are the most commonly used cathode materials. However, due to the strong tendency of Ba, Ca, and low work function metals in general to release electrons, they are strongly reactive and prone to oxidation. The devices require robust encapsulation to avoid contact with air, moisture, or oxygen (Jørgensen et al., 2008), which cause oxidation and consequent degradation of the metal contact by modification of both its energy levels and conductivity (Bröms et al., 1997). Wide bandgap metal oxides, such as TiO_2 and ZnO, serve as excellent electron acceptors with good electron-conducting and hole-blocking properties.

Contact engineering allows one to tune the device for hole or electron transport, as shown for quantum dot (QD) films in the scheme of Figure 10.5 (Oh et al., 2014). Furthermore, the rate of bulk transport can be tuned, and hence switch the device between bulk or injection-limited conduction, by modifying the length of the ligands between QDs, which determine the tunneling distance, as explained in Section 19.4.

It is also frequently observed that the injection barrier does not correlate with the metal work function. This situation was discussed in Chapters 2 and 4; an interfacial dipole is formed by the deposition of a very thin organic layer on a metal and changes the vacuum level (VL) at the interface. The insensitivity of the barrier to the metal work function is called the *pinning of the Fermi level*; see Section 4.6. For example, for the Alq_3 barrier, a widely investigated small molecule for OLED applications, the Fermi level is pinned at about 3.6 eV for low work function materials, whereas a normal Schottky barrier is formed for higher work function metals as shown in Figure 10.6. Tuning the energy level of the cathode to the organic material can be achieved by interfacial dipole layers that reduce the effective work function of a metal such as Al (Hung et al., 1997; Ding and Gao, 2007). Inserting a very thin layer of a wide bandgap material between the conducting cathode and the active organic material is often an effective approach. Specifically, inorganic insulators such as LiF, Li_2O, MgF_2, and MgO under aluminum metal have been widely used to reduce the effective work function. As an example, Figure 10.7 shows the effect of a LiF layer at the Alq_3–Al interface. The ultraviolet photoelectron spectroscopy (UPS) measurement shows that the LiF interlayer produces a large (0.6-eV) interfacial dipole that considerably reduces the barrier to electron injection from Al to Alq_3.

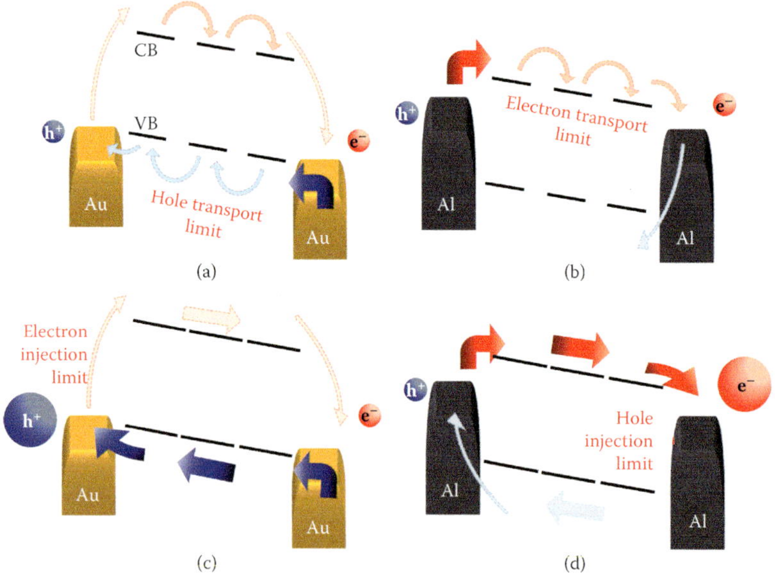

FIGURE 10.5 Band diagram depicting charge injection and transport in PbSe nanocrystal thin films with long organic ligands for field-effect transistors (FETs) fabricated with (a) high work function Au contacts and (b) low work function Al contacts and with short inorganic ligands for FETs fabricated with (c) Au and (d) Al contacts. (Reproduced with permission from Oh et al., 2014.)

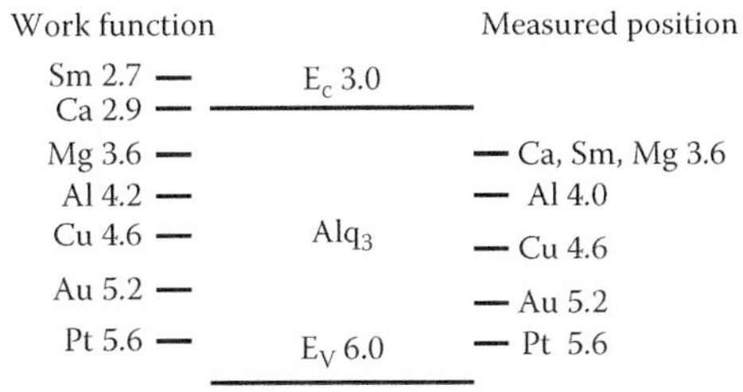

FIGURE 10.6 Metal work functions and electron energy diagram for metal contacts with Alq_3 derived from the internal photoemission and built-in potential measurements. (Reproduced with permission from Campbell and Smith, 1999.

On the other hand, contact materials with a large work function provide an efficient contact to holes. PEDOT:PSS, a conjugated polymer formed by a mixture of two ionomers, poly(3,4-ethylenedioxythiophene) and poly(styrenesulfonate), is a good hole conductor that is often deposited on a transparent conducting oxide (TCO) such as indium-doped tin oxide (ITO) to form a hole-injection contact. PEDOT and similar materials also serve to control the wettability and compatibility of the organic active layer with the contact layer. The work function of ITO is 4.8 eV, whereas that of PEDOT:PSS is 5.1 eV. These two materials readily form an ohmic contact, and PEDOT matches well with the ordinary values of the Fermi level of holes in organic devices. It was found that the higher work function of PEDOT:PSS compared to ITO facilitates hole injection into poly(2-methoxy-5-(2′-ethylhexyloxy)-1,4-phenylene vinylene (MEH-PPV; with ionization energy at 5.1 eV) (Cao et al., 1997; Brown et al., 1999). Transition metal oxides, such as molybdenum oxide (MoO_3), tungsten oxide (WO_3), and vanadium oxide (V_2O_5), easily exchange charges with several organic semiconductors and produce efficient hole-injection layers in OLEDs and organic solar cells as discussed further in Section 20.5.

Injection models have been developed to consider the essential physical mechanisms that take part in the charge injection from a metal into a disordered organic semiconductor viewed as a random hopping system. One detailed model was proposed by Arkhipov and collaborators that describes the injection as follows; see Figure 10.8 (Arkhipov et al., 1998; Arkhipov et al., 1999; Arkhipov et al., 2003). First, carriers jump from the Fermi level of the metal contact into localized states of a Gaussian density of states (DOS) that forms a distribution of hopping sites that are sufficiently close to the metal–organic interface. Every injected carrier creates an image charge of the opposite sign at the contact. The superposition of the external electric field at the contact and the Coulomb field of the image charge form the potential barrier at the metal–organic interface, which will restrict the charge

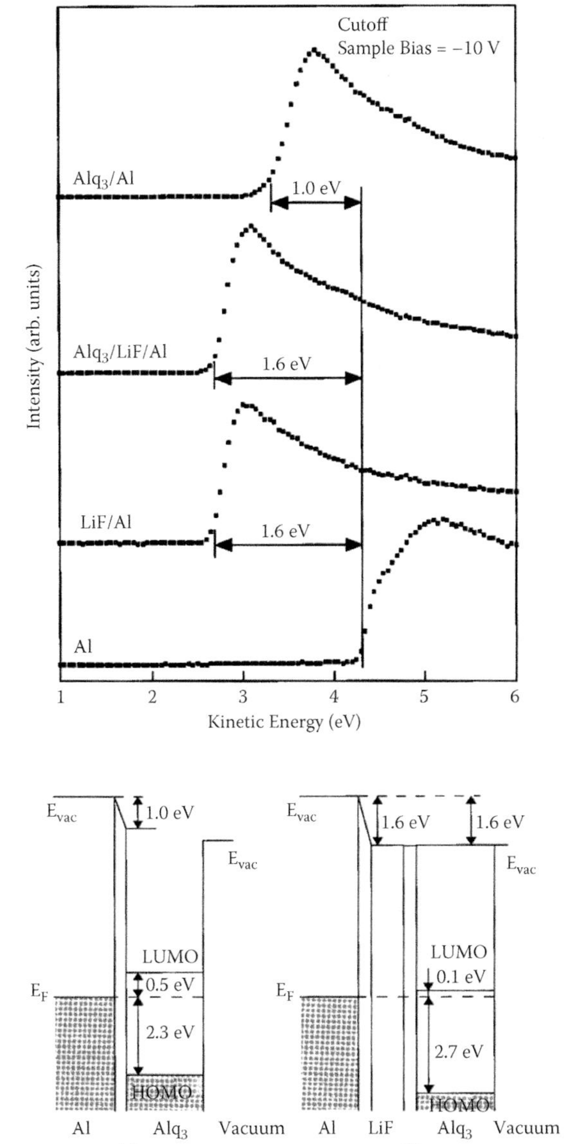

FIGURE 10.7 Top: UPS spectra (low kinetic energy region) measured immediately after each layer deposited in Alq_3/LiF/Al and Alq_3/Al multilayers. Bottom: Energy diagrams of the Alq_3–Al interface and the Alq_3–LiF–Al interface. (Reproduced with permission from Mori et al., 1998.)

injection. In this way, the injection is described in terms of thermally and field-assisted charge transfer from the metal to the organic semiconductor. Part of this injected charge is followed by a return to the metallic contact. At this point, the model evaluates the probability of crossing the potential barrier for a carrier that has made an initial jump at a distance x close to the metal–organic interface. This probability is formulated by the one-dimensional Onsager theory to avoid geminate recombination for a pair of a carrier in the organic and its image twin at the metal. Further discussion of this approach is given in Figures 6.22 and 6.23.

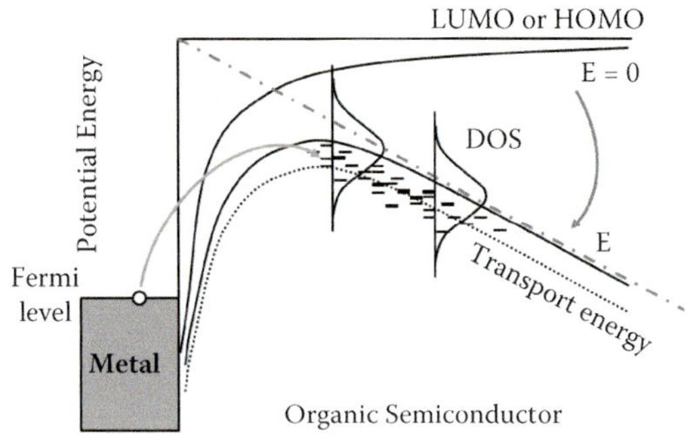

FIGURE 10.8 Potential energy distribution at the metal–organic contact in Arkhipov's model considering charge carrier injection into a Gaussian DOS. The small lines in Gaussian curves represent the distribution of available energy states in the semiconductor. Solid lines show the barrier potential energy of a charge carrier after considering the Coulombic field of the image charge and in the external electric field. Dotted lines represent the energy of the transport level for the regime of upward carrier jumps.

10.3 THE METAL-INSULATOR-METAL MODEL

Let us consider a material slab with electronic contacts on both sides. If the material has intrinsic carriers, at low applied voltage, conduction current flows steadily, governed by drift transport, according to Equation 10.15. If the number of carriers is low, at large applied voltage, the injected carriers must develop a space charge that contributes to maintaining the electrical field, as discussed later in Chapter 6 of this volume. Both the ohmic and space-charge limited current are indicated in Figure 10.3. In the metal-insulator-metal (MIM) model, an insulator layer is contacted by two metals of very different work function. As a result, the bands of the insulator obtain a slope in equilibrium due to the built-in potential. This model with simplified characteristics of drift transport offers us a good example to introduce many aspects of electrical device operation (Simmons, 1971). We will discuss the general physical properties and implications of this model, considering first the energetic picture at equilibrium and then the effect of applied voltage. This approach has been amply applied in OLEDs and also as a primitive model for organic solar cells. The low-voltage regime of this device has been studied by Kim et al. (2011).

In Figure 10.9, we consider the physical basis for the equilibrium distribution of the MIM model. Figure 10.9a shows the separate components of the device. The central layer is an insulator with electron energy level E_c (conduction band) and hole energy level E_v (valence band). The corresponding levels in organic semiconductors are the LUMO and the highest occupied molecular orbital (HOMO), respectively. The layer is contacted by two metals, one of low work function Φ_c (cathode) and another one of high work function Φ_a (anode). When the system comes to equilibrium, the Fermi level must be homogeneous. Equalizing the Fermi levels of the metals requires the construction of a difference of potential, which is described by the slant of the bands. Under the vacuum level alignment (VLA) rule at the contacts, the situation of Figure 10.9b is obtained. Because we assumed a low charge quantity, we neglect any space-charge formation, and therefore the electrical field in the insulating layer, which we call the *drift field*, is constant and has the value

$$\mathbf{E}_{dr} = -\Delta\varphi_{dr}/d, \tag{10.19}$$

where $\Delta\varphi$ is defined as the difference of potentials across the insulator

$$\Delta_{dr}\varphi = \varphi(d) - \varphi(0). \tag{10.20}$$

It is important to remark that even in the presence of an electric field, the carriers are in equilibrium and no conduction takes place. The built-in potential between the metals $V_{bi}^{a,c}$ coincides with the contact potential difference of the metals

$$\Delta_{bi}^{a,c} = \Delta_{dr}\varphi^{eq} = \varphi^{eq}(d) - \varphi^{eq}(0) = \frac{\Phi_a - \Phi_c}{q}. \tag{10.21}$$

The application of the VLA rule in Figure 10.9b has the consequence that the injection barrier for electrons is fixed at the value given by Equation 4.18, $\Phi_{B,n} = \Phi_c - \chi_n$, and the value for holes by Equation 4.19, $\Phi_{B,p} = \chi_p - \Phi_a$. Thus, the density of the respective carrier is fixed at the boundary at the equilibrium value as follows (Haney, 2011):

$$p(0) = p_0 \tag{10.22}$$

$$n(L) = n_0. \tag{10.23}$$

This type of model contact has been used in the past in the discussion of photoconductivity in insulators (Goodman and Rose, 1971). The model of Sokel and Hughes (1982) sets all carrier densities to zero at the boundary.

The VLA property need not be generally satisfied. When the contact forms a dipole, as mentioned in the previous section, the built-in voltage is not invested fully in the drift field. As a result, the barrier for electron injection increases and the drift field decreases as shown in Figure 10.9c.

In Figure 10.10, we observe different situations caused by an applied voltage V_{app} in the MIM model

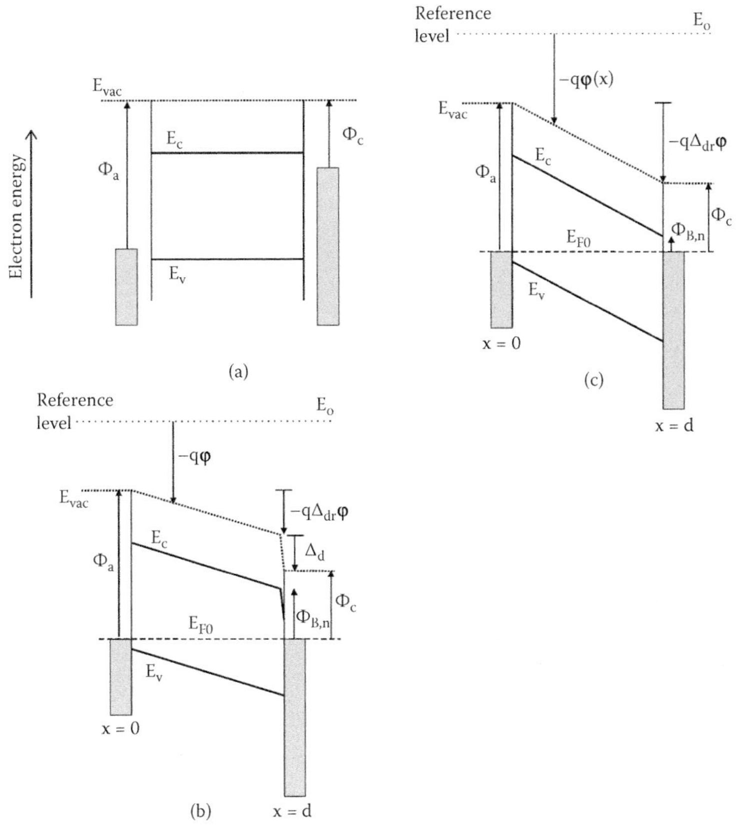

FIGURE 10.9 Energy diagram of an insulator layer that is contacted by two metals of dissimilar work function. (a) The energies of the separate materials. (b) The equilibrium condition with equilibrium Fermi level E_{F0} under vacuum level alignment. (c) The equilibrium condition with formation of the dipole layer at the cathode, which reduces the drift field $\Delta_{dr}\varphi$.

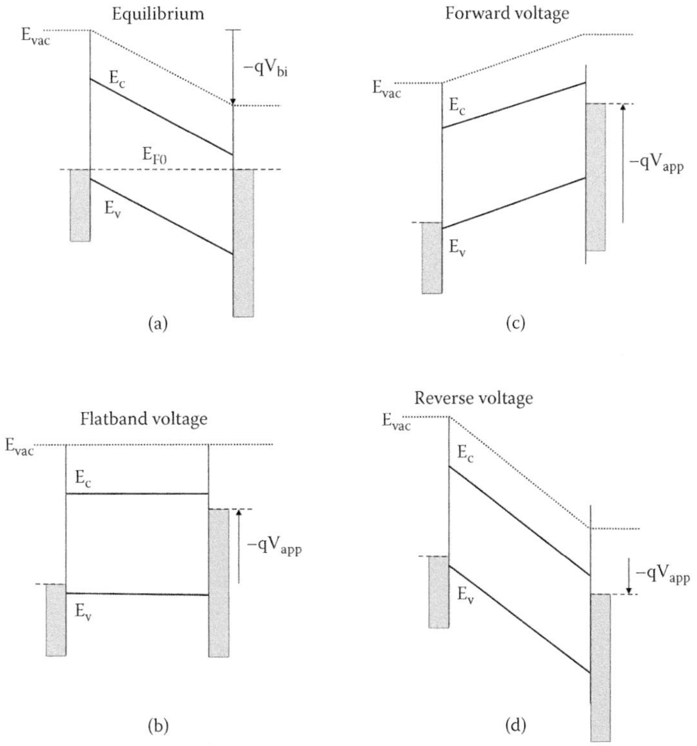

FIGURE 10.10 Energy diagram of an insulator layer that is contacted by two metals of dissimilar work function, under different cases of applied voltage as indicated.

with straight bands. The drift field changes according to the following expression

$$\Delta_{dr}\varphi = V_{bi}^{a,c} + V_{app}. \tag{10.24}$$

The drift field vanishes when the applied voltage equals the built-in potential, i.e., at $V_{app} = -V_{bi}^{a,c}$, the flat-band condition. Under forward bias, injected carriers drift in the electrical field and current increases with the voltage. Under reverse bias, the field also increases but the large injection barrier prevents current flow. Thus, the model shows diode characteristics. The diagrams of Figure 10.10 constitute only a first approximation, since they neglect Schottky barriers and band bending produced by space charge.

In practice, the built-in voltage $V_{bi}^{a,c}$ can be estimated from the current-voltage measurements or from the onset of photocurrent under illumination. The rationale for this method is that the drift force on carriers is zero at the flat-band condition and the drift current starts only when the voltage causes the bands to be sloped. Therefore, it is common in the literature to represent the voltage scale as $(V - V_0)$, where V_0 is the compensation voltage at the point where the current changes sign in the current-voltage curve, identified as the built-in voltage $V_{bi}^{a,c}$ Another method to determine the built-in voltage is the electro-absorption measurement (Campbell et al., 1996). This technique involves the measurement of the modulation of the reflection coefficient for monochromatic light, resulting from the modulated ac voltage imposed over the steady-state voltage. The detection of the field is possible because the optical absorption coefficient α changes with the square of the electric field due to a Stark shift in the allowed optical transition. The energy levels of PPV and the built-in voltage obtained in devices with different metal contacts are shown in Figure 10.11. The presence of some native carrier density explains the divergence observed between the built-in voltage determined by electro-absorption and current-voltage curves (de Vries et al., 2010).

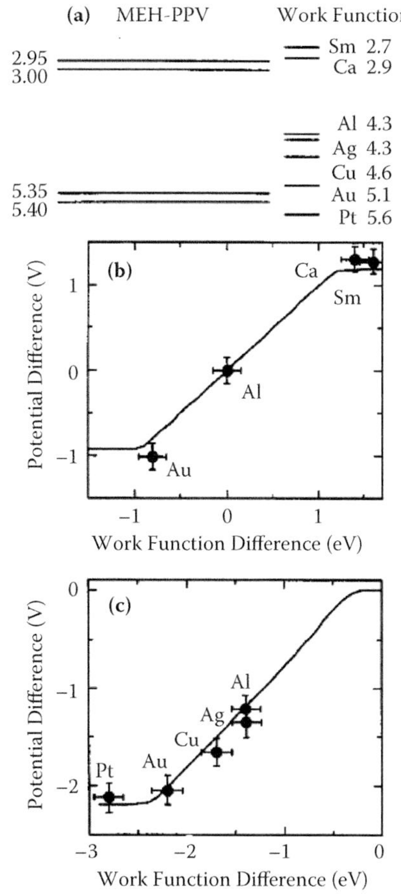

FIGURE 10.11 (a) Energy-level diagram for MEH-PPV deduced from the electro-absorption and internal photoemission measurements. The line at 2.95 eV (5.40 eV) corresponds to electron (hole) polarons and that at 3.00 eV (5.35 eV) corresponds to electron (hole) bipolarons. Calculated (solid line) and experimental (points) potential difference across (b) metal/MEH-PPV/Al structures and (c) metal/MEH-PPV/Ca structures as a function of the work function difference of the contacts. (Reproduced with permission from Campbell et al., 1996.)

Let us describe more realistically the distribution of carriers in the MIM model including space-charge formation (Simmons, 1971). Figure 10.12 shows the formation of an ohmic contact with the insulator. The electron charge is injected into the insulator to achieve equilibration of Fermi levels and produces a bending of the conduction band in the insulator by establishing an accumulation region. The amount of space charge fixes the width of the accumulation region λ. The relationship of the size of band bending to the size of the insulator layer gives rise to different situations. In Figure 10.13b, the two contacts are separately equilibrated and a field-free region of the flat conduction band is obtained in the central region. If the amount of space charge decreases, the width λ increases until the accumulation regions overlap; see Figure 10.13c. But if the amount of space charge is negligible, no significant band bending occurs, and equilibration of the Fermi levels produces a tilted straight band, corresponding to Figure 10.9b. Finally, we describe another situation, which is the equilibration between a p-type semiconductor and two metal contacts. If the work functions of the metals are lower than those of the semiconductor, the contacts form depletion layers, consisting of Schottky barriers. It should therefore be noted that the model of Figure 10.9b, in which the built-in potential translates into a constant drift field, is

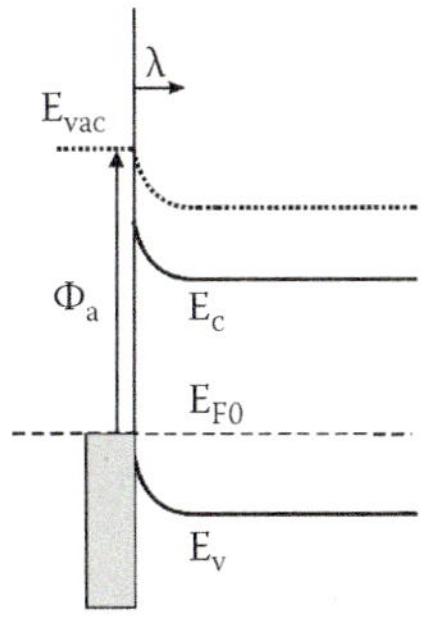

FIGURE 10.12 Energy diagrams for different situations of the metalinsulator-metal contacts. The top line is the local vacuum level. (a) Formation of an ohmic contact by electron transfer from the metal to the insulator, which forms an accumulation region of width λ in the insulator.

restricted to situations in which the intrinsic density of carriers of the semiconductor is very low.

10.4 THE TIME-OF-FLIGHT METHOD

The time-of-flight (TOF) method is a widely used experimental method to measure mobility in organic semiconductors. It has its origin in the xerographic copying process. In the TOF setup, as shown in Figure 10.14a, the sample is contacted by at least one transparent electrode and is biased to a large extent with applied voltage in reverse mode, so that only a photogenerated charge can conduct electrical current. A short pulse of strongly

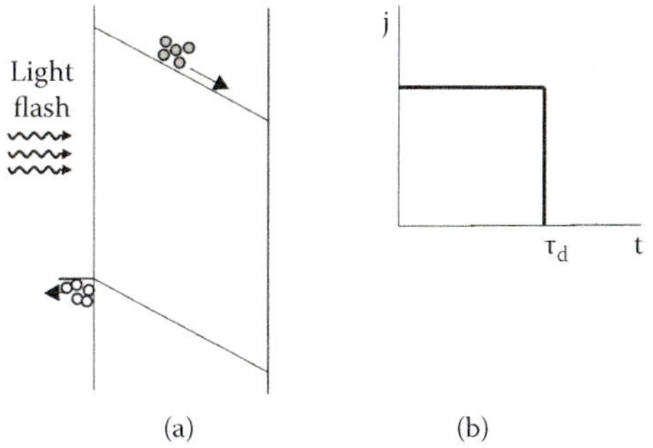

FIGURE 10.14 (a) Scheme of the time-of-flight experimental setup. (b) Schematic representation of drift current after a light pulse in the absence of dispersive transport.

absorbed light generates electrons and holes close to the transparent contact and one species is extracted while the other drifts to the opposite contact. In a trap-free material, the photocurrent transient should exhibit a plateau during which the photoexcited carriers move with constant velocity; see Figure 10.14b. The photocurrent drops to zero suddenly when the carriers arrive at the opposite electrode. This *transit* time τ_d for carriers to cross the sample is related to their velocity by the expression

$$v\tau_d = d \quad (10.25)$$

$$\tau_d = \frac{d}{u_n \mathbf{E}} = \frac{d^2}{u_n V}. \quad (10.26)$$

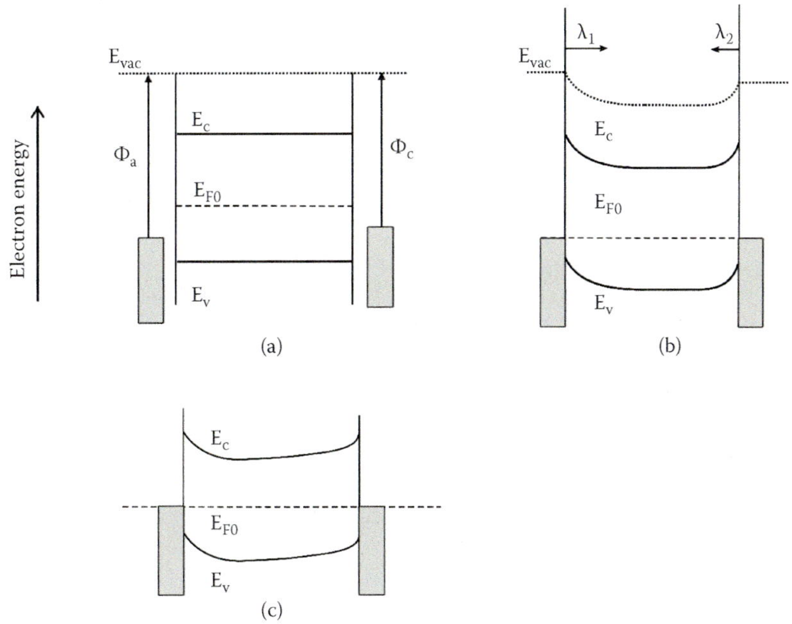

FIGURE 10.13 Energy diagram of a p-type semiconductor layer that is contacted by two metals of dissimilar work function, lower than the work function of the semiconductor. (a) Before contact. (b) Formation of two separate accumulation regions in the case λ > *L*/2 The insulator contains a field-free region in which the conduction band is flat. (c) The accumulation regions overlap.

Note that in Figure 10.14b, the drift current can be detected immediately without waiting for the arrival of the electron at the electrode. The induced charge created at the electrodes produces a voltage (as in the dielectric capacitor) and the modification of the position of the carriers in the layer produces electrical current that is associated with variation of the charge at the electrodes. Such current can be determined by calculation of the image charge at the contacts induced by the electron in motion. A method derived by Shockley (1938) and Ramo (1939) allows simple calculation of the current.

The ideal characteristic indicated in Figure 10.14b is not realized in the presence of dispersive transport, which produces a slow decrease of the current due to the spread of carriers and time-dependent mobility (Scher and Montroll, 1975; Marshall, 1983); see Section 13.7.

GENERAL REFERENCES

Carrier injection: Henisch, 1984; Arkhipov et al., 1999; Malliaras and Scott, 1999; Scott and Malliaras, 1999; Arkhipov et al., 2003; Scott, 2003; Lonergan, 2004; Shen et al., 2004; Walzer et al., 2007; Ratcliff et al., 2011; and Lopez Varo et al., 2012.

Ionic transport in electrolytes: Kontturi et al., 2008.

Metal-insulator-metal model: Brütting et al., 2001; Brabec et al., 2002; Mihailetchi et al., 2003; Mihailetchi et al., 2004; Waldauf et al., 2004; Koster et al., 2005; Deibel and Dyakonov, 2010.

Method of Shockley and Ramo: Shockley, 1938; Ramo, 1939; Sirkis and Holonya, 1966; He, 2001.

REFERENCES

Abkowitz, M.; Facci, J. S.; Rehm, J. Direct evaluation of contact injection efficiency into small molecule based transport layers: Influence of extrinsic factors. *Journal of Applied Physics* 1998, *83*, 2670–2676.

Arkhipov, V. I.; Emelianova, E. V.; Tak, Y. H.; Bässler, H. Charge injection into light-emitting diodes: Theory and experiment. *Journal of Applied Physics* 1998, *84*, 848–856.

Arkhipov, V. I.; Seggern, H. v.; Emelianova, E. V. Charge injection versus space-chargelimited current in organic light-emitting diodes. *Applied Physics Letters* 2003, *83*, 5074–5076.

Arkhipov, V. I.; Wolf, K.; Bässler, H. Current injection from a metal to a disordered hopping system. II. Comparison between analytic theory and simulation. *Physical Review B* 1999, *59*, 7514–7520.

Brabec, C. J.; Shaheen, S. E.; Winder, C.; Sariciftci, N. S.; Denk, P. Effect of LiF/metal electrodes on the performance of plastic solar cells. *Applied Physics Letters* 2002, *80*, 1288–1290.

Bröms, P.; Birgerson, J.; Salaneck, W. R. Magnesium as electrode in polymer LEDs. *Synthetic Metals* 1997, *88*, 255–258.

Brown, T. M.; Kim, J. S.; Friend, R. H.; Cacialli, F.; Daik, R.; Feast, W. J. Built-in field electroabsorption spectroscopy of polymer light-emitting diodes incorporating a doped poly(3,4-ethylene dioxythiophene) hole injection layer. *Applied Physics Letters* 1999, *75*, 1679–1681.

Brütting, W.; Berleb, S.; Mücki, A. G. Device physics of organic light-emitting diodes based on molecular materials. *Organic Electronics* 2001, *2*, 1–36.

Campbell, I. H.; Hagler, T. W.; Smith, D. L.; Ferraris, J. P. Direct measurement of conjugated polymer electronic excitation energies using metal/polymer/metal structures. *Physical Review Letters* 1996, *1900*, 76.

Campbell, I. H.; Smith, D. L. Schottky energy barriers and charge injection in metal/Alq/ metal structures. *Applied Physics Letters* 1999, *74*, 561–563.

Cao, Y.; Yu, G.; Zhang, C.; Menon, R.; Heeger, A. J. Polymer light-emitting diodes with polyethylene dioxythiophene:polystyrene sulfonate as the transparent anode. *Synthetic Metals* 1997, *87*, 171–174.

Davids, P. S.; Campbell, I. H.; Smith, D. L. Device model for single carrier organic diodes. *Journal of Applied Physics* 1997, *82*, 6319.

de Vries, R. J.; van Mensfoort, S. L. M.; Janssen, R. A. J.; Coehoorn, R. Relation between the built-in voltage in organic light-emitting diodes and the zero-field voltage as measured by electroabsorption. *Physical Review B* 2010, *81*, 125203.

Deibel, C.; Dyakonov, V. Polymer–fullerene bulk heterojunction solar cells. *Reports in Progress in Physics* 2010, *73*, 096401.

Ding, H.; Gao, Y. Au/LiF/tris(8-hydroxyquinoline) aluminum interfaces. *Applied Physics Letters* 2007, *91*, 172107.

Fishchuk, I. I.; Kadashchuk, A.; Bässler, H.; Nespurek, S. Nondispersive polaron transport in disordered organic solids. *Physical Review B* 2003, *67*, 224303.

Gerischer, H. The impact of semiconductors on the concepts of electrochemistry. *Electrochimica Acta* 1990, *35*, 1677–1699.

Goodman, A. M.; Rose, A. Double extraction of uniformly generated electron-hole pairs from insulators with noninjecting contacts. *Journal of Applied Physics* 1971, *52*, 2823–2830.

Haney, P. M. Organic photovoltaic bulk heterojunctions with spatially varying composition. *Journal of Applied Physics* 2011, *110*, 024305.

He, Z. Review of the Shockley-Ramo theorem and its application in semiconductor gammaray detectors. *Nuclear Instruments and Methods in Physics Research Section A* 2001, *463*, 250–267.

Henisch, H. K. *Semiconductor Contacts*; Clarendon Press: Oxford, 1984.

Hung, L. S.; Tang, C. W.; Mason, M. G. Enhanced electron injection in organic electroluminescence devices using an Al/LiF electrode. *Applied Physics Letters* 1997, *70*, 152–154.

Jørgensen, M.; Norrman, K.; Krebs, F. C. Stability/degradation of polymer solar cells. *Solar Energy Materials and Solar Cells* 2008, *92*, 686–714.

Kim, C. H.; Yaghmazadeh, O.; Bonnassieux, Y.; Horowitz, G. Modeling the low-voltage regime of organic diodes: Origin of the ideality factor. *Journal of Applied Physics* 2011, *110*, 093722.

Kontturi, K.; Murtomäki, L.; Manzanares, J. A. *Ionic Transport Processes in Electrochemistry and Membrane Science*; Oxford University Press: Oxford, 2008.

Koster, L. J. A.; Smits, E. C. P.; Mihailetchi, V. D.; Blom, P. W. M. Device model for the operation of polymer/fullerene bulk heterojunction solar cells. *Physical Review B* 2005, *72*, 085205.

Lonergan, M. Charge transport at conjugated polymer-inorganic semiconductor and conjugated polymer-metal interfaces. *Annual Review of Physical Chemistry* 2004, *55*, 257–298.

Lopez-Varo, P.; Jimenez-Tejada, J. A.; Lopez-Villanueva, J. A.; Carceller, J. E.; Deen, M. J. Modeling the transition from ohmic to space charge limited current in organic semiconductors. *Organic Electronics* 2012, *13*, 1700–1709.

Malliaras, G. G.; Scott, J. C. Numerical simulation of the electrical characteristics and the efficiencies of single-layer organic light emitting diodes. *Journal of Applied Physics* 1999, *85*, 7426.

Marshall, J. M. Carrier diffusion in amorphous semiconductors. *Reports in Progress in Physics* 1983, *46*, 1235–1282.

Mihailetchi, V. D.; Blom, P. W. M.; Hummelen, J. C.; Rispens, M. T. Cathode dependence of the open-circuit voltage of polymer:fullerene bulk heterojunction solar cells. *Journal of Applied Physics* 2003, *94*, 6849.

Mihailetchi, V. D.; Koster, L. J. A.; Blom, P. W. M. Effect of metal electrodes on the performance of polymer:fullerene bulk heterojunction solar cells. *Applied Physics Letters* 2004, *85*, 970–972.

Mori, T.; Fujikawa, H.; Tokito, S.; Taga, Y. Electronic structure of 8-hydroxyquinoline aluminum/LiF/Al interface for organic electroluminescent device studied by ultraviolet photoelectron spectroscopy. *Applied Physics Letters* 1998, *73*, 2763–2765.

Oh, S. J.; Wang, Z.; Berry, N. E.; Choi, J.-H.; Zhao, T.; Gaulding, E. A.; Paik, T.; et al. Engineering charge injection and charge transport for high performance PBSE nanocrystal thin film devices and circuits. *Nano Letters* 2014, *14*, 6210–6216.

Pai, D. M. Transient photoconductivity in poly(N-vinylcarbazole). *The Journal of Physical Chemistry* 1970, *52*, 2285.

Ramo, S. Current induced by electron motion. *Proceedings of the I.R.E.* 1939, *27*, 584–585.

Ratcliff, E. L.; Zacher, B.; Armstrong, N. R. Selective interlayers and contacts in organic photovoltaic cells. *The Journal of Physical Chemistry Letters* 2011, *2*, 1337–1350.

Scher, H.; Montroll, E. W. Anomalous transit-time dispersion in amorphous solids. *Physical Review B* 1975, *12*, 2455–2477.

Scott, J. C. Metal–organic interface and charge injection in organic electronic devices. *Journal of Vacuum Science & Technology A: Vacuum, Surfaces, and Films* 2003, *21*, 521–531.

Scott, J. C.; Malliaras, G. G. Charge injection and recombination at the metal-organic interface. *Chemical Physics Letters* 1999, *299*, 115.

Shen, Y.; Hosseini, A. R.; Wong, M. H.; Malliaras, G. G. How to make ohmic contacts to organic semiconductors. *ChemPhysChem* 2004, *5*, 16–25.

Shockley, W. Current to conductors induced by a moving point charge. *Journal of Applied Physics* 1938, *9*, 635–636.

Simmons, J. G. Theory of metallic contacts on high resistivity solids. I. Shallow traps. *Journal of Physics and Chemistry of Solids* 1971, *32*, 1987–1999.

Sirkis, M. D.; Holonya, N. Currents induced by moving charges. *American Journal of Physics* 1966, *34*, 943–946.

Sokel, R.; Hughes, R. C. Numerical analysis of transient photoconductivity in insulators. *Journal of Applied Physics* 1982, *53*, 7414–7424.

Waldauf, C.; Schilinsky, P.; Hauch, J.; Brabec, C. J. Material and device concepts for organic photovoltaics: Towards competitive efficiencies. *Thin Solid Films* 2004, *451–452*, 503–507.

Walzer, K.; Maennig, B.; Pfeiffer, M.; Leo, K. Highly efficient organic devices based on electrically doped transport layers. *Chemical Reviews* 2007, *107*, 1233–1271.

11 Diffusion Transport

Diffusion is a net displacement of particles related to differences in concentration, or more precisely, to differences of the chemical potential. In a homogeneous environment, a variety of mechanisms, such as collisions with neighbors or thermally activated hops, produce a displacement by frequent steps in random directions, in a type of motion known as a *random walk*. When a large number of particles are considered, distribution functions can be derived to provide the most likely distribution of particles in space, which relates to their concentration. At the macroscopic level, diffusive motion takes a definite direction established by the gradient of the concentration. Thus, there are two types of approaches to diffusion, one is single-particle random motion and the other is collective motion depending on the distribution of carrier density over the available space. Both approaches lead to kinetic coefficients with distinct meanings that are nonetheless related by thermodynamic quantities. The statistics of the carriers play a dominant role in establishing the connection between single carrier and collective diffusion, and determining the thermodynamic factor that relates mobility to the chemical diffusion coefficient.

11.1 DIFFUSION IN THE RANDOM WALK MODEL

Free or quasifree carriers (electrons, holes, ions, molecules) in a solid or liquid medium possess an average kinetic energy given by k_BT, and hence a thermal velocity $v_{th} = \sqrt{2k_BT/m}$. The carriers thermalized to the medium realize a random motion with no definite average direction (Figure 11.1a). To analyze this type of motion, we discuss the random walk model, which assumes that diffusion takes place by successive hops of the carrier from site to site through a lattice. This model also incorporates the transport of trapped carriers by hopping events.

Let us consider a one-dimensional regular lattice composed of sites separated by a distance a (Figure 11.2a). If the particle randomly jumps the distance a between neighboring sites with a transition rate ν, after N_τ jumps, it can be at any of the sites between $-N_\tau$ and $+N_\tau$. By statistical arguments, it can be shown that the probability function that the particle is situated at site m has the form (Chandrasekaran, 1943)

$$p_m(N_\tau) = \left(\frac{2}{\pi N_\tau}\right)^{1/2} \exp\left(-\frac{m^2}{2N_\tau}\right) \tag{11.1}$$

If there are a large number of particles that do not interfere with each other, $p_m(N_\tau)$ can be viewed as the probability of occupation of site m after a time $t = N_\tau/\nu$. We can denote the distance from the origin as $x = ma$. We consider an interval $\Delta x >> a$, containing many lattice points, as in Figure 11.2b. We identify the probability that the particle doing a random walk is situated between x and $x + \Delta x$ after N_τ displacements. This function is a probability density of the occupancy $f(x)$, related to p as

$$f(x, N_\tau)\Delta x = p_m(N_\tau)\frac{\Delta x}{2a}. \tag{11.2}$$

From Equations 11.1 and 11.2,

$$f(x, N) = \left(\frac{1}{2\pi N_\tau a^2}\right)^{1/2} \exp\left(-\frac{m^2}{2N_\tau}\right), \tag{11.3}$$

or alternatively, with respect to time t,

$$f(x,t)\Delta x = \frac{1}{2(\pi D_J t)^{1/2}} \exp\left(-\frac{x^2}{4D_J t}\right), \tag{11.4}$$

Where

$$D_J = \frac{1}{2}\nu a^2. \tag{11.5}$$

In random walks of a carrier in d dimensions, the *jump diffusion coefficient*, D_J (or kinetic diffusion coefficient), has the form

$$D_J = \frac{1}{2dt}\left\langle\left(\frac{1}{N}\sum_{i=1}^{N}\Delta r_i\right)^2\right\rangle, \tag{11.6}$$

where Δr_i is the displacement of the ith particle at time t and $\langle\ \rangle$ denotes a statistical average. More precisely, the jump (or kinetic) diffusion coefficient defined by Equation 11.6 reflects diffusion of the *center of mass* of N particles, while the tracer diffusion coefficient D^* reflects random walks of a particle

$$D^* = \frac{1}{2dNt}\left\langle\sum_{i=1}^{N}(\Delta r_i)^2\right\rangle. \tag{11.7}$$

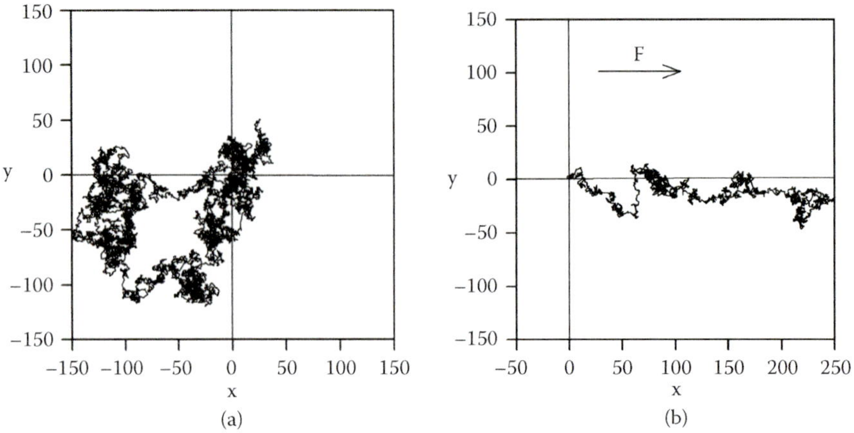

FIGURE 11.1 (a) Brownian motion of a particle starting at the origin. (b) Brownian motion of a particle starting at the origin biased by a force in the x direction.

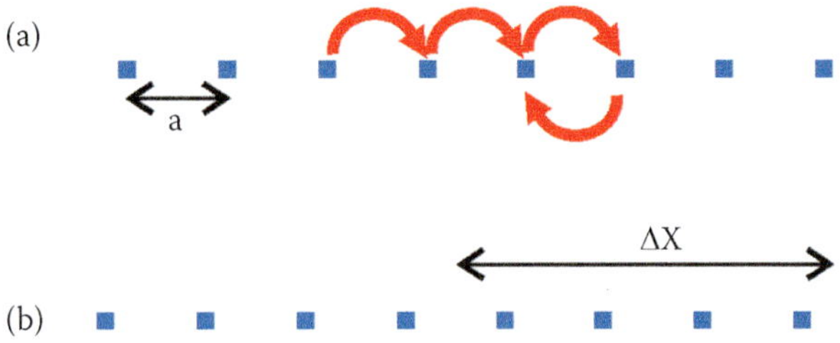

FIGURE 11.2 (a) Random walk on a one-dimensional lattice. The spacing between neighbor sites is a. The length element $\Delta x > a$ allows us to define the concentration.

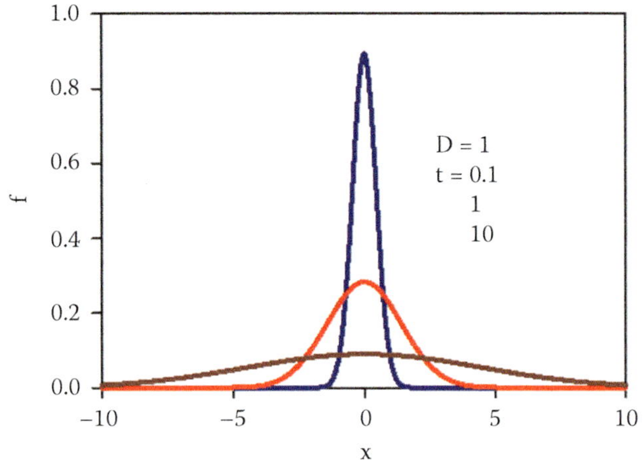

FIGURE 11.3 Distribution function for diffusion in one dimension for particles initially injected at $x = 0$.

The difference between these two coefficients (the *Haven ratio*) depends on cross correlations between displacements $\Delta r_i(t)$ of different particles at different times (Ala-Nissila et al., 2002). If on average there are no cross correlations, D_J and D^* become equivalent. Monte Carlo simulations show that jump and tracer diffusion coefficients are practically identical in many conditions (Uebing and Gomer, 1994). The jump diffusion coefficient can often be expressed in a simplified form

$$D_J = \nu r^2 \quad (11.8)$$

in terms of a mean effective jump frequency $\langle \nu \rangle$ and the square of effective jump length $\langle r^2 \rangle$. For the diffusion of ions in solid media, the dependence of transition rates on the crystal characteristics has been described in Figure 5.22 and in Section 6.2.

The probability density in Equation 11.4 has the form of a Gaussian packet that spreads from the starting point, Figure 11.3. If N number of particles are injected at $x = 0$, the local concentration at x is

$$n(x,t) = Nf(x,t). \quad (11.9)$$

As time increases, the concentration wave becomes wider and consequently, the probability that a particle has travelled a larger distance from the origin increases. This feature is computed by the mean-square displacement from the starting point

$$x^2(t) = \int_0^{\infty} x^2 f(x,t)\,dx. \quad (11.10)$$

The calculation from Equation 11.4 gives the result

$$x^2 = 2D_J t. \quad (11.11)$$

This means that the distance travelled by diffusing particles increases as the square root of time. A diffusion process over a distance L has the characteristic time

$$\tau_{diff} = \frac{L^2}{D}. \quad (11.12)$$

This is the time scale in which an accumulation of solute concentrated in a small region will be dispersed. Furthermore, if there is a gradient of concentration, the composition is equalized by diffusion on a time scale τ_{diff}.

For diffusion in three dimensions

$$D_J = \frac{1}{6} \nu a^2 \tag{11.13}$$

$$r^2 = 6 D_J t. \tag{11.14}$$

More generally, in a cubic coordination lattice of dimension d, with coordination number Z,

$$D_J = \frac{1}{2d} Z \nu a^2. \tag{11.15}$$

This last expression is, however, restricted to the case of homogeneous transition rates across the whole diffusion space. The elementary hops of ions in a solid lattice were described in Chapter 5 and 6; see Figure 5.22. The transition rate is constant in the case of identical barriers as shown in Figure 11.4a. However, the transition rates can depend on position or neighbor occupation, as discussed in Section 6.2. For example, in the random barrier model, the barrier to be overcome, the activation energy E_m, is assumed to vary randomly (see Figure 11.4b). In general, in a one-dimensional system of size N_L with arbitrary transition rates $\nu_{i,i+1}$, a general result by Zwanzig (1982) states that the diffusion coefficient is given by the average inverse hopping rate

$$D_J = \left\{ \frac{1}{N_L} \sum_{i=0}^{N_L - 1} \frac{1}{y_{i,i+1}} \right\}^{-1} a^2. \tag{11.16}$$

Clearly, the slowest transition rates provide the main contribution to the long-range diffusion process, such as hops that form a bottleneck for the displacement of a particle along a chain of localized sites. When the directions of successive atom jumps are related to each other, it is necessary to include the correlation between successive jumps. This can be done by multiplying the jump frequency ν by a correlation factor f, with νf being the "effective" jump frequency (Manning, 1959). Note that Equation 11.16 is in contrast to the simplified Equation 11.8, where the diffusion coefficient is proportional to $\langle \nu \rangle$. In general, $\langle \nu \rangle \neq \langle 1/\nu \rangle^{-1}$.

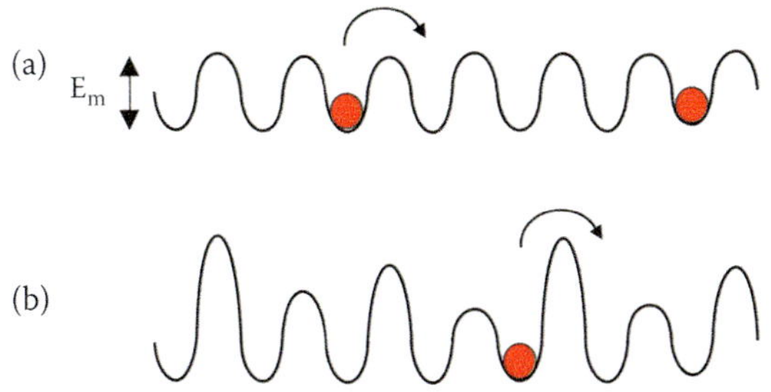

FIGURE 11.4 (a) Diffusion on a one-dimensional lattice by hopping between sites overcoming an energy barrier E_m. (b) The random barrier model.

If there is a net force F acting on the carriers that effect a random walk, it will produce a net displacement in the direction of the force as indicated in Figure 11.1b. Note that the average velocity gained by the action of the force v_F may be much less than the thermal velocity, however v_F gives rise to a systematic drift of the carriers alongside F. In a typical conductor, the drift velocity is about 10^{-4} cm/s, which is significantly lower than the random velocity of 10^8 cm/s. Considering a large ensemble of carriers with density n, the flux J (per unit area) in the direction of F depends on the velocity v as $J = nv$ (see Section 1.1 of this volume), and the current density is $j = Jq = nqv$. This effect is usually produced by an electrical field. Therefore, for carriers that do not interact strongly, it is usual to separate carrier transport into diffusion and drift mechanisms.

11.2 MACROSCOPIC DIFFUSION EQUATION

For applications in device models, it is important to formulate the diffusion process in terms of differential equations. Consider that the microscopic transport is a hopping process between sites separated by a distance a, Figure 11.2, with a symmetric transition rate ν. We assume that each site can contain one carrier at most. Although the occupation probabilities should take only values 0 or 1, we can view p_i as the normalized average occupancy of the site i and it then takes fractional values. The master equation for the probability p_i is established in terms of hopping probabilities with a transition rate ν_0, taking into account the occupancy of the starting site and the vacancies in the destination site, as discussed in Equation 6.8:

$$\frac{dp_i}{dt} = \nu_0 \left[p_{i+1}(1 - p_i) + p_{i-1}(1 - p_i) - p_i(1 - p_{i+1}) - p_i(1 - p_{i-1}) \right] \tag{11.17}$$

To formulate an equation for the macroscopic diffusion flux, known as Fick's law, we start with the flux between two points in Equation 11.17 and we proceed to an average flux in terms of the gradient of the concentration (Reed and Ehrlich, 1981; Guidoni and Aldao, 2002).

A carrier flux is a product of carrier density and velocity. The flux of carriers of number density n moving from site 1 to site 2 is given by

$$J_1 = a\nu_0 n1\left(1 - p_2\right). \quad (11.18)$$

The term $a\nu_0$ is the velocity associated with the elementary jump, and $(1 - p_2)$ is the probability that the arrival site is empty. The flux in the opposite direction is $J_2 = a\nu_0 n_2(1 - p_1)$. Because the concentration n is related to the probability of site occupancy by $n = p/N_V$, where N_V is the number of sites per unit volume, the terms np cancel out in the net flux between two neighbor sites and the result is

$$J_n = -\nu_0 a^2 \frac{\left(n_2 - n_1\right)}{a}. \quad (11.19)$$

The fraction on the right side is a gradient of the concentration. Averaging over Δx, we get Fick's law:

$$J_n = -D_n \frac{\partial n}{\partial x}. \quad (11.20)$$

Here, D_n is the *chemical diffusion coefficient* that is defined more generally below. Jump and chemical diffusion coefficient are identical if the carriers obey ideal statistics but the difference in more general situations will be clarified later on.

If we assume that the different sites are far from fully occupied, Equation 11.17 reduces to

$$\frac{dp_m}{dt} = \nu\left(p_{m+1} + p_{m-1} - 2p_m\right). \quad (11.21)$$

As before, the position can be written as $x = ma$ and we introduce a probability density of occupancy $f(x)$. From Equation 11.21, we obtain

$$\frac{\partial f\left(x,t\right)}{\partial t} = \nu\left\{f\left(x+a,t\right) + f\left(x-a,t\right) - 2f\left(x,t\right)\right\}. \quad (11.22)$$

Expanding around x, we arrive at

$$\frac{\partial f}{\partial t} = D_n \frac{\partial^2 f}{\partial x^2}. \quad (11.23)$$

Equation 11.23 is a conservation equation for the evolution of the probability density. Using Equation 11.20, the conservation equation can be expressed as

$$\frac{\partial n}{\partial t} = -\frac{\partial J_n}{\partial x}. \quad (11.24)$$

If G is the generation rate per unit volume and U_n is the recombination or reaction rate, the conservation equation has the form

$$\frac{\partial n}{\partial t} = -\frac{\partial J_n}{\partial x} + G - U_n. \quad (11.25)$$

11.3 THE DIFFUSION LENGTH

Our next task is to provide a description of the combined processes of diffusion and recombination. We remarked in Figure 11.3 that a number of particles N injected at $x = 0$ diffuse away from the origin with a mean-square displacement in one dimension given by Equation 11.11. We consider an important additional physical effect in which diffusing particles have a probability of disappearing during their random walk displacement. This is a central feature in a number of situations, for example, the transport and recombination of a minority electronic species in semiconductor devices or the diffusion of an ionic species coupled with homogeneous reaction in electrochemical systems.

When recombination of the injected carriers occurs, the spread of the packet is not well described by Figure 11.3 because the number of particles is decreasing as time advances. The distance that can be covered by a random walker before it disappears is accounted for by the *diffusion length* L_n. This quantity was first introduced by Amaldi and Fermi (1936) in the context of neutron diffusion in paraffin samples, as the distance that "a neutron will diffuse before it gets captured by a proton."

The excess carriers with respect to an equilibrium situation are $\Delta n = n - n_0$. We assume that carriers are injected at $x = 0$ at a continuous rate. That is, we arbitrarily fix the concentration of carriers at two points:

1 $\Delta n(0)$ is the concentration injected at $x = 0$ over the background.
2 $\Delta n = 0$ at $x = +\infty$.

We define the diffusion length L_n as the average distance traveled by the injected carrier flux from $x = 0$.

$$L_n = \frac{\int_0^{+\infty} x\Delta n\left(x\right)dx}{\int_0^{+\infty} \Delta n\left(x\right)dx} \quad (11.26)$$

Let us consider diffusion–recombination of injected electrons with chemical diffusion coeffi ient D_n and recombination rate U_n, formulated in Equation 11.25. The steady-state distribution of carriers is determined by the equation

$$D_n \frac{\partial^2 n}{\partial x^2} - U_n\left(x\right) = 0. \quad (11.27)$$

We discuss the simplest recombination model, which is that of linear recombination introduced in Equation 18.35 in terms of the electron lifetime τ_n, Equation 11.27 can be written as

$$D_n \frac{\partial^2 n}{\partial x^2} - \frac{1}{\tau_n}(n - n_0) = 0. \tag{11.28}$$

We calculate the concentration profile by solving Equation 11.28 with the boundary conditions (i) and (ii):

$$n(x) = n_0 + (n(0) - n_0) e^{-x/(D_n \tau_n)^{1/2}}. \tag{11.29}$$

By calculating of the right-hand side of Equation 11.26, we obtain

$$L_n = (D_n \tau_n)^{1/2}. \tag{11.30}$$

Thus, the diffusion length is determined by the combination of the diffusion coefficient and carrier lifetime.

Equation 11.28 governing the carrier distribution can be written

$$\frac{\partial^2 n}{\partial x^2} - \frac{n - n_0}{L_n^2} = 0. \tag{11.31}$$

Therefore, in *steady state*, the physical constant L_n determines the distribution of electrons. However, transient time measurements are governed by the equation

$$\frac{\partial n}{\partial t} = +D_n \frac{\partial^2 n}{\partial x^2} - \frac{1}{\tau_n}(n - n_0). \tag{11.32}$$

Thus, it is possible to determine the time constants D_n and τ_n separately. These constants are also decoupled if the generation term G is present. Note that if the carrier diffusion rate is very fast and can be neglected, we get back to the decay controlled by lifetime described in Section 18.5.

11.4 CHEMICAL DIFFUSION COEFFICIENT AND THE THERMODYNAMIC FACTOR

So far, we have introduced two different diffusion coefficients, the jump diffusion coefficient D_J (similar to D^*), which is related to random walk of the particles (Equation 11.6), and the chemical diffusion coefficient D_n, which connects the flux to the gradient of the concentration (Equation 11.20). In the simplest situations, both coincide, for example, in a cubic lattice for dilute statistics,

$$D_n = D_J = \frac{1}{6} v a^2. \tag{11.33}$$

But in general, the diffusion of carriers is affected by several factors: the available states for hopping, the interactions between carriers, the extent of shielding of Coulomb interactions by surrounding media, etc. The relationship between the microscopic hopping mechanisms and the macroscopic transport coefficients must be carefully defined. We discuss a useful framework that dates back to the statement by Onsager and Fuoss (1932) of the thermodynamic component of the diffusion coefficient in a nonideal solution. These ideas have been vastly applied in surface diffusion (Reed and Ehrlich, 1981) and in the simulation of diffusion systems consisting of interacting particles on the lattice (Gomer, 1990; Uebing and Gomer, 1991). Table 11.1 summarizes the main parameters of diffusion including general relationships derived in the following chapters.

When an electronic or ionic species with concentration n diffuses in a material, the true driving force for diffusion is the gradient of its chemical potential, μ_n, defined in Section 2.6. In the Onsager form of the diffusion law, a linear relationship is assumed between the diffusive flux and the gradient of the chemical potential,

$$J_n = -M_n \frac{\partial \mu_n}{\partial x}. \tag{11.34}$$

The prefactor M_n is a phenomenological coefficient known as the Onsager coefficient and depends on the carrier density n and the mobility u_n

$$M_n = \frac{n u_n}{q}. \tag{11.35}$$

The mobility has already been defined in Equation 10.10. Equation 11.34 can be re-stated as

$$J_n = -\frac{n u_n}{q} \frac{\partial \mu_n}{\partial x}. \tag{11.36}$$

On the other hand, diffusion is also formulated in terms of the *concentration* gradient in Fick's form, Equation 11.20. The coefficient D_n in Equation 11.20 is called the *chemical diffusion coefficient*. It is given by

$$D_n = -\frac{n u_n}{q} \frac{\partial \mu_n}{\partial x}. \tag{11.37}$$

D_n contains two components:

1. The phenomenological coefficient u_n (the mobility; see Section 10.1).
2. The term $n\partial\mu_n/\partial n$ that accounts for the difference between a gradient in concentration and a

TABLE 11.1
Diffusion Parameters Generally Valid for Nonideal Statistics

Relation	Significance
D_J	Jump diffusion coeffcient
D_n	Chemical diffusion coeffcient
$T_n = \frac{n}{k_B T}\frac{\partial \mu_n}{\partial n}$	Thermodynamic factor
$D_n = T_n D_J$	Relation of chemical and jump diffusion coeffcients
$D_J = \frac{k_B T}{q} u_n$	Relation of jump diffusion coefficient and mobility
$D_n = T_n \frac{k_B T}{q} u_n$	Generalized Einstein relation
$j_n = m r_n \frac{\partial E_{Fn}}{\partial x}$	Electron transport current in terms of the gradient of the Fermi level
$j_n = \frac{\sigma_n}{q}\frac{\partial E_{Fn}}{\partial x}$	Electron transport current in terms of the gradient of the Fermi level

gradient in chemical potential. This last term is expressed in dimensionless form as the *thermodynamic factor* introduced by Darken (1948).

$$T_n = \frac{n}{k_B T}\frac{\partial \mu_n}{\partial n} \tag{11.38}$$

Note that $T_n = 1$ in the case of Boltzmann distribution, that is, when the chemical potential of electrons satisfies $\mu_n = k_B T \ln (n/n_0)$; see Equation 2.32. The thermodynamic factor can be expressed with respect to the chemical capacitance C_μ defined in Equation 7.6 as

$$T_n = \frac{q^2 n}{k_B T}\frac{1}{C_\mu}. \tag{11.39}$$

Furthermore, the chemical capacitance is related to the mean-square fluctuation of the particle number N in a volume V by the formula (Landsberg, 1978)

$$\left(\delta N\right)^2 = \frac{k_B T V}{q^2} C_\mu. \tag{11.40}$$

Therefore, the thermodynamic factor is

$$T_n = \frac{N}{\left(\delta N\right)^2}. \tag{11.41}$$

This last equation is correct only in the absence of cross correlations between velocities. Equation 11.37 can be written as

$$D_n = T_n \frac{k_B T}{q} u_n. \tag{11.42}$$

The chemical diffusion coefficient takes the form

$$D_n = T_n D_J, \tag{11.43}$$

where the jump (or kinetic) diffusion coefficient defined in Equation 11.6 is simply proportional to mobility:

$$D_J = \frac{k_B T}{q} u_n. \tag{11.44}$$

Equation 11.43 indicates the two components of the chemical diffusion coefficient. All the dynamic components of D_n that are included in D_J. D_n in Equation 11.44 can be derived from a microscopic approach using the Green-Kubo theory (Gomer, 1990; Uebing and Gomer, 1991). In numerical simulation methods, the quantities T_n and D_J are evaluated separately and then combined to form D_n by the so-called indirect method (Argyrakis et al., 2001). The term T_n can be determined in grand canonical Monte Carlo simulations by use of the fluctuation formula in Equation 11.42.

In electrochemistry, it is customary to write the chemical potential function in terms of the activity of the species a_i; see Equation 3.44. Therefore, the thermodynamic factor has the expression

$$T_i = \frac{c_i}{k_B T}\frac{\partial \mu_i}{\partial c_i} = \frac{\partial \ln a_i}{\partial \ln c_i}. \tag{11.45}$$

The activity relates to the concentration of the species c_i through the activity coefficient as $\gamma_i = a_i / c_i$, hence

$$T_i = 1 + \frac{d \ln \gamma_i}{d \ln c_i} \quad (11.46)$$

and the chemical diffusion coefficient is given as (Weppner and Huggins, 1977)

$$D_{ch(i)} = \left(1 + \frac{d \ln \gamma_i}{d \ln c_i}\right) D_{J(i)}. \quad (11.47)$$

GENERAL REFERENCES

Diffusion fundamental concepts: Chandrasekhar, 1943; Wang and Uhlenbeck, 1945; Crank, 1956; Alexander et al., 1981; Cussler, 1984; Savéant, 1988; Philibert, 1991; Yu et al., 2000; Guidoni and Aldao, 2002; Kontturi et al., 2008.

Random barrier model: Alexander et al., 1981; Teitel, 1988; Kehr et al., 1998; Dyre and Schroder, 2000.

Diffusion length: Lindmayer and Wrigley, 1965; Bisquert and Mora-Seró, 2010.

Jump and chemical diffusion coefficient: Reed and Ehrlich, 1981; Gomer, 1990; Uebing and Gomer, 1991; Myshlyavtsev et al., 1995; van der Ven et al., 2001; Ansari-Rad et al., 2013.

Einstein relation: Butcher, 1972; Landsberg, 1981; Ando et al., 1982; Lee, 1982; Marshak, 1989.

REFERENCES

Ala-Nissila, T.; Ferrando, R.; Ying, S. C. Collective and single particle diffusion on surfaces. *Advances in Physics* 2002, *51*, 949–1078.

Alexander, S.; Bernasconi, J.; Schneider, W. R.; Orbach, R. Excitation dynamics in random one-dimensional systems. *Reviews of Modern Physics* 1981, *53*, 175–198.

Amaldi, E.; Fermi, E. On the absorption and the diffusion of slow neutrons. *Physical Review* 1936, *50*, 899–928.

Ando, T.; Fowler, A. B.; Stern, F. Electronic properties of two-dimensional systems. *Reviews of Modern Physics* 1982, *54*, 437–672.

Ansari-Rad, M.; Anta, J. A.; Bisquert, J. Interpretation of diffusion and recombination in nanostructured and energy disordered materials by stochastic quasiequilibrium simulation. *The Journal of Physical Chemistry C* 2013, *117*, 16275–16289.

Argyrakis, P.; Groda, Y. G.; Bokun, G. S.; Vikhrenko, V. S. Thermodynamics and diffusion of a lattice gas on a simple cubic lattice. *Physical Review E* , *64*, 2001, 066108.

Bisquert, J.; Mora-Seró, I. Simulation of steady-state characteristics of dye-sensitized solar cells and the interpretation of the diffusion length. *Journal of Physical Chemistry Letters* 2010, *1*, 450–456.

Butcher, P. N. On the definition of energy dependent mobility and diffusivity. *Journal of Physics C: Solid State Physics* 1972, *5*, 3164–3167.

Chandrasekaran, S. Stochastic problems in physics and astronomy. *Reviews of Modern Physics* 1943, *15*, 1–89.

Crank, J. *The Mathematics of Diffusion*; Oxford University Press: London, 1956.

Cussler, E. L. *Diffusion. Mass Transfer in Fluid Systems*; Cambridge University Press: Cambridge, 1984.

Darken, L. S. Diffusion, mobility and their interrelation through free energy in binary metallic systems. *Transactions of the American Institute of Mining and Metallurgical Engineers* 1948, *175*, 184.

Dyre, J. C.; Schroder, T. B. Universality of ac conduction in disordered solids. *Reviews of Modern Physics* 2000, *72*, 873.

Gomer, R. Diffusion of adsorbates on metal surfaces. *Reports in Progress in Physics* 1990, *53*, 917–1002.

Guidoni, S. E.; Aldao, C. M. On diffusion, drift and the Einstein relation. *European Journal of Physics* 2002, *23*, 395–402.

Kehr, K. W.; Mussawisade, K.; Wichmann T. *Diffusion in Condensed Matter*; Kärger, J., Heitjans, P., Haberlandt, R., Eds.; Vieweg: Wiesbaden, 1998, pp. 265–305.

Kontturi, K.; Murtomäki, L.; Manzanares, J. A. *Ionic Transport Processes in Electrochemistry and Membrane Science*; Oxford University Press: Oxford, 2008.

Landsberg, P. T. *Thermodynamics and Statistical Mechanics*; Dover: New York, 1978.

Landsberg, P. T. Einstein and Statistical thermodynamics III: The diffusion-mobility rela-tion in semiconductors. *European Journal of Physics* 1981, *2*, 213–219.

Lee, P. A. Density of states and screening near the mobility edge. *Physical Review B* 1982, *26*, 5882.

Lindmayer, J.; Wrigley, C. Y. *Fundamentals of Semiconductor Devices*; Van Nostrand: New York, 1965.

Manning, J. R. Correlation effects in impurity diffusion. *Physical Review* 1959, *116*, 819–827.

Marshak, A. H. Modeling semiconductor devices with position-dependent material parameters. *IEEE Transactions on Electron Devices* 1989, *36*, 1764–1772.

Myshlyavtsev, A. V.; Stepanov, A. A.; Uebing, C.; Zhdanov, V. P. Surface diffusion and continuous phase transitions in absorbed overlayers. *Physical Review B* 1995, *52*, 5977.

Onsager, L.; Fuoss, R. M. Irreversible processes in electrolytes, diffusion, conductance, and viscous flow. *The Journal of Physical Chemistry* 1932, *36*, 2689–2778.

Philibert, J. *Atom Movements Diffusion and Mass Transport in Solids*; Les Editions de Physique: Les Ulis, 1991.

Reed, D. A.; Ehrlich, G. Surface diffusion, atomic jump rates and thermodynamics. *Surface Science* 1981, *102*, 588–609.

Savéant, J. M. Electron hopping between localized sites. Effect of ion pairing on diffusion and migration. General rate laws and steady-state responses. *The Journal of Physical Chemistry* 1988, *92*, 4526.

Teitel, S. Transition to anomalous relaxation. *Physical Review Letters* 1988, *60*, 1154.

Uebing, C.; Gomer, R. A Monte Carlo study of surface diffusion coefficients in the presence of adsorbate-adsorbate interactions. I. Repulsive interactions. *Journal of Chemical Physics* 1991, *95*, 7626–7652.

Uebing, C.; Gomer, R. Determination of surface diffusion coefficients by Monte Carlo methods: Comparison of fluctuation and Kubo-Green methods. *Journal of Chemical Physics* 1994, *100*, 7759–7766.

van der Ven, A.; Ceder, G.; Asta, M.; Tepesch, P. D. First-principles theory of ionic diffusion with nondilute carriers. *Physical Review B* 2001, *64*, 184307.

Wang, M. C.; Uhlenbeck, G. E. On the theory of the Brownian motion II. *Reviews of Modern Physics* 1945, *17*, 323–342.

Weppner, W.; Huggins, R. A. Determination of the kinetics parameters of mixed-conducting electrodes and application to the system Li_3Sb. *Journal of the Electrochemical Society* 1977, *124*, 1569.

Yu, Z. G.; Smith, D. L.; Saxena, A.; Martin, R. L.; Bishop, A. R. Molecular geometry fluctuation model for the mobility of conjugated polymers. *Physical Review Letters* 2000, *84*, 721–724.

Zwanzig, R. Non-Markoffian diffusion in a one-dimensional disordered lattice. *Journal of Statistical Physics* 1982, *28*, 127–133.

12 Drift-Diffusion Transport

In previous chapters, the frameworks of drift and diffusion have been established and the effect of nonideal statistics on the kinetic transport coefficients was introduced. Here, we take more general steps, formulating the general problem of combined diffusion and drift, the experimental measurement of transport coefficients, and the Einstein relation between mobility and diffusivity in nonideal conditions. Thereafter, we establish the conditions of transport for a set of coupled carriers, considering also the Poisson equation. In particular, we analyze ambipolar transport where charge transport rates are linked by an electroneutrality condition. We discuss the effects of relaxation toward charge compensation and the competition between neutralization and recombination of injected charges in semiconductors. We also describe one particular model that is controlled by displacement current. We finish with a general discussion of simulation of transport problems.

12.1 GENERAL TRANSPORT EQUATION IN TERMS OF ELECTROCHEMICAL POTENTIAL

The current density due to electronic carrier transport in a semiconductor or in ionic transport in condensed media is proportional to the thermodynamic driving force, the gradient of the electrochemical potential η_i (see Equation 2.26), provided that the carrier distribution is not far from equilibrium, so that nonlinear effects can be neglected. For the transport of a charge carrier with concentration c_i and a charge number z_i, with electrical charge $Q_i = z_i q$, the conduction current is

$$j_i = -\frac{z_i}{|z_i|} c_i u_i \frac{\partial \eta_i}{\partial x}. \quad (12.1)$$

Equation 12.1 can be derived fundamentally for electron transport by solving the Boltzmann transport equation in the relaxation time approximation (Marshak and van Vliet, 1978). This derivation assumes a distribution of carriers in the energy space, which is close to an equilibrium distribution. The same result is obtained by the standard approach in irreversible thermodynamics on the assumptions that Onsager's cross-coefficients are negligible, isothermal, and adiabatic conditions for the diffusion-drift displacement in quasi-equilibrium. Moreover, Equation 12.1 is valid even in situations in which the carriers interact and the distinction between chemical potential and the electrostatic term (Equation 2.29) is not satisfied.

If we split the electrochemical potential as in Equation 3.41, Equation 12.1 takes the form

$$j_i = |z_i| q c_i u_i \mathbf{E} - \frac{z_i}{|z_i|} c_i u_i \frac{\partial \mu_i}{\partial c_i} \frac{\partial c_i}{\partial x}. \quad (12.2)$$

We introduce an extension of Equation 11.44 for the jump diffusion coefficient:

$$D_{Ji} = \frac{k_B T}{|z_i| q} u_i. \quad (12.3)$$

Then, Equation 12.2 can be written using D_i, the chemical diffusion coefficient of the species c_i, as follows:

$$j_i = q|z_i| c_i u_i \mathbf{E} - z_i q D_i \frac{\partial c_i}{\partial x}. \quad (12.4)$$

Equation 12.4 is the drift-diffusion equation for an ionic or electronic species. The first term is the current by drift in the electrical field and the second term is the diffusion current. In the field of ionic transport, it is called the Nernst-Planck equation. These equations can be generalized to include the displacement under a temperature gradient in the system (the Seebeck effect).

12.2 THE TRANSPORT RESISTANCE

Let us consider the particular case of electrons and holes in a semiconductor. For electrons with density n, we can write Equation 12.1 in terms of the electron Fermi level using Equation 2.26:

$$j_n = n u_n \frac{\partial E_{Fn}}{\partial x}. \quad (12.5)$$

We apply the expression 2.30 to give Equation 12.5 in terms of chemical and electrostatic potential

$$j_n = -nqu_n \frac{\partial \varphi}{\partial x} + n u_n \frac{\partial \mu_n}{\partial x}. \quad (12.6)$$

We introduce the chemical diffusion coefficient of electrons by Equation 11.37 and we obtain

$$j_n = qnu_n\mathbf{E} + qD_n\frac{\partial n}{\partial x}. \tag{12.7}$$

Taking into account Equation 2.41, the current density resulting from Equation 12.1 for the transport of holes is

$$j_p = pu_p\frac{\partial E_{Fp}}{\partial x}. \tag{12.8}$$

The drift-diffusion current equation for holes is

$$j_p = qpu_p\mathbf{E} - qD_p\frac{\partial p}{\partial x}. \tag{12.9}$$

We recall from Equation 10.18 that the conductivity has the form

$$\sigma = \sigma_n + \sigma_p, \tag{12.10}$$

where the electron and hole conductivities are

$$\sigma_n = qnu_n \tag{12.11}$$

$$\sigma_p = qpu_p. \tag{12.12}$$

In terms of conductivities, we may write Equation 12.1 as

$$j_n = \frac{\sigma_n}{q}\frac{\partial E_{Fn}}{\partial x} \tag{12.13}$$

$$j_p = \frac{\sigma_p}{q}\frac{\partial E_{Fp}}{\partial x}. \tag{12.14}$$

In the drift mechanism, although the electric field tends to accelerate the carriers, they quickly reach a constant velocity because of the presence of resistive forces. However, under a gradient of compositions, a displacement of carriers occurs in the lower composition region by diffusion, independent of their charge sign or number. The two transport mechanisms indicated in Equations 12.7 and 12.9 can be separately viewed in the energy diagram. Figure 12.1a (as Figure 10.1 earlier) shows the drift transport of electrons (Vanmaekelbergh and de Jongh, 1999). We observe a tilt of the band due to the electric field across the transport layer and we also note that the Fermi level follows the same tilt as that of the band, implying there is no gradient of concentration. The carrier density is constant and the electrons move in the electric field. Conversely, in Figure 12.1b,

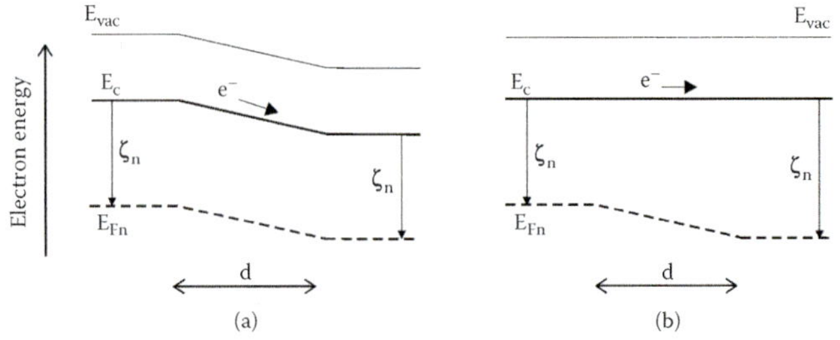

FIGURE 12.1 A flux of carriers induced by a gradient of the Fermi level: (a) transport by drift and (b) transport by diffusion. In both cases, the gradient $\Delta E_{Fn}/d$ is the same so that the electron flux is the same.

there is a gradient of the Fermi level and the band is flat, then the gradient represents a change of the chemical potential and hence the transport is diffusive. Both systems show a difference of Fermi level at the two external sides. Therefore, these are nonequilibrium situations in which a local voltage is applied, equal to the difference of the Fermi level between two points, $qV = \Delta E_{Fn}$. The gradient of the electrochemical potential is the same in both cases. According to Equation 12.13, if the voltage is small, in the sense that local differences of conductivities can be neglected, the flux of electrons will be the same in both systems.

However, once the transport distance becomes large and band-bending or space-charge regions are formed, the variations of conductivity become rather significant to determine transport rates. To deal with the local variations of conductivity in the description of transport problems, it is useful to define local impedances related to carrier transport that can be introduced in the larger framework of transmission line models (Bisquert, 2002; Bisquert et al., 2006; Bisquert et al., 2014). We thus need to obtain the conduction current in response to a small perturbation of voltage. Consider in Figure 12.1 the part of thickness d where the resistive losses occur. The electrical current is

$$j_n = \frac{\sigma_n V}{d}. \tag{12.15}$$

Equation 12.15 describes an ohmic system in which the current is linear with the voltage. A small perturbation of voltage and current provide the impedance

$$Z = \frac{\hat{V}}{\hat{j}} = \frac{d}{\sigma_n}. \tag{12.16}$$

This impedance is just the transport resistance (per area)

$$R_{tr} = \frac{d}{\sigma_n}. \tag{12.17}$$

When carrier density, electric field, and so on are position-dependent, the local transport resistance per unit length is defined as

$$r_{tr} = \lim_{d\to 0} \frac{R_{tr}}{d} = \frac{1}{\sigma_n}. \tag{12.18}$$

r_{tr} coincides with the reciprocal of the conductivity (Bisquert and Fabregat-Santiago, 2010). The origin of the resistance is a drop in the Fermi level, which is required to drive the current by transport. It should be noted that transport resistance is therefore not necessarily associated with the difference in electrostatic potential. The conductivity can be associated with a *diffusion* process as well.

Let us summarize the measured quantities that can be obtained in a system that is close to equilibrium. It is therefore assumed that carriers are thermalized to the steady-state Fermi level. The measurements are made by small perturbation over the steady state and provide the following information:

1 The chemical capacitance is measured as a relation of charge ΔQ to voltage ΔV when the voltage displaces the Fermi level (see Chapter 7). The carrier density n can be found readily by integrating the chemical capacitance with respect to voltage.
2 The chemical diffusion coefficient of electrons, D_n, is directly measured by transient methods (in either the time or frequency domain) such as impedance spectroscopy (IS). The determination of D_n consists of inducing a disequilibrium by a voltage step ΔV and measuring the time constant for equilibration, which relates to the transit time for diffusion across the sample. These methods are very common in the electrochemistry of ionic conductors. Routine electrochemical methods, based on a step of the voltage (in either the time or frequency domain) measure the chemical diffusion coefficient. Thermodynamic factors play an important role in Li diffusion in intercalation materials (Weppner and Huggins, 1977) and a correlation of chemical capacitance and the Li chemical diffusion coefficient is often observed in such materials (Xia et al., 2006).
3 In organic semiconductors, the mobility is often measured in transistor configuration (see Chapter 14), or in the space-charge-limited conduction regime in planar diode configuration (see Chapter 15). Mobility can also be measured by the time-of-flight method (Section 10.4), but in dispersive transport (as in multiple trapping), time-transient techniques do not establish quasi-equilibrium and the resulting mobility may be undefined (see Section 13.7). In electrochemical systems, the mobility is found from the conductivity and the carrier density (Patil et al., 2004).
4 In random walk numerical simulations, the diffusion coefficient obtained is the jump diffusion coefficient. As mentioned previously, the thermodynamic factor T_n and jump diffusion coefficient D_J are evaluated directly and then combined to form D_n. One can also evaluate the transport parameters by simulating the relaxation from a nonequilibrium state (De et al., 2002).
5 The conductivity is measured as a relation of electrical current ΔI to voltage ΔV. This can be obtained directly from the ohmic region of current-voltage curve, as the low-frequency resistance in IS, or by electrochemical gating (Paul et al., 1985, Vanmaekelbergh et al., 2007); see Section 14.6.
6 The recombination lifetime is derived by small perturbation methods, as discussed in Section 18.5.
7 In the case of mixed ionic electronic conductors, the polarization at interfaces causes important effects commented in Section 15.7 and special techniques are required (Hebb, 1952; Yokota, 1961; Huggins, 2002).

12.3 THE EINSTEIN RELATION

Consider the band-bending region of a semiconductor in equilibrium, for example, Figure 2.1a. There is both a curvature of the band, implying an electric field and a change of the chemical potential that provides diffusive force, but both quantities are adjusted to establish a flat Fermi level, thus $\partial E_{Fn}/\partial x = 0$ and consequently

$$q\frac{\partial \varphi}{\partial x} = \frac{\partial \mu_n}{\partial x}. \tag{12.19}$$

This is the equilibrium condition in which there is no flux of the carrier. We assume that we start from separate drift and diffusion current terms and write Equation 3.6 as

$$j_n = n\left(-u_n + \frac{qD_n}{n}\frac{\partial n}{\partial \mu_n}\right)\frac{\partial \mu_n}{\partial x}. \tag{12.20}$$

Since $j_n = 0$, the parenthesis in Equation 3.20 is zero. This imposes a relationship between the mobility and the chemical diffusion coefficient that is termed the *generalized Einstein relationship*:

$$\frac{D_n}{u_n} = T_n \frac{k_B T}{q}. \qquad (12.21)$$

For electron transport in nondegenerate band-conduction materials, a single transport level consisting of extended states is well defined. Then the thermodynamic factor takes the value $T_n = 1$ and the mobility u_n and the diffusion coefficient D_n are constant quantities that satisfy the Einstein relation

$$\frac{D_n}{u_n} = \frac{k_B T}{q}. \qquad (12.22)$$

However, Equation 12.22 has important limitations. For band transport in semiconductors, it holds only under the Boltzmann distribution. The generally valid Einstein relationship is stated in Equation 12.21 and can also be written in the form

$$\frac{D_n}{u_n} = \frac{1}{q} \frac{n}{q(dn/d\mu_n)}. \qquad (12.23)$$

A number of examples of the generalized Einstein equation for transport in disordered materials will be discussed in Chapter 13. The conductivity can be written as

$$\sigma_n = q^2 D_n \frac{dn}{d\mu_n}. \qquad (12.24)$$

The conductivity can also be stated in terms of the chemical capacitance (Bisquert, 2008):

$$\sigma_n = C_\mu D_n. \qquad (12.25)$$

The conductivity–diffusivity relationship 12.25 was stated earlier in Equation 7.24 and now can be viewed as a direct expression of the generalized Einstein relation. In accumulation conditions, one cannot rely on simplified approximations of the Fermi–Dirac function. In this case, the Fermi level of carriers enters the respective band edge and the electron (n) and hole (p) densities must be evaluated in degenerate conditions with the Fermi–Dirac 1/2 integral $\Im_{1/2}$ as follows:

$$n = N_c 2/\sqrt{\pi}\Im_{1/2}\left(-(E_c - E_{Fn})/k_B T\right) \qquad (12.26)$$

$$p = N_v 2/\sqrt{\pi}\Im_{1/2}\left((E_v - E_{Fp})/k_B T\right) \qquad (12.27)$$

$$\Im_{1/2}(x_F) \equiv \int_0^\infty \frac{x^{1/2}}{\exp(x - x_F) + 1} dx. \qquad (12.28)$$

The Einstein relationship between diffusion coefficient D_n and mobility u_n (or D_p and u_p) takes the form (Landsberg, 1981)

$$D_n = u_n \frac{k_B T \Im_{1/2}(E_c - E_{Fn})}{q \Im_{-1/2}(E_c - E_{Fn})}. \qquad (12.29)$$

Equation 12.29 can be simplified with the approximation (Kroemer, 1978; Sze, 1981)

$$D_n = u_n \frac{kT}{q}\left[1 + 0.35355\left(\frac{n}{N_c}\right) - 9.9\times10^{-3}\left(\frac{n}{N_c}\right)^2 + 4.45\times10^{-4}\left(\frac{n}{N_c}\right)^3 + \ldots\right]. \qquad (12.30)$$

12.4 DRIFT-DIFFUSION EQUATIONS

Let us summarize the equations that determine the transport of electrons and holes in a semiconductor. The current carried by each of the species is obtained by the drift-diffusion Equations 12.6 and 12.9, so that total current is given by

$$j = j_n + j_p \qquad (12.31)$$

where

$$j_n = -qJ_n = qnu_n\mathbf{E} + qD_n \frac{\partial n}{\partial x} \qquad (12.32)$$

$$j_p = +qJ_p = qpu_p\mathbf{E} - qD_p \frac{\partial p}{\partial x}. \qquad (12.33)$$

The drift-diffusion equations can be written in terms of the respective fluxes as

$$J_n = -nu_n\mathbf{E} - D_n \frac{\partial n}{\partial x}. \qquad (12.34)$$

$$J_p = pu_p\mathbf{E} - D_p \frac{\partial p}{\partial x}. \qquad (12.35)$$

Since the generation G and recombination rate U_{np} of electrons and holes are the same for either species, the continuity equations related to Equation 11.25 can be stated as

$$\frac{\partial n}{\partial t} = +\frac{1}{q}\frac{\partial j_n}{\partial x} + G - U_{np} \tag{12.36}$$

$$\frac{\partial p}{\partial t} = -\frac{1}{q}\frac{\partial j_p}{\partial x} + G - U_{np}. \tag{12.37}$$

The problem of transport of electrons and holes is completed by stating the Poisson equation that determines the connection between spatial distribution of charge and the local electrical field. The Poisson equation will include every fixed (immobile) or free charge in the system. Assuming fixed ionic charge-density N_{ionic}, we have

$$\frac{\partial \mathbf{E}}{\partial x} = +\frac{q}{\varepsilon}\left(p - n + N_{ionic}\right). \qquad 12.38$$

In a neutral semiconductor, the system is balanced in equilibrium, hence

$$p_0 - n_0 + N_{ionic} = 0 \tag{12.39}$$

$$\Delta n \equiv n - n_0 \tag{12.40}$$

$$\Delta p \equiv p - p_0 \tag{12.41}$$

$$\frac{\partial \mathbf{E}}{\partial x} = +\frac{q}{\varepsilon}\left(\Delta p - \Delta n\right). \tag{12.42}$$

The transport rate can also be expressed in terms of the Fermi levels as shown before and the intrinsic level E_i that differs by a constant from the vacuum level (VL) and therefore indicates the local electric field. The expression of the intrinsic level can be written as

$$E_i = -E_g - q\varphi + \frac{1}{2}k_B T \ln\left(\frac{N_v}{N_c}\right). \tag{12.43}$$

The densities of electrons and holes and the Poisson equation are formulated as follows:

$$n = n_i e^{(E_{Fn} - E_i)/k_B T} \tag{12.44}$$

$$p = n_i e^{-(E_{Fp} - E_i)/k_B T}. \tag{12.45}$$

$$\frac{1}{q}\frac{\partial^2 E_i}{\partial x^2} = +\frac{q}{\varepsilon}\left(p - n + N_{ionic}\right). \tag{12.46}$$

Finally, the boundary conditions indicating current and charge densities at the edge of the transport layers must be specified. Detailed analysis of boundary conditions for solar cell modeling is presented in Section 26.4.

12.5 AMBIPOLAR DIFFUSION TRANSPORT

Ambipolar transport is a specific phenomenon in solid-state physics that describes the motion of excess electronic charge carriers in a quasi-neutral region of a semiconductor. A similar model is utilized to describe the transport of the ion components of a salt in electrochemistry, making use of the local electroneutrality assumption in transport problems as first suggested by Planck (1890). This mechanism has been validated in a wide variety of situations including membranes, conducting polymer films (Vorotyntsev et al., 1994), and polymer light-emitting electrochemical cells (Manzanares et al., 1998).

We assume that the injected or photogenerated electrons and holes compensate each other to maintain electroneutrality. Then we have that Equation 12.40 and 12.41 for the excess carriers satisfy

$$\Delta p = \Delta n. \tag{12.47}$$

It follows that the gradients of electrons and holes are the same:

$$\frac{\partial n}{\partial x} = \frac{\partial p}{\partial x}, \tag{12.48}$$

so that diffusion displaces the carrier densities of both species in the same direction. However, to maintain electroneutrality, the injected electrons and holes must move as a single entity. Figure 12.2 illustrates the motion of excess neutral injected charge in a semiconductor.

By summation of Equations 12.34 and 12.35 and using the condition 12.48, we can obtain the following expression for the electrical field

$$\mathbf{E} = \frac{j}{\sigma} - q\frac{D_n - D_p}{\sigma}\frac{\partial n}{\partial x}. \tag{12.49}$$

The field is established to maintain the solidary motion of the two kinds of carriers. The first term in Equation 12.49 is for ohmic transport. The second term is adjusted to establish the drift of the slowest carrier. A small amount of space charge may adjust itself to produce macroscopic currents and quasi-neutrality still prevails, while the residual space charge remains a property of the steady-state carrier density.

Inserting the field expression in the equations of the current density, we have

$$j_n = \frac{\sigma_n}{\sigma} j + qD^* \frac{\partial n}{\partial x} \tag{12.50}$$

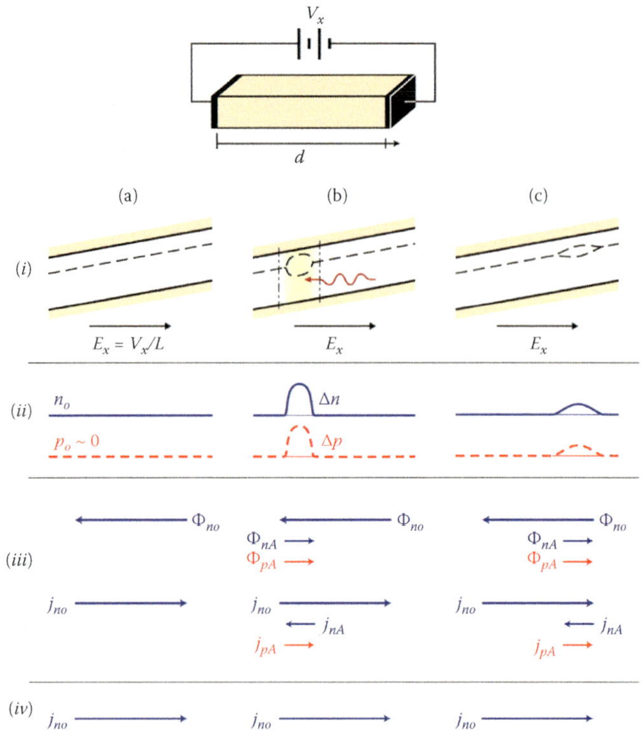

FIGURE 12.2 Illustration of ambipolar transport. Top of figure: a bulk slab of n-type semiconductor with a potential applied across its length. (a) Steady-state conditions prior to excess charge injection. (b) The initial excess charge injection, illustrated by electron-hole-pair ($\Delta n \approx \Delta p$) generation via photon absorption. (c) The subsequent motion of the excess charge by ambipolar transport. (i) Band diagram and applied electric field in each case; (ii) electron and hole carrier concentrations; (iii) majority equilibrium electron and excess electron and hole fluxes (Φ) and currents (j), and (iv) the total (net) current. (Reproduced from Champlain, 2011, with permission.)

$$j_p = \frac{\sigma_p}{\sigma} j - qD^* \frac{\partial p}{\partial x}, \quad (12.51)$$

where the ambipolar diffusion coefficient is

$$D^* = \frac{nu_n D_p + pu_p D_n}{nu_n + pu_p}. \quad (12.52)$$

Using the generalized Einstein relation, we can write

$$D^* = \frac{\dfrac{n}{T_n} + \dfrac{p}{T_p}}{\dfrac{n}{T_n D_p} + \dfrac{p}{T_p D_n}}. \quad (12.53)$$

In particular, for Boltzmann statistics in the case $n = p$, the diffusion coefficient is

$$D^* = \frac{2D_n D_p}{D_n + D_p}. \quad (12.54)$$

12.6 RELAXATION OF INJECTED CHARGE

A central point to establish the dominant aspects of the physics of a particular device is the clarification of the presence or absence of extended quasi-neutral regions. In the absence of charge compensation, the device is forced to enter space-charge limited transport, which is highly energy demanding. This model will be treated in detail in Chapter 15. In the presence of electrolytes or charge-compensating media in nanostructured or semiconducting materials, the formation of space charge is prevented by local electroneutrality and transport flows mainly by diffusion. In general, if quasi-electroneutrality is invoked, space-charge regions related to excess carriers cannot be built up even in the case of current flow and the treatment of the transport characteristics in the device is considerably simplified. In Section 5.7, we observed that the mobile charge of both signs can be neutralized but only to the scale of the Debye length. We also introduced a characteristic time, the dielectric relaxation time, related to the material permittivity ε and its conductivity σ_n as

$$\tau_{die} = \frac{\varepsilon}{\sigma_n}. \quad (12.55)$$

Now we analyze the dynamics of charge screening to show the meaning of Equation 12.55. We consider that positive and negative space-charge regions of size a and density ρ_1 have been formed corresponding to an excess of mobile electron carrier density that leaves an equivalent deficiency (positive charge) in a medium of background density n_0; see Figure 12.3. An electric field is created between the positive and negative regions, and the Poisson equation states that

$$\frac{\partial \mathbf{E}}{\partial x} = \frac{\rho_1}{\varepsilon}. \quad (12.56)$$

The charge in each separate region is $\rho_1 a$ per unit area, and therefore the field between both regions is

$$\mathbf{E} = \frac{\rho_1 a}{\varepsilon}. \quad (12.57)$$

The field creates a current density

$$j = \sigma_n \mathbf{E} = \frac{\sigma_n \rho_1 a}{\varepsilon} \quad (12.58)$$

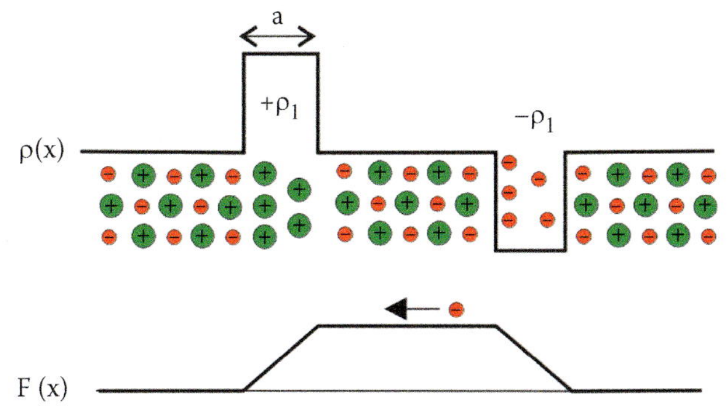

FIGURE 12.3 Two regions of positive and negative space charge in a neutral background, creating an electric field that tends to neutralize the separation of charge.

that rapidly displaces the electrons toward the positive charge region and the space charge collapses. To evaluate the time scale for the neutralization of the space charge, we state the continuity Equation 12.36 as

$$\frac{\partial \rho_1}{\partial t} = -\frac{\partial j}{\partial x}. \quad (12.59)$$

Combining with Equation 12.58, we obtain

$$\frac{\partial \rho_1}{\partial t} = -\frac{\sigma_n}{\varepsilon}\rho_1. \quad (12.60)$$

Equation 12.60 states that the time constant in which an excess mobile charge is swept away is the dielectric relaxation time. If we use the Einstein relation in Equation 12.55, we can write

$$\tau_{die} = \frac{\varepsilon k_B T}{q^2 n_0 D_n}. \quad (12.61)$$

Comparing Equation 5.42, we find

$$\lambda_D = \left(D_n \tau_{die}\right)^{1/2}. \quad (12.62)$$

We realize that the Debye length is the characteristic length for diffusion on the time scale of the dielectric relaxation time.

When considering the net injection of the charge in a semiconductor layer, in addition to the dielectric relaxation of local charge imbalance, the lifetime of carriers (recombination time, τ_n, discussed in Section 18.5) also plays a central role in the charge distribution (van Roosbroek and Casey, 1972). The diffusion length is defined as the average distance in which mobile carriers can diffuse before recombination, as described by Equation 11.30:

$$L_n = \left(D_n \tau_n\right)^{1/2}. \quad (12.63)$$

The occurrence of quasi-neutral regions relies on the property that the carrier lifetime is much larger than the dielectric relaxation time, $\tau_n > \tau_{die}$, so that mobile carriers can exist long enough to neutralize the charge. Depending on whether $\tau_n > \tau_{die}$ (*lifetime* semiconductor regime) or $\tau_n < \tau_{die}$ (*relaxation* semiconductor regime), the physics governing the device operation changes drastically. In terms of a length scale comparison, the lifetime regime is determined by $L_n > \lambda_D$, whereas the opposite holds true in the relaxation regime with typical values of τ_{die} of the order 10^{-14} s.

Charge screening is justified in all conditions in the quasi-neutral regions in the lifetime regime. An injection of a small quantity of minority carriers in the quasi-neutral region is rapidly neutralized by the same amount of majorities and then progressively disappears by recombination. In the relaxation semiconductor regime, electroneutrality cannot be achieved, and hence regions of near-zero net local recombination may occur, implying spatially separated excess electron and hole concentrations. This will enhance the amount of space charge in dielectric relaxation-dependent decay. For instance, low-conductivity amorphous *p-i-n* silicon solar cells were modeled as developing photogenerated hole space-charge regions near the *p* contact, which concentrate the voltage drop (Schiff, 2003). In the extreme case, currents should be space-charge limited as occurring in organic light-emitting diodes based on low-mobility polymers or molecules (Bozano et al., 1999).

12.7 TRANSIENT CURRENT IN INSULATOR LAYERS

So far, most of the examples we have discussed rely on the conduction current. We analyze a particular device in which the application of a voltage produces both conduction and displacement current. The model is formed by a conducting and an insulating layer in series, as shown in Figure 12.4 (Hu et al., 2010). Let d_a, ε_a, $\mathbf{E}_a$, σ_a denote the thickness, relative dielectric constant, electric field, and electron conductivity in the conducting region. Similarly, we have d_b, ε_b, $\mathbf{E}_b$ in the insulating region where $\sigma_b = 0$. The total voltage is

$$V = d_a \mathbf{E}_a + d_b \mathbf{E}_b. \quad (12.64)$$

The electrical current density is given by Equation 10.1 and since the current is continuous, it is the same in both regions. Thus,

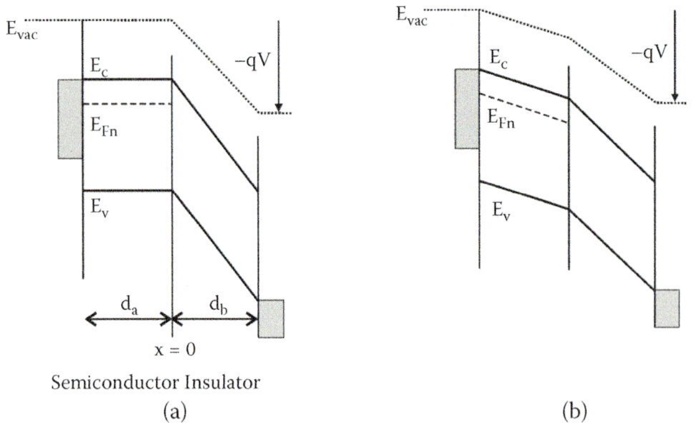

FIGURE 12.4 Model device consisting of an electron conducting region a in contact with an insulating layer b. (a) In the equilibrium situation under a voltage, the current is zero in b and the potential drops across the insulating layer. (b) After application of a voltage, V, drift current flows in region a and displacement current in region b.

$$j = \sigma_a \mathbf{E}_a + \varepsilon_0 \varepsilon_a \frac{d\mathbf{E}_a}{dt} = \varepsilon_0 \varepsilon_b \frac{d\mathbf{E}_b}{dt}. \qquad (12.65)$$

In equilibrium at applied voltage V, $j = 0$, hence the entire field drops in the insulating region as shown in Figure 12.4a. When the voltage V is first applied, it is distributed as determined by the equation that establishes the continuity of the dielectric displacement at the boundary,

$$\varepsilon_a \mathbf{E}_a (t = 0) = \varepsilon_b \mathbf{E}_b (t = 0). \qquad (12.66)$$

Each region takes a voltage according to the respective dielectric capacitance, so that

$$\mathbf{E}_a (t = 0) = \frac{V_a}{d_a} = \frac{\varepsilon_b}{d_a \varepsilon_b + d_b \varepsilon_a} V, \qquad (12.67)$$

and similarly for $\mathbf{E}_b(t = 0)$; see Figure 3.4b. Then a current drifts in region a, which charges the interface with the insulating medium. Note in Equation 3.65 that in region b, the current is a displacement current. The accumulation charge makes the polarization increase in the region b until all the applied potential drops in this region, as was indicated in Figure 12.4a. To determine the time dependence of the electric fields and the current, we combine these equations

$$\frac{d\mathbf{E}_a}{dt} = -\frac{1}{\tau_a} \mathbf{E}_a, \qquad (12.68)$$

where

$$\tau_a = \tau_{die,a} \left(1 + \frac{\varepsilon_b d_a}{\varepsilon_a d_b} \right). \qquad (12.69)$$

Here, $\tau_{die,a}$ is the dielectric relaxation time in region a. Hence, the field responds as

$$\frac{d\mathbf{E}_a}{dt} = \mathbf{E}_a (0) e^{-t/\tau}, \qquad (12.70)$$

where the initial value is given in Equation 12.67. We can obtain the time dependence of the electric fields

$$\mathbf{E}_a = \frac{\varepsilon_b}{d_a \varepsilon_b + d_b \varepsilon_a} V e^{-t/\tau} \qquad (12.71)$$

$$\mathbf{E}_b = \left(1 - \frac{d_a \varepsilon_b}{d_a \varepsilon_b + d_b \varepsilon_a} e^{-t/\tau} \right) \frac{V}{d_b} \qquad (12.72)$$

and the current is

$$j = \frac{\sigma_a d_a \varepsilon_b^2}{\left(d_a \varepsilon_b + d_b \varepsilon_a \right)^2} V e^{-t/\tau}. \qquad (12.73)$$

12.8 MODELING TRANSPORT PROBLEMS

Let us consider the general demands for the modeling of complex devices. It is required to consider four main aspects:

1 The number of types of transport species: charged and neutral, ionic and electronic.
2 The spatial distribution of the charge carriers that relates to important properties such as shielding, space charge, macroscopic electrical fields, recombination, and the main transport mechanisms.
3 The structure of interfaces, especially at the contacts, including a description of interfacial capacitances and charge transfer mechanisms. Particularly important is the description of carrier injection at contacts that determine how the device exchanges current with the external world. In multiple phase devices, the conditions of charge transfer at heterojunctions constitute a key feature of the model.
4 The energy axis for each carrier, governed by disorder at each material (see Figure 13.2), and by energy level alignment at interfaces (Jaegermann, 1996).

The device is therefore formed by some geometry and morphology that sets the first constraint in establishing regions where the carriers can be distributed, either in motion or stationary. Such regions have

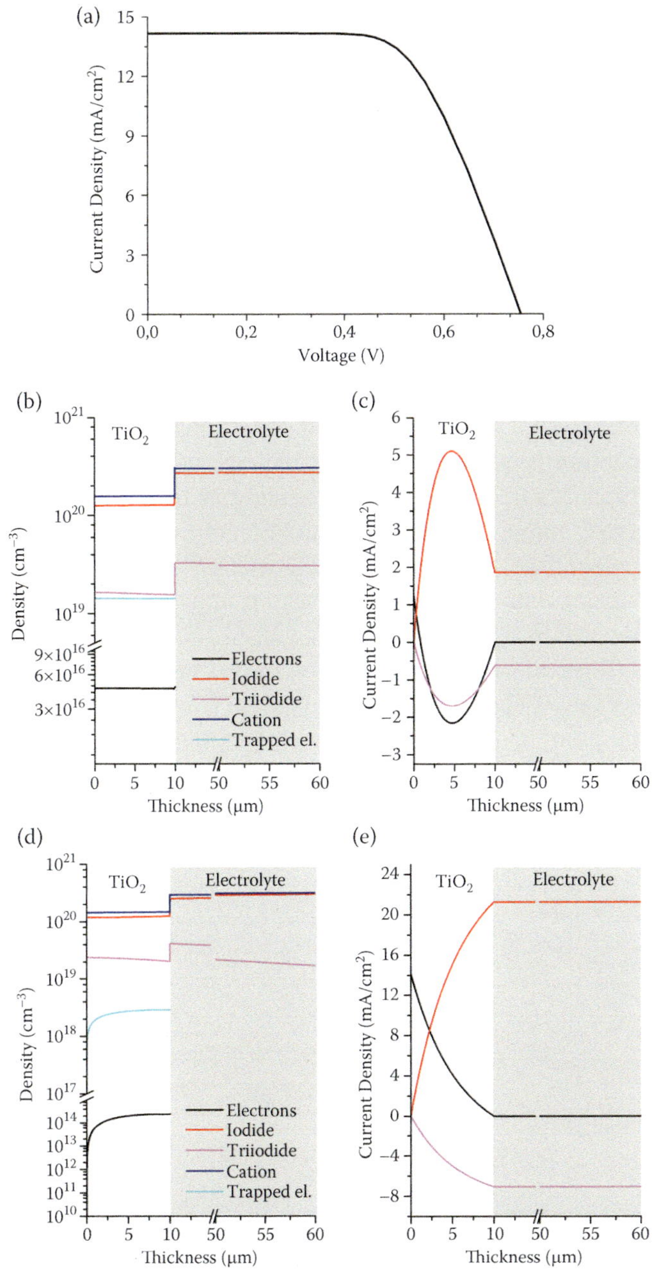

FIGURE 12.5 (a) Simulation of current-voltage characteristics for a typical dye-sensitized solar cells (DSC; 10 mm of mesoporous TiO_2/electrolyte, 50 mm of pure electrolyte) with N719 dye. (b) The density distribution and current density (c) within the cell at 737 mV (close to open-circuit condition). (b) Shows all of the charged species in the system: electrons (free and trapped), iodide and triiodide, positive counter-ion. For the current density in (c), only the charged species that contribute to the total current are shown: free electrons, iodide, and triiodide ions. (d) The density distribution and current density (e) within the cell at 100 mV (close to short-circuit condition). (Simulation performed using TiberCAD software, courtesy of Alessio Gagliardi and Aldo di Carlo.)

boundaries that are described by suitable boundary conditions. Having set morphologies and boundaries, there are two different dimensionalities—the *spatial space*, which can be described with one or more dimensions, and the *energy space*. At each point, an energy diagram gives the allowed energy levels for different types of carriers. Energy levels for electronic carriers may be stationary states, also called localized states or traps, or extended states that allow fast transport. Based on these general properties, one formulates a series of macroscopic equations and boundary conditions that provide, as a result, carrier densities and carrier dynamics, expressed as output current densities, either for steady state or any desired transient condition.

The modeling of a solar cell formed by the stack of a specific combination of absorber and contact materials is described in detail in Chapter 26. It requires a prior knowledge of parameters as carrier mobility and energy levels, and consists on the application of: Drift-diffusion equations that combine the continuity Equations 12.36 and 12.37, including the generation terms in light-absorbing layers and the recombination rates, and the current density Equations 12.32 and 12.33; the Poisson Equation 12.38; and the boundary conditions that establish the rate of transference (or blocking, or recombination) of the carriers at the interface between two materials. This procedure can be applied to a broad variety of materials and devices is generally termed "drift-diffusion modeling"; see Figure 26.18, 26.19, and 26.20 for examples of solar cell simulation.

As an illustration, two examples of simulation of the drift-diffusion-recombination model for different types of devices are shown in Figure 12.5, Figure 12.6, and Figure 12.7. Figure 12.5 is the simulation of carrier density and currents in a dye-sensitized solar cell like that shown in Figure 1.5. The panels (b), (c) correspond to an open-circuit condition in the current density-voltage curve (a), while panels (d), (e) show the short-circuit condition. The calculation is performed by taking into account the motion of electronic species in the semiconductor layer and ionic species across the whole device (Gagliardi et al., 2010), and the boundary condition whereby only electrons are extracted at the left outer contact and only ions discharge at the right contact. Note that the different charged species add up for local electroneutrality in the left column and the constant global electrical current in the right panels.

Figure 12.6 represents a model for organic light-emitting diodes (OLEDs), where the ITO and LiF/Al

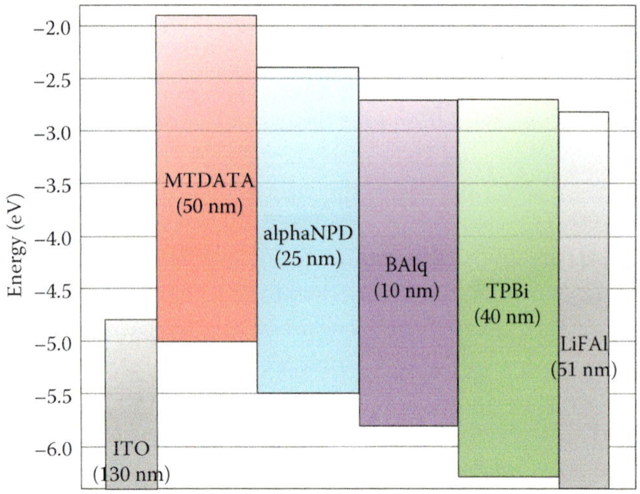

FIGURE 12.6 Energy levels of a red-emitting OLED.

layers are considered as electrodes. Thermal equilibrium is assumed at the interface between organic layers and electrodes. The results of the simulation of the layers (MTDATA, α-NPD, BAlq and TPBi) under applied bias by the drift-diffusion equations where the charge transport is described by a field-dependent mobility are shown in Figure 12.7, indicating carrier distribution, current distribution for electron and hole current, and local recombination rate, which depends on the product of the carrier concentrations. The light-emitting layer is α-NPD, which is doped with $Ir(MDQ)_2(acac)$. High barriers occur for both charges at the MTDATA–α-NPD and α-NPD–BAlq interfaces. Holes are injected at the anode (position 0 nm) and electrons are injected at the cathode (position 125 nm). The hole current is decreased by the MTDATA–α-NPD, α-NPD–BAlq, and BAlq–TPBi interfaces, whereas the electron current is primarily blocked by the MTDATA–α-NPD interface. Note the continuity of total current, while electron and hole densities are not compensated, leading to the formation of space charge. A large accumulation of hole and electron pairs at the MTDATA–α-NPD and α-NPD–BAlq interfaces occurs. Diffusion allows excitons to diffuse into the α-NPD.

Instead of the phenomenological transport and conservation equations, one may adopt a more fundamental point of view in which electronic states are modeled

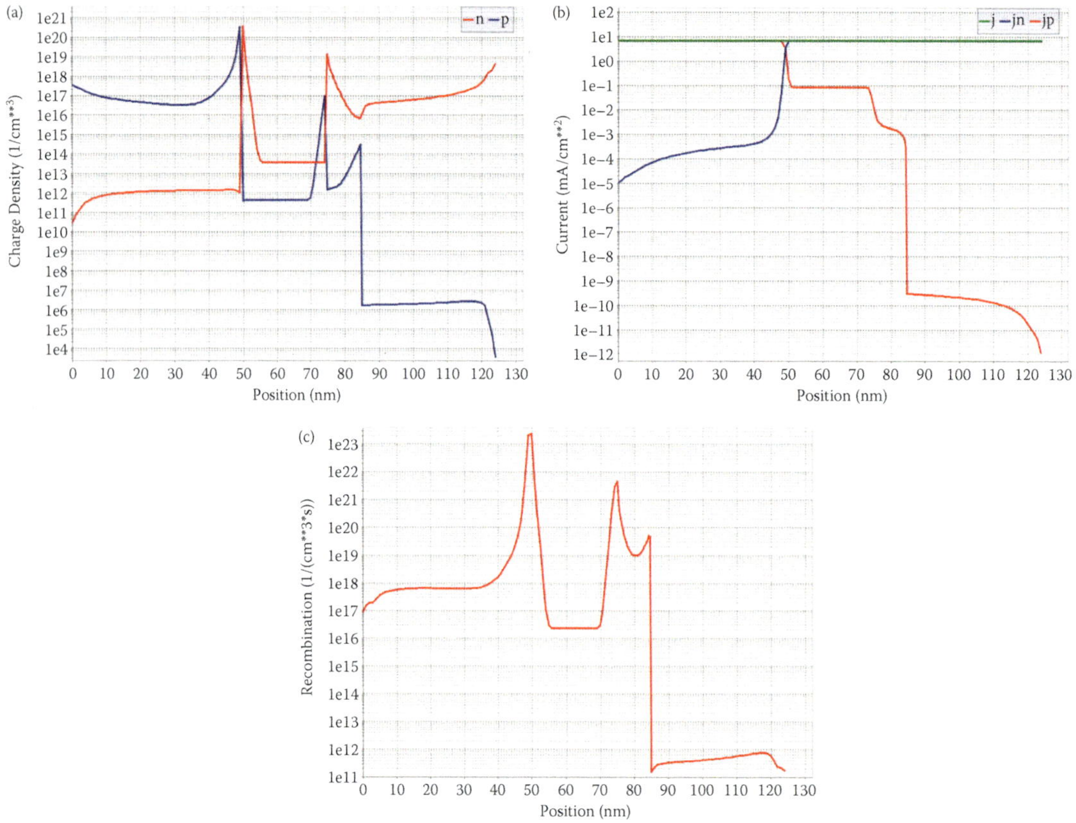

FIGURE 12.7 (a) Calculated charge-density distribution of electrons (n) and holes (p). The electrically simulated domain is MTDATA (50 nm)/α-NPD (25 nm)/BAlq (10 nm)/TPBi (40 nm). The blocking of holes and electrons occurs mainly at the MTDATA–α-NPD and α-NPD–BAlq interfaces. The calculation of charges across an organic–organic interface is based on a quasi-Fermi level continuity. (b) Current distribution at applied voltage 7 V. (c) Recombination profile. The electrons and holes recombine according to Langevin recombination. (Courtesy of Benjamin Perucco, Stéphane Altazin, and Beat Ruhstaller. All calculations were made with SETFOS 3.4 by Fluxim AG (http://www.fluxim.com).)

individually and the transfer rates between states allow us to analyze the global dynamics by Monte Carlo simulations (Ansari-Rad et al., 2013); see Section 13.6. These methods permit us to establish arbitrary morphologies and to investigate complex effects such as energy disorder, as further explained in Section 13.5. A still more fundamental approach consists of a simulation of the molecular details of the components of the system. These methodologies can produce very valuable knowledge about the structure of interfaces and the origin of the observed energy distribution features and also the nature of excited or intermediate states for charge transfer phenomena (Pastore and de Angelis, 2013). Microscopic simulation methods have been widely explored in the field of organic materials and they have important applications for the investigation of carrier and energy transport in organic solids (Beljonne and Cornil, 2014). In summary, there are four main approaches: the mean field (solving microscopic rate equations), kinetic Monte Carlo simulation, molecular dynamics, and the density functional theory (DFT).

GENERAL REFERENCES

Fundamental transport equation: Guggenheim, 1929; de Groot and Mazur, 1962; Weppner and Huggins, 1977; Marshak and van Vliet, 1978; Marshak and Van Vliet, 1984; Marshak, 1989.

Ambipolar diffusion: van Roosbroeck, 1953; Nussbaum, 1981; Champlain, 2011.

Measurement of chemical diffusion coefficient: Weppner and Huggins, 1977; Levi et al., 2005; Niklasson and Granqvist, 2007.

Lifetime and relaxation regimes: van Roosbroeck and Casey, 1972; Santana and Jones, 1998.

Simulation of percolation and nonthermalized electron transport: van de Lagemaat and Frank, 2001; Benkstein et al., 2003; Anta, 2009; Ansari-Rad et al., 2012; Mendels and Tessler, 2013.

Microscopic simulation of transport: Ruhle et al., 2011; Baumeier et al., 2012; Coehoorn and Bobbert, 2012.

REFERENCES

Ansari-Rad, M.; Abdi, Y.; Arzi, E. Simulation of non-linear recombination of charge carriers in sensitized nanocrystalline solar cells. *Journal of Applied Physics* 2012, *112*, 074319

Ansari-Rad, M.; Anta, J. A.; Bisquert, J. Interpretation of diffusion and recombination in nanostructured and energy disordered materials by stochastic quasiequilibrium simulation. *The Journal of Physical Chemistry C* 2013, *117*, 16275–16289.

Anta, J. A. Random walk numerical simulation for solar cell applications. *Energy and Environmental Science* 2009, *2*, 387–392.

Baumeier, B.; May, F.; Lennartz, C.; Andrienko, D. Challenges for in silico design of organic semiconductors. *Journal of Materials Chemistry* 2012, *22*, 10971–10976.

Beljonne, D.; Cornil, J., Eds., *Multiscale Modelling of Organic and Hybrid Photovoltaics*; Springer: Berlin, 2014.

Benkstein, K. D.; Kopidakis, N.; Van de Lagemaat, J.; Frank, A. J. Influence of the percolation network geometry on the electron transport in dye-sensitized titanium dioxide solar cells. *The Journal of Physical Chemistry B* 2003, *107*, 7759–7767.

Bisquert, J. Theory of the impedance of electron diffusion and recombination in a thin layer. *The Journal of Physical Chemistry B* 2002, *106*, 325–333.

Bisquert, J. Interpretation of electron diffusion coefficient in organic and inorganic semiconductors with broad distributions of states. *Physical Chemistry Chemical Physics* 2008, *10*, 3175–3194.

Bisquert, J.; Bertoluzzi, L.; Mora-Sero, I.; Garcia-Belmonte, G. Theory of impedance and capacitance spectroscopy of solar cells with dielectric relaxation, drift-diffusion transport, and recombination. *The Journal of Physical Chemistry C* 2014, *118*, 18983–18991.

Bisquert, J.; Fabregat-Santiago, F. Impedance spectroscopy: A general introduction and application to dye-sensitized solar cells. In *Dye-Sensitized Solar Cells*; Kalyanasundaram, K., Ed.; CRC Press: Boca Raton, 2010.

Bisquert, J.; Grätzel, M.; Wang, Q.; Fabregat-Santiago, F. Three-channel transmission line impedance model for mesoscopic oxide electrodes functionalized with a conductive coating. *The Journal of Physical Chemistry B* 2006, *110*, 11284–11290.

Bozano, L.; Carter, S. A.; Scott, J. C.; Malliaras, G. G.; Brock, P. J. Temperature and field dependent electron and hole mobilities in polymer light-emitting diodes. *Applied Physics Letters* 1999, *74*, 1132.

Champlain, J. G. On the use of the term "ambipolar." *Applied Physics Letters* 2011, *99*, 123502.

Coehoorn, R.; Bobbert, P. A. Effects of Gaussian disorder on charge carrier transport and recombination in organic semiconductors. *Physica Status Solidi (a)* 2012, 209, 2354–2377.

de Groot, S. R.; Mazur, P. *Non-Equilibrium Thermodynamics*; North Holland: Amsterdam, 1962.

De, S.; Teitel, S.; Shapir, Y.; Chimowitz, E. H. Monte Carlo simulation of Fickian diffusion in the critical region. *The Journal of Chemical Physics* 2002, *116*, 3012–3017.

Gagliardi, A.; Mastroianni, S.; Gentilini, D.; Giordano, F.; Reale, A.; Brown, T. M.; Di Carlo, A. Multiscale modeling of dye solar cells and comparison with experimental data. *IEEE Journal of Selected Topics in Quantum Electronics* 2010, *16*, 1611–1618.

Guggenheim, E. A. On the conception of electrical potential difference between two phases. II. *The Journal of Physical Chemistry* 1929, *34*, 1540–1543.

Hebb, M. H. Electrical conductivity of silver sulfide. *The Journal of Chemical Physics* 1952, *20*, 185.

Hu, L.; Noda, Y.; Ito, H.; Kishida, H.; Nakamura, A.; Awaga, K. Optoelectronic conversion by polarization current, triggered by space charges at organic-based interfaces. *Applied Physics Letters* 2010, *96*, 243303.

Huggins, R. A. Simple method to determine electronic and ionic components of the conductivity in mixed conductors a review. *Ionics* 2002, *8*, 300–313.

Jaegermann, W. The semiconductor/electrolyte interface: A surface science approach. *Modern Aspects of Electrochemistry*, 1996, *30*.

Kroemer, H. The Einstein relation for degenerate carrier concentrations. *IEEE Transactions on Electron Devices* 1978, *25*, 850–850.

Landsberg, P. T. Einstein and statistical thermodynamics. III. The diffusion-mobility relation in semiconductors. *European Journal of Physics* 1981, *2*, 213.

Levi, M. D.; Markevich, E.; Aurbach, D. Comparison between Cottrell diffusion and moving boundary models for determination of the chemical diffusion coefficients in ioninsertion electrodes. *Electrochimica Acta* 2005, *51*, 98–110.

Manzanares, J. A.; Reiss, H.; Heeger, A. J. Polymer light-emitting electrochemical cells: A theoretical study of junction formation under steady-state conditions. *The Journal of Physical Chemistry B* 1998, *102*, 4327–4336.

Marshak, A. H. Modeling semiconductor devices with position-dependent material parameters. *IEEE Transactions on Electron Devices* 1989, *36*, 1764–1772.

Marshak, A. H.; van Vliet, C. M. Electrical current and carrier density in degenerate materials with nonuniform band structure. *Proceedings of the IEEE* 1984, *72*, 148–164.

Marshak, A. H.; van Vliet, K. M. Electrical current in solids with position-dependent band structure. *Solid-State Electronics* 1978, *21*, 417–427.

Mendels, D.; Tessler, N. Drift and diffusion in disordered organic semiconductors: The role of charge density and charge energy transport. *The Journal of Physical Chemistry C* 2013, *117*, 3287–3293.

Niklasson, C. A.; Granqvist, C.-G. Electrochromics for smart windows: Thin films of tungsten oxide and nickel oxide, and devices based on these. *Journal of Materials Chemistry* 2007, *17*, 127–156.

Nussbaum, A. Semiconductors and semimetals. In *The Theory of Semiconducting Junctions*; Willardson, R. K., Beer, A. C., Eds.; Academic Press: New York, 1981; Vol. 15; pp. 39–194.

Pastore, M.; de Angelis, F. Intermolecular interactions in dye-sensitized solar cells: A computational modeling perspective. *Journal of Physical Chemistry Letters* 2013, *4*, 956–974.

Patil, R.; Harima, Y.; Jiang, X. Mobilities of charge carriers in pol(o-methylaniline) and poly(o-methoxyaniline). *Electrochimica Acta* 2004, *49*, 4687.

Paul, E. W.; Ricco, A. J.; Wrighton, M. S. Resistance of polyaniline films as a function of electrochemical potential and the fabrication of polyaniline based microelectronic devices. *The Journal of Physical Chemistry* 1985, *89*, 1441.

Planck, M. *Ann. Phys. Chem. N.F.* 1890, *39*, 161.

Ruhle, V.; Lukyanov, A.; May, F.; Schrader, M.; Vehoff, T.; Kirkpatrick, J.; Baumeier, B. et al. Microscopic simulations of charge transport in disordered organic semiconductors. *Journal of Chemical Theory and Computation* 2011, *7*, 3335–3345.

Santana, J.; Jones, B. K. Semi-insulating GaAs as a relaxation semiconductor. *Journal of Applied Physics* 1998, *83*, 7699–7705.

Schiff, E. A. Low-mobility solar cells: A device physics primer with application to amorphous silicon. *Solar Energy Materials and Solar Cells* 2003, *78*, 567–595.

Sze, S. M. *Physics of Semiconductor Devizes*, 2nd ed.; John Wiley and Sons: New York, 1981.

van de Lagemaat, J.; Frank, A. J. Nonthermalized electron transport in dye-sensitized nanocrystalline TiO_2 films: Transient photocurrent and random-walk modeling studies. *The Journal of Physical Chemistry B* 2001, *105*, 11194–11205.

van Roosbroeck, W. The transport of added current carriers in a homogeneous semiconductor. *Physical Review* 1953, *91*, 282–289.

van Roosbroeck, W.; Casey, H. C., Jr. Transport in relaxation semiconductors. *Physical Review B* 1972, *5*, 2154–2175.

Vanmaekelbergh, D.; de Jongh, P. E. Driving force for electron transport in porous nanostructured photoelectrodes. *The Journal of Physical Chemistry B* 1999, *103*, 747–750.

Vanmaekelbergh, D.; Houtepen, A. J.; Kelly, J. J. Electrochemical gating: A method to tune and monitor the (opto)electronic properties of functional materials. *Electrochimica Acta* 2007, *53*, 1140–1149.

Vorotyntsev, M. A.; Daikhin, L. I.; Levi, M. D. Modelling the impedance properties of electrodes coated with electroactive polymer films. *Journal of Electroanalytical Chemistry* 1994, *364*, 37.

Weppner, W.; Huggins, R. A. Determination of the kinetics parameters of mixed-conducting electrodes and application to the system Li_3Sb. *Journal of the Electrochemical Society* 1977, *124*, 1569.

Xia, H.; Lu, L.; Ceder, G. Li diffusion in $LiCoO_2$ thin films prepared by pulsed laser deposition. *Journal of Power Sources* 2006, *159*, 1422–1427.

Yokota, I. On the theory of mixed conduction with special reference to conduction in silver sulfide group semiconductors. *Journal of the Physical Society of Japan* 1961, *16*, 2213–2223.

13 Transport in Disordered Media

Localized electronic states frequently dominate the transport in disordered semiconductors. The hopping transitions between localized states and the exchange of electrons between localized and extended states form a variety of complex phenomena that determine the carrier transport mechanisms. In this chapter, we discuss a variety of transport phenomena either by multiple trapping or hopping transport. The multiple trapping model is amply used for the description of photoinduced transport phenomena in disordered solids. This model is usually valid at temperatures where the activated transitions are dominant. The occupation of the bandgap states determined by the position of the Fermi level is then a main factor governing the different kinetic coefficients. We will also analyze a number of important aspects that pertain to the problem of hopping conduction in a spatially disordered medium, including the models of the transport energy and variable range hopping.

13.1 MULTIPLE TRAPPING AND HOPPING TRANSPORT

In Chapters 7 and 8, we described many of the properties caused by a broad distribution of localized states in the energy axis such as the exponential or the Gaussian distribution. In a semiconductor that contains a broad distribution of localized states in addition to extended states such as the conduction band, the transport of carriers occurs by displacement in the band of extended states, affected by trapping events, or directly by transitions between localized states. In Section 7.2, we introduced the idea of a mobility edge. This corresponds to a region in the bandgap that separates extended states from traps or denominates a region where localized states allow fast transitions that produce long-range transport, as shown in Figure 13.1a. Below the mobility edge, additional traps exist that maintain the localized nature of the carriers in the long time regime.

Two main formalisms have been developed to describe transport in electron localizing systems with a broad distribution of states: the multiple trapping transport and the hopping transport. The multiple trapping transport of electrons introduces a division between transport states above the mobility edge and localized states in the bandgap (Rose, 1963; Tiedje and Rose, 1981). The localized states do not communicate electronically with each other; only the extended states in the conduction band allow for long-range transport. The traps may capture a carrier and retain it until its release by thermal activation as discussed in Section 6.3. The process of trap-limited transport is indicated in Figure 13.1b for a single-trap level, and in Figure 13.1c and in Figure 7.7 for electron transport in extended states (E_c) affected by trapping in a wide distribution of states in the energy axis. The second approach is the hopping transport. The term *hopping* refers to a sudden displacement of a charge carrier from one position to another (Böttger and Bryksin, 1985). Carriers remain localized in specific states and realize random hops to nearby vacant electronic states. These hopping steps give rise to long-range motion by diffusion or field-oriented (drift) displacement.

Hopping conductivity theory has been widely applied to doped inorganic semiconductors and to insulating amorphous organic materials. The main factors governing the transition probability between two localized states are the distance and their energetic difference. Two models are usually adopted for the description of hopping rates (Tessler et al., 2009). The first is thermally assisted quantum mechanical tunneling according to the Miller–Abrahams (MA) upward and downward jump rates, as was shown in Equation 6.2. The second type of transition rate, usually applied in "soft" condensed media, is also a thermally assisted quantum mechanical tunneling but, in addition, it contains polaronic effects as given in Equation 6.63 that are determined by two main aspects: the electronic coupling (transfer integral) between adjacent molecules and the reorganization energy λ, corresponding to the Marcus model for electron transfer rate. When hopping conduction is assisted by barrier lowering by an electrical field, it is termed the *Poole-Frenkel mechanism.*

The long-range transport in a model two-level hopping system is shown in Figure 13.1d. However, in a realistic situation, disorder is spatially and energetically random and it is described by a particular density of states (DOS) as discussed in Chapter 8. Then a large variety of hopping transitions become possible as suggested in Figure 13.1e. The semiconductors used in applications possess specific morphological features such as molecular dispersion and aggregation, phase segregation, and meso-porosity. In the case of organic bulk heterojunctions, transport of each carrier occurs

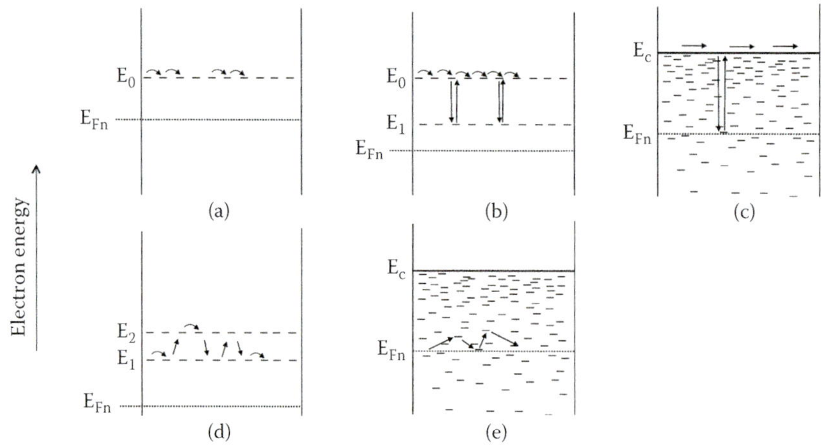

FIGURE 13.1 Schematic representation of several transport models in a semiconductor with a distribution of electronic states. (a) Hopping transport through a unique level. (b) Transport in a single level affected by trapping in a deeper level. (c) Conduction band transport affected by trapping in a wide distribution of states in the bandgap. (d) Hopping transport in a two-level system. (e) Hopping transport in a wide distribution of states in the bandgap. E_{Fn} is the Fermi level, E_c is a transport level by hopping, E_c represents the level of extended electronic states such as the conduction band energy and E_i are the energies of discrete levels.

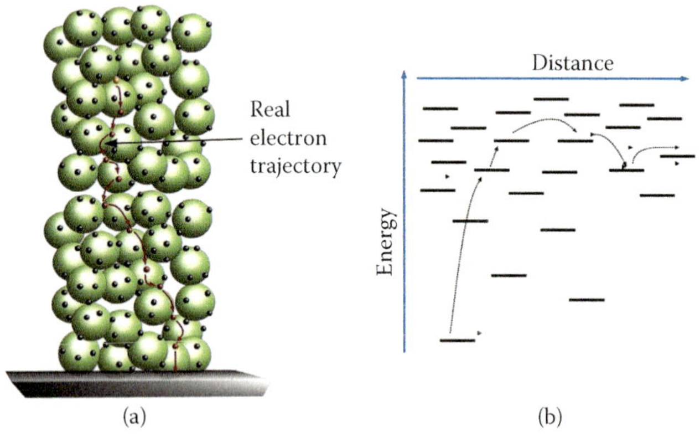

FIGURE 13.2 (a) Transport of electrons in a nanostructured semiconductor up to a current collector situated at the bottom. (b) The hopping transitions in an exponential DOS in energy space. (Reproduced with permission from Anta, 2012.)

in a meso-structured, phase-separated donor–acceptor morphology. Representation of the conduction network in heterogeneous media are suggested in Figure 13.2 for nanostructured inorganic semiconductor electrodes and in Figure 8.23 for organic conductors.

The combination of spatial disorder and energy disorder, as shown in Figure 13.3, makes a general analysis of hopping conductivity quite laborious and prevents a generally valid solution. To obtain the conduction properties of a given type of disordered system, it is required first to establish assumptions on the spatial and energy distribution of localized sites (the DOS), the hopping rates, and then to develop a procedure for solving the conduction properties in the desired conditions of the carrier concentration, applied electrical field, and boundary conditions. If hopping transitions are governed by MA hopping rates or similar, the transition rate from a trap depends on the occupancy, distance, and energy difference of the surrounding localized sites. The exponential dependence on both distance and energy difference implies that small differences become amplified in the resulting transition rates, which display an enormous variability that cannot be easily smoothed. As a result, transport in diluted trap systems will be more favorable in a chain of sites connected by large hopping rates, surrounded by many other pairs with a very low hopping rate, which become irrelevant for long-range conduction. Inside one such pathway, the contribution of the least favorable hops is dominant, as indicated by Equation 11.16. The problem can be viewed as a connection of resistors where the highest resistance in the conduction pathway determines the conductivity. Then the percolation theory establishes the least-resistant pathway and hence the hopping conductivity is governed by a percolation criterion (Shklovskii and Efros, 1984). More details on this topic are given in Section 13.8. The analytical solution to this type of problem has been predominantly attempted in lattice systems in which the localized sites for transport occupy spatially regular positions. The more realistic spatially and energetically disordered distributions have been mainly studied by numerical computation (Anta, 2009). Many results have been developed by empirical fitting of numerical calculations (Coehoorn and Bobbert, 2012).

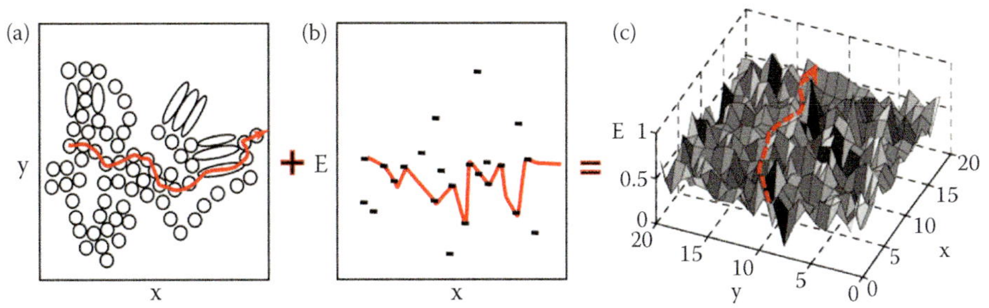

FIGURE 13.3 Schematic description showing that the transport sites are distributed both (a) in space and (b) in energy. (c) A schematic 2D map of a rough energy surface. The line describes a hypothetical path a charge might take while hopping from one side of the sample to the other (under applied electric field). (Reproduced with permission from Tessler et al., 2009.)

The theoretical analysis of electronic transport in disordered solids is considerably simplified in a system with a steep distribution of localized states. For carriers situated deep enough energetically, a particular level, called the transport energy, E_{tr}, determines the dominant hopping events. Whenever carrier transport is governed by thermal activation to some type of transport level within a broad distribution of localized states, the localized states become trapping levels. Then the problem of determining transport coefficients can be treated with spatially homogeneous equations of conservation, so that one can employ the multiple trapping model methodology. We saw in Equation 6.29, and also will see in Equation 18.61, that the energy for thermal promotion of an electron to the transport level becomes larger as the localized states are deeper in the bandgap. Hence, the localized states at the Fermi level normally play a dominant role in the time and temperature dependence of the kinetic coefficients, as indicated in Figure 13.1d (although the case of the Gaussian distribution is somewhat different, as will be discussed in Section 13.6). For example, in Figure 13.1c, the traps below E_{Fn} are occupied, so they do not capture any carriers from the transport states. In addition, the fastest transition for an electron to be delocalized again is from the traps at the Fermi level to the conduction band minimum. Thus, the transport coefficients show a large variation depending on the specific position of the Fermi level.

In the studies of transport in disordered systems, the main focus has been placed on the analysis of highly diluted systems as a function of temperature and electric field. However, in energy devices such as solar cells and organic light-emitting diodes (OLEDs), the carrier density can be varied over many orders of magnitude up to high levels of 10^{21} cm^{-3} and the density dependence strongly influences different aspects of transport in the presence of disorder. For example, in organic conductors used in organic light-emitting diodes (LEDs) and field effect transistors (FETs), mobility is largely dependent on carrier density (Tanase et al., 2003; Coehoorn and Bobbert, 2012); see Figure 14.16. In nanostructured TiO_2, variations of electron diffusion coefficient were observed (Fisher et al., 2000), and are related to trap-limited transport (Bisquert, 2004). Furthermore, the interesting part of the operational regimes normally occurs at room temperature and even higher due to heating effects. Thus, the electronic transport in disordered solids normally occurs in a thermally activated regime and the dominant changes are the diffusion coefficient, mobility, and recombination lifetime dependence on carrier density. These aspects will be explored in depth in the next sections using the multiple trapping model.

13.2 TRANSPORT BY HOPPING IN A SINGLE LEVEL

The first application we discuss is the transport of electrons by hopping between neighboring localized sites of a unique energy level E_0 with a volume density N_L; see Figure 13.1a. This model includes spatial disorder but not energetic disorder effects. In terms of the occupancy $f = n/N_L$, the equilibrium occupancy is given by the Fermi–Dirac distribution. The chemical capacitance has the form indicated in Equation 7.12:

$$c_\mu = \frac{N_L q^2}{k_B T} f\left(1-f\right). \tag{13.1}$$

The general shape of the capacitance, shown in Figure 13.4b, forms a peak at the value $E_F = E_0$, at which the occupancy $f = 1/2$ (Equation 7.9). In a first approximation (neglecting distance dependence), the mean effective jump frequency is

$$\nu = \nu_1\left(1-f\right), \tag{13.2}$$

where ν_1 is the rate constant for hopping from an occupied site to an empty neighbor site. In a cubic lattice,

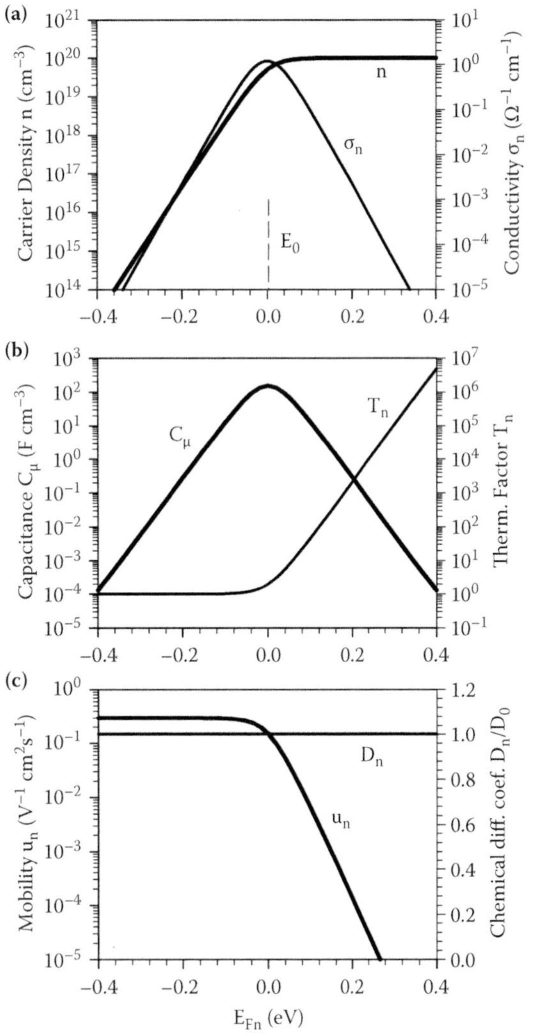

FIGURE 13.4 Representation of several quantities for charge accumulation and transport in a single energy level at $E_0 = 0$ eV with respect to E_{Fn}, the Fermi level. (a) Carrier density and conductivity. (b) Chemical capacitance and thermodynamic factor. (c) Mobility and chemical diffusion coefficient. The following parameters were used in the calculation: $N_0 = 1.0 \times 10^{20}$ cm^{-3}, $T = 300$ K, $\nu_0 = 10^{12}$ s^{-1}, $a = 10^{-7}$ cm.

the distance between sites is $r = (N_L)^{-1/3}$. In a spatially disordered system in three dimensions, the average site separation is

$$r = \left[\frac{4\pi N_L}{3}\right]^{-1/3} \tag{13.3}$$

The jump diffusion coefficient is

$$D_J = \frac{1}{6} r^2 \nu \tag{13.4}$$

and the mobility has the form

$$u_n(E_F) = \frac{qr^2\nu_1}{6k_BT}(1-f). \tag{13.5}$$

The chemical diffusion coefficient is a constant,

$$D_n = D_0 = \frac{1}{6}\nu_1 r^2 \tag{13.6}$$

There is a strong difference between mobility and diffusivity due to the exclusion of occupied sites for hopping targets; see Figure 13.4c. Since D_J and u_n relate to the random walk displacement of electrons, they become affected by the lack of target sites for hopping when the electronic level becomes occupied. In contrast to this, the chemical diffusion coefficient describes the net flux under a gradient of the concentration and in this case, the exclusion effects of forward and backward jumps between two neighbor sites compensate, as mentioned in Equation 11.19, implying that D_n is a constant (Reed and Ehrlich, 1981). The difference between mobility and the diffusion coefficient is described by the thermodynamic factor, as indicated in Equation 11.43. In the present example, it is

$$T_n = 1/(1-f); \tag{13.7}$$

(see Figure 13.4b).

A relevant instance of the difference between mobility and the chemical diffusion coefficient is found in the study of electron transport in an array of quantum dots with a series of discrete energy levels by van de Lagemaat (2005). By filling the first quantum $1S_0$ level, the diffusion coefficient decreases by a factor of 10, while the mobility shows a much stronger decay by three orders of magnitude, which is observed in measurements (Yu et al., 2003).

A peak of the conductivity shown in Figure 13.4a can be explained by the combined behaviors of carrier density and mobility. At a low Fermi level, the mobility is constant and the conductivity increases by the increase of the carrier density. Above $E_{Fn} = E_0$, the density of electrons is $\approx N_L$, but the mobility starts to decrease because most of the transport states have been occupied. As a result, the conductivity has the same shape as the chemical capacitance, which was summarily expressed in Equation 11.60.

Including nearest neighbor interactions in the lattice gas model provokes a major departure from the Fermi–Dirac statistics, and more so if the system is below the critical temperature and may form an ordered phase, as was discussed in Figures 5.27 and 5.28. The effect of interactions on the thermodynamic factor and on the jump and chemical diffusion coefficient is shown in Figure 13.5.

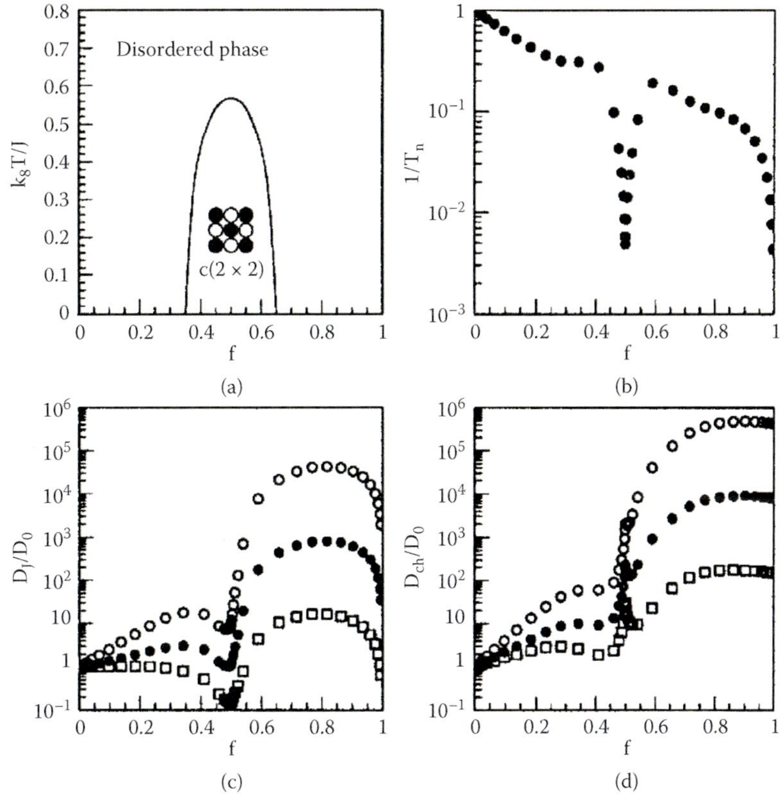

FIGURE 13.5 A square lattice gas model with repulsive nearest neighbor interactions. (a) Schematic picture of the phase diagram. Numerical results for (b) $1/T_n$ (reciprocal thermodynamic factor), (c) jump, and (d) chemical diffusion coefficient as a function of the fractional occupation of the lattice. The results are calculated well below the critical temperature or order–disorder transition, at $k_BT = J/3$ (J is the strength of the near-neighbor repulsion). The full dots correspond to no saddle-point interactions, the open squares and circles correspond to positive and negative saddle-point interactions J', respectively ($J' = \pm k_BT$). Positive values of J' raise the saddle-point energy, thus depressing the jump rate and D_J; negative values of J' have the opposite effect. Both T_n^{-1} and D_J show deep minima at half occupation $f = 0.5$, corresponding to the perfect ordered phase. (Reproduced with permission from Ala-Nissila et al., 2002.)

13.3 TRAPPING FACTORS IN THE KINETIC CONSTANTS

As explained in Section 13.1, the multiple trapping transport describes the effect of trap levels over the rate of displacement through transport states. Here, we formulate the general form of the chemical diffusion coefficient and electron lifetime in a multiple trapping framework with a general localized DOS. We distinguish between *free electrons* in the extended transport states (conduction band), n_c, and *trapped electrons* in localized band gap states with density n_L. The total carrier density is

$$n = n_c + n_L \tag{13.8}$$

Associated with each type of electronic state, we have the corresponding chemical capacitance, as discussed in Chapter 7. The conduction band capacitance c_μ^{cb} is given by Equation 7.14, while the localized states capacitance c_μ^L takes the general form in Equation 7.19, which depends on the localized DOS, $g(E)$. We have the following relationships between the chemical capacitances:

$$\begin{aligned} c_\mu &= c_\mu^{cb} + c_\mu^L \\ &= c_\mu^{cb}\left(1 + \frac{\partial n_L}{\partial n_c}\right) \end{aligned} \tag{13.9}$$

In the multiple trapping model, the jump diffusion coefficient is defined for the average displacement of all carriers, and it is lower than the intrinsic diffusion coefficient of the transport level, D_0. The rate of transport of the total carrier density can be related to that of the free carrier density because the time constants for trapping and detrapping depend on the free and total carrier density to satisfy the detailed balance principle. The jump diffusion coefficient of total carrier density is given by Bisquert (2008) and Ansari-Rad et al. (2013):

$$D_J = \frac{n_c}{n} D_0. \tag{13.10}$$

By Equation 11.39, the thermodynamic factor takes the form

$$T_n = \frac{n}{n_c}\left(1 + \frac{\partial n_L}{\partial n_c}\right)^{-1} = \frac{n}{n_c}\frac{c_\mu^{cb}}{c_\mu}. \tag{13.11}$$

By combining Equations 11.43, 13.10, and 13.11, the chemical diffusion coefficient of electrons has the general expression

$$D_n = \left(1 + \frac{\partial n_L}{\partial n_c}\right)^{-1} D_0. \tag{13.12}$$

This result shows that the chemical diffusion coefficient will always be smaller than the free electrons diffusion coefficient due to the effect of trapping and detrapping. The presence of traps also influences the observed rate of recombination as further discussed in Section 18.3. The electron lifetime τ_n is defined in terms of the recombination rate $U_n(n)$ and total carrier density n, as indicated in Equation 18.52. It is also interesting to introduce a separate quantity, the *free carrier lifetime* τ_f, which corresponds to the recombination rate of carriers in the transport state (without trapping effects):

$$\tau_f = \left(\frac{\partial U_n}{\partial n_c}\right)_{n_c}^{-1}. \tag{13.13}$$

Then we have

$$\tau_n = \left(1 + \frac{\partial n_L}{\partial n_c}\right)\tau_f. \tag{13.14}$$

Equation 13.14 now gives the measured lifetime τ_n as a combination of two effects: trapping–detrapping and subsequently recombination. Further details about interpretation of lifetimes are described by Bisquert et al. (2009) and Ansari-Rad et al. (2013).

In summary, the traps shorten the diffusion coefficient and increase the lifetime. We now consider the different kinetic effects together, combined in the continuity equation for free carrier density and flux $J = D_0 \partial n_c/\partial x$, Equation 11.25, modified with an additional term $\partial n_L/\partial t$, due to the net capture and release of free carriers by traps

$$\frac{\partial n_c}{\partial t} = -\frac{\partial J}{\partial x} + G(x) - U_n(n_c) - \frac{\partial n_L}{\partial t}. \tag{13.15}$$

Let us use the small perturbation approach discussed in Section 18.5. Equation 13.15 is split in two parts. The first is the *steady-state equation*:

$$\frac{\partial J}{\partial x} + G(x) - U_n(n_c) = 0. \tag{13.16}$$

Note that the localized states do not introduce any new effect in the steady-state conservation equation. Therefore, quantities that can be measured at steady state, such as the electron conductivity, are independent of the number and occupation of traps.

The second equation is for the small perturbed density:

$$\frac{\partial \hat{n}_c}{\partial t} = -\frac{\partial \hat{J}}{\partial x} - \frac{1}{\tau_f}\hat{n}_c - \frac{\partial \hat{n}_L}{\partial t}, \tag{13.17}$$

where we have included the free carrier lifetime as defined by Equation 13.13. The model system may be completed by an additional kinetic equation for the traps that defines the variation $\partial n_L/\partial t$, Equation 6.30. However, if the trapping kinetics is fast with respect to the time scale of the transient measurement, we may assume that the traps follow the equilibrium relation with the free carriers:

$$\frac{\partial n_L}{\partial t} = \frac{\partial n_L}{\partial n_c}\frac{\partial n_c}{\partial t}. \tag{13.18}$$

Equation 13.18 states that equilibrium between free and trapped electrons (with a common Fermi level) will be maintained for any time variation during kinetic measurements. Equation 13.18 is termed *the quasi-static approximation* (Bisquert and Vikhrenko, 2004). On the other hand, if the trapping times are slow with respect to measurement times, then the trapping kinetics must be described explicitly (Bisquert, 2008a). Applying the quasi-static approximation and using Fick's law, Equation 13.17 becomes

$$\left(1 + \frac{\partial n_L}{\partial n_c}\right)\frac{\partial \hat{n}_c}{\partial t} = \frac{\partial}{\partial x}\left(D_0 \frac{\partial \hat{n}_c}{\partial x}\right) - \frac{1}{\tau_f}\hat{n}_c \tag{13.19}$$

and therefore

$$\frac{\partial \hat{n}_c}{\partial t} = D_n \frac{\partial^2 \hat{n}_c}{\partial x^2} - \frac{1}{\tau_n}\hat{n}_c. \tag{13.20}$$

Equation 13.20 indicates that the diffusion-recombination of electrons in the presence of traps can be treated with the dynamic equations of a single level but with kinetic coefficients that depend on the steady state: the

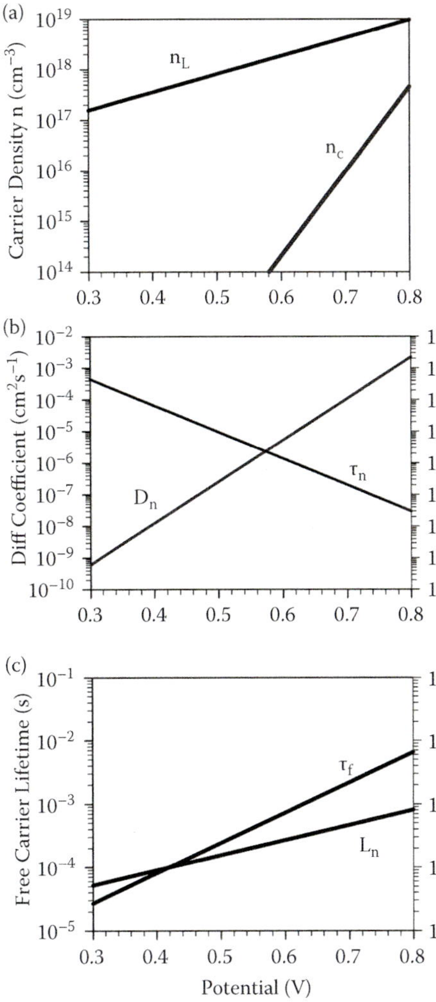

FIGURE 13.6 (a) Representation of the free and localized carrier density as a function of potential (Fermi level position) for an exponential distribution of localized states with a recombination rate $U_n = k_{rec} n_c^{\beta} \left(T = 300\,\mathrm{K}, T_0 = 1400\,\mathrm{K}, \beta = 0.71\right)$ (b) Electron lifetime, τ, and the diffusion coefficient, D_n, measured by small perturbation. (c) The free carrier lifetime, τ_f, and diffusion length $L_n = \sqrt{D_n \tau_n} = F D_0 \tau_f$.

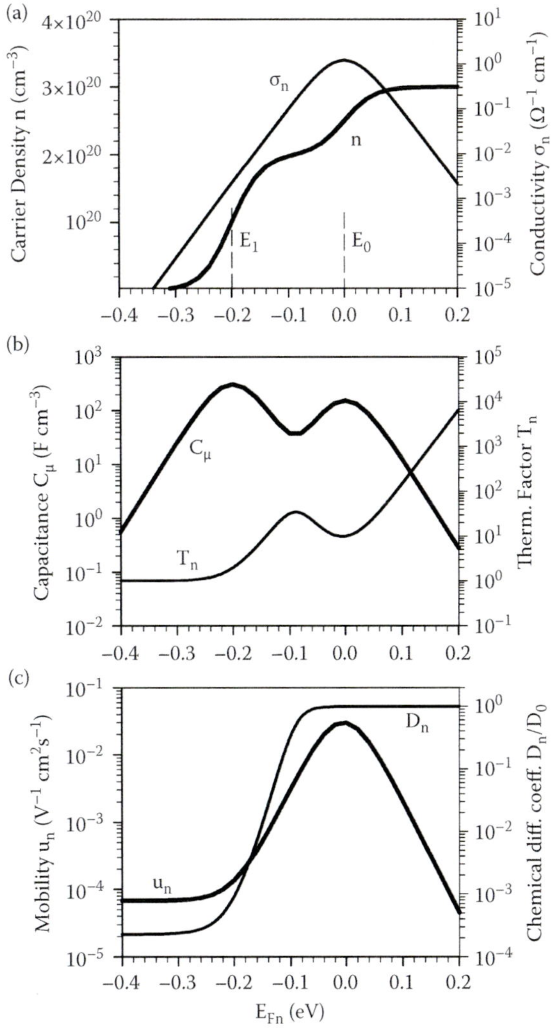

FIGURE 13.7 Representation of several quantities for charge accumulation and transport with respect to E_F, the Fermi level, or chemical potential μ. The model assumes transport by hopping between localized states in a material with a single band gap state of energy $E_0 = 0$ eV and a trap level at $E_1 = -0.2$ eV. (a) Carrier density and conductivity. (b) Chemical capacitance and thermodynamic factor. (c) Mobility and chemical diffusion coefficient. The following parameters were used in the calculation: $N_0 = 1.0 \times 10^{20}$ cm^{-3}, $N_1 = 2N_0$, $T = 300$ K, $\nu_0 = 10^{12}$ s^{-1}, $a = 10^{-7}$ cm.

electron lifetime $\tau_n(E_{Fn})$ and the chemical diffusion coefficient $D_n(E_{Fn})$. The dependence of these quantities on the voltage is illustrated in Figure 13.6 for the case of an exponential distribution of states (Bisquert and Marcus, 2014) as discussed in detail later in Section 13.5 (Figure 13.6).

13.4 TWO-LEVEL (SINGLE-TRAP) MODEL

The simplest model of trap-limited transport is composed of a transport level at the energy E_0, Section 13.2, and a trap level at energy E_1 with volume density $N_1 = \delta N_0$, where δ is a constant; see Figure 13.1b (Bisquert and Vikhrenko, 2002). In equilibrium, the occupancies of the two levels are determined by the equalization of their respective Fermi levels:

$$E_F = E_0 + k_B T \ln\left(\frac{f_0}{1 - f_0}\right) = E_1 + k_B T \ln\left(\frac{f_1}{1 - f_1}\right). \quad (13.21)$$

When the electron Fermi level increases, the two states are consecutively filled up with carriers, Figure 13.7a. Consequently, there are two peaks in the chemical capacitance that correspond to the addition of two terms as that in Equation 13.1, one for each level; see Figure 13.7b.

The chemical diffusion coefficient is obtained from Equation 13.12:

$$D_n = \frac{D_0}{1+\delta\dfrac{f_1(1-f_1)}{f_0}} \quad (13.22)$$

When the Fermi level lies deep below the trap state E_1, both E_0 and E_1 are populated following the ideal statistics ($f_0, f_1 << 1$). Hence, the thermodynamic factor is 1 and the mobility and chemical diffusion coefficient in Equation 13.22 take constant values

$$D_n = \frac{D_0}{1+\delta\exp\left[(E_0-E_1)/k_BT\right]}. \quad (13.23)$$

This result was obtained by Hoesterey and Letson (1963) for doped anthracene crystals. Changes in D_n and u_n appear when the deep states begin to be more heavily occupied. Filling the deep traps reduces their slowing effect, hence the chemical diffusion coefficient increases rapidly (Figure 13.7c) until the deep state is filled completely, at which point the chemical diffusion coefficient becomes a constant identical to the single-level case. The conductivity, shown in Figure 13.7a, reflects only the hopping along the shallow level and is not changed with respect to Figure 13.4a.

Let us consider the effect of interactions between particles. A complete calculation of one-dimensional diffusion in two-level systems with interactions is given by Bulnes et al. (2006). A calculation of Monte Carlo simulations of a two-dimensional two-level lattice gas with repulsive interactions between nearest neighbors is shown in Figure 13.8 (Groda et al., 2005; Bisquert, 2008). With respect to the noninteracting two-level system discussed earlier in Figure 13.7, a new feature in Figure 13.8 is a strong decrease of the occupancy of the deep site when the Fermi level lies between the energies of the two levels (Figure 13.8b). This is reflected in a new peak of the chemical capacitance (Figure 13.8c), or, equivalently, a dip in the thermodynamic factor. The jump diffusion coefficient is characterized by a minimum value at $c = 1/3$ because at this concentration, almost all the deep sites are occupied. When the concentration is approaching this value and a particle jumps to a nearest shallow site, it is repelled by its neighbors on the nearest deep sites. At $c > 1/3$, some particles must be on the shallow sublattice and thus the mobility sharply increases.

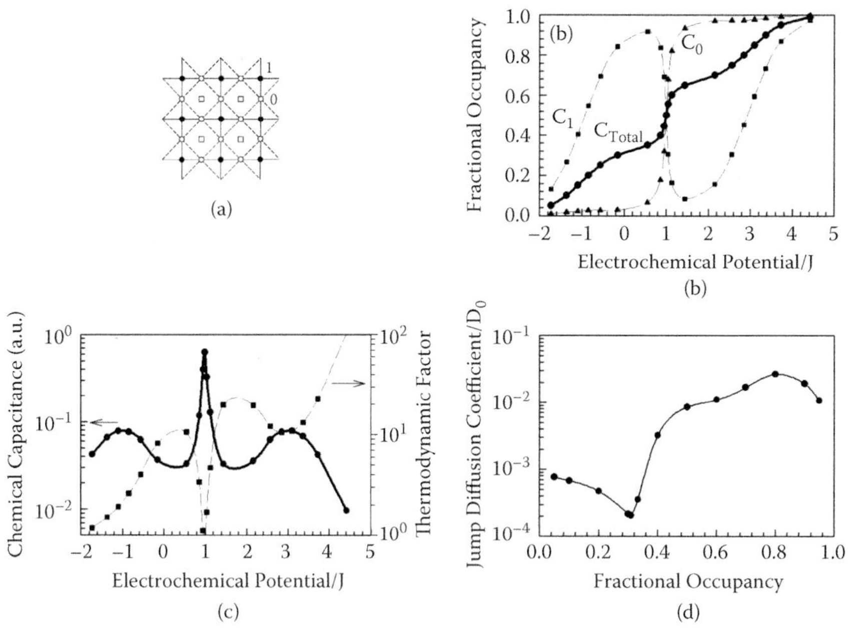

FIGURE 13.8 Calculation of thermodynamic and transport coefficients in a two-level system with interactions. $J = E_2 - E_1$ is the interaction energy between carriers on nearest neighbor sites, where E_1 and E_2 are site energies ($N_1 = 1/3N$, $N_2 = 2/3N$). The critical temperature of this system is given by $J/k_BT_c = 3.064$. The separation between the two energy levels is of the order of the thermal energy, $E_2 - E_1 = 2.55\ k_BT$. (a) Arrangement of sites in the two-level system. Shallow 0 (open circles) and deep 1 (full circles) lattice sites form two square sublattices, open squares designate the substrate particles. Results of Monte Carlo simulations for this system at $T = 1/2T_c$, above the critical value: (b) Fractional occupancies of the two levels and total number of particles, (c) chemical capacitance and thermodynamic factor, and (d) jump diffusion coefficient. (Adapted with permission from Groda et al., 2005.)

13.5 MULTIPLE TRAPPING IN EXPONENTIAL DOS

Properties of the exponential distribution of localized states in the band gap were described in Sections 8.2 and 8.3. Figure 7.8 shows the thermalized occupation of the exponential DOS with tail parameter T_0 that consists of a full occupancy below E_F and a Boltzmann tail of occupied states above the Fermi level.

Using the zero-temperature approximation, the upper tail is neglected. The localized electron carrier density is then given by the integration of DOS in Equation 8.19 up to the Fermi level,

$$n_L = \int_{-\infty}^{E_{Fn}} g(E)dE = n_{L0}e^{(E_{Fn}-E_{F0})/k_BT_0}, \quad (13.24)$$

where the equilibrium value is given by

$$n_{L0} = N_L e^{(E_{F0}-E_c)/k_BT_0} \quad (13.25)$$

Using Equations 5.53 and 13.24, we obtain the relationship between localized and free carrier density in equilibrium

$$\frac{n_L}{n_{L0}} = \left(\frac{n_c}{n_{c0}}\right)^{\alpha} \quad (13.26)$$

with the coefficient

$$\alpha = \frac{T}{T_0}. \quad (13.27)$$

The chemical capacitance in the zero-temperature approximation is given in Equation 8.20. Note the following property

$$n_L = \frac{k_BT_0}{q^2}c_\mu^L. \quad (13.28)$$

The chemical diffusion coefficient from Equation 13.12 is then given by

$$D_n = \frac{N_0T_0}{N_LT}\exp\left[\left(E_{Fn}-E_c\right)\left(\frac{1}{k_BT}-\frac{1}{k_BT_0}\right)\right]D_0$$
$$= \frac{T_0}{T}\frac{n_{c0}}{n_{L0}^{1/\alpha}}n_L^{(1-\alpha)/\alpha}D_0. \quad (13.29)$$

A representation of the free and trapped carrier density and the chemical diffusion coefficient as a function of the Fermi level is shown in Figure 13.6. Also shown in Figure 13.6 are the recombination lifetime and diffusion length (Bisquert et al., 2009). The jump diffusion coefficient and the mobility dependence on carrier concentration can be found from Equations 13.10 and 13.26:

$$u_n = \frac{n_{c0}u_0}{1/\alpha}n^{1/\alpha-1} \quad (13.30)$$

with respect to the mobility of the transport state, u_0. The thermodynamic factor in Equation 13.11 is a constant

$$T_n = \frac{T_0}{T}. \quad (13.31)$$

For typical values of T_0, $T_n \approx 2-5$ at room temperature. The diffusion-mobility ratio is independent of temperature:

$$\frac{D_n}{u_n} = \frac{k_BT_0}{q}. \quad (13.32)$$

Figure 13.9b shows an additional feature of the multiple trapping model. When approaching the conduction band, the chemical capacitance is dominated by the free carriers, $c_\mu^{cb} \gg c_\mu^L$

Then, the thermodynamic factor decays to the ideal value 1 and the transport coeffcients take a constant value for the free electrons; see Figure 13.9c The conductivity shown in Figure 13.9b is governed only by the free carrier density (Abayev et al., 2003; Wang et al., 2006) (Figure 13.9).

13.6 ACTIVATED TRANSPORT IN A GAUSSIAN DOS

As remarked in Section 8.6, the transport of carriers in disordered organic semiconductors and insulators is usually described within the framework of a Gaussian disorder model. Here, we discuss the properties of the multiple trapping model with the Gaussian distribution. We will consider the connection between single particle and collective diffusion properties based on numerical simulation. The analysis is also useful for describing transport in organic DOS in quasi-equilibrium situations in which the carrier transport is determined by thermally activated events, as in the concept of the transport energy further discussed in Section 13.8.

There is a large difference in the equilibrium population of the exponential and Gaussian DOS (Tessler and Roichman, 2005). In the exponential DOS, the density of electrons is mainly located at the Fermi level; see Figure 7.8. The density of electrons decays upward with a scale k_BT and downward with the scale k_BT_0. In contrast

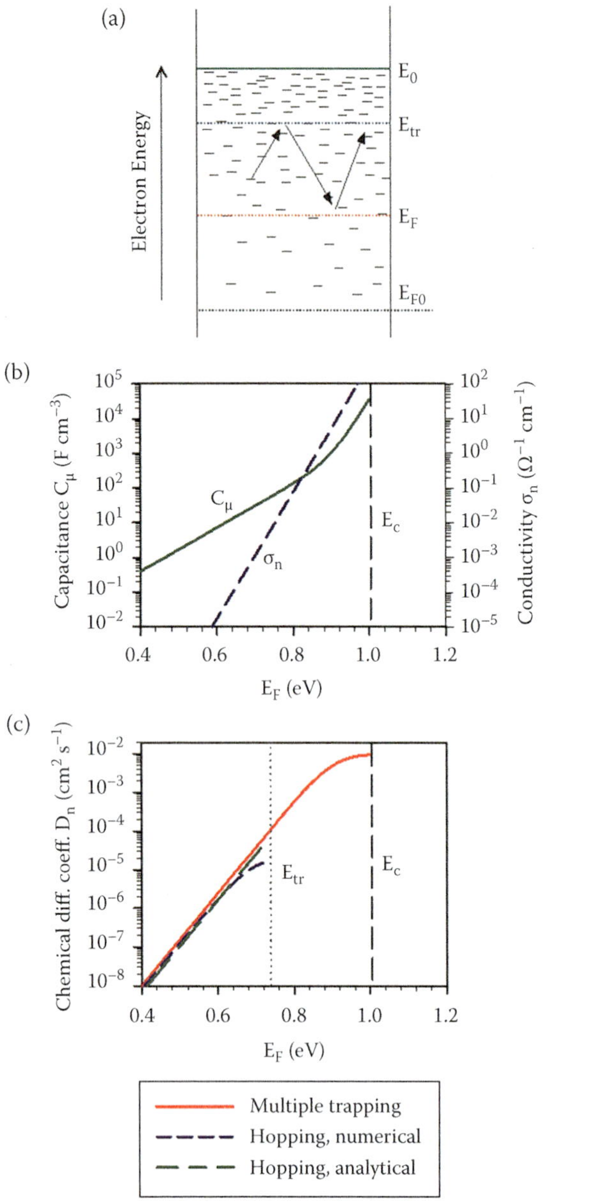

FIGURE 13.9 (a) The scheme shows the trajectory for electron transport by hopping between localized states in the bandgap of a semiconductor. Upward hops predominantly occur to the transport level (E_c). Representation of several quantities for charge accumulation and transport in an exponential DOS with the conduction band edge at the energy E_c = 1 eV. E_F is the Fermi level potential: (b) Chemical capacitance and conductivity. (c) Chemical diffusion coefficient, calculated in multiple trapping approximation, and in the hopping model with the transport energy concept, both with the numerical integration of the average jump frequency and with the analytical expression. The following parameters were used in the calculation: $N_0 = 5.0 \times 10^{21}$ cm^{-3}, $N_L = 10^{21}$ cm^{-3}, T = 275 K, T_0 = 800 K, $D_0 = 10^{-2}$ cm^{2}s^{-1} (for multiple trapping), localization length α_l = 0.5 nm and $\nu_0 = 5 \times 10^{12}$ s^{-1} (for hopping transport).

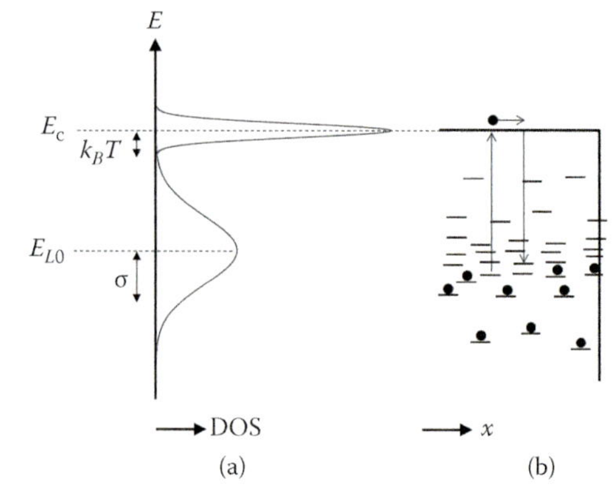

FIGURE 13.10 (a) Multiple trapping transport and recombination in the Gaussian DOS. Trapping into the localized Gaussian DOS and then detrapping to the transport level E_c (right) is essentially the same as the multiple trapping in a bi-Gaussian DOS (left), in which the upper Gaussian has a narrow width in the order of thermal energy. (Reproduced with permission from Ansari-Rad et al., 2013.)

to this, in the Gaussian distribution, the electron carriers are centered at the level $E_m = E_{L0} - \sigma^2/k_BT$ as stated in Equation 8.26, which lies above E_{Fn} at low concentration; see Figure 8.27. The total carrier density follows a Boltzmann distribution at low concentration (Equation 8.25), and the DOS can become fully occupied as discussed in Figure 7.10.

The kinetic Monte Carlo (KMC) simulation is a stochastic computational procedure that allows for a flexible description of transport of charge carriers in a network of traps without huge computational demands (Anta et al., 2008). In the KMC calculation, a certain number of carriers are allowed to jump between neighboring traps. The hopping time between two traps labeled i and j is

$$t_{ij} = -\ln(R)(\nu_{ij})^{-1}, \tag{13.33}$$

where ν_{ij} is the chosen energy-dependent hopping rate and R is a random number distributed uniformly between 0 and 1. The carriers are dispersed in the specific DOS and for each carrier, hopping times to neighboring traps are computed via Equation 13.33. The process is repeated in such a way that for each simulation step, the carrier that happens to have the minimum hopping time moves along the network and the simulation is advanced by time intervals of variable size t_{ij}.

To simulate multiple trapping transport in the Gaussian DOS, we use the model in Figure 13.10 (Ansari-Rad et al., 2013), in which the upper Gaussian is narrow and plays the role of the transport level. Thus, the jump rate

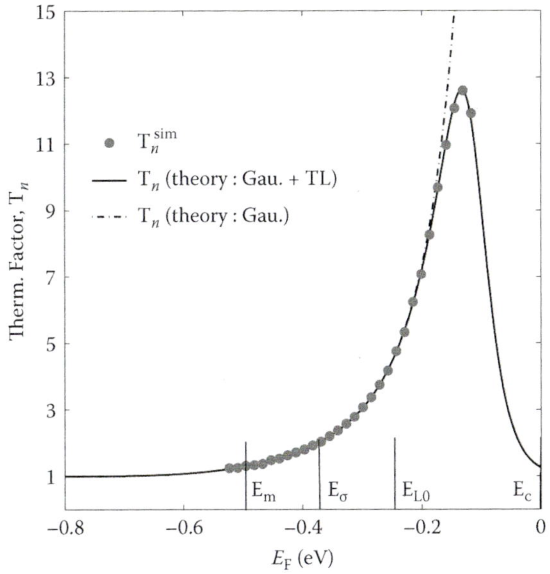

FIGURE 13.11 Thermodynamic factor vs. Fermi level for the energy-level alignment of Figure 13.10. The points are the Monte Carlo simulation results and the solid line is the theoretical value for thermodynamic factor calculated with Equation 13.11. Also shown (dashed line) is the thermodynamic factor for the Gaussian distribution without the transport level (TL). In this case, the thermodynamic factor diverges at high Fermi level. Parameters: $T = 300$ K, $E_c = 0$, $E_{L0} = -0.25$ eV, $\sigma = 0.08$ eV, $N_L/N_{tr} = 0.25$. Reproduced with permission from (Ansari-Rad et al., 2013).

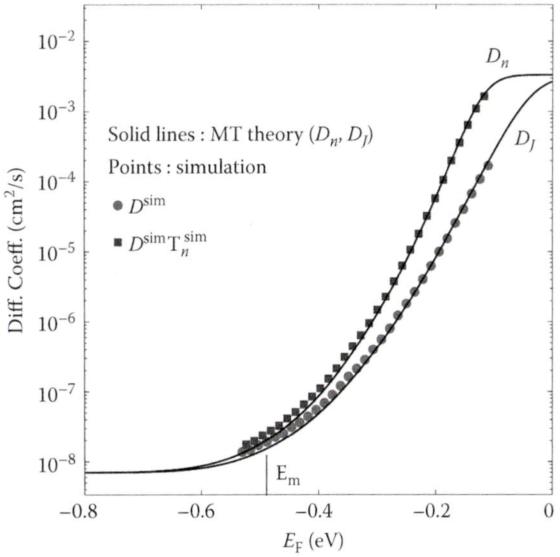

FIGURE 13.12 Diffusion coefficient vs. Fermi level for multiple trapping in the Gaussian DOS. The lines show the theoretical predictions for D_n and D_J, calculated with Equations 4.12 and 4.10, respectively. The filled circles (D^{sim}) show the random walk diffusion coefficient obtained in the simulation. As can be seen, D^{sim} coincides with the jump coefficient D_J. Filled squares show the quantity $D^{sim}T_n^{sim}$. The results are in agreement with the chemical diffusion coefficient D_n. (Reproduced with permission from Ansari-Rad et al., 2013.)

from a trap in Equation 13.33 is thermally activated to the transport level E_c. A bimodal Gaussian DOS has been used in related problems to describe the effects of traps induced by impurities or doping (Arkhipov et al., 2003; Peng et al., 2006; Coehoorn, 2007). Figure 13.11 shows the calculated thermodynamic factor in comparison with the theoretical expression of Equation 13.11. Due to the properties of the carrier density in the Gaussian distribution, the thermodynamic factor at low concentration has the ideal value 1, whereas it increases with the Fermi level due to the full occupation of the finite DOS (Roichman and Tessler, 2002; Bisquert, 2008). Although a certain number of carriers are placed in the simulation box in this type of simulation, the dynamics of only one of them is tracked during the simulation. Figure 13.12 shows the numerical simulation of the random walk diffusion coefficient defined in Equation 11.6, D_J, and the confirmation that it coincides with the jump diffusion coefficient given in Equation 13.10. The features of the mobility (proportional to D_J) can be well understood from the values of the thermodynamic factor. When $E_{Fn} < E_m$, the trapped carrier density follows Boltzmann statistics so that $T_n = 1$, implying that D_J and the mobility are constant. Compare the case of the exponential distribution where D_J is density dependent at an arbitrary low concentration as indicated in Equation 13.30. We noted in Section 8.6 that the carriers trapped in the tail of the Gaussian have the same density as that in a monoenergetic level at $E_\sigma = E_{L0} - \sigma^2/2k_BT$. Thus, the thermodynamic factor in Figure 13.11 when the Fermi level approaches E_σ behaves as that in Figure 13.7 when approaching E_1. These features explain the Fermi level dependence of jump and the chemical diffusion coefficient observed in Figure 13.12. The figure also confirms the statement that the random walk diffusion coefficient is different from the one that appears in Fick's law. Figure 13.13 shows a separate calculation of the diffusion coefficient and the mobility, and confirms the Einstein relationship that was stated in Equation 12.21.

13.7 MULTIPLE TRAPPING IN THE TIME DOMAIN

When a light pulse impacts a semiconductor with localized states in the bandgap, at first, optically

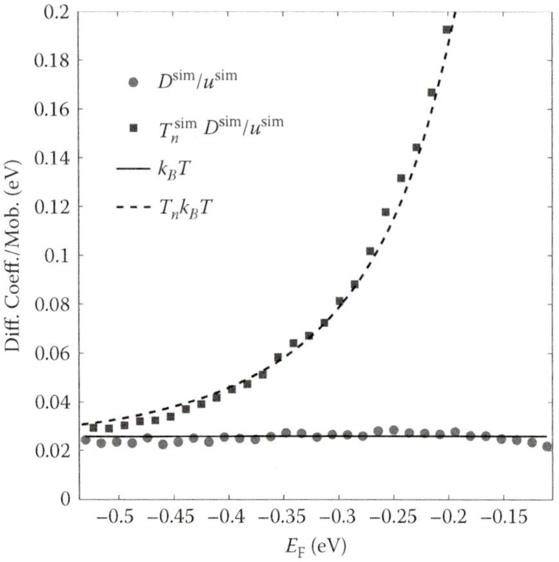

FIGURE 13.13 Einstein relation for multiple trapping in the Gaussian DOS. The filled circles show the Monte Carlo simulation results. For each point we have made two separate simulations, one without the electric field to calculate the diffusion coefficient D^{sim} and the other with the electric field to compute the mobility u^{sim}. As can be seen, results of the simulation are in good agreement with the prediction $D_n/u_n = k_BT/q = 0.258$ eV, Equation 12.21. The dashed line and the filled squares show the other statement of the generalized Einstein relation, that is, Equation 12.22. Note that *both* sets of data justify the generalized relationship. (Reproduced with permission from Ansari-Rad et al., 2013.)

excited electrons are generated in the conduction band, followed by the carriers undergoing trapping and thermalization processes, which make the initial free carrier concentration decrease. The free carrier distribution dependence on time can be probed by the experimental technique of transient photocurrents (Orenstein and Kastner, 1981; Göbel and Graudszus, 1982). We assume that any intrinsic carrier concentration and recombination can be neglected. The optically excited carriers in the conduction band will be trapped in the localized states for a time of the order of one trapping time ($\approx 10^{-13} - 10^{-12}$ s), as indicated in Figure 13.14. For an energy-independent capture cross section, the distribution of trapped carriers is proportional to the density of states (Moon et al., 1994). The energy depth below the transport band becomes a key factor for the frequency of the transitions between a given trap and the conduction band because the probability of detrapping to the conduction band is proportional to $\exp[E_t - E_c)/k_BT]$; see Equation 18.61. After a time t, the shallower states release and retrap electrons many times, so that these states effectively obtain a thermal distribution, as described in Figure 7.8. In contrast, the electrons in deeper levels remain frozen, as indicated in Figure 13.14b. The demarcation energy level $E_d(t)$ above which electrons are released at time t is determined by

$$t = \frac{1}{\nu_0} e^{(E_c - E_d)/k_BT}, \qquad (13.34)$$

where $\nu_0 = \beta_n N_c$; see Equation 6.29. The demarcation level sinks with time into the distribution of localized states, depopulating the states above E_d and adding more trapped charge to the states below E_d in addition to the already present frozen charge, so that the total density of excess carriers remains constant, as indicated in Figure 13.14a. This $E_d(t)$ acts in the form of a quasi-Fermi level and the bulk of the injected charge

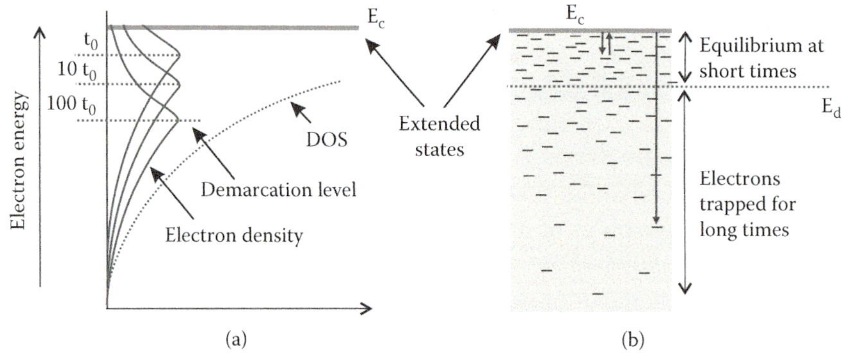

FIGURE 13.14 (a) Energy diagram of a semiconductor showing the exponential DOS and the energy distribution of carriers. The peak of the distribution is at the demarcation level E_d and sinks deeper in the bandgap with time. (b) The energy diagram shows the transport level (indicated by the energy of the conduction band, E_c) and localized states in the bandgap. The trapping–detrapping kinetics is indicated as a function of the depth of the localized state in the bandgap. At time t after injection, states shallower than the demarcation level E_d are able to frequently detrap and retrap electrons and establish thermal equilibrium, while electrons trapped at states deeper than E_d have a negligible probability of being released. (Reproduced with permission from Bisquert, 2005.)

will be concentrated near $E_d(t)$. The distribution of carriers in the band gap is described by Schiff (1981):

$$n_t(E,t) = \phi(t) g(E) F\left[E, E_d(t)\right] \qquad (13.35)$$

in terms of a time-dependent occupancy factor $\phi(t)$ that ensures that the initial excitation is conserved and the Fermi–Dirac distribution function, Equation 5.14. The conservation of carriers can be stated as

$$\phi(t) \int_{-\infty}^{E_c} g(E) F\left[E, E_d(t)\right] dE = n_0 \qquad (13.36)$$

For an exponential DOS, Equation 13.36 can be transformed to

$$\phi(t)(\nu_0 t)^{-a} N_t \alpha \int_0^{\nu_0 t} \frac{x^{\alpha-1}}{1+x} dx = n_0. \qquad (13.37)$$

Provided that $\nu_0 t >> 1$, Equation 13.37 gives

$$\phi(t) = \frac{n_0}{N_t \alpha \Gamma(\alpha) \Gamma(1-\alpha)} (V_0 t)^{\alpha}. \qquad (13.38)$$

The occupancy factor $\phi(t)$ in Equation 13.38 increases with time to compensate for the sinking of the quasi-Fermi level $E_d(t)$. The density of electrons in the conduction band is a small tail of the total distribution,

$$n_c(t) = N_c \phi(t) e^{(E_d(t) - E_c)/k_B T} \qquad (13.39)$$

such that $n_c << n_0$; see Figure 13.14a. The occupancy factor of electrons in extended states $f_c = n_c / N_c$ decreases with time as

$$f_c(t) = \frac{n_0}{N_t \alpha \Gamma(\alpha) \Gamma(1-\alpha)} (V_0 t)^{\alpha-1}. \qquad (13.40)$$

If the sample is weakly absorbing, the carriers will be homogeneously generated. Equation 13.40 describes the decay law of photoconductivity as shown in Figure 13.15. The method allows us to derive drift mobilities (Michiel et al., 1986).

On the other hand, in the TOF technique explained in Section 10.4, the carriers are generated close to one contact of a carrier-free sample and drift to the other contact. In the presence of localized states, the trap-free result, Equation 10.26, is strongly modified (Spear, 1969; Tiedje et al., 1981; Tyutnev et al., 2014). For multiple trapping in an exponential DOS, the mean displacement of

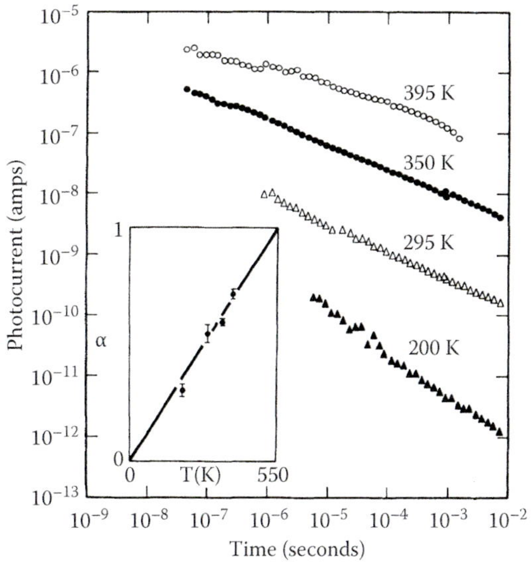

FIGURE 13.15 Time decay of the photocurrent in amorphous As_2Se_3 after pulsed light excitation at different temperatures. At each temperature, the decay is accurately described by $\Delta j_{ph} \propto t^{-1+\alpha}$. The temperature dependence of α is shown in the inset, in excellent agreement with Equation 13.27. (Reproduced with permission from Orenstein and Kastner, 1981.)

the photocarrier distribution $x(t)$ is given by the expression (Schiff, 2004)

$$x(t) = \frac{\sin(\alpha\pi)}{\alpha\pi(1-\alpha)} \frac{\mathbf{E} u_0}{\nu_0} (\nu_0 t)^{\alpha}. \qquad (13.41)$$

where $\mathbf{E}$ is the electric field, provided that $\alpha = T/T_0 < 1$.

13.8 HOPPING CONDUCTIVITY

As explained in Section 13.1, the long-range transport that occurs by hopping between localized states is denoted the hopping conductivity. It is often assumed that the carriers hop from site to site as determined by the MA hopping rate indicated in Equation 6.2:

$$\nu_{ij} = \nu_0 \exp\left[-2\frac{r_{ij}}{\alpha_l} - \frac{E_j - E_i}{k_B T} \theta(E_j - E_i)\right], \qquad (13.42)$$

where θ is a unit step function. Since the hopping process is assisted by phonons, ν_0 is usually considered a phonon frequency of the order 10^{12} s^{-1}. The localization length is often estimated as $\alpha_l \approx 10^{-8}$ cm.

The model of nearest neighbor hopping (NNH) applies for conduction that occurs by hopping transitions between dopant impurities, provided that their average distance r, Equation 13.3, is much larger than the Bohr radius of one localized electron (Mott and

Twose, 1961; Yildiz et al., 2009). The average hopping rate is given by

$$\nu_{hop} = \nu_0 \exp -\left[2\frac{r}{\alpha_l} - \frac{W}{k_B T}\right], \quad (13.43)$$

where W is the activation energy. The diffusion coefficient is established by Equation 13.4 and the conductivity can be formed by Equation 12.25

$$\sigma_n = \frac{1}{6} c_\mu r^2 \nu_{hop}. \quad (13.44)$$

Since the chemical capacitance is given by the DOS, Equation 7.23, Equation 13.44 provides the result

$$\sigma_n = \frac{1}{6} g(E_F) \nu_0 q^2 r^2 \exp\left[-2\frac{r}{\alpha_l} - \frac{W}{k_B T}\right], \quad (13.45)$$

which is usually stated as

$$\sigma_n = \sigma_1 \exp\left[-\frac{W}{k_B T}\right]. \quad (13.46)$$

When the DOS is very broad, a large variety of possible transitions appears. The local conductivity has the form

$$\sigma_{ij} = \frac{q^2 r_{ij}^2 \nu_0}{6} e^{\frac{2r_{ij}}{\alpha_l} - \theta\frac{E_j - E_i}{k_B T}}. \quad (13.47)$$

The calculation of hopping conductivity can be reduced to the calculation of the conductivity of an MA network of random resistances with the elementary resistance given by Miller and Abrahams (1960) and Shklovskii and Efros (1984):

$$R_{\lambda\lambda'} = \frac{k_B T}{q^2} \frac{1}{\nu_{\lambda\lambda'} f(E_\lambda)(1 - f(E_{\lambda'}))}. \quad (13.48)$$

The conductivity of the network can be calculated in the framework of the percolation theory and expressed as a function of the critical exponent ξ_χ, which is determined by the percolation criterion (Shklovskii and Efros, 1984; Böttger and Bryksin, 1985):

$$\sigma = \sigma_0 \exp(-\xi_c). \quad (13.49)$$

The results of this calculation are shown in Figure 13.16, indicating the approximations of multiple trapping and variable range hopping. Frequency-dependent conduction in disordered systems is further discussed in Section 16.5. (Figure 13.16).

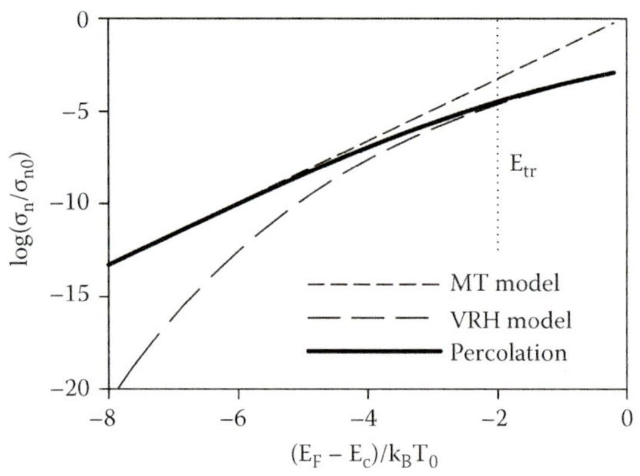

FIGURE 13.16 Dependence of normalized conductivity on Fermi energy (with respect to conduction band level E_c, calculated by a numerical solution of percolation conductivity in the multiple trapping model and VRH). Transport energy is shown by vertical lines. Parameters $N_L^{1/3}\alpha_l = 0.34$, $T = 0.26\,T_0$. (Calculation courtesy of Vladimir G. Kytin.)

13.9 THE TRANSPORT ENERGY

The concept of transport energy was originally formulated for amorphous inorganic semiconductors (Grünewald and Thomas, 1979; Monroe, 1985). Let us consider the hopping transport in an exponential distribution of states. The following derivation aims to find the fastest hop of a charge carrier that determines transport in equilibrium conditions.

The most probable upward jump corresponds to an optimized combination of the distance and energy difference, Equation 13.47. Let $a = N_L^{-1/3}$ be the mean distance between localized sites. The average distance for states below the energy E_1 is

$$r(E_1) = \left[\frac{4\pi}{3}\int_{-\infty}^{E_1} g(E)\,dE\right]^{-1/3} = \left(\frac{4\pi}{3}\right)^{-1/3} \exp\left(-\frac{E_1 - E_0}{3k_B T_0}\right) a. \quad (13.50)$$

Now one can find the energy that optimizes the upward jump rate $\nu_\uparrow$, and the result is that the fastest hops occur in the vicinity of a certain level, independent of the energy of the starting site, called the *transport energy* E_{tr}, given by (Baranovskii et al., 1995; Hartenstein and Bässler, 1995)

$$E_{tr} = E_0 - \Delta E_{tr}. \quad (13.51)$$

where

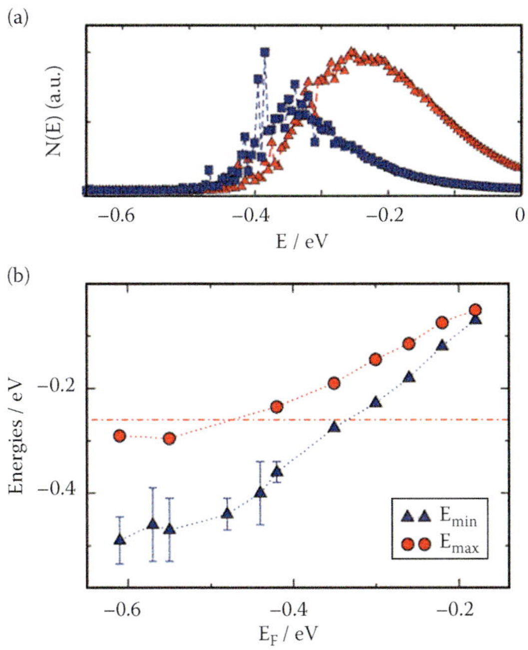

FIGURE 13.17 (a) Histograms of the target energies of hopping in an exponential DOS (T = 275 K, T_0 = 800 K and α_l = 0.5 nm) with spatial disorder. Shown are the target sites (squares) and the same without considering backward jumps between pairs of sites. (b) Energy of the most probable jump (triangles) and estimation of the effective transport energy, E_{tr} (circles) as a function of the Fermi level. The former are extracted from the maxima of the energy histograms, whereas the latter are extracted from the maxima of the corrected histograms with backward jumps between pairs of sites removed. The horizontal line represents the classical analytical value of the transport energy. (Reproduced with permission from González-Vázquez et al., 2009.)

$$\Delta E_{tr} = 3k_BT_0 \ln\left[\frac{3\alpha_l T_0}{2aT}\left(\frac{4\pi}{3}\right)^{1/3}\right]. \tag{13.52}$$

Including Equation 13.52 in Equation 13.50 yields the average jump distance

$$r\left(E_{tr}\right) = \frac{3T_0}{2T}\alpha_l. \tag{13.53}$$

The occurrence of the effective transport level effectively reduces the hopping transport to multiple trapping, with E_{tr} playing the role of the mobility edge. Stochastic simulations to check the validity of the approximation are shown in Figure 13.17.

13.10 VARIABLE RANGE HOPPING

An important class of hopping models makes an evaluation of the competition between hopping at different energies and at different distances. These types of models consider that the hopping from a particular site takes place over a variable distance to the most favorable energy site. This class is generally termed *variable range hopping* (VRH) and was first suggested by Mott (1968). The calculation optimizes the hopping rate of a single hop from one site to another under the constraint that at least one such hop is possible. Later, more systematic approaches were developed based on percolation arguments, the so-called critical path analysis (Shklovskii and Efros, 1984; Böttger and Bryksin, 1985), presented in Equation 13.49. All of these models lead to a dependence of conductivity on the temperature of the type

$$\sigma_n = \sigma_0 e^{-(T_1/T)_p} \tag{13.54}$$

where the parameter $p = (d+1)^{-1}$ takes the values ¼, ⅓, ½, depending on the dimensionality d of the problem, and

$$k_BT_1 = \frac{c_d}{g\left(E_F\right)}\left(\frac{2}{\alpha_l}\right)^d, \tag{13.55}$$

where $g(E_F)$ is the (constant) density of states around the Fermi level and c_d is a constant.

To derive the above expressions, let us consider stationary transport of electrons in a wide distribution of states at low temperatures. Since the energy for thermal activation is scarce, the localized states that contribute the most to the hopping transport are those close to the Fermi level. The VRH model makes the simple assumption that the conductivity is realized by the localized states in a stripe of width $2E_0$ centered at E_F, and thus it is essentially proportional to an average hopping rate. The effective concentration of states for hopping is

$$N\left(\varepsilon_0\right) = 2g\left(E_F\right)E_0 \tag{13.56}$$

and the average distance between states is given by

$$r_{ij} = \left[N\left(E_0\right)\right]^{-1/3}. \tag{13.57}$$

The resistivity $\rho = \sigma^{-1}$ adopts the following form

$$\rho = \rho_0 \exp\left[2\frac{1}{\alpha_l N\left(E_0\right)^{1/3}} + \frac{E_0}{k_BT}\right]. \tag{13.58}$$

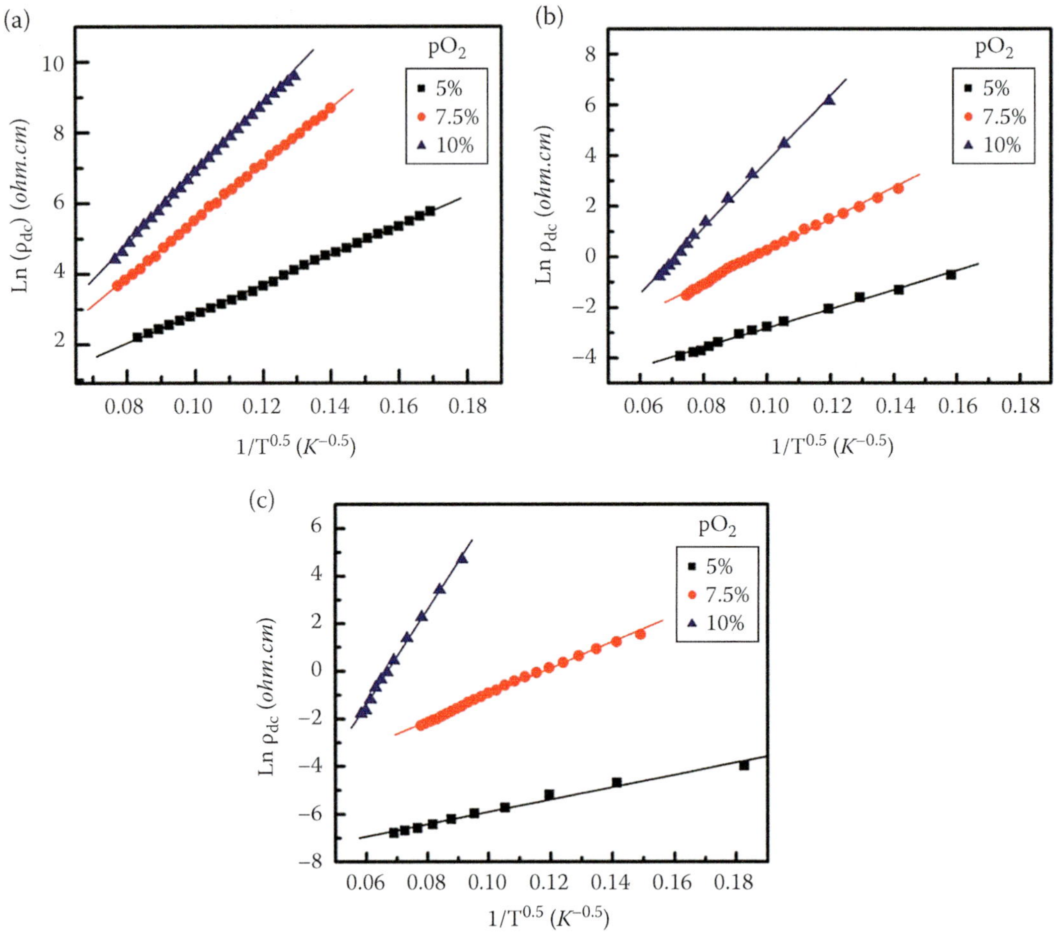

FIGURE 13.18 Resistivity (ρ) dependence on temperature (T) for VO_x thin films, as a function of oxygen partial pressure during deposition at different temperatures (a) 40°C, (b) 200°C, and (c) 300°C. Solid lines are least-square fits to the data. (Reproduced with permission from Bharadwaja et al., 2009.)

We observe that the hopping rate becomes strongly dependent on the size of the stripe E_0 where hopping occurs. When E_0 becomes large, thermal activation becomes more difficult and the conductivity decreases. But if E_0 increases, there are more states for tunneling. There is a competition between the two terms in the exponential of Equation 13.58, and the electron hopping rate is optimized for

$$E_0 = \frac{(k_B T)^{3/4}}{\left[g(E_F)\alpha_l^3\right]^{1/4}}. \tag{13.59}$$

Including this result in Equation 13.58 gives the expression of the resistivity in VRH

$$\rho = \rho_0 \exp\left\{\frac{2}{\left[g(E_F)\alpha_l^3\right]^{1/4}} \frac{1}{(k_B T)^{1/4}}\right\}. \tag{13.60}$$

VRH has been observed in a wide variety of systems such as Siand Ge-based inorganic semiconductors (Pollak and Shklovskii, 1991), conducting polymers, and assemblies of quantum dots (Yu et al., 2003; Liu et al., 2010). (Efros and Shklovskii, 1975) showed that the Coulomb interaction would open a soft gap in the DOS, which leads to the power $p = 1/2$ in the conductivity. Figure 13.16 shows that VRH provides a good approximation to the hopping conductivity at high carrier density. The data of resistivity dependence on temperature in Figure 13.18 show the VRH dependence following the Efros and Shklovskii model with the value $p = 1/2$. Figure 13.19 shows the validity of different approximations to the hopping conduction for a two-dimensional solid formed by CdSe colloidal quantum dots. For a classification of the transport regimes in the Gaussian distribution, see Baranovskii (2014).

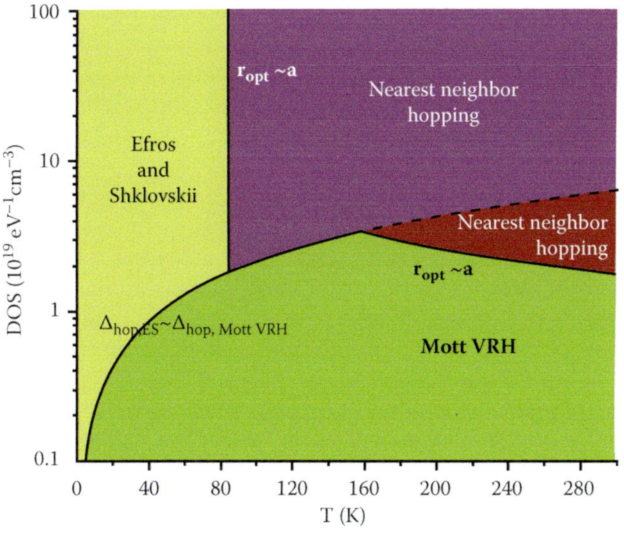

FIGURE 13.19 Phase diagram of electron transport in monodispersed CdSe colloidal quantum dot 3D solids. Indicated are the domains of the Efros and Shklovskii (ES) Coulomb gap model, Mott VRH, and the nearest neighbor hopping that takes place when the optimized hopping distance is of the order of the nearest neighbor distance. (Reproduced with permission from Liu et al., 2010.)

GENERAL REFERENCES

Multiple trapping transport: Hoesterey and Letson, 1963; Rose, 1963; Orenstein and Kastner, 1981; Tiedje and Rose, 1981; Bisquert, 2008; Bisquert, 2008; Tyutnev et al., 2014.

Generalized Einstein relation: Baranovskii et al., 1998; Guidoni and Aldao, 2002; Roichman and Tessler, 2002; Nguyen and O'Leary, 2003; Peng et al., 2005; van de Lagemaat, 2005; Neumann et al., 2006.

Einstein relation in exponential distribution: Ritter et al., 1988; Arkhipov and Adriaenssens, 1996; Baranovskii et al., 1996; Gu et al., 1996; Nguyen and O'Leary, 2003; Nguyen and O'Leary, 2005.

Hopping transport: Shklovskii and Efros, 1984; Böttger and Bryksin, 1985; Monroe, 1985; Shapiro and Adler, 1985; Bässler, 1993; Vissenberg and Matters, 1998; Coehoorn et al., 2005; Bisquert, 2008; Tessler et al., 2009; Coehoorn and Bobbert, 2012.

Hopping in organic materials: Bässler, 1993; Vissenberg and Matters, 1998; Gruhn et al., 2002; Arkhipov et al., 2003; Arkhipov et al., 2003; Coehoorn et al., 2005; Hutchison et al., 2005; Fishchuk et al., 2007; Nagata and Lennartz, 2008; Baranovskii, 2014.

The percolation criteria in hopping transsport: Miller and Abrahams, 1960; Ambegaokar et al., 1971; Shklovskii and Efros, 1971; Pollak, 1972; Shklovskii and Efros, 1984; Böttger and Bryksin, 1985; Zvyagin, 2008.

Multiple trapping in time domain: Noolandi, 1977; Orenstein and Kastner, 1981; Tiedje et al., 1981; Tiedje and Rose, 1981; Michiel et al., 1986; Monroe and Kastner, 1986.

The transport energy concept and simulations: Monroe, 1985; Baranovskii et al., 1997; Arkhipov et al., 2001; Arkhipov et al., 2001; Nguyen et al., 2007; Nikitenko and von Seggern, 2007; Iacchetti et al., 2012; Mendels and Tessler, 2013.

REFERENCES

Abayev, I.; Zaban, A.; Fabregat-Santiago, F.; Bisquert, J. Electronic conductivity in nanostructured TiO_2 films permeated with electrolyte. *Physics Status Solidii (a)* 2003, *196*, R4–R6.

Ala-Nissila, T.; Ferrando, R.; Ying, S. C. Collective and single particle diffusion on surfaces. *Advances in Physics* 2002, *51*, 949–1078.

Ambegaokar, V.; Halperin, B. I.; Langer, J. S. Hopping conductivity in disordered systems. *Physical Review B* 1971, *4*, 2612–2620.

Ansari-Rad, M.; Anta, J. A.; Bisquert, J. Interpretation of diffusion and recombination in nanostructured and energy disordered materials by stochastic quasiequilibrium simulation. *The Journal of Physical Chemistry C* 2013, *117*, 16275–16289.

Anta, J. A. Random walk numerical simulation for solar cell applications. *Energy & Environmental Science* 2009, *2*, 387–392.

Anta, J. A. Electron transport in nanostructured metal-oxide semiconductors. *Current Opinion in Colloid & Interface Science* 2012, 17, 124–131.

Anta, J. A.; Mora-Seró, I.; Dittrich, T.; Bisquert, J. Interpretation of diffusion coefficients in nanostructured materials from random walk numerical simulation. *Physical Chemistry Chemical Physics* 2008, *10*, 4478–4485.

Arkhipov, V. I.; Adriaenssens, G. J. Low-temperature relaxations of charge carriers in disordered hopping systems. *Journal of Physics: Condensed Matter* 1996, *8*, 7909–7916.

Arkhipov, V. I.; Emelianova, E. V.; Adriaenssens, G. J. Effective transport energy versus the energy of most probable jumps in disordered hopping systems. *Physical Review B* 2001a, *64*, 125125.

Arkhipov, V. I.; Heremans, P.; Emelianova, E. V.; Adriaenssens, G. J. SCLC in materials with Gaussian energy distribution of localized states. *Applied Physics Letters* 2001b, *79*, 4154.

Arkhipov, V. I.; Heremans, P.; Emelianova, E. V.; Adriaenssens, G. J.; Bässler, H. Charge carrier mobility in doped semiconducting polymers. *Applied Physics Letters* 2003a, *82*, 3245.

Arkhipov, V. I.; Heremans, P.; Emelianova, E. V.; Adriaenssens, G. J.; Bässler, H. Equilibrium trap-controlled and hopping transport of polarons in disordered materials. *Chemical Physics* 2003b, *288*, 51.

Baranovskii, S. D. Theoretical description of charge transport in disordered organic semiconductors. *Physica Status Solidi (b)* 2014, *251*, 487–525.

Baranovskii, S. D.; Faber, T.; Hensel, F.; Thomas, P. The applicability of the transport-energy concept to various disordered materials. *Journal of Physics: Condensed Matter* 1997, *9*, 2699.

Baranovskii, S. D.; Faber, T.; Hensel, T.; Thomas, P. On the Einstein relation for hopping electrons. *Journal of Non-Crystalline Solids* 1998, *227–230*, 158–161.

Baranovskii, S. D.; Faber, T.; Hensel, T.; Thomas, P.; Adriaenssens, G. J. Einstein's relationship for hopping electrons. *Journal of Non-Crystalline Solids* 1996, *198–200*, 214–217.

Baranovskii, S. D.; Thomas, P.; Adriaenssens, G. J. The concept of transport energy and its application to steady-state photoconductivity in amorphous silicon. *Journal of Non- Crystalline Solids* 1995, *190*, 283–287.

Bässler, H. Charge transport in disordered organic photoconductors. *Physics Status Solidii (b)* 1993, *175*, 15–56.

Bharadwaja, S. S. N.; Venkatasubramanian, C.; Fieldhouse, N.; Ashok, S.; Horn, M. W.; Jackson, T. N. Low temperature charge carrier hopping transport mechanism in vanadium oxide thin films grown using pulsed dc sputtering. *Applied Physics Letters* 2009, *94*, 222110.

Bisquert, J. Chemical diffusion coefficient in nanostructured semiconductor electrodes and dye-sensitized solar cells. *The Journal of Physical Chemistry B* 2004, *108*, 2323–2332.

Bisquert, J. Interpretation of a fractional diffusion equation with nonconserved probability density in terms of experimental systems with trapping or recombination. *Physical Review E* 2005, *72*, 011109.

Bisquert, J. Beyond the quasi-static approximation: Impedance and capacitance of an exponential distribution of traps. *Physical Review B* 2008a, *77*, 235203.

Bisquert, J. Interpretation of electron diffusion coefficient in organic and inorganic semiconductors with broad distributions of states. *Physical Chemistry Chemical Physics* 2008b, *10*, 3175–3194.

Bisquert, J. Physical electrochemistry of nanostructured devices. *Physical Chemistry Chemical Physics* 2008c, *10*, 49–72.

Bisquert, J.; Fabregat-Santiago, F.; Mora-Seró, I.; Garcia-Belmonte, G.; Giménez, S. Electron lifetime in dye-sensitized solar cells: Theory and interpretation of measurements. *The Journal of Physical Chemistry C* 2009, *113*, 17278–17290.

Bisquert, J.; Marcus, R. A. Device modeling of dye-sensitized solar cells. *Topics in Current Chemistry* 2014, *352*, 325–396.

Bisquert, J.; Vikhrenko, V. S. Analysis of the kinetics of ion intercalation. Two state model describing the coupling of solid state ion diffusion and ion binding processes. *Electrochimica Acta* 2002, *47*, 3977–3988.

Bisquert, J.; Vikhrenko, V. S. Interpretation of the time constants measured by kinetic techniques in nanostructured semiconductor electrodes and dye-sensitized solar cells. *The Journal of Physical Chemistry B* 2004, *108*, 2313–2322.

Böttger, H.; Bryksin, V. V. *Hopping Conduction in Solids*; Akademie Verlag: Berlin, 1985.

Bulnes, F.; Ramirez-Pastor, A. J.; Riccardo, J. L.; Zgrablich, G. Diffusion of interacting particles in one-dimensional heterogeneous systems. *Surface Science* 2006, *600*, 1917–1923.

Coehoorn, R. Hopping mobility of charge carriers in disordered organic host-guest systems: Dependence on the carrier concentration. *Physical Review B* 2007, *75*, 155203.

Coehoorn, R.; Bobbert, P. A. Effects of Gaussian disorder on charge carrier transport and recombination in organic semiconductors. *Physica Status Solidi (a)* 2012, *209*, 2354–2377.

Coehoorn, R.; Pasveer, W. F.; Bobbert, P. A.; Michels, C. J. Charge-carrier concentration dependence of the hopping mobility in organic materials with Gaussian disorder. *Physical Review B* 2005, *72*, 155206.

Efros, A. L.; Shklovskii, B. I. Coulomb gap and low temperature conductivity of disordered systems. *Journal of Physics C* 1975, *8*, L49–L51.

Fishchuk, I. I.; Arkhipov, V. I.; Kadashchuk, A.; Heremans, P.; Bässler, H. Analytic model of hopping mobility at large charge carrier concentration in disordered organic semiconductors: Polarons versus bare charge carriers. *Physical Review B* 2007, *76*, 045210.

Fisher, A. C.; Peter, L. M.; Ponomarev, E. A.; Walker, A. B.; Wijayantha, K. G. U. Intensity dependence of the back reaction and transport of electrons in dye-sensitized nanocrystalline TiO_2 solar cells. *The Journal of Physical Chemistry B* 2000, *104*, 949–958.

Göbel, E. O.; Graudszus, W. Optical detection of multiple-trapping relaxation in disordered crystalline semiconductors. *Physical Review Letters* 1982, *48*, 1277–1280.

González-Vázquez, J. P.; Anta, J. A.; Bisquert, J. Random walk numerical simulation for hopping transport at finite carrier concentrations: Diffusion coefficient and transport energy concept. *Physical Chemistry Chemical Physics* 2009, *11*, 10359–10367.

Groda, Y. G.; Lasovsky, R. N.; Vikhrenko, V. S. Equilibrium and diffusion properties of two-level lattice systems: Quasi-chemical and diagrammatic approximation versus Monte Carlo simulation results. *Solid State Ionics* 2005, *176*, 1675–1680.

Gruhn, N. E.; da Silva Filho, D. A.; Bill, T. G.; Malagoli, M.; Coropceanu, V.; Kahn, A.; Bredas, J.-L. The vibrational reorganization energy in pentacene: Molecular influences on charge transport. *Journal of the American Chemical Society* 2002, *124*, 7918–7919.

Grünewald, M.; Thomas, P. A hopping model for activated charge transport in amorphous silicon. *Physics Status Solidii (b)* 1979, *94*, 125–133.

Gu, Q.; Schiff, E. A.; Grebner, S.; Wang, F.; Schwartz, R. Non-Gaussian transport measurements and the Einstein relation in amorphous silicon. *Physical Review Letters* 1996, *76*, 3196.

Guidoni, S. E.; Aldao, C. M. On diffusion, drift and the Einstein relation. *European Journal of Physics* 2002, *23*, 395–402.

Hartenstein, B.; Bässler, H. Transport for hopping in a Gaussian density of states distribution. *Journal of Non-Crystalline Solids* 1995, *190*, 112–116.

Hoesterey, D. C.; Letson, G. M. The trapping of photocarriers in anthracene. *Journal of Physical Chemistry of Solids* 1963, *24*, 1609.

Hutchison, G. R.; Ratner, M. A.; Marks, T. J. Hopping transport in conductive heterocyclic oligomers: Reorganization energies and substituent effects. *Journal of the American Chemical Society* 2005, *127*, 2339–2350.

Iacchetti, A.; Natali, D.; Binda, M.; Beverina, L.; Sampietro, M. Hopping photoconductivity in an exponential density of states. *Applied Physics Letters* 2012, *101*, 103307.

Liu, H.; Pourret, A.; Guyot-Sionnest, P. Mott and Efros-Shklovskii variable range hopping in CdSe quantum dots films. *ACS Nano* 2010, *4*, 5211–5216.

Mendels, D.; Tessler, N. The topology of hopping in the energy domain of systems with rapidly decaying density of states. *The Journal of Physical Chemistry C* 2013, *117*, 24740–24745.

Michiel, H.; Adriaenssens, G. J.; Davis, E. A. Extended-state mobility and its relation to the tail-state distribution in a-Si:H. *Physical Review B* 1986, *34*, 2486–2499.

Miller, A.; Abrahams, S. Impurity conduction at low concentrations. *Physical Review* 1960, *120*, 745.

Monroe, D. Hopping in exponential band tails. *Physical Review Letters* 1985, *54*, 146–149.

Monroe, D.; Kastner, M. A. Exactly exponential band tail in a glassy semiconductor. *Physical Review B* 1986, *33*, 8881–8884.

Moon, J. A.; Tauc, J.; Lee, J. K.; Schiff, E. A.; Wickboldt, P.; Paul, W. Femtosecond photomodulation spectroscopy of a-Si: H and a-Si:Ge: H alloys in the midinfrared. *Physical Review B* 1994, *50*, 10608–10618.

Mott, N. F. Conduction in glasses containing transition metal ions. *Journal of Non- Crystalline Solids* 1968, *1*, 1–17.

Mott, N. F.; Twose, W. D. The theory of impurity conduction. *Advances in Physics* 1961, *10*, 107–163.

Nagata, Y.; Lennartz, C. Atomistic simulation on charge mobility of amorphous tris(8- hydroxyquinoline) aluminum (Alq3): Origin of Poole-Frenkel-type behavior. *The Journal of Chemical Physics* 2008, *129*, 034709.

Neumann, F.; Genenko, Y. A.; von Seggem, H. The Einstein relation in systems with trap-controlled transport. *Journal of Applied Physics* 2006, *99*, 013704.

Nguyen, T. H.; O'Leary, S. K. Generalized Einstein relation for disordered semiconductors with exponential distributions of tail states and square-root distributions of band states. *Applied Physics Letters* 2003, *83*, 1998–2000.

Nguyen, T. H.; O'Leary, S. K. Einstein relation for disordered semiconductors: A dimensionless analysis. *Journal of Applied Physics* 2005, *98*, 076102.

Nguyen, T. H.; Schmeits, M.; Loebl, H. P. Determination of charge-carrier transport in organic devices by admittance spectroscopy: Application to hole mobility in α-NPD. *Physical Review B* 2007, *75*, 075307.

Nikitenko, V. R.; von Seggern, H. Nonequilibrium transport of charge carriers and transient electroluminescence in organic light-emitting diodes. *Journal of Applied Physics* 2007, *102*, 103708.

Noolandi, J. Multiple-trapping model of anomalous transit-time dispersion in a-Se. *Physical Review B* 1977, *16*, 4466.

Orenstein, J.; Kastner, M. Photocurrent transient spectroscopy: Measurement of the density of localized states in a-As_2Se_3. *Physical Review Letters* 1981, *46*, 1421–1424.

Peng, Y. Q.; Sun, S.; Song, C. A. Generalization of Einstein relation for organic semiconductor thin films. *Materials Science in Semiconductor Processing* 2005, *8*, 525–530.

Peng, Y. Q.; Yang, J.-H.; Lu, F.-P. Generalization of Einstein relation for doped organic semi-conductors. *Applied Physics A* 2006, *83*, 305–311.

Pollak, M. A percolation treatment of dc hopping conduction. *Journal of Non-Crystalline Solids* 1972, *11*, 1–24.

Pollak, M.; Shklovskii, B. *Hopping Transport in Solids.* Holland: Amsterdam, 1991.

Reed, D. A.; Ehrlich, G. Surface diffusion, atomic jump rates and thermodynamics. *Surface Science* 1981, *102*, 588–609.

Ritter, D.; Zeldov, E.; Weiser, K. Ambipolar transport in amorphous semiconductors in the lifetime and relaxation-time regimes investigated by the steady-state photocarrier grating technique. *Physical Review B* 1988, *38*, 8296.

Roichman, Y.; Tessler, N. Generalized Einstein relation for disordered semiconductors: Implications for device performance. *Applied Physics Letters* 2002, *80*, 1948.

Rose, A. *Concepts in Photoconductivity and Allied Problems.* Interscience: New York, 1963.

Schiff, E. A. Trap-controlled dispersive transport and exponential band tails in amorphous silicon. *Physical Review B* 1981, *24*, R6189–R6192.

Schiff, E. A. Drift-mobility measurements and mobility edges in disordered silicons. *Journal of Physics: Condensed Matter* 2004, *16*, S5265.

Shapiro, F. R.; Adler, D. Equilibrium transport in amorphous semiconductors. *Journal of Non-Crystalline Solids* 1985, *74*, 189–194.

Shklovskii, B. I.; Efros, A. L. Impurity band and conductivity of compensated semiconductors. *Soviet Physics JETP* 1971, *33*, 468–474.

Shklovskii, B. I.; Efros, A. L. *Electronic Properties of Doped Semiconductors*; Springer: Heidelberg, 1984.

Spear, W. E. Drift mobility techniques for the study of electrical transport properties in insulating solids. *Journal of Non-Crystalline Solids* 1969, *1*, 197.

Tanase, C.; Meijer, E. J.; Blom, P. W. M.; de Leeuw, D. M. Unification of the hole transport in polymeric field effect-transistors and LED. *Physical Review Letters* 2003, *91*, 216601.

Tessler, N.; Preezant, Y.; Rappaport, N.; Roichman, Y. Charge transport in disordered organic materials and its relevance to thin-film devices: A tutorial review. *Advanced Materials* 2009, *21*, 2741–2761.

Tessler, N.; Roichman, Y. Amorphous organic molecule/polymer diodes and transistors: Comparison between predictions based on Gaussian or exponential density of states. *Organic Electronics* 2005, *6*, 200–210.

Tiedje, T.; Cebulka, J. M.; Morel, D. L.; Abeles, B. Evidence for exponential bandtails in amorphous hydrogenated silicon. *Physical Review Letters* 1981, *46*, 1425–1428.

Tiedje, T.; Rose, A. A physical interpretation of dispersive transport in disordered semiconductors. *Solid State Communications* 1981, *37*, 49.

Tyutnev, A. P.; Ikhsanov, R. S.; Novikov, S. V. Comparison of the time of flight current shapes predicted by hopping and multiple trapping models. *Chemical Physics* 2014, *440*, 1–7.

van de Lagemaat, J. Einstein relation for electron diffusion on arrays of weakly coupled quantum dots. *Physical Review B* 2005, *72*, 235319.

Vissenberg, M. C. J. M.; Matters, M. Theory of the field-effect mobility in amorphous organic transistors. *Physical Review B* 1998, *57*, 12964–12967.

Wang, Q.; Ito, S.; Grätzel, M.; Fabregat-Santiago, F.; Mora-Seró, I.; Bisquert, J.; Bessho, T.; et al. Characteristics of high efficiency dye-sensitized solar cells. *The Journal of Physical Chemistry B* 2006, *110*, 19406–19411.

Yildiz, A.; Serin, N.; Serin, T.; Kasap, M. Crossover from nearest-neighbor hopping conduction to Efros-Shklovskii variable-range hopping conduction in hydrogenated amorphous silicon films. *Japanese Journal of Applied Physics* 2009, *48*, 111203.

Yu, D.; Wang, C.; Guyot-Sionnest, P. n-Type conducting CdSe nanocrystal solids. *Science* 2003, *300*, 1277.

Yu, D.; Wang, C.; Wehrenberg, B. L.; Guyot-Sionnest, P. Variable range hopping conduction in semiconductor nanocrystal solids. *Physical Review Letters* 2003, *92*, 216802.

Zvyagin, I. P. A percolation approach to the temperature and charge carrier concentration dependence of the hopping conductivity in organic materials. *Physica Status Solidi (c)* 2008, *5*, 725–729.

14 Thin Film Transistors

Field effect transistors are the main components of electronic circuits and are used in many applications such as sensors and light-emitting diodes (LEDs). In this chapter, we review the basic characteristics of thin film transistors based on organic materials that form a key part of organic electronics. We first describe their general structure and operating principles. The transistor is a three-terminal device in which the current that flows for two terminals is controlled by the third terminal. This configuration is also used for the determination of the electronic properties of organic conductors such as the mobility and density of states. Supported by energy diagrams that show the effects of manipulation of voltages, we will examine in detail the determination of the charge density and the mobility of carriers in the organics. We analyze specific models of concentration-dependent mobility and we discuss a special kind of electrochemical gating useful for nanostructured materials.

14.1 ORGANIC THIN FILM TRANSISTORS

A thin film transistor (TFT) is a three-terminal device, formed by a thin active film deposited on top of an insulating (normally a metal-oxide) layer that separates the bottom electrode (the *gate* G) from the active film, which is denoted as the *channel*; see Figure 14.1. The active film has two contacts, the source (S) and the drain (D). The application of bias voltage V_{ds} between these contacts allows us to measure the current along the channel. The fundamental advantage of the three-contact structure is that the carrier density in the film is controlled by the voltage applied below the substrate, the gate voltage, with respect to the source contact, V_{gs}, denoted simply V_g; see Figure 14.1. The conductivity of the channel can be changed by orders of magnitude, from an "off" state below a given *threshold voltage* V_t in which there is no conduction at all, to a highly conducting state. We have already observed in Section 8.5 that the properties of organic conductors can vary widely, from intrinsic and insulating to highly doped, from crystalline to disordered, or formed by a polymer and even a blend of electron and hole conductor materials. To analyze the main features of transistor operation, we will consider two types of situations.

In the first case, the organic film has properties similar to an inorganic semiconductor, with an established degree of doping and a well-developed band bending, as indicated in Figure 14.2 and in Figure 9.18. This behavior corresponds to a conventional inorganic transistor such as the metal-insulator-semiconductor field effect transistor (MISFET). In this model, the energy diagram is similar to a standard metal-insulator-semiconductor (MIS) capacitor (Sze, 1981), see Figure 9.20, although the TFT is a variable resistor from D to S. In the case of an n-type semiconductor material, a positive gate voltage produces an accumulation of electrons in the space charge region at the semiconductor–insulator contact, while a negative gate voltage produces depletion of electrons and further negative bias will lead to the appearance of holes at the contact by the formation of an inversion layer. A more detailed analysis must consider the electrostatics of the semiconductor space charge, depending on V_g (Yong et al., 2011). The three possible regimes of gate voltage in a standard p/type metal-oxide-semiconductor (MOS) transistor are shown in Figure 9.18. The accumulation of carriers near the interface and the depletion characteristics can be detected by capacitance techniques (Hyuk-Ryeol et al., 1998). The normal mode of operation of a conventional MISFET results from the formation of a minority-carrier channel in the strong inversion regime; see Figures 9.19 and 9.20. The threshold voltage required for the onset of current conduction in the channel is that at which the Fermi level at the insulator–semiconductor interface crosses the middle of the gap.

The second type of transistor that we will discuss is a very thin layer, commonly denoted as a thin film transistor (TFT) (see Figures 14.3 and 14.4). Now, there is no band-bending structure and the carrier density changes homogeneously in the active film by the gate voltage as shown in Figure 14.4, which assumes perfect ohmic S and D contacts. This configuration occurs in nm-thin organic layers deposited by evaporation or a 2D layer of one atom thickness (Figure 4.3), and also in thick, porous layers measured in the electrochemical transistor that is explained in Section 14.6. The gate voltage modifies the position of the energy levels of the organic layer with respect to the Fermi level (Figure 14.3), which remains at the same level at S and D electrodes. A positive V_g displaces the energy level downward and increases the electron density in the organic layer because the conduction band or lowest unoccupied molecular orbital (LUMO) level approaches the Fermi level (Figure 14.3b). Conversely, a negative gate voltage shifts the organic levels up and favors the increase

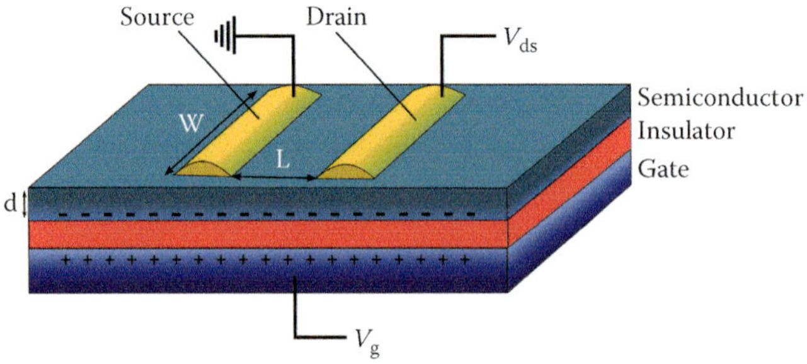

FIGURE 14.1 Schematic geometry and voltages of a thin film transistor. (Adapted from Newman et al., 2004.)

of positive electronic carriers, or holes (Figure 14.3c). Figure 14.3d and 14.3e show, in addition, the application of source–drain voltage that drives electrical current along the channel.

In the TFT formed with organic materials or amorphous semiconductors, the enhanced current is due to majority carrier formation. Although there is no depletion layer, a threshold voltage is also observed due to effects such as trap filling and Fermi level mismatch between contacts and the organic layer (Horowitz et al., 1998). We can determine the density of states (DOS) based on the fact that the change of the gate voltage induces a displacement of the Fermi level in the gap of the organic conductor, as discussed in Section 14.3. This configuration has been employed to explore the remarkable properties correlated with 2D ultrathin atomic layer structure materials like graphene or the family of monolayer transition-metal dichalcogenides. These materials offer excellent electrostatic control, which makes them candidates for the next generation of transistors or similar devices. Graphene is the most well-known material of this kind and the measurements have been discussed in Section 7.7 in connection with the chemical capacitance. Figure 14.5 shows the dynamics of electrons and hole carriers in a single-layer MoS_2 transistor under illumination. There are two main types of measurements for material characterization in the transistor configuration. The first one is capacitive and allows determination of the DOS of the thin film material in the channel. No current between S and D is needed, so that $V_{ds} = 0$ V is maintained. Second, the conductivity of the semiconductor can be varied widely by the gate voltage and measured by applying voltage $|V_{ds}| > 0$ that makes a current flow along the channel. Since the carrier density can also be established, mobilities can be obtained, as discussed in Section 14.5. Both methods require control over the carrier density in the channel that we will discuss in the following section (Figure 14.5).

14.2 CARRIER DENSITY IN THE CHANNEL

We analyze the increase of charge in the channel by the application of a gate voltage while $V_{ds} = 0$ V. The gate voltage is distributed in two contributions that are both normal to the substrate as shown in Figure 14.4b. First, a voltage drop V_{ox} occurs across the dielectric capacitor formed by the insulating oxide layer with a capacitance C_{ox} per unit area. The second part of the voltage V_L is a change of the vacuum level (VL) in the organic

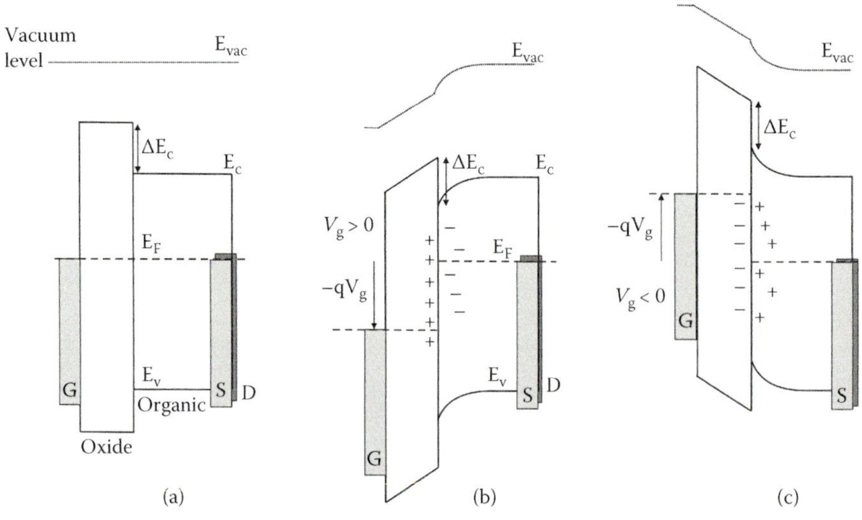

FIGURE 14.2 Energy scheme of a transistor viewed from the side, showing the energy levels in the insulator and channel, the gate electrode (G), and the source (S) and drain (D) electrodes. (a) Equilibrium without bias. (b) Positive gate voltage. (c) Negative gate voltage.

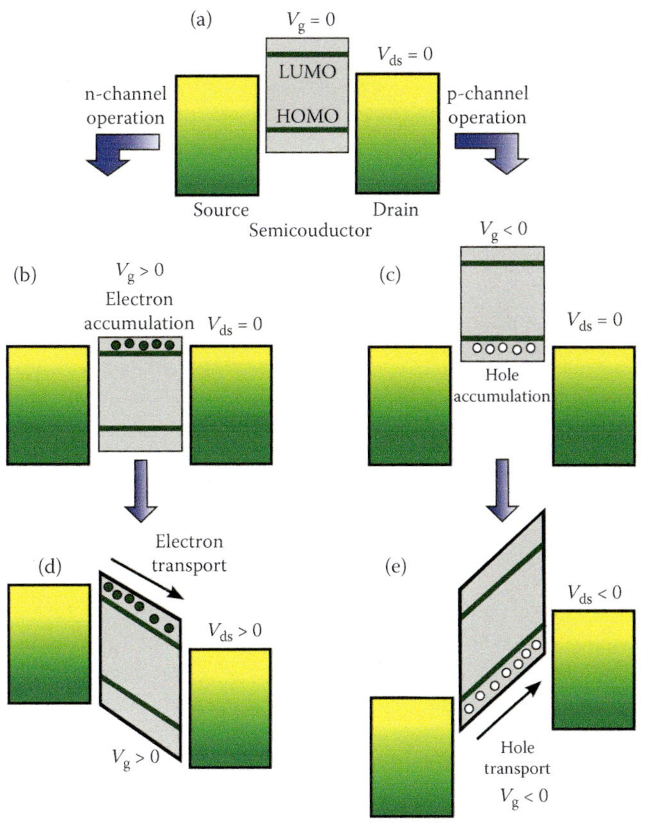

FIGURE 14.3 (a) Idealized energy level diagram of an organic TFT at $V_g = 0$ and $V_{ds} = 0$ (with a threshold voltage $V_t = 0$). (b–e) Demonstrate the principle of field effect transistor operation for the case of electron accumulation (b) and transport (d) and hole accumulation (c) and transport (e). (Adapted from Newman et al., 2004.)

layer. The actual introduction of charge in the channel usually requires overcoming a threshold voltage, V_t, as mentioned earlier. Therefore

$$V_g = V_{ox} + V_L + V_t. \tag{14.1}$$

As remarked in Figure 14.3b and 14.4b, biasing the gate positively ($V_g - V_t > 0$) is equivalent to applying a negative bias in the active film that raises the Fermi level of electrons, increasing the electron density. If n is the density of carriers and d_n is the thickness of the charged layer in the channel, the total charge is

$$Q_{ch} = nqd_n. \tag{14.2}$$

On the other hand, the charge in the capacitor plates is

$$Q_{ox} = C_{ox}V_{ox}. \tag{14.3}$$

In the absence of surface traps and other related effects, we have

$$n = \frac{C_{ox}V_{ox}}{qd_n}. \tag{14.4}$$

Therefore

$$n = \frac{C_{ox}}{qd_n}\left(V_g - V_t - V_L\right). \tag{14.5}$$

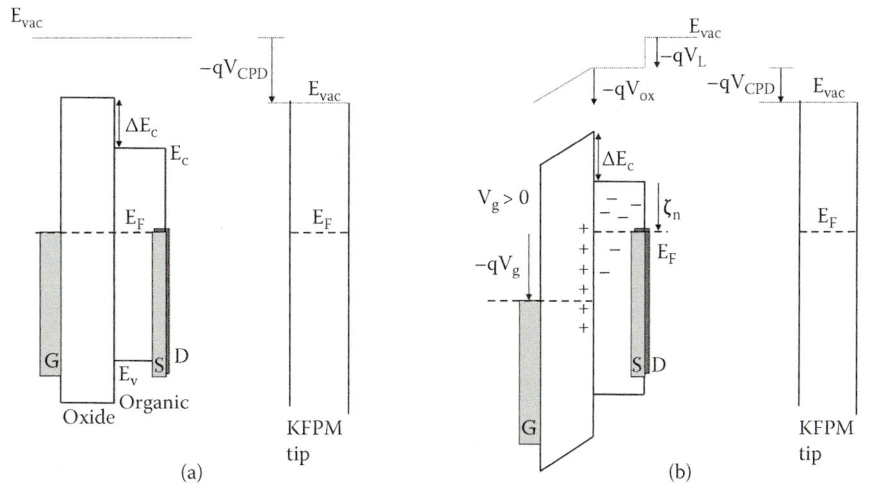

FIGURE 14.4 Energy scheme of a TFT viewed from the side, showing the energy levels in the insulator and channel, the gate electrode (G), and the source (S) and drain (D) electrodes. Also shown are the vacuum level and the Fermi level of the tip material in Kelvin probe force microscopy (KFPM), which measures the contact potential difference (CPD). The organic layer (channel) is considered very thin and its energy levels shift homogeneously. (a) Equilibrium without bias. (b) Positive gate voltage.

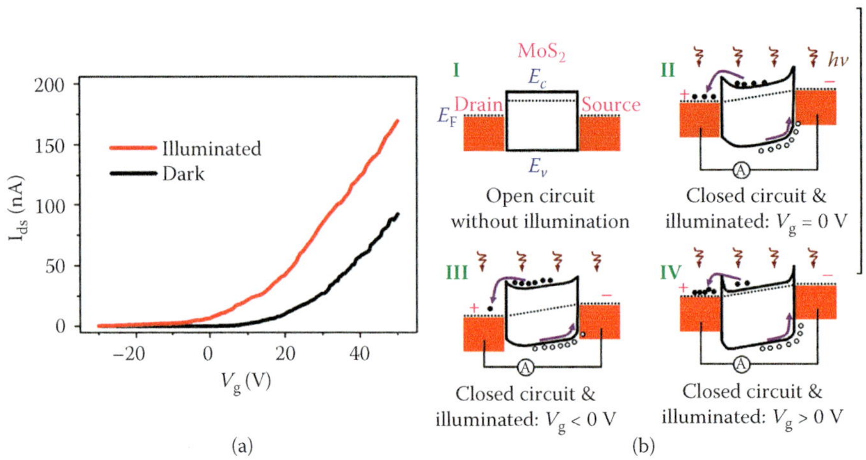

FIGURE 14.5 (a) Typical output characteristics of a single-layer MoS_2 phototransistor at gate voltage varied from −30 to 50 V. (b) Diagram of a single-layer MoS2 phototransistor circuit: Diagram I represents the initial state of the device under an open circuit without illumination. Diagrams II, III, and IV represent the device under a closed circuit (source–drain voltage V_{ds} = 1 V) and illumination while different gate voltages (V_g) were applied. The full dots at the conduction band energy level and the empty dots at the valence band energy level in MoS_2 represent the photoexcited electrons and holes respectively. At negative gate voltage, the conduction band of MoS_2 is shifted away from the Fermi level and the photoconductivity is reduced. On the contrary, at positive gate voltage, the conduction band is taken close to the Fermi level, strongly increasing the electron density and consequently the conductivity and the drift current. At the same time, the voltage V_{ds} causes current flow. (Reproduced with permission from Yin et al., 2011.)

14.3 DETERMINATION OF THE DOS IN THIN FILM TRANSISTOR CONFIGURATION

Let us assume that the band bending produced by the gate voltage is negligible. Since the D and S are grounded, the Fermi level effectively approaches the LUMO for $V_g > 0$. The shift of the Fermi level is homogeneous in the whole thickness of the organic layer d as indicated in Figure 14.4b.

In disordered or amorphous layers, the displacement of the Fermi level fills the localized states in the material (Shur and Hack, 1984). For example, the DOS derived for amorphous silicon transistors are shown in Figure 7.4. In Chapter 7, we explained in detail the significance of the determination of the chemical capacitance and in particular the relation of C_μ with the density of states of disordered materials (see Equation 7.23). Let us analyze the measurement of the chemical capacitance in the transistor configuration (Tal and Rosenwaks, 2006) (see Figure 14.6).

We recall from Section 4.3 that the Kelvin Probe (KP) methods measure the difference of the VL between the tip and the sample surface. The measurement therefore gives the contact potential difference (CPD), V_{CPD}, as indicated in Figure 14.4a. When the gate is biased, there is a change of the V_{CPD} with respect to the threshold value (defined as the onset of conduction in the channel,

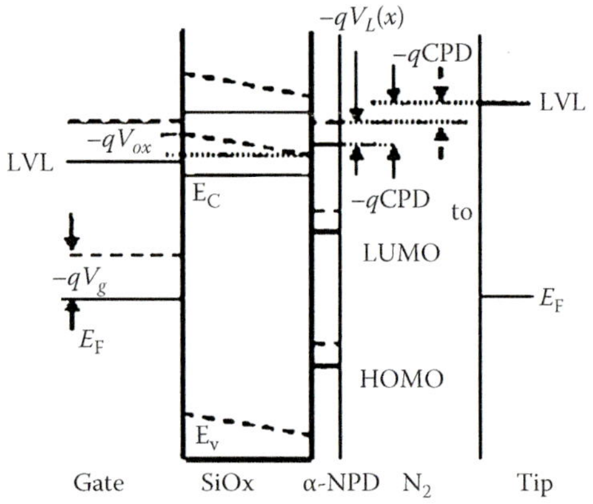

FIGURE 14.6 Qualitative energy-level scheme showing the changes of the local vacuum level (LVL) across a TFT measured by KPFM under nitrogen atmosphere far from the source and drain contacts for $V_g = V_t$ and V_{ds} = 0 V (solid curves) and $V_g < V_t$ and V_{ds} = 0 V (dashed curves) in the case of negligible band bending. (Reproduced with permission from Tal and Rosenwaks, 2006.)

V_{to}), so that V_L can be directly measured as shown in Figure 14.6:

$$V_L = V_{CPD}\left(V_g\right) - V_{CPD,to}. \tag{14.6}$$

Figure 14.7 shows the displacement of the Fermi level with respect to the Gaussian distribution under negative gate voltage, which produces increasing hole occupation

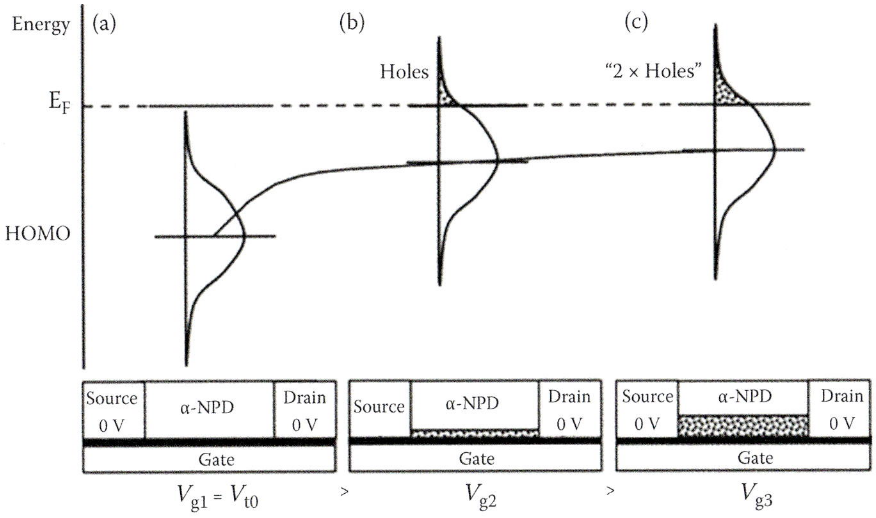

FIGURE 14.7 Top: Qualitative scheme of the hole occupation of a Gaussian DOS in α-NPD at different positions with respect to E_F. Bottom: TFT cartoons in which the relevant charge concentration is represented by a dotted region. (a) $V_g = V_t$, zero gate-induced hole concentration; (b) $V_{g2} < V_t$, thus holes accumulate to form a conducting channel; (c) $V_{g3} < V_{g2}$ such that the hole concentration is twice that in part (b). (Reproduced with permission from Tal and Rosenwaks, 2006.)

at the tail of the DOS. To obtain the chemical capacitance, we note that V_g displaces E_F in the organic layer, but the levels also shift by V_{ox}. For determination of the DOS, it is therefore necessary to obtain the amount that the Fermi level shifts with respect to the conduction band/LUMO, that is, the change of the chemical potential $\zeta_n = E_{Fn} - E_c$ (Equation 2.35) with respect to threshold value. According to Figure 14.4:

$$qV_L = \zeta_n - \zeta_{n,to}. \tag{14.7}$$

The Fermi level variation (with respect to E_c) correlates to V_L as $d\zeta_n = qdV_L$. We calculate the chemical capacitance in Equation 7.6 corresponding to the channel thickness d and we have

$$C_\mu = q^2 d \frac{dn}{d\zeta_n}. \tag{14.8}$$

Using Equation 14.5, we obtain the expression

$$C_\mu = C_{ox}\left(\frac{dV_g}{dV_L} - 1\right). \tag{14.9}$$

Therefore, a measurement of V_L with respect to V_g allows us to obtain C_μ. The results of the application of this method to a thin layer of organic conductor α-NPD are shown in Figure 14.8.

Another approach to the determination of the chemical capacitance in the transistor does not involve the

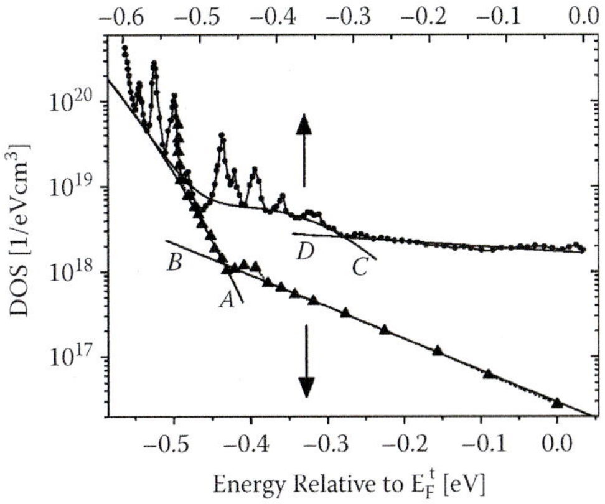

FIGURE 14.8 DOS vs. energy relative to $E_F^t\left(E_F \text{ at } V_g = V_t\right)$ for undoped (solid triangles) and doped (solid circles) α-NPD. The solid curves are fttings of a Gaussian function (curve A) and an exponential function (curve B) to given ranges in the undoped sample DOS curve and similar fttings (curve *Ck* and D) to the doped sample DOS curve. (Reproduced with permission from Tal and Rosenwaks, 2006.)

contact potential difference (CPD) but rather measures the gate capacitance directly, given by

$$C_g = qd\frac{dn}{dV_g}. \tag{14.10}$$

From Equation 14.8, we obtain

$$C_\mu = C_g \frac{q}{d\zeta_n/dV_g}. \tag{14.11}$$

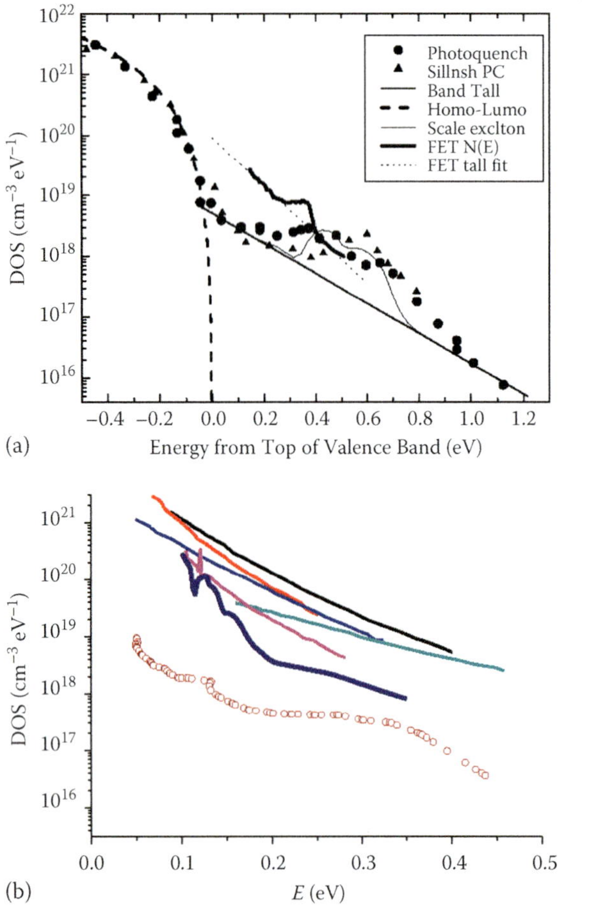

FIGURE 14.9 (a) Density of states in amorphous silicon derived by various methods. Reproduced with permission from (Lang et al., 2004) (top). (b) Density of states in a pentacene single crystal derived by various methods. (Reproduced with permission from Yogev et al., 2011.)

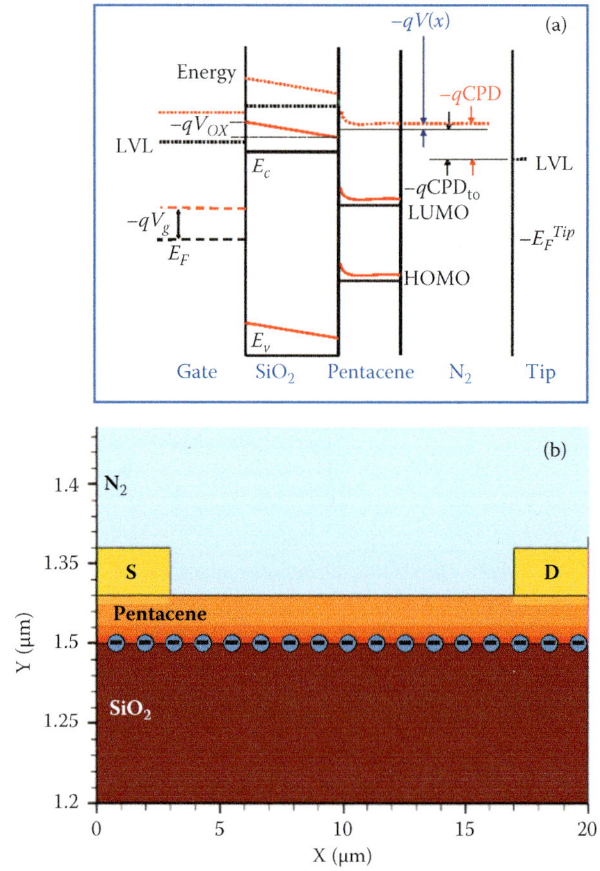

FIGURE 14.10 (a) Qualitative energy-level scheme across a pentacene FET far from the source and drain contacts for $V_g = V_t, V_{ds} = 0$ V (black lines) and for $V_g < V_t$, $V_{ds} = 0$ V (red lines). (b) Schematic device layout. (Reproduced with permission from Yogev et al., 2011.)

The oxide capacitance has the form

$$C_{ox} = qd\frac{dn}{dV_{ox}}. \tag{14.12}$$

Given the relationship $dV_g = dV_{ox} + dV_L$, the oxide capacitance and chemical capacitance are connected in series and we obtain

$$C_g = \frac{C_{ox}C_\mu}{C_{ox} + C_\mu}. \tag{14.13}$$

As mentioned earlier, resolving the chemical capacitance requires establishment of the displacement of the Fermi level with respect to the conduction band. One method to apply Equation 14.11 is to derive ζ_n from the activation energy of the conductivity in the channel (Lang et al., 2004; Puigdollers et al., 2010). Other approaches obtain the chemical capacitance directly from the measured gate capacitance

$$C_\mu = \frac{C_{ox}C_g}{C_{ox} - C_g} \tag{14.14}$$

using various techniques to determine the potential V_L (Das et al., 2008; Kichan et al., 2008; Xu et al., 2011). The application of Equation 14.14 to determine the DOS of graphene is shown in Figure 7.15. The DOS measured for amorphous silicon and organic crystals of pentacene using a variety of methods summarized by Kalb and Batlogg (2010) including Equation 14.11 is shown in Figure 14.9. Roelofs et al. (2012) determined the DOS of a self-assembled monolayer field effect transistor. Figure 14.10 shows the energy diagram and charge distribution of a pentacene TFT indicating the flat-band condition $V_g = V_{to}$ and also the negative gate bias in which holes are injected into the organic film channel from the grounded source and drain electrodes. The gate voltage produces a significant amount of band bending in the organic layer, which makes V_L smaller than in the homogeneous case. Figure 14.11 shows a hydrogenated

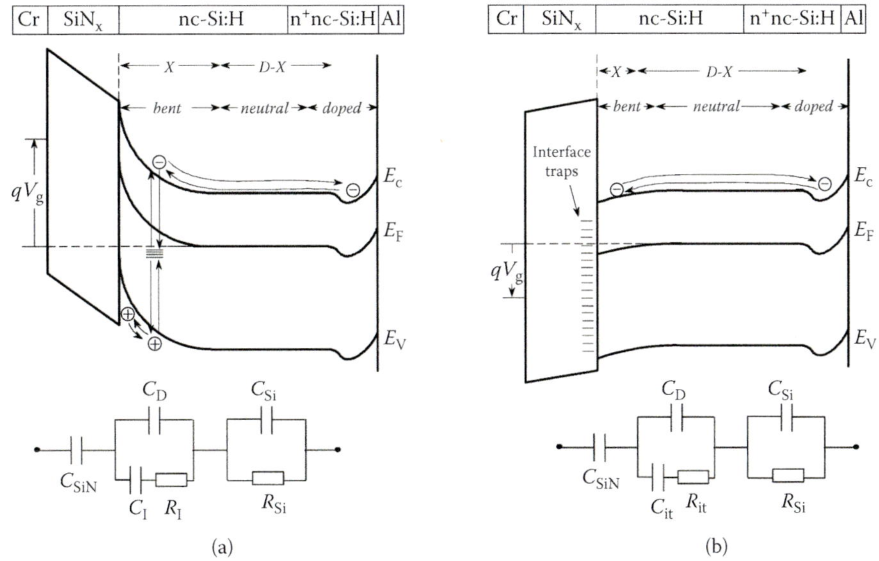

FIGURE 14.11 The energy band diagram and the equivalent circuit model of the MIS structure for the (a) inversion–depletion and (b) accumulation regimes. C_{SiN} is the insulator capacitance. C_{Si} and R_{Si} stand for the nc–Si:H layer bulk capacitance and resistance, respectively. C_D is the depletion capacitance in the depletion regime and also corresponds to the Debye capacitance in the accumulation regime. The inversion capacitance C_I (or interface state capacitance C_{it}) is parallel to C_D. R_I is the resistance due to the generation–recombination processes in the inversion regime. R_{it} is the resistance due to the capture (or emission) of electrons by (or from) the interface traps and can be determined in the inversion-free depletion regime. (Reproduced with permission from Anutgan et al., 2011.)

nanocrystalline silicon bottom-gate TFT and metal-insulator amorphous silicon (MIS) structure (Anutgan et al., 2011). The nanocrystalline–Si:H layer is thick and remains quasi-neutral so that the gate voltage changes the carrier density at the semiconductor surface, either in depletion (a) or accumulation (b).

14.4 CURRENT-VOLTAGE CHARACTERISTICS

The source–drain voltage V_{ds} induces a variable carrier distribution along the channel. Figure 14.12 schematically shows the carrier distribution in the organic layer from source to drain, which depends on the relative magnitude of drain and gate voltage. The voltage $V_L(x)$ in Equation 14.1 changes as a function of position x. If V_{ds} is much lower than the gate voltage, there is a small gradient of charge in the organic layer that drives the current from drain to source. In this domain, the current is *linear* with V_{ds}, as shown in Figure 14.13a. On the other hand, at large $V_{ds} \approx V_g - V_t$, the carrier concentration is exhausted close to the drain contact and an increase of drain voltage does not cause an increase of the current in the channel. The current enters a saturation regime. The different domains of current are illustrated in Figure 14.14. The voltage changes along the channel can be resolved by monitoring the VL by means of the contactless method of Kelvin probe force microscopy (KPFM), which provides a high-resolution determination of the surface potential, as shown previously in Figure 4.3 (see also Zhang et al., 2015).

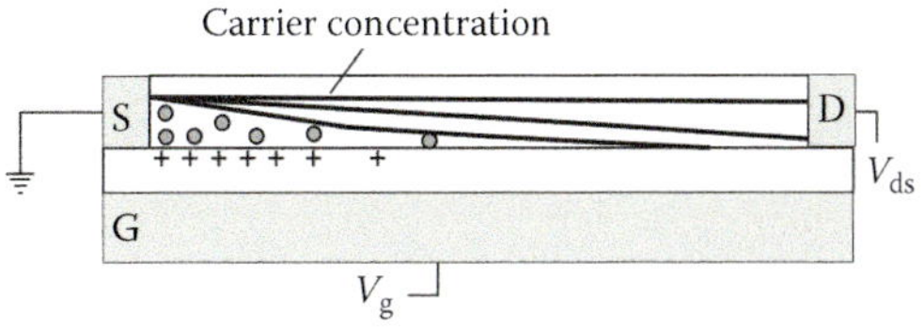

FIGURE 14.12 Carrier concentration in a TFT. When $V_g - V_t >> V_{ds}$, the carrier profile is nearly uniform. A small source–drain bias voltage V_{ds} allows measurement of the current. However, by increasing V_{ds}, the concentration becomes progressively inhomogeneous, decreasing from the source to the drain. When the source–drain voltage is comparable to the gate voltage, $V_{ds} \approx V_g - V_t$, the region close to D becomes depleted of electrons and the channel is pinched. Increasing the source–drain voltage does not increase the channel current, which attains a saturation value.

We now derive expressions for the conductance and the mobility. The channel is described by the dimensions W, d, L; see Figure 14.1. Assuming that the carrier profile is linear as indicated in Figure 14.12, we use the average value of $V_L(x)$ in Equation 14.5 and the average value of the charge is

$$n = \frac{C_{ox}}{qd}\left(V_g - V_t - \frac{V_{ds}}{2}\right). \tag{14.15}$$

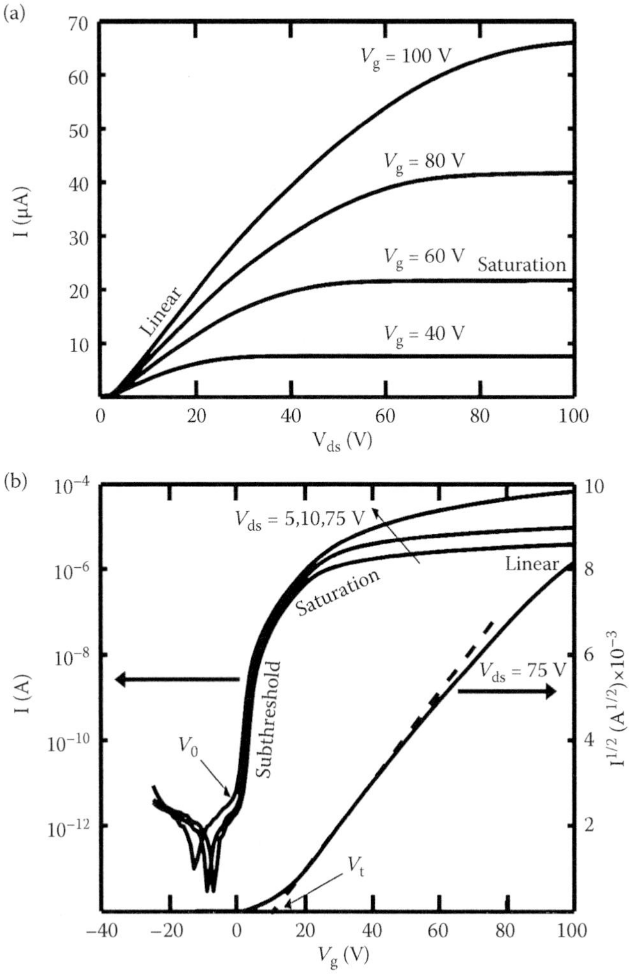

FIGURE 14.13 (a) $I_{ds} - V_{ds}$ curves for a PTCDI-C8 TFT for various values of V_g. (b) $I_d - V_g$ curves plotted on semilogarithmic axes for the same device for various values of V_{ds}. The $I^{1/2}$ vs. V_g curve for V_{ds} = 75 V is shown on the right-hand axis. (Reproduced with permission from Newman et al., 2004.)

The current density in the channel relates to the current as

$$j = \frac{I_{ds}}{dW}. \tag{14.16}$$

Applying the drift current expression

$$I_{ds} = \frac{W}{L} nqdu_n V_{ds} \tag{14.17}$$

and combining with Equation 14.15, we have the result

$$I_{ds} = \frac{W}{L} C_{ox} u_n \left[\left(V_g - V_t\right) V_{ds} - \frac{V_{ds}^2}{2} \right]. \tag{14.18}$$

In the linear regime, in which $V_{ds} << V_g$, we defne two quantities, the *conductance*

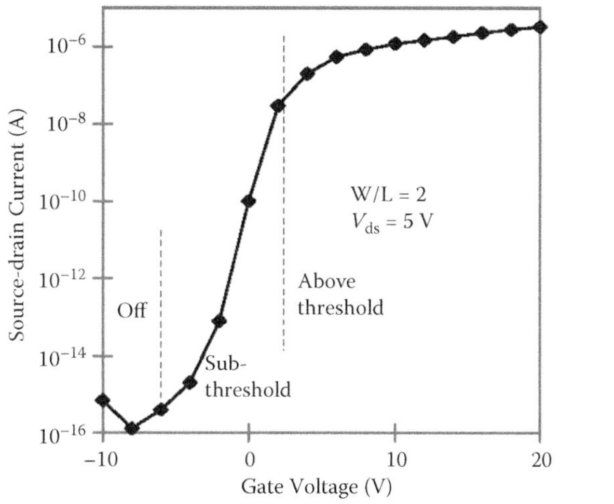

FIGURE 14.14 Current/voltage characteristics of an amorphous silicon transistor. (Reproduced with permission from Street, 2010.)

$$g_d = \frac{\partial I_{ds}}{\partial V_{ds}} = \frac{W}{L} C_{ox} u_n \left(V_g - V_t\right) \tag{14.19}$$

and the *transconductance*

$$g_m = \frac{\partial I_{ds}}{\partial V_g} = \frac{W}{L} C_{ox} u_n V_{ds}. \tag{14.20}$$

The field effect mobility is directly calculated from the transconductance

$$u_{FE,lin} = \frac{Lg_m}{WC_{ox}} \frac{1}{V_{ds}}. \tag{14.21}$$

Substituting saturation regime condition $V_{ds} = V_g - V_t$ into Equation 14.18, the channel current is given by

$$I_{ds,sat} = \frac{W}{2L} C_{ox} u_{FE} \left(V_g - V_t\right)^2. \tag{14.22}$$

Therefore, there is a linear relationship between $I_{ds,sat}^{1/2}$ and V_g, as shown in Figure 14.13*b*. The mobility can be obtained from the slope of this line as

$$u_{FE,lin} = \frac{2L}{WC_{ox}} \left(\frac{\partial \sqrt{I_{ds}}}{\partial V_g} \right)^2. \tag{14.23}$$

The channel resistance is just the reciprocal of the conductance

$$R_{ch} = \left(\frac{\partial I_{ds}}{\partial V_{ds}} \right)^{-1} = g_{d^{-1}} = \frac{L}{Wu_n C_{ox} \left(V_g - V_t\right)}. \tag{14.24}$$

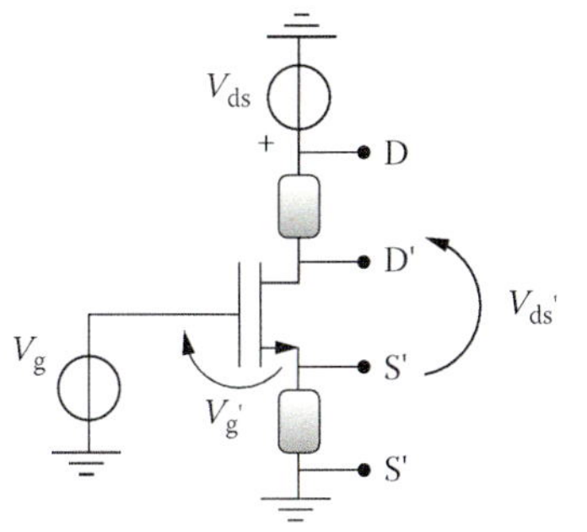

FIGURE 14.15 Model of a transistor with charge injection (and extraction) limitations at the contacts. The externally applied voltages V_g and V_{ds} do not drop entirely in the actual transistor; the voltage drops driving the transistor are V_g' and V_{ds}'. The shaded two-terminal boxes between the physically accessible terminals, S and D, and the physically inaccessible transistor terminals, S' and D', represent nonideal contacts. (Reproduced with permission from Natali and Caironi, 2012.)

An additional resistance R_p is associated with injection of charge at the source and drain contacts as observed in Figure 14.15. The total resistance is usually normalized to the channel width as follows (Yong et al., 2010):

$$R_{tot} \times W = \frac{L}{u_n C_{ox}\left(V_g - V_t\right)} + R_p \times W \tag{14.25}$$

The contact contribution can be separated from the total resistance by measuring the total resistance at varying channel length L (Gundlach et al., 2006).

If the active layer has the geometry of a cylinder, as in nanowires (Weng et al., 2011), the gate capacitance per unit length is obtained from a model that consists of a cylinder on a conducting plate (Lind et al., 2006)

$$\frac{C_g}{L} = \frac{2\pi\varepsilon_{ox}}{\cosh^{-1}\left(\frac{r+h}{r}\right)}, \tag{14.26}$$

where r is the nanowire radius, L is the source–drain distance, h is the insulator thickness, and ε_{ox} the insulator dielectric constant. For this geometry, the mobility in the low field region is given by

$$u_n = \frac{\partial I_{ds}}{\partial V_g}\frac{L^2}{V_{ds}C_g}. \tag{14.27}$$

14.5 THE MOBILITY IN DISORDERED SEMICONDUCTORS

As mentioned in previous sections, many of the materials used in transistors show a broad DOS of exponential or Gaussian shape and the effects of trapping exert different types of influences on the carrier dynamics and response times of the transistor (Zhang et al., 2015). Here, we describe the effects of traps on the conductivity and field effect mobility employing the frequently used model of multiple trapping that was described in detail in Chapter 14. We have also observed that carrier transport in the multiple trapping regime introduces a density-dependent mobility that is given by the combination of Equation 11.44 and Equation 13.10,

$$u_n = \frac{qD_J}{k_B T} = \frac{n_c}{n}\frac{qD_0}{k_B T}. \tag{14.28}$$

For an exponential density of states, the following dependence of mobility on total carrier density n was obtained in Equation 13.30:

$$u_n(n) = An^{\tau_0/T-1}. \tag{14.29}$$

Using Equation 12.11, the conductivity shows the dependence

$$\sigma_n(n) = Bn^{\tau_0/\tau}, \tag{14.30}$$

where A and B are constants dependent on the temperature. If the charge distribution is not uniform in the channel (Figure 14.2), then the carrier density $n(x)$ at a distance x from the insulator depends on the gate-induced potential, as was indicated in Equation 13.24:

$$n(x) = n_0 \exp\left(\frac{qV(x)}{k_B T_0}\right). \tag{14.31}$$

Here n_0 is the occupation far away from the interface where $V(x) = 0$. For an accumulation layer $n(x) >> n_0$, the carrier density is related to the electrical field $\mathbf{E}(x)$ as

$$\mathbf{E}^2(x) = \frac{2K_b t}{\varepsilon} n(x). \tag{14.32}$$

The field at the interface depends on the gate voltage as

$$\mathbf{E}(0) = \frac{c_{0x}V_g}{\varepsilon}, \tag{14.33}$$

where the current in the channel is given by

$$I_{ds} = \frac{WV_{ds}}{L}\int_0^t \sigma\left[n(x), T\right]dx. \tag{14.34}$$

The calculation of the field effect mobility using Equation 14.23 provides the mobility dependence

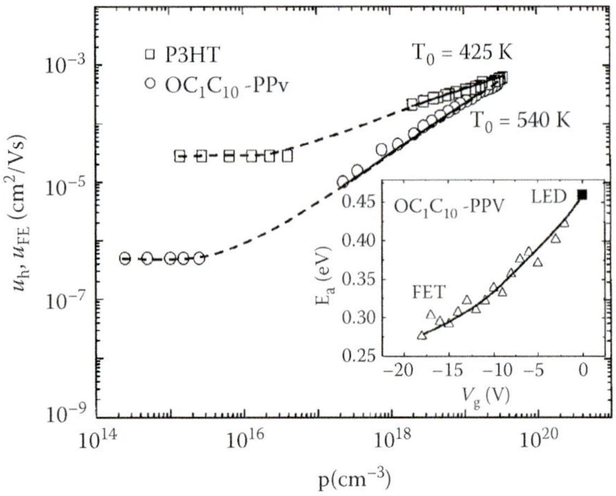

FIGURE 14.16 Mobility as a function of hole density p in a diode and field effect transistor for P3HT and OC_1C_{10}-PPV. The values at low carrier density are obtained with a hole-only diode. At high carrier density they represent the measured field effect mobility in a TFT. The dashed line is a guide to the eye. Inset: The activation energy of the mobility. (Reproduced with permission from Tanase et al., 2003.)

on voltage as follows (Vissenberg and Matters, 1998; Sungsik et al., 2011):

$$u_n = H\left(C_{ox}V_g\right)^{2(T_0/T-1)}, \tag{14.35}$$

where H is a temperature-dependent constant. In summary, for the exponential density of states, we have Equations 14.29, 14.30, and 14.35, which state the behavior of the mobility and conductivity.

The model of Vissenberg and Matters (1998) determines a more specific prefactor to Equation 14.29 using the hopping model with the percolation criterion, in which $B_c \approx 2.8$ is the critical number of bonds in the percolation network. The following expression of the conductivity dependence on carrier density and temperature is obtained

$$\sigma(n,T) = \sigma_0\left(\frac{\pi\left(\alpha T_0/T\right)^3}{8B_c\Gamma\left(1-T/T_0\right)\Gamma\left(1+T/T_0\right)}\right)^{T_0/T} n^{T_0/T}, \tag{14.36}$$

where σ_0 is a prefactor of the conductivity. For the mobility

$$u_{FE} = \frac{\sigma_0}{q}\left(\frac{\pi\left(\alpha T_0/T\right)^3}{8B_c\Gamma\left(1-T/T_0\right)\Gamma\left(1+T/T_0\right)}\right)^{T_0/T}\left[\frac{C_{ox}V_g}{\left(2k_BT_0\varepsilon\right)^{1/2}}\right]^{2(T_0/T-1)}. \tag{14.37}$$

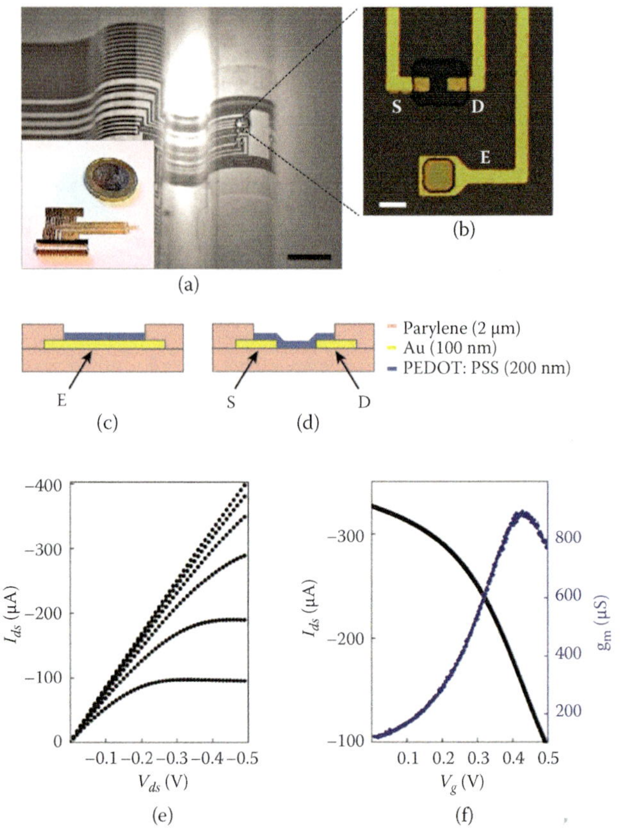

FIGURE 14.17 (a) Optical micrograph of an electrocorticography probe conforming to a curvilinear surface. Scale bar, 1 mm. The inset shows an image of the whole probe, in which the transistor–electrode arrays are on the right-hand side, whereas the external connections, onto which a zero insertion force connector is attached, are on the left-hand side. (b) Optical micrograph of the channel of a transistor and a surface electrode, in which the Au films that act as source (S), drain (D), and electrode pad (E) are identified. Scale bar, 10 mm. (c,d) Layouts of the surface electrode and of the transistor channel respectively (not to scale). (e) In vitro characterization of the transistor. Output characteristics showing the drain current, I_{ds}, as a function of drain voltage, V_{ds}, for a gate voltage V_g varying from 0 V to 0.5 V (with a step of 0.1 V) of a PEDOT:PSS transistor in Ringer's solution and with a stainless-steel gate electrode. (f) Transfer curve and resulting transconductance at $V_{ds} = -0.4$ V. (Reproduced with permission from Khodagholy et al., 2013.)

Results of the measurement of the mobility of two organic conductors as a function of the hole density are shown in Figure 14.16. The combined dependence of the mobility on carrier density and electrical field is discussed in Pasveer et al. (2005).

14.6 ELECTROCHEMICAL TRANSISTOR

Electrochemical gating was introduced by Kittlesen et al. (1984) and White et al. (1984) for the measurement of

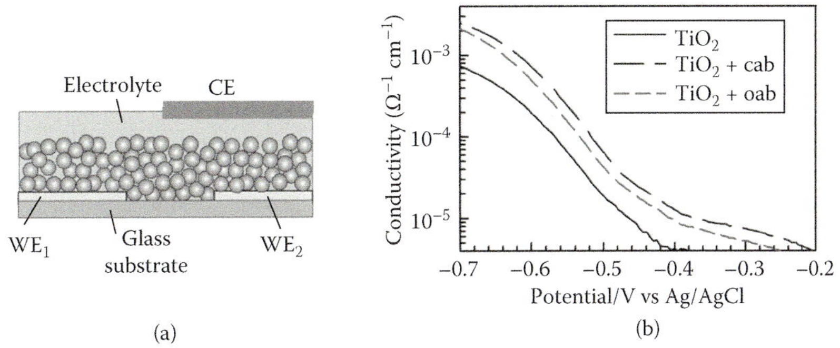

FIGURE 14.18 (a) Electrochemical transistor measurement configuration. The conducting substrate over which the film is deposited is divided in two regions. The separated regions serve as two working electrodes, WE_1 and WE_2 that are operated with a potential against a counterelectrode (CE), and S and D with a difference of potential V_{ds}. (b) Conductivity plot of a bare, mesoporous TiO2 film and molecular modified films with electrochemically deposited 4-methoxybenzenediazonium tetrafluoroborate (oab) and 4-cyanobenzenediazonium tetrafluoroborate (cab). (Reprinted with permission from Ruhle et al., 2005.)

organic conductors that operate in wet conditions. These materials and methods have been amply developed, for example in biomedical applications; see Figure 14.17 (Khodagholy et al., 2013). The technique has been applied to determine the DOS of graphene as shown in Figure 7.17. In general, the electrochemical transistor uses an electrolyte as the electronic insulator instead of the oxide layer gate; see Figure 14.18. Thus, the insulator is able to penetrate any nanostructured or porous layer while still forming an equipotential at the material surface. This has the advantage that the Fermi level in the active layer can be modified homogeneously, even in thick layers. The two coplanar regions can be controlled as independent working electrodes that are biased against the electrode in solution. Then, it is possible to govern the Fermi level in the film with a bias $V_1 \cong V_2$ while maintaining a small potential difference between the two sides, $V_{ds} = \Delta V = V_2 - V_1$, which causes a current flow I_{ds} that enables measurement of the electronic conductivity σ. An example of the application of this technique is shown in Figure 14.18 for measurements of nanostructured TiO_2 films with different surface treatments. Changes in the conductivity are caused by a shift of the conduction band of TiO_2 due to the different dipole moments of the molecular absorbed species, see Figure 9.32. Electrolyte gating has been investigated in a variety of thin film materials including organic conductors and metal oxides (see General References).

GENERAL REFERENCES

Thin film transistors: Horowitz, 1998; Newman et al., 2004; Mas-Torrent and Rovira, 2008; Bain et al., 2009; Kumar et al., 2014.

Transport models of TFT: Horowitz and Delannoy, 1991; Shengwen and Gerold, 1992; Vissenberg and Matters, 1998; Horowitz et al., 2000; Necliudov et al., 2000; Estrada et al., 2005; Torricelli et al., 2012.

Measurement of the DOS: Shur and Hack, 1984; Khakzar and Lueder, 1992; Paasch and Scheinert, 2007; Kalb and Batlogg, 2010.

Carrier density dependence of mobility: Shur and Hack, 1984; Shaw and Hack, 1988; Horowitz and Delannoy, 1991; Khakzar and Lueder, 1992; Servati et al., 2003.

Electrochemical transistor: Laibinis et al., 1994; Meulenkamp, 1999; Abayev et al., 2003; Roest et al., 2003; Agrell et al., 2004; Ruhle et al., 2005; Gstrein et al., 2007; Vanmaekelbergh et al., 2007; Paulsen and Frisbie, 2012, Leng et al. 2016, Leng et al. 2017

REFERENCES

Abayev, I.; Zaban, A.; Fabregat-Santiago, F.; Bisquert, J. Electronic conductivity in nanostructured TiO_2 films permeated with electrolyte. *Physica Status Solidii (a)* 2003, *196*, R4–R6.

Agrell, H. G.; Boschloo, G.; Hagfeldt, A. Conductivity studies of nanostructured TiO_2 films permeated with electrolyte. *The Journal of Physical Chemistry B* 2004, *108*, 12388.

Anutgan, T.; Anutgan, M.; Atilgan, I.; Katircioglu, B. Capacitance analyses of hydrogenated nanocrystalline silicon based thin film transistor. *Thin Solid Films* 2011, *519*, 3914–3921.

Bain, S.; Smith, D. C.; Wilson, N. R.; Carrasco-Orozco, M. Kelvin force gradient microscopy of pBTTT transistors in both the linear and saturation electrical regimes. *Applied Physics Letters* 2009, *95*, 143304.

Das, A.; Pisana, S.; Chakraborty, B.; Piscanec, S.; Saha, S. K.; Waghmare, U. V.; Novoselov, K. S. et al. Monitoring dopants by Raman scattering in an electrochemically top-gated graphene transistor. *Nature Nano* 2008, *3*, 210–215.

Estrada, M.; Cerdeira, A.; Puigdollers, J.; Resandiz, L.; Pallares, J.; Marsal, L. F.; Voz, C. et al. Accurate modeling and parameter extraction method for organic TFTs. *Solid-State Electronics* 2005, *49*, 1009–1016.

Gstrein, F.; Michalak, D. J.; Knapp, D. W.; Lewis, N. S. Near-surface channel impedance measurements, open-circuit impedance spectra, and differential capacitance vs potential measurements of the Fermi level position at Si/CH_3CN contacts. *The Journal of Physical Chemistry C* 2007, *111*, 8120–8127.

Gundlach, D. J.; Zhou, L.; Nichols, J. A.; Jackson, T. N.; Necliudov, P. V.; Shur, M. S. An experimental study of contact effects in organic thin film transistors. *Journal of Applied Physics* 2006, *100*, 024509.

Horowitz, G. Organic field-effect transistors. *Advanced Materials* 1998, *10*, 365–377.

Horowitz, G.; Delannoy, P. An analytical model for organic-based thin-film transistors. *Journal of Applied Physics* 1991, *70*, 469–475.

Horowitz, G.; Hajlaoui, M. E.; Hajlaoui, R. Temperature and gate voltage dependence of hole mobility in polycrystalline oligothiophene thin film transistors. *Journal of Applied Physics* 2000, *87*, 4456–4463.

Horowitz, G.; Hajlaoui, R.; Bouchriha, H.; Bourguiga, R.; Hajlaoui, M. The concept of "threshold voltage" in organic field-effect transistors. *Advanced Materials* 1998, *10*, 923–927.

Hyuk-Ryeol, P.; Daewon, K.; Cohen, J. D. Electrode interdependence and hole capacitance in capacitance-voltage characteristics of hydrogenated amorphous silicon thin-film transistor. *Journal of Applied Physics* 1998, *83*, 8051–8056.

Kalb, W. L.; Batlogg, B. Calculating the trap density of states in organic field-effect transistors from experiment: A comparison of different methods. *Physical Review B* 2010, *81*, 035327.

Khakzar, K.; Lueder, E. H. Modeling of amorphous-silicon thin-film transistors for circuit simulations with SPICE. *IEEE Transactions on Electron Devices* 1992, *39*, 1428–1434.

Khodagholy, D.; Doublet, T.; Quilichini, P.; Gurfinkel, M.; Leleux, P.; Ghestem, A.; Ismailova, E. et al. In vivo recordings of brain activity using organic transistors. *Nature Communicatons* 2013, *4*, 1575.

Kichan, J.; Changjung, K.; Ihun, S.; Jaechul, P.; Sunil, K.; Sangwook, K.; Youngsoo, P. et al. Modeling of amorphous InGaZnO thin-film transistors based on the density of states extracted from the optical response of capacitance-voltage characteristics. *Applied Physics Letters* 2008, *93*, 182102.

Kittlesen, G. P.; White, H. S.; Wrighton, M. S. Chemical derivatization of microelectrode arrays by oxidation of pyrrole and N-methylpyrrole: Fabrication of molecule-based electronic devices. *Journal of the American Chemical Society* 1984, *106*, 7389–7396.

Kumar, B.; Kaushik, B. K.; Negi, Y. S. Organic thin film transistors: Structures, models, materials, fabrication, and applications: A review. *Polymer Reviews* 2014, *54*, 33–111.

Laibinis, P. E.; Stanton, C. E.; Lewis, N. S. Measurement of barrier heights of semiconductor/liquid junctions using a transconductance method: Evidence for inversion at n-Si/CH_3OH-1,1′-dimethylferrocene+/0 junctions. *The Journal of Physical Chemistry* 1994, *98*, 8765–8774.

Lang, D. V.; Chi, X.; Siegrist, T.; Sergent, A. M.; Ramirez, A. P. Amorphous like density of gap states in single-crystal pentacene. *Physical Review Letters* 2004, *93*, 086802.

Leng, X., Bollinger, A. T.; Božović, I. Purely electronic mechanism of electrolyte gating of indium tin oxide thin films. *Scientific Reports* 2016, *6*, 31239.

X. Leng, J. Pereiro, J. Strle, A.T. Bollinger, G. Dubuis, A. Gozar, N. Litombe, D. Pavuna, and I. Bozovic. Insulator to metal transition in WO_3 induced by electrolyte gating. *Quantum Materials* 2017, *2*, 35.

Lind, E.; Persson, A. I.; Samuelson, L.; Wernersson, L.-E. Improved subthreshold slope in an InAs nanowire heterostructure field-effect transistor. *Nano Letters* 2006, *6*, 1842–1846.

Mas-Torrent, M.; Rovira, C. Novel small molecules for organic field-effect transistors: Towards processability and high performance. *Chemical Society Reviews* 2008, *37*, 827–838.

Meulenkamp, E. A. Electron transport in nanoparticle ZnO films. *The Journal of Physical Chemistry B* 1999, *103*, 7831–7838.

Natali, D.; Caironi, M. Charge injection in solution-processed organic field-effect transistors: Physics, models and characterization methods. *Advanced Materials* 2012, *24*, 1357–1387.

Necliudov, P. V.; Shur, M. S.; Gundlach, D. J.; Jackson, T. N. Modeling of organic thin film transistors of different designs. *Journal of Applied Physics* 2000, *88*, 6594–6597.

Newman, C. R.; Frisbie, C. D.; da Silva Filho, D. A.; Bredas, J.-L.; Ewbank, P. C.; Mann, K. R. Introduction to organic thin film transistors and design of n-channel organic semiconductors. *Chemistry of Materials* 2004, *16*, 4436–4451.

Paasch, G.; Scheinert, S. Space charge layers in organic field-effect transistors with Gaussian or exponential semiconductor density of states. *Journal of Applied Physics* 2007, *101*, 024514.

Pasveer, W. F.; Cottar, J.; Tanase, C.; Coehoorn, R.; Bobbert, P. A.; Blom, P. W. M.; de Leeuw, D. M. et al. Unified description of charge-carrier mobilities in disordered semiconducting polymers. *Physical Review Letters* 2005, *94*, 206601.

Paulsen, B. D.; Frisbie, C. D. Dependence of conductivity on charge density and electrochemical potential in polymer semiconductors gated with ionic liquids. *The Journal of Physical Chemistry C* 2012, *116*, 3132–3141.

Puigdollers, J.; Marsal, A.; Cheylan, S.; Voz, C.; Alcubilla, R. Density-of-states in pentacene from the electrical characteristics of thin-film transistors. *Organic Electronics* 2010, *11*, 1333–1337.

Roelofs, W. S. C.; Mathijssen, S. G. J.; Janssen, R. A. J.; de Leeuw, D. M.; Kemerink, M. Accurate description of charge transport in organic field effect transistors using an experimentally extracted density of states. *Physical Review B* 2012, *85*, 085202.

Roest, A. L.; Kelly, J. J.; Vanmaekelbergh, D. Coulomb blockade of electron transport in a ZnO quantum-dot solid. *Applied Physics Letters* 2003, *83*, 5530.

Ruhle, S.; Greenshtein, M.; Chen, S.-G.; Merson, A.; Pizem, H.; Sukenik, C. S.; Cahen, D. et al. Molecular adjustment of the electronic properties of nanoporous electrodes in dye-sensitized solar cells. *The Journal of Physical Chemistry B* 2005, *109*, 18907–18913.

Servati, P.; Striakhilev, D.; Nathan, A. Above-threshold parameter extraction and modeling for amorphous silicon thin-film transistors. *IEEE Transactions on Electron Devices* 2003, *50*, 2227–2235.

Shaw, J. G.; Hack, M. An analytic model for calculating trapped charge in amorphous silicon. *Journal of Applied Physics* 1988, *64*, 4562–4566.

Shengwen, L.; Gerold, W. N. An experimental study of the source/drain parasitic resistance effects in amorphous silicon thin film transistors. *Journal of Applied Physics* 1992, *72*, 766–772.

Shur, M.; Hack, M. Physics of amorphous silicon based alloy field-effect transistors. *Journal of Applied Physics* 1984, *55*, 3831–3842.

Street, G. B. Amorphous silicon transistors and photodiodes. In *Solar Cells and Their Applications*; Fraas, L., Partain, L., Eds.; Wiley: New York, 2010.

Sungsik, L.; Khashayar, G.; Arokia, N.; John, R.; Sanghun, J.; Changjung, K.; Song, I. H. et al. Trap-limited and percolation conduction mechanisms in amorphous oxide semiconductor thin film transistors. *Applied Physics Letters* 2011, *98*, 203508.

Sze, S. M. *Physics of Semiconductor Devices*, 2nd ed.; John Wiley and Sons: New York, 1981.

Tal, O.; Rosenwaks, Y. Electronic properties of doped molecular thin films measured by Kelvin Probe Force Microscopy. *The Journal of Physical Chemistry B* 2006, *110*, 25521–25524.

Tanase, C.; Meijer, E. J.; Blom, P. W. M.; de Leeuw, D. M. Unification of the hole transport in polymeric field effect-transistors and LED. *Physical Review Letters* 2003, *91*, 216601.

Torricelli, F.; O'Neill, K.; Gelinck, G. H.; Myny, K.; Genoe, J.; Cantatore, E. Charge transport in organic transistors accounting for a wide distribution of carrier energies; Part II: TFT modeling. *IEEE Transactions on Electron Devices* 2012, *59*, 1520–1528.

Vanmaekelbergh, D.; Houtepen, A. J.; Kelly, J. J. Electrochemical gating: A method to tune and monitor the (opto)electronic properties of functional materials. *Electrochimica Acta* 2007, *53*, 1140–1149.

Vissenberg, M. C. J. M.; Matters, M. Theory of the field-effect mobility in amorphous organic transistors. *Physical Review B* 1998, *57*, 12964–12967.

Weng, W. Y.; Chang, S. J.; Hsu, C. L.; Hsueh, T. J. A ZnO-nanowire phototransistor prepared on glass substrates. *ACS Applied Materials & Interfaces* 2011, *3*, 162–166.

White, H. S.; Kittlesen, G. P.; Wrighton, M. S. Chemical derivatization of an array of three gold microelectrodes with polypyrrole: Fabrication of a molecule-based transistor. *Journal of the American Chemical Society* 1984, *106*, 5375–5377.

Xu, H.; Zhang, Z.; Peng, L.-M. Measurements and microscopic model of quantum capacitance in graphene. *Applied Physics Letters* 2011, *98*, 133122.

Yin, Z.; Li, H.; Li, H.; Jiang, L.; Shi, Y.; Sun, Y.; Lu, G. et al. Single-layer MoS_2 phototransistors. *ACS Nano* 2011, *6*, 74–80.

Yogev, S.; Halpern, E.; Matsubara; R., Nakamura; M.; Rosenwaks, Y. Direct measurement of density of states in pentacene thin film transistors. *Physical Review B* 2011, *84*, 165124.

Yong, X.; Gwoziecki, R.; Chartier, I.; Coppard, R.; Balestra, F.; Ghibaudo, G. Modified transmission-line method for contact resistance extraction in organic field-effect transistors. *Applied Physics Letters* 2010, *97*, 063302.

Yong, X.; Takeo, M.; Kazuhito, T.; Romain, G.; Romain, C.; Mohamed, B.; Jan, C. et al. Modeling of static electrical properties in organic field-effect transistors. *Journal of Applied Physics* 2011, *110*, 014510.

Zhang, Y.; Chen, Q.; Alivisatos, A. P.; Salmeron, M. Dynamic charge carrier trapping in quantum dot field effect transistors. *Nano Letters* 2015, *15*, 4657–4663.

15 Space-Charge-Limited Transport

In insulator materials, or in a semiconductor with large imposed current, the carriers injected from the contacts determine the distribution of the electrical field. This is termed the *space-charge-limited conduction regime*. In this chapter, we discuss the characteristics of carrier and field distribution, the current-voltage characteristics, and the dynamic properties of this regime by calculation of the capacitance.

15.1 SPACE-CHARGE-LIMITED CURRENT

Consider a semiconductor layer in which the cathode is able to supply a large current density by injection of electrons. When the cathode is negatively biased (forward bias), the layer first shows a domain of ohmic conduction determined by the intrinsic conductivity as indicated by Equation 10.16, that we may write as

$$j = qnu_n\mathbf{E}, \tag{15.1}$$

where $\mathbf{E} = V/d$ and $n = n_0$, the equilibrium carrier density. However, when a large density of electrons enters the sample, the carrier density $n >> n_0$ modifies the internal field distribution and a new conduction regime occurs (see Figure 10.3). Assuming that the current is dominated by drift, in the steady-state situation, the charge density and local field must be adjusted to provide the same value of current along all points in the conduction pathway (Equation 10.4), so that the local electric field increases in inverse proportion to the increase of the carrier density. These quantities must also satisfy Poisson's equation

$$\frac{d\mathbf{E}}{dx} = -\frac{q}{\varepsilon}n. \tag{15.2}$$

Such a domain of conduction is termed the space-charge-limited current (SCLC). By combining these equations, we obtain

$$\mathbf{E}\frac{d\mathbf{E}}{dx} = -\frac{j}{\varepsilon u_n}. \tag{15.3}$$

If we assume that the field at the surface of the cathode is zero, we get the solution

$$\mathbf{E}(x) = \left(\frac{2j}{\varepsilon u_n}\right)^{1/2} x^{1/2} \text{ for } \mathbf{E}(0) = 0. \tag{15.4}$$

The electric field increases from the cathode toward the anode and the carrier density decreases as we move away from the cathode. The distribution of carrier density takes the form

$$n(x) = \frac{3}{4}\frac{\varepsilon V}{qd^2}\left(\frac{x}{d}\right)^{-1/2}. \tag{15.5}$$

Figure 4.1 indicates the carrier injection and the distribution of voltage between the electrodes. Figure 15.1 shows the spatial dependence of the electrical field and carrier density. The total voltage can be obtained from Equation 15.4:

$$V = -\int_0^d \mathbf{E}dx = \left(\frac{8j}{9\varepsilon u_n}\right)^{1/2} d^{3/2}. \tag{15.6}$$

Hence, current density has the expression

$$j = \frac{9}{8}\varepsilon u_n \frac{V^2}{d^3}. \tag{15.7}$$

Equation 15.7 shows a quadratic dependence of the SCLC on voltage, which is termed the *Mott-Gourney square law* or *Child's law*. Current voltage curves of regio-regular P3HT displaying this dependence are shown in Figure 15.2. Note that it is straightforward to extract the mobility using Equation 15.7.

The dependence of the local electric field on voltage is

$$\mathbf{E} = \frac{3}{2}\frac{V}{d}\left(\frac{x}{d}\right)^{1/2}. \tag{15.8}$$

This model shows an anomaly in that the carrier density at the injecting contact is infinite. Taking into account the electric field at the injection boundary $\mathbf{E}_0$, the current-voltage characteristic is given by the equation that can be derived from Equations 15.1 and 15.2:

$$V = \frac{u_n\varepsilon}{3j}\left[\left(\frac{2j}{u_n\varepsilon}d + \mathbf{E}_0^2\right)^{3/2} - \mathbf{E}_0^3\right]; \tag{15.9}$$

If the injection field vanishes, the maximum current and Child's law are obtained. Thus, the increase of current requires a decrease of the injection field.

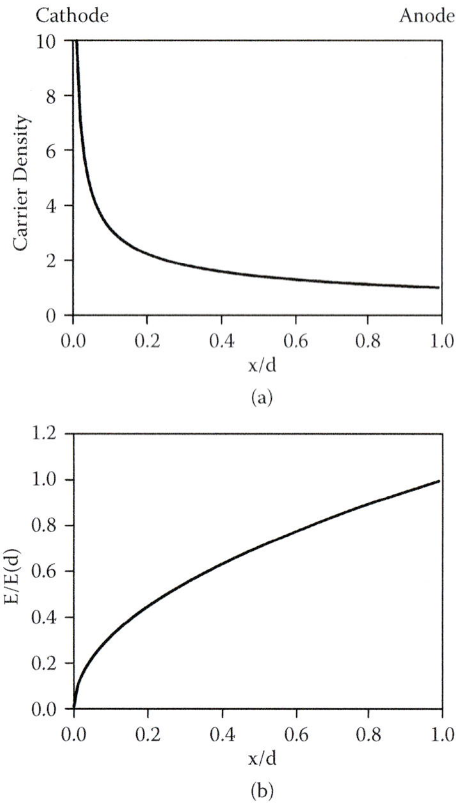

FIGURE 15.1 Spatial dependence of (a) carrier density and (b) the electric field in SCLC transport.

As a matter of fact, for the application of SCLC in problems of carrier injection in insulators, like in organic light-emitting diodes (OLEDs), the carrier density at the injecting contacts cannot grow in an unlimited fashion because the maximum is given by the local density of states (DOS), which is the statement in Equations 10.22 and 10.23, where $n_0 = N_c$, $p_0 = N_v$ (Blom et al., 1997). A more complete solution by Wright (1961) of the drift-diffusion transport and Poisson equations showed that current occurs predominantly by diffusion at low forward bias while at higher voltages, the dominant mechanism is carrier drift where the current follows the space-charge-limited power law. More recent models describe a transition from injection to bulk-limited transport (Bullejos et al., 2008).

Equation 15.7 is valid when the semiconductor presents a constant mobility or the diode operates at a moderate electric field. However, the transport in organic layers often shows a mobility that depends on the electric field (see Equation 10.12 as well as the example in Figure 15.2c). An extension of the current-voltage expression for a field-dependent mobility was given by Murgatroyd (1970).

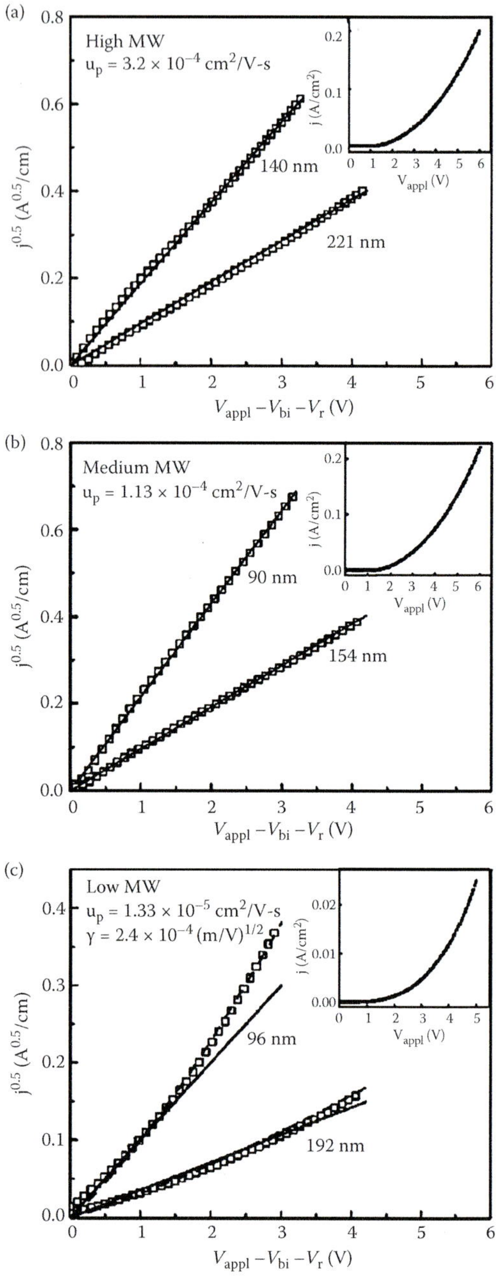

FIGURE 15.2 $j^{0.5}$ vs V plots of regioregular P3HT of different molecular weights in the direction perpendicular to the substrate. (a) High-molecular-weight film (31.1 kg/mol), (b) mediummolecular-weight film (9.72 kg/mol), and (c) low-molecular-weight film (2.89 kg/mol) at room temperature. The thickness of the films is indicated in the plots. The inset in each figure shows the j vs. V_{app} plot of the thicker film device before any correction for the voltage. (Reproduced with permission from Goh et al., 2005.)

The presence of traps exerts a great influence on the SCLC characteristics (Rose, 1955; Lampert and Mark, 1970). Although the trapped carriers do not participate in conduction current, they modify the field distribution via the Poisson equation. For an exponential

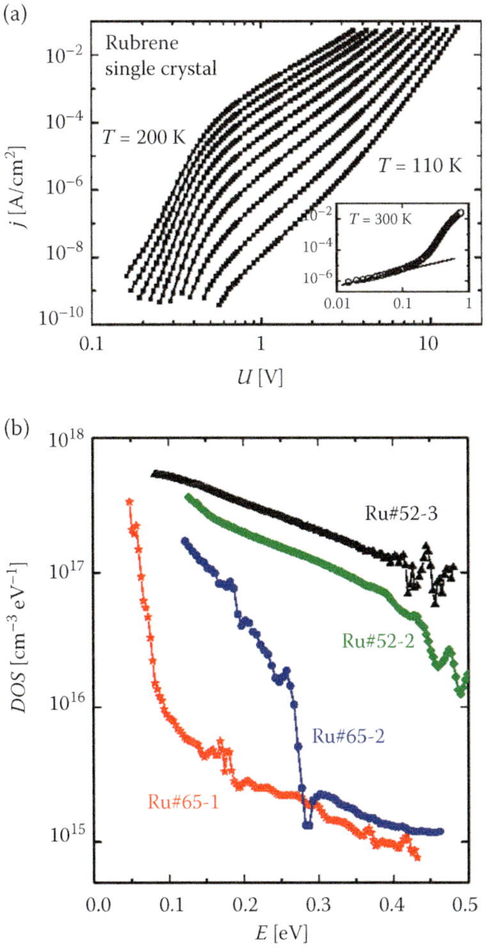

FIGURE 15.3 (a) Space-charge-limited current density vs. applied voltage curves at different temperatures for a rubrene single crystal, d = 0.6 μ m. The temperature step is 10 K. The inset shows j(V) at 300 K. The straight line indicates the ohmic behavior of thermally generated charge carriers. (b) Density of states above the valence band for four rubrene single crystals. (Reproduced with permission from Krellner et al., 2007.)

distribution of traps with parameter T_0, Equation 8.19, the Mott–Gurney law is modified as follows (Mark and Helfrich, 1962)

$$j = qN_L u_n \left(\frac{\varepsilon}{qN_L} \frac{l}{l+1} \right)^l \left(\frac{2l+1}{l+1} \right)^{l+1} \frac{V^{l+1}}{d^{2l+1}}, \quad (15.10)$$

where $l = T_0 / T > 1$.

In general, for other trap distributions, the current can be related to the increment of the space charge with respect to the shift of the Fermi energy, dn/dE_{Fn}, thus it is possible to develop a spectroscopy of the DOS of an insulator by measuring the SCLC at different temperatures (Krellner et al., 2007). The results of this method for semiconducting organic single crystals of rubrene are shown in Figure 15.3. The observed DOS shows a low density of deep trap states of 10^{15} cm^{-3}, an exponentially varying shallow trap density near the band edge, and an oxygen-related sharp hole bulk trap state at 0.27 eV above the valence band. Note that the inset in Figure 15.3a shows the transition of ohmic to SCLC conduction.

15.2 INJECTED CARRIER CAPACITANCE IN SCLC

In general, we must bear in mind that there are three main effects that can govern the observed current: ohmic current, injection current, and the SCLC. In many situations, distinguishing between them is far from trivial, as discussed earlier in Section 10.2, especially when traps lead to exponential dependencies like those in Equation 15.10. The main criterion to distinguish injection and bulk current is the thickness dependence of the bulk SCLC current for trap-free transport as d^{-3} at constant voltage, indicated in Equation 15.7, or as d^{-3} at constant field (Brütting et al. 2001). For trap-limited transport, it is d^{-2l-1} in Equation 15.10. As shown in Figure 15.4, measurement of Alq_3 samples of varying thickness indicate injection limitations for the thinner films (< 200 nm), whereas for samples with thickness greater than 200 nm, the current is space-charge-limited including effects of trap states. Nevertheless, this approach is not readily feasible in many organic and hybrid materials.

An alternate route to characterize the SCLC is to apply the small perturbation method of impedance spectroscopy (IS), which provides the response of the device over a broad variety of frequencies. In this method, the response of the contacts and traps normally occurs at low frequencies, so that one can determine the transport mechanism and parameters unambiguously by observing rather specific patterns in the high-frequency part of spectra. The analysis of the frequency-dependent capacitance is more generally explained in Chapter 16. If there is no carrier injection into an insulating layer, we can only observe the dielectric relaxation phenomena. If the contacts can inject carriers, then the dielectric can be filled with mobile charge and we obtain the SCLC transport domain that was described earlier.

Some preliminary values of characteristic parameters can be obtained by simple intuitive reasoning. Let us assume that the dielectric constant ε is independent of frequency in order to evaluate the capacitance in SCLC. If electrons are injected from the cathode, they are distributed along the layer as indicated in Equation 15.5 while positive compensating charge is at the cathode

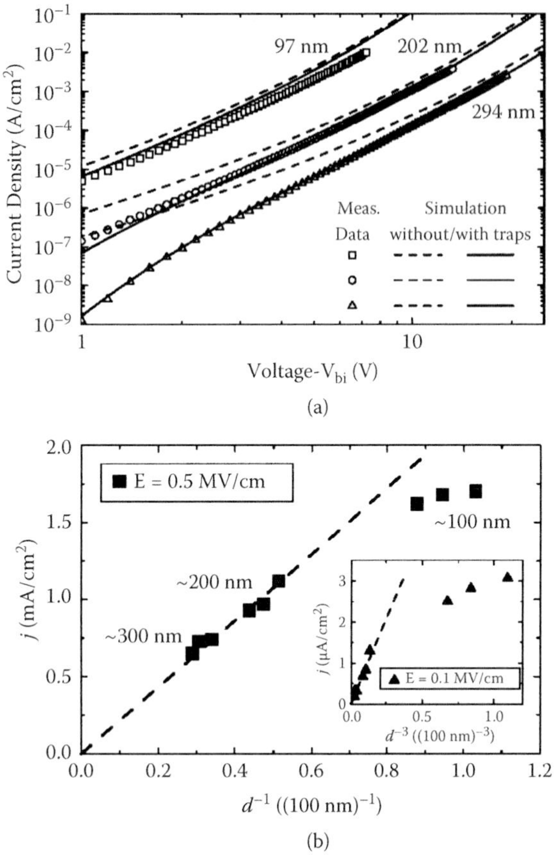

FIGURE 15.4 (a) Current-voltage characteristics of Al/Alq/Ca electron-only devices with different Alq layer thickness in double-logarithmic representation. (b) Thickness dependence of the current (j versus d^{-1}) at a constant electric field of 0.5 MV/cm for Al/Alq/Ca electron-only devices. The inset shows a plot of (j versus d^{-3}) at an electric field of 0.1 MV/cm. (Reproduced with permission from Brütting et al., 2001.)

surface. The amount of charge at voltage V can be estimated by the basic relationship $Q = C_g V$ in terms of the geometric capacitance $C_g = \varepsilon/d$, so that the current is $j = Q/\tau_{tr}$. The transit time τ_{tr} is taken from Equation 10.26, hence the current is $j = \varepsilon u_n V^2/d^3$. This result is close to the numerically correct value of Equation 15.7. The failure of the prefactor is due to the inhomogeneous distribution of charge caused by the position-dependent electric field.

For a quantitative evaluation of parameters, we note first of all that the transit time can be calculated averaging the reciprocal velocity of carriers

$$\tau_{tr} = \int_0^d \frac{1}{u_n \mathbf{E}(x)} dx \tag{15.11}$$

with the result, from Equation 15.8,

$$\tau_{tr} = \frac{4}{3} \frac{d^2}{u_n V}. \tag{15.12}$$

The total charge can be evaluated using Gauss's law across the whole film. Note in Figure 15.1b that $\mathbf{E}(0) = 0$, therefore, taking the two electrodes as the Gaussian surfaces, we have $Q = \varepsilon \mathbf{E}(d)$, and hence

$$Q = \frac{3}{2} \frac{\varepsilon}{d} V. \tag{15.13}$$

The static capacitance, calculated as the voltage derivative from Equation 7.2, has the value

$$C = \frac{3}{2} C_g. \tag{15.14}$$

However, Equation 15.14 neglects the dynamic properties of the injected carriers in the SCLC regime. To account for such properties, we use IS at varying angular frequency ω. The insulator layer has fixed j and V values given by Equation 15.7 and is additionally perturbed by a small, frequency-dependent voltage. Using the equations outlined in Section 15.1, we obtain the impedance of the SCLC regime for a trap-free semiconductor that has the expression (van der Ziel, 1976):

$$Z(\omega) = \frac{6}{g_0 (i\omega\tau_{tr})^3} \left[1 - i\omega\tau_{tr} + \frac{1}{2}(i\omega\tau_{tr})^2 - e^{-i\omega\tau_{tr}} \right], \tag{15.15}$$

where dc conductance is

$$g_0 = \frac{d\bar{j}}{d\bar{V}} = \frac{9}{4} \varepsilon u_n \frac{\bar{V}}{d^3}. \tag{15.16}$$

For low frequency, the admittance is

$$Y(\omega) = g_0 + i\omega \frac{3}{4} C_g \tag{15.17}$$

and for high frequency, it is

$$Y(\omega) = \frac{2}{3} g_0 + i\omega C_g. \tag{15.18}$$

It is observed that the low-frequency dynamic capacitance is $3C_g/4$, which is considerably less than the static capacitance in Equation 15.14. The difference has been explained by Kassing (1975) in terms of three separate contributions: velocity modulation, density modulation, and a displacement current term.

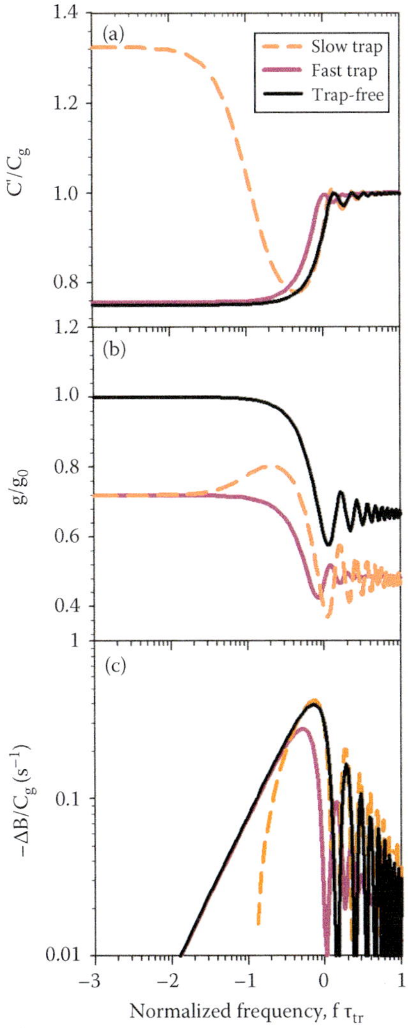

FIGURE 15.5 Simulation of the frequency-dependent quantities in SCLC for a trap-free layer, with slow traps and with fast traps, with respect to the frequency normalized to transit time. (a) Capacitance spectra normalized to C_g. (b) Conductance spectra normalized to g_0. (c) Negative differential susceptance divided by C_g. (Reproduced with permission from Montero and Bisquert, 2012.)

When the modulation frequency is larger than the reciprocal transit time, the charge traveling in the layer cannot respond to the ac signal and the capacitance corresponds to the geometrical value as indicated in Equation 15.18. The step from a low-frequency to high-frequency value occurs approximately at a frequency $f = \tau_{tr}^{-1}$, allows a determination of the transit time from the IS data. These characteristics are illustrated in Figure 15.5, also showing the effects of traps obtained from a general transport-trap model. If the traps respond slowly with respect to the transit time across the layer, they produce a large capacitive contribution at low frequency. Actual experimental measurements of

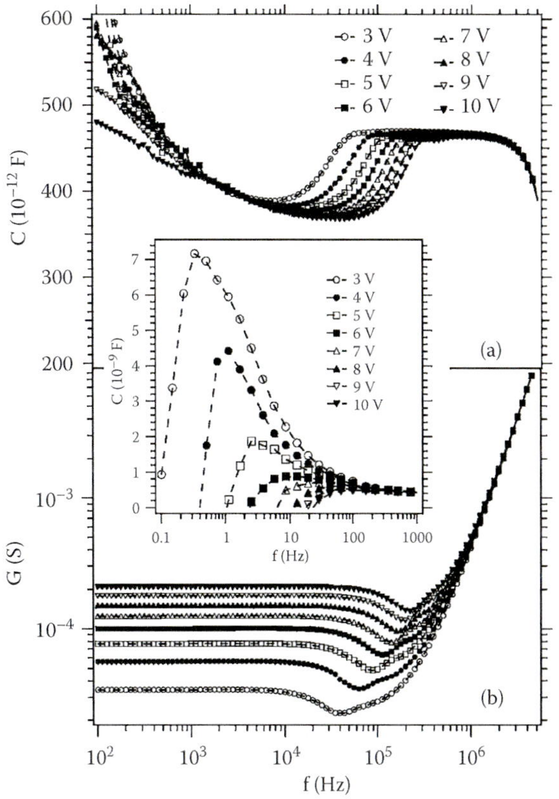

FIGURE 15.6 Experimental (a) capacitance C and (b) conductance G of a hole-only α-NPD device as a function of the modulation frequency f for applied dc voltages V_0 between 3 V and 10 V at room temperature. The inset shows the capacitance C as a function of frequency f in the low-frequency range (0.1–10^3 Hz) for the same applied dc voltage values. (Reproduced with permission from Nguyen et al., 2007.)

the transition of the real part of the capacitance at $f = \tau_{tr}^{-1}$ in an organic layer are shown in Figure 15.6. For a quantitative evaluation of τ_{tr}, the representation of the susceptance $B(\omega) = \mathrm{Im}\left[Y(\omega)\right]$ is used (Martens et al., 2000). The frequency f_{max} of the peak of the negative excess susceptance

$$-\Delta B(\omega) = -\omega\left[C(\omega) - C_g\right] \quad (15.19)$$

is related to the transit time as $\tau = 0.72 \cdot f^{-1}$; see Figure 15.5c.

15.3 SPACE CHARGE IN DOUBLE INJECTION

Let us now consider a two-carrier diode model where electrons and holes are voltage injected at the respective electrodes. Recombination occurs at the bimolecular rate indicated in Equation 18.32 with rate constant B_{rec}. Based on the equations given earlier in Section 12.4, with transport governed by the drift component, the total current is

$$j = q\left[n(x)u_n + p(x)u_p\right]E(x). \quad (15.20)$$

The conservation Equation 12.36 gives

$$\frac{\partial}{\partial x}(nu_n\mathbf{E}) = -\frac{\partial}{\partial x}(pu_p\mathbf{E}) = B_{rec}np \quad (15.21)$$

and the Poisson equation

$$\frac{\partial \mathbf{E}}{\partial x} = \frac{q}{\varepsilon}(n-p). \quad (15.22)$$

solved by Parmenter and Ruppel (1959) with the zero electric field boundary conditions at the electrodes. The full expression is rather complicated because all of the variables are spatially mixed,

$$j = \frac{9}{8}\varepsilon \cdot u_{eff}\frac{V^2}{d^3} \quad (15.23)$$

$$u_{eff} = \frac{4}{9}u_{rec}v_n v_p\left[\frac{\Gamma\left(\frac{3}{2}(v_n+v_p)\right)}{\Gamma\left(\frac{3}{2}v_n\right)\Gamma\left(\frac{3}{2}v_p\right)}\right]^2\left[\frac{\Gamma(v_n)\Gamma(v_p)}{\Gamma((v_n+v_p))}\right]^2 \quad (15.24)$$

with the mobility ratios $v_n = u_n/u_{rec}$, $v_n = u_p/u_{rec}$ and the recombination mobility being defined as

$$u_{rec} = \frac{\varepsilon B_{rec}}{2q}. \quad (15.25)$$

The general solution may be simplified in special situations. For instance, under the approximation of a very strong recombination rate, the carriers only meet at a recombination front situated closer to the slow carrier as shown in Figure 15.7. The total current simplifies into the expression

$$j = \frac{9}{8}\varepsilon(u_n+u_p)\frac{V^2}{d^3}. \quad (15.26)$$

In the opposite case, under the approximation of the so-called plasma limit, i.e., similar concentrations of electrons and holes are present within the layer ($p \approx n$), the mutual compensation of charge increases significantly the current within the device, given by

$$j = \left(\frac{9\pi}{8}\right)^{1/2}\varepsilon\left(\frac{2qu_nu_p(u_n+u_p)}{\varepsilon B_{rec}}\right)^{1/2}\frac{V^2}{d^3}. \quad (15.27)$$

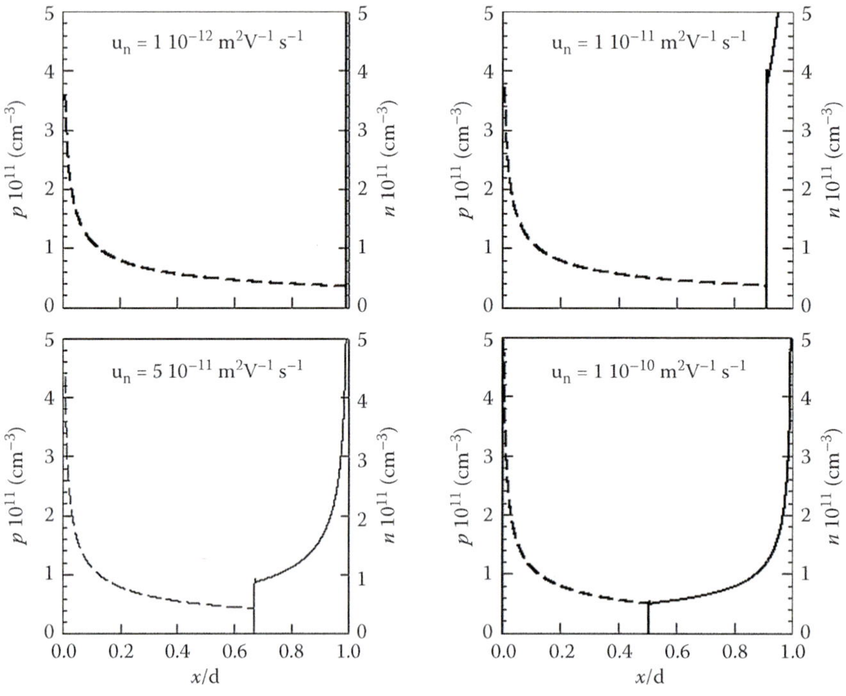

FIGURE 15.7 Carrier distribution in a double-injection diode with infinite recombination at the front where electrons and holes meet. Hole mobility is $u_p = 10^{-10}$ m2V^{-1}s^{-1} and electron mobility as indicated.

GENERAL REFERENCES

Space-charge-limited current: Lampert, 1956; Parmenter and Ruppel, 1959; Lampert and Mark, 1970; Rosenberg and Lampert, 1970; Umstattd et al., 2005; Lopez Varo et al., 2012.

Capacitance in space-charge-limited current: Shao and Wright, 1961; Dascalu, 1969; Kassing and Kähler, 1974; van der Ziel, 1976; Ramachandhran et al., 2006.

REFERENCES

Blom, P. W. M.; de Jong, M. J. M.; van Munster, M. G. Electric-field and temperature dependence of the hole mobility in poly(p-phenylene vinylene). *Physical Review B* 1997, *55*, R656.

Brütting, W.; Berleb, S.; Mücki, A. G. Device physics of organic light-emitting diodes based on molecular materials. *Organic Electronics* 2001, *2*, 1–36.

Bullejos, P. L.; Tejada, J. A. J.; Deen, M. J.; Marinov, O.; Datars, W. R. Unified model for the injection and transport of charge in organic diodes. *Journal of Applied Physics* 2008, *103*, 064504.

Dascalu, D. High-frequency impedance of silicon SCL diode. *Solid-State Electronics* 1969, *12*, 444–446.

Goh, C.; Kline, R. J.; McGehee, M. D.; Kadnikova, E. N.; Frechet, J. M. J. Molecular-weightdependent mobilities in regioregular poly(3-hexyl-thiophene) diodes. *Applied Physics Letters* 2005, *86*, 122110.

Kassing, R. Calculation of the frequency dependence of the admittance of SCLC diodes. *Physics Status Solidii (a)* 1975, *28*, 107.

Kassing, R.; Kähler, E. The small signal behavior of SCLC-diodes with deep traps. *Solid State Communications* 1974, *15*, 673–677.

Krellner, C.; Haas, S.; Goldmann, C.; Pernstich, K. P.; Gundlach, D. J.; Batlogg, B. Density of bulk trap states in organic semiconductor crystals: Discrete levels induced by oxygen in rubrene. *Physical Review B* 2007, *75*, 245115.

Lampert, M. A. Simplified theory of space-charge-limited currents in an insulator with traps. *Physical Review* 1956, *103*, 1648–1656.

Lampert, M. A.; Mark, P. *Current Injection in Solids*; Academic: New York, 1970.

Lopez Varo, P.; Jimenez Tejada, J. A.; Lopez Villanueva, J. A.; Carceller, J. E.; Deen, M. J. Modeling the transition from ohmic to space charge limited current in organic semiconductors. *Organic Electronics* 2012, *13*, 1700–1709.

Mark, P.; Helfrich, W. Space-charge-limited currents in organic crystals. *Journal of Applied Physics* 1962, *33*, 205–215.

Martens, H. C. F.; Huiberts, J. N.; Blom, P. W. M. Simultaneous measurement of electron and hole mobilities in polymer light-emitting diodes. *Applied Physics Letters* 2000, *77*, 1852.

Montero, J. M.; Bisquert, J. Features of capacitance and mobility of injected carriers in organic layers measured by impedance spectroscopy. *Israel Journal of Chemistry* 2012, *52*, 519–528.

Murgatroyd, P. N. Theory of space-charge-limited current enhanced by Frenkel effect. *Journal of Physics D* 1970, *3*, 151.

Nguyen, T. H.; Schmeits, M.; Loebl, H. P. Determination of charge-carrier transport in organic devices by admittance spectroscopy: Application to hole mobility in α-NPD. *Physical Review B* 2007, *75*, 075307.

Parmenter, R. H.; Ruppel, W. Two-carrier-space-charge-limited current in a trap-free insulator. *Journal of Applied Physics* 1959, *30*, 1548–1558.

Ramachandhran, B.; Huizing, H. G. A.; Coehoorn, R. Charge transport in metal/semiconductor/metal devices based on organic semiconductors with an exponential density of states. *Physical Review B* 2006, *73*, 233306.

Rose, A. Space-charge-limited currents in solids. *Physical Review* 1955, *97*, 1538–1544.

Rosenberg, L. M.; Lampert, M. A. Double injection in the perfect insulator: Further analytic results. *Journal of Applied Physics* 1970, *41*, 508.

Shao, J.; Wright, G. T. Characteristics of the space-charge-limited dielectric diode at very high frequencies. *Solid-State Electronics* 1961, *3*, 291–303.

Umstattd, R. J.; Carr, C. G.; Frenzen, C. L.; Luginsland, J. W.; Lau, Y. Y. A simple physical derivation of Child—Langmuir space-charge-limited emission using vacuum capacitance. *American Journal of Physics* 2005, *73*, 160–163.

van der Ziel, A. *Solid-State Electronics*; Prentice-Hall: Englewood Cliffs, 1976.

Wright, G. T. Mechanisms of space-charge-limited current in solids. *Solid-State Electronics* 1961, *2*, 165–189.

16 Impedance and Capacitance Spectroscopies

We address different aspects of the characterization of materials and devices by small perturbation methods in the frequency domain. Due to the fact that conduction and polarization mechanisms are highly sensitive to the frequency of the external perturbation, the impedance and capacitance spectroscopies provide a great deal of information on carrier accumulation and transport. We describe the main characteristics of the frequency-dependent electrical relaxation and transport phenomena that can be observed in organic and inorganic materials. We first summarize the principal functions of data representation and analysis. Then we introduce the main types of relaxation functions that are used to discuss different phenomena such as dielectric relaxation data, conducting or insulating materials, and devices with a combination of internal processes. The electrical response of a system in the frequency domain can be synthetically described by an equivalent circuit model that combines resistances, capacitances, and inductors. The kinetic behavior can also be analyzed by time transients and the response will be determined by the relaxation functions that represent the decay of an applied perturbation in the time domain. The case of disordered solids, such as ion conducting glasses, introduces a universal behavior of the frequency-dependent conductivity. At low frequency, blocking of ions provides the peculiar response associated with electrode polarization phenomena.

16.1 FREQUENCY DOMAIN MEASUREMENTS

In Section 3.5, we discussed some basic aspects of the measurements of capacitance and impedance of a material layer with metal contacts. The impedance is defined as the relationship of small perturbation voltage $(\hat{V})$ to current $(\hat{I})$ for a sinusoidal perturbation at the angular frequency ω

$$Z(\omega) = \frac{\hat{V}(\omega)}{\hat{I}(\omega)} \tag{16.1}$$

Depending on the convention for voltage and current, Equation 16.1 can carry a negative sign. The *impedance spectroscopy* (IS) technique consists of the measurements of the electrical impedance and the tools of interpretation by frequency domain models and equivalent circuits. More generally, beyond the electrical technique, one can measure the system by the modulated perturbation of other physical quantities such as illumination. The complex function that relates input to output quantities is called a *transfer function*. One important technique is the intensity-modulated photocurrent spectroscopy (IMPS), which is widely used in photelectrochemical and solar cell devices (Peter, 1990; Ravisshankar, 2019b).

In an IS measurement, the system is kept at a fixed steady state by imposing stationary constraints such as dc current and illumination intensity, and the $Z(\omega)$ is measured by scanning the frequency $f = \omega/2\pi$ at many values, typically over several decades that comprise the phenomena of interest, for example, from mHz to 10 MHz, with 5–10 points per decade. At each frequency, the impedance meter must check that the $Z(\omega)$ is stable. Below 10^{-5} Hz, the measurement is not practical and at frequencies in excess of THz, the system is dominated by inertial, quantum, and photonic phenomena, which are commented on in Section 19.2.

The measured impedance contains real and imaginary parts that can be expressed as

$$Z(\omega) = Z'(\omega) - iZ''(\omega). \tag{16.2}$$

It is useful to display the measured impedance in different representations. For a sample of electrode area A and thickness d, the complex impedance $Z(\omega)$ allows us to define the complex conductivity

$$\sigma(\omega) = \frac{d}{A} Z(\omega)^{-1}. \tag{16.3}$$

The complex capacitance $C^*(\omega)$ is defined from the impedance as

$$C^*(\omega) = \frac{1}{i\omega Z(\omega)}. \tag{16.4}$$

The capacitance is often written as

$$C^*(\omega) = C'(\omega) - iC''(\omega). \tag{16.5}$$

The real part of the capacitance can be written simply as $\mathrm{Re}[C^*(\omega)] = C'(\omega) \equiv C(\omega)$. The complex capacitance is also called $C(\omega)$, dropping the asterisk if there is no risk of confusion. The complex dielectric constant is obtained from the complex capacitance

$$C^*(\omega) = A\frac{\varepsilon_0 \varepsilon(\omega)}{d}. \quad (16.6)$$

The complex dielectric constant can be separated into real and imaginary parts as

$$\varepsilon(\omega) = \varepsilon'(\omega) - i\varepsilon''(\omega). \quad (16.7)$$

The real part indicates the increase of charge accumulated by polarization, while ε'' is denoted as *dielectric loss* because it determines the dissipation of energy in excess of the dc dissipation. According to Equations 16.3 and 16.6, there is a direct connection between the complex conductivity and complex dielectric constant as follows:

$$\sigma(\omega) = i\omega\varepsilon_0\varepsilon(\omega). \quad (16.8)$$

The ac conductivity can be separated into real and imaginary parts as

$$\sigma(\omega) = \sigma'(\omega) + io'(\omega) \quad (16.9)$$

We have the relationships

$$\sigma'(\omega) = \omega\varepsilon_0\varepsilon''(\omega) \quad (16.10)$$

$$\sigma''(\omega) = \omega\varepsilon_0\varepsilon'(\omega) \quad (16.11)$$

All the previous different representations are just a rearrangement of the originally measured data given by the instrument, the complex impedance. The representations are thus neutral with respect to the physical interpretation, as no model has been formulated so far.

Sometimes we are interested in systems that show a constant or nearly constant conductivity in the low-frequency range; here, a special form of the conductivity is employed, which consists of the sum of dc conduction and dipolar mechanisms that only actuate in the ac regime. Equation 16.9 is modified as follows:

$$\sigma(\omega) = o(0) + i\omega\varepsilon_0\varepsilon_d(\omega) \quad (16.12)$$

where ε_d (ω) is the dipolar polarization. Note that this last equation corresponds to adding a drift current due to mobile carriers to the displacement current. In the time domain, the correspondence of Equation 16.12 is

$$I = Ao(0)\mathbf{E} + A\varepsilon_0\varepsilon_d\frac{\partial \mathbf{E}}{\partial t}. \quad (16.13)$$

Combining Equations 16.7, 16.8, and 16.12, the dielectric loss takes the form

$$\varepsilon'(\omega) = \frac{\sigma(0)}{\omega\varepsilon_0} + \varepsilon_d'(\omega). \quad (16.14)$$

The dc conduction is characterized by a rise of the dielectric loss that increases as f^{-1} toward low frequency. The loss of the dipolar mechanism is related to the real part of the conductivity as follows:

$$\varepsilon_d''(\omega) = \frac{\sigma'(\omega) - \sigma(0)}{\omega\varepsilon_0}. \quad (16.15)$$

16.2 DIELECTRIC RELAXATION FUNCTIONS

To start the analysis of specific applications, let us consider the behavior of a material layer with contacts. If the layer is an insulator that lacks free carriers for conduction, then the capacitance is a dielectric capacitance. However, the dielectric material has dynamic features due to dipolar oscillation and short-range carrier motion that cause time- or frequency-dependent polarization phenomena generally termed *dielectric relaxation*. Dielectric relaxation in insulators is part of a wider class of *relaxation* phenomena that generally describe the recovery toward an equilibrium state and include the structural relaxation of glasses and the decay of photoconductivity and photoluminescence.

The relaxation behavior produces specific features of the complex permittivity $\varepsilon(\omega)$ that are due to the response of different physical elements to the external perturbation. Under a time-varying local electrical field, both the intrinsic dipoles in the material and the induced dipoles react by changing their orientation. These phenomena were first represented by the "ideal" Debye model of freely floating dipoles in a fluid medium (Fröhlich, 1958). Now assume that the material layer is an insulator or conductor that allows the transport of free and localized charge carriers. A number of phenomena associated with the short-range displacement of charge carriers can also provide polarization and relaxation: the dipole moments of complex defects and the short-range hopping of electronic carriers. In addition, under an external bias, carriers may be injected forming a conduction current that flows across the device and charges the chemical capacitances in the system. Therefore, rotating dipoles and conduction mechanisms may provide similar features in the complex permittivity, so that polarization and conduction appear as different facets of a single mechanism. We will focus first on the dielectric response of an insulator that is often associated with the relaxation of permanent dipoles. The conduction mechanisms that also produce components of the complex dielectric constant according to Equation 16.8 are discussed in Section 16.5.

The dipolar units in the bulk of an insulating layer produce a polarization vector $P(\omega)$ that influences the dielectric displacement field in response to the applied field $\mathbf{E}(\omega)$ as follows:

$$D(\omega) = P(\omega) + \varepsilon_0 \mathbf{E}(\omega). \tag{16.16}$$

For a linear dielectric, we obtain the relation

$$D(\omega) = \varepsilon_0 \varepsilon(\omega) \mathbf{E}(\omega). \tag{16.17}$$

The two components of the permittivity are indicated in Equation 16.7. The features of the frequency dependence of the real part of the permittivity and the dielectric loss are closely related. A peak of the dielectric loss that corresponds to a specific dielectric relaxation mechanism produces an increase of the real part of the permittivity. This general property can be rigorously determined by the Kramers-Kronig relationship (Böttcher and Bordewijk, 1978; Jonscher, 1983). For a given mechanism, the polarization increment of the complex permittivity from its high-frequency value can be expressed as

$$\Delta\varepsilon = \varepsilon_s - \varepsilon_\infty. \tag{16.18}$$

Here, ε_s is the "static" value of the real part of the dielectric constant, whereas ε_∞ is the high-frequency part. The latter can be associated with an "instantaneous" relaxation due to the displacement of the electrons with respect to the nuclei, and of ions with respect to their normal equilibrium positions, yielding the *geometrical capacitance*

$$C_g = \frac{\varepsilon_0 \varepsilon_\infty}{d}. \tag{16.19}$$

However, the components of the permittivity that should be included in ε_∞ are not defined in absolute terms. Rather, they depend on the frequency range of interest of the measurement.

The simplest relaxation type is described by the Debye expression

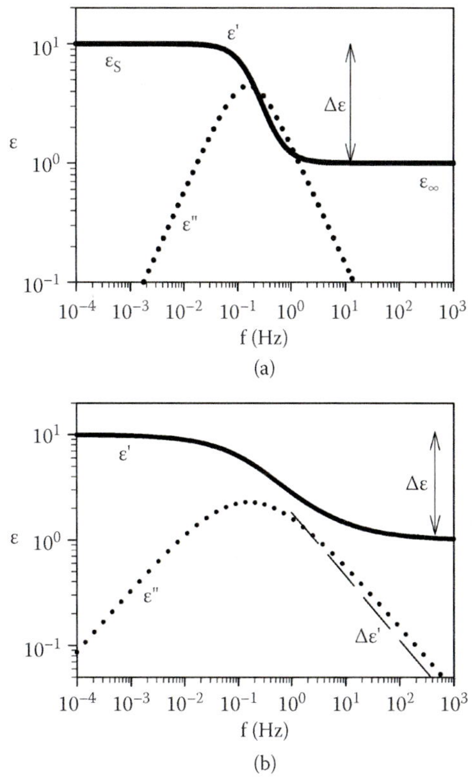

FIGURE 16.1 Real (ε') and imaginary part (ε'') of the complex permittivity for an elementary relaxation with increment $\Delta\varepsilon = 10$. (a) Debye relaxation. (b) Cole-Cole relaxation ($\alpha = 0.4$). The thin dashed line is the asymptote $\Delta\varepsilon'(\omega)$, which has the same slope as $\varepsilon''(\omega)$ in the high-frequency wing.

$$\varepsilon(\omega) = \varepsilon_\infty + \frac{\varepsilon_s - \varepsilon_\infty}{1 + i\omega\tau_d} \tag{16.20}$$

The shape of this last dielectric function is shown in Figure 16.1. The imaginary part of the permittivity makes a symmetric peak centered at the frequency that corresponds to the angular frequency $\omega_d = \tau_d^{-1}$, where τ_d is the relaxation time. At frequencies higher than $\omega_d/2\pi$, the dipole units cannot follow the ac perturbation and the polarization effect vanishes. Thus, the real part of the permittivity ε' changes from the static value ε_s to the high-frequency value by a step indicated in Equation 16.18. The dielectric relaxation of a material over a very broad range of frequencies usually shows a variety of steps of ε' coincident with dielectric loss peaks that represent different types of polarization and relaxation mechanisms as indicated in Figure 16.2a. In the presence of nonoverlapping loss processes, a constant value of the dielectric permittivity is observed between the successive peaks of the dielectric loss (Jonscher, 1983) a. The separated high-frequency dipolar relaxation and low-frequency electrode polarization are shown later in Figure 16.7, where $\Delta\varepsilon$ is observed as a capacitance step.

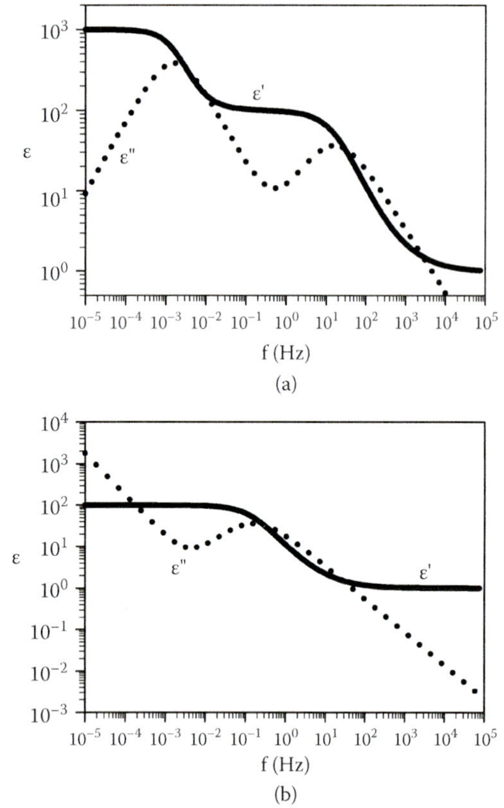

FIGURE 16.2 Real (ε′) and imaginary part (ε″) of the complex permittivity for (a) two consecutive relaxations with values $\varepsilon_\infty = 1$, $\varepsilon_1 = 100$ for the high-frequency process, $\varepsilon_s = 1000$ for the low-frequency process, and (b) a single dielectric relaxation peak combined with a small dc conduction component.

A number of phenomenological relaxation functions are practically used to describe the dielectric relaxation in disordered or entangled systems (Böttcher and Bordewijk, 1978). These models usually contain one or two adding parameters that recover the equation for a single relaxation time at certain parameter values. Among the more widely used is the Cole-Cole relaxation function,

$$\varepsilon(\omega) = \varepsilon_\infty + \frac{\varepsilon_s - \varepsilon_\infty}{1 + (i\omega\tau_C)^{1-\alpha}}. \tag{16.21}$$

The Cole-Cole function is characterized by a broader dielectric loss peak and longer transition range of the step Δε than Debye relaxation; see Figure 16.1b. The Cole-Cole model and similar phenomenological functions can be considered as a combination of elementary Debye relaxations with a broad distribution of relaxation times. τ_C represents the central characteristic time in the corresponding time distribution. The parameter α ($0 < \alpha < 1$) states the broadening of the distribution and in the limiting case of $\alpha = 0$, a peaked delta function is obtained for the time distribution so that Equation 16.21 reduces to a single relaxation time process.

Historically, the generalized dielectric relaxation was observed in the form of a ubiquitous power law decay in time $j(t) \propto t^{-n}$ of depolarization currents, denoted the Curie-von Schweidler law, instead of the exponential decay predicted by the Debye relaxation; see Section 16.4. The occurrence of broad relaxation and dispersive exponents in the relaxation laws is not generally due to a simple combination of independent processes, but rather due to cooperative and entangled relaxation phenomena (Jonscher, 1977; Halpern, 1992). Indeed, in organic and inorganic solids, and in softmatter formulations such as ionic liquids, one encounters the general phenomenon of *frequency dispersion* in which the limiting tails of the relaxation function depart from the idealized case of a Debye relaxation associated with independently relaxing units.

For Equation 16.21, the asymptotes of the real and imaginary parts of the permittivity for $\omega \to 0$ are related to the dispersive exponent as follows:

$$\varepsilon'(\omega) \sim \varepsilon_s \tag{16.22}$$

$$\varepsilon''(\omega) \sim \omega^{1-\alpha} \tag{16.23}$$

and in the high-frequency part, both real and imaginary parts show the same slope as indicated in Figure 16.1b

$$\Delta\varepsilon'(\omega) = \varepsilon'(\omega) - \varepsilon_\infty \sim \omega^{\alpha-1} \tag{16.24}$$

$$\varepsilon''(\omega) \sim \omega^{\alpha-1}. \tag{16.25}$$

16.3 RESISTANCE AND CAPACITANCE IN EQUIVALENT CIRCUIT MODELS

The electrical ac perturbation of a device stimulates internal physical processes that respond with specific impedances. The resistance R is a real and positive Z. The resistances are associated with different processes of carrier flux, either in the bulk materials or at interfaces. We have previously analyzed specific examples such as the charge transfer resistance at the electrochemical contact in Section 6.3 and the charge transport resistance in Section 12.2. For any type of energy conversion device, the resistance is a central quantity that determines the performance and particularly the power conversion efficiency in steady-state operation, as discussed below. However, the impedance response also contains information on the kinetics of the phenomena that are measured. As explained in Section 3.5, any kind of relaxation introduces a time delay in the response that brings in an imaginary component in the impedance, which is associated with capacitive behavior. The capacitances represent a charge storage mechanism and they can be divided into two main kinds, as discussed in Chapter 7. The dielectric capacitances are associated with an internal electric field that is produced by spatial charge separation, and they can be of many kinds, due to bulk relaxation phenomena or to contact interfaces. The other type is the chemical capacitance due to the variation of chemical potential or carrier concentration in one type or a distribution of electronic states. Kinetic phenomena also cause inductive responses (Bisquert, 2011).

In the measurement of material systems, it is rather frequent that the impedance response is composed of the combination of several processes. These phenomena can be included in a physicochemical model based on transport and conservation equations, as discussed in Section 12.8. Solving the full model for a small perturbation condition, we obtain the analytical impedance model. However, in most cases, one can observe that the impedance model, which corresponds to a linearized form of the transport equations, consists of a combination of resistances and capacitors. The time constants and the connection of the elements describing such processes depend on the internal structure of the system. Two elements are arranged in series if they carry the same current, or in parallel if they respond to the same voltage. Therefore, the full model can be translated into the scheme of an equivalent circuit (EC) that is very useful for data analysis (Bertoluzzi and Bisquert, 2012). Furthermore, the EC enables a rapid connection between the detailed information included in the model and the actual experimental response. Usually, the model needs to be simplified to adapt it to those processes that can actually be observed in the experimental data.

It should be recognized that the EC approach presents two main issues. First, we must recall that IS is a *two-contact measurement*. Different internal elements may not be separated in the frequency response function, so that the information of different internal regions in the sample may provide a single EC element. In this type of situation, there is not an attribution of different representative circuit elements in correspondence to different internal phenomena. The impedance measures only the impediment to current flow, not the internal geometrical or material composition. Second, the expression of a given impedance model may be interpreted in terms of several equivalent circuit representations. To tackle both issues, it is therefore necessary to perform further experiments and verify the physical meaning of the extracted parameters. One useful approach to remove ambiguity consists of plotting the voltage variations of the capacitances and resistances obtained with a given EC. The progression of capacitances and resistances along the voltage variations gives valuable information about the meaning of each element and the overall behavior of the system. In the example shown in Figure 16.3, two different models for water oxidation at the illuminated semiconductor surface, involving distinct charge transfer mechanisms, provide the same EC model but the physical characteristics of the parameters can be distinguished by tracking the capacitance characteristics, which allows the identification of the physical model of the system (Bertoluzzi et al., 2015). In IS measurements, we obtain the data and such data may be transformed as desired between the different representations of Section 16.1 to find a suitable physical interpretation. In the analysis of energy devices, it is often particularly important to distinguish the contact/interfacial and bulk phenomena, as the former depend on material combinations while the latter depend on a material's property. For example, the discussion in Section 16.3 focused on polarization mechanisms for a dielectric capacitance that is distributed homogeneously in the bulk of the sample. In some fields of study, the dielectric relaxation phenomena are the central topic of interest so that $\varepsilon(\omega)$ and $\sigma(\omega)$ are preferred quantities to present the data. Then, the permittivity and conductivity are intensive bulk properties of the material. They are independent of sample thickness, and contact effects should be identified and removed from the relevant data analysis. However, electrical measurements contain different types of interfacial effects, especially those related to the presence of the outer contacts. If the main interest is in the value of resistances, one adopts the representation $Z(\omega)$. But the capacitances also play a crucial role for the interpretation, since different elements with similar resistance provide very distinct spectral features if their associated capacitances sufficiently differ in magnitude. The capacitance is, therefore, a key to the understanding of the origin of the measured resistances. To this end, representations such as $C(\omega)$ become very useful to analyze the data.

The main goal for understanding the operation of a photoelectrode or a solar cell is to explain the steady-state $j - V$ behavior (Bisquert et al., 2016c). This is the curve used to derive the power conversion efficiency, considering the number of photons impinging on the semiconductor (Murphy et al., 2006), so that ultimately, this curve provides the effectiveness of the device for energy production from sunlight. Similarly, for different types of devices such as light-emitting diodes (LEDs), the $j - V$ characteristic is a central feature of the electrical operation. Steady-state $j - V$ curves can be treated with analytical models, but since the operation of the devices involves a large number of internal steps, it is not conclusive to analyze the system only in the steady state, even if this is the only information that we ultimately wish to obtain. The frequency-resolved small perturbation methods provide detailed dynamic information about a given point of the $j - V$ curve. As we vary the value of the steady-state voltage, the internal characteristics of the device can become very different and this is reflected in the impedance and capacitance spectra that can be interpreted with suitable equivalent circuits.

Let us choose a certain point of bias voltage V_0 with the associated current density j_0. At this point, a small displacement of voltage $\hat{V}(0)$ implies a change of current $\hat{j}(0)$. The value $\omega = 0$ in parentheses indicates that the displacement is infinitely slow. The connection between the small perturbation values is

$$\hat{j} = \frac{\partial j}{\partial V}\hat{V}. \tag{16.26}$$

The impedance in Equation 16.1 is

$$Z(0) = \frac{\hat{V}}{A\hat{j}} = R_{tot}. \tag{16.27}$$

In the limit $\omega = 0$, all kinetic parts of the impedance vanish and the impedance is a resistance, called the *total resistance*, with the value

$$R_{tot} = \left(A\frac{\partial j}{\partial V}\right)^{-1}. \tag{16.28}$$

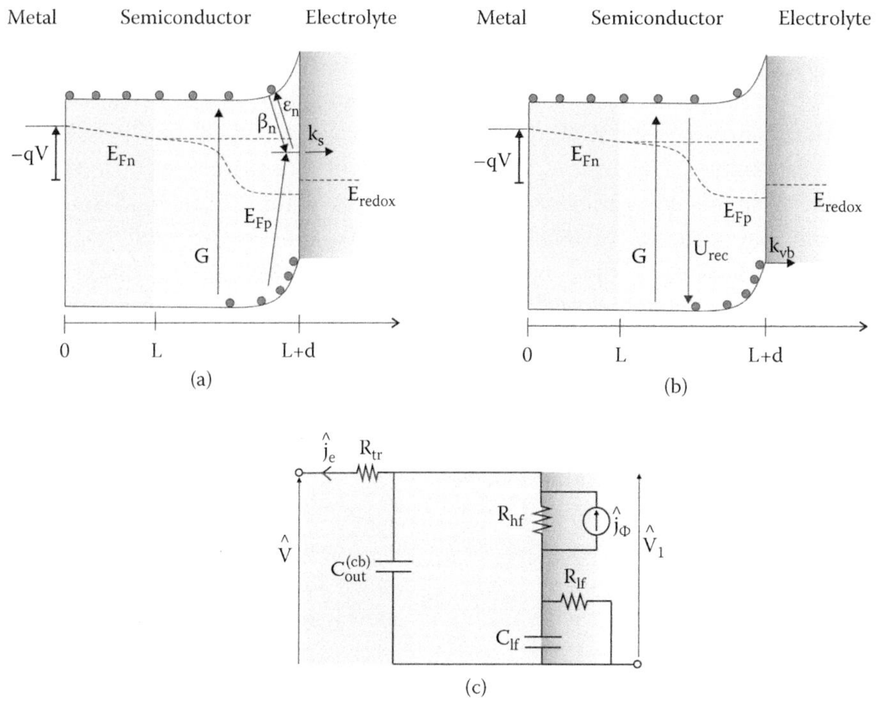

FIGURE 16.3 (a) Indirect charge transfer model for solar fuel production by a photoanode of length $L + d$. Electrons and holes are photogenerated at a rate G. Electrons can be trapped (kinetic constant: β_n) and detrapped (ε_n) in surface states at the semiconductor–electrolyte interface. All the photogenerated holes are trapped and can either recombine with trapped electrons or be transferred at a rate k_s to the redox couple in solution with energy E_{redox}. The concentration of free electrons and holes follow the quasi-Fermi levels, E_{Fn} and E_{Fp}. At the metal–semiconductor interface, E_{Fn} is directly modulated by the applied voltage V. (b) Direct charge transfer model. Photogenerated holes can either recombine with electrons at a rate U_{rec} or be transferred to the electrolyte from the valence band at a rate k_{vb}. (c) The general EC for both models, calculated for a small perturbation of light intensity ($\hat{\Phi}$) and a small perturbation of voltage ($\hat{V}$). The small perturbation of light intensity is modeled by a current source. Note that the voltage perturbation can arise from the light intensity perturbation, as for intensity-modulated photocurrent spectroscopy, or can be imposed electrically, as for IS. In those circuits, R_{tr} is the transport resistance, C_μ is the conduction band capacitance, R_{lf} and C_{lf} are the low-frequency resistance and capacitance and R_{hf} is the high-frequency resistance. (Reproduced with permission from Bertoluzzi and Bisquert, 2017.)

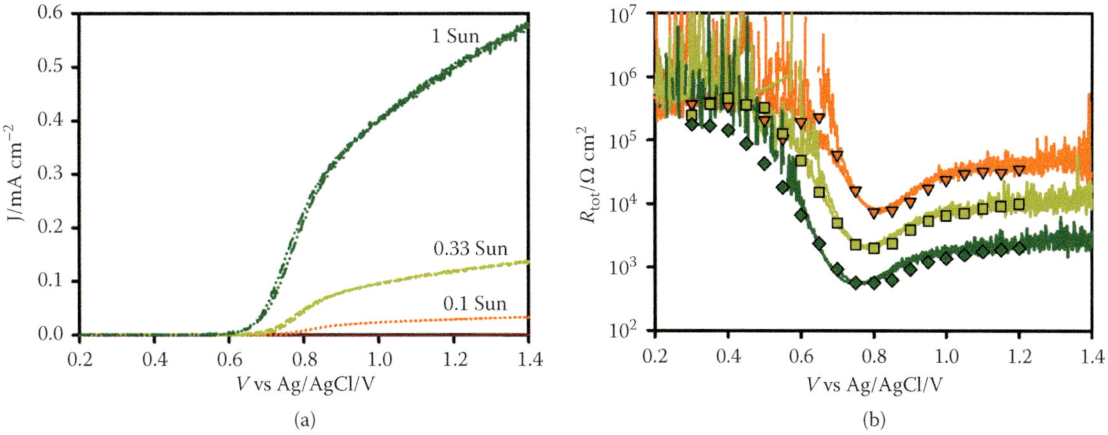

FIGURE 16.4 (a) Current density-voltage ($j(V)$) curves for a 60-nm-thick Fe_2O_3 electrode at pH 6.9 under different illumination intensities. (b) R_{tot} values determined by two methods: (1) calculated dV/dj from the $j(V)$ curves in upper pannel (lines) and (2) the calculated R_{tot} values by adding the resistances associated with charge transfer from IS (symbols). (Reprinted with permission from Klahr et al., 2012.)

We conclude that the small perturbation quantities give a *derivative* of voltage with respect to current. This is the reciprocal of the slope of the $j - V$ curve, which is in turn the dc resistance of the solar cell or photoelectrode in those particular conditions. Figure 16.4 illustrates the point by giving perfect agreement between the R_{tot} measured by impedance spectroscopy and that derived from the local slope of the $j - V$ curve in photoelectrodes for water splitting.

A more general analysis includes the incident photon flux Φ_{ph}, forming the (j_e, V, Φ_{ph}) space of external variables. The steady-state performance forms a surface (called Λ_0 surface) as shown in Figure 16.5. The modulation of the different variables produces different transfer functions as explained by Bertoluzzi and Bisquert (2017). It was shown that all different small perturbation techniques must be described by a single unique equivalent circuit. The combination of methods such as IS and Intensity-Modulated Photocurrent Spectroscopy (IMPS) allows one to establish a consistent model, and furthermore it is possible that IMPS reveals features that remain hidden in IS spectra. This has been shown experimentally for some types of hybrid perovskite solar cells (Ravishankar et al. 2019a, 2019b).

In addition to the general formulation based on Figure 16.5, still an additional magnitude can be modulated, namely the outgoing radiation flux produced by luminescence; see Chapter 18. The Light-Modulated Emission Spectroscopy is an impedance of light that may provide significant information on the coupling of electronic and optical phenomena in a semiconductor sample, provided that the material shows significant emission (Ansari-Rad and Bisquert, 2017).

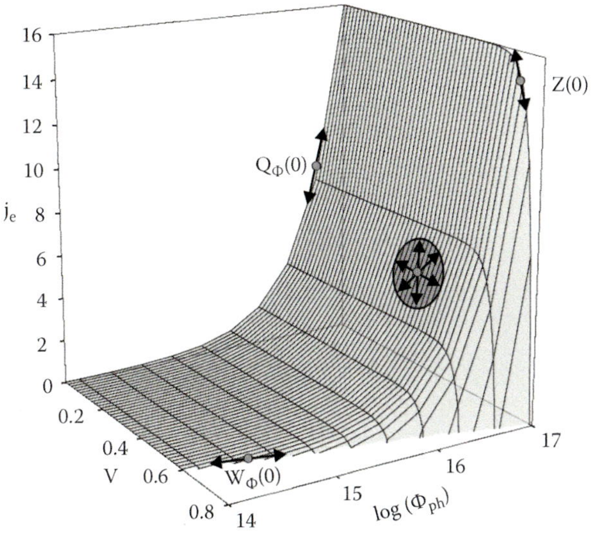

FIGURE 16.5 Steady-state performance of a solar energy conversion device, the light Λ_0 surface in terms of electrical current j_e, voltage V, and illumination flux Φ_{ph}. When these variables are taken in pairs, the slope of the curve is given by the zero frequency value of the transfer functions, Z, Q_Φ and W_Φ. The slopes are shown in the main 2D planes but the same analysis can be performed at any point of the surface (when the three variables are not zero) as indicated by the circular tangent plane. (Reproduced with permission from Bertoluzzi and Bisquert, 2017.)

A first example of an equivalent circuit is the R_1C_1 series combination. From the impedance

$$Z(\omega) = R_1 + \frac{1}{i\omega C_1}, \tag{16.29}$$

we obtain the complex capacitance

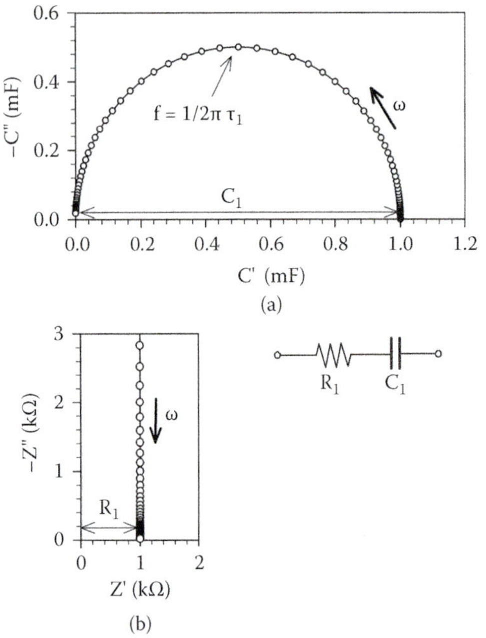

FIGURE 16.6 Representations of the impedance of an equivalent circuit. R_1 = 1 kΩ, C_1 = 1 mF, τ_1 = 1 s. The thick arrows indicate the direction of increasing angular frequency ω.

$$C^*(\omega) = \frac{C_1}{1 + i\omega\tau_1}, \tag{16.30}$$

where the characteristic frequency of the relaxation ω_1 and the relaxation time τ_1 are defi ed as

$$\omega_1 = \frac{1}{\tau_1} = \frac{1}{R_1 C_1}. \tag{16.31}$$

When Equation 16.30 represents a bulk capacitance, it is equivalent to the model of Equation 16.20. The plot of the capacitance in the complex plane is shown in Figure 16.6a. The capacitance displays an arc from the dc value $C^*(0) = C_1$ to the high-frequency value. The top of the arc occurs at the characteristic frequency of the relaxation ω_1. The impedance, shown in the complex plane in Figure 16.6b, forms a vertical line.

In a device composed of several layers, the EC will show corresponding equivalent circuit elements. Examples of thin film transistors are presented in Figure 14.11. Note that the capacitance of the channel corresponds to interfacial effects such as depletion, inversion, or traps. Another important example of an equivalent circuit is the *RC* parallel combination (Figure 16.7), often applied for the electrochemical interface with the assumption of independent charging and transfer currents (Randles, 1947; Grahame, 1952); see Chapter 6. In this model, R_1C is the combination of the charge transfer resistance and the double-layer capacitance. The admittance of the combination is

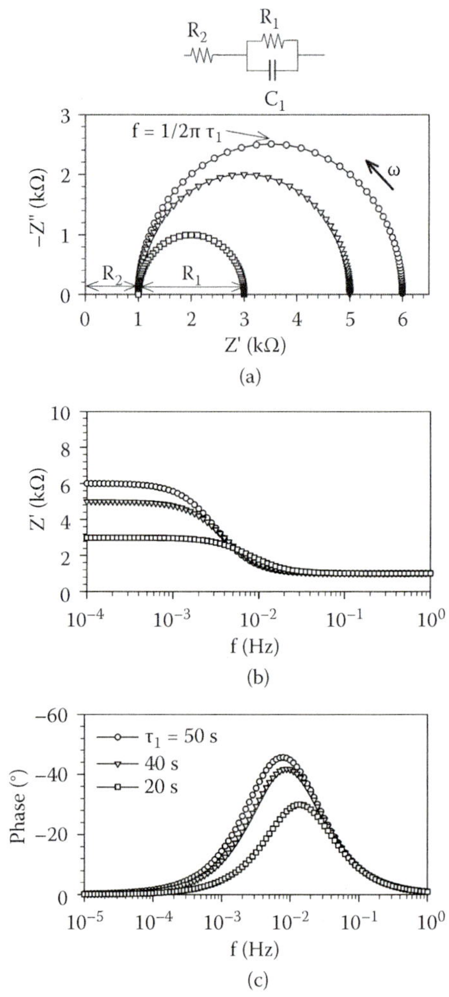

FIGURE 16.7 Representations of the impedance of an equivalent circuit. R_1 takes values 5,4,2 kΩ, C_1 = 10 mF, τ_1 = 50,40,20 s, R_2 = 1 kΩ. (a) Complex plane impedance plot. (b) Real part of the impedance vs. frequency. (c) The phase angle of the impedance vs. frequency.

$$Y_1(\omega) = \frac{1}{R_1} + i\omega C_1. \tag{16.32}$$

With the addition of a series resistance R_2, the impedance is

$$Z(\omega) = R_2 + Y_1 = R_2 + \frac{R_1}{1 + i\omega\tau_1}. \tag{16.33}$$

The complex plane impedance plot is shown in Figure 16.7a. The parallel *RC* forms an arc in the complex plane that is shifted positively along the real axis by the series resistance R_2. In this circuit, dc conduction is determined by the low-frequency intercept, that is, the total resistance $Z(0) = R_{dc} = R_1 + R_2$.

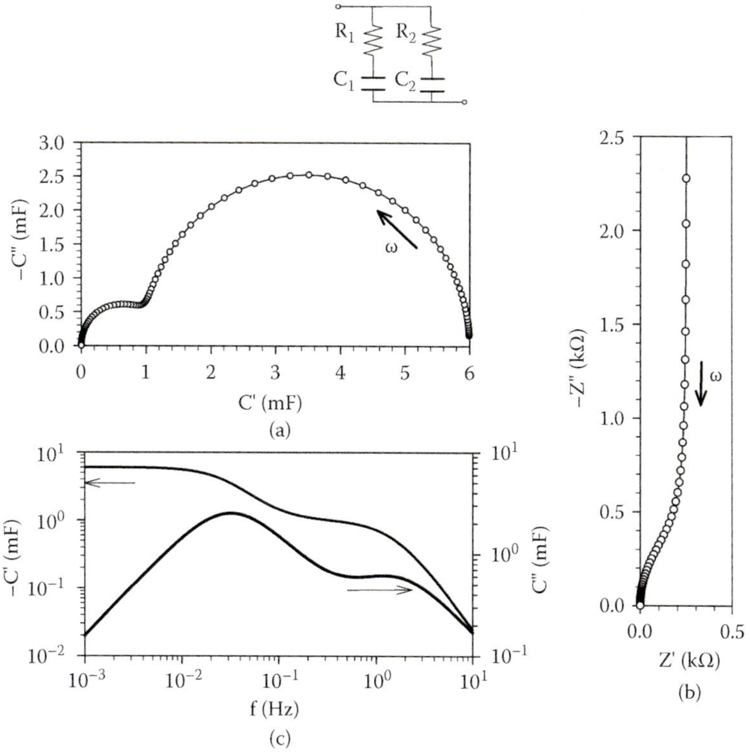

FIGURE 16.8 Representations of the capacitance of an equivalent circuit. R_1 = 1 kΩ, C_1 = 5 mF, τ_1 = 5 s, R_2 = 0.1 kΩ, C_2 = 1 mF, τ_2 = 0.1 s. (a) Complex plane capacitance plot. (b) Complex plane impedance plot. (c) Real and imaginary part of the capacitance vs. frequency.

Figure 16.8 shows an example of a system composed of several relaxation phenomena represented by two series *RC* circuits connected in parallel. This circuit allows no passage of dc current because the low-frequency current is blocked by the capacitors. The representation of the data in the complex impedance plane in Figure 16.8b only shows the blocking response at low frequencies and an additional feature at high frequency. If we analyze the capacitance in the complex plane (Figure 16.8a), we see the two relaxation processes clearly resolved. The representation of the capacitance vs. frequency, *C′(f)*, is also very useful, indicating two plateaus of the real part of the capacitance between the loss peaks of the capacitance, C″, associated with distinct relaxation processes. Note the equivalence of this EC model to the model of Figure 16.2a. The experimental results in Figure 16.9 summarize the capacitance versus frequency behavior of a broad class of lead halide perovskite solar cells. By measurements of samples of different thicknesses, it has been possible to conclude that the high-frequency plateau relates to the bulk polarization of the sample, while the low-frequency rise of the capacitance is associated with the response of the contacts (Almora et al., 2015; Zarazua et al., 2016).

16.4 RELAXATION IN TIME DOMAIN

The relaxation time τ in Equation 16.31 has been shown to play a major role in the interpretation of dielectric relaxation phenomena measured in the frequency domain. Let us look at the response of the system in the time domain. We consider the evolution of the electrical current in the circuit of Figure 16.6 after a step of voltage ΔV is applied at time $t = 0$. This is described as $V(t) = \Delta V \cdot \theta(t)$, where $\theta(t)$ is the unit step function. In the frequency domain, the step voltage has the expression

$$\hat{V}(s) = \frac{\Delta V}{s} \tag{16.34}$$

in terms of the variable $s = i\omega$. The electrical current is

$$\hat{I}(s) = \frac{\hat{V}(s)}{Z(s)} = \frac{\Delta V}{sZ(s)} = \frac{\tau_1 \Delta V}{R_1(1+s\tau_1)}. \tag{16.35}$$

Inverting Equation 16.35 to the time domain, we obtain the exponential decay function

$$I(t) = \frac{\Delta V}{R_1} e^{-t/\tau_1}. \tag{16.36}$$

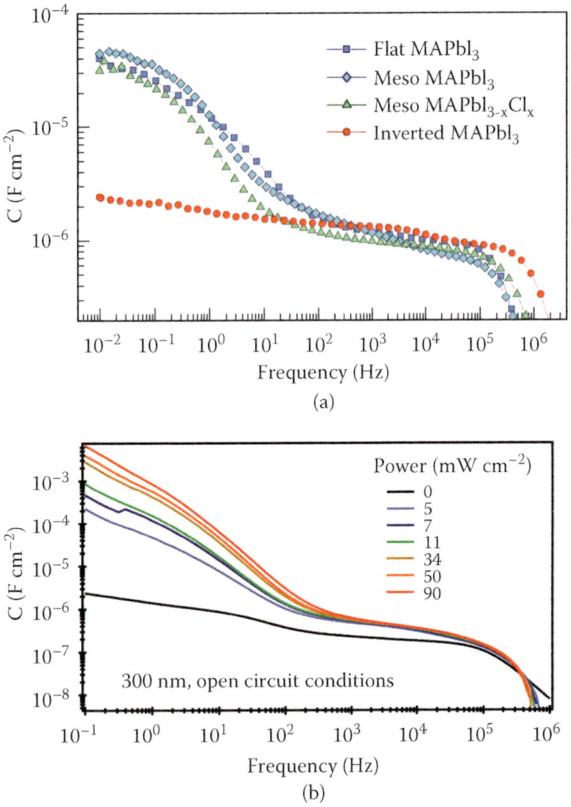

FIGURE 16.9 (a) Comparison of the capacitance spectra of a variety of perovskite solar cells with compact (c) layer and mesoporous (m) or planar TiO_2 contacts (cTiO_2/mTiO_2/$MAPbI_{3-x}Cl_x$/*spiro*-OMeTAD, cTiO_2/mTiO_2/$MAPbI_3$/*spiro*-OMeTAD, and cTiO2/$MAPbI_3$/*spiro*- OMeTAD) and inverted (PEDOT:PSS/$MAPbI_3$/$PC_{70}BM$) structures measured in the dark at room temperature. (b) Effect of the illumination intensity on the low-frequency capacitance of $CH(NH_2)_2PbI_{3-x}Cl_x$-based planar solar cell. (Reproduced with permission from Garcia-Belmonte and Bisquert, 2016.)

The process described by Equations 16.20, 16.30, and 16.36 is an elementary relaxation defined by the constants in Equation 16.31.

We now consider the Cole-Cole relaxation function indicated in Equation 16.21, which is the simplest example of a non-Debye response. The Cole-Cole function is defined in terms of a suitably normalized dielectric permittivity function $\phi(\omega)$ as

$$\phi(\omega)=\frac{\varepsilon(\omega)-\varepsilon(\infty)}{\varepsilon(0)-\varepsilon(\infty)}=\frac{1}{1+(i\omega\tau_C)^a}, \qquad (16.37)$$

where $a = 1 - \alpha$. The pulse-response function is obtained by writing Equation 16.37 as a series expansion and taking the Laplace transform (Cole and Cole, 1942). This leads to an expansion as a power series in the reduced time $\xi = t/\tau_C$

$$f_+(t)=\frac{1}{\tau_C}\sum_{n=0}^{\infty}\frac{(-1)^n}{\Gamma[(n+1)a]}\xi^{(n+1)a-1} \qquad (16.38)$$

and also to an expansion in inverse powers

$$f_-(t)=\frac{1}{\tau_C}\sum_{n=1}^{\infty}\frac{(-1)^n}{\Gamma(-na)}\xi^{-na-1} \qquad (16.39)$$

For $t << \tau_c$, the current is described by the leading term of Equation 16.38

$$f_+(t)=\frac{1}{\tau_C}\frac{1}{\Gamma(a)}\xi^{a-1}, \xi \ll 1 \qquad (16.40)$$

and similarly for long time, one finds from Equation 16.39 that

$$f_-(t)=\frac{1}{\tau_C}\frac{1}{\Gamma(-a)}\xi^{-(a+1)}, \xi \ll 1. \qquad (16.41)$$

The last two equations are obviously related by Fourier transform to Equations 16.23 and 16.25. We conclude that frequency dispersion phenomena produce a power law response in both frequency and time domains, corresponding to the Curie-von Scweidler law. In fact, the expression for the current at short times can be related to the high frequency loss by the formula

$$f(t)=\omega_{eq}\phi'(\omega_{eq}). \qquad (16.42)$$

where the equivalent angular frequency is given by

$$\omega_{eq}=\frac{c}{\xi\tau_C} \qquad (16.43)$$

and c is a factor of order unity. The current at long times can be related to the low frequency loss by the same formula in Equation 16.42, but with a different value of c (Baird, 1968).

The isothermal decay of the polarization of glass-forming and polymeric systems is well described in a wide variety of cases by the *stretched exponential* or Kohlrausch-Williams-Watts (KWW) function

$$F(t)=\exp\left[-(t/\tau_{se})^{\beta}\right]. \qquad (16.44)$$

A stretched exponential decay in the time domain can correspond in practice to a number of different forms of response in the frequency domain. Indeed, a fit to a stretched exponential is usually only performed for

0.99 > $F(t)$ > 0.01 so that a variety of types of behavior at very low and very high frequencies will fit the same stretched exponential. The Fourier transform in Equation 16.44 can be expressed in several ways, for example, as the series (Williams, 1985)

$$\phi(\omega) = \sum_{n=1}^{\infty} \frac{\Gamma(n\beta+1)}{\Gamma(n+1)} \frac{(-1)^{n-1}}{(i\omega\tau_{se})^{n\beta}} \tag{16.45}$$

but for its practical application, this and related expressions present problems of slow convergence, and consequently, the KWW function as such is not well suited for treating frequency-resolved measurements of dielectric relaxation. Several procedures have been suggested to relate the results of a fit to Equation 16.44 to the corresponding analysis in the frequency domain. Alvarez et al. (1991) used a distribution of relaxation times to examine the connection between the KWW function and the Havrilak-Negami function,

$$\phi(\omega) = \frac{1}{\left[1+(i\omega\tau_{HN})^{\alpha}\right]^{\gamma}}. \tag{16.46}$$

This last relaxation model has been widely used in the analysis of the α relaxation in polymers.

16.5 UNIVERSAL PROPERTIES OF THE FREQUENCY-DEPENDENT CONDUCTIVITY

It has been discussed in Section 5.11 that the transport of ions in a solid-state environment occurs by hopping between localized states. This picture includes fast transitions between energetically favorable sites in the interstices of the solid network by overcoming free-energy barriers between adjacent sites. A consequence of this is that the transport cannot be described by a single constant conductivity at all the frequencies. An enormous variety of ion conducting glasses have been investigated and a series of quite general or even "universal" patterns of conduction and polarization in these materials has been determined (Dyre and Schroder, 2000). In the explanation of these patterns, the structural disorder that is always present in glass usually plays a key role. The real part of the conductivity, taken over a frequency range from mHz to GHz, shows a constant domain at low frequencies σ_0 and increases rapidly at high frequencies as described by the expression (Jonscher, 1977; Almond et al., 1985)

$$\sigma'(\omega) = \sigma_0\left[1+(\omega/\omega_m)^n\right] \tag{16.47}$$

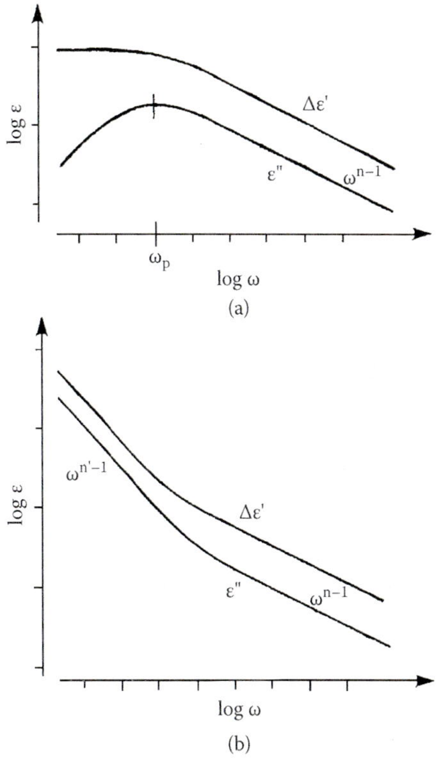

FIGURE 16.10 Increment of the real part of the complex permittivity (ε′) and the imaginary part (ε″) for two types of dielectric response, showing the universal behavior with ω^{n-1} at high frequencies. (Reproduced with permission from Jonscher, 1977.)

where the frequency ω_m marks the onset of the dispersion, which takes the form of a power law $\sigma' \propto \omega^n$, with the exponent $0 < n \le 1$ usually lying in the range 0.6–0.8. Note that Equation 16.47 is consistent with both types of "universal" relaxation behavior of the complex permittivity pointed out by Jonscher, shown in Figure 16.10. Indeed, by Equation 16.10, Equation 16.47 corresponds to $\varepsilon'' \approx \omega^{n-1}$ at large frequencies. The frequency dependence of the conductivity in Equation 16.47 was first explained from models that treat dc and ac conduction in solids as independent macroscopic phenomena. For example, in amorphous semiconductors, the dc conductivity was described by models for hopping of electrons near the Fermi level, such as the variable-range hopping (VRH) described earlier in Section 13.10. The acconduction was typically evaluated using the pair approximation (PA) introduced by Pollak and Geballe (1961) in which the ac response was given by the sum of the individual responses of pairs of sites as first described by Fröhlich (1958), distributed at random through the material. There are, however, some limitations in combining independent dc and ac contributions to the conductivity. This procedure neglects the contributions to

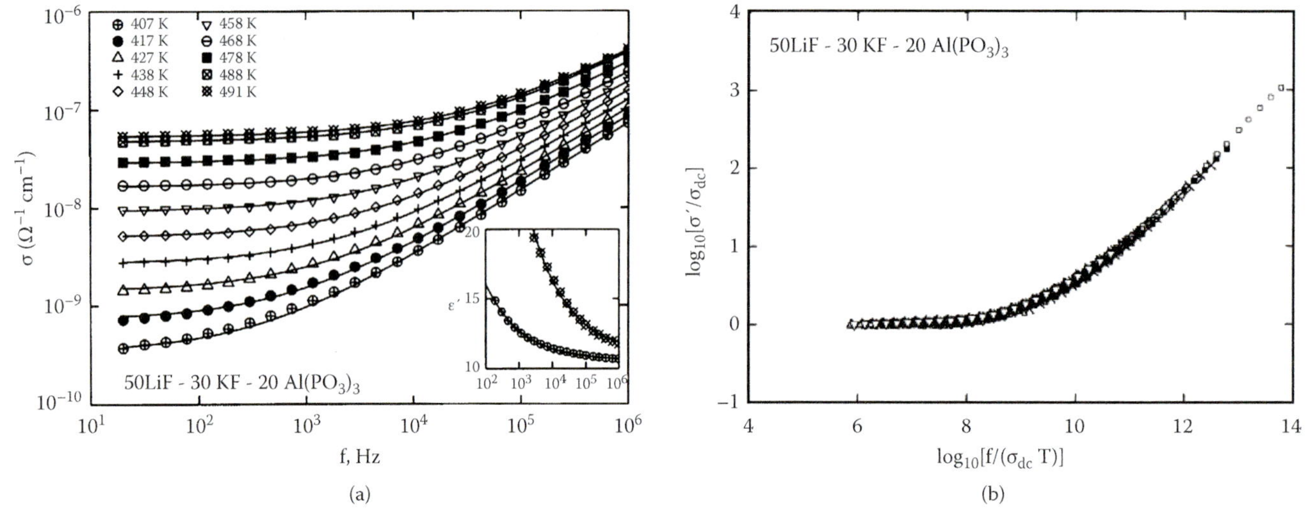

FIGURE 16.11 (a) Variation of the ac conductivity with frequency of the mixed alkali glass with composition 50 LiF–30 KF–20 $Al(PO_3)_3$ over the temperature range 407–491 K. The lines are fits to the data with the sum of a power law and $\sigma(0)$. The inset shows the variation of the real part of the dielectric constant with frequency for the lowest (407 K) and the highest (491 K) temperature. The lines are the fits to the data with a sum of a power law and ε_∞. The dielectric data below 100 Hz, due to electrode polarization, are not included in the inset. (b) Master curve for the data in (a). (Reproduced with permission from Kulkarni et al., 1998.)

the conductivity from clusters containing more than two states, apart from the infinite percolation cluster, which is treated separately. Consequently, the PA fails in the low-frequency region (Zvyagin, 1980).

Additional universal properties of the ac conductivity of disordered systems were shown first by Taylor (1956) and Isard (1961). Figure 16.11a shows the characteristic behavior of ac conductivity of a mixed alkali glass at different temperatures. The shape of the relaxation curve is shifted in frequency due to the temperature dependence of the onset frequency ω_m and in the vertical axis by the changes of dc conductivity $\sigma(0)$. To remove these dependencies, a representation in terms of a scaled frequency and conductivity in terms of a universal function F as follows

$$\tilde{\sigma} \equiv \sigma'(\omega)/\sigma(0) = F\left(C\frac{\omega}{\sigma(0)}\right) \tag{16.48}$$

allows us to construct a single master plot into which the curves at different temperatures collapse (Dyre and Schroder, 2000), as shown in Figure 16.11b, implying that the shape of the conductivity curve is nearly independent of the temperature. It has also been observed that the frequency of onset of ac conduction, which corresponds to the peak of the dielectric loss, satisfies the Barton-Nakajima-Namikawa relation (Dyre and Schroder, 2000)

$$\sigma(0) = p\Delta\varepsilon\varepsilon_0\omega_m, \tag{16.49}$$

where the step of the dielectric constant is given in Equation 16.18 and p is a numerical constant of order 1. Equation 16.49 can be obtained if the constant in Equation 16.48 satisfies the relation $C = \Delta\varepsilon\varepsilon_0$, as suggested by scaling plots presented by (Sidebottom, 2009).

The connection between dc conduction and onset of relaxation features indicated in Equation 16.48 shows that that dc and ac conduction should be described by the same mechanism, which should also account also the intermediate frequency range at the onset of ac conduction. Several approaches have been developed to achieve this goal. The characteristics of ac conduction in solids can be obtained from the effects of a static disordered energy landscape of the ion dynamics. At high frequencies, the ions can jump back and forth between two sites connected by a low-energy barrier. This type of localized hopping gives rise to a relaxation process. In a larger time scale, the ions realize long-range excursions over many sites, which explains a dc conduction process. To contribute to the long-range transport, the ions have to surmount the percolation barriers, which are the highest free-energy barriers through the network, and the dispersion is caused by localized motions in regions limited by these barriers. Therefore, conduction and polarization are two manifestations of a single mechanism, and so the onset of relaxation by short scale motion at a certain frequency provides an increase of the ac conductivity. For instance, Bryksin (1980) used the effective medium approximation to calculate the conductivity for electrons tunneling between nearest neighbors in a solid with electron sites randomly located in space. A similar result was later derived by Dyre (1988) using the coherent medium approximation for a model in which particles hop over barriers with randomly distributed energies. Despite advances in the picture of ion conduction in glasses, the process is not completely understood and there remain open questions such as the characteristic scale of the relaxation at the onset frequency, the relationship of percolation to the time–temperature superposition principle, the role of coulombic and exclusion interactions, and the dependence of diffusivity on concentration (Sidebottom, 2009).

16.6 ELECTRODE POLARIZATION

Different types of polarization occur at electrodes and device contacts. The Helmholtz, or more generally the double-layer capacitance in electrochemical systems, and the Schottky barrier leading to the Mott-Schottky plot in semiconductor barriers are familiar instances of polarization at contacts. In ionic conducting systems, one finds the phenomenon of electrode polarization, which is related to the fact that ions cannot penetrate the metal collector contact, so that conductivity is blocked at low frequency. Therefore, the general trend of Equation 16.47, due to the bulk conductivity of the material, will be supplemented by a decrease of the conductivity toward low frequency as shown in Figure 16.12. In a pure ionic conductor, the conductivity must actually decrease to null value as the frequency tends to zero.

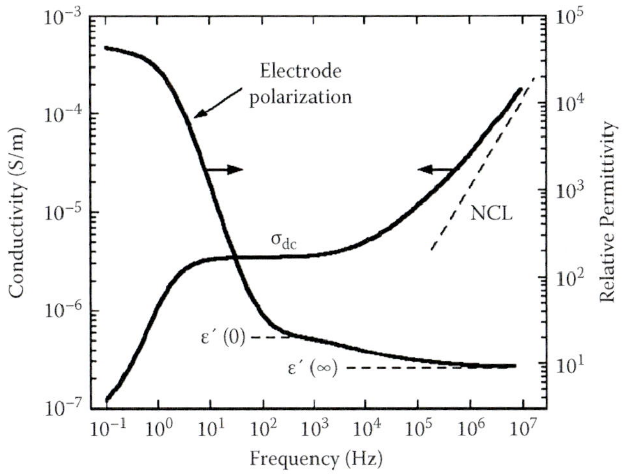

FIGURE 16.12 Representation of ac conductivity and the real part of the permittivity for a lithium-phosphate glass, indicating the step Δε, the electrode polarization, and the "nearly constant loss" (p = 1) regime of the power law dependence of frequency. (Reproduced with permission from Dyre et al., 2009.)

The accumulation of ions at the contact interface produces a capacitance due to surface space charge. This capacitance occurs at a short distance from the contact and it is independent of film thickness d if the film is thick enough. In a permittivity plot, electrode polarization gives rise to very large values of the apparent permittivity as shown in Figure 16.12, which occurs in addition to the conductivity polarization that is indicated as $\varepsilon'(0)$. The large rise of the ε' is a typical feature of many types of ionic systems; see Figure 16.13 (Lunkenheimer et al., 2002). Electrode polarization gives rise to capacitances of the order 10 μF cm^{-2} and can be modeled following the classical Gouy-Chapman double-layer model. If the applied potential is very small, the interfacial capacitance is well described by a Debye capacitance

$$C = A\frac{\varepsilon_0\varepsilon_s}{\lambda_D} \qquad (16.50)$$

where λ_D is the Debye length given in Equation 5.42. In a more general formalism, ρ is a dimensionless

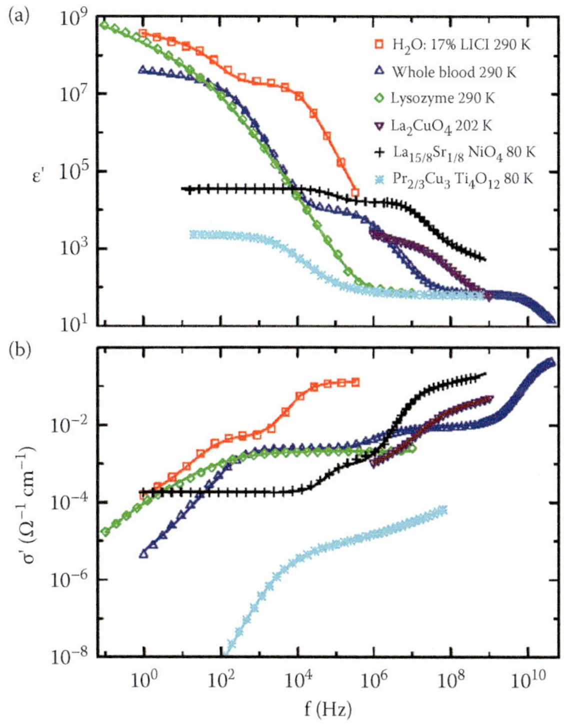

FIGURE 16.13 (a) Dielectric constant and (b) real part of the conductivity of different samples that are affected by electrode polarization effects: 17% LiCl–/water solution at 290 K (squares), human whole blood at 290 K (triangles), 0.009% Lysozym–/water solution (diamonds), single-crystalline La_2CuO_4 (down triangles), single crystalline $La_{15/8}Sr_{1/8}NiO_4$ (crosses), and polycrystalline $Pr_{2/3}Cu_3Ti_4O_{12}$ (stars). The lines are fits to an equivalent circuit model. (Reproduced with permission from Emmert et al., 2011.)

parameter that indicates the extent of charge transfer ($\rho = 0$ for blocking electrodes) (Beaumont and Jacobs, 1967; Tomozawa and Shin, 1998). The apparent increase of the dielectric constant is given by

$$\varepsilon_s = \varepsilon_\infty + \frac{2\lambda_D^{-1} d}{(2+\rho)^2} \varepsilon_\infty. \tag{16.51}$$

Therefore, the low-frequency dielectric constant depends on the concentration of mobile ions n as follows:

$$\varepsilon_s = \varepsilon_\infty + \left(\frac{n\varepsilon_\infty}{k_B T \varepsilon_0}\right)^{1/2} \frac{2}{(2+p)^2} qd. \tag{16.52}$$

The result of applying the above expressions often provides a Debye length that is rather short, about 0.1 nm, and the correspondent density of charge associated with electrode polarization is approximately 10^{-3} times the total number of mobile ions, which is of the range of 10^{22} cm^{-3}. The associated capacitance is similar to a Helmholtz capacitance and has a value around 30 μF cm^{-2}. To describe much larger capacitances caused by electrode polarization, it is necessary to take into account specific absorption of ions at the metal surface (Mariappan et al., 2010). If the surface concentration is governed by a Langmuir isotherm, then we observe a chemical capacitance that is given by the expression

$$C_\mu = \frac{A}{A_{eff}} \frac{q^2}{k_B T} f(1-f), \tag{16.53}$$

where f is the fractional occupation of ions at the surface and A_{eff} is the effective area available for an ion at the surface.

GENERAL REFERENCES

Impedance functions: Macdonald, 1987; Bisquert and Fabregat-Santiago, 2010.

Impedance spectroscopy of nanostructured energy devices: Bisquert and Fabregat-Santiago, 2010; Fabregat-Santiago et al., 2011; Bisquert et al., 2016; Bisquert et al., 2016; Bisquert et al., 2016.

Dielectric relaxation: Frohlich, 1958; Bottcher and Bordewijk, 1978; Jonscher, 1983.

Capacitance involving localized states: Bisquert et al., 2006; Garcia-Belmonte et al., 2007; Bisquert et al., 2008; Garcia-Belmonte et al., 2008; Guerrero et al., 2012; Montero and Bisquert, 2012.

Multiple trapping impedance models: Hens, 1999; Bisquert, 2002; Bisquert, 2008; Bisquert, 2010.

Universality of ac conduction: Bryksin, 1980; Zvyagin, 1980; Dyre, 1988.

Electrode polarization: Chang and Jaffe, 1952; Macdonald, 1953; Beaumont and Jacobs, 1967; Tomozawa and Shin, 1998; Lunkenheimer et al., 2002; Klein et al., 2006; Mariappan et al., 2010; Emmert et al., 2011.

REFERENCES

Almond, D. P.; Duncan, G. K.; West, A. R. Analysis of conductivity prefactors and ion hopping rates in AgI-Ag_2MoO_4 glass. *Journal of Non-Crystalline Solids* 1985, *74*, 285–301.

Almora, O.; Zarazua, I.; Mas-Marza, E.; Mora-Sero, I.; Bisquert, J.; Garcia-Belmonte, G. Capacitive dark currents, hysteresis, and electrode polarization in lead halide perovskite solar cells. *The Journal of Physical Chemistry Letters* 2015, *6*, 1645–1652.

Alvarez, F.; Alegria, A.; Colmenero, J. Relationship between the time-domain Kohlrausch- Williams-Watts and frequency-domain Havriliak-Negami relaxation functions. *Physical Review B* 1991, *44*, 7306.

Ansari-Rad, M.; Bisquert, J. Theory of light-modulated emission spectroscopy. *The Journal of Physical Chemistry Letters* 2017, *8*, 3673–3677.

Baird, M. E. Determination of dielectric behaviour at low frequencies from measurements af anomalous charging and discharging currents. *Reviews of Modern Physics* 1968, *40*, 219–227.

Beaumont, J. H.; Jacobs, P. W. M. Polarization in potassium chloride crystals. *Journal of Physical Chemistry of Solids* 1967, *28*, 657.

Bertoluzzi, L.; Bisquert, J. Equivalent circuit of electrons and holes in thin semiconductor films for photoelectrochemical water splitting applications. *The Journal of Physical Chemistry Letters* 2012, *3*, 2517–2522.

Bertoluzzi, L.; Bisquert, J. Investigating the consistency of models for water splitting systems by light and voltage modulated techniques. *The Journal of Physical Chemistry Letters* 2017, *8*, 172–180.

Bertoluzzi, L.; Lopez-Varo, P.; Jimenez Tejada, J. A.; Bisquert, J. Charge transfer processes at the semiconductor/electrolyte interface for solar fuel production: Insight from impedance spectroscopy. *Journal of Materials Chemistry A* 2015, *4*, 2873–2879.

Bisquert, J. Analysis of the kinetics of ion intercalation. Ion trapping approach to solid-state relaxation processes. *Electrochimica Acta* 2002, *47*, 2435–2449.

Bisquert, J. Beyond the quasi-static approximation: Impedance and capacitance of an exponential distribution of traps. *Physical Review B* 2008, *77*, 235203.

Bisquert, J. Theory of the impedance of charge transfer via surface states in dye-sensitized solar cells. *Journal of Electroanalytical Chemistry* 2010, *646*, 43–51.

Bisquert, J. Variable series resistance mechanism to explain the negative capacitance observed in impedance spectroscopy measurements of nanostructured solar cells. *Physical Chemistry Chemical Physics* 2011, *13*, 4679–4685.

Bisquert, J.; Fabregat-Santiago, F. Impedance spectroscopy: A general introduction and application to dye-sensitized solar cells. In *Dye-Sensitized Solar Cells*; Kalyanasundaram, K., Ed.; CRC Press: Boca Raton, 2010.

Bisquert, J.; Garcia-Belmonte, G.; Guerrero, A. Impedance characteristics of hybrid organometal halide perovskite solar cells. In *Organic-Inorganic Halide Perovskite Photovoltaics: From Fundamentals to Device Architectures*; Park, N.-G., Grätzel, M., Miyasaka, T., Eds.; Springer, Berlin, 2016a.

Bisquert, J.; Garcia-Belmonte, G.; Mora-Sero, I. Characterization of capacitance, transport and recombination parameters in hybrid perovskite and organic solar cells. In *Unconventional Thin Film Photovoltaics*; Como, E. D., Angelis, F. D., Snaith, H., Walker, A., Eds.; The Royal Society of Chemistry: London, 2016b.

Bisquert, J.; Garcia-Belmonte, G.; Munar, A.; Sessolo, M.; Soriano, A.; Bolink, H. J. Band unpinning and photovoltaic model for P3HT:PCBM organic bulk heterojunctions under illumination. *Chemical Physics Letters* 2008, *465*, 57–62.

Bisquert, J.; Garcia-Belmonte, G.; Pitarch, A.; Bolink, H. Negative capacitance caused by electron injection through interfacial states in organic light-emitting diodes. *Chemical Physics Letters* 2006, *422*, 184–191.

Bisquert, J.; Gimenez, S.; Bertoluzzi, L.; Herraiz-Cardona, I. Analysis of photoelectrochemical systems by impedance spectroscopy. In *Photoelectrochemical Solar Fuel Production. From Basic Principles to Advanced Devices*; Gimenez, S., Bisquert, J., Eds.; Springer, Berlin, 2016c.

Böttcher, C. J. F.; Bordewijk, P. *Theory of Electric Polarization*; Elsevier: Amsterdam, 1978; Vol. II.

Bryksin, V. V. Frequency dependence of the hopping conductivity for three dimensional systems in the framework of the effective-medium method. *Soviet Physics Solid State* 1980, *22*, 1421–1426.

Chang, H. C.; Jaffé, G. Polarization in electrolytic solutions. Part I. Theory. *The Journal of Chemical Physics* 1952, *20*, 1071.

Cole, K. S.; Cole, R. H. Dispersion and absorption in dielectrics II. Direct current characteristics. *The Journal of Chemical Physics* 1942, *10*, 98–105.

Dyre, J. C. The random free-energy barrier model for ac conductivity in disordered solids. *Journal of Applied Physics* 1988, *64*, 2456.

Dyre, J. C.; Maass, P.; Roling, B.; Sidebottom, D. L. Fundamental questions relating to ion conduction in disordered solids. *Reports on Progress on Physics* 2009, *72*, 046501.

Dyre, J. C.; Schroder, T. B. Universality of ac conduction in disordered solids. *Reviews of Modern Physics* 2000, *72*, 873.

Emmert, S.; Wolf, M.; Gulich, R.; Krohns, S.; Kastner, S.; Lunkenheimer, P.; Loidl, A. Electrode polarization effects in broadband dielectric spectroscopy. *European Physics Journal B* 2011, *83*, 157–165.

Fabregat-Santiago, F.; Garcia-Belmonte, G.; Mora-Seró, I.; Bisquert, J. Characterization of nanostructured hybrid and organic solar cells by impedance spectroscopy. *Physical Chemistry Chemical Physics* 2011, *13*, 9083–9118.

Fröhlich, H. *Theory of Dielectrics*, 2nd ed.; Oxford University Press: Oxford, 1958.

Garcia-Belmonte, G.; Bisquert, J. Distinction between capacitive and noncapacitive hysteretic currents in operation and degradation of perovskite solar cells. *ACS Energy Letters* 2016, *1*, 683–688.

Garcia-Belmonte, G.; Bisquert, J.; Bueno, P. R.; Graeff, C. F. O. Impedance of carrier injection at the metal–organic interface mediated by surface states in electron-only tris(8-hydroxyquinoline) aluminium (Alq3) thin layers. *Chemical Physics Letters* 2008, *455*, 242–248.

Garcia-Belmonte, G.; Bolink, H.; Bisquert, J. Capacitance-voltage characteristics of organic light-emitting diodes varying the cathode metal: Implications for interfacial states. *Physical Review B* 2007, *75*, 085316.

Grahame, D. C. Fiftieth anniversary: Mathematical theory of the faradaic admittance. *Journal of the Electrochemical Society* 1952, *99*, 370C–385C.

Guerrero, A.; Marchesi, L. F.; Boix, P. P.; Ruiz-Raga, S.; Ripolles-Sanchis, T.; Garcia-Belmonte, G.; Bisquert, J. How the charge-neutrality level of interface states controls energy level alignment in cathode contacts of organic bulk-heterojunction solar cells. *ACS Nano* 2012, *6*, 3453–3460.

Halpern, V. Dielectric relaxation, the superposition principle, and age-dependent transition rates. *Journal of Physics D: Applied Physics* 1992, *25*, 1533–1537.

Hens, Z. The electrochemical impedance on one-equivalent electrode processes at dark semiconductor redox electrodes involving charge transfer through surface states. 1. Theory. *The Journal of Physical Chemistry B* 1999, *103*, 122–129.

Isard, J. O. A study of migration loss in glass and a generalized method of calculating the rise of dielectric loss with temperature. *Proceedings Institute Electrical Engineers* 1961, *109B*, Suppl. No. 22, 440–447.

Jonscher, A. K. The 'universal' dielectric response. *Nature* 1977, *267*, 673.

Jonscher, A. K. *Dielectric Relaxation in Solids*; Chelsea Dielectrics Press: London, 1983.

Klahr, B.; Gimenez, S.; Fabregat-Santiago, F.; Hamann, T.; Bisquert, J. Water oxidation at hematite photoelectrodes: The role of surface states. *Journal of the American Chemical Society* 2012, *134*, 4294–4302.

Klein, R. J.; Zhang, S.; Dou, S.; Jones, B. H.; Colby, R. H.; Runt, J. Modeling electrode polarization in dielectric spectroscopy: Ion mobility and mobile ion concentration of single-ion polymer electrolytes. *The Journal of Chemical Physics* 2006, *124*, 144903.

Kulkarni, A. R.; Lunkenheimer, P.; Loidl, A. Scaling behaviour in the frequency dependent conductivity of mixed alkali glasses. *Solid State Ionics* 1998, *112*, 69–74.

Lunkenheimer, P.; Bobnar, V.; Pronin, A. V.; Ritus, A. I.; Volkov, A. A.; Loidl, A. Origin of apparent colossal dielectric constants. *Physical Review B* 2002, *66*, 052105.

Macdonald, J. R. Theory of ac space-charge polarization effects in photoconductors, semiconductors, and electrolytes. *Physical Review* 1953, *92*, 4–17.

Macdonald, J. R. *Impedance Spectroscopy*; John Wiley and Sons: New York, 1987.

Mariappan, C. R.; Heins, T. P.; Roling, B. Electrode polarization in glassy electrolytes: Large interfacial capacitance values and indication for pseudocapacitive charge storage. *Solid State Ionics* 2010, *181*, 859–863.

Montero, J. M.; Bisquert, J. Features of capacitance and mobility of injected carriers in organic layers measured by impedance spectroscopy. *Israel Journal of Chemistry* 2012, *52*, 519–528.

Murphy, A. B.; Barnes, P. R. F.; Randeniya, L. K.; Plumb, I. C.; Grey, I. E.; Horne, M. D.; Glasscock, J. A. Efficiency of solar water splitting using semiconductor electrodes. *International Journal of Hydrogen Energy* 2006, *31*, 1999–2017.

Peter, L. M. Dynamic aspects of semiconductor photoelectrochemistry. *Chemical Reviews* 1990, *90*, 753–769.

Pollak, M.; Geballe, T. H. Low-frequency conductivity due to hopping processes in silicon. *Physical Review* 1961, *122*, 1742.

Randles, E. B. *Faraday Discussion Chemical Society* 1947, *1*, 11.

Sidebottom, D. L. Understanding ion motion in disordered solids from impedance spectroscopy scaling. *Reviews of Modern Physics* 2009, *81*, 999–1014.

Ravishankar, S.; Aranda, C.; Sanchez, S.; Bisquert, J.; Saliba, M.; Garcia-Belmonte, G. Perovskite solar cell modeling using light and voltage modulated techniques. *The Journal of Physical Chemistry C* 2019a, *123*, 6444–6449.

Ravishankar, S.; Riquelme, A.; Sarkar, S. K.; Garcia-Batlle, M.; Garcia-Belmonte, G.; Bisquert, J. Intensity-modulated photocurrent spectroscopy and its application to perovskite solar cells. *The Journal of Physical Chemistry C* 2019b, *123*, 24995–25014.

Taylor, H. E. The dielectric spectrum relaxation of glass. *Transactions on Faraday Society* 1956, *52*, 873–881.

Tomozawa, M.; Shin, D.-W. Charge carrier concentration and mobility of ions in a silica glass. *Journal of Non-Crystalline Solids* 1998, *241*, 140–148.

Williams, G. Dielectric relaxation behavior of amorphous polymers and related materials. *IEEE Transactions on Electrical Insulation* 1985, *EI-20*, 843–857.

Zarazua, I.; Bisquert, J.; Garcia-Belmonte, G. Light-induced space-charge accumulation zone as photovoltaic mechanism in perovskite solar cells. *The Journal of Physical Chemistry Letters* 2016, *7*, 525–528.

Zvyagin, I. P. A percolative approach to the theory of the AC hopping conductivity. *Physica Status Solidii (b)* 1980, *97*, 143–149.

Part III

Radiation, Light, and Semiconductors

17 Blackbody Radiation and Light

The fundamental properties of radiation are a central tool for the design and utilization of energy devices. Radiation from the sun is the principal source of energy exploited in solar cell devices. Even in the dark, ambient thermal radiation causes electronic processes in a photoactive device, and this fact is used in arguments based on detailed balance that provide fundamental information of semiconductors interaction with radiation. The properties of light and color are necessary for the characterization of light sources. This chapter provides a brief revision of the main features of the electromagnetic spectrum, and some fundamental quantities that are employed to quantify the radiation. The main part of the chapter describes the blackbody radiation for later reference, starting with the radiative properties of a blackbody, and formulating the properties of thermal radiation, consisting of the spectral distribution of number of photons and their energy. The next step is to characterize the flux of photons and the energy flux. Finally, we examine the main aspects of the solar spectrum, which is the energy source that we wish to utilize for the production of useful energy, and how it relates to the idealized blackbody spectrum.

17.1 PHOTONS AND LIGHT

Electromagnetic radiation can be viewed as being composed of *electromagnetic waves*, or as consisting of massless energy quanta called *photons*. Radiation, in either view, can be classified according to its wavelength, λ, or frequency ν. These quantities hold the relationship

$$\lambda = \frac{c_\gamma}{\nu} \tag{17.1}$$

where c_γ is the speed of light, which depends on the *refractive index* of the medium, n_r, through which the radiation travels, as

$$c_\gamma = \frac{c}{n_r} \tag{17.2}$$

Here, $c = 2.998 \times 10^8$ ms^{-1} is the speed of light in vacuum. Common units of wavelength are the micrometer, μm, and the nanometer, nm.

The energy of the photon is given by the expression

$$E = h\nu = \hbar\omega \tag{17.3}$$

The angular frequency is $\omega = 2\pi\nu$. h is Planck's constant and $\hbar = h/2\pi$. The photon energy can be converted to wavelength using the formula

$$\lambda = \frac{1240}{h\nu(\text{eV})}\text{nm} \tag{17.4}$$

The parts of the electromagnetic spectrum are indicated in Figure 17.1. The *optical region*, which is the wavelength region of interest for solar energy conversion and lighting, is the long-wave portion comprising the ultraviolet region (UV), the visible light region extending from approximately $\lambda = 0.4$–0.7 μm, and the infrared region (IR) from beyond the red end of the visible spectrum to about $\lambda = 1000$ nm. We denote *light* the part of the electromagnetic spectrum that provokes a visual response in humans. Figure 17.2 shows the solar spectrum in two different fashions: the number of photons arriving at the top of the earth's atmosphere, and at the earth's surface, per interval of wavelength. Approximately 50% of solar irradiation occurs outside the visible range, especially in the IR; therefore this part provides radiant energy and heat but not light. The spectral differences above and below the atmosphere are due to the filtering of certain wavelengths, as will be discussed in Section 17.8.

17.2 SPREAD AND DIRECTION OF RADIATION

Light emitted at a point source propagates in the three dimensions of space, so that the intensity decreases with distance. The spread of light is defined by two factors: the area and the angle.

To quantify the direction of propagation, it is necessary to use a two-dimensional angle: the solid angle Ω, defined in terms of size and distance of a distant object. It corresponds to a fragment of a sphere that is invariant with the sphere's radius. In SI units the arc is measured in radians, $\theta = s/R$, where s is the arc and R is the radius of the circle. The solid angle is measured in *steradians*. A steradian (sr) is defined as the solid angle subtended at the center of a sphere by an area on its surface numerically equal to the square of its radius. As shown in Figure 17.3a, for a surface element dS seen from a distance R the solid angle is given by

$$d\Omega = \frac{dS}{R^2} \tag{17.5}$$

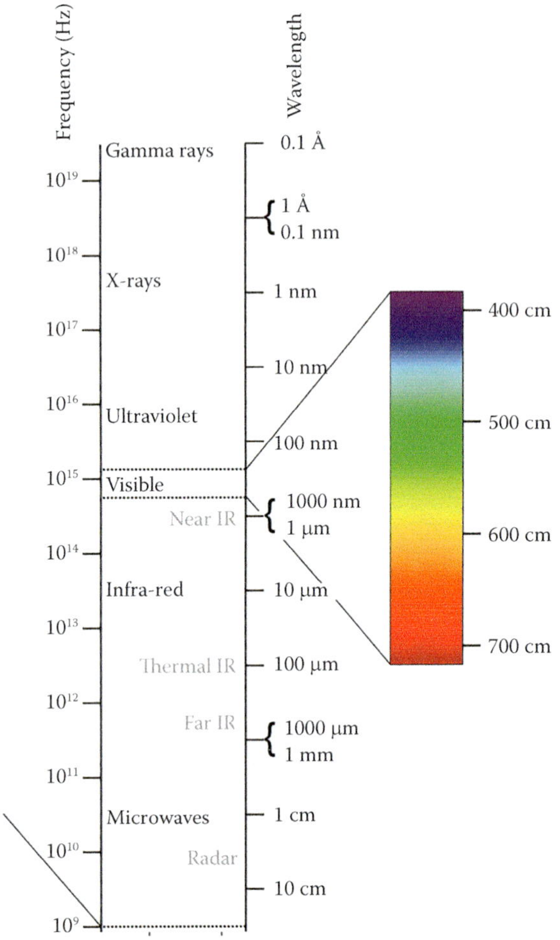

FIGURE 17.1 Electromagnetic spectrum in the range between microwaves and gamma rays. (Reprinted from the public domain from http://en.wikipedia.org/wiki/File:Electromagnetic-Spectrum.svg.)

The element of solid angle can be written in terms of angular coordinates as

$$d\Omega = \sin\theta d\phi d\theta \qquad (17.6)$$

The total solid angle is 4π, and the solid angle of a hemisphere is 2π.

In geometrical optics étendue is a convenient concept to describe the propagation of light through an optical system. Consider a beam of radiation in a direction **s** in a medium of refractive index n_r shown in Figure 17.3b. dA is a cross-sectional element of the beam, and **n** is the normal to dA. The *étendue* ε is the product of the solid angle of the radiation and the projected area (Markvart, 2008).

$$d\varepsilon = n_r^2 \cos\theta dA d\Omega \qquad (17.7)$$

An important property is that the étendue of a beam propagating in a clear and transparent medium is conserved. However, as we are mainly interested in radiance emission from a solar cell where the optical aperture is unique—the front surface—rather than multiple optical systems, we drop the area from the expression in Equation 17.7. We will find situations in which a cone of radiation of half angle θ_m, impacts, or is emitted from, a surface as shown in Figure 17.3c. The étendue of this light is

$$\varepsilon = n_r^2 \int_0^{\theta_m} \cos\theta \sin\theta d\theta \int_0^{2\pi} d\varphi = n_r^2 \pi \sin^2\theta_m \qquad (17.8)$$

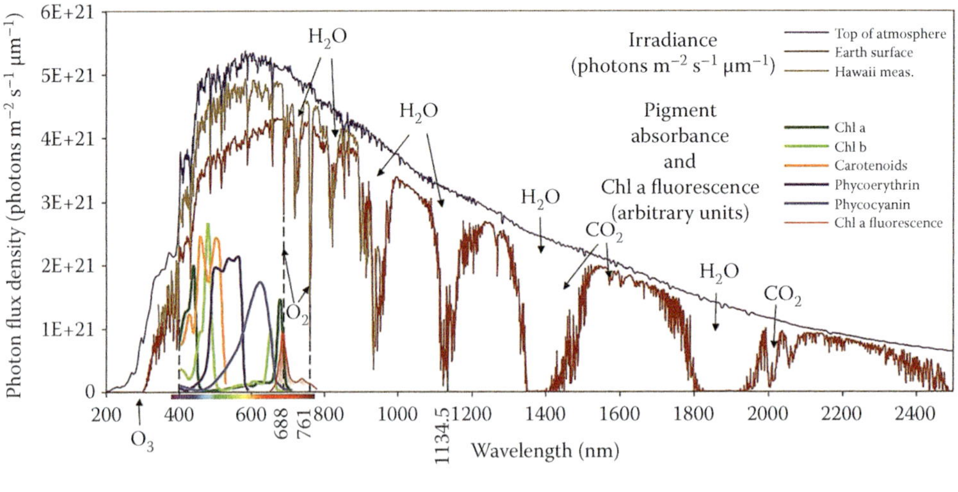

FIGURE 17.2 Solar spectral photon flux densities at the top of the earth's atmosphere and at the earth's surface, and estimated *in vivo* absorption spectra of photosynthetic pigments of plants and algae: Chl a and Chl b, carotenoid, phycoerythrin, phycocyanin absorption spectra. Chl a fluorescence spectrum, from spinach chloroplasts. (Reprinted with permission from Kiang, N. Y. et al. *Astrobiology* 2007, 7, 222–251.)

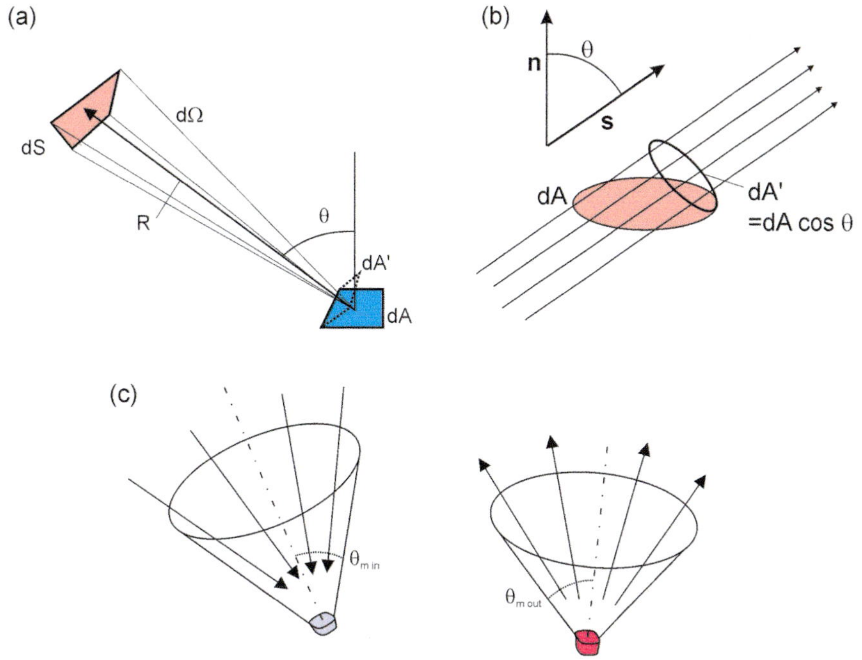

FIGURE 17.3 (a) Emission of radiation along a solid angle $d\Omega = dO/R^2$ from a surface element of area dA. The angle between the radiation and the normal to the surface is θ. (b) Representation of the flux in the direction **s** across an area dA with unit normal vector **n**. (c) Conical solid angles of incident and outgoing radiation on a solid object.

We consider the following particular cases:

- The radiation emitted from a planar solid surface into a hemisphere ($\theta_m = \pi/2$) has $\varepsilon = n_r^2\pi$. This is the étendue for emission of radiation from the solar cell through the front aperture.
- The radiation incident on a solar cell carries the solid angle of the sun. The half angle subtended by the sun is θ_S = 0.265°, so that $\sin^2\theta_s = 2.139\times10^{-5}$ and $\varepsilon_S = n_r^2\ 6.8\times10^{-5}$ sr.
- The radiation incident on the solar cell under maximal concentration on a planar black absorber is again $\varepsilon = n_r^2\pi$.

For an ideal diffusely reflecting surface or ideal diffuse radiator, the number of photons and the energy flux Φ across the small solid angle $d\Omega$ depends on the projected solid angle (the étendue)

$$\Phi(\theta) \propto (\cos\theta)\, d\Omega \qquad (17.9)$$

Such a dependence is called *Lambert's* law. The factor $\cos\theta$ gives the reduction of apparent area of the emitting element dA seen from the receiving point, and the radiation is reduced by the same factor $\cos\theta$ with respect to the normal. Therefore, for the Lambertian reflector or radiator, the same number of photons is perceived when looking at dA from any angle. Conventional LEDs are approximately Lambertian. If we consider a curved surface that radiates in our direction, we obtain that for each element of the surface the emitted power is reduced by the same factor that the area is. Thus all elements of the Lambertian surface appear equally bright, and this is actually observed looking at the sun. Note that the Lambertian diffusor for light emitted from a surface is the reciprocal of a maximal concentration for the incoming light.

17.3 COLOR AND PHOTOMETRY

The visible spectrum ranges from 390 nm (violet) to 780 nm (red), see Figure 17.4. Our perception of color results from the composition of the light (the energy spectrum of the photons) that enters the eye. Cone cells in human eye are of three different types, sensible to three different ranges of frequency, which the eye interprets as blue (with a peak close to 419 nm), green (peaking at 531 nm), and red (with a peak close to 558 nm, which is more yellowish) (Figure 17.4). Naturally occurring colors are composed of a broad range of wavelengths. The wavelength that appears to be the most dominant in a color is the color's *hue*. The *saturation* is a measure of the purity of the color and indicates the amount of distribution in wavelengths in the color. A highly saturated color will contain a very narrow set of wavelengths.

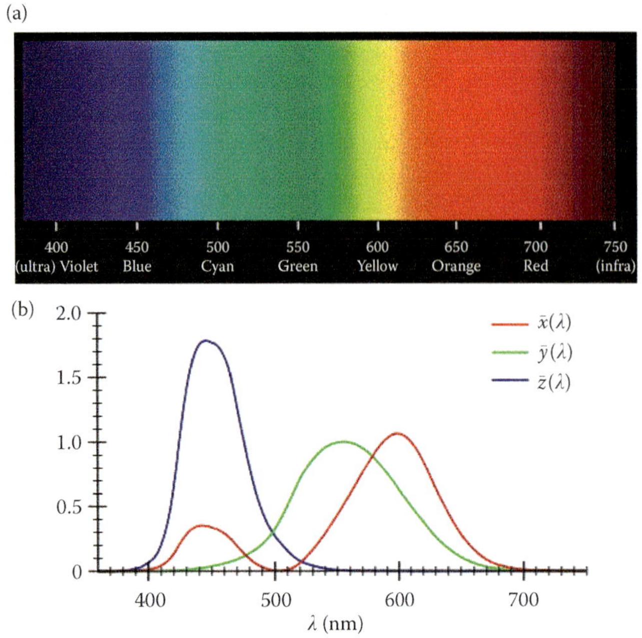

FIGURE 17.4 (a) Approximated wavelengths associated with the colors perceived in the visible spectrum. (b) CIE 1931 Standard Colorimetric Observer functions used to map blackbody spectra to XYZ coordinates. (Reprinted from the public domain from https://en.wikipedia.org/wiki/File:CIE_1931_XYZ_Color_Matching_Functions.svg.)

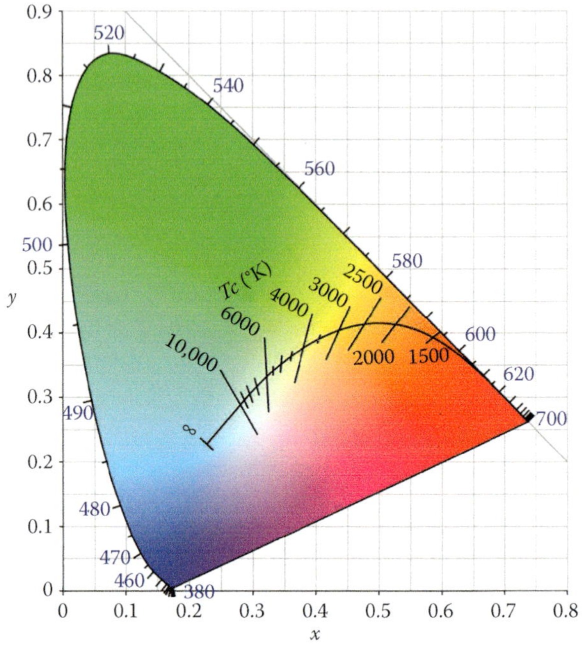

FIGURE 17.5 CIE *xy* 1931 chromaticity diagram including the Planckian locus. The Planckian locus is the path that a blackbody color will take through the diagram as the blackbody temperature changes. Lines crossing the locus indicate lines of constant CCT. Monochromatic wavelengths are shown at the boundary in units of nanometers. (Reprinted from the public domain from http://en.wikipedia.org/wiki/File:PlanckianLocus.png.)

The perception of color by the human eye is quantified by tristimulus values X, Y, Z defined by Commission Internationale de l'Éclairage (CIE). These parameters are derived from the spectral power distribution of the light emitted by a colored object, weighted by sensitivity curves, the standard colorimetric functions in Figure 17.4, that have been determined by actual measurement of the average human eye. Tristimulus values uniquely represent a perceivable hue. The two lowercase coordinates *xy* in the chromaticity diagram, Figure 17.5, are derived from the tristimulus values, and represent the relative contribution of the three primaries. The boundaries indicate maximum saturation, that is, the spectral colors. The diagram forms the boundary for all perceivable hues.

The emission of hot objects, described in Section 17.4, is called *incandescence*. Incandescent light sources emit a broad set of frequencies that covers a wide range of the visible spectrum. We normally denote white light, the light of the sun. One could also consider white light that of the tungsten filament bulb, which is similar to blackbody radiation at about 2900 K. In general, one can parametrize the white light emitted by a blackbody radiator by using the correlated color temperature (CCT). The trace of the points of the irradiating blackbodies in the chromaticity diagram, Figure 17.5, forms the Planckian locus. From low to very high temperatures, the emission changes from deep red, through orange, yellowish white, white, and finally bluish white.

There are three classes of lamps for lighting: incandescent, discharge, and solid state. Incandescent lamps heat a filament that glows. Discharge lamps and white LEDs (light-emitting diodes) are based on the emission of UV or blue photons, either by gas ionization or by electroluminescence. These high-energy photons are then used to excite phosphorescent atoms or molecules that emit lower energy photons in red or green. The downconversion process, further discussed in Section 27.2, displaces the wavelength and allows the production of different colors with a single lamp.

White light can be created by a combination of red, green, and blue monochromatic sources. The emission of two complementary colors, at opposite sides of the Planckian line, also gives rise to white light. When we look at illuminated objects, we only see the reflected light. The spectrum of the light source affects the appearance of objects according to the color rendering. Incandescent light sources allow our visual system to easily distinguish the colors of objects. Discharge and fluorescent lamps may produce light that peaks strongly

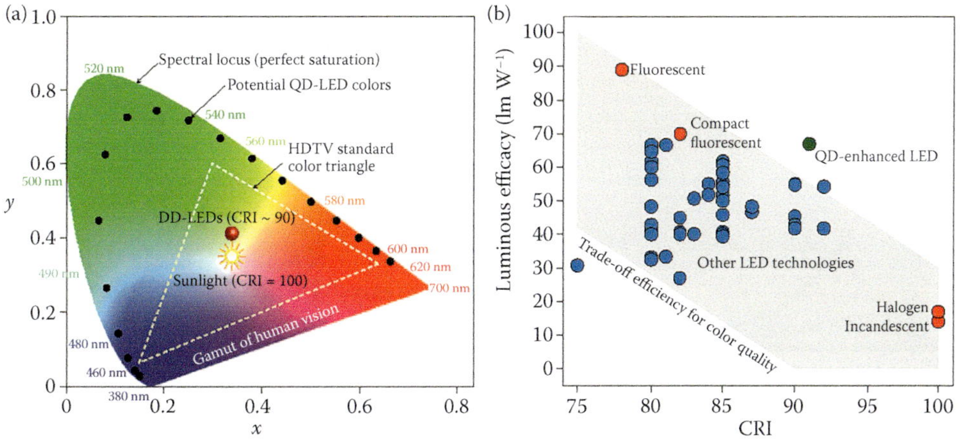

FIGURE 17.6 (a) CIE chromaticity diagram showing that the spectral purity of quantum dots (QDs) enables a color gamut (*dotted line*) larger than the high-definition television (HDTV) standard (*dashed line*). (b) Plot showing the luminous efficacy and CRI of various commercially available lighting solutions. (Adapted with permission from Shirasaki, Y. et al. *Nature Photonics* 2012, 7, 13–23.)

at certain emission wavelengths and lacks a significant part of the visible spectrum causing the perceived color of an object to be very different from that under natural light. An index denoted *Color Rendering Index* (CRI) compares the ability of a light source to reproduce the colors in comparison with a natural source. The reference is a perfect blackbody radiator at the same nominal temperature, which is assigned the value 100. General lighting requires a CRI of 70 and some applications demand 80 or higher.

The intensity of the light is measured in SI in candela (cd). The *luminous intensity* per unit area of light, travelling in a given direction, is measured in cd m^{-2}. *Radiant intensity* is the amount of power radiated per unit solid angle, measured in W sr^{-1}. *Radiance* (in W m^{-2} sr^{-2}) is the energy flux per unit area per solid angle received by a surface.

The vision system of humans does not detect all wavelengths equally: UV and infrared light is not useful for illumination. The effect of production of light in a source for human vision therefore depends both on the efficacy of production of radiation and on how the produced spectrum adapts to the human eye. The response of vision of the average eye, in the visible spectrum, in conditions of bright illumination, is called the *photopic response*. The eye's response is maximum for green light of wavelength 555 nm and 1 W of irradiated power at this wavelength is defined as 683 lumen (lm). Solid state lighting and display applications require, respectively, a brightness of 10^3–10^4 and 10^2–10^3 cd m^{-2}. *Illuminance* measures the photometric flux per unit area, or visible flux density, in lux (lm m^{-2}). *Luminance* is the illuminance per unit solid angle, in lm m^{-2} sr^{-2}. Luminance is the density of visible radiation in a given direction. The spectrum of any light source can be measured and decomposed with the photopic response to provide a total production of lumens by the lamp. The luminous efficacy of a source is then the number of lumens produced per watt of electrical power supplied (lm W^{-1}).

High CCT sources (>5000 K) produce a bluish emission and low CRI in the range 80–85. For low CCT lights, in the range of the incandescent emitters (about 2700 K), higher CRI can be achieved but it is difficult to maintain high luminous efficiency due to losses in the infrared. Figure 17.6 compares the luminous efficiency and the CRI of different types of lighting solutions (Shirasaki et al., 2012).

17.4 BLACKBODY RADIATION

Sources of electromagnetic radiation can emit discrete wavelengths, as in phosphors, LEDs, or gases, by specific quantum transitions that radiate photons of energy equal to the difference between initial and final energy levels, Equation 17.3. On the other hand, many solids contain a near-continuum of energy levels that allow for a very broad scope of excitations and consequently radiate a continuous electromagnetic spectrum extending over a wide range. The electromagnetic energy emitted from the surface of a heated body is called *thermal* radiation. Many solids radiate a spectrum that can be well approximated by *blackbody radiation*.

The spectrum of a blackbody is shown in Figure 17.7. Note that Figure 17.7 displays the power radiated while the solar spectrum in Figure 17.2 is the number of photons, per wavelength interval. The spectra in Figure 17.7

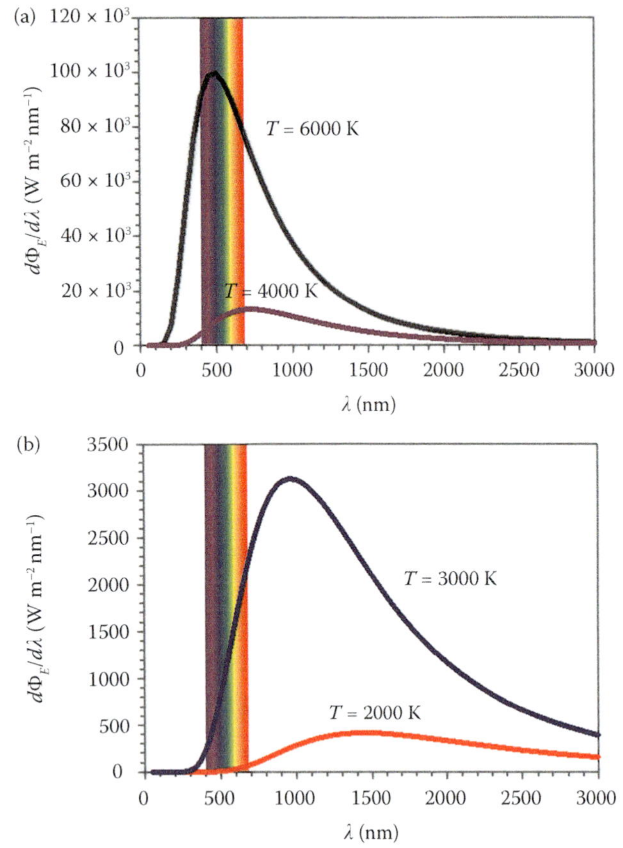

FIGURE 17.7 The spectrum of a blackbody at different temperatures with respect to wavelength. (a) and (b) show the spectra at different temperatures as indicated.

contain power at all wavelengths but display a characteristic distribution, consisting of a peak at the wavelength of maximum emission. Each spectrum shows a sharp decay toward the shorter wavelength region, while there is a longer and slower decay toward longer frequencies in the infrared. The wavelength of maximum emission is a function of temperature according to the expression

$$\lambda T = 2.9 \cdot 10^{-3}\ \text{mK} \tag{17.10}$$

This rule is known as Wien's displacement law. According to Equation 17.10, the wavelength of maximum emission moves to shorter wavelength when the temperature of the radiator rises. The total radiated power increases with the body temperature according to the Stefan–Boltzmann law discussed in Equation 17.39.

At room temperature, an object emits radiation in the infrared, and no visible light is generated. When an object is heated, the radiation spectrum approaches the visible wavelengths and the emission will change from dark red at low temperatures, through orange red, to white at the very high temperatures, as remarked before in the chromaticity diagram of Figure 17.5. To emit red light, a body must be heated above 850 K and at 3000 K it appears yellowish white, Figure 17.7b. The emission of a tungsten filament in a light bulb (at a power of 150 W) occurs at 2900 K. It is observed in Figure 17.7b that only a small fraction of the radiated energy (11%) is visible light, thus while the incandescent light bulb produces light containing all the frequencies of the visible spectrum, which is comfortable for human vision, the efficiency of conversion of electricity to light is small. Figure 17.7a shows the Planck spectrum at 6000 K, which is similar to the solar spectrum further discussed in Section 17.8.

17.5 THE PLANCK SPECTRUM

The *Planck spectrum* (*blackbody spectrum*) is a spectral distribution of energy per unit volume per unit frequency range $d\nu$. It is given by the form

$$\rho_{bb}(\nu) = \frac{8\pi h\nu^3}{c_\gamma^3} \frac{1}{e^{h\nu/k_B T} - 1} \tag{17.11}$$

A justification of Equation 17.11 will be presented below. It is derived from the concept of a blackbody, which is a body that absorbs all incoming radiation, so that it does not return, by reflection or scattering, any of the radiation falling on it. For a blackbody, the absorptivity, defined in Section 18.1, is $a = 1$. Such a body appears *black* if its temperature is low enough such that it is not luminous by virtue of the radiation emitted on its own. In fact, the blackbody emits electromagnetic radiation, and such radiation, covering a broad spectral range, is the topic of interest here.

Although the blackbody is an idealized object, it is possible to build a radiation source with an emissivity close to that of the blackbody. Cavity radiation is a paradigmatic example of blackbody radiation and has actually been used for experiments in thermal radiation for more than a century. The cavity blackbody is an enclosure that has black internal walls, with a small hole drilled on it. Even though the internal wall may not be perfectly absorbing, photons entering the hole have a very low probability to bounce back out before being absorbed. The body is held at temperature T and the walls emit radiation so that the cavity is filled with isotropic radiation with the spectral distribution corresponding to the temperature T. In consequence, the hole in the cavity absorbs all incoming radiation and emits thermalized photons, so that *the hole* (not the cavity) can be considered a blackbody according to the above definition.

It can be rigorously shown that the radiation in equilibrium with the blackbody has a unique spectral distribution that is a function only of the absolute temperature of the blackbody, and not of its composition or the nature of the radiation enclosure (Landsberg, 1978). In addition, an ideal blackbody emits more energy than any other material at the same temperature, see Equation 22.29. In general, however, blackbody radiation refers to the spectral distribution of the radiation, as given by Planck's formula, and we do not really need to think about cavities in most of the applications discussed here. The essential point is that all radiation coming from a blackbody must be regarded as *emitted* radiation.

The blackbody spectrum is a very important tool for a number of aspects of devices, physics and chemistry. First, it describes thermal radiation, which is the radiation of hot objects such as the sun, a hot filament or incandescent materials in general. Since our eyes are used to seeing objects under sunlight, the Planckian radiation is the landmark of white light. The radiation of artificial light sources is compared to the blackbody spectrum by attributing a temperature to the obtained light by comparison of its actual spectral distribution with the exact blackbody spectrum, which has a CRI of 100, as mentioned in Section 17.3. Historically, the quest for the Planckian spectrum was propelled at the end of the 19th century by the need to characterize the light emanating from a hot filament. For solar energy conversion, we need to know the spectral distribution of solar photons, their number, and energy in order to devise the transduction of the photon energy to electricity or chemical fuel. The blackbody radiation is a very good model for the spectrum of the sun, as already mentioned, although solar radiation contains additional features that are discussed in Section 17.8. Finally, thermalized radiation is also used to derive transition rates for light absorption and emission in semiconductors and molecules using the detailed balance arguments (Kennard, 1918). Fundamental rates of emission are obtained from the hypothesis of equilibrium with the ambient radiation, which is blackbody radiation at temperature T = 300 K in the normal environment of the earth surface.

17.6 THE ENERGY DENSITY OF THE DISTRIBUTION OF PHOTONS IN BLACKBODY RADIATION

In Chapter 5, we discussed the thermalization of electrons in a semiconductor to a Fermi–Dirac distribution. We consider a similar situation concerning the *distribution of photons* that form electromagnetic radiation. Let us analyze a set of photons in an enclosement with perfect reflecting walls. These walls neither absorb nor emit radiation. Since the photons do not interact among themselves, the initial spectral distribution that was produced by the source of the radiation will be maintained. But, as Planck (1914) said,

> "as soon as an arbitrarily small quantity of matter is introduced into the vacuum, a stationary state of radiation is gradually established. If the substance introduced is not diathermanous for any color, for example, a piece of carbon, however small, there exists at the stationary state of radiation in the whole vacuum for all colors the intensity of black radiation corresponding to the temperature of the substance."

Just a small piece of material that can absorb and emit all frequencies in the cavity (like a piece of carbon dust) will absorb and release photons which ultimately come to a thermal distribution, formed by isotropic radiation in equilibrium with the small body at temperature T. Such a thermal distribution of the photons is the Planck spectrum.

We analyze the spectral distribution of isotropic radiation, moving in all directions. This is composed of photons of all wavelengths (frequencies) that occupy the available states consisting of the volume density of electromagnetic modes, $D_{ph}(E)$, according to a thermal distribution defined by the occupation function $f_{ph}(h\nu)$.

The density of photons n_{ph} in a small range of energies around the energy $h\nu$ is given by

$$dn_{ph}(h\nu) = D_{ph}(h\nu) f_{ph}(h\nu) d(h\nu) \quad (17.12)$$

The Bose-Einstein distribution gives the occupancy of boson modes

$$f_{ph}(E) = \frac{1}{e^{(E-\eta_{ph})/k_BT} - 1} \quad (17.13)$$

Here η_{ph} is the electrochemical potential of the photons that coincides with their chemical potential $\mu_{ph} = \eta_{ph}$. The distribution of photons as function of their energy is shown in Figure 17.8. In contrast to the Fermi–Dirac function, there is no limit to the number of bosons that occupy an available mode so that the photons accumulate in the lowest energy available state (the ground state). For $E - \mu_{ph} \gg k_BT$, the distribution takes the form of the Boltzmann expression

$$f(E,T) \approx e^{-(E-\mu_{ph})/k_BT} \quad (17.14)$$

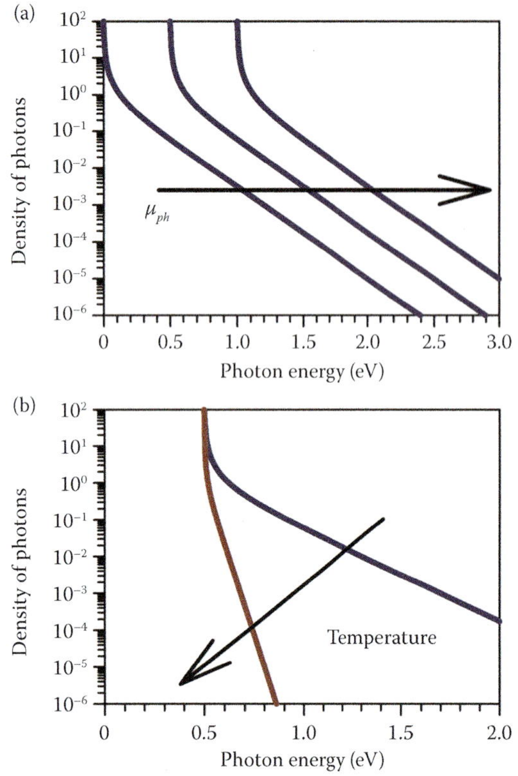

FIGURE 17.8 Number of photons at different electrochemical potentials (μ_{ph} = 0, 0.5, 1.0 eV) (a) and temperatures (b). The case $\mu_{ph} = 0$ is the thermal radiation.

This term dominates the spectral luminescence as discussed later in Section 22.4. The change of chemical potential of the photons (later to be identified with the voltage of the light-emitting device) produces a horizontal shift of the spectra, Figure 17.8a, while the temperature determines the slope, Figure 17.8b.

Thermal radiation is characterized by $\eta_{ph} = 0$ and the distribution takes the form

$$f_{ph}(E) = \frac{1}{e^{E/k_BT} - 1} \tag{17.15}$$

The number of electromagnetic modes having energy lower than $h\nu$, per unit volume, is (Fowles, 1989)

$$N_{ph}(\nu) = \frac{(8\pi/3)(h\nu)^3}{(h^3 c_\gamma^3)} \tag{17.16}$$

Therefore, the density of states per volume and energy interval is

$$D_{ph}(h\nu) = \frac{dN_{ph}(\nu)}{d(h\nu)} = \frac{8\pi\nu^2}{hc_\gamma^3} \tag{17.17}$$

These states are isotropically distributed in a solid angle 4π, so that the density of states in a direction specified by the solid angle $d\Omega$ is $d\Omega/4\pi$. We include the index of refraction n_r, which causes an enhanced light intensity when the radiation is trapped in a medium (Yablonovitch and Cody, 1982), and we give the density of states per solid angle as follows:

$$D_{ph}(h\nu) = \frac{2n_r^3 d\Omega \nu^2}{hc^3} \tag{17.18}$$

Now *Planck's law of radiation* describes the number of photons, dn_{ph}, per volume per frequency interval. From Equation 17.12, we obtain

$$\frac{dn_{ph}(\nu)}{d\nu} = \frac{8\pi n_r^3 \nu^2}{c^3} \frac{1}{e^{h\nu/k_BT} - 1} \tag{17.19}$$

The photon density, per volume per frequency interval per solid angle, is

$$\frac{dn_{ph}(\nu)}{d\nu} = \frac{2n_r^3 \nu^2}{c^3} \frac{1}{e^{h\nu/k_BT} - 1} d\Omega \tag{17.20}$$

Figure 17.9a shows the distribution function for the number of photons at different temperatures for blackbody radiation. The energy per unit volume in a small range of wavelengths, de_{ph}, is obtained by the product of the photon density and their energy

$$de_{ph}(\nu) = (h\nu) dn_{ph}(\nu) \tag{17.21}$$

As already anticipated in Equation 17.11, the Planck spectrum (blackbody spectrum) is normally given in terms of the radiant energy, that is, the spectral distribution of energy per unit volume per unit frequency range $d\nu$

$$\rho_{bb}(\nu) = \frac{de_{ph}(\nu)}{d\nu} = \frac{dn_{ph}(\nu)}{d\nu} h\nu = \frac{2n_r^3 h\nu^3 d\Omega}{c^3} \frac{1}{e^{h\nu/k_BT} - 1} \tag{17.22}$$

This function is shown in Figure 17.9b. We can also write the spectral distribution in terms of wavelength, noting that

$$\rho_{bb}(\nu) d\nu = \rho_{bb}(\lambda) d\lambda \tag{17.23}$$

Since

$$|d\nu| = \frac{c}{\lambda^2} |d\lambda| \tag{17.24}$$

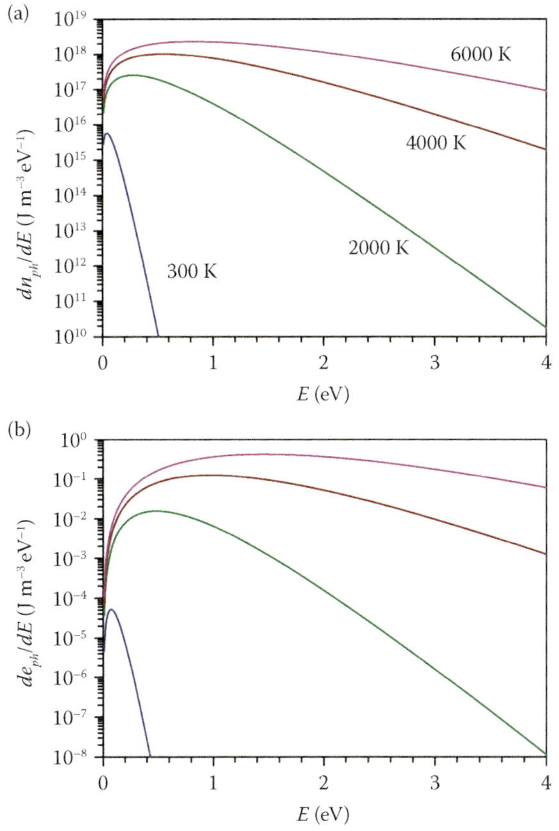

FIGURE 17.9 Representation of the photon and energy density distributions in the Planck radiation, as a function of the energy of the photons: (a) the number of photons per unit volume and (b) the energy density per unit volume.

we obtain

$$\rho_{bb}(\lambda) = \frac{de_{ph}(\lambda)}{d\lambda} = \rho_{bb}(\nu)\frac{d\nu}{d\lambda} = \frac{2n_r^3 hcd\Omega}{\lambda^5}\frac{1}{e^{hc/k_B\lambda T} - 1} \quad (17.25)$$

From Equation 17.25, it can be shown that the maximum of the spectrum is displaced to shorter wavelengths (higher energies) as the temperature of the radiation is raised, which is Wien's displacement law, discussed earlier in Equation 17.10.

17.7 THE PHOTON AND ENERGY FLUXES IN BLACKBODY RADIATION

Radiation in a cavity-type environment is isotropic but the radiation emitted from a source is directed, from the source to the observation point. For a radiating surface element dA with unit normal vector **n**, Figure 17.3b, the radiation can be emitted in any direction in the hemisphere characterized by the unit vector **s** and the solid angle $d\Omega$. The incoming or outgoing radiation is characterized by a number of photons $\Phi_{ph}(\nu_1, \nu_2, \mathbf{s})$ in the frequency interval (ν_1, ν_2) crossing the unit surface per second and they carry a flux of energy $\Phi_E(\nu_1, \nu_2, \mathbf{s})$. The variables used for the description of radiation are indicated in Table 17.1.

TABLE 17.1
Physical Variables for Blackbody Radiation

Quantity	Symbol
Total photon flux	Φ_{ph}
Total energy flux	Φ_E
Spectral photon flux	ϕ_{ph}
Spectral energy flux	ϕ_E

However, the *spectral flux* refers to the number of photons or photon energy per second in an interval in a small range of wavelengths $(\nu, \nu + d\nu)$ around a single frequency ν. The spectral quantity

$$\phi_E(\nu, s) = \frac{d\Phi_E(0, \nu, s)}{d\nu} \quad (17.26)$$

describes the energy flux (power). The radiant flux density is called the spectral *irradiance* E_ν (Sizmann et al., 1991). The spectral photon flux is

$$\phi_{ph}(\nu, s) = \frac{d\Phi_{ph}(0, \nu, s)}{d\nu} \quad (17.27)$$

The two previous magnitudes are connected by the relationship $\phi_E = h\nu\phi_{ph}$.

The total flux of photons in a frequency interval is obtained as

$$\Phi_{ph}(\nu_1, \nu_2, s) = \int_{\nu_1}^{\nu_2} \frac{d\Phi_{ph}(\nu, s)}{d\nu} d\nu = \int_{\nu_1}^{\nu_2} \phi_{ph}^{bb}(\nu, s) d\nu \quad (17.28)$$

Similarly, we can obtain

$$\phi_E^{bb}(\lambda, s) = \frac{d\Phi_E^{bb}(\lambda, s)}{d\lambda} \quad (17.29)$$

which is a very different quantity compared to Equation 17.26, due to the varying relationship between frequency and wavelength intervals, Equation 17.24. However, the integral of Equation 17.29 gives the same result as Equation 17.26 for the corresponding wavelength interval. The different flux distributions derived below are assembled in Table 17.2. Some of the distribution

TABLE 17.2
Formulas of Thermal Blackbody Radiation ($n_r = 1$)

Quantity as Function of	Energy/Frequency	Wavelength
Photons per volume	$\frac{dn_{ph}(\nu)}{d\nu}=\frac{8\pi\nu^2}{c^3}\frac{1}{e^{h\nu/k_BT}-1}$	
Energy per volume	$\rho_{bb}(\nu)=\frac{de_{ph}(\nu)}{d\nu}=\frac{8\pi h\nu^3}{c^3}\frac{1}{e^{h\nu/k_BT}-1}$	$\rho_{bb}(\lambda)=\frac{de_{ph}(\lambda)}{d\lambda}=\frac{8\pi hc}{\lambda^5}\frac{1}{e^{hc/k_B\lambda T}-1}$
Spectral photon flux	$\phi_{ph}^{bb}(E,\Omega)=\frac{\varepsilon}{\pi}b_\pi\frac{E^2}{e^{E/k_BT}-1}$	$\phi_{ph}^{bb}(\lambda,\Omega)=\frac{2\varepsilon c}{\lambda^4}\frac{1}{e^{hc/k_B\lambda T}-1}$
	$\phi_{ph}^{bb,\text{hemi}}(\nu)=\frac{2\pi\nu^2}{c^2}\frac{1}{e^{h\nu/k_BT}-1}$	
Spectral energy flux	$\phi_E^{bb}(E,\Omega)=\frac{\varepsilon}{\pi}b_\pi\frac{E^3}{e^{E/k_BT}-1}$	$\phi_E^{bb}(\lambda,\Omega)=\frac{2\varepsilon hc^2}{\lambda^5}\frac{1}{e^{hc/k_B\lambda T}-1}$
	$\phi_E^{bb}(\nu,\Omega)=\frac{2\varepsilon h\nu^3}{c^2}\frac{1}{e^{h\nu/k_BT}-1}$	$\phi_E^{bb,\text{hemi}}(\lambda)=\frac{C_1}{\lambda^5}\frac{1}{e^{C_2/\lambda T}-1}$

$$b_\pi=\frac{2\pi}{h^3c^2}=9.883\times10^{26}\,\text{eV}^{-3}\,\text{m}^{-2}\,\text{s}^{-1}$$

$$C_1=2\pi hc^2=3.742\times10^{-16}\,\text{Wm}^2$$

$$C_2=\frac{hc}{k_B}=1.439\times10^{-2}\,\text{mK}$$

functions are represented in Figure 17.10, in the case of radiation at T = 5800 K with a total energy flux $\Phi_{E,tot}^{bb}=1000\,\text{Wm}^{-2}$ corresponding to that of the terrestrial solar spectrum, as was discussed in Section 17.8.

In many cases, when analyzing solar cells we will be interested in the radiation flux of a Lambertian surface. Accordingly, we need to describe the flux of photons in blackbody radiation that pass through a surface element, θ being the angle between **s** and **n**, as indicated in Figure 17.3b. It is given by

$$d\Phi_{ph}^{bb}(\nu,n)=c_\gamma\frac{dn_{ph}(\nu)}{d\nu}\cos\theta d\Omega d\nu \quad (17.30)$$

We now will consider a cone of radiation incoming or outgoing from a solid object as shown in Figure 17.3c. The projected solid angle in Equation 17.30 is integrated around the normal as indicated in Equation 17.8. Hence, the spectral photon number flux density in terms of étendue per unit area is

$$\phi_{ph}^{bb}(\nu,n)=\frac{d\Phi_{ph}^{bb}(\nu)}{d\nu}=\frac{2\varepsilon\nu^2}{c^2}\frac{1}{e^{h\nu/k_BT}-1} \quad (17.31)$$

If the spectral variable is the photon energy, the spectral distribution of photon flux in blackbody radiation at temperature T per unit energy interval is

$$\phi_{ph}^{bb}(E,\Omega)=\frac{\varepsilon}{\pi}b_\pi\frac{E^2}{e^{E/k_BT}-1} \quad (17.32)$$

Note that étendue in Equation 17.32 includes the refractive index n_r, as indicated in Equation 17.8. The most frequent situation for a solar cell study is the spectral radiant flux emitted at the front face, considered a Lambertian surface, into a hemisphere in the vacuum ($n_r = 1$). For reducing the expression of many formulas, in Equation 17.32 we introduce the constant prefactor adapted to this situation ($\varepsilon/\pi = 1$):

$$b_\pi=\frac{2\pi}{h^3c^2}=9.883\times10^{26}\,\text{eV}^{-3}\,\text{m}^{-2}\,\text{s}^{-1} \quad (17.33)$$

On the other hand, for isotropic radiation of a volume element emitted in all directions (see Figure 22.3), we use in Equation 17.32– the total solid angle and it will be $\varepsilon/\pi=4n_r^2$.

In the range of photon energy $h\nu\gg k_BT$, the term -1 in the denominator of Equation 17.32 can be

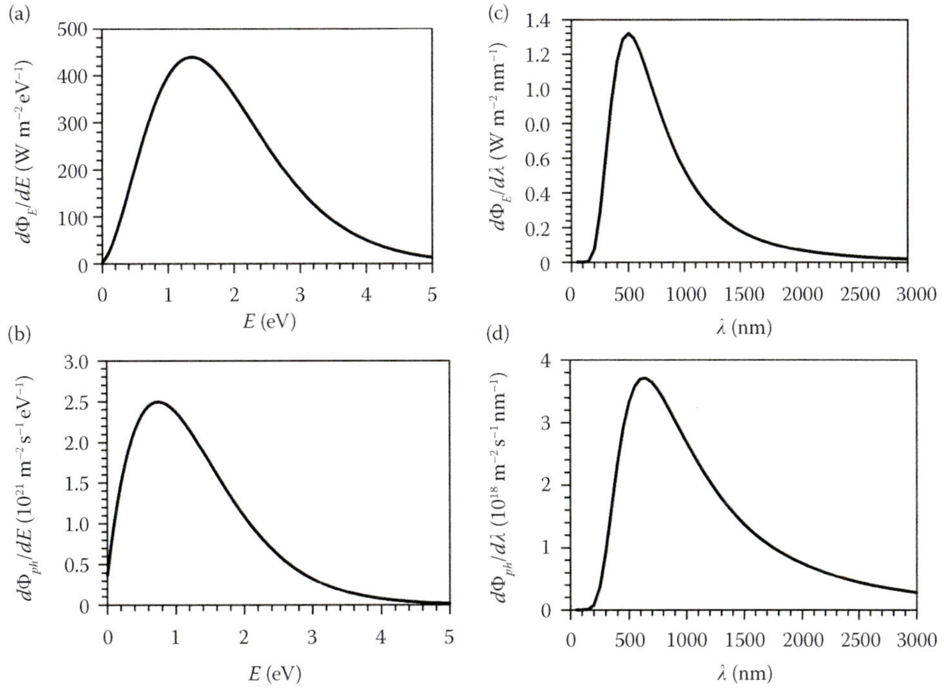

FIGURE 17.10 Energy and photon flux emitted by blackbody at $T = 5800$ K, as a function of energy (a, b) and wavelength (c, d). The energy flux is normalized to a total power density $\phi_E^{\mathrm{AM1.5G}} = 1000\,\mathrm{Wm}^{-2}$.

neglected and the resulting distribution is Wien's approximation.

$$\phi_{ph}^{bb}(E,\Omega) = \frac{\varepsilon}{\pi} b_\pi E^2 e^{-E/k_BT} \tag{17.34}$$

The spectral distribution of energy flux per second, per energy interval dE, is

$$\phi_E^{bb}(E) = \frac{\varepsilon}{\pi} b_\pi \frac{E^3}{e^{E/k_BT} - 1} \tag{17.35}$$

This quantity is related to the *spectral radiance* L_ν (Sizmann et al., 1991)

$$L_\nu(\nu) = \frac{d\Phi_E(\nu)}{\cos\theta\, d\Omega\, d\nu} \tag{17.36}$$

as $\phi_E^{bb}(\nu) = (\varepsilon/\pi) L_\nu(\nu)$. The total energy flux is

$$\Phi_E^{bb} = \frac{\varepsilon}{\pi} b_\pi \int_0^\infty \frac{E^3}{e^{E/k_BT} - 1} dE \tag{17.37}$$

With the change $x = E/k_BT$ and applying

$$\int_0^\infty \frac{x^3}{e^x - 1} dx = \frac{\pi^4}{15} \tag{17.38}$$

we obtain

$$\Phi_{E,tot}^{bb} = \frac{\varepsilon}{\pi} b_\pi \frac{\pi^4}{15} (k_BT)^4 = \frac{\varepsilon}{\pi} \sigma T^4 \tag{17.39}$$

where the Stefan–Boltzmann constant is defined as

$$\sigma = \frac{2\pi^5 k_B^4}{15 h^3 c^2} = 5.6704 \times 10^{-8}\ \mathrm{Wm}^{-2}\,\mathrm{K}^{-4} \tag{17.40}$$

The *Stefan–Boltzmann law* $\Phi_{E,tot}^{bb} = \sigma T^4$ gives the energy flux emitted by the blackbody radiator into a hemisphere, per unit area of the emitting surface.

17.8 THE SOLAR SPECTRUM

The solar spectrum is distributed over a range of wavelengths from 280 to 4000 nm. The total irradiation at the surface of the earth depends on the length of the path through the atmosphere, which is determined by the orientation of the sun with respect to the normal to the earth's surface at a given location. The length of the path is quantified by a coefficient denoted *air mass* (AM) that has the value 1 for normal incidence. The solar irradiation arriving at the earth fluctuates with the seasons and varies over extended periods of time. However, for energy conversion applications, it is convenient to refer the operation of devices and processes to a standardized

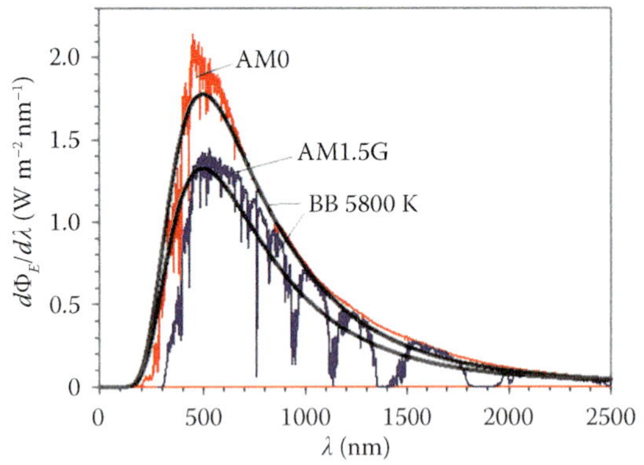

FIGURE 17.11 The spectral irradiance (energy current density, per wavelength interval) from the sun just outside the atmosphere (AM0 reference spectrum) and (AM1.5G) terrestrial solar spectrum. The lines are reference spectra of a blackbody at $T = 5800$ K, normalized to a total power density of $\phi_E^{AM0} = 1366\,\mathrm{Wm}^{-2}$ and $\phi_E^{AM1.5G} = 1000\,\mathrm{Wm}^{-2}$.

spectrum. Reference spectra are the AM0 corresponding to the spectrum outside the atmosphere and Air Mass 1.5 Global (AM1.5G) that describes the radiation arriving at the earth's surface after passing through a standard air mass 1.5 times, with the sun at 48.2°. Both are shown in Figure 17.11.

The AM0 reference spectrum, representing the typical spectral solar irradiance measured outside the atmosphere, is given by the international standard ASTM E-490 spectrum of the American Society for Testing and Materials. Two reference spectra are defined for the terrestrial irradiance under absolute air mass of 1.5. The AM1.5G (ASTM G173) has an integrated power of

$$\Phi_E^{AM1.5} = 1000\,\mathrm{Wm}^{-2} = 100\,\mathrm{m\,Wcm}^{-2} \qquad (17.41)$$

and an integrated photon flux of

$$\Phi_{ph}^{AM1.5} = 4.31\times10^{21}\,\mathrm{s}^{-1}\,\mathrm{m}^{-2} \qquad (17.42)$$

The AM1.5 Direct (+circumsolar) spectrum is defined for applications of solar concentrator and has an integrated power density of 900 W m^{-2}.

Consider the sun as a blackbody radiator of radius R_S and temperature T_S. The rate of energy emission from the whole surface is $\sigma T_S^4 4\pi R_S^2$. The flow of energy across a sphere of radius R, centered at the sun, is $S4\pi R$ where the *solar constant* S is a standard measure of the average energy received from sunlight and is defined as the energy received per unit time per unit area (perpendicular to the radiation) at the earth's mean distance from the sun. Therefore, the solar constant has the value

$$S = \left(\frac{R_S}{R}\right)^2 \sigma T_S^4 \qquad (17.43)$$

Using $R = 1.5 \times 10^{11}$ m and the outer radius of the sun $R_S = 6.95 \times 10^8$ m, we have $(R_S/R)^2 = 2.147 \times 10^{-5}$. From the effective temperature $T_S = 5760$ K, one obtains the value $S = 1349$ W m^{-2} (Landsberg, 1978). The actual value of the solar constant is obtained by an average of satellite measurements, and the integrated spectral irradiance of ASTM E-490 is made to conform to the accepted value of the solar constant which is

$$S = \Phi_E^{AM0} = 1366.1\,\mathrm{Wm}^{-2} \qquad (17.44)$$

The terrestrial solar spectral irradiance on the surface actually applied in solar energy conversion differs from the extraterrestrial irradiation due to the effect of the filtering by the atmosphere. The scattering by the atmosphere disperses the blue light more than the red part of the spectrum, which is the cause of the red color of sunset. Therefore, the atmospheric extinction affects predominantly the shorter frequencies of the incoming radiation. In addition, selective absorption by low concentration gases causes a strong decrease or even full extinction of the radiation in certain specific ranges of wavelengths that are indicated in Figure 17.2. In the infrared, the terrestrial solar spectrum is especially highly structured by the principal agents of opacity that are carbon dioxide and water vapor.

Most bodies on the earth's surface emit thermal infrared radiation, and the interception by the atmosphere is a cause of the greenhouse effect. The emission and absorption properties of the atmosphere are observed in detail in Figure 17.12. By looking at space from the surface, the atmosphere shows a transparency window (8–13 mm) that can be used to dissipate heat from the earth into outer space for different applications such as passive building cooling, renewable energy harvesting, and passive refrigeration in arid regions (Chen et al., 2016). For thermophotovoltaic applications, it is important to develop materials in which the normal radiative emission in the infrared is suppressed (Dyachenko et al., 2016).

It is sometimes practical to use an analytical function that approximates solar AM1.5G spectrum for evaluation of the efficiency of solar energy conversion. If the receiver measures only the spectral band of the visible and the close infrared, a blackbody spectrum of 5800 K approximately describes the spectrum in this wavelength range (Zanetti, 1984), see Figure 17.11. Although no single function can approach all details of the standard spectrum, we choose a blackbody spectrum at $T_S = 5800$ K (with the thermal factor $k_BT_S = 0.5$ eV) as shown in Figure 17.13,

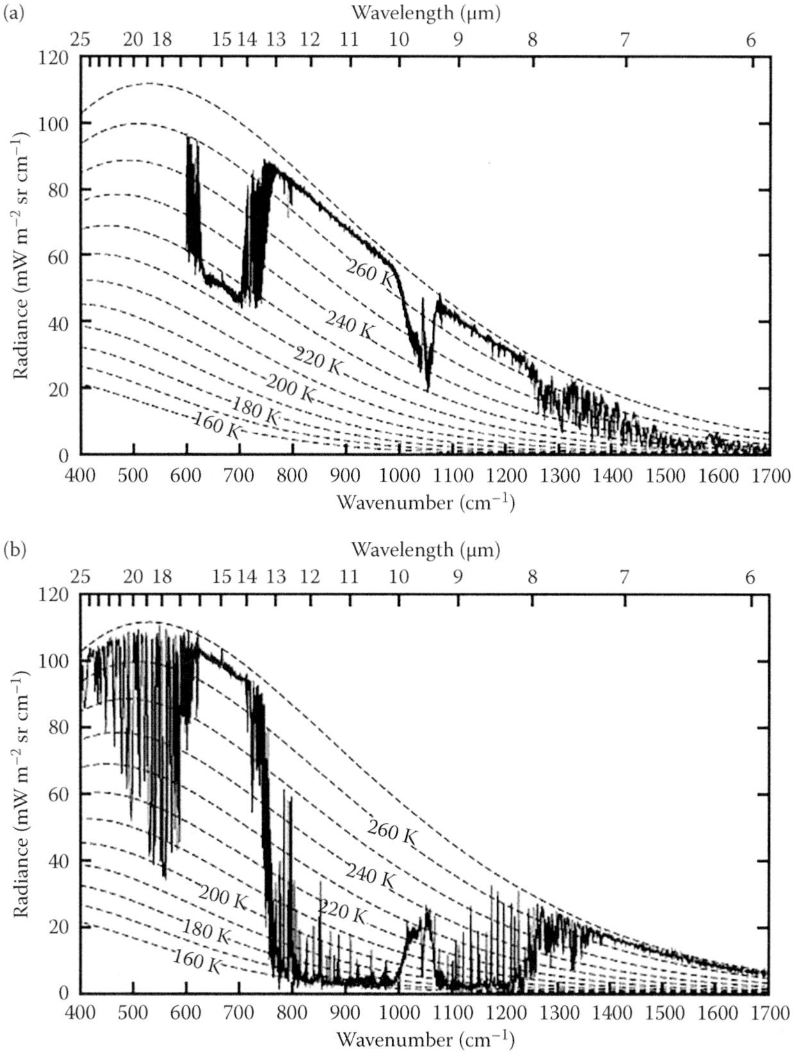

FIGURE 17.12 (a) Top of the atmosphere from 20 km and (b) bottom of the atmosphere from surface in the Arctic. Dashed curves represent the Planck function at different temperatures. The CO_2*bite* is between the 14 and 17 μm, and the *bite* in the 9.5–10 μm area is due to O_2 and O_3 absorption spectra. In these bands, the atmosphere is opaque to outgoing radiation. Emission observed when looking up is from the lowest levels of the atmosphere, closest to the ground. However, the emission of the ground, at ≈270 K, is observed when looking down from a plane 20 km high. In (a) the emission from CO_2 and O bands comes from higher levels in the atmosphere and corresponds to the Planck spectrum of lower temperatures, ≈270 K. (Reproduced with permission from Petty, G. W. *Atmospheric Radiation*, 2nd edition; Sundog Publisher: Madison, 2006.)

which indicates the quality of the approximation. To use the blackbody radiation expression, the photon flux at the earth's surface is reduced with respect to the emission at the sun's surface by the projected solid angle $\varepsilon_S^{AM1.5}$

$$\phi_{ph}^{bb,1sun}\left(E,T_S\right)=\frac{\varepsilon_S^{AM1.5}}{\pi}b_\pi\frac{E^2}{e^{E/k_BT_S}-1} \quad (17.45)$$

The total energy flux is given in Equation 17.38. Comparing the total irradiated power of the blackbody at $T_S = 5800$ K to Air Mass 1.5 Global (AM1.5G) conditions ($\Phi_{E,tot} = 1000$ W m^{-2}), we obtain the value of the dilution factor (Smestad and Ries, 1992)

$$D=\frac{\varepsilon_S^{1.5AM}}{\pi}=1.56\times10^{-5} \quad (17.46)$$

and the corresponding *equivalent* projected solid angle is $\varepsilon_S^{1.5AM} = D\pi = 4.8\times10^{-5}$. This is slightly different from the actual solid angle of the sun since $\varepsilon_S^{1.5AM}$ takes into account the extinction of the atmosphere.

The Standard Test Conditions (STC) for testing and verification of photovoltaic cell performance consist of air mass, sunlight intensity, and cell temperature at AM1.5G, 100 mW cm^{-2}, and 25°C, respectively. To test solar conversion devices, it is necessary to use a

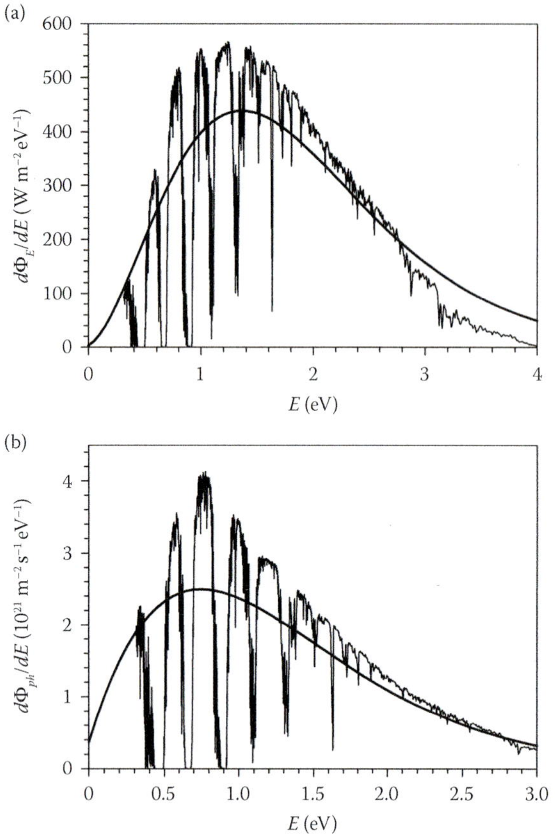

FIGURE 17.13 Energy and photon flux for the terrestrial solar spectrum (AM1.5G) as a function of photon energy. The lines are reference spectra of a blackbody at $T = 5800$ K, normalized to a total power density of $\phi_E^{AM1.5G} = 1000\,\mathrm{Wm^{-2}}$.

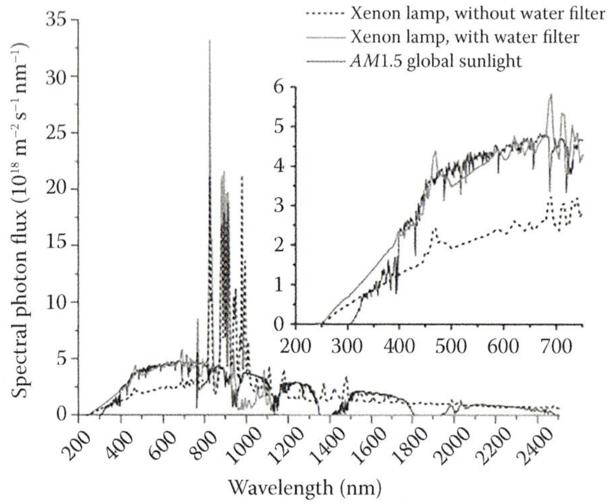

FIGURE 17.14 Spectral photon flux for the xenon lamp, with and without the water filter, compared to the AM1.5G solar spectrum. The lamp had logged 1250 h of operation when the spectrum was recorded. The inset shows detail of short wavelengths. In all cases, the total irradiance is normalized to 1000 W m^{-2}. (Reproduced from Murphy, A. B. et al. *International Journal of Hydrogen Energy* 2006, 31, 1999–2017.)

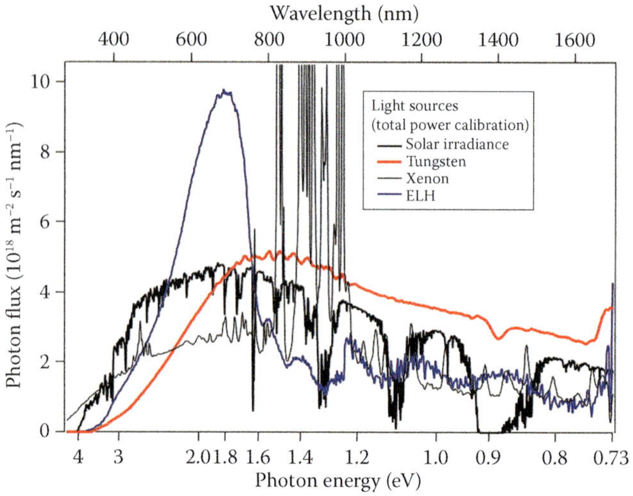

FIGURE 17.15 Spectral distribution of flux for AM1.5G irradiance compared to typical laboratory white light sources (adjusted to provide solar equivalent illumination power). ELH source is a tungsten halogen lamp with dichroic reflector. (Reproduced with permission from Doscher, H. et al. *Energy & Environmental Science* 2016, 9, 74–80.)

radiation source that approaches the AM1.5G spectrum as much as possible. Artificial light sources have important limitations to mimic the solar spectrum. It should be checked that the calibrated photon flux and energy flux matches the standard spectrum in the absorption range of the device (Murphy et al., 2006; Doscher et al., 2016). The spectra of widely used xenon lamp solar simulators are shown in Figure 17.14. Figure 17.15 shows the spectra of typical laboratory sources of white light, compared to the AM1.5G irradiance. It is also interesting to consider the spectral characteristic of warm while lamps used in ambient lighting since they can be used for energy supply in indoor photovoltaics (Ho, 2020).

GENERAL REFERENCES

Blackbody radiation and the Planck spectrum: Kangro (1976), Waldman (1983), Sizmann et al. (1991), MacIsaac et al. (1999), Robitaille (2008), Castrejón-García et al. (2010), and Howell et al. (2010).

Distribution of photons in blackbody radiation: Ross (1967), Landsberg (1978), Yablonovitch (1982), Ries and McEvoy (1991), Overduin (2003), Markvart (2007), Würfel (2009), and Mooney and Kambhampati (2013).

REFERENCES

Castrejón-García, R.; Castrejón-Pita, J. R.; Castrejón-Pita, A. A. Design, development, and evaluation of a simple blackbody radiative source. *Review of Scientific Instruments* 2010, *81*, 055106.

Chen, Z.; Zhu, L.; Raman, A.; Fan, S. Radiative cooling to deep sub-freezing temperatures through a 24-h day–night cycle. *Nature Communications* 2016, *7*, 13729.

Doscher, H.; Young, J. L.; Geisz, J. F.; Turner, J. A.; Deutsch, T. G. Solar-to-hydrogen efficiency: Shining light on photoelectrochemical device performance. *Energy & Environmental Science* 2016, *9*, 74–80.

Dyachenko, P. N.; Molesky, S.; Petrov, A. Y.; Störmer, M.; Krekeler, T.; Lang, S.; Ritter, M.; et al. Controlling thermal emission with refractory epsilon-near-zero metamaterials via topological transitions. *Nature Communications* 2016, *7*, 11809.

Fowles, G. R. *Introduction to Modern Optics*; Dover: New York, 1989.

Ho J. K. W., Yin H., So S. K. From 33% to 57% – an elevated potential of efficiency limit for indoor photovoltaics. *Journal of Materials Chemistry A*. 2020, 8, 1717–23.

Howell, J. R.; Siegel, R.; Menguc, M. P. *Thermal Radiation Heat Transfer*; CRC Press: New York, 2010.

Kangro, H. *Early History of Planck's Radiation Law*; Taylor & Francis: London, 1976.

Kennard, E. H. On the thermodynamics of fluorescence. *Physical Review* 1918, *11*, 29–38.

Kiang, N. Y.; Siefert, J.; Govindjee, B.; Blankenship, R. E. Spectral signatures of photosynthesis. I. Review of Earth organisms. *Astrobiology* 2007, *7*, 222–251.

Landsberg, P. T. *Thermodynamics and Statistical Mechanics*; Dover: New York, 1978.

MacIsaac, D.; Kanner, G.; Anderson, G. Basic physics of the incandescent lamp (lightbulb). *The Physics Teacher* 1999, *37*, 520–523.

Markvart, T. Thermodynamics of losses in photovoltaic conversion. *Applied Physics Letters* 2007, *91*, 064102.

Markvart, T. The thermodynamics of optical étendue. *Journal of Optics A: Pure and Applied Optics* 2008, *10*, 015008.

Mooney, J.; Kambhampati, P. Get the basics right: Jacobian conversion of wavelength and energy scales for quantitative analysis of emission spectra. *The Journal of Physical Chemistry Letters* 2013, *4*, 3316–3318.

Murphy, A. B.; Barnes, P. R. F.; Randeniya, L. K.; Plumb, I. C.; Grey, I. E.; Horne, M. D.; Glasscock, J. A. Efficiency of solar water splitting using semiconductor electrodes. *International Journal of Hydrogen Energy* 2006, *31*, 1999–2017.

Overduin, J. M. Eyesight and the solar Wien peak. *American Journal of Physics* 2003, *71*, 219–219.

Petty, G. W. *Atmospheric Radiation*, 2nd ed.; Sundog Publisher: Madison, 2006.

Planck, M. *The Theory of Heat Radiation*; P. Blakiston's Son & Co: Philadelphia, 1914.

Ries, H.; McEvoy, A. J. Chemical potential and temperature of light. *Journal of Photochemistry and Photobiology A: Chemistry* 1991, *59*, 11–18.

Robitaille, P.-M. Blackbody radiation and the carbon particle. *Progress in Physics* 2008, *3*, 36–55.

Ross, R. T. Some thermodynamics of photochemical systems. *The Journal of Chemical Physics* 1967, *46*, 4590–4593.

Shirasaki, Y.; Supran, G. J.; Bawendi, M. G.; Bulovic, V. Emergence of colloidal quantum-dot light-emitting technologies. *Nature Photonics* 2012, *7*, 13–23.

Sizmann, R.; Köpke, P.; Busen, R. Solar radiation conversion. In *Solar Power Plants. Fundamentals, Technology, Systems, Economics*; Winter, C.-J., Sizmann, R. L., Vant-Hull, L. L., Eds.; Springer: Berlin, 1991.

Smestad, G.; Ries, H. Luminescence and current-voltage characteristics of solar cells and optoelectronic devices. *Solar Energy Materials and Solar Cells* 1992, *25*, 51–71.

Waldman, G. *Introduction to Light*; Dover Publications: New York, 1983.

Würfel, P. *Physics of Solar Cells. From Principles to New Concepts*, 2nd ed.; Wiley: Weinheim, 2009.

Yablonovitch, E. Statistical ray optics. *Journal of the Optical Society of America* 1982, *72*, 899–907.

Yablonovitch, E.; Cody, G. D. Intensity enhancement in textured optical sheets for solar cells. *IEEE Transactions on Electron Devices* 1982, *29*, 300–305.

Zanetti, V. Sun and lamps. *American Journal of Physics* 1984, *52*, 1127–1130.

18 Light Absorption, Carrier Recombination, and Luminescence

The interaction of photons with molecules and materials, resulting in the absorption and production of radiation, is a fundamental component of many energy devices. A range of spectral techniques based on light absorption and emission shows the physical properties of ions, molecules, and materials. Light is used to generate carriers on a large scale in a solar cell and conversely, radiative recombination produces light emission in chromophores and semiconductor LEDs. In this chapter, we discuss the general properties of light absorption and luminescence in solids, nanostructured materials, and molecules. We address the general concepts of quantum efficiency (QE) and quantum yield (QY) that govern the conversion between electricity and light. Finally, we review basic recombination mechanisms in semiconductors and the methods of determination of carrier lifetime.

18.1 ABSORPTION OF INCIDENT RADIATION

Incident radiation on a material results in different types of interactions. The incident light can be reflected at any surface. A part of the radiation is transmitted through the back surface. The *coefficient of reflection* R is defined as the ratio of reflected power to the incident power of the electromagnetic radiation and similarly, the *transmittance* T is the ratio of transmitted power to the incident power. For a beam of energy E propagating in the x direction, with an intensity (power per unit area) $\Phi_E(E, x)$, we have

$$T = \frac{\Phi_E(E,x)}{\Phi_E(E,0)} \tag{18.1}$$

In the absence of absorption, the conservation of energy states that

$$R + T = 1. \tag{18.2}$$

The laws of reflection and refraction can be deduced from Huygens' principle. The refractive index of a medium is related to the relative dielectric constant as

$$n_r = \sqrt{\varepsilon_r} \tag{18.3}$$

For normal incidence, the coefficient of reflection is

$$R = \left[\frac{n_r - 1}{n_r + 1}\right]^2 \tag{18.4}$$

The light passing through the material can also be absorbed or scattered. In addition, some of the absorbed light can be reemitted. In this section, we consider the description of the absorption processes. The light-emitting processes will be described in the next section. Different effects of light propagation via various reflection modes in multilayer organic light-emitted diodes (OLEDs) are shown in Figure 18.1.

The absorption of light by a medium is quantified by its linear absorption coefficient α (m^{-1}), which is defined as the fraction of the power absorbed per unit volume. In terms of the decrease of intensity in an incremental slice of thickness dx, we have

$$\alpha(E) = -\frac{1}{\Phi_E(E,x)}\frac{d\Phi_E}{dx} \tag{18.5}$$

The integral of Equation 18.5 gives Beer-Lambert's law that describes the attenuation of light in a medium

$$\Phi_E(x) = \Phi_E(0)e^{-\alpha x} \tag{18.6}$$

In solar cells, efficiently absorbing most of the solar photons is a crucial aspect of materials and device design. The quantity $L_\alpha = \alpha^{-1}$ is denominated as the absorption length. It indicates the characteristic size of the absorber layer needed to collect all the solar spectral photons. Increasing the thickness of the active layer improves the device absorptivity but involves other drawbacks in device operation. To enhance the probability to convert photons to electricity, the light in the solar cell should be internally reflected until it is absorbed.

Consider the passage of the incident radiation through a medium across a length z. The dimensionless product

$$A = \alpha z \tag{18.7}$$

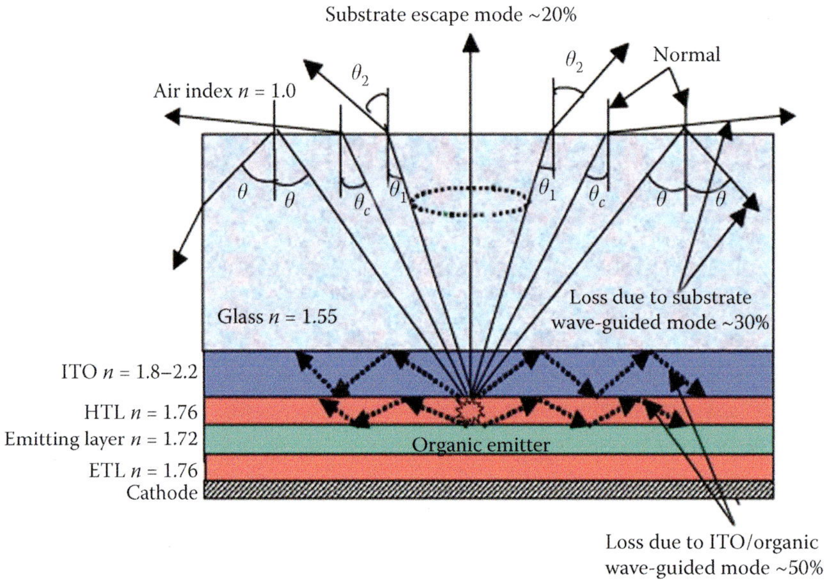

FIGURE 18.1 Schematic of multilayer OLED structure and optical ray diagram of light propagation via various modes, that is, substrate escape, substrate wave-guided mode, and ITO/organic wave-guided modes. (Reproduced with permission from Saxena, K. et al. *Optical Materials* 2009, 32, 221–233.)

is called the *absorbance* or the *optical density*. The intensity decreases as

$$\Phi_E = \Phi_E(0)e^{-A} \tag{18.8}$$

and we have

$$\log T = -A \tag{18.9}$$

When the absorption is a function of the absorbing centers in the medium, the absorbance is expressed as

$$\Phi_E(z) = \Phi_E(0)10^{-\varepsilon c z} \tag{18.10}$$

where c is the molar concentration of absorbing species and the quantity ε, in units $M^{-1}\ cm^{-1}$, is called the decadic *molecular extinction coefficient*.

We now consider the light absorption property of a slab of thickness d. The fraction of incoming radiation absorbed at a given wavelength or photon energy E and not reflected or emitted is described by the function $a(E)$, the spectral absorptivity or *absorptance* ($0 \le a \le 1$). It depends on the absorption coefficient $\alpha(E)$, the thickness of the layer, and the reflection properties of the surfaces. If reflection can be neglected, the absorptance relates to the absorption coefficient as

$$a(E) = \frac{\Phi_E(0) - \Phi_E(d)}{\Phi_E(0)} = 1 - e^{-\alpha d} \tag{18.11}$$

For an absorber layer with a backside mirror, the optical path length is doubled, and we have

$$a(E) = 1 - e^{-2\alpha d} \tag{18.12}$$

Considering the reflectivity at the front and back surface $R_f(E)$ and $R_b(E)$, a more general expression is obtained (Trupke et al., 1998):

$$a(E) = \frac{\left(1 - R_f(E)\right)\cdot\left(1 - e^{-\alpha(E)d}\right)\left(1 + R_b(E)e^{-\alpha(E)d}\right)}{1 - R_f(E)R_b(E)e^{-2\alpha(E)d}}. \tag{18.13}$$

If both reflectivities are identical and the absorption is very weak, $\alpha d \ll 1$, then

$$a(E) = \alpha d \tag{18.14}$$

However, if $d \gg L_\alpha$, all photons are absorbed except those reflected at the front surface

$$a(E) = 1 - R_f(E) \tag{18.15}$$

We note that if we neglect reflection effects, the absorptance has two important limits. When the film is optically thick, it is always $a = 1$ and when it is thin, the absorptance is proportional to absorption coefficient as in Equation 18.14.

Scattering refers to the interaction of a light beam with small particles in the medium. The scattering coefficient α_{sc} depends on the ratio of the size of the scattering particles to the wavelength of the light. The reduction in the intensity of a beam of light, which has traversed a medium containing scattering particles, is given by an expression identical to Beer's law:

$$\Phi_E(x) = \Phi_E(0)e^{-\alpha_{sc}x} \tag{18.16}$$

Scattering is maximized when the particle size is somewhat less than the wavelength of the light. It also depends on the ratio of the refractive indices of the particle and the surrounding medium. Scattering layers can play a large role to enhance photon collection in nanostructured solar cells (Usami, 1997).

In a solar cell with planar surfaces, the absorptance is essentially given by Equation 18.11, and if a reflective coating is used in the backside surface the optical pathway is only doubled. Weakly absorbed light in the long wavelength region, therefore, has a low chance to contribute to carrier generation by absorption. The light generated or scattered inside the solar cell that impacts the surface will leave the sample if it is included in the *escape cone* of the surface, Ω_c. The solid angle subtended by the escape cone is

$$\Omega_c = \frac{1}{2n_r^2}4\pi \tag{18.17}$$

Thus, for a planar surface, a randomly generated ray has a probability $p_e = \Omega_c/4\pi = 1/2n_r^2$ to escape from the active zone in an encounter with the surface (Yablonovitch, 1982). The high-refraction index of the semiconductor reduces the escape cone for the emission of the generated radiation. For silicon with a refractive index $n_r \approx 3.6$ in the wavelength range around $\lambda = 1150$ nm (Trupke et al., 2003a, b), a high-reduction factor $2n_r^2 \approx 26$ is obtained.

Randomly textured surfaces or regularly textured surfaces enhance the light absorption providing a light trapping effect due to the randomizing of the internal reflection angle avoiding the escape of internally reflected rays. This effect results in a much longer propagation distance of the light ray and, hence, a substantial increase of absorptance. A randomly textured surface also increases the probability of light emission, since the photons that hit the surface out of the narrow escape cone can be scattered into the cone by the random orientation. A Lambertian scatterer at the front surface disperses the light according to the law in Equation 17.9, which gives the highest degree of randomization. The upper limit of light path enhancement for random scattering textures has been shown to be the quantity $4n_r^2$ (Yablonovitch, 1982), and this is called the Yablonovitch limit (or the Lambertian limit). In a textured cell, in the weak absorption limit, the absorptance is well described by the expression given by Tiedje et al. (1984):

$$a(E) = \frac{\alpha}{\alpha + \dfrac{1}{4n_r^2 d}} \tag{18.18}$$

For a number of analytical solutions see (Green, 2002). Figure 18.2a shows the higher light emission of a textured silicon diode compared to planar diodes. The light emission properties of the textured cell compared with the blackbody radiation will be further discussed in section 18.3, 23.4, and in Figure 25.14.

In solar cells made of very thin layers that use high-absorption materials, the active layers may be of the same order of size as the wavelength of the light, which leads to interference effects. The cell design must take these effects into account in order to optimize the light harvesting properties as well as the spatial distribution of photogeneration of charge.

18.2 LUMINESCENCE AND ENERGY TRANSFER

In Chapter 17, we described blackbody radiation, which is the emission of radiation resulting from heat, also called as incandescence. *Luminescence* generally denotes the emission of light by a material after it has absorbed energy. Luminescence requires the promotion of an atom, a molecule, or a solid to an excited state that is subsequently demoted, emitting a photon.

Depending on the nature of the excitation source, there are different types of luminescence. If the excited state is obtained by photoexcitation, then the subsequent light emission is called photoluminescence (PL), Figure 18.3. The major forms of PL are *fluorescence* and *phosphorescence*. If the carriers are created by voltage injection to a solid, then the light emission process is called *electroluminescence* (EL), Figure 18.4.

In Section 19.5, we will show that the main PL emission in organic molecules is by spontaneous photon emission from the first singlet excited state and is called *fluorescence*. In general, fluorescence is a luminescence that occurs only during excitation or shows a very short decay time ($\tau < 10$ ms). On the other hand, *phosphorescence* persists for some time after the excitation has been disconnected, due to the long decay time ($\tau > 0.1$ s).

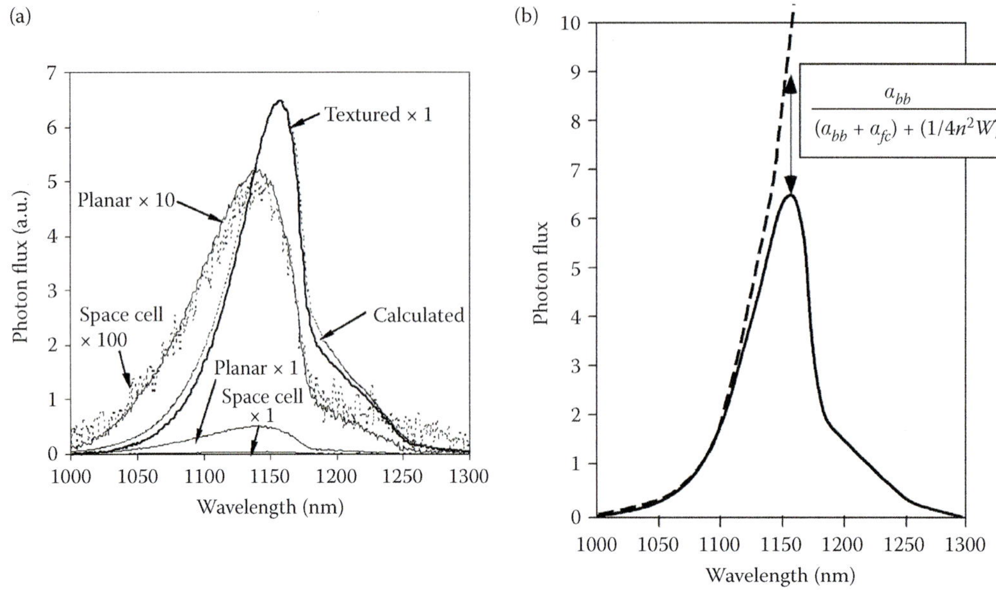

FIGURE 18.2 (a) Photon flux versus wavelength emitted by three silicon diodes. The lowest curve (shown multiplied 100 times) is for a high-performance silicon space cell. Planar cells perform 10 times better, due to reduction of parasitic non-radiative recombination. Textured cells perform 10 times higher again due to better optical properties, particularly much higher absorptance and hence emittance at the wavelengths shown. (b) Luminescence of a front-textured high-performance silicon solar cell compared to the blackbody emission (*dashed line*) for energies above the silicon bandgap (wavelengths shorter than 1102 nm). The expression in the inset (Equation 18.18) gives the ratio of these values, α_{bb} being the band-to-band absorption, α_{fc} the free carrier absorption, n the refraction index, and W the device thickness. (Reproduced with permission from Green, M. A. et al. *Physica E: Low-dimensional Systems and Nanostructures* 2003, 16, 351–358.)

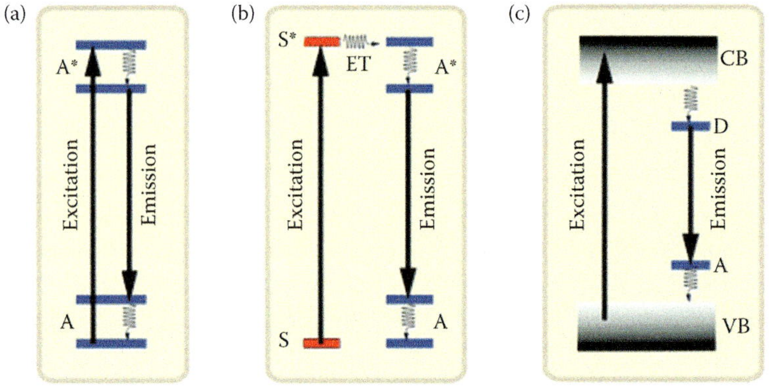

FIGURE 18.3 Luminescence processes. (a) Emission from a luminescence activator upon excitation. (b) Sensitized emission from an activator through energy transfer from a sensitizer to the activator upon excitation of the sensitizer. (c) Emission from a semiconductor after band-to-band excitation. A and A* represent the ground and excited states of the activator, respectively. S and S* represent the ground and excited states of the sensitizer, respectively. VB and CB represent the valence and conduction bands of the semiconductor while D and A represent the donor and acceptor energy levels, respectively. (Reproduced with permission from Huang, X. et al. *Chemical Society Reviews* 2013, 42, 173–201.)

Photon absorption and emission in an ion or molecule are physically reciprocal processes. The respective probability coefficients are connected by a detailed balance relationship by Einstein (Strickler and Berg, 1962). If the absorption coefficient is large, so is the emission coefficient, and the radiative lifetime is short as shown in Equation 22.23. However, in organic chromophores and in inorganic quantum dots, absorption and emission do not occur at the same wavelength due to the vibrational or thermal relaxation losses that occur after the absorption process, as shown in Figure 18.3a. This general principle is known as Stokes' law. The displacement of the peak of the emission to longer wavelengths from the absorption peak is generally known as *Stokes' shift*. It is

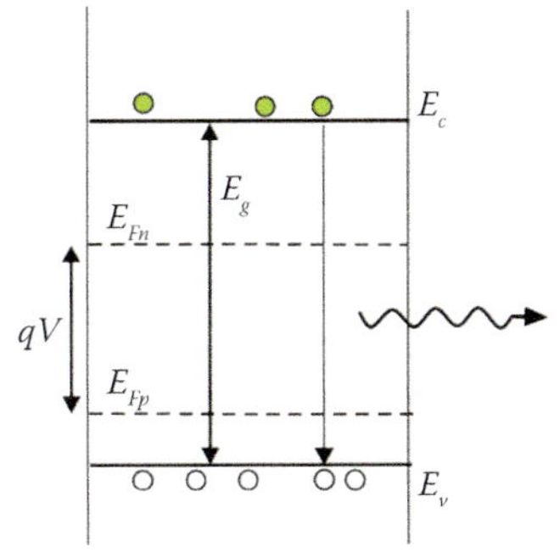

FIGURE 18.4 Emission of a photon by radiative recombination in an LED.

normally measured as the difference between the maximum wavelengths in the excitation and emission spectra. A Stokes' shift due to vibrational or orientational relaxation is discussed in Section 19.5.

Inorganic solids that give rise to luminescence are called *phosphors* or luminescent materials. The luminescence can be obtained from two different mechanisms. One is the emission of localized centers or activators embedded in a larger host crystalline structure that constitutes the bulk of the phosphors. The second is recombination in semiconductors, which may be band to band or via recombination centers. In semiconductors, most electrons and holes undergo rapid thermalization on a ps time scale, as discussed in Figure 5.8, so that band-to-band recombination occurs between carriers close to the respective band edges, Figures 18.3 and 18.4. Therefore, the emitted photon spectrum peaks sharply at energy $h\nu \approx E_g$. A variety of recombination mechanisms are discussed in Section 18.4. EL can also be influenced by the free-exciton, bound-exciton, free-to-bound, and donor–acceptor recombination mechanisms.

The main metal ions used in inorganic host materials to form phosphors are lanthanide ions, which are a family of 15 chemically similar elements from lanthanum (La) to lutetium (Lu). The lanthanides and the yttrium (Y) are generally denoted as "rare earths." The lanthanides show rich optical properties due to a large number of possible transitions between diverse $4f^n$ states ($0 < n < 14$). The transitions between the different $4f^n$ states are parity forbidden and become allowed through mixing with opposite parity states of crystal field components of vibrations. The absorption lines are very narrow because the coupling with the vibrations is weak. There is no Stokes' shift from absorption to emission, which allows for avoiding thermal losses by vibrations, enhancing the luminescent QY.

The decay from an excited molecular state or the recombination of an electron-hole pair does not necessarily imply the emission of a photon, because the electron in the excited state may decay via other *nonradiative* recombination pathways. In this case, the energy of the excitation is dissipated in the lattice of the material as heat. Nonradiative recombination processes are mediated by phonons and are largely suppressed at low temperatures. Another way to prevent radiationless recombination is to use a dielectric environment in which the phonon frequency is small. This is because multiphonon processes become less likely when the number of phonons that is involved in the decay of the excited state is larger.

The mechanism of sensitization separates the absorption and emission processes in PL into two different ions or molecules, as shown in Figure 18.3b. In the terminology of phosphors, the *sensitizer* absorbs light and transfers the excitation to an *activator*, which emits a photon. The acceptor can also be termed the *annihilator* or *emitter*. Conservation of energy in the transfer process and energy relaxation in the vibrational modes of the activator implies that Stokes' shift will normally occur.

There are several ways to transfer the energy from a sensitizer to a target molecule. In photovoltaic applications, one can extract electrons and holes from the sensitizer to metal oxide and electrolytic contacts (Gerischer et al., 1968), as in a dye-sensitized solar cell. For the transfer from one molecule to another, one normally distinguishes three mechanisms: radiative transfer, nonradiative transfer, and multiphonon-assisted energy transfer. In the *radiative transfer* mechanism, a real photon is emitted by the sensitizer and reabsorbed in the acceptor. *Resonant energy transfer involves* the excitation of a donor molecule that decays and passes its energy to the acceptor molecule by a suitable interaction, before the sensitizer is able to emit a quantum of fluorescence. *Förster resonant energy transfer* (FRET) occurs by Coulombic dipole–dipole interaction (Scholes, 2003). The rate constant for energy transfer is given by the expression

$$k_{ET} = k_D \left(\frac{R_0}{R} \right)^6 = \frac{1}{\tau_D^0} \left(\frac{R_0}{R} \right)^6 \tag{18.19}$$

Here $k_D = 1/\tau_D^0$ is the decay rate constant of the excited donor in the absence of energy transfer, R the distance between donor and acceptor, and R_0 is the *critical quenching radius* or *Förster radius*, that is, the distance at which the rate constants $k_{ET} = k_D$, so that energy transfer and spontaneous decay of the excited donor are equally probable (Braslavsky et al., 2008). This mechanism operates for large separations up to 2 nm. FRET gives rise to diffusion of excitons and is applied in antenna complexes in the natural photosystem, in which the excitation travels until it arrives to the reaction center.

Energy transfer based on higher multipole and exchange interaction is termed *Dexter energy transfer*. It depends on the spatial overlap of wave functions and operates only at very short distances (< 0.5 nm). FRET has become a general denomination for energy transfer that does not involve fluorescence in the donor. Different types of excitonic states are discussed in Chapter 19.

18.3 THE QUANTUM EFFICIENCY

Starting from the QY, we define several quantum efficiencies for LEDs and solar cells. QY of a photo-induced process is the number of defined events (electron-hole generation, fluorescence, photochemical reaction) occurring per photon absorbed. The QY is defined for monochromatic radiation of frequency ν, hence the denominator contains the incident spectral photon flux $\phi_{ph}^{in}(\nu)$. Assuming $a(\nu) = 1$, we have

$$\eta_{QY}(\nu) = \frac{\text{Number of events}}{\phi_{ph}^{in}(\nu)} \tag{18.20}$$

The *QE* of a luminescent material is the number of photons emitted per input. The input can be an incident photon flux, as in PL, or the (current density)/q, in EL, as in an LED. The QE is defined for monochromatic radiation of frequency ν.

The *external quantum efficiency* (EQE) in PL is given by the ratio of incoming and outgoing photon flux

$$EQE_{PL}(n) = \frac{\phi_{ph}^{out}(\nu)}{\phi_{ph}^{in}(\nu)} \tag{18.21}$$

The EQE is a combination of several factors:

$$EQE_{PL} = \eta_{LHE} IQE \eta_{outco} \tag{18.22}$$

The light harvesting efficiency (LHE) is the number of photons that are absorbed by the converter, determined by absorption coefficient, thickness, and other optical features such as reflection, texturing, scattering layers, etc., which were discussed in Section 18.1, so that, in terms of the absorptivity, $\eta_{LHE} = a(\nu)$. The *internal quantum efficiency* (IQE) is the probability that an excited molecule or an electron-hole pair recombines radiatively. Extraction or outcoupling efficiency η_{outco} is the fraction of internally generated photons that escape out of the material and contribute to the radiated emission. It depends strongly on the geometry and optical characteristics of the material, as indicated in Figure 18.1.

A semiconductor LED operates on the principle of injection of electrons and holes from separate contacts that recombine radiatively in the active layer. The EQE is defined as the ratio of output-emitted photons to electron-hole pairs injected by the current density j

$$EQE_{LED} = \frac{\Phi_{ph}^{out}}{j/q}. \tag{18.23}$$

The EQE is related to two efficiency factors:

$$EQE_{LED} = IQE_{LED} \eta_{outco}^{LED} \tag{18.24}$$

The external voltage produces total recombination current j_{rec} that contains both radiative j_{rad} and nonradiative current j_{nrad}, hence the LED IQE is given by the relationship

$$IQE_{LED}(V) = \frac{j_{rad}(V)}{j_{rec}(V)} = \frac{j_{rad}(V)}{j_{rad}(V) + j_{nrad}(V)} \tag{18.25}$$

In organic LEDs, it is often necessary to establish the process of formation of excitons that subsequently produce the radiative emission. Hence, the EQE contains additional terms as follows:

$$EQE_{LED} = \eta_{inject}^{LED} \chi_{op} \eta_{rad}^{LED} \eta_{outco}^{LED} \tag{18.26}$$

The injection efficiency, η_{inject}^{LED}, is the fraction of injected carriers that become excited electron-hole pairs, or excitons. χ_{op} is the fraction of excitons whose states have spin-allowed optical transitions. For thermalized triplet and singlet states in organic LEDs, $\chi_{op} = 0.25$. η_{rad}^{LED} is the ratio of radiative to total recombination. The first three factors in Equation 18.26 form the IQE of the LED.

Internally trapped photons can be absorbed and reemitted in a cyclic fashion. *Photon recycling*, which will be described in more detail in Section 23.4, refers to the process of reabsorption and reemission of photons that are not able to escape the LED in the first emission cycle produced by the voltage. Photon recycling allows photons to be reemitted several times before eventually escaping the LED provided that the nonradiative recombination rate is low. Under total internal reflection and photon recycling processes, the existence of nonradiative recombination pathways enhances the probability that multiple regenerated photons are finally recombined nonradiatively, decreasing the external quantum efficiency of the LED. In the presence of photon recycling and reabsorption effects, for a large IQE value the product in Equation 18.24 is not valid and a more general expression is given in Equation 23.32.

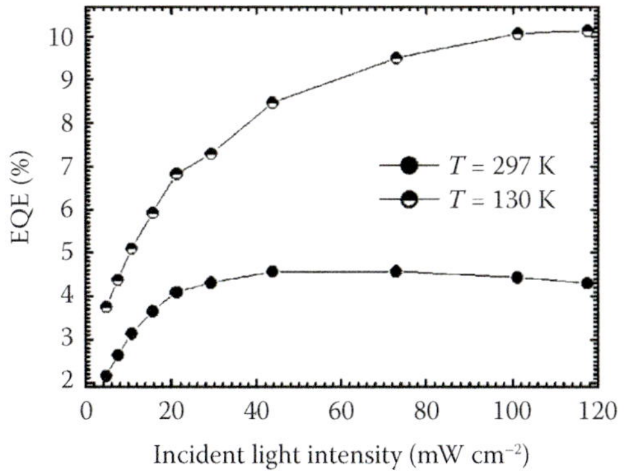

FIGURE 18.5 PL *EQE* of a textured 500 μm thick 30 Ω cm n-type silicon sample as a function of the incident light intensity. (Reproduced with permission from Trupke, T. et al. *Journal of Applied Physics* 2003a, 94, 4930–4937. Trupke, T. et al. *Applied Physics Letters* 2003b, 82, 2996–2998.)

The rate of external emission of either direct light radiation or regenerated photons in an LED also depends on the optical characteristics of the device as shown in Figure 18.1. As commented earlier in Equation 18.17, the escape probability of a photon is considerably reduced when the refractive index of the medium is high. Determination of η_{outco} can be made at low temperatures where nonradiative recombination is reduced and the IQE is close to unity. In practice, separating all the different factors involved in the EQE of an LED is a difficult task (Matioli and Weisbuch, 2011; Kivisaari et al., 2012).

The PL EQE of a textured 500 μm thick n-type silicon sample is shown in Figure 18.5. The EQE increases at low temperature because the absorption coefficient of silicon for band-to-band transitions decreases strongly in the spectral range 1000–1250 nm (1.0–1.2 eV), where significant luminescence is emitted from bulk silicon, see Figure 18.2. At lower temperatures the PL is enhanced, as the reabsorption of internally generated photons by band-to-band transitions is considerably reduced (Trupke et al., 2003a, b).

In a solar cell, the photovoltaic EQE is the ratio of electron flux, in the form of electrical current (at short circuit) to the incoming photon flux:

$$EQE_{PV}(\nu) = \frac{j(\nu)}{q\phi_{ph}(\nu)} \tag{18.27}$$

The EQE_{PV} is obtained by measuring the short-circuit current under monochromatic light as a function of the frequency or wavelength. The photovoltaic EQE depends on the conditions of conductivity and recombination, which may depend on background illumination conditions, which will be discussed in Section 26.10.

The EQE_{PV} can be separated into its optical and electrical parts.

$$\begin{aligned} EQE_{PV}(\nu) &= \eta_{LHE} IQE_{PV} \\ &= a(\nu) IQE_{PV}(\nu) \end{aligned} \tag{18.28}$$

IQE_{PV} is the flux of collected electrons per photon absorbed, given by

$$IQE_{PV} = \eta_{sep}\eta_{col} \tag{18.29}$$

The efficiency of charge separation of a generated electron-hole pair to free carriers is denoted η_{sep} (or injection efficiency, η_{inj}, in heterogeneous solar cells). This process must be considered in systems in which photogeneration creates a spatially localized electron-hole pair, or an exciton, that has a probability to recombine (see the full discussion in Chapter 24). The charge collection efficiency η_{col} is the probability that the separated electron and hole carriers are collected at the contacts and not lost by recombination or other processes, as will be discussed in Chapter 26. In summary, the EQE can be expressed as a product

$$EQE_{PV}(\nu) = \eta_{LHE}\eta_{sep}\eta_{col} \tag{18.30}$$

The EQE_{PV} is also termed incident-photon-to-current-col lected-electron-efficiency (η_{IPCE}) and IQE_{PV} is denoted as absorbed-photon-to-collected-electron-efficiency (η_{APCE}).

For a layer of reflectivity R of the front surface, absorption coefficient α, and no back layer reflection, the EQE relates to IQE as

$$EQE_{PV} = (1-R)\left(1-e^{-\alpha d}\right) IQE_{PV} \tag{18.31}$$

The normalized EQE of record photovoltaic devices of several technologies is shown in Figure 26.2a.

It is important to remark that EQE_{LED} is a scalar quantity, which describes the whole outgoing photon flux, while EQE_{PV} is a function of photon energy or wavelength.

18.4 THE RECOMBINATION OF CARRIERS IN SEMICONDUCTORS

The excitation of a semiconductor by supra-bandgap light or the voltage injection process creates excess carrier densities with respect to equilibrium values n_0, p_0. In the recombination process, an electron in the conduction band of the semiconductor makes a downward transition to a valence band state resulting in annihilation of the

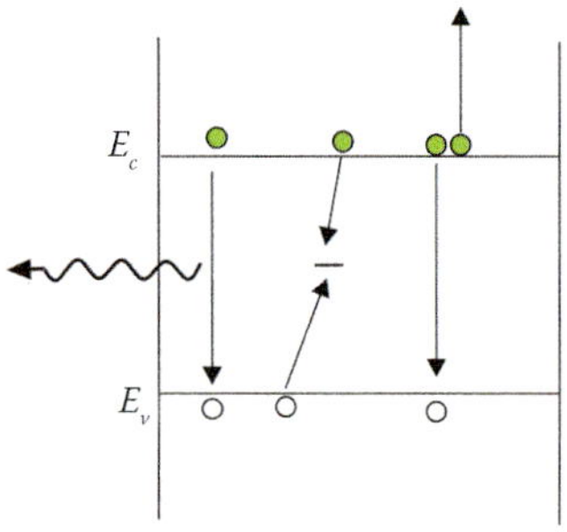

FIGURE 18.6 From left to right: Band-to-band, SRH, and Auger recombination.

electron-hole pair (Figure 18.6). In crystalline inorganic semiconductors such as silicon or GaAs, there are three principal recombination mechanisms: band to band, via defect levels, and Auger.

In band-to-band recombination, the electron makes a single transition from a state in the conduction band to an empty state in the valence band. Most of the energy of the excited carrier is released via photon emission, and this process is hence called radiative recombination. The density dependence of the recombination rate from band to band is well described by the expression

$$U_{np} = Bnp \tag{18.32}$$

where B is a recombination coefficient that depends on the temperature and other factors. The value for radiative recombination in Si at 300 K is $B \approx 1 \times 10^{-14}$ cm^3 s^{-1} (Michaelis and Pilkuhn, 1969; Gerlach et al., 1972; Trupke et al., 2003a, b), and for GaAs it is $B = 7.2 \times 10^{-10}$ cm^3 s^{-1} (Varshni, 1967a, b). More generally, the radiative recombination coefficient can be extracted from the absorption coefficient by reciprocity arguments as described in Section 22.2. From Equation 5.56, we may re-write Equation 18.32 showing the dependence of recombination rate on applied voltage as

$$U_{np} = Bn_i^2 e^{qV/k_BT} \tag{18.33}$$

The recombination of excess carriers is

$$U_{np} = B\left(np - n_i^2\right) \tag{18.34}$$

In a p-doped semiconductor, the majority carrier density p is basically constant, $p \approx p_0$ (except at very high injection levels), and the recombination rate (18.34) depends only on the local electron density as follows:

$$U_n = k_{rec}\left(n - n_0\right) \tag{18.35}$$

where

$$k_{rec} = Bp_0 \tag{18.36}$$

The recombination rate is often described in terms of the *electron lifetime*. This quantity requires a careful definition as commented in Section 18.5, but for a linear recombination rate indicated in Equation 18.35, it is given simply by

$$\tau_n = \frac{1}{k_{rec}} \tag{18.37}$$

(see also Equation 18.46). If the recombination event occurs immediately after the generation of an electron-hole pair, before any charge separation occurs, it is called geminate recombination.

Recombination may occur as well via trap levels in the band gap. Therefore, the drop of an electron from the conduction band to a band gap level can produce the emission of a photon that is redshifted with respect to the band gap photons. Recombination via a mid-gap state more often consists of a multiphonon process in which the band gap energy is released to a number of phonons rather than by the emission of photons, hence called radiationless or *nonradiative* recombination. Nonradiative recombination can be associated with defects located in the bulk or the surface of the semiconductor. The surface of the semiconductor where the contacts are located is often a source of defects and recombination sites. Recombination at these surface sites is called surface recombination.

Recombination via mid-gap states is often an important effect in semiconductor devices. For a nonradiative recombination process, the larger the energy between initial and final electron states, the greater the number of phonons required and the process becomes more unlikely. Thus, a mid-gap state, that is readily able to capture both electrons and holes, greatly increases the probability of radiationless recombination with respect to band-to-band recombination. A frequently used recombination model is the Shockley-Read-Hall (SRH) recombination that was described in Section 6.3. It occurs at a localized state in the bandgap that receives both an electron from the conduction band and a hole from the valence band (Figure 18.6). The recombination rate of excess carriers is given by Equation 6.35, often expressed as

$$U_{\text{SRH}} = \frac{\left(np - n_i^2\right)}{\tau_p\left(n + n_1\right) + \tau_n\left(p + p_1\right)} \tag{18.38}$$

where τ_n is the electron lifetime for a large density of holes, τ_p is the hole lifetime for a large density of electrons, and n_1, $p_1 = n_i^2/n_1$ are the electron and hole densities when the Fermi level coincides with the energy of the trap through which the recombination takes place:

$$n_1 = N_c \exp\left(\left(E_t - E_c\right)/k_B T\right)$$
$$p_1 = N_v \exp\left(\left(E_v - E_t\right)/k_B T\right) \qquad (18.39)$$

A value τ_n of 4 ms is typical for the low injection SRH electron lifetime measured in lightly doped silicon (Stephens et al., 1994).

Another important effect in heavily doped inorganic semiconductors is the Auger recombination, in which the energy released in band-to-band recombination of an electron and a hole is transmitted to another electron or hole (Figure 18.6). The resultant gain of kinetic energy of the third particle (electron) is normally lost by relaxation of the carrier to the band edge energy. It is, therefore, a three carrier nonradiative process usually expressed as

$$U_{\mathrm{Auger}} = \left(C_n n + C_p p\right)\left(np - n_i^2\right) \qquad (18.40)$$

where C_n and C_p are the Auger coefficients.

Figure 18.7 shows the minority carrier lifetime and luminescent IQE of different semiconductor solar cells. Auger recombination becomes the dominant effect at high-carrier density (Vossier et al., 2010).

Recombination in semiconductors is often investigated by observing luminescence dependence on incident light intensity. The identification of the dominant recombination processes as a function of the excitation rate can be established using an expression of the type

$$U_n = An + Bn^2 + Cn^3 \qquad (18.41)$$

The first term is for monomolecular recombination, when the recombination is dominated by minority carriers under low-carrier injection, or for SRH when the capture of one carrier to traps governs the overall rate in Equation 18.38. The second term is for bimolecular recombination, which involves the densities of both electrons and holes. This is the case in band-to-band radiative recombination under high injection, when the condition $n = p$ holds. The third term is trimolecular recombination that corresponds to Auger processes. An illustration is shown in Figure 18.8 for the lead halide perovskite solar cells, indicating first the PL intensity as a function of the excitation intensity, and then the IQE that allows to recognize the dominant recombination

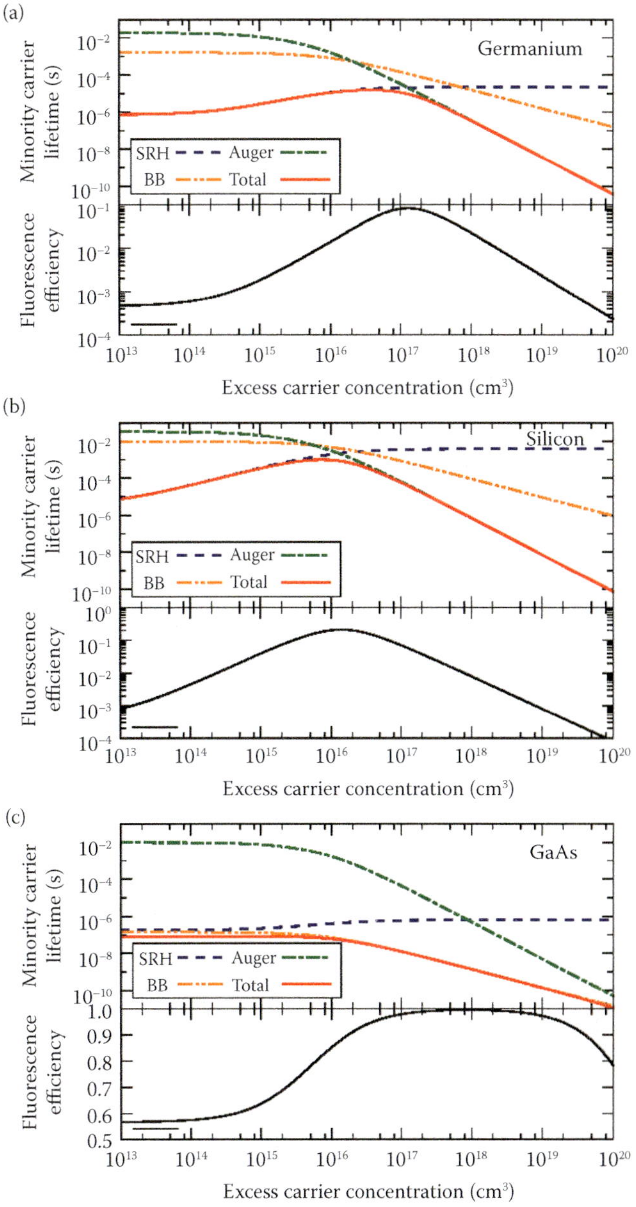

FIGURE 18.7 Minority carrier lifetime and internal fluorescence efficiency (luminescent IQE) of Ge, Si, and GaAs solar cells as a function of excess carrier concentration. (Reproduced with permission from Vossier, A. et al. *Journal of Applied Physics* 2015, 117, 015102.)

mechanism. At lower intensities of illumination, the PL is controlled by recombination via trap states. When the traps become saturated, the PL is dominated by band-to-band mechanism, and finally the Auger recombination sets in at very high-generation levels. The kinetic model based on Equation 18.41 allows to find the recombination lifetimes (Staub et al., 2016).

In nanostructured semiconductors and organic blends, recombination occurs by the transference of a carrier between two different phases. In these systems, the

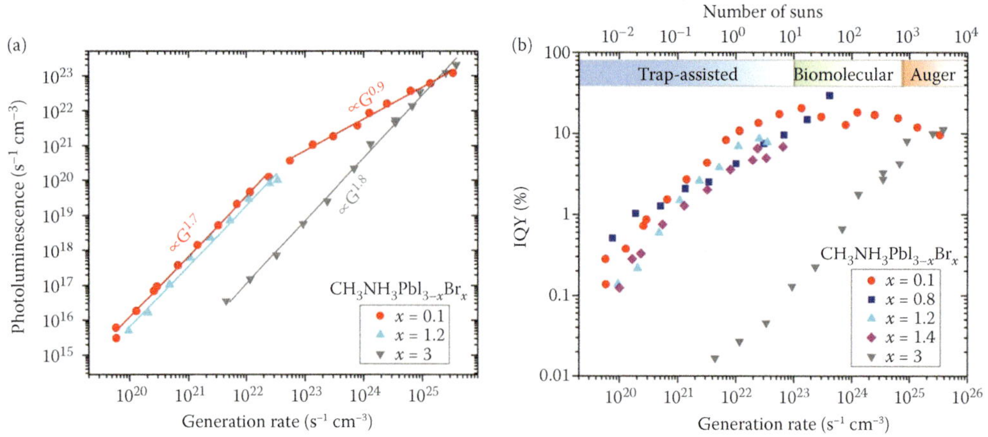

FIGURE 18.8 Steady-state PL of lead halide perovskite solar cells. (a) Pump-power dependence of the integrated PL signal for three Br concentrations. (b) Pump-power dependence of the IQE. (Reproduced with permission from Sutter-Fella, C. M. et al. *Nano Letters* 2016, 16, 800–806.)

charge transfer process is often influenced by disorder and a combination of surface states. The electron transfer rate can often be expressed using a phenomenological model as follows (Bisquert and Marcus, 2014):

$$U_n = k_{rec}\left(n^{\beta} - n_0^{\beta}\right) \tag{18.42}$$

where $0 < \beta \leq 1$ is a recombination order.

18.5 RECOMBINATION LIFETIME

Let us consider the question of the probability of survival of a quantity of excess electrons injected to a bulk or nanostructured semiconductor. We assume that the electrons are minority carriers and that electro-neutrality is maintained by fast displacement of the majorities. The decay of a population of electrons is governed by their rate of recombination according to the equation

$$\frac{dn}{dt} = -U_n(n) \tag{18.43}$$

We take first the simplest recombination model, which is that of linear recombination introduced in Equation 18.35. Excess electrons injected can be written as $\Delta n = n - n_0$, and their decay is determined by

$$\frac{d(\Delta n)}{dt} = -k_{rec}\Delta n \tag{18.44}$$

Therefore, the decay with time takes the form

$$\Delta n(t) = \Delta n(0)e^{-t/\tau_n} \tag{18.45}$$

where the lifetime, τ_n, is

$$\tau_n = k_{rec}^{-1} \tag{18.46}$$

In general, we define the lifetime as the constant in the denominator of the exponential decay law of Equation 18.45. However, we observe that such decay law depends critically on the fact that our starting recombination law in Equation 18.35 is linear, which is far from being the general case, as discussed in Section 18.4. It is useful to develop a procedure whereby a lifetime can be established in any type of decay process, so that the results of experiments can be well defined. Here, we describe a method that is widely used in device characterization, in which the determination of the lifetime is based on small perturbation measurement. The general rationale for small perturbation method was introduced earlier, in Section 3.10. It has the advantage that the measured quantity is independent of the amplitude of the measurement signal, since all the equations become linear in the small perturbation domain. Small perturbation recombination lifetime has been amply adopted in the field of dye-sensitized and organic solar cells (Bisquert et al., 2009), and it has also been developed in amorphous silicon (Ritter et al., 1988) and crystalline silicon solar cells (Brendel, 1995; Schmidt, 1999).

Let us take a system that is determined by any general recombination law that has the form $U_n(n)$ in terms of the carrier concentration n. The decay of injected carriers by recombination is outlined in Figure 18.9, which emphasizes that the measurement of lifetime basically consists of a perturbation of the Fermi level that induces the recombination toward a certain equilibrium value. Experimentally, the shift of the Fermi level will be

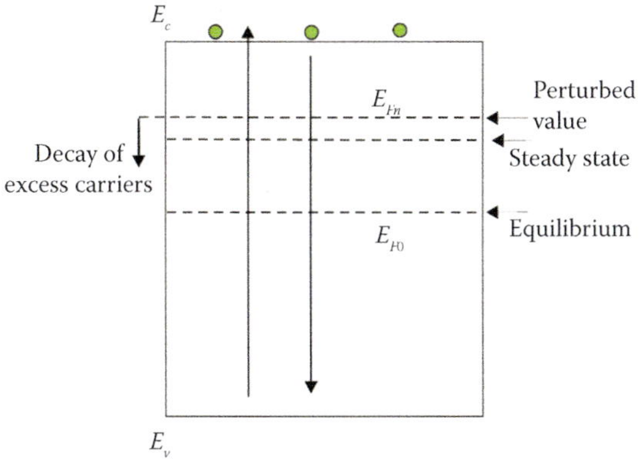

FIGURE 18.9 Scheme of the measurement of a lifetime. The semiconductor has an equilibrium density of electrons n_0 and a steady-state value $\bar{n}$, established by equilibrium of generation-recombination. The corresponding Fermi levels are indicated. An excess carrier density is injected, $\Delta n = n - n_0$, which decays in a characteristic time τ_n as the Fermi level E_{Fn} returns to the steady-state value.

recorded; for instance, measuring the transient of photovoltage. We assume that the stable carrier density $\bar{n}$ (the steady state) is maintained by background photogeneration. Considering the balance of generation and recombination, we have

$$\frac{dn}{dt} = G_\Phi - U_n(n) \tag{18.47}$$

which in equilibrium sets the background carrier density by the equation

$$U_n(\bar{n}) = G_\Phi \tag{18.48}$$

Now a small perturbation $\hat{n}$ that is induced on top of the steady state provides the density dependence on time as

$$n(t) = \bar{n} + n(t) \tag{18.49}$$

Let us determine the transient behavior of $\hat{n}$. With an expansion

$$U(n) = U(\bar{n}) + \frac{\partial U}{\partial n} n \tag{18.50}$$

we obtain from Equation 18.47

$$\frac{d\hat{n}}{dt} = -\frac{\partial U}{\partial n}\hat{n} \tag{18.51}$$

The result we obtain in Equation 18.51 is that the linearization procedure always takes the evolution equation to a form of Equation 18.44, which will provide an exponential decay of the small perturbation excess density. Equation 18.51 defines a lifetime in terms of the recombination rate $U_n(n)$ as

$$\tau_n = \left(\frac{\partial U_n}{\partial n}\right)_{\bar{n}}^{-1} \tag{18.52}$$

As an example of the small perturbation procedure, consider the nonlinear recombination law introduced in Equation 18.42

$$U_n = k_{rec}\left(n^\beta - n_0^\beta\right) \tag{18.53}$$

Equation 18.47 can be written as

$$\frac{d\hat{n}}{dt} = G_\Phi - k_{rec}\left[\left(\bar{n} + \hat{n}\right)^\beta - n_0^\beta\right] \tag{18.54}$$

Expanding the sum to first order in $\hat{n}$ and removing the steady-state terms (that cancel out), we have

$$\frac{d\hat{n}}{dt} = -k_{rec}\beta\bar{n}^{\beta-1}\hat{n} \tag{18.55}$$

Therefore, the lifetime is

$$\tau_n(\bar{n}) = \left(k_{rec}\beta\bar{n}^{\beta-1}\right)^{-1} \tag{18.56}$$

Note that the lifetime is a function of the steady state; this is a general feature of nonlinear systems.

The decay of a small population that serves as a probe of the kinetics of the system has, therefore, been characterized for any recombination rate law and steady-state condition. More generally, the decay may require a number of sequential processes coupled to recombination, for example, when prior de-trapping of localized carriers is required. The linearizing method toward an exponential decay can easily be generalized and the characteristic decay time in general is denoted a *response time* (Rose, 1963).

In general, the response time τ of a carrier in a given type of electronic state with concentration c is determined by the decay of a small variation of the concentration $\hat{c}$ to equilibrium (Rose, 1963; Bisquert et al., 2009):

$$\frac{\partial \hat{c}}{\partial t} = -\frac{1}{\tau}\hat{c} \tag{18.57}$$

Let us take Equation 6.26 for capture and release of electrons in a trap. We recall that f is trap occupancy, determined by kinetic exchange with electrons from the conduction band with density n_c as indicated in

Figure 6.4. For a small perturbation of the trap occupancy (holding $\bar{n}_c$ fixed), we get the following equation that regulates the transient behavior:

$$\frac{\partial \hat{f}}{\partial t} = -\left(\beta_n \bar{n}_c + \varepsilon_n\right)\hat{f} \tag{18.58}$$

Comparing Equation 18.57, we observe that the response time for trapping is $\tau_c = \left(\beta_n \bar{n}_c\right)^{-1}$ and the response time for release is $\tau_r = \varepsilon_n^{-1}$. The time for decay is

$$\tau_t = \left(\tau_c^{-1} + \tau_r^{-1}\right)^{-1} \tag{18.59}$$

Using the definitions in Section 6.3, Equation 18.59 can be written as

$$\tau_t = \frac{1}{\beta_n N_c e^{-E_c/k_BT}} \frac{1}{e^{E_t/k_BT} + e^{E_{Fn}/k_BT}} \tag{18.60}$$

If we assume that $E_{Fn} < E_t$ (so that the trap is unoccupied and available to capture electrons), we obtain

$$\tau_t = \frac{1}{\beta_n N_c} e^{(E_c - E_t)/k_BT} \tag{18.61}$$

Therefore, according to Equation 18.61, the release time from a trap to the conduction band becomes increasingly long the deeper the trap is with respect to E_c.

It is possible to adopt a different approach to define a recombination lifetime, which is to define an apparent lifetime as (Hornbeck and Haynes, 1955)

$$\tau_{app} = -\frac{n}{(dn/dt)} = -\frac{n}{U_n(n)} \tag{18.62}$$

It should be emphasized that Equation 18.62 uses total density and recombination rate instead of the small perturbation. Equation 18.62 is commonly used in silicon solar cells, as one can measure separately the recombination rate (equal to photogeneration) and the minority carrier density by the photovoltage transient or by other contactless methods, such as photoconductance (Sinton and Cuevas, 1996). In silicon solar cells, there are several mechanisms that lead to recombination of minority carriers, including radiative, SRH, Auger, and surface recombination (Richter et al., 2012), see Figure 18.7.

GENERAL REFERENCES

Optimization of light harvesting in solar cells: Redfield (1974), Yablonovitch (1982), Deckman et al. (1983), Campbell and Green (1987), Pettersson et al. (1999), Schueppel et al. (2010), and Wehrspohn et al. (2015).

Luminescence and phosphorescence: Pankove (1971), Zacks and Halperin (1972), Schmidt et al. (1992), Auzel (2003), Feldmann et al. (2003), de Wild et al. (2011), and Huang et al. (2013).

The quantum efficiency: Boroditsky et al. (2000), Schubert (2003), Heikkilä et al. (2009), Matioli and Weisbuch (2011), and Kivisaari et al. (2012).

Recombination in semiconductors: Hall (1952), Shockley and Read (1952), Blackmore (1962), Haug (1983), Wang et al. (2006), and Bisquert and Mora-Seró (2010).

Recombination lifetime: van Roosbroeck (1953), van Roosbroeck and Shockley (1954), Rose (1963), Schmidt (1999), Zaban et al. (2003), and Bisquert et al. (2009).

REFERENCES

Auzel, F. Upconversion and anti-Stokes processes with f and d ions in solids. *Chemical Reviews* 2003, *104*, 139–174.

Bisquert, J.; Fabregat-Santiago, F.; Mora-Seró, I.; Garcia-Belmonte, G.; Giménez, S. Electron lifetime in dye-sensitized solar cells: Theory and interpretation of measurements. *The Journal of Physical Chemistry C* 2009, *113*, 17278–17290.

Bisquert, J.; Marcus, R. A. Device modeling of dye-sensitized solar cells. *Topics in Current Chemistry* 2014, *352*, 325–396.

Bisquert, J.; Mora-Seró, I. Simulation of steady-state characteristics of dye-sensitized solar cells and the interpretation of the diffusion length. *Journal of Physical Chemistry Letters* 2010, *1*, 450–456.

Blackmore, J. S. *Semiconductor Statistics*; Dover Publications: New York, 1962.

Boroditsky, M.; Gontijo, I.; Jackson, M.; Vrijen, R.; Yablonovitch, E.; Krauss, T.; Chuan-Cheng, C.; et al. Surface recombination measurements on III–V candidate materials for nanostructure light-emitting diodes. *Journal of Applied Physics* 2000, *87*, 3497–3504.

Braslavsky, S. E.; Fron, E.; Rodriguez, H. B.; Roman, E. S.; Scholes, G. D.; Schweitzer, G.; Valeur, B.; et al. Pitfalls and limitations in the practical use of Forster's theory of resonance energy transfer. *Photochemical & Photobiological Sciences* 2008, *7*, 1444–1448.

Brendel, R. Note on the interpretation of injection-level-dependent surface recombination velocities. *Applied Physics A* 1995, *60*, 523.

Campbell, P.; Green, M. A. Light trapping properties of pyramidally textured surfaces. *Journal of Applied Physics* 1987, *62*, 243–249.

de Wild, J.; Meijerink, A.; Rath, J. K.; van Sark, W. G. J. H. M.; Schropp, R. E. I. Upconverter solar cells: Materials and applications. *Energy & Environmental Science* 2011, *4*, 4835–4848.

Deckman, H. W.; Wronski, C. R.; Witzke, H.; Yablonovitch, E. Optically enhanced amorphous silicon solar cells. *Applied Physics Letters* 1983, *42*, 968–970.

Feldmann, C.; Jüstel, T.; Ronda, C. R.; Schmidt, P. J. Inorganic luminescent materials: 100 years of research and application. *Advanced Functional Materials* 2003, *13*, 511–516.

Gerischer, H.; Michel-Beyerle, M. E.; Rebentrost, F.; Tributsch, H. Sensitization of charge injection into semiconductors with large band gap. *Electrochimica Acta* 1968, *13*, 1509–1515.

Gerlach, W.; Schlangenotto, H.; Maeder, H. On the radiative recombination rate in silicon. *Physica Status Solidi (a)* 1972, *13*, 277–283.

Green, M. A. Lambertian light trapping in textured solar cells and light-emitting diodes: Analytical solutions. *Progress in Photovoltaics: Research and Applications* 2002, *10*, 235–241.

Green, M. A.; Zhao, J.; Wang, A.; Trupke, T. High-efficiency silicon light emitting diodes. *Physica E: Low-Dimensional Systems and Nanostructures* 2003, *16*, 351–358.

Hall, R. N. Electron-hole recombination in germanium. *Physical Review* 1952, *87*, 387.

Haug, A. Auger recombination in direct-gap semiconductors: Band-structure effects. *Journal of Physics C: Solid State Physics* 1983, *16*, 4159–4172.

Heikkilä, O.; Oksanen, J.; Tulkki, J. Ultimate efficiency limit and temperature dependency of light-emitting diode efficiency. *Journal of Applied Physics* 2009, *105*, 093119.

Hornbeck, J. A.; Haynes, J. R. Trapping of minority carriers in silicon. I. p-Type silicon. *Physical Review* 1955, *97*, 311.

Huang, X.; Han, S.; Huang, W.; Liu, X. Enhancing solar cell efficiency: The search for luminescent materials as spectral converters. *Chemical Society Reviews* 2013, *42*, 173–201.

Kivisaari, P.; Riuttanen, L.; Oksanen, J.; Suihkonen, S.; Ali, M.; Lipsanen, H.; Tulkki, J. Electrical measurement of internal quantum efficiency and extraction efficiency of III-N light-emitting diodes. *Applied Physics Letters* 2012, *101*, 021113.

Matioli, E.; Weisbuch, C. Direct measurement of internal quantum efficiency in light emitting diodes under electrical injection. *Journal of Applied Physics* 2011, *109*, 073114.

Michaelis, W.; Pilkuhn, M. H. Radiative recombination in silicon p-n junctions. *Physica Status Solidi (b)* 1969, *36*, 311–319.

Pankove, J. I. *Optical Processes in Semiconductors*; Prentice-Hall: Englewood Cliffs, 1971.

Pettersson, L. A. A.; Roman, L. S.; Inganas, O. Modeling photocurrent action spectra of photovoltaic devices based on organic thin films. *Journal of Applied Physics* 1999, *86*, 487–496.

Redfield, D. Multiple-pass thin-film silicon solar cell. *Applied Physics Letters* 1974, *25*, 647–648.

Richter, A.; Glunz, S. W.; Werner, F.; Schmidt, J.; Cuevas, A. Improved quantitative description of Auger recombination in crystalline silicon. *Physical Review B* 2012, *86*, 165202.

Ritter, D.; Zeldov, E.; Weiser, K. Ambipolar transport in amorphous semiconductors in the lifetime and relaxation-time regimes investigated by the steady-state photocarrier grating technique. *Physical Review B* 1988, *38*, 8296.

Rose, A. *Concepts in Photoconductivity and Allied Problems*; Interscience: New York, 1963.

Saxena, K.; Jain, V. K.; Mehta, D. S. A review on the light extraction techniques in organic electroluminescent devices. *Optical Materials* 2009, *32*, 221–233.

Schmidt, J. Measurement of differential and actual recombination parameters in crystalline silicon wafers. *IEEE Transactions on Electron Devices* 1999, *46*, 2018–2025.

Schmidt, T.; Lischka, K.; Zulehner, W. Excitation-power dependence of the near-band-edge photoluminescence of semiconductors. *Physical Review B* 1992, *45*, 8989–8994.

Scholes, G. D. Long-range resonance energy transfer in molecular systems. *Annual Review of Physical Chemistry* 2003, *54*, 57–87.

Schubert, E. F. *Light-Emitting Diodes*; Cambridge University Press: New York, 2003.

Schueppel, R.; Timmreck, R.; Allinger, N.; Mueller, T.; Furno, M.; Uhrich, C.; Leo, K.; et al. Controlled current matching in small molecule organic tandem solar cells using doped spacer layers. *Journal of Applied Physics* 2010, *107*, 044503.

Shockley, W.; Read, W. T. Statistics of the recombinations of holes and electrons. *Physical Review* 1952, *87*, 835–842.

Sinton, R. A.; Cuevas, A. Contactless determination of current–voltage characteristics and minority-carrier lifetimes in semiconductors from quasi-steady-state photoconductance data. *Applied Physics Letters* 1996, *69*, 2510.

Staub, F.; Hempel, H.; Hebig, J.-C.; Mock, J.; Paetzold, U. W.; Rau, U.; Unold, T.; et al. Beyond bulk lifetimes: Insights into lead halide perovskite films from time-resolved photoluminescence. *Physical Review Applied* 2016, *6*, 044017.

Stephens, A. W.; Aberle, A. G.; Green, M. A. Surface recombination velocity measurements at the silicon–silicon dioxide interface by microwave-detected photoconductance decay. *Journal of Applied Physics* 1994, *76*, 363–370.

Strickler, S. J.; Berg, R. A. Relationship between absorption intensity and fluorescence lifetime of molecules. *Journal of Chemical Physics* 1962, *37*, 814–820.

Sutter-Fella, C. M.; Li, Y.; Amani, M.; Ager, J. W.; Toma, F. M.; Yablonovitch, E.; Sharp, I. D.; et al. High photoluminescence quantum yield in band gap tunable bromide containing mixed halide perovskites. *Nano Letters* 2016, *16*, 800–806.

Tiedje, T.; Yablonovitch, E.; Cody, G. D.; Brooks, B. G. Limiting efficiency of silicon solar cells. *IEEE Transactions on Electron Devices* 1984, *31*, 711–716.

Trupke, T.; Daub, E.; Würfel, P. Absorptivity of silicon solar cells obtained from luminescence. *Solar Energy Materials and Solar Cells* 1998, *53*, 103–114.

Trupke, T.; Green, M. A.; Wurfel, P.; Altermatt, P. P.; Wang, A.; Zhao, J.; Corkish, R. Temperature dependence of the radiative recombination coefficient of intrinsic crystalline silicon. *Journal of Applied Physics* 2003a, *94*, 4930–4937.

Trupke, T.; Zhao, J.; Wang, A.; Corkish, R.; Green, M. A. Very efficient light emission from bulk crystalline silicon. *Applied Physics Letters* 2003b, *82*, 2996–2998.

Usami, A. Theoretical study of application of multiple scattering of light to a dye-sensitized nanocrystalline photoelectrochemical cell. *Chemical Physics Letters* 1997, *277*, 105–108.

van Roosbroeck, W. The transport of added current carriers in a homogeneous semiconductor. *Physical Review* 1953, *91*, 282–289.

van Roosbroeck, W.; Shockley, W. Photon-radiative recombination of electrons and holes in germanium. *Physical Review* 1954, *94*, 1558–1560.

Varshni, Y. P. Band-to-band radiative recombination in groups IV, VI, and III-V semiconductors (I). *Physica Status Solidi (b)* 1967a, *19*, 459–514.

Varshni, Y. P. Band-to-band radiative recombination in groups IV, VI, and III–V semiconductors (II). *Physica Status Solidi (b)* 1967b, *20*, 9–36.

Vossier, A.; Gualdi, F.; Dollet, A.; Ares, R.; Aimez, V. Approaching the Shockley-Queisser limit: General assessment of the main limiting mechanisms in photovoltaic cells. *Journal of Applied Physics* 2015, *117*, 015102.

Vossier, A.; Hirsch, B.; Gordon, J. M. Is Auger recombination the ultimate performance limiter in concentrator solar cells? *Applied Physics Letters* 2010, *97*, 193509.

Wang, Q.; Ito, S.; Grätzel, M.; Fabregat-Santiago, F.; Mora-Seró, I.; Bisquert, J.; Bessho, T.; et al. Characteristics of high efficiency dye-sensitized solar cells. *The Journal of Physical Chemistry* 2006, *110*, 19406–19411.

Wehrspohn, R. B.; Rau, U.; Gombert, A. *Photon Management in Solar Cells*; Wiley: Weinheim, 2015.

Yablonovitch, E. Statistical ray optics. *Journal of the Optical Society of America* 1982, *72*, 899–907.

Zaban, A.; Greenshtein, M.; Bisquert, J. Determination of the electron lifetime in nanocrystalline dye solar cells by open-circuit voltage decay measurements. *ChemPhysChem* 2003, *4*, 859–864.

Zacks, E.; Halperin, A. Dependence of the peak energy of the pair-photoluminescence band on excitation intensity. *Physical Review B* 1972, *6*, 3072–3075.

19 Optical Transitions in Organic and Inorganic Semiconductors

This chapter provides a summary of the light absorption properties in a range of materials relevant to energy conversion including bulk inorganic semiconductors, semiconductor quantum dots, and organic molecules and materials. Emphasis is placed on the properties that determine the capabilities of solar energy harvesting when these materials are used as the light absorber in energy conversion devices, such as the spectral distribution of absorption features. A number of effects that are significant in the study of optical properties of semiconductors are revised, such as plasmonic absorption, the Burstein-Moss shift, and excitonic absorption. We also address the optical absorption features due to charge-transfer complexes at heterojunctions and some of their applications.

19.1 LIGHT ABSORPTION IN INORGANIC SOLIDS

Let us consider the light absorption process in an inorganic solid material as indicated in Figure 19.1a. In the interband transition process, an electron jumps from a state of energy E_i in the lower band to a state E_f in the upper band by absorption of a photon. By the law of conservation of energy,

$$E_f = E_i + h\nu \tag{19.1}$$

Figure 19.2a shows a semiconductor or insulator material with conduction and valence bands separated by the gap energy E_g. Furthermore, the VB maximum and CB minimum occur at the same value of the crystal momentum axis, so that the optical transition occurs satisfying the conservation of momentum. This is called a *direct transition*. The *joint density of states* of the valence band and conduction band states, D_{vc}, for the transition of an electron by an energy $h\nu$, is the probability that a state in the valence band is occupied by an electron (unoccupied by a hole) combined with the probability that a state in the conduction band, separated by the energy of the photon, is unoccupied by an electron. Taking the DOS for the parabolic-band model described by Equation 2.10, the joint DOS is given by the expression

$$D_{vc}(h\nu) = 0 \text{ for } h\nu < E_g \tag{19.2}$$

$$D_{vc}(h\nu)d(h\nu) = \frac{(2m^*_{eh})^{3/2}}{2\pi^2\hbar^3}(h\nu - E)^{1/2}\, d(h\nu) \text{for } h\nu \geq E_g \tag{19.3}$$

where m^*_{eh} is the reduced mass for electrons and holes,

$$m^*_{eh} = \frac{m^*_e m^*_h}{m^*_e + m^*_h} \tag{19.4}$$

Obviously, photons of energy lower than E_g cannot be absorbed. At energies $h\nu$ larger than the band gap, the joint density of states increases with the square root of the energy as a result of the parabolic (effective mass) approximation for $E(k)$ when $k \to 0$. As a consequence of this, photoexcited electrons arrive in the conduction band with energies in excess of E_c, as indicated in Figure 19.1a. As described in Section 5.5, an electron in a high-energy state in Figure 19.1a is termed a hot electron, as it is a very short-lived state (about 1 ps). In this time scale, the electron loses the kinetic energy and cools down to the bottom of the band by collisions with phonons.

The bonding character of a semiconductor largely determines the molecular orbitals that are involved in optical electronic transitions. For ionic solids, the valence band is usually formed from the highest occupied p orbitals of the anions, with a mixture of d levels to a certain extent. The conduction band is formed by the s levels of the cation. In the case of covalently bound solids, the valence and conduction band come, respectively, from bonding and antibonding states of sp^3 hybrid orbitals.

The optical absorption coefficient α is determined by the quantum mechanical transition rate $W_{i\to f}$ from the initial to the final quantum state by absorption of a photon of frequency ν. The transition rate is given by Fermi's golden rule

$$W_{i\to f} = \frac{2\pi}{\hbar}|M|^2 D_{vc}(h\nu) \tag{19.5}$$

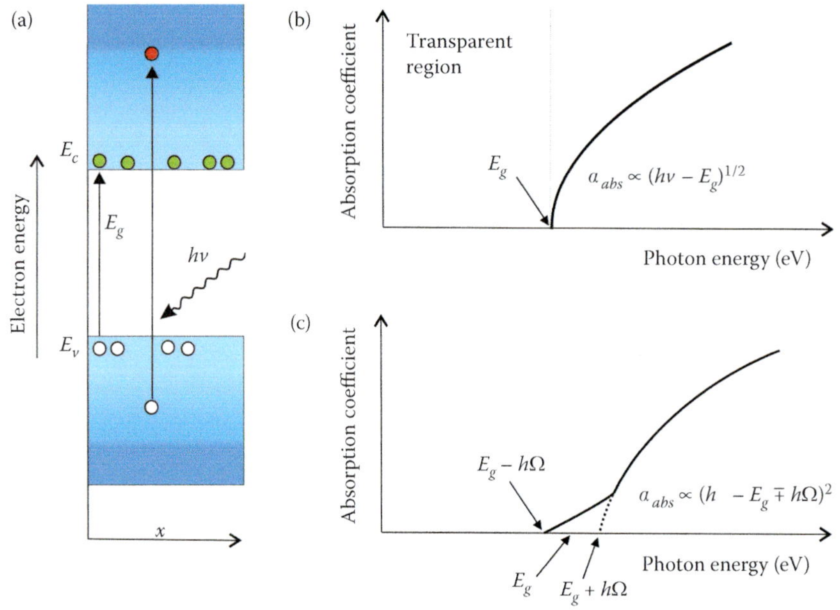

FIGURE 19.1 (a) Illustration of a photoinduced electronic transition across the semiconductor band gap, (b) the absorption coefficient as a function of photon energy for a direct gap semiconductor, and (c) for an indirect band gap semiconductor.

The matrix element M represents the electric dipole moment for the transition. Furthermore, the transition must be allowed by the selection rules that depend on the crystal symmetry (Klingshirn, 1995). According to Equation 19.5, the absorption rate is proportional to the joint density of states given by Equation 19.3. Therefore, we obtain the following spectral dependence:

$$\alpha(h\nu) \propto (h\nu - E)^{1/2} \qquad (19.6)$$

Note that this relatively simple relation for band-to-band transitions in direct absorption materials is in many cases obeyed only approximately due to defects in the band gap, Coulomb interaction (exciton formation), and other factors, as discussed below.

In semiconductors where the VB maximum and CB minimum do not have the same crystal momentum, the direct transition excited by only a photon cannot occur, since the photon has a very small linear momentum. However an *indirect transition* is possible that is assisted by phonons, which take the difference of momentum of the electron in the final and initial states. This process is indicated in Figure 19.2b. The spectral dependence of the absorption coefficient is

$$\alpha(h\nu) \propto (h\nu - E_g \mp h\Omega)^2 \qquad (19.7)$$

Here, E_g is the indirect band gap and $h\Omega$ is the phonon energy. The signs $\mp$ depend on whether the phonon is absorbed or emitted. The spectral dependence of the absorption coefficient is shown in Figure 19.1c. Since the absorption event is a two-particle quantum transition, the absorption coefficient for indirect band gap material is usually much smaller than for direct band gap semiconductors.

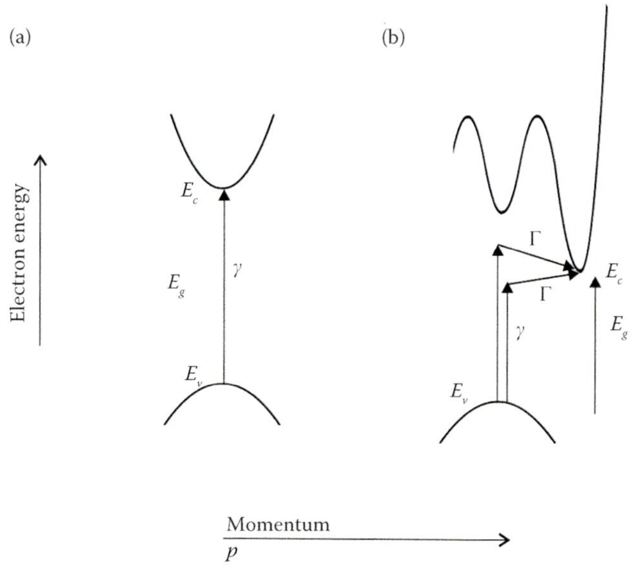

FIGURE 19.2 (a) Lowest energy electronic transition from the valence band to the conduction band for a direct band gap semiconductor by absorption of a photon γ. (b) Electronic transitions in a semiconductor of indirect band gap by the absorption of a photon γ and simultaneous absorption or release of a phonon Γ.

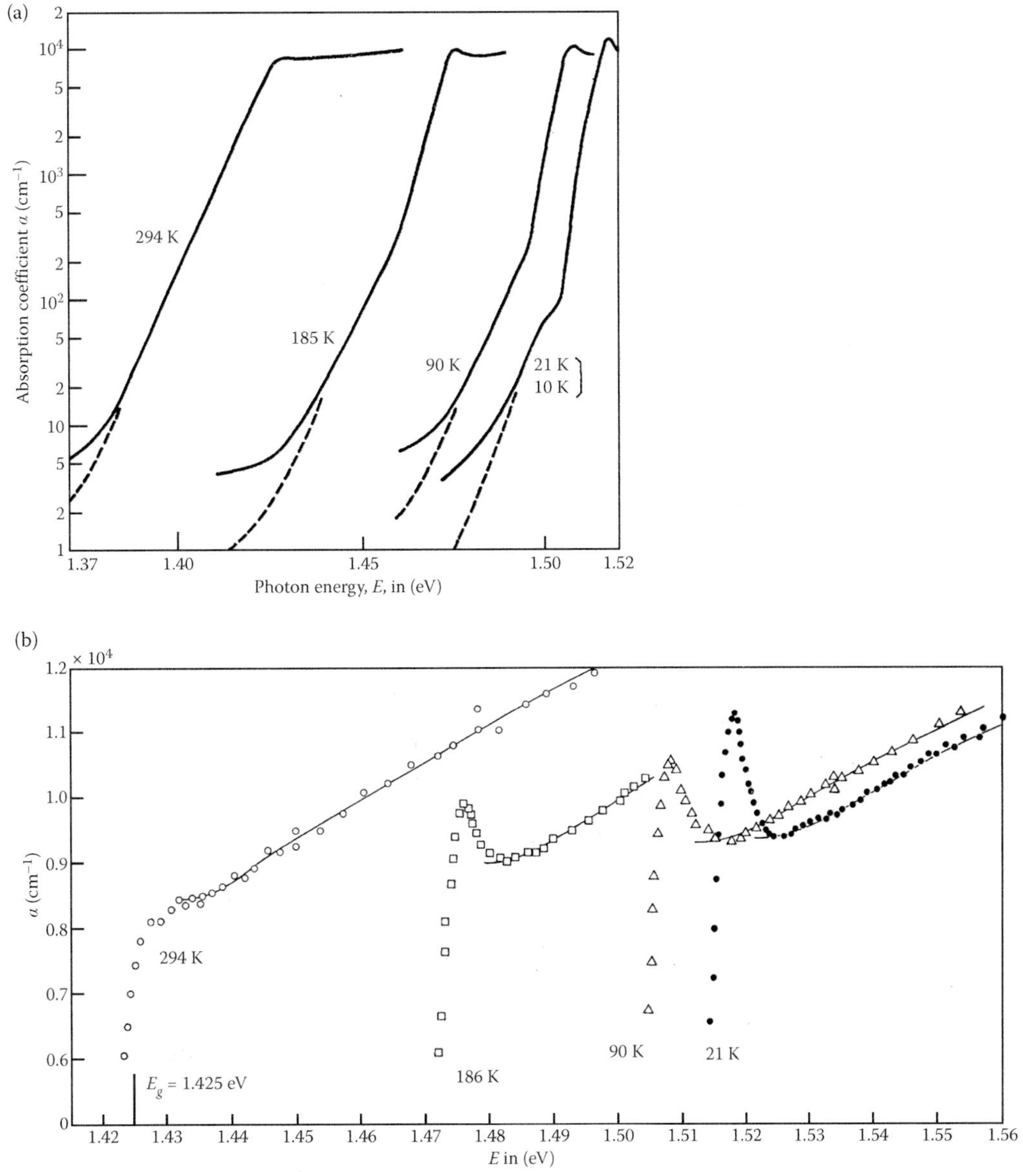

FIGURE 19.3 (a) Absorption of GaAs in a broad photon energy range and (b) in the exciton region. (Reproduced with permission from Sturge, M. D. *Physical Review* 1962, 127, 768–773.)

In summary, direct semiconductors are characterized by an absorption edge and their band gap E_g. Photons with energies less than E_g are not absorbed. Photons with energies larger than E_g are absorbed with increasing absorbance as their energy increases. However, in practice, the absorption (and luminescence) spectral features of semiconductors can be influenced by different factors that modify the ideal features for an abrupt band gap. Phonon-assisted transitions in indirect absorption produce a tail of absorption below the band gap energies. In addition, any symmetry-breaking disorder in the lattice disrupts the perfect crystalline structure and causes the appearance of localized states close to the band edge that generate the Urbach tail at energies below the main absorption edge (Urbach, 1953). Spatial fluctuations of the fundamental band gap also cause tailing absorption and emission (Mattheis et al., 2007). The precise shape of sub-band-gap absorption is an important feature for studies of PL and photovoltaic properties of a semiconductor. A general model for absorption coefficient including different physical effects for sub-band-gap tails is reported by Katahara and Hillhouse (2014). As an example, Figure 19.3a shows the main absorption edge of GaAs, which is a direct absorption semiconductor. The high absorption coefficient in combination with a band gap of 1.42 eV constitutes excellent features for solar energy harvesting that make GaAs the highest efficiency solar cell material at this time.

The absorption of GaAs is nearly constant at $E > E_g$ and it is well described by the expression (Miller et al., 2012)

$$\alpha = \alpha_0 \left(1 + \frac{E - E_g}{E_1}\right) \qquad E > E_g \tag{19.8}$$

where E_1 is a constant. In common to many semiconductors, the optical absorption coefficient for photons around the absorption edge shows a linear-exponential shape termed the Urbach tail, that is described by the expression

$$\alpha(E) = \alpha_0 \exp\left(\frac{E - E_C}{E_U}\right) \tag{19.9}$$

Here, α_0 and E_C are material constants and E_U is a tailing parameter, associated to structural and thermal disorder, as previously discussed in Figure 8.9, see also (John et al., 1986; Studenyak et al., 2014). The exponential tail for GaAs is shown in Figure 22.7b.

In many cases the single exponential 19.9 is not sufficient to describe accurately the bandtail. An alternative expression for the absorption coefficient in a direct band gap semiconductor with an extent of disorder is a Gaussian distribution for the local band gap (Mattheis et al., 2007; Gokmena et al., 2013). In addition, the hump at the absorption edge is the excitonic absorption, shown in Figure 19.3b and discussed in Section 19.3.

Figure 19.4 shows the absorption coefficient of several semiconductors with respect to wavelength. Figure 19.5 shows the absorption coefficient as a function of energy highlighting the Urbach tail. As noted before, direct semiconductors show a sharp increase of the absorption coefficient above the band gap energy, with values $E_U \approx k_B T$ near room temperature. In contrast to this, in indirect semiconductors the bandtail broadens considerably, and the absorption coefficient is much lower and rises very slowly. Further analysis of the Urbach tail in relation to photovoltaic performance is presented in Figure 25.13.

Silicon has a direct band gap of 3.4 eV and an indirect band gap of 1.1 eV at 300 K. The latter is relevant for the conversion of solar energy but the absorption coefficient at the band-edge energy is very small (Green, 2008; Wang et al., 2013a). To measure the entire range of the absorption coefficient below E_g in silicon, many samples with increasing thickness would be needed. Alternatively, techniques based on reciprocity of luminescence and

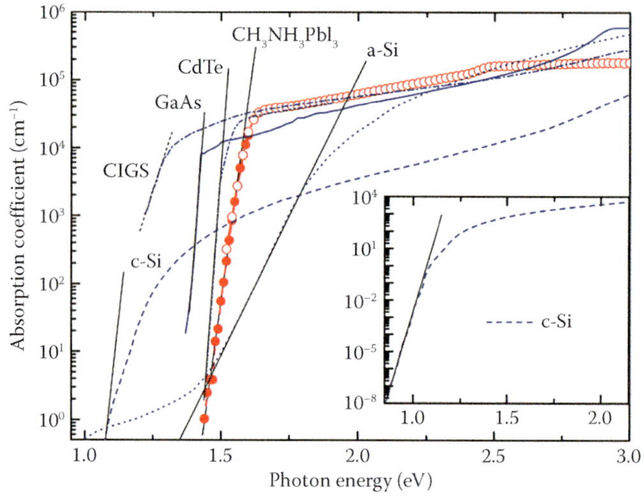

FIGURE 19.5 Effective absorption coefficient of several photovoltaic materials with respect to photon energy, all measured at room temperature. For each material, the slope of the Urbach tail is shown. The inset shows the data for c-Si down to low-absorption values. (Reproduced with permission from De Wolf, S. et al. *The Journal of Physical Chemistry Letters* 2014, 5, 1035–1039.)

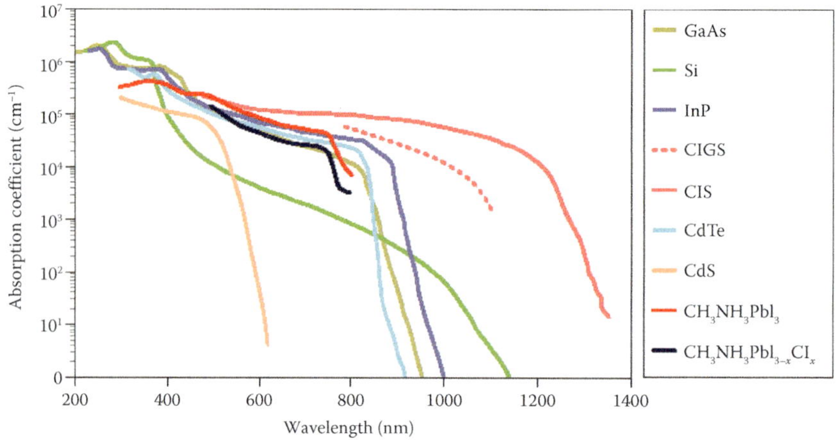

FIGURE 19.4 Effective absorption coefficient with respect to wavelength of several photovoltaic materials. (Reproduced with permission from Green, M. A. et al. *Nature Photonics* 2014, 8, 506–514.)

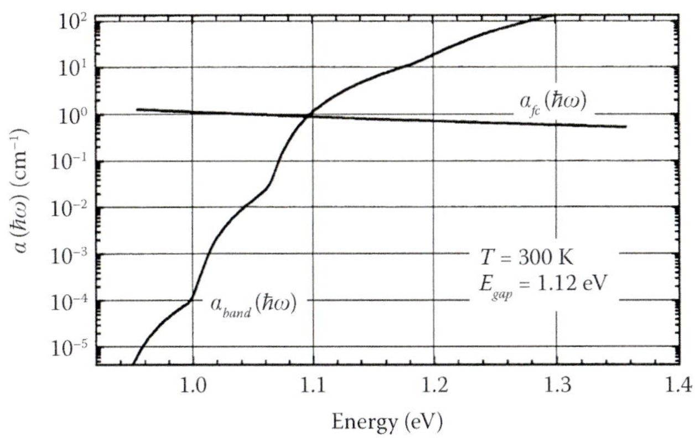

FIGURE 19.6 The band-to-band and free carrier absorption coefficients of silicon. (Reproduced with permission from Trupke, T. et al. *Solar Energy Materials and Solar Cells* 1998, 53, 103–114.)

absorption that are developed in Chapter 22 allow for extraordinary accuracy at very long wavelengths down to about 10^{-7} cm^{-1} (Daub and Würfel, 1995; Trupke et al., 1998; Barugkin et al., 2015) and provide absorption coefficient values shown in Figure 19.6. The staircase processes observed in this figure correspond to successive phonon absorptions of the indirect transition.

As mentioned in Section 18.1, the absorption length L_α of the absorber material is a central characteristic determining the necessary solar cell thickness. If we take as a reference the absorption coefficient at 775 nm wavelength (1.6 eV photon energy) for silicon, we observe in Figure 19.4 that the absorption length is in the range of L_α = 10 μm, while GaAs solar cells with $\alpha = 10^4$ cm^{-1} have values L_α = 1 μm that require considerably thinner films. The $CH_3NH_3PbI_3$ perovskite requires an even thinner film of 300 nm to absorb all the incoming radiation at this wavelength.

19.2 FREE CARRIER PHENOMENA

The presence of a large density of free carriers in metals, or in the CB, and/or VB of semiconductors, produces significant effects in their optical properties. We discuss briefly several important effects: the plasma reflectivity, the Burstein shift, and the techniques based on the absorption of radiation by free carriers.

A *plasma* is a neutral gas of positive or negative carriers, as electrons or holes. In metals and heavily doped semiconductors, the plasma is hence formed by an electron gas electrically compensated by fixed ions. The plasma absorption is due to a collective excitation of free carriers by the electrical field of the electromagnetic waves. The quantization leads to quasiparticles termed *plasmons* that obey the Bose statistics. Nonetheless, the oscillations of the electrons in the metal can be well described as a classical effect by combining Maxwell's equations, the Drude model of free electron conductivity, and the Lorentz model of dipole oscillators. For an undamped displacement of the electrons, the relative dielectric constant depends on the angular frequency of the incoming light as follows:

$$\varepsilon_r(\omega) = 1 - \frac{\omega_p^2}{\omega^2} \tag{19.10}$$

where the *plasmon frequency* is given by

$$\omega_p = \left(\frac{nq^2}{\varepsilon_0 m_e}\right)^{1/2} \tag{19.11}$$

Here, n is the density of free electrons. Using the relationships (18.3) and (18.4), it is obtained that the reflectivity R is near unity at frequencies $\omega \leq \omega_p$, and decreases from $\omega = \omega_p$ approaching zero at $\omega \gg \omega_p$. Due to electrical field screening, all light of frequencies below the plasmon frequency is reflected. Equation 19.11 indicates that the plasmon frequency is proportional to the square root of the carrier concentration, n. In metals, the plasmon energy $\hbar\omega_p$ is around 5 eV. Thus, metals reflect all light below this energy and become transparent in the ultraviolet. The transparent conducting oxides (TCO, Section 5.5) usually have a lower electron concentration than metals by 1–2 orders of magnitude, hence plasmon energies lie around 1 eV for the highest doped TCOs. Consequently, the TCOs do not transmit infrared radiation and are used as low-emissivity window coatings (Klein, 2013). Figure 19.7 shows the reduction of the transmittance at long wavelengths by the decrease of the plasmon wavelength λ_p with increasing carrier concentration as indicated by the conductivity of the oxide.

Considering a metal nanoparticle, the plasmon causes an enhancement of the local field intensity in the nearby space by redistribution of the optical field. A localized surface plasmon resonance induces sharp spectral absorption and scattering peaks so that chromophores situated in the region of enhanced field effectively absorb more light (Munechika et al., 2010). Plasmonic nanostructures can thus redistribute the optical field, increasing the fluorescence intensity. The shift of the plasmon resonance spectral peak allows the detection of molecular interactions near the metal nanoparticle surface (Mayer and Hafner, 2011). These effects have been utilized in applications such as biological labeling, sensing and imaging, LEDs, and single photon sources. The plasmonic nanostructures can maximize the light absorption in solar cells (Ming et al., 2012).

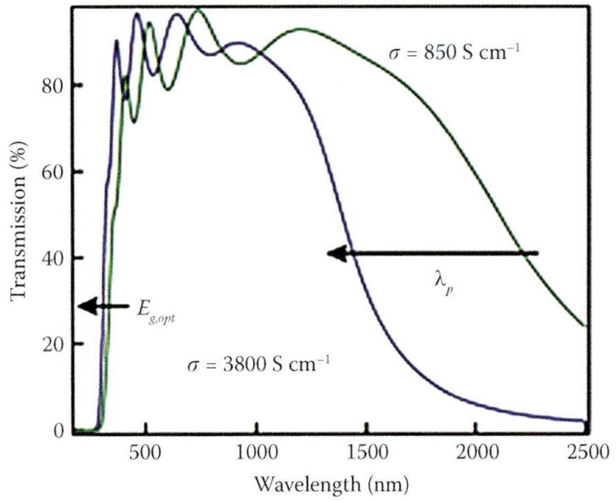

FIGURE 19.7 The optical transmission spectra of two ITO films with different electrical conductivity that indicate the free carrier density. The increase in the optical band gap is due to the Burstein shift and the reduction in the plasmon wavelength λ_p with increasing carrier concentration changes the absorption at long wavelength. (Reproduced with permission from Klein, A. *Journal of the American Ceramic Society* 2013, 96, 331–345.)

The semiconductor optical gap is defined as the energy of the lowest electronic transition accessible via absorption of a single photon (Bredas, 2014). The effective increase of the optical gap due to the creation of a very large density of electrons (holes) in the conduction (valence) band, which fills up the energy levels near the edge of the band, causes the blue shift of the optical absorption edge known as the *Burstein-Moss shift*. This effect is illustrated in Figure 19.8a. In the short wavelength region of Figure 19.7, the shift of optical absorption edge of differently doped TCOs should be noted. It is observed that the transmission onset shifts to shorter wavelength for the degenerately doped, high-conductivity material.

Changes in the occupation of the CB can be spectrally detected by the Burstein-Moss shift. The transient absorption spectra (TAS) shown in Figure 19.9a for a $CH_3NH_3PbI_3$ perovskite film consist of the difference of the semiconductor absorption at different, short times after photogeneration by a pump signal, with respect to the dark spectrum (Manser and Kamat, 2014). The negative parts are termed bleaches and indicate enhanced transient absorption due to specific photoinduced transitions. According to Equation 17.4, the bleach at 760 nm is associated with an optical transition at 1.63 eV that corresponds to the direct transition across the band gap of the $CH_3NH_3PbI_3$ perovskite. The change of absorption is incremented when the number of photo-generated carriers is increased, as shown in Figure 19.8b. Using the absorbance of the film, the absorption coefficient of $CH_3NH_3PbI_3$, and the excitation energy density of the laser, it is possible to determine the initial photogenerated carrier density. The result of this calculation at various pump powers enables to fit the difference in absorbance using the expression that relates the shift of absorption to the carrier number density in parabolic-band theory (Muñoz et al., 2001):

$$\Delta E_g^{BM} = \frac{\hbar^2}{2m^*_{eh}}\left(3\pi^2 n\right)^{2/3} \tag{19.12}$$

This model provides a very good agreement with the results of photoexcitation of $CH_3NH_3PbI_3$ at 760 nm, as shown in Figure 19.8c. The results indicate that the perovskite becomes degenerate at relatively low carrier densities of order 5×10^{17} cm^{-3}, which points to a low-effective density of states in the conduction band, associated with a small effective mass (sharp $E(k)$ curvature), cf. Table 2.1.

Free carrier absorption involves the absorption of a photon by excitation of an electron in the CB to a higher-energy level. The conservation of momentum is satisfied by phonons or by impurity scattering. Free carrier absorption usually decreases monotonically at higher photon energies as shown for silicon crystal in Figure 19.6. The Burstein-Moss shift reveals the change of electronic occupation of the CB of nanostructured semiconductors such as TiO_2 by large negative voltage that induces strong electron accumulation (Figure 19.9b). The spectra may be divided into two regions separated by the band gap energy, which lies approximately at 400 nm. At long wavelengths, the intensity increases because of the absorbance of free electrons in the CB. The bleaching peak below 400 nm appears due to the increase of excitation energy across the gap.

Another technique to detect transient phenomena associated with free carriers is the time-resolved microwave conductivity (TRMC) method. This technique is based on the measurement of the relative change of the microwave power reflected from a semiconductor that is caused by a small increase in the conductivity (Savenije et al., 2013). However, TRMC operates in the GHz frequencies and the trap states in the band gap of nanocrystalline semiconductors may contribute to the observed response. Carriers in band gap localized states, discussed in Section 8.4, also induce light absorption in the long wavelength region of the spectrum. Figure 19.10 illustrates different mechanisms whereby the electrons in traps in TiO_2 provide transitions that contribute to the optical absorption. These effects result in absorption features in the middle infrared at energies much lower than

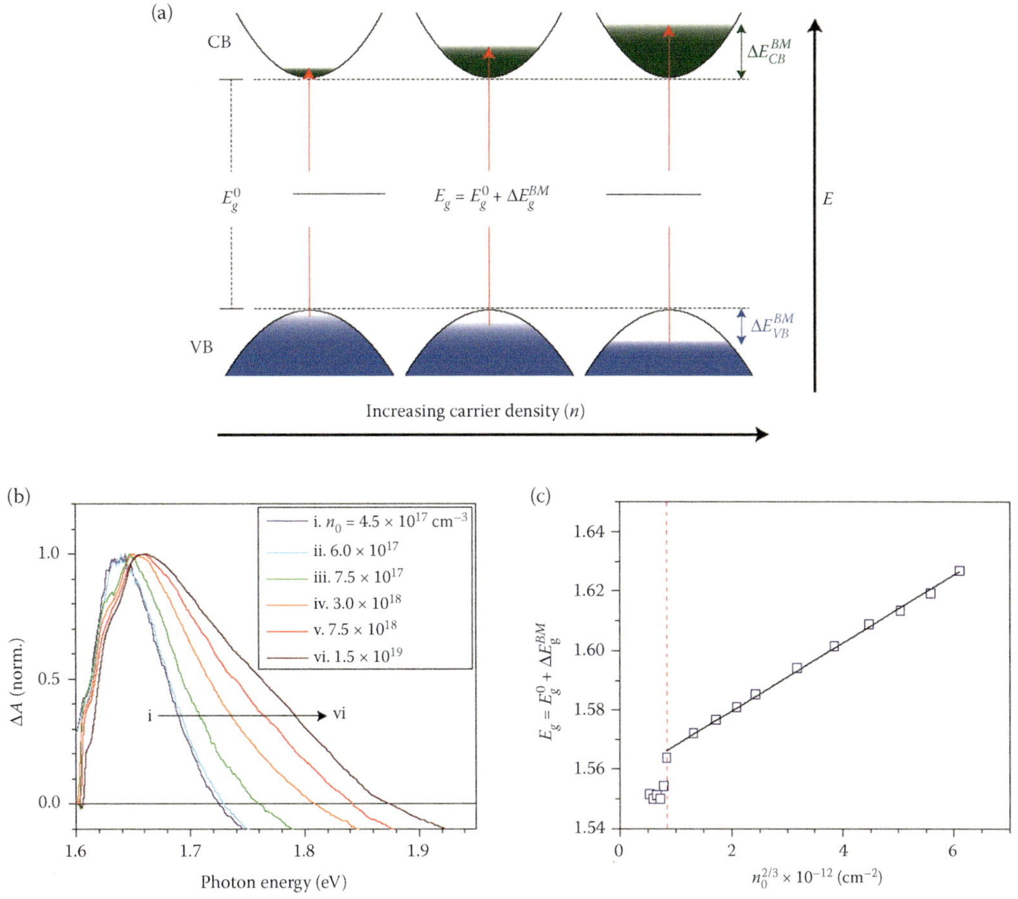

FIGURE 19.8 (a) Schematic representation of the Burstein-Moss effect showing the contribution from both electrons in the conduction band and holes in the valence band in the case of similar effective masses. (b) Normalized TAS of the band-edge transition in $CH_3NH_3PbI_3$ recorded at the maximum bleach signal (5 ps) after 387 nm pump excitation of varying intensity. The corresponding carrier densities are indicated in the legend. (c) Modulation of the intrinsic band gap of $CH_3NH_3PbI_3$ according to the Burstein-Moss model. The vertical dashed line marks the onset of band gap broadening. The solid line is a linear fit to the data after the onset threshold. The linear trend indicates agreement with band filling by free charge carriers. (Reproduced with permission from Manser, J. S.; Kamat, P. V. *Nature Photonics* 2014, 8, 737–743.)

the band gap, 0.2 eV, corresponding to a wavelength of 1 μm, as shown in Figure 19.11. The same difference in spectra can be obtained either from UV generation of carriers or from the electrons injected at negative bias. In the case of ZnO nanostructures, electrons accumulated within the conduction band cause a bleach of the excitonic band (Subramanian et al., 2003).

The method of time-resolved Terahertz spectroscopy (TRTS) makes use of sub-picosecond pulses of freely propagating electromagnetic radiation in the Terahertz range, that is characterized by sub-mm wavelengths (300 μm for 1 THz), and very low-photon energies (4.2 meV at 1 THz) (Ulbricht et al., 2011; Canovas et al., 2013). As the detection energy corresponds to less-than-thermal energies at room temperature (1 THz corresponds to 48 K), this method directly investigates the dynamics of free carriers and provides information about the complex dielectric function of the material, yielding the complex conductivity. It is a powerful tool for studying ultrafast charge carrier dynamics and carrier transfer processes in semiconductor nanostructures.

19.3 EXCITONS

Free electrons in the conduction band and free holes in the valence band of a semiconductor interact by the Coulomb attraction and their relative motion is quantized, provided that this interaction is stronger than the random thermal fluctuation (caused by phonon collisions). The stable quasiparticle state associated with this electron-hole interaction is an *exciton*, Figure 19.12a. In a first approximation, the exciton can be treated as a hydrogenic atom of reduced mass m_{eh} in a medium of

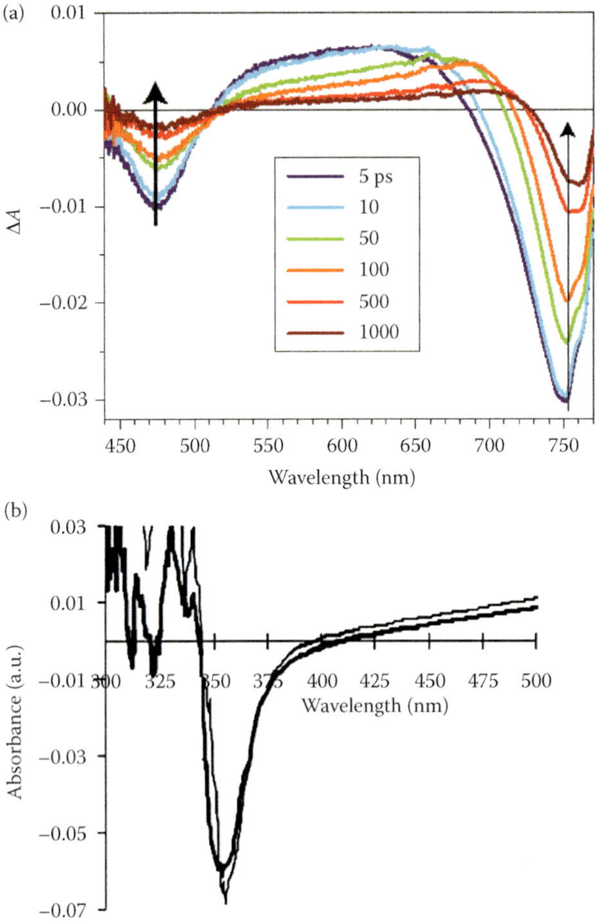

FIGURE 19.9 (a) Time-resolved TAS of $CH_3NH_3PbI_3$ at different probe delay times following 387 nm laser excitation with an energy density of 8 μJ cm^{-2}. Arrows indicate low- and high-energy bleach recovery. (Reproduced with permission from Manser and Kamat, 2014). (b) Spectral changes of the bare TiO_2 (thin line) and $SrTiO_3$-coated TiO_2 electrodes (thick line) at applied -0.9 V versus SCE in $HClO_4$ aqueous solution (pH 1.8). The short wavelength bleach that is related to the electron accumulation in TiO_2 is not affected by the coating. (Reproduced with permission from Diamant, Y. et al. *The Journal of Physical Chemistry B* 2003, 107, 1977–1981.)

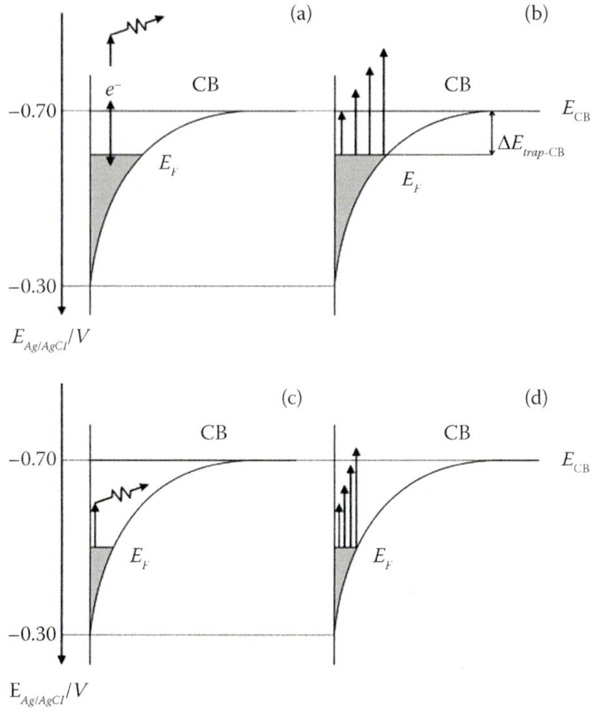

FIGURE 19.10 Possible transitions in the middle infrared of accumulated electrons in anatase TiO_2 nanocrystal electrodes. (a) Intra-conduction band transitions associated with phonon absorption. (b) Excitation of shallow trapped electrons to the conduction band. (c) Polaron excitations by coupling with two-dimensional surface phonons. (d) Intra-band-gap transitions. (Reproduced with permission from Berger, T. et al. *The Journal of Physical Chemistry C* 2012, 116, 11444–11455.)

dielectric constant ε_r. Using the Bohr model, the binding energy of the n_l level relative to the ionization limit is

$$E_B(n_l) = -\frac{m_{eh}^*}{m_0\varepsilon_r^2}\frac{R_H}{n_l^2} = -\frac{R_X}{n_l^2} \tag{19.13}$$

Here, m_0 is the mass of the free, noninteracting electron; R_H = 13.6 eV is the Rydberg constant of the hydrogen atom; and

$$R_X = \frac{m_{eh}^*}{m_0\varepsilon_r^2}R_H \tag{19.14}$$

is known as the effective Rydberg of the exciton. The average separation between an electron and a hole in the $n_l = 1$ exciton or effective Bohr radius is hence defined as

$$a_X = \frac{m_{eh}^*}{m_0}\frac{1}{\varepsilon_r}a_B \tag{19.15}$$

where a_B = 0.053 nm is the Bohr radius of the hydrogen atom.

Here, we describe briefly the excitonic characteristics of inorganic solids with direct transitions. The strong optical coupling of the exciton with the radiation field exerts a great influence on the optical absorption features as outlined in Figure 19.12b. The photon energy required to create the electron-hole pair is the band gap energy minus the binding energy due to the Coulomb interaction, given in Equation 19.13. Therefore, a strong optical absorption is expected below the band-edge energy, so that the optical gap is lower than the transport gap. In addition, higher-excitonic levels produce a flat shape at $E = E_g$ instead of the smooth rise of the parabolic band-to-band absorption, as predicted by the theory of

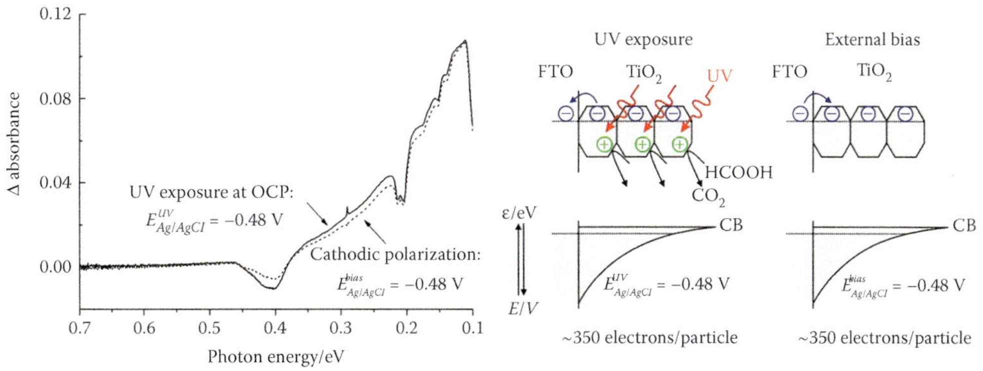

FIGURE 19.11 IR spectra of an anatase TiO_2 nanocrystal electrode during UV exposure at open circuit potential and after polarization to $V_{Ag/AgCl} = -0.48$ V. (Reproduced with permission from Berger, T.; Anta, J. A. *Analytical Chemistry* 2011, 84, 3053–3057.)

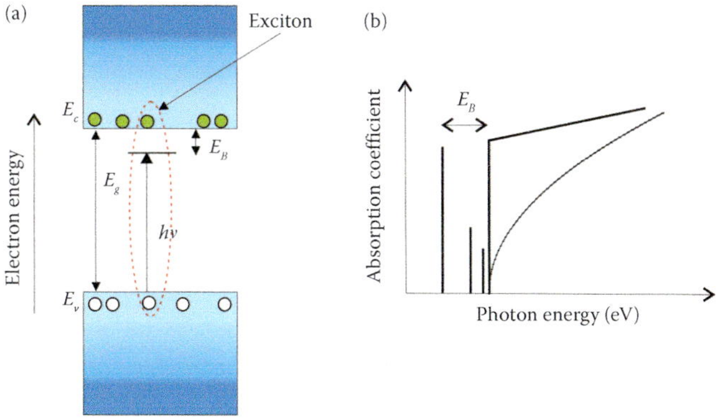

FIGURE 19.12 (a) Illustration of a photoinduced excitonic transition indicating the binding energy of the exciton. (b) Characteristic absorption associated with the excitonic electron-hole correlations. The smooth spectrum of the direct band gap semiconductor is also shown in thin line.

Elliott (1957). For GaAs, the binding energy of the $n_l = 1$ exciton calculated by values in Table 2.1 is $E_B = 5.6$ meV, which is very weak with respect to k_BT at room temperature. Therefore, the exciton, with an effective Bohr radius $a_X = 10.6$ nm, can remain bound only at very low temperatures. However, the *Sommerfeld enhancement* due to the exciton continuum of the absorption edge is observed from low-to-high temperatures (as explained by the theory of Elliot) instead of the parabolic absorption. The excitonic absorption features of GaAs are shown in Figure 19.3 and for InP in Figure 19.13. GaN is a large band gap semiconductor, $E_g = 3.4$ eV, where $m_e^* = 0.067m_e$, $m_h^* = 0.51m_e$, and $\varepsilon_r = 12$. The exciton-binding energy and radius for this material are $E_B = 26$ meV and $a_X = 2.8$ nm, respectively, so that the exciton effects are more significant than in the former case. In general, for semiconductors with a small band gap, the dielectric constant is large and the effective masses are small. Hence, binding energy tends to increase and the radius decreases as the band gap E_g of the semiconductor increases. In lead halide perovskite materials, exciton populations are not stable at room temperature at moderate photoexcited carrier densities but excitonic resonances dominate the absorption onset (Chen et al., 2018).

Self-consistency of Equation 19.14 requires the exciton radius to be much larger than the lattice spacing, so that the dielectric screening applies. Excitons with small-binding energy and a large radius are called the *Wannier–Mott excitons*. These are delocalized states that can move freely through the crystal (represented by the exciton center of mass). In organic materials, the dielectric constant tends to be low and the exciton-binding energy E_0 is larger than in inorganic semiconductors, so that excitons are often stable in organic materials. In insulators and some organic crystals, the tightly bound electron-hole pair wave function is confined to one crystal unit and then it is called a *Frenkel exciton*.

Free carriers present in heavily doped samples shield the Coulomb interaction and reduce the binding forces by screening, impeding the formation of excitons (Mahan, 1967). Simple arguments based on the Debye–Hückel theory show that the exciton-binding energy decreases strongly at concentrations of $n \approx 10^{17}$ cm^{-3} (Gay, 1971). If the Debye length is λ_D, the exciton-binding energy E_X is reduced from the ground state energy E_0 as (Sachenko and Kryuchenko, 2000)

$$E_X \approx E_0 \left(1 - \frac{a_X}{\lambda_D}\right)^2 \tag{19.16}$$

This expression is plotted in Figure 19.14 with respect to the free carrier concentration. It is observed that when the Debye length becomes of similar size as the exciton radius, $\lambda_D \approx 4a_X/3$, the exciton-binding energy is reduced to half the ground state value. If $a_X = 3$–5 nm

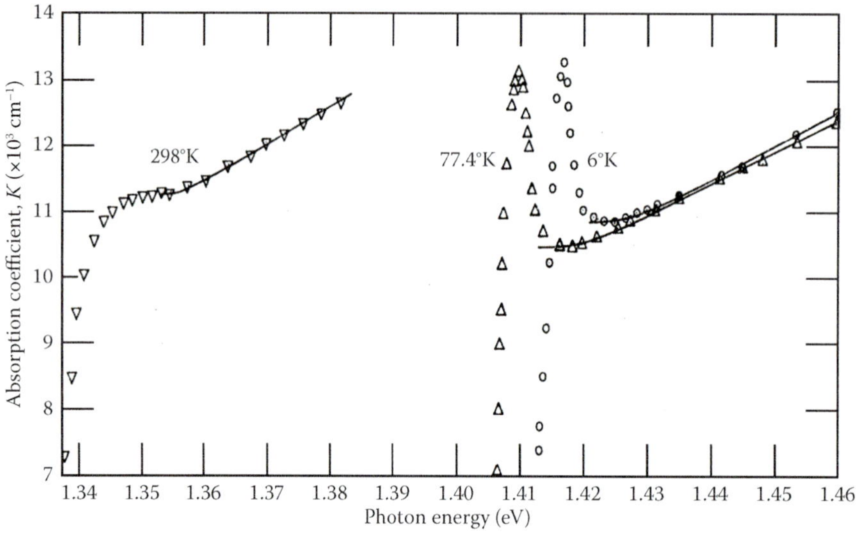

FIGURE 19.13 Absorption of InP in the exciton region. (Reproduced with permission from Turner, W. J. et al. *Physical Review* 1964, 136, A1467.)

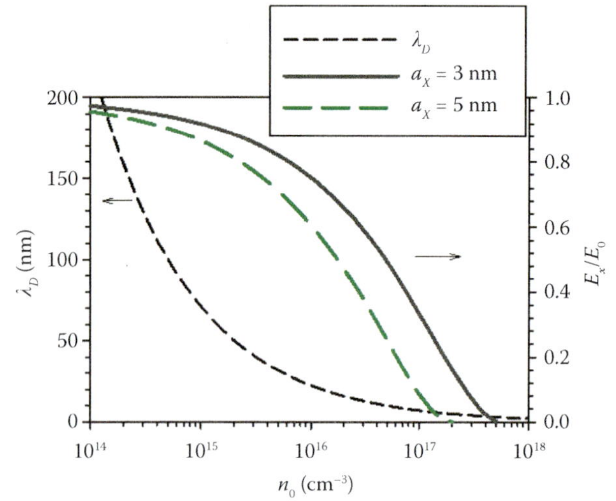

FIGURE 19.14 Reduction of exciton-binding energy E_X/E_0 by screening for a semiconductor with ε = 3.5 and two values of exciton Bohr radius as indicated. The Debye screening length λ_D is also shown.

and ε = 3.5, the reduction of binding energy happens at $n_0 \approx 5 \times 10^{16}$ cm^{-3} with the consequence of increased dissociation probability. In addition, at $a_X \approx \lambda_D$, the exciton Mott transition occurs, and the bound electron-hole pairs do not occur anymore.

19.4 QUANTUM DOTS

Quantum dots (QDs) are regular fragments of a material whose size is smaller than the exciton radius, usually in the range of a few nm. QDs have fully developed crystalline structure but the conduction and valence bands are replaced by discrete levels due to the confinement effect, while some of the semiconductor bulk properties such as high-extinction coefficient are preserved. In semiconductor QDs, the band gap can be easily tuned by the control of their size and shape, providing an excellent tool for nanoscale design of light absorber materials. The induced dipole moment by the electromagnetic field, as represented by the oscillator strength, is superior to their bulk counterparts, so that the absorption coefficient is larger. The increased specific surface area also plays an important role in the photophysical properties of QDs.

Colloidal QDs are prepared by low-temperature wet chemical routes that provide a very precise control of shape and size and also high monodispersity that is especially important for light emission applications, see Figure 17.6. The resulting QDs are freestanding in colloidal solutions, which is made possible by their surrounding organic ligands that provide solubility in many nonpolar solvents, thus further facilitating the formation of films for application in devices. Figures 19.15 and 19.16 illustrate main absorption and PL properties of semiconductor QDs. Figure 19.15 shows the room temperature absorption and emission spectra of InP QDs with a mean diameter of 3.2 nm. The absorption spectrum shows a broad excitonic peak at about 590 nm due to the inhomogeneous size distribution of the QDs. The PL spectrum shows two emission bands. The weak emission near the band edge with a peak at 655 nm is attributed to the ground exciton recombination. The

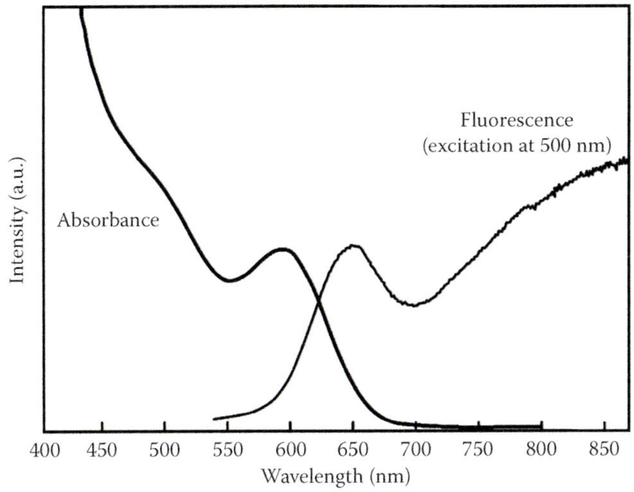

FIGURE 19.15 Absorption and emission spectra at 298 K of untreated 32 Å InP QDs. (Reprinted with permission from Micic, O. I. et al. *Applied Physics Letters* 1996, 68, 3150–3152.)

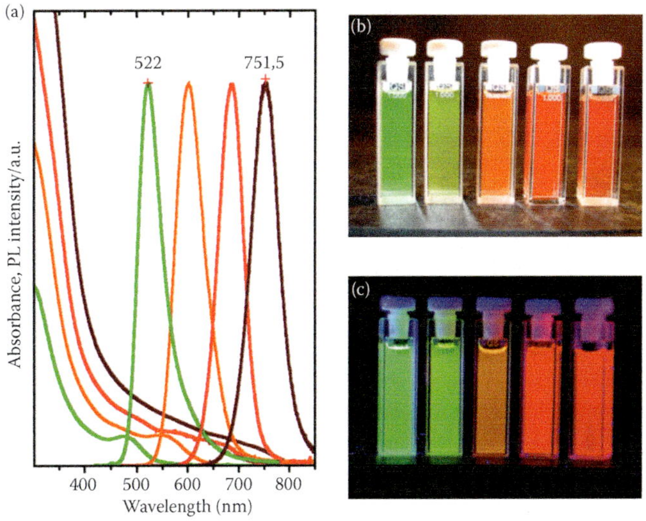

FIGURE 19.16 (a) Absorption and PL spectra of HF-photoetched InP nanocrystals of different size. (b, c) Visualization of size-dependent change of the PL color of HF-photoetched InP nanocrystals. The smallest (1.7 nm) particles emit green, whereas 4 nm particles emit deep red. Larger InP nanocrystals emit in near-IR (not shown). The high PL quantum yield makes the colloidal solutions "glow" in room light (b). Photo (c) shows the luminescence of the etched InP nanocrystals placed under UV-lamp. (Reprinted with permission from Talapin, D. V. et al. *The Journal of Physical Chemistry B* 2002, 106, 12659–12663.)

larger band above 850 nm is due to radiative surface states. It has been widely observed that recombination at surface states produces broadband and very slow PL in QDs (Harruff and Bunker, 2003; Shea-Rohwer and Martin, 2007). The room temperature absorption spectra of InP nanocrystals as a function of size between 1.7 and 4 nm are shown in Figure 19.16. The spectra shift to higher energy as the QD size decreases due to the change of the band gap by the quantum size effect.

The properties of semiconductor nanocrystals can be modified by the formation of nanoheterostructures consisting of QDs with a combination of core and shell materials, or embedding a single material QD in a solid framework. The combination of sizes, composition, and shapes provides a large versatility for the engineering of specific properties and physical effects. In core–shell QDs, the electronic and optical properties of the combined material are largely determined by the band alignment type, as indicated in Figure 19.17. All these structures produce a modification of the original core QDs. The type I heterojunction occurs when the band gap of the shell layer encloses that of the central core. This structure causes the confinement of both the electron and the hole in the core, away from surface traps, which reduces surface recombination and improves the radiative quantum yield (Hines and Guyot-Sionnest, 1996). In type II, the band alignment is staggered, which causes light absorption and emission at lower energies than both core and shell original materials (Kim et al., 2003), as further discussed in Section 19.6. Finally, in the intermediate case, either conduction or valence band is aligned producing a larger delocalization of either the electron or the hole.

The assembly of semiconductor QDs into QD films is significantly cheaper compared to their bulk semiconductor counterparts since their synthesis takes place at lower temperatures and with solution-based approaches (Vanmaekelbergh, 2011; Guyot-Sionnest, 2012). Unlike sintered nanoparticulate networks studied in Chapters 8 and 9, one may form films in which QDs are held together by organic linkers so that each QD maintains its electronic identity and the conductivity is governed by transitions between dots, Figure 10.5. Solar cells based on QD layers have been developed based on PbS, which has a bulk band gap in the IR. High-power conversion efficiencies have been obtained employing the bidentate organic linkers, which bring the nanoparticles into close packing while achieving the best-available surface passivation (Luther et al., 2008; Tang et al., 2011; Ip et al., 2012). On the other hand, for LED applications, the proximity of QDs causes a quenching of the luminescence by FRET, and the EQEs are greatly decreased with respect to the colloidal suspension in solution (Shirasaki et al., 2012).

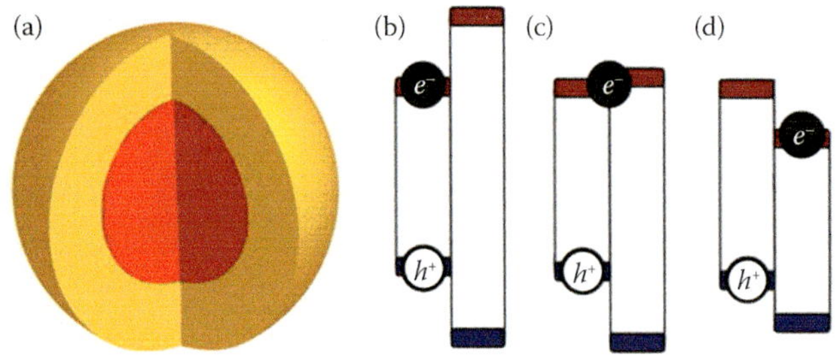

FIGURE 19.17 (a) Schematic of a core/shell nanocrystal. Differing alignments of core (*left* in each case) and shell (*right* in each case) conduction (*red*) and valence (*blue*) bands afford distinct electronic structures: (b) Type I, (c) Quasi-type II, and (d) Type II. Location of an excited-state electron or hole in the core–shell structure is indicated by e^- or h^+, respectively, and determined by the particular band alignments. (Reprinted with permission from Hollingsworth, J. A. *Coordination Chemistry Reviews* 2014, 263–264, 197–216.)

19.5 ORGANIC MOLECULES AND MATERIALS

Organic materials and molecules consist of carbon and hydrogen with a few heteroatoms such as sulfur or nitrogen, see Section 8.5. The light absorption features of organic materials and chromophores are of high importance for artificial photosynthesis schemes as well as for all solar energy conversion devices based on organic molecules and materials.

Light absorption in natural organic pigments provides the basic fuel for life on the earth. In the process of oxygenic photosynthesis, pigment molecules absorb sunlight and use its energy to synthesize carbohydrates from CO_2 and water. The absorption of the main pigments such as chlorophylls and carotenoids is shown in Figure 17.2, in comparison with the solar spectral photon flux at the top of the earth's atmosphere and on the earth's surface. We note that oxygenic photosynthesis utilizes photons in a restricted range of wavelengths between 400 nm and 700 nm. A photon having a wavelength of 700 nm has an energy of 1.8 eV. This energy is required to drive the required reactions in the photosynthetic apparatus (Milo, 2009).

As discussed in the preceding sections of this chapter, in inorganic materials at room temperature the excited electron-hole pair is formed by very weakly bound carriers. Therefore, the excited carriers are delocalized in their bands with small exchange energy. Furthermore, there is a continuum of electronic states in the CB and VB over a broad energy range. Consequently, absorption bands extend all the way to very short wavelengths in the energy range above the band gap.

In contrast to these characteristic features of inorganic crystalline semiconductors, the main factors determining the interaction of photons with electrons in organic molecules are caused by the strong influence of the correlation of spin states of the electrons in the ground and excited states and by the fact that molecules have many vibrational degrees of freedom. The absorption features of organic molecules show specific bands centered at the wavelengths that correspond to transitions between quantum states having a large spatial overlap. The exchange energy corresponding to the excited and unexcited electrons can have a large value 0.7–1.0 eV (Köhler and Bässler, 2009).

In organic materials, the highest molecular level filled with electrons is the HOMO (highest occupied molecular state), which for a conjugated molecule will be a π orbital and the first available level above the HOMO is the LUMO (lowest unoccupied molecular state), which is an excited configuration of the π orbital termed a π^* state. The main transition for optical absorption in organic molecules is the π–π^* transition due to its high-molar absorptivity. The π–π^* transition involves electrons that are strongly correlated and it is important to consider in detail the possible pairing of the spin of the electron in the excited state with that remaining in the ground state (Figure 19.18). Electrons in the ground state are paired off with their spins antiparallel and the net spin angular momentum is zero, according to the Pauli exclusion principle. This is called a *singlet state*, S_0. In the excited state, the electrons can have their spin states either antiparallel or parallel. The first case is the excited singlet state S_1. In the second case, the total spin angular momentum is 1 and there are three possible spin wave functions according to the z-component of the angular momentum; hence these are called *triplet states*, T_1. Since photons carry no spin, a photon-induced transition $S_0 \rightarrow T_1$ is not allowed and the main optical absorption route corresponds to the transition $S_0 \rightarrow S_1$.

In an organic crystal or chain, the promotion of an electron occurs predominantly in one molecule or to a neighboring molecule in a charge-transfer (CT) event. Accordingly, distinct classes of excitons exist in organic crystals depending on the extent of charge separation (Bardeen, 2014). In the Frenkel exciton, the electron and hole associated with the excited state reside in the same molecule (Figure 19.18b). Two electrons are localized on the LUMO and HOMO level of the same site, implying that the exchange interaction between them can be large. This effect separates the Frenkel exciton into two singlet and triplet bands. In a CT event, the electron is transferred to the LUMO of another molecule, leaving a hole in the HOMO, see Figure 19.18c, and the resulting CT exciton has ionic character. Long-range transport of excitons in the spin singlet state by hopping occurs

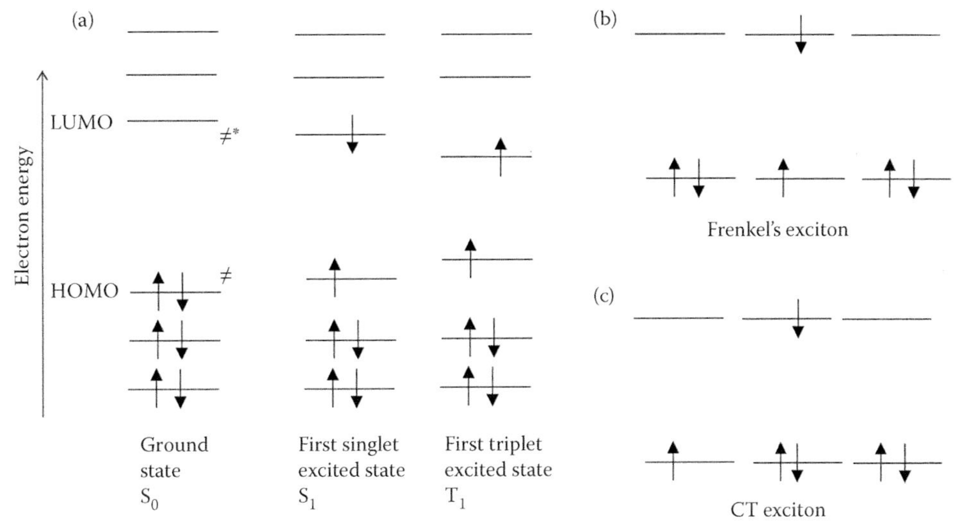

FIGURE 19.18 (a) The unpaired electrons in the excited state of a molecule can have their spins parallel or antiparallel, forming, respectively, a singlet or triplet state. The ground state is a singlet. The diagram also indicates the effect of the Coulomb and exchange energy in the energy levels of the frontier orbitals involved in the first excited singlet and triplet states. (b) Frenkel exciton. (c) CT exciton.

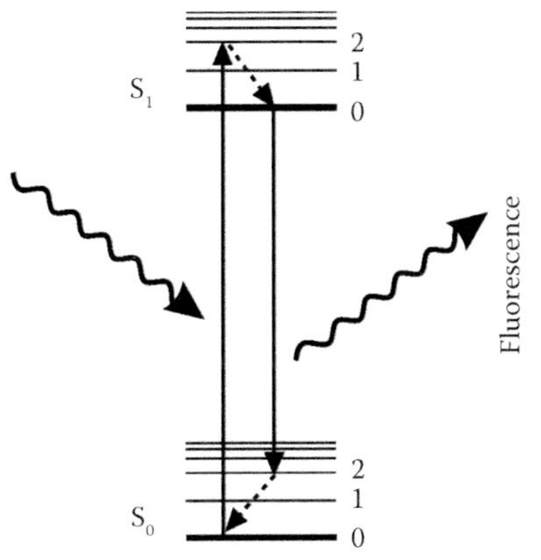

FIGURE 19.19 The photon-induced transition from the ground singlet to the first excited singlet states. The excited state relaxes rapidly in the manifold of vibronic states and decays back to the ground singlet state emitting a photon in the process of fluorescence.

by the Forster energy transfer. Triplet exciton hopping is dominated by the Dexter energy transfer, since excitons in the triplet state are spin forbidden from emitting.

Because of the vibrational degrees of freedom of the molecule or polymer, there are many singlet excited states, above the ground state S_1, that are termed S_2, S_3, etc. and similarly for the triplet states T_2, T_3, etc. In Figure 19.19, the ground state of the vibronic ensemble of a class of spin states is indicated with a thick line while the upper energy states are indicated by thinner lines. The vibronic levels in a given type of spin-correlated system correspond to the quantum number of the harmonic oscillator $v = 0, 1, 2, \ldots$. The transitions involving photon absorption or emission will start from the lowest energy state $v = 0$ of the ground or excited singlet ensemble, respectively. The excited state will rapidly relax to the lowest energy state, in a process similar to that explained in Figure 6.9 and shown in Figure 19.20 for the case of a vibronic manifold. Then, the excited state S_1 may decay back to one of the energy states in the ground singlet, emitting a photon in the process termed *fluorescence*.

While the photon absorption and emission process starts from the lowest energy state of the respective singlet vibronic manifold, the intensity of the different transitions will determine the lineshape of absorption and fluorescence spectra. In Figure 19.20a, the vibronic manifolds of the singlet states are represented by the quantum states of oscillators. The oscillators are displaced in the horizontal axis that indicate the nuclear coordinates of the molecule, due to the fact that equilibrium position of excited and ground states is not the same. The nuclear coordinates can be taken as normal coordinates of the vibration, Q, since rotations and translations can be neglected. The process of the electronic transition is so fast that the nuclear coordinates do not change their values during the transition. This general rule is termed the *Franck–Condon principle*, as discussed in Chapter 6. Therefore, the probability of a specific transition is largely determined by the spatial overlap of the initial and final states of the electronic vibrational wave functions.

If the ground and excited states have a similar distribution of oscillator quantum states, then the upward

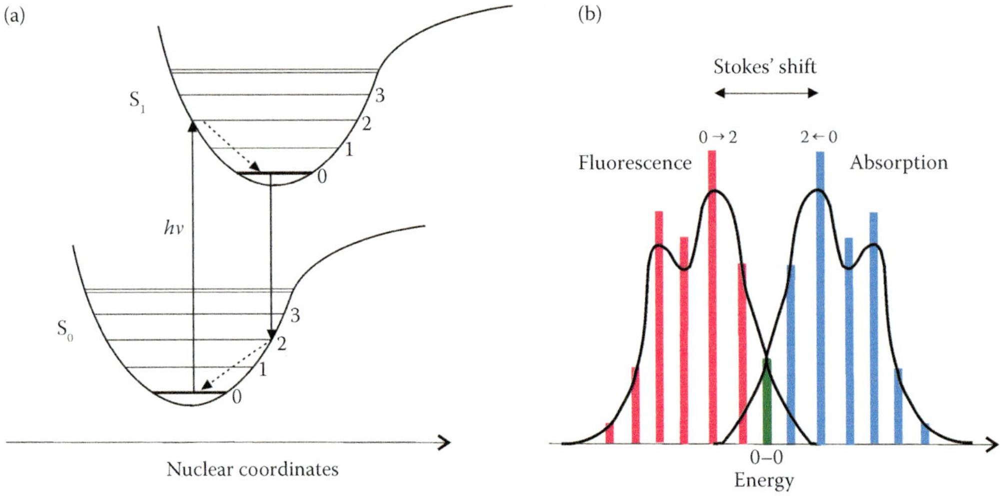

FIGURE 19.20 (a) The photon-induced transition from the 0-ground singlet to the 2-level of the excited singlet state in an oscillator model. The initial excitation in 2-level excited state relaxes rapidly to the 0-level of the excited state and decays by fluorescence to the 2-level of the ground singlet state, which then relaxes to 0-level of the ground singlet state. (b) The intensity of individual transitions is determined by the spatial overlap of the electronic states involved in the transition. The smoother lines correspond to the actual lineshape expected in the observation of absorption and fluorescence.

transitions from the 0 level of the ground singlet state will have the same probabilities as the downward transitions from the 0 level of the excited singlet state. However, the fluorescent transitions will occur at lower energies (larger wavelengths) than the absorption transitions, thus causing Stokes shift. The fluorescent peak is often a mirror image of the absorption peak and the intercept occurs at the 0–0 (zero–zero) transition that has the same energy in both absorption and fluorescence. Some examples are shown in Figure 19.21. The fluorescent luminescence shows the vibronic splitting. The vibrational states can be closely spaced and are coupled with normal thermal motion. Therefore, the emission and absorption spectra occur over a band of wavelengths rather than in sharp lines. The vibronic features of oligothiophenes are shown in several peaks that are clearly resolved at low temperature in the left column of Figure 19.22, while at higher temperatures, those features are smoothed out in the right column of Figure 19.22.

The direct transition from the ground singlet state to a triplet state requires a spin flip that is forbidden on the basis of the conservation of the spin angular momentum alone. However, the transition in which an electron is moved between orbitals of differing symmetry becomes possible if assisted by the spin–orbit coupling, since the change in spin angular momentum compensates for the change in orbital angular momentum. The most effective way to increase the efficiency of these transitions is the presence of heavy atoms incorporated in the molecule that provoke a very large, metal-induced spin–orbit coupling. The excited triplet state is usually more stable than the excited singlet manifold due to larger stabilization by the exchange interaction between the electrons. Therefore, T_1 is lower in energy than S_1. The transition $S_1 \rightarrow T_1$ is termed *intersystem crossing* (ISC) and depends on spin–orbit coupling and on the vibrational overlap between the singlet and triplet states, see Figure 19.23. From the relaxed state, a luminescent transition $T_1 \rightarrow S_0$ is weakly allowed. This process occurs at lower energy and longer wavelengths than fluorescence, and with much longer radiative lifetime (microseconds to miliseconds), and is called *phosphorescence.* Since ISC is facilitated if the states have similar energy, the transition from S_1 may preferentially occur to a higher-lying triplet state T_n that subsequently relaxes vibronically to T_1.

The metal-to-ligand charge transfer (MLCT) in excited states of $d\pi^6$ coordination compounds has been widely applied in solar energy harvesting using dye-sensitized solar cells (Ardo and Meyer, 2009). In this type of transition, light absorption promotes an electron from the metal d orbitals to the ligand π^* orbitals. The processes following the vertical excitation of [Ru(bpy)3]2+ to the Franck–Condon state are shown in Figure 19.24. Absorption of visible light (450 nm) by $Ru(bpy)_3^{2+}$ leads to a $d - \pi^*$ MLCT transition that proceeds with unit quantum efficiency. The lifetime of the excited state, $Ru(bpy)_3^{2+*}$, is quite long (~600 ns) (Henry et al., 2008). The excited electron undergoes ISC to the lowest triplet states of $Ru(bpy)_3^{2+}$ from where phosphorescent

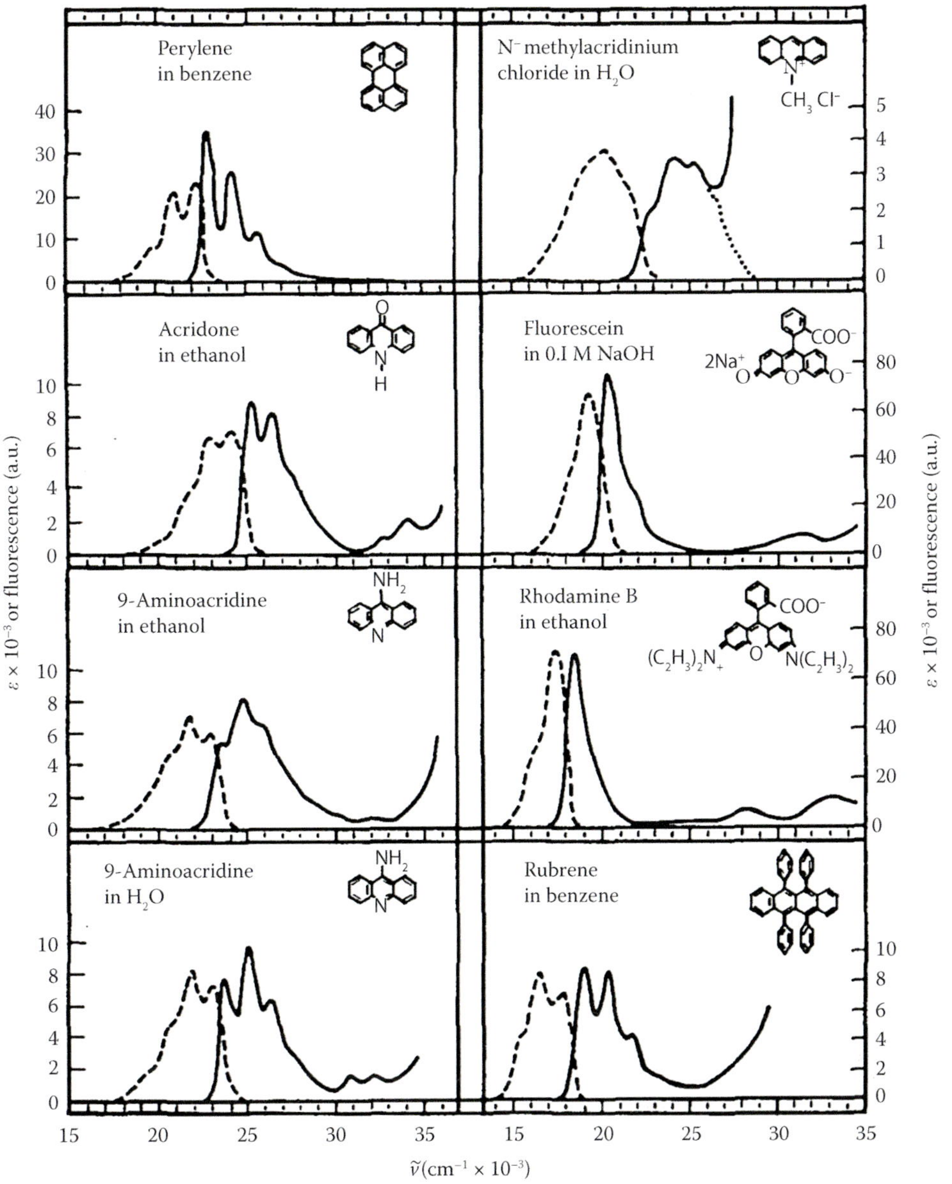

FIGURE 19.21 Absorption spectra (*solid lines*) and fluorescence spectra (*dashed lines*) of organic compounds. (Reproduced with permission from Strickler, S. J.; Berg, R. A. *Journal of Chemical Physics* 1962, 37, 814–820.)

emission of a photon of energy $E = 2.1$ eV (610 nm) occurs.

The size of a conjugated system greatly influences all the photophysical quantities because the confinement effect increases when the number of repetition units decreases. The energy gap, ionization potential, and electron affinity of oligomers show a reciprocal dependence on the number of repeat unit (or degree of polymerization) (Wegner et al., 2008). When the molecules are condensed to form an organic solid, the absorption properties of the solid may either remain similar to those of the molecule in solution or else may be strongly modified. An example is shown in Figure 19.25. In a random molecular distributed solid, the molecular absorption still predominates. However, if the solid is a crystal or forms ordered aggregates, the electronic transition obtains a predominant CT character, which induces efficient coupling of intermolecular vibrational modes (Gierschner et al., 2005; Oksana et al., 2005).

19.6 THE CT BAND IN ORGANIC BLENDS AND HETEROJUNCTIONS

The combination of a donor and acceptor pair of molecules, or the formation of an intimate heterojunction between semiconductors opens new optical transitions

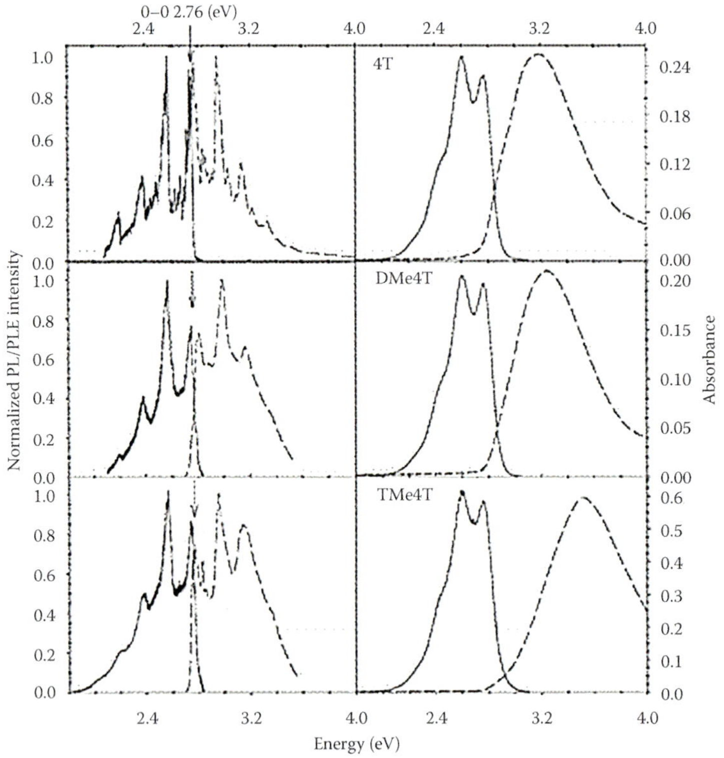

FIGURE 19.22 (b) Normalized PL (*solid line*) and absorption spectra (*dashed line*) collected at 4 K (*left*) and 300 K (*right*) of several oligothiophenes. (Reproduced with permission from Macchi, G. et al. *Physical Chemistry Chemical Physics* 2009, 11, 984–990.)

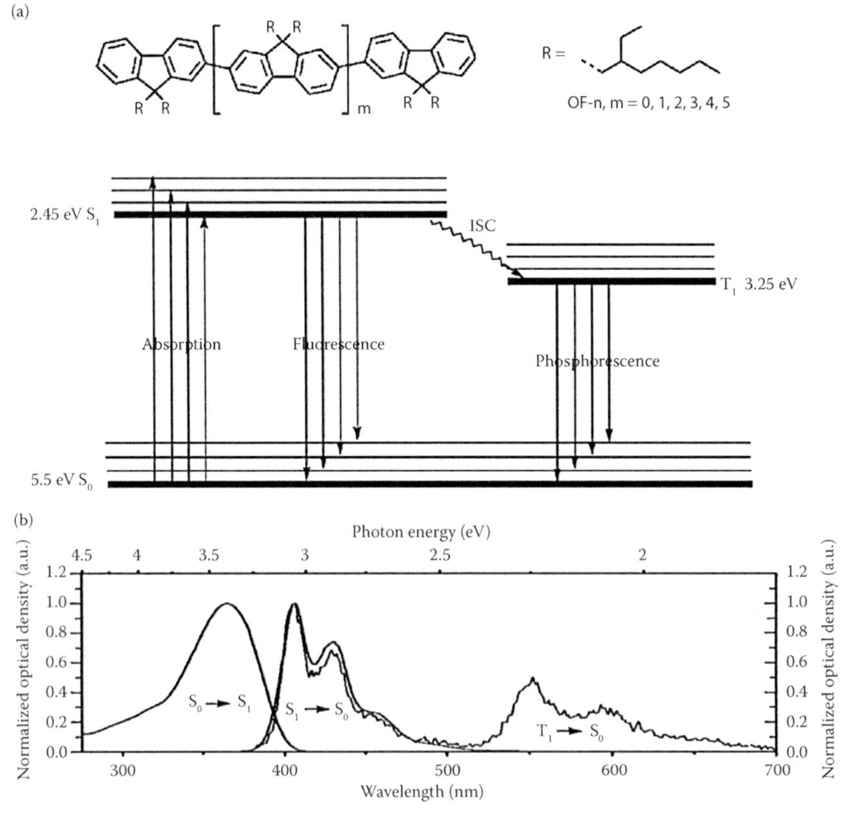

FIGURE 19.23 Structure of oligofluorenes with 2-ethylhexyl side chains (OFn). (a) Energy diagram and (b) absorption and luminescence of OF5 in a matrix of methyltetrahydrofurane (m-THF). (Reproduced with permission from Wegner, G. et al. *Macromolecular Symposia* 2008, 268, 1–8.)

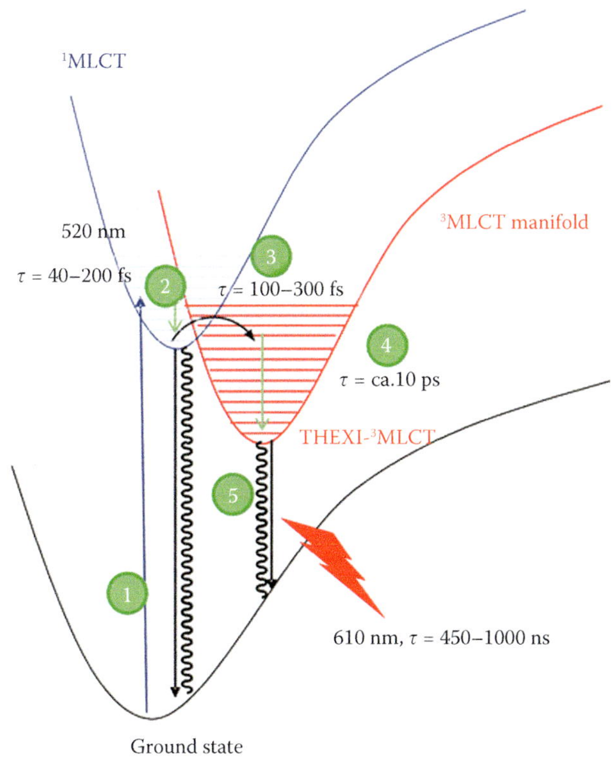

FIGURE 19.24 Overview of processes following the excitation of [Ru(bpy)3]2+ to the Franck–Condon (FC) state. (1) Excitation from the ground state to the FC state, (2) relaxation of the FC state and fluorescence (520 nm), the lifetime of which is determined by the rate of ISC, ca. <300 fs, (3) to the vibrationally hot ^{3}MLCT state followed by (4) vibrational cooling to the THEXI–^{3}MLCT state (complete by 20 ps), which itself undergoes both (5) nonradiative and radiative relaxation with a lifetime of 400–1000 ns. (Reproduced with permission from Henry, W. et al. *The Journal of Physical Chemistry A* 2008, 112, 4537–4544.)

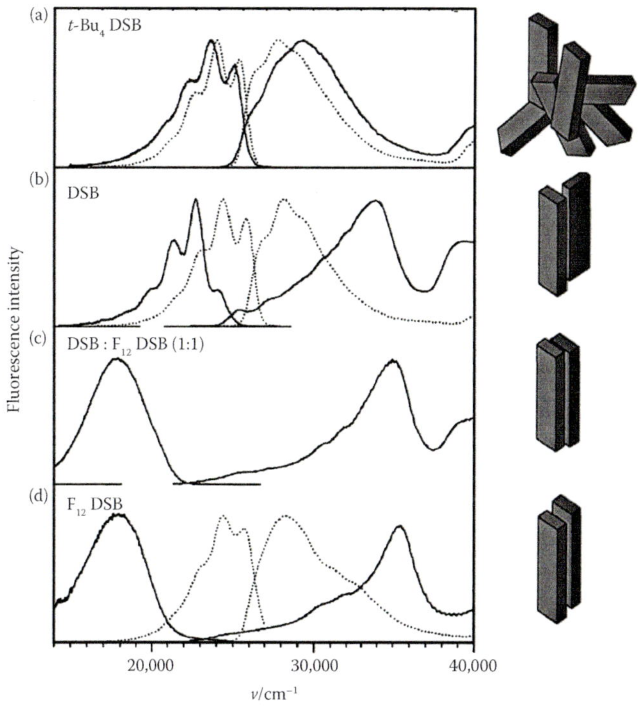

FIGURE 19.25 Fluorescence (*left*) and absorption spectra (*right*) of distyrylbenzene nanoparticles: (a) *t*-Bu$_4$DSB, (b) DSB, (c) cocrystallized DSB/F$_{12}$DSB, and (d) F$_{12}$DSB. Spectra in hexane solution (*dashed lines*) are shown for comparison. A schematic representation of the respective condensed phase structures is given on the right. (Reproduced with permission from Gierschner, J. et al. *The Journal of Chemical Physics* 2005, 123, 144914.)

with respect to those of the separate materials by excitation of an electron from the VB (HOMO) of the acceptor to the CB (LUMO) of the donor, as shown in Figure 19.26. The heterojunction state formed in this type of transition is termed the *exciplex* (shorthand for excited states complex formation) or *charge-transfer complex* (CTC). The CTC can be strongly excitonic in which case the opposite carriers remain localized at the interface as indicated in Figure 19.26, or it may give rise to fast charge separation, and the optical gap will be smaller than the transport gap of the organic blend (Bredas, 2014).

CT absorption or emission processes in donor–acceptor molecular systems have been amply investigated comparing the absorption properties of the blend with respect to the parent materials. CT absorption bands of π-conjugated polymers and oligomers combined with electron acceptors correspond to the transition from the ground state to the charge-separated state. The peak absorbance energy is seen to correlate with the difference between oxidation potential of the donor and reduction potential of the acceptor, as shown in Figure 19.27. The CTC usually provides featureless, redshifted emission spectra and long-radiative decay times (Gebler et al., 1997; Morteani et al., 2004). The CTC can also be realized in core–shell quantum dots with type II band alignment, indicated in Figure 19.17d. Figure 19.28 shows that the absorptivity of CdTe/CdSe and CdSe/ZnTe core–shell QDs is extended to the infrared and the PL occurs at energies that are smaller than the band gap of both materials that form the QD (Kim et al., 2003). The absorptivity in the infrared is weak since the QD behaves as an indirect semiconductor, nonetheless it provides a significant increase of light harvesting for photovoltaic applications (Wang et al., 2013a, b). On the other hand, the application of the exciplex emission for LEDs requires donor–acceptor blends or bilayer systems where the CTC emission is highly efficient (Cocchi et al., 2002).

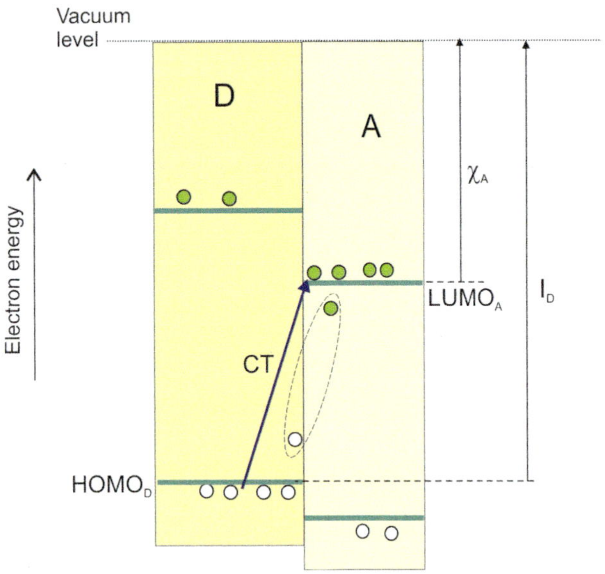

FIGURE 19.26 Interface between an electron donor (D) and electron acceptor (A) material showing optical excitation across the interface known as exciplex or charge-transfer complex. The energy of the CT state may be less than $I_D - \chi_A$ due to Coulomb-binding energy.

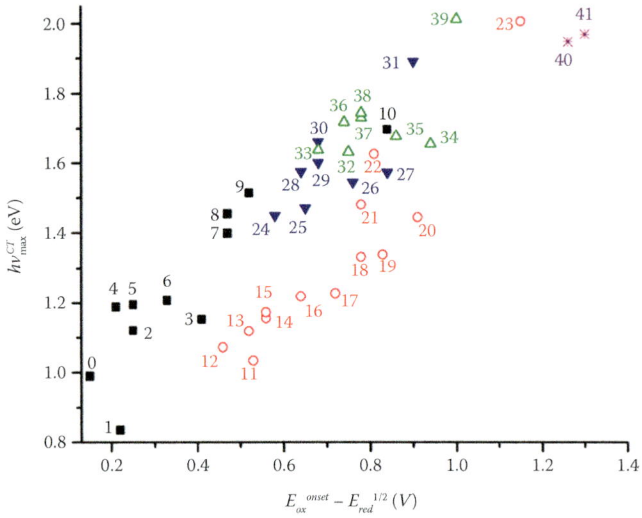

FIGURE 19.27 Correlation of the photon energy of maximum absorbance for the first CT band ($h\nu_{\max}^{CT}$) with the difference between oxidation potential of the donor and reduction potential of the acceptor for various donors mixed with different acceptors (see Table 2 of the original publication). (Reproduced with permission from Panda, P. et al. *The Journal of Physical Chemistry B* 2007, 111, 5076–5081.)

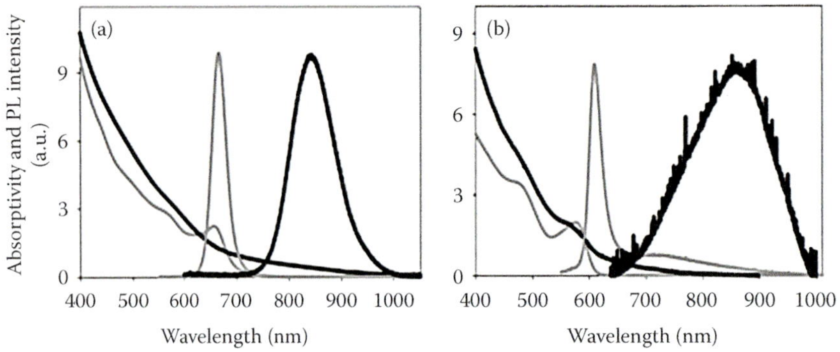

FIGURE 19.28 Absorptivity and normalized PL spectra of 3.2 nm radius CdTe QD (gray lines on (a)), CdTe/CdSe (3.2 nm radius core/thickness of 1.1 nm shell) QD (black lines on (a)), 2.2 nm radius CdSe QD (gray lines on (b)), and CdSe/ZnTe (2.2 nm radius core/thickness of 1.8 nm shell) QD (black lines on (b)). (Reproduced with permission from Kim, S. et al. *Journal of the American Chemical Society* 2003, 125, 11466–11467.)

GENERAL REFERENCES

Light absorption in inorganic solids: Pankove (1971), Klingshirn (1995), and Fox (2001).

Free carrier absorption: Burstein (1952), Liu and Bard (1989), and Fox (2001).

Burstein shift: Burstein (1952), Moss (1954), Kamat et al. (1989), Muñoz et al. (2001), Kawamura et al. (2003), and Manser and Kamat (2014).

Time-resolved microwave conductivity: Friedrich and Kunst (2011), Dunn et al. (2012), Fravventura et al. (2013), and Savenije et al. (2013).

Excitons: Pankove (1971), Fox (2001), Rosencher and Vinter (2002), and Bardeen (2014).

Exciton screening: Gay (1971) and Amo et al. (2006).

Quantum dots: Alivisatos (1996), Yu et al. (2003), Kamat (2008), and de Mello (2010).

REFERENCES

Alivisatos, A. P. Perspectives on the physical chemistry of semiconductor nanocrystals. *The Journal of Physical Chemistry* 1996, *100*, 13226–13239.

Amo, A.; Martin, M.; Viña, L.; Toropov, A.; Zhuravlev, K. Interplay of exciton and electron-hole plasma recombination on the photoluminescence dynamics in bulk GaAs. *Physical Review B* 2006, *73*, 035205.

Ardo, S.; Meyer, G. J. Photodriven heterogeneous charge transfer with transition-metal compounds anchored to TiO_2 semiconductor surfaces. *Chemical Society Reviews* 2009, *38*, 115–164.

Bardeen, C. J. The structure and dynamics of molecular excitons. *Annual Review of Physical Chemistry* 2014, *65*, 127–148.

Barugkin, C.; Cong, J.; Duong, T.; Rahman, S.; Nguyen, H. T.; Macdonald, D.; White, T. P.; et al. Ultralow absorption coefficient and temperature dependence of radiative recombination of $CH_3NH_3PbI_3$ perovskite from photoluminescence. *The Journal of Physical Chemistry Letters* 2015, *6*, 767–772.

Berger, T.; Anta, J. A. IR and spectrophotoelectrochemical characterization of mesoporous semiconductor films. *Analytical Chemistry* 2011, *84*, 3053–3057.

Berger, T.; Anta, J. A.; Morales-Florez, V. Electrons in the band gap: Spectroscopic characterization of anatase TiO_2 nanocrystal electrodes under Fermi level control. *The Journal of Physical Chemistry C* 2012, *116*, 11444–11455.

Bredas, J.-L. Mind the gap! *Materials Horizons* 2014, *1*, 17–19.

Burstein, E. Anomalous optical absorption limit in InSb. *Physical Review* 1952, *93*, 632–633.

Canovas, E.; Pijpers, J.; Ulbricht, R.; Bonn, M. Carrier dynamics in photovoltaic structures and materials studied by time-resolved terahertz spectroscopy, Chapter 11. In *Solar Energy Conversion: Dynamics of Interfacial Electron and Excitation Transfer*; The Royal Society of Chemistry: London, 2013, pp. 301–336.

Chen, X.; Lu, H.; Yang, Y.; Beard, M. C. Excitonic effects in methylammonium lead halide perovskites. *Journal of Physical Chemistry Letters* 2018, *9*, 2595–2603.

Cocchi, M.; Virgili, D.; Giro, G.; Fattori, V.; Marco, P. D.; Kalinowski, J.; Shirota, Y. Efficient exciplex emitting organic electroluminescent devices. *Applied Physics Letters* 2002, *80*, 2401–2403.

Daub, E.; Würfel, P. Ultralow values of the absorption coefficient of Si obtained from luminescence. *Physical Review Letters* 1995, *74*, 1020–1023.

de Mello, C. Synthesis and properties of colloidal heteronanocrystals. *Chemical Society Reviews* 2010, *40*, 1512–1546.

De Wolf, S.; Holovsky, J.; Moon, S.-J.; Löper, P.; Niesen, B.; Ledinsky, M.; Haug, F.-J.; et al. Organometallic halide perovskites: Sharp optical absorption edge and its relation to photovoltaic performance. *The Journal of Physical Chemistry Letters* 2014, *5*, 1035–1039.

Diamant, Y.; Chen, S. G.; Melamed, O.; Zaban, A. Core-shell nanoporous electrode for dye sensitized solar cells: The effect of the $SrTiO_3$ shell on the electronic properties of the TiO_2 core. *The Journal of Physical Chemistry B* 2003, *107*, 1977–1981.

Dunn, H. K.; Peter, L. M.; Bingham, S. J.; Maluta, E.; Walker, A. B. In situ detection of free and trapped electrons in dye-sensitized solar cells by photo-induced microwave reflectance measurement. *The Journal of Physical Chemistry C* 2012, *116*, 22063–22072.

Elliott, R. J. Intensity of optical absorption by excitons. *Physical Review* 1957, *108*, 1384–1389.

Fox, M. *Optical Properties of Solids*; Oxford University Press: Oxford, 2001.

Fravventura, M. C.; Deligiannis, D.; Schins, J. M.; Siebbeles, L. D. A.; Savenije, T. J. What limits photoconductance in anatase TiO_2 nanostructures? A real and imaginary microwave conductance study. *The Journal of Physical Chemistry C* 2013, *117*, 8032–8040.

Friedrich, D.; Kunst, M. Analysis of charge carrier kinetics in nanoporous systems by time resolved photoconductance measurements. *The Journal of Physical Chemistry C* 2011, *115*, 16657–16663.

Gay, J. G. Screening of excitons in semiconductors. *Physical Review B* 1971, *4*, 2567–2575.

Gebler, D. D.; Wang, Y. Z.; Blatchford, J. W.; Jessen, S. W.; Fu, D. K.; Swager, T. M.; MacDiarmid, A. G.; et al. Exciplex emission in bilayer polymer light-emitting devices. *Applied Physics Letters* 1997, *70*, 1644–1646.

Gierschner, J.; Ehni, M.; Egelhaaf, H.-J.; Medina, B. M.; Beljonne, D.; Benmansour, H.; Bazan, G. C. Solid-state optical properties of linear polyconjugated molecules: Pi-Stack contra herringbone. *The Journal of Chemical Physics* 2005, *123*, 144914.

Gokmena, T.; Gunawan, O.; Todorov, T. K.; Mitzi, D. B. Band tailing and efficiency limitation in kesterite solar cells. *Applied Physics Letters* 2013, *103*, 103506.

Green, M. A. Self-consistent optical parameters of intrinsic silicon at 300 K including temperature coefficients. *Solar Energy Materials and Solar Cells* 2008, *92*, 1305–1310.

Green, M. A.; Ho-Baillie, A.; Snaith, H. J. The emergence of perovskite solar cells. *Nature Photonics* 2014, *8*, 506–514.

Guyot-Sionnest, P. Electrical transport in colloidal quantum dot films. *The Journal of Physical Chemistry Letters* 2012, *3*, 1169–1175.

Harruff, B. A.; Bunker, C. E. Spectral properties of AOT-protected CdS nanoparticles: Quantum yield enhancement by photolysis. *Langmuir* 2003, *19*, 893–897.

Henry, W.; Coates, C. G.; Brady, C.; Ronayne, K. L.; Matousek, P.; Towrie, M.; Botchway, S. W. The early picosecond photophysics of ru(ii) polypyridyl complexes: A tale of two timescales. *The Journal of Physical Chemistry A* 2008, *112*, 4537–4544.

Hines, M. A.; Guyot-Sionnest, P. Synthesis and characterization of strongly luminescing ZnS-Capped CdSe nanocrystals. *The Journal of Physical Chemistry* 1996, *100*, 468–471.

Hollingsworth, J. A. Nanoscale engineering facilitated by controlled synthesis: From structure to function. *Coordination Chemistry Reviews* 2014, *263–264*, 197–216.

Ip, A. H.; Thon, S. M.; Hoogland, S.; Voznyy, O.; Zhitomirsky, D.; Debnath, R.; Levina, L. Hybrid passivated colloidal quantum dot solids. *Nature Nanotechnology* 2012, *7*, 577–582.

John, S.; Soukoulis, C.; Cohen, M. H.; Economou, E. N. Theory of electron band tails and the urbach optical-absorption edge. *Physical Review Letters* 1986, *57*, 1777–1780.

Kamat, P. V. Quantum dot solar cells. Semiconductor nanocrystals as light harvesters. *The Journal of Physical Chemistry C* 2008, *112*, 18737–18753.

Kamat, P. V.; Dimitrijevic, N. M.; Nozik, A. J. Dynamic Burstein-Moss shift in semiconductor colloids. *The Journal of Physical Chemistry* 1989, *93*, 2873–2875.

Katahara, J. K.; Hillhouse, H. W. Quasi-Fermi level splitting and sub-bandgap absorptivity from semiconductor photoluminescence. *Journal of Applied Physics* 2014, *116*, 173504.

Kawamura, K.-i.; Maekawa, K.; Yanagi, H.; Hirano, M.; Hosono, H. Observation of carrier dynamics in CdO thin films by excitation with femtosecond laser pulse. *Thin Solid Films* 2003, *445*, 182–185.

Kim, S.; Fisher, B.; Eisler, H.-J.; Bawendi, M. Type-II quantum dots: CdTe/CdSe(core/shell) and CdSe/ZnTe(core/shell) heterostructures. *Journal of the American Chemical Society* 2003, *125*, 11466–11467.

Klein, A. Transparent conducting oxides: Electronic structure–property relationship from photoelectron spectroscopy with in situ sample preparation. *Journal of the American Ceramic Society* 2013, *96*, 331–345.

Klingshirn, C. F. *Semiconductor Optics*; Springer-Verlag: Berlin, 1995.

Köhler, A.; Bässler, H. Triplet states in organic semiconductors. *Materials Science and Engineering: R: Reports* 2009, *66*, 71–109.

Liu, C. Y.; Bard, A. J. Effect of excess charge on band energetics (optical absorption edge and carrier redox potentials) in small semiconductor particles. *The Journal of Physical Chemistry* 1989, *93*, 3232–3237.

Luther, J. M.; Law, M.; Beard, M. C.; Song, Q.; Reese, M. O.; Ellingson, R. J.; Nozik, A. J. Schottky solar cells based on colloidal nanocrystal films. *Nano Letters* 2008, *8*, 3488–3492.

Macchi, G.; Medina, B. M.; Zambianchi, M.; Tubino, R.; Cornil, J.; Barbarella, G.; Gierschner, J.; et al. Spectroscopic signatures for planar equilibrium geometries in methyl-substituted oligothiophenes. *Physical Chemistry Chemical Physics* 2009, *11*, 984–990.

Mahan, G. D. Excitons in degenerate semiconductors. *Physical Review* 1967, *153*, 882–889.

Manser, J. S.; Kamat, P. V. Band filling with free charge carriers in organometal halide perovskites. *Nature Photonics* 2014, *8*, 737–743.

Mattheis, J.; Rau, U.; Werner, J. H. Light absorption and emission in semiconductors with band gap fluctuations—A study on Cu(In,Ga)Se_2 thin films. *Journal of Applied Physics* 2007, *101*, 113519.

Mayer, K. M.; Hafner, J. H. Localized surface plasmon resonance sensors. *Chemical Reviews* 2011, *111*, 3828–3857.

Micic, O. I.; Sprague, J.; Lu, Z.; Nozik, A. J. Highly efficient band-edge emission from InP quantum dots. *Applied Physics Letters* 1996, *68*, 3150–3152.

Miller, O. D.; Yablonovitch, E.; Kurtz, S. R. Strong internal and external fluorescence as solar cells approach the Shockley-Queisser limit. *IEEE Journal of Photovoltaics* 2012, *2*, 303–311.

Milo, R. What governs the reaction center excitation wavelength of photosystems I and II? *Photosynthesis Research* 2009, *101*, 59–67.

Ming, T.; Chen, H.; Jiang, R.; Li, Q.; Wang, J. Plasmon-controlled fluorescence: Beyond the intensity enhancement. *The Journal of Physical Chemistry Letters* 2012, *3*, 191–202.

Morteani, A. C.; Sreearunothai, P.; Herz, L. M.; Friend, R. H.; Silva, C. Exciton regeneration at polymeric semiconductor heterojunctions. *Physical Review Letters* 2004, *92*, 247402.

Moss, T. S. The interpretation of the properties of indium antimonide. *Proceedings of the Physical Society* 1954, *76*, 775.

Munechika, K.; Chen, Y.; Tillack, A. F.; Kulkarni, A. P.; Plante, I. J.-L.; Munro, A. M.; Ginger, D. S. Spectral control of plasmonic emission enhancement from quantum dots near single silver nanoprisms. *Nano Letters* 2010, *10*, 2598–2603.

Muñoz, M.; Pollak, F.; Kahn, M.; Ritter, D.; Kronik, L.; Cohen, G. Burstein-Moss shift of n-doped In0.53Ga0.47As/InP. *Physical Review B* 2001, *63*, 233302.

Oksana, O.; Svitlana, S.; David, G. C.; Ray, F. E.; Frank, A. H.; Rik, R. T.; Sean, R. P.; et al. Optical and transient photoconductive properties of pentacene and functionalized pentacene thin films: Dependence on film morphology. *Journal of Applied Physics* 2005, *98*, 033701.

Panda, P.; Veldman, D.; Sweelssen, J.; Bastiaansen, J. J. A. M.; Langeveld-Voss, B. M. W.; Meskers, S. C. J. Charge transfer absorption for pi-conjugated polymers and oligomers mixed with electron acceptors. *The Journal of Physical Chemistry B* 2007, *111*, 5076–5081.

Pankove, J. I. *Optical Processes in Semiconductors*; Prentice-Hall: Englewood Cliffs, 1971.

Rosencher, E.; Vinter, B. *Optoelectronics*; Cambridge University Press: Cambridge, 2002.

Sachenko, A. V.; Kryuchenko, Y. V. Excitonic effects in band-edge luminescence of semiconductors at room temperatures. *Semiconductor Physics, Quantum Electronics & Optoelectronics* 2000, *3*, 150–156.

Savenije, T. J.; Ferguson, A. J.; Kopidakis, N.; Rumbles, G. Revealing the dynamics of charge carriers in polymer: Fullerene blends using photoinduced time-resolved microwave conductivity. *The Journal of Physical Chemistry C* 2013, *117*, 24085–24103.

Shea-Rohwer, L. E.; Martin, J. E. Luminescence decay of broadband emission from CdS quantum dots. *Journal of Luminescence* 2007, *127*, 499–507.

Shirasaki, Y.; Supran, G. J.; Bawendi, M. G.; Bulovic, V. Emergence of colloidal quantum-dot light-emitting technologies. *Nature Photonics* 2012, *7*, 13–23.

Strickler, S. J.; Berg, R. A. Relationship between absorption intensity and fluorescence lifetime of molecules. *Journal of Chemical Physics* 1962, *37*, 814–820.

Studenyak, I.; Kranjec, M.; Kurik, M. Urbach rule in solid state physics. *International Journal Optical Applications* 2014, *4*, 76–83.

Sturge, M. D. Optical absorption of Gallium Arsenide between 0.6 and 2.75 eV. *Physical Review* 1962, *127*, 768–773.

Subramanian, V.; Wolf, E. E.; Kamat, P. V. Green emission to probe photoinduced charging events in ZnO/Au nanoparticles. Charge distribution and Fermi-level equilibration. *The Journal of Physical Chemistry B* 2003, *107*, 7479–7485.

Talapin, D. V.; Gaponik, N.; Borchert, H.; Rogach, A. L.; Haase, M.; Weller, H. Etching of colloidal InP nanocrystals with fluorides: Photochemical nature of the process resulting in high photoluminescence efficiency. *The Journal of Physical Chemistry B* 2002, *106*, 12659–12663.

Tang, J.; Wang, X.; Brzozowski, L.; Barkhouse, D. A. R.; Debnath, R.; Levina, L.; Sargent, E. H. Schottky quantum dot solar cells stable in air under solar illumination. *Advanced Materials* 2011, *22*, 1398–1402.

Trupke, T.; Daub, E.; Würfel, P. Absorptivity of silicon solar cells obtained from luminescence. *Solar Energy Materials and Solar Cells* 1998, *53*, 103–114.

Turner, W. J.; Reese, W. E.; Pettit, G. D. Exciton absorption and emission in InP. *Physical Review* 1964, *136*, A1467.

Ulbricht, R.; Hendry, E.; Shan, J.; Heinz, T. F.; Bonn, M. Carrier dynamics in semiconductors studied with time-resolved terahertz spectroscopy. *Reviews of Modern Physics* 2011, *83*, 543–586.

Urbach, F. The long-wavelength edge of photographic sensitivity and of the electronic absorption of solids. *Physical Review* 1953, *92*, 1324–1324.

Vanmaekelbergh, D. Self-assembly of colloidal nanocrystals as route to novel classes of nanostructured materials. *Nano Today* 2011, *6*, 419–437.

Wang, H.; Liu, X.; Zhang, Z. Absorption coefficients of crystalline silicon at wavelengths from 500 nm to 1000 nm. *International Journal of Thermophysics* 2013a, *34*, 213–225.

Wang, J.; Mora-Sero, I.; Pan, Z.; Zhao, K.; Zhang, H.; Feng, Y.; Yang, G.; et al. Core/shell colloidal quantum dot exciplex states for the development of highly efficient quantum dot sensitized solar cells. *Journal of the American Chemical Society* 2013b, *135*, 15913–15922.

Wegner, G.; Baluschev, S.; Laquai, F.; Chi, C. Managing photoexcited states in conjugated polymers. *Macromolecular Symposia* 2008, *268*, 1–8.

Yu, W.; Qu, L. H.; Guo, W. Z.; Peng, X. G. Experimental determination of the extinction coefficient of CdTe, CdSe, and CdS nanocrystals. *Chemistry of Materials* 2003, *15*, 2854–2860.

Part IV

Photovoltaic Principles and Solar Energy Conversion

20 Fundamental Model of a Solar Cell

In this chapter, we study the models and physical features that explain those general properties of diodes that are central to their application for energy conversion. We first examine the operation of realistic semiconductor device models, starting with majority carrier diodes controlled by charge transference at the contact. Then we discuss in detail the recombination diode formed by a semiconductor layer of long diffusion length with asymmetric contacts to either electrons or holes. The latter model is the basis for formulating a fundamental model of a solar cell, which consists of a semiconductor light absorber with selective contacts. We discuss the conversion of light-generated carriers to voltage, and we describe the general function of the selective contacts as well as their specific materials and interfacial properties.

20.1 MAJORITY CARRIER INJECTION MECHANISMS

In Section 5.8, we introduced the basic properties of a diode. We defined forward and reverse voltage, and we remarked the current-voltage characteristic associated with rectification:

$$j = j_0\left(e^{qV/mk_BT} - 1\right) \tag{20.1}$$

In the case $m = 1$, Equation 20.1 is the *Shockley ideal diode equation* or the *diode law*. The parameter m is the *ideality factor*, also known as the *quality factor*, and it describes the specific exponential voltage dependence of the current. Often, the current density–voltage curve depends on a combination of processes and m describes the best exponential approximation. For example, ideality factors approaching $m = 2$ occur by the generation/recombination in the space-charge region (SCR) shown later in Figure 26.16. j_0 is the *reverse saturation current*. In this chapter, we aim to describe the basic models for solar cells based on operation of semiconductor devices; therefore, we start with a detailed view of semiconductor devices, which lead to the diode structure that realizes Equation 20.1. There are two main kinds of diodes: based on either majority carrier injection at one interface, or those controlled by minority carrier injection and recombination. We begin with the diodes of the first class.

The properties of a Schottky barrier (SB) were discussed in Chapter 9. We examine the types of diodes that can be formed with an SB. Figure 20.1a shows an SB at the contact of an n-type semiconductor with a metal, see also Figure 9.3. The semiconductor has an *ohmic contact* to majority carriers at the right side. This is a convenient condition in order to focus our attention on only one junction, as it means that the polarization by applied voltage V translates into a difference of Fermi levels across the SB junction. As described in Section 5.8, the forward bias occurs when the voltage is made negative at the ohmic contact of the semiconductor, which means that positive voltage applied to the SB side decreases the size of the barrier (Figure 20.1b and Figure 9.2). The interfacial barriers for injection of electrons and holes, $\Phi_{B,n}$ and $\Phi_{B,p}$, are defined as the energy levels at the contact measured from the Fermi level, as discussed in Section 4.5. Figure 20.2 shows the rectification at the semiconductor-electrolyte interface discussed in Section 6.6, where the voltage governs the surface concentration of electronic carriers, and the rate of transfer is determined by the interfacial charge transfer rates.

As shown in Figure 20.1b, a change of voltage modifies the Fermi level of the majority carrier with respect to the conduction band edge level E_c at the contact. Therefore, the voltage produces a change in the concentration of electrons at the semiconductor surface, n_s. This property allows to govern the rate of electron transfer by the voltage applied to the barrier as further discussed below, or alternatively the hole injection as shown in Figure 20.2. The rates will be increased or decreased according to the number of electrons available at the surface.

In the equilibrium situation of Figure 20.1a, the same current j_0 flows in both directions, from the metal to the semiconductor and vice versa. The electron transfer rate is quantified by the electron current density *in equilibrium*, j_0, which plays a similar role to the *exchange current density* in electrochemistry.

The fundamental assumption about a biased SB is that the Fermi levels in the two phases in contact (the metal and the semiconductor) are homogeneous up to the interface where the step of the Fermi level occurs as was discussed in Section 9.2. The details of the transition of the Fermi level across the interface have important influence on the majority carrier rate of transfer. The location of the Fermi level step actually determines whether a disturbance with respect to equilibrium affects either the concentration of electrons in the semiconductor or in

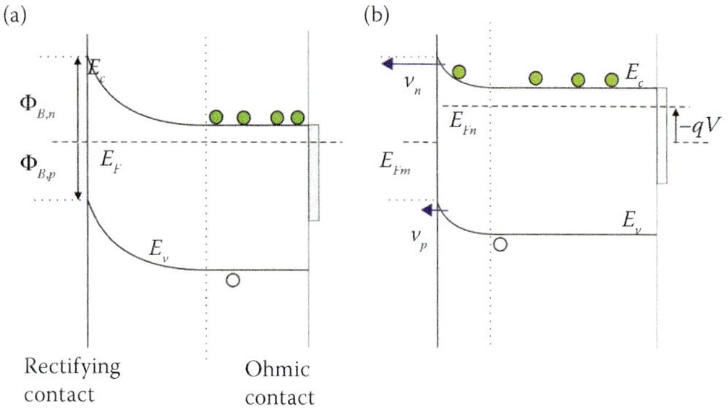

FIGURE 20.1 Model for an SB with a metal contact at (a) equilibrium and (b) at forward bias, showing the majority and minority carrier injection velocities from semiconductor to metal. The injection of majorities predominates at the contact, and the diode is a majority carrier device.

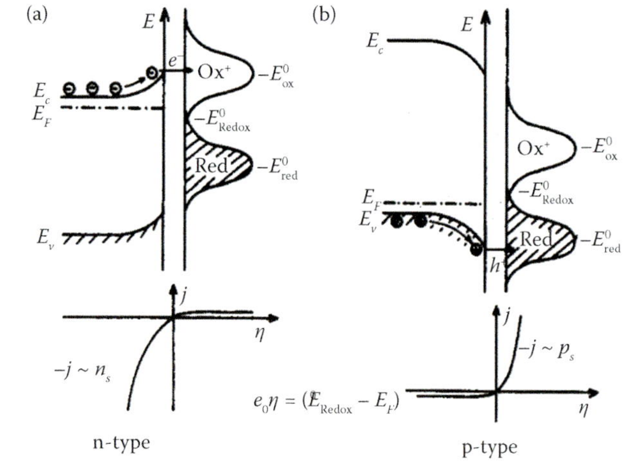

FIGURE 20.2 Redox reactions at n-type (a) and p-type (b) semiconductor electrodes, showing the energy diagrams and the current-voltage curves. (Reproduced with permission from Gerischer, H. *Electrochimica Acta* 1990, 35, 1677–1699.)

the metal, which sets the mechanism of electron transfer. These assumptions correspond respectively to the *diffusion theory*, which sets the gradient of the Fermi level in the semiconductor side, and the *Bethe thermionic-emission theory*, which states the drop at the metal side. A synthesis of these mechanisms is formulated in the combined *thermionic emission–diffusion theory*, which considers the two Fermi level drops in series (Crowell and Sze, 1966). These questions have been amply discussed in the literature of crystalline semiconductors and are summarized by Rhoderick and Williams (1988). It has been concluded that in the Schottky diodes with high mobility semiconductors, the forward current is limited by thermionic emission provided the forward bias is not too large. In the following, we apply the generalized thermionic-emission theory discussed in Section 6.7 to derive the current across the SB.

20.2 MAJORITY CARRIER DEVICES

The surface concentration of majority carriers is a crucial quantity to determine the current flowing in an SB biased by a voltage V. In the derivation of the properties of the SB in Chapter 9, we have assumed that the only significant electric charge in the barrier is the donor density implying $n \ll n_0 = N_D$, Equation 9.9. As we move from the quasi-neutral region toward the surface, the electron density decreases to a value n_s related to the concentration in the bulk by the expression $n_s = n_0 e^{qV_{sc}/k_BT}$, where V_{sc} is the potential difference across the barrier in the semiconductor surface, Equation 9.4. Therefore, the surface concentration dependence on voltage is

$$n_s = n_{s0} e^{qV/k_BT} \tag{20.2}$$

where n_{s0} is the carrier concentration at the surface in the unbiased SB. In the equilibrium situation, electron flow across the barrier, between the metal and the semiconductor conduction band, is balanced in both directions. When a forward bias is applied to the diode, as in Figure 20.1b, the surface concentration of electrons n_s increases as indicated in Equation 20.2; hence, the thermionic flow of electrons from the semiconductor to the metal increases, while the thermal emission from the metal remains the same. Therefore, the resulting current is given by the diode law Equation 4.1 as follows:

$$\begin{aligned} j &= qv_n\left(n_s - n_{s0}\right) \\ &= qv_n n_{s0}\left(e^{qV/k_BT} - 1\right) \end{aligned} \tag{20.3}$$

where v_n is a transfer velocity (Crowell and Sze, 1966); see Figure 20.1b. Since the rectification is due to the injection or suppression of majority carriers, the diode is called a *Schottky diode* or more generally a *majority carrier device* (Shannon, 1979). The reverse saturation current j_0 is given by Equation 6.92:

$$j_0 = qn_{s0}v_n \tag{20.4}$$

The barrier height in equilibrium, $\Phi_{B,n}$, is drawn in Figure 20.1 and defined in Equation 4.24. We can write Equation 20.4 as

$$j_0 = qv_n N_c e^{-\Phi_{B,n}/k_BT} \tag{20.5}$$

where N_c is the effective density of states of the conduction band. Therefore, a fit of log j_0 with respect

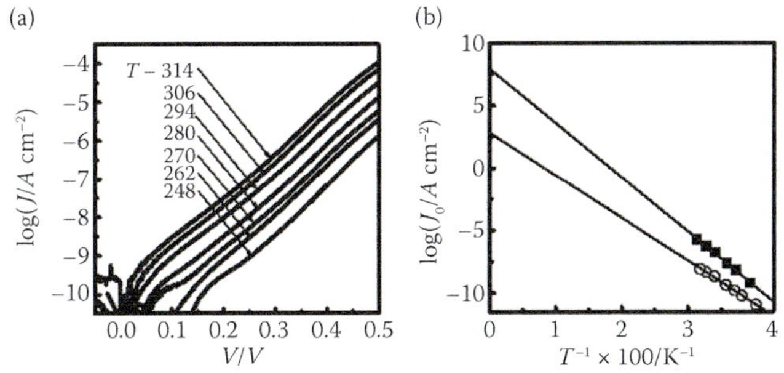

FIGURE 20.3 (a) Forward bias current density-voltage response for an n-Si/PEDOT:PSS device measured at several temperatures (K). (b) Temperature dependence of j_0 for (■) n-Si/Au and (○) n-Si/PEDOT:PSS contacts. (Reproduced with permission from Price, M. J. et al. *Applied Physics Letters* 2010, 97, 083503.)

to reciprocal temperature provides the barrier height (Sze, 1981).

Figure 20.3 compares the parameters of majority carrier diodes formed on n-type Si by Au and PEDOT:PSS contacts; two materials that have the same work function of 5.1 eV and hence similar $\Phi_{B,n}$ (Price et al., 2010). The current-voltage characteristic of n-Si/PEDOT:PSS at different temperatures is shown in Figure 20.3a, and it follows well the model of Equation 20.3. Figure 20.3b shows the temperature dependence of j_0. The slopes are well described by the exponential dependence of Equation 20.5 with a large offset, which indicates that the carrier charge transfer velocity v_n is much larger for n-Si/Au contact than for n-Si/PEDOT:PSS contact. This result shows that v_n is dramatically influenced by carrier transference properties at the interface between the two materials. Figure 20.4 shows the jV characteristics and the derived parameters of a Cr/n-GaAs/In Schottky contact.

20.3 MINORITY CARRIER DEVICES

Based on the SB with an n-type semiconductor, we discuss a different operation mechanism of a semiconductor diode, governed by injection and recombination of minority carriers as shown in Figure 20.5. As explained in Section 20.2, the voltage modulates the size of the depletion layer, so that the forward voltage makes the Fermi level of majority carrier electrons, E_{Fn}, move toward the conduction band in the depletion zone and consequently reduces the barrier width. In the device shown in Figure 20.5, the contact facilitates the exchange of *minority carrier* (holes) over that of majority carrier (electrons), in contrast to the majority carrier exchange, seen in Figure 20.1. Minority carrier injection predominates because of higher hole transfer velocity at the interface $v_p \gg v_n$, or the barrier for injection of holes $\Phi_{B,p}$ is small (Green and Shewchun, 1973). In Figure 20.5b, the dominant current flows by hole transfer from the metal to the valence band. The injected holes travel toward the interior of the semiconductor and eventually recombine in the quasi-neutral region. This SB diode is a *minority carrier device.*

Another example is shown in Figure 20.6. In these two figures, we observe that the Fermi levels across the

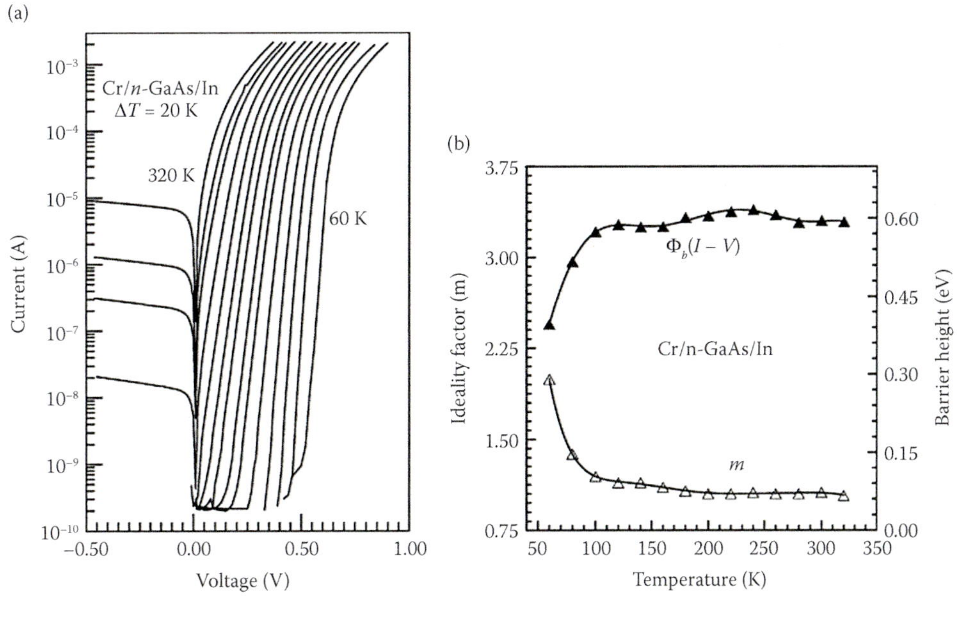

FIGURE 20.4 (a) Experimental forward bias current–voltage characteristics of Cr/n-GaAs/In Schottky contact at various temperatures. (b) Temperature dependence of the ideality factor (*the open triangles*) and barrier height (*filled triangles*). (Reproduced with permission from Korkut, H. et al. *Microelectronic Engineering* 2009, 86, 111–116.)

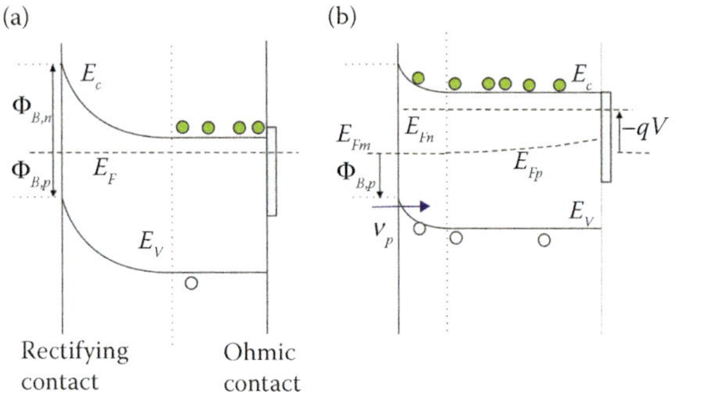

FIGURE 20.5 Model for an SB with a metal contact at (a) equilibrium and (b) at forward bias. The injection of minorities (holes) predominates at the contact and the diode is a minority carrier device.

space-charge region (SCR) are flat. This is due to the common assumption that neglects recombination in the SCR, which will be discussed in Chapter 26.

In general, in the SB diode, we must take into account two effects due to the modification of surface concentrations by the forward voltage:

1 Increased flux of majorities to the metal
2 Decreased injection of minorities to the metal that results in an increased minority flux from the metal to the semiconductor

The dominant condition, according to the interfacial kinetics, will determine the type of diode, either majority or minority carrier type.

We return to Figures 5.11 and 5.12, and we observe that the devices in these figures are minority carrier diodes operating in the same mechanism as that of Figure 20.5. One obvious difference is that Figure 5.12 shows minority electrons device and Figure 20.5 a minority holes device. Another difference is the minority carrier distribution inside the diode at forward bias. In Figure 5.12b, the Fermi level of minorities remains homogeneous across the whole bulk semiconductor. Meanwhile, in Figures 20.5b and 20.6, the Fermi level of minorities progressively decays to the equilibrium value when the injected holes travel further away from the selective contact for holes. These differences are controlled by the minority carrier diffusion length, as shown in Section 26.2. Another common characteristic is illustrated in Figure 20.6: the Fermi level of holes remains in equilibrium with a metal and across a dielectric layer that forms an interfacial dipole at the contact. In the bulk, there is full recombination of injected minorities, before they reach the back contact. This is a diode with a short diffusion length, which will be analyzed in Chapter 26.

20.4 FUNDAMENTAL PROPERTIES OF A SOLAR CELL

The solar cell is a device capable of absorbing photons of the sunlight, as a result of which free electron-hole pairs are produced. These pairs can be extracted and made to circulate through an external load that supports a

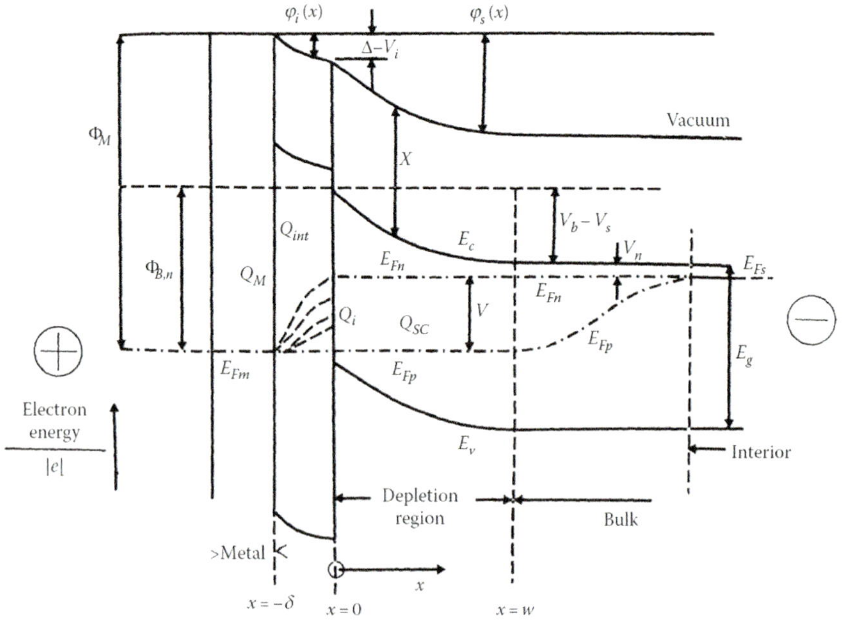

FIGURE 20.6 Model for an SB with a contact to the minority carrier (holes) at forward bias. There is a barrier formed by a thin metal oxide at the metal-semiconductor contact. This is a model for an SB solar cell. (Reproduced with permission from Landsberg, P. T.; Klimpke, C. *Proceedings Royal Society (London) A* 1977, 354, 101–118.)

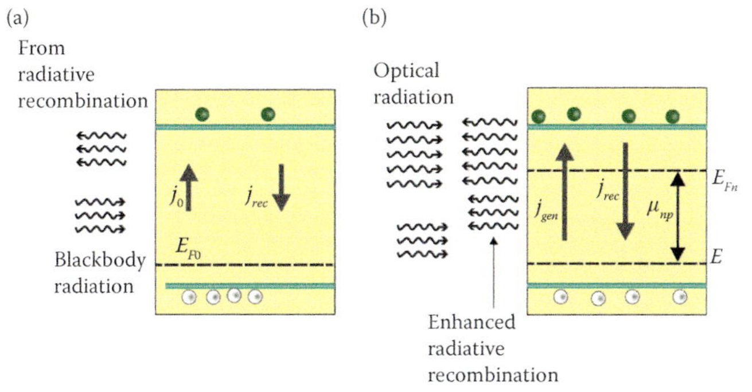

FIGURE 20.7 (a) Scheme of p-type semiconductor material in dark equilibrium, showing the balance of incoming and outgoing photons absorbed or produced in the material, and the equilibrium Fermi level E_{F0}. The full dots at the conduction band energy level and the empty dots at the valence band energy level represent electrons and holes, respectively. The generation by incoming thermal radiation produces a generation flux j_0/q and radiative recombination internal flux j_{rec}/q. (b) Under excess optical radiation, the rate of generation increases to j_{gen}/q, the minority carrier density increases producing the larger internal recombination flux and the splitting of the Fermi levels of electrons and holes, expressed as an internal chemical potential μ_{np}. The increased radiative recombination produces excess radiation so that external fluxes of photons can be equilibrated. Eventually, the radiation field spectral flux becomes that of photons at nonzero chemical potential with $\eta_{ph} = \mu_{np}$, as will be shown in Section 22.4.

positive voltage, thereby resulting in net delivery of electrical power to the outer circuit just from the consumption of photons and with no other change in the device (Araújo and Martí, 1994).

This process of conversion of light to electricity is called the *photovoltaic effect* and it will be progressively explained in the following chapters. The interaction of radiation with the electron-hole system in the semiconductor light absorber results in the transference of energy from the incoming photons to the electron-hole gas, causing the modification of the Fermi levels that can be used to produce useful work on an external system, and vice versa: the electrons and holes system may lose energy by creating photons that are emitted outward.

These processes are outlined in Figure 20.7, where a model semiconductor without contacts is indicated. In Chapter 5, we described the properties of an electron-hole gas in a semiconductor, consisting of the set of electrons in the extended states of the conduction band and set of holes in the valence band that thermalize to the temperature of the absorber, T_A. In equilibrium, the two sets of carriers share a common electrochemical potential that determines the equilibrium Fermi level E_{F0}, as indicated in Figure 20.7a. When the semiconductor film is illuminated in excess of the ambient radiation, it is displaced to a different steady state of recombination-generation equilibrium with excess populations of generated carriers (Equation 24.5) that cause a separation of the Fermi levels, as shown in Figure 20.7b. The difference of electrochemical potentials of electrons and holes μ_{np} is stated in Equation 2.46:

$$\mu_{np} = \eta_n + \eta_p = \mu_n + \mu_p = E_{Fn} - E_{Fp} \qquad (20.6)$$

When the free energy of the photons has been captured in the form of a difference of electrochemical potentials of electronic carriers, additional steps are still necessary for the conversion into useful electrical energy. Electronic devices for energy and light production require a transformation of the electronic carrier concentration to a voltage. This functionality is realized by the diode structure, which imparts directionality to the carrier flow with respect to the external contacts. When trying to exit the absorber, electrons move preferentially in one direction and are impeded to move in the opposite one, as in a valve, while holes experience a similar effect in the contrary direction.

We have discussed so far two classes of diodes. In the majority carrier device, the diode operation is controlled by the transference rate across the rectifying contact. This device is called *unipolar*, and it is regulated by injection. However, the minority carrier device is called *bipolar*. It is controlled by the flow of minority carriers whose distribution is established in the active layer by a competition between transport and recombination. We will use the shorthand *recombination diode* for this type of diode. The minority carrier device requires good selective contacts (as discussed below), especially if the device is thin and the minorities arrive to the selective contact of the majority carrier.

In the analysis of realistic solar cell technologies, there are a wide variety of detailed considerations concerning the material properties and operation mechanisms. At this stage, we are interested in a fundamental model that allows us to establish the dominant mechanisms that set the limits to efficiency and a benchmark for evaluation of actual solar cells. The basic structure of a solar cell that we focus our attention on in the following chapters consists of a semiconductor diode composed of a good light absorber material with selective contacts to electrons and holes, forming a recombination diode.

In order to discuss the properties of this model, we show in Figure 20.8 the formation of a recombination diode from the separate materials by application of contacts with different work function to the semiconductor light absorber layer. The process of equilibration of the separate materials to a common Fermi level has been

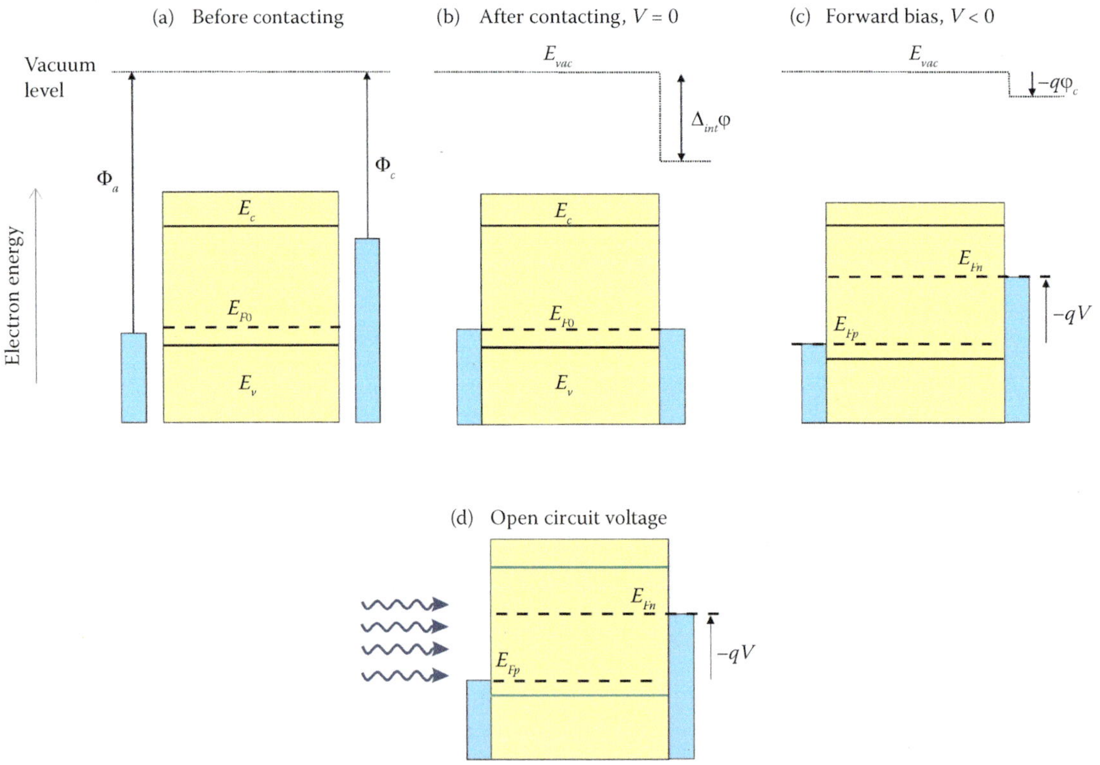

FIGURE 20.8 Schematic representation energy diagram of a p-type layer with conduction band for electrons E_c and valence band for holes E_v and two contacts: an electron selective contact (ESC) with work function Φ_c and hole selective contact (HSC) with work function Φ_a. The contacts are considered as metals in which the carrier energy level is at the Fermi level. The voltage produces a modification of the Fermi level of minority carriers (electrons) with respect to E_c at the right contact, while holes remain at equilibrium at the left contact. (a) Energies of the separate materials. (b, c) Different situations of bias voltage V indicating the Fermi levels of electrons (E_{Fn}) and holes (E_{Fp}). In (d), the semiconductor is illuminated and minority carrier generation raises the Fermi level of electrons consequently producing a photovoltage.

shown in Figure 10.9, and it was noted that the main question is the resulting distribution of the original difference of work functions. In Chapter 10, we noted two fundamentally different possibilities. One is the distribution of the band bending in the absorber layer, like in the SB model of Figure 20.6. In the other possibility, the modification of the vacuum level (VL) when the materials achieve equilibrium is absorbed by the minority carrier selective contact as a dipole layer; this was described in Section 4.6 and in Figure 10.9b. Any potential drop (change of VL) thereafter due to nonzero voltage appears just at this interface, as in the case of forward voltage (Figure 20.8c). This is just one possibility among many types of device behavior, as discussed in Chapter 26. However, Figure 20.8 is probably the simplest diode model in that the bands remain flat in the absorber and transport occurs entirely by diffusion. We also assume low-device thickness, high mobilities, and neglect surface recombination. These properties imply that the Fermi levels are flat as shown in Figure 20.8. The internal chemical potential μ_{np} is extracted and a measurable voltage V occurs, of value given by the expression (Bisquert et al., 2004; Luque et al., 2012)

$$qV = E_{Fn} - E_{Fp} \tag{20.7}$$

Instead of providing a voltage, the outer contacts of the cell can be adapted to produce the synthesis of chemical substances that store the energy in the form of chemical bonds, and then it is called a solar fuel converter; this will be discussed further in Section 27.3.

20.5 PHYSICAL PROPERTIES OF SELECTIVE CONTACTS IN SOLAR CELLS

To form a solar cell, it is necessary to obtain selective contacts with electrons and holes at each side of the semiconductor absorber layer (Bisquert et al., 2004). The use of contacts for facile injection of one specific carrier was discussed in Section 10.2. When the contact barrier to one specific carrier is reduced, it becomes progressively more *ohmic* and hence better suited for injection,

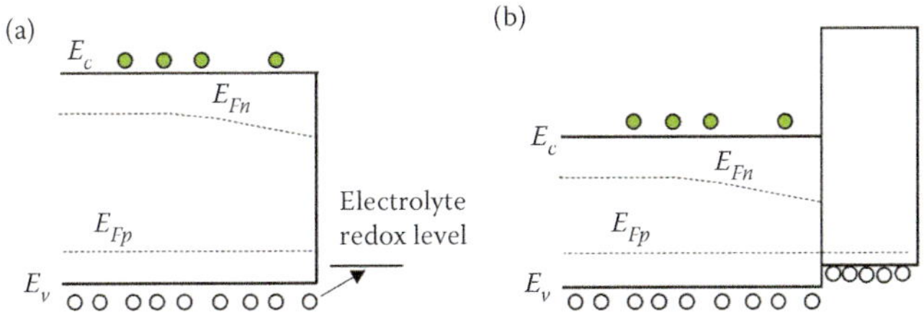

FIGURE 20.9 Mechanisms of selective contacts. (a) Preferential kinetic exchange at one semiconductor energy level at the semiconductor-electrolyte contact. (b) Equilibration of Fermi levels for holes at the semiconductor–hole transport layer contact.

as shown in Figure 10.4 (Abkowitz et al., 1998). Now we introduce the idea of a selective contact, which is more specific and demanding than the injection ohmic contact, especially in the case of photovoltaic devices. An ideally selective contact is one that is transparent to one carrier type and blocks completely the other. The selective contact must tend to ohmic for one carrier, or at least have a negligible impedance, and reject the opposite charge carrier type. The ideal selective contact may operate reversibly for injection and extraction, which is a central property of the contact in solar cells type devices.

In a solar cell, the selective contacts perform two important functions. They introduce the asymmetry of extraction and injection of electrons and holes at the two contacts which is required to obtain a diode structure. Thereafter, the device is directional by this construction. Each side of the device equilibrates to a separate carrier, as outlined in Figure 20.8, see also Figure 26.6. The electrons can be extracted from one contact and retrieved at the other one at a lower electrochemical potential, having delivered the difference of free energy to the external load. In addition, the selective contacts have the function of taking the carriers from a state in which the Fermi level in the absorber is displaced from thermal dark equilibrium, to a material (normally a metal) in which the carrier is at the equilibrium Fermi level (Honsberg et al., 2002). The light only creates carriers in a nonequilibrium state, and the selective contact converts such nonequilibrium Fermi levels into stable Fermi levels in the wires, and their difference is properly a voltage, as was discussed in Chapter 3.

There is a variety of ways to realize the asymmetry of the contacts usually required in devices for energy production, storage, or lighting. Selective contacts may be formed by materials that have the required kinetic properties to extract only one kind of carrier at the interface with the absorber. In electrochemical systems, the contact selectivity can be obtained by the kinetic asymmetry (for electrons and holes) of the interfacial reaction at the semiconductor-electrolyte junction, previously discussed in Section 6.8 and also shown in Figure 20.2. A kinetic preference for extraction of holes at the semiconductor electrochemical contact, caused by the match of energy levels at both sides of the interface, is shown in Figure 20.9a. An extensive analysis of electrochemical diodes will be presented in Chapter 21. In a solution that contains different redox species formed by photochemical reaction, a contact that allows one reaction and blocks another species allows us to form a photogalvanic solar cell (Albery, 1982). The selectivity to chemical reaction by biological molecules is used to build fuel cells that provide energy from a glucose-rich environment without the need to separate anode and cathode compartments (Heller, 2004).

Quantum dot films can be regulated for selective charge exchange at the contact as shown in Figure 10.5, which indicates the preferential injection of electrons or holes by contact engineering. Hodes et al. (1992) first observed that a nanostructured solar cell, formed by an array of quantum dots in contact with electrolyte, operates by kinetic selectivity. Figure 20.10 shows a quantum dot film immersed in solution with only one oxide or organic contact. The sign of photovoltage can be inverted by the type of contact, which indicates that the contact layer may adapt itself either to the extraction of electrons or holes, as shown in the scheme in Figure 20.10c. Solar cells composed of a layer of quantum dots have provided promising conversion efficiency (Luther et al., 2008; Barkhouse et al., 2011; Ning et al., 2012).

In inorganic semiconductors, the selectivity is often formed by properties of the junction, involving SBs and space-charge regions, as discussed earlier in this chapter. A very important method to address one specific carrier selectively, by regulating the injection barrier at the interface, employs a semiconductor with adequate energy level for alignment to that of the absorber layer. In addition, the contact material has preferential doping of one type, and it is called a transport layer. This method to build a selective contact by match of the energy levels is indicated in the scheme of Figure 20.9b, and it has been applied in the models in Figure 10.9 and Figure 20.8.

In order to impart selectivity to electrons and holes, the transport layers at the two sides of the absorber need to have a substantial difference of work function of the order of the absorber bandgap. From the difference of work functions occurs another function of the contacts, which is to create an initial built-in potential in the device. This topic was described in Section 10.3 for the metal-insulator-metal (MIM) model (see Figure 10.10), consisting of an insulator layer contacted by two metals of low and high-work function. Based on this

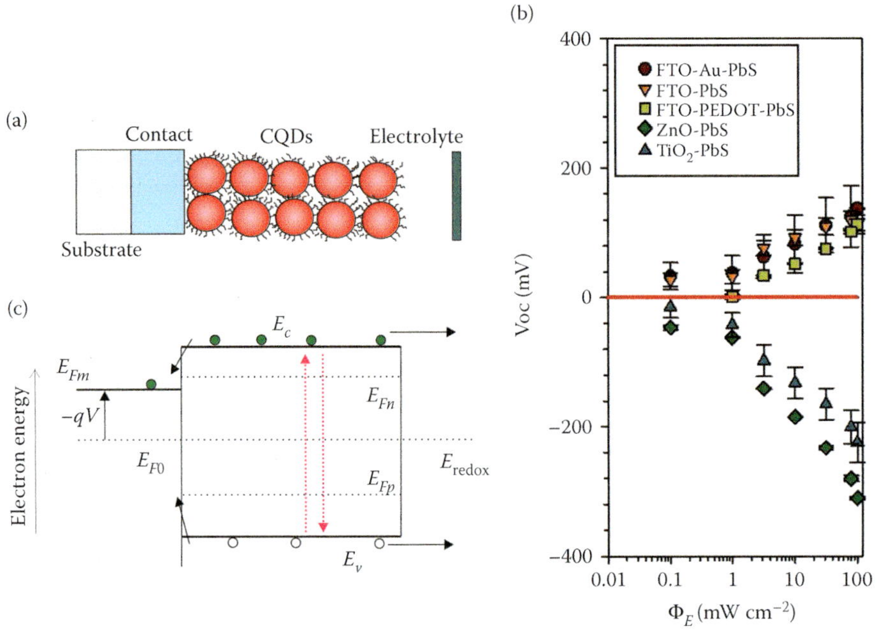

FIGURE 20.10 (a) Scheme of a QD film immersed in electrolyte and measured versus a counterelectrode. (b) Evolution of the open-circuit voltage V_{oc} with illumination intensity obtained experimentally for different metal or semiconductor-QD contacts. (c) Model for an SB with a contact to the minority carrier (holes) at forward bias. There is a barrier formed by a thin metal oxide at the metal-semiconductor contact. This is a model for an SB solar cell. (Adapted with permission from Mora-Sero, I. et al. *Nature Communications* 2013, 2, 2272.)

previously discussed fundamental device, we can use as in Figure 20.11 an absorber "intrinsic" layer sandwiched between an n-doped layer that functions as the electron selective contact of the solar cell and a p-doped layer as the hole selective contact. Here, there are two main properties of the contacts. First, the difference of their work functions produces the slant of the bands at zero bias as in the MIM device. The second property is the conductivity to only one carrier that supports the selectivity function. This structure is usually termed a p–i–n solar cell. In contrast to Figure 20.8, the presence of a majority carrier, or charge accumulation in the absorber layer leading to band bending, is ignored. The external voltage affects the inclination of the straight bands (see Figure 10.10), so that charge collection is strongly influenced by an electrical field. In this class of models, the mechanism of selectivity is not associated with the properties of the contacts, rather the contacts have the function to produce the starting slant of the bands that impart directionality to carrier flow. The current-voltage curve is generated by the modification of charge collection induced by the electrical field, while recombination is neglected (Schilinsky et al., 2004). The open-circuit condition under illumination is achieved when the bands are flat (horizontal) and charge extraction ceases, as shown in Figure 20.11b. This approach was used in early analysis of polymer solar cells and also in ferroelectric solar cells, formed by materials that are very poor conductors (Lopez-Varo et al., 2016). In general, this type of model treats the device as a capacitor, that is it consists of an insulator that extracts photogenerated charges as in early models of photoconductors (Goodman and Rose, 1971; Sokel and Hughes, 1982). Although this solar cell model is conceptually useful, the application to realistic high-quality photovoltaic devices is rather limited.

The main method to form a selective contact for holes is to introduce a hole transport layer that readily conducts hole carriers and blocks electrons. Devices based on organic materials exploit specifically doped layers with adequate thermodynamic and kinetic properties of charge transfer at the interface. Often, an electron layer for good injection at the cathode cannot function as effective hole blocking layer to confine the holes in the absorber layer region, hence different layers are used to perform the separate functions of ohmic carrier extraction of one type and blocking the complementary carrier. These *blocking layers* suppress injection or extraction of the undesired carrier at each contact and in addition, they conduct the carrier that is extracted with low impedance to avoid losses by the transport resistance. This structure has been amply developed in organic light emitted diodes (OLEDS), and it will be described in more detail in Section 21.4. The selective contact in solid solar cells should provide low rate of recombination resulting from

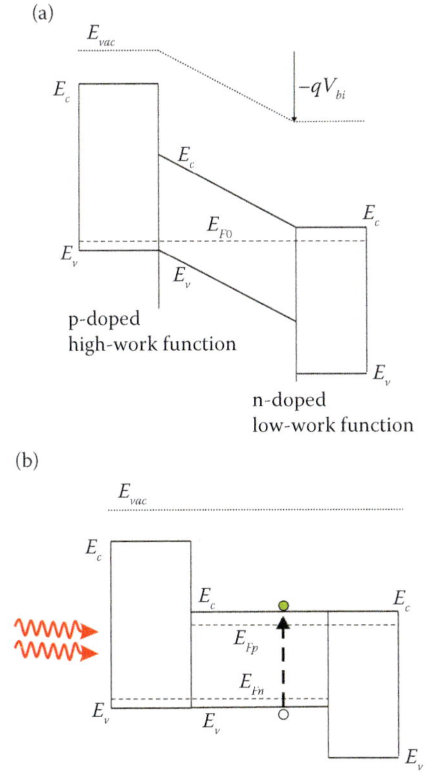

FIGURE 20.11 The model of a p–i–n solar cell, showing the slanted bands when the system is in equilibrium (a) and the flat bands when charge carrier extraction compensates the internal electrical field, in open-circuit condition (b).

collection of the undesired carrier and annihilation of the desired carrier. Buffer layers often improve the operation of inorganic solar cells by passivation of surface states and reducing recombination of carriers (Green et al., 2003). In silicon solar cell technology, the treatment of the majority carrier contact is very important for obtaining high efficiency because the back surface of the crystalline material produces a large recombination rate. To prevent the carrier loss, a back surface field is formed, which reflects the majority carrier or a SiO_2 passivation layer is introduced. These treatments facilitate large diffusion lengths that yield significant gains of sunlight conversion efficiency (Fossum, 1977; Bai et al., 1998).

GENERAL REFERENCES

Rectification at semiconductor/electrolyte junction: Memming (2001).

The Schottky diode: Sharma (1984) and Rhoderick and Williams (1988).

Fundamental model of the solar cell with selective contacts: Green (2002), Bisquert et al. (2004), and Würfel (2009).

Transition metal oxide contacts: Meyer et al. (2007), Toshinori et al. (2007), Meyer and Kahn (2011), Greiner et al. (2012), and Jasieniak et al. (2012).

REFERENCES

Abkowitz, M.; Facci, J. S.; Rehm, J. Direct evaluation of contact injection efficiency into small molecule based transport layers: Influence of extrinsic factors. *Journal of Applied Physics* 1998, *83*, 2670–2676.

Albery, W. J. Development of photogalvanic cells for solar energy conservation. *Accounts of Chemical Research* 1982, *15*, 142–148.

Araújo, G. L.; Martí, A. Absolute limiting efficiencies for photovoltaic energy conversion. *Solar Energy Materials and Solar Cells* 1994, *33*, 213–240.

Bai, Y.; Phillips, J. E.; Barnett, A. M. The roles of electric fields and illumination levels in passivating the surface of silicon solar cells. *IEEE Transactions on Electron Devices* 1998, *45*, 1784–1790.

Barkhouse, D. A. R.; Debnath, R.; Kramer, I. J.; Zhitomirsky, D.; Pattantyus-Abraham, A. G.; Levina, L.; Etgar, L.; et al. Depleted bulk heterojunction colloidal quantum dot photovoltaics. *Advanced Materials* 2011, *23*, 3134–3138.

Bisquert, J.; Cahen, D.; Rühle, S.; Hodes, G.; Zaban, A. Physical chemical principles of photovoltaic conversion with nanoparticulate, mesoporous dye-sensitized solar cells. *The Journal of Physical Chemistry B* 2004, *108*, 8106–8118.

Crowell, C. R.; Sze, S. M. Current transport in metal semiconductor barriers. *Solid-State Electronics* 1966, *9*, 1035–1048.

Fossum, J. G. Physical operation of back-surface-field silicon solar cells. *IEEE Transactions on Electron Devices* 1977, *24*, 322–325.

Gerischer, H. The impact of semiconductors on the concepts of electrochemistry. *Electrochimica Acta* 1990, *35*, 1677–1699.

Goodman, A. M.; Rose, A. Double extraction of uniformly generated electron-hole pairs from insulators with noninjecting contacts. *Journal of Applied Physics* 1971, *52*, 2823–2830.

Green, M. A. Photovoltaic principles. *Physica E* 2002, *14*, 11–17.

Green, M. A.; Shewchun, J. Minority carrier effects upon the small signal and steady state properties of Schottky diodes. *Solid-State Electronics* 1973, *16*, 1141–1150.

Green, M. A.; Zhao, J.; Wang, A.; Trupke, T. High-efficiency silicon light emitting diodes. *Physica E: Low-dimensional Systems and Nanostructures* 2003, *16*, 351–358.

Greiner, M. T.; Chai, L.; Helander, M. G.; Tang, W.-M.; Lu, Z.-H. Metal/metal-oxide interfaces: How metal contacts affect the work function and band structure of MoO_3. *Advanced Functional Materials* 2012, *23*, 215–226.

Heller, A. Miniature biofuel cells. *Physical Chemistry Chemical Physics* 2004, *6*, 209–216.

Hodes, G.; Howell, I. D. J.; Peter, L. M. Nanocristallyne photoelectrochemical cells. A new concept in photovoltaic cells. *Journal of the Electrochemical Society* 1992, *139*, 3136–3140.

Honsberg, C. B.; Bremmer, S. P.; Corkish, R. Design trade-offs and rules for multiple energy levels solar cells. *Physica E* 2002, *14*, 136.

Jasieniak, J. J.; Seifter, J.; Jo, J.; Mates, T.; Heeger, A. J. A solution-processed MoOx anode interlayer for use within organic photovoltaic devices. *Advanced Functional Materials* 2012, *22*, 2594–2605.

Korkut, H.; Yildirim, N.; Turut, A. Temperature-dependent current–voltage characteristics of Cr/n-GaAs Schottky diodes. *Microelectronic Engineering* 2009, *86*, 111–116.

Landsberg, P. T.; Klimpke, C. Theory of the Schottky barrier solar cell. *Proceedings Royal Society (London) A* 1977, *354*, 101–118.

Lopez-Varo, P.; Bertoluzzi, L.; Bisquert, J.; Alexe, M.; Coll, M.; Huang, J.; Jimenez-Tejada, J. A. Physical aspects of ferroelectric semiconductors for photovoltaic solar energy conversion. *Physics Reports* 2016, *653*, 1–40.

Luque, A.; Marti, A.; Stanley, C. Understanding intermediate-band solar cells. *Nature Photonics* 2012, *6*, 146–152.

Luther, J. M.; Law, M.; Beard, M. C.; Song, Q.; Reese, M. O.; Ellingson, R. J.; Nozik, A. J. Schottky solar cells based on colloidal nanocrystal films. *Nano Letters* 2008, *8*, 3488–3492.

Memming, R. *Semiconductor Electrochemistry*; Wiley-VCH: Weinheim, 2001.

Meyer, J.; Hamwi, S.; Bulow, T.; Johannes, H. H.; Riedl, T.; Kowalsky, W. Highly efficient simplified organic light emitting diodes. *Applied Physics Letters* 2007, *91*, 113506.

Meyer, J.; Kahn, A. Electronic structure of molybdenum-oxide films and associated charge injection mechanisms in organic devices. *Journal of Photonics for Energy* 2011, *1*, 011109.

Mora-Sero, I.; Bertoluzzi, L.; Gonzalez-Pedro, V.; Gimenez, S.; Fabregat-Santiago, F.; Kemp, K. W.; Sargent, E. H.; et al. Selective contacts drive charge extraction in quantum dot solids via asymmetry in carrier transfer kinetics. *Nature Communications* 2013, *2*, 2272.

Ning, Z.; Ren, Y.; Hoogland, S.; Voznyy, O.; Levina, L.; Stadler, P.; Lan, X.; et al. All-inorganic colloidal quantum dot photovoltaics employing solution-phase halide passivation. *Advanced Materials* 2012, *24*, 6295–6299.

Price, M. J.; Foley, J. M.; May, R. A.; Maldonado, S. Comparison of majority carrier charge transfer velocities at Si/polymer and Si/metal photovoltaic heterojunctions. *Applied Physics Letters* 2010, *97*, 083503.

Rhoderick, E. H.; Williams, R. H. *Metal-Semiconductor Contacts*, 2nd ed.; Clarendon Press: Oxford, 1988.

Schilinsky, P.; Waldauf, C.; Hauch, J.; Brabec, C. J. Simulation of light intensity dependent current characteristics of polymer solar cells. *Journal of Applied Physics* 2004, *95*, 2816–2819.

Shannon, J. M. A majority-carrier camel diode. *Applied Physics Letters* 1979, *35*, 63–65.

Sharma, B. L. *Metal-Semiconductor Schottky Barrier Junctions and Their Applications*; Plenum: New York, 1984.

Sokel, R.; Hughes, R. C. Numerical analysis of transient photoconductivity in insulators. *Journal of Applied Physics* 1982, *53*, 7414–7424.

Sze, S. M. *Physics of Semiconductor Devizes*, 2nd ed.; John Wiley and Sons: New York, 1981.

Toshinori, M.; Yoshiki, K.; Hideyuki, M. Formation of ohmic hole injection by inserting an ultrathin layer of molybdenum trioxide between indium tin oxide and organic hole-transporting layers. *Applied Physics Letters* 2007, *91*, 253504.

Würfel, P. *Physics of Solar Cells. From Principles to New Concepts*, 2nd ed.; Wiley: Weinheim, 2009.

21 Recombination Current in the Semiconductor Diode

We continue with the study of the fundamental model of the solar cell, based on the assumption of radiative recombination that produces the maximal theoretical efficiency. We analyze the behavior of the diode in forward bias, introducing the parameters that determine the recombination rates. Then, we observe the main features of current density-voltage curves. We study specific types of diodes like the LEDs, the molecular diodes represented by dye-sensitized solar cells (DSCs).

21.1 DARK EQUILIBRIUM OF ABSORPTION AND EMISSION OF RADIATION

In this chapter, we analyze the behavior of the fundamental solar cell model that was introduced in Chapter 20. Here we investigate the features of the model in dark conditions, as shown in the upper row of Figure 21.1. In the subsequent chapters, the operation of the solar cell under illumination, shown in the bottom row of Figure 21.1, will be discussed step by step.

To analyze the interplay between light absorption, carrier densities, and photocurrents in a solar cell, it is important to first establish some basic properties that are settled in the dark equilibrium by detailed balance arguments that dictate recombination rates. Detailed balance of light absorption and emission is a rather fundamental physical restriction that must be realized by recombination processes in a semiconductor. Radiative recombination of an excitation is a necessary process that cannot be suppressed. As a matter of fact, at the molecular and atomic level, the light absorption event is subject to microscopic reversibility. Thus, if a molecule absorbs light it must also radiate. Therefore, the fundamental radiation rate of an atom or a molecule can be determined by the detailed balance of incoming blackbody radiation and outgoing radiation as was formulated by Einstein and applied to fluorescence by Kennard (1926). The ensuing relationships connect the linear optical spectra of luminescence and absorbance for homogeneous luminescent materials under thermal equilibrium (Strickler and Berg, 1962).

The traffic of electronic carriers in the operation of a solar cell is governed by three elementary processes: light absorption (leading to carrier generation), carrier recombination, and charge extraction, as shown in the bottom row of Figure 21.1. In the absence of any source of optical radiation, that is in the "dark" (the upper row of Figure 21.1), the semiconductor material is bathed in the Lambertian blackbody radiation coming from all directions that generates a flux of generated electrons termed j_0^{th}/q, where j_0^{th} is a current density. The voltage in the solar cell device can be controlled independently of the incident light, either by connecting the solar cell to a load resistor, or using a potentiostat. In Figure 21.1b, we show a situation in which the solar cell diode in the dark is forward biased so that carriers are injected to the semiconductor. In *steady state*, the net current extracted from the diode is

$$j_{inj} = j_0^{th} - j_{rec} \tag{21.1}$$

The recombination current density j_{rec} is defined as the integration of recombination rate across the film thickness

$$j_{rec} = \int_0^d U_{rec}(n, p, x)\, dx \tag{21.2}$$

Note that the fluxes j_0^{th}/q and j_{rec}/q correspond to electron transitions between the semiconductor quantum states, the valence band, and the conduction band, and not to spatial flux of carriers. However, by virtue of transport in extended states, these local fluxes are able to produce real output currents at the contacts of the solar cell, quantified by j_{inj}, as indicated in Equation 21.1 and Figure 21.1b. The effect of realistic finite mobilities on the transport features of solar cells will be discussed in Chapter 26.

When a semiconductor attains thermal equilibrium in the dark with the surrounding blackbody radiation, electrons are continuously promoted to the conduction band by the absorption of the radiation. The detailed balance principle requires that the amount of absorbed external radiation is emitted at the same rate by photons resulting from radiative recombination of electrons and holes. The role of radiative recombination is therefore essential for expelling the energy without

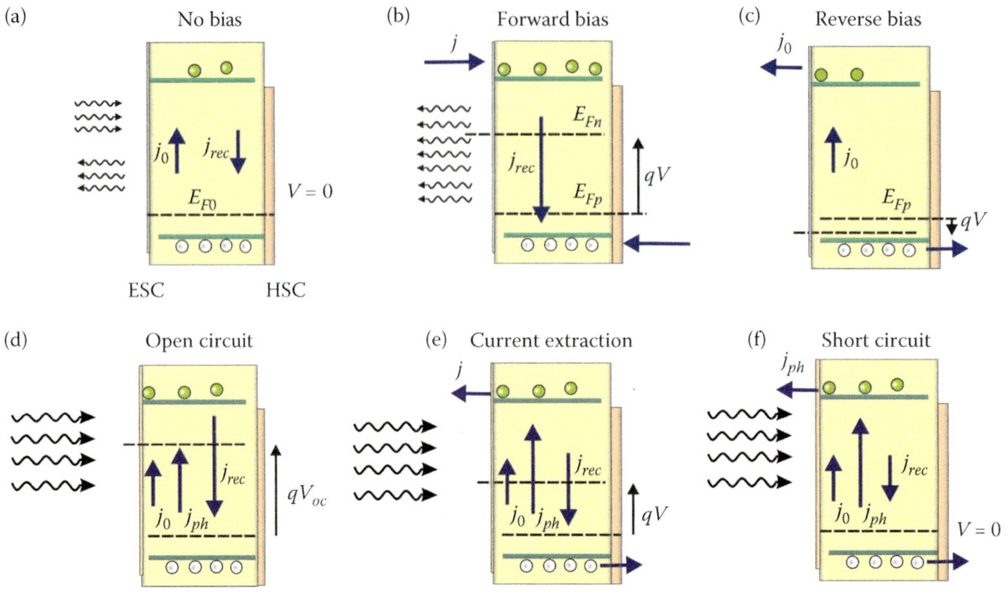

FIGURE 21.1 (a) Scheme of p-type semiconductor material with selective contacts to electrons and holes, ESC and HSC, a model of a solar cell (a–c) in dark. j_0 is the dark generation current (by incoming photons) that coincides with the diode reverse saturation current. j_{rec} is the recombination current that generates outgoing photons (neglecting photon-recycling effects). (a) In dark equilibrium, showing the balance of incoming and outgoing photons absorbed or produced in the material, and the equilibrium Fermi level E_{F0}. (b) At forward bias, the minority carrier electrons Fermi level E_{Fn} increases and current is injected at the contacts. (c) At reverse bias, the carriers are extracted determining the reverse saturation current j_0. (d–f) Under illumination, absorbed photons promote excitation of electrons from the valence band to the conduction band. (d) Open circuit: The generation current achieves equilibrium with recombination. This raises the electron Fermi level and causes the production of the photovoltage. No current is extracted from the solar cell. (e) At voltage lower than V_{oc}, the recombination rate is less than the total generation, electrons and holes are extracted in the external circuit and this constitutes a photocurrent. (f) Short circuit: The voltage is zero, all the current created by photogeneration from the illumination in excess of thermal radiation is extracted and constitutes the short-circuit current j_{ph}.

thermal losses. The fundamental solar cell model established in Section 20.4 is complemented by the assumption that all recombination is purely radiative. The local rate of recombination is given by the expression of band-to-band radiative decay, as was indicated in Equation 18.32:

$$U_{rad} = B_{rad} np \tag{21.3}$$

From the principle of detailed balance, in the situation of equilibrium at $V = 0$ shown in Figure 20.7a and Figure 21.1a, the incoming and outgoing photon fluxes must be the same, otherwise a net flux of energy would be produced, contradicting the condition of thermodynamic equilibrium. Therefore,

$$j_0^{th} = j_{rec}\left(V = 0\right) \tag{21.4}$$

In consequence, if the semiconductor does absorb light, it also has to emit light at the rate fixed by the dark equilibrium balance. In Figure 21.1c, the diode is reverse biased so that recombination is suppressed and all generated carriers are extracted; hence, $j_{inj} = j_0^{th}$. It must be noted that j_0^{th} is the same parameter as the *reverse saturation current* mentioned in the diode Equation 20.1, as it is expected that the current in the diode will be saturated to this value in reverse voltage, see Figure 5.13. We observe in Equation 21.4 that the thermal generation parameter is also a measure of the recombination rate in the solar cell. Hence, we drop the superscript that labels it as a generation quantity, and hereafter we simply write j_0. In addition, since the radiative recombination is unavoidable, we write $j_{0,rad}$ for the special case that establishes the minimum possible recombination rate in the solar cell.

As we will discuss later on, recombination is a loss process in the overall energy conversion scheme, hence radiative recombination parameter $j_{0,rad}$ forms the basis for the determination of the maximum optimized performance of a solar cell. This parameter is finally calculated in Equation 23.3 and it gives the physical limit to the efficiency of performance of a semiconductor material as an energy converter from light to electricity.

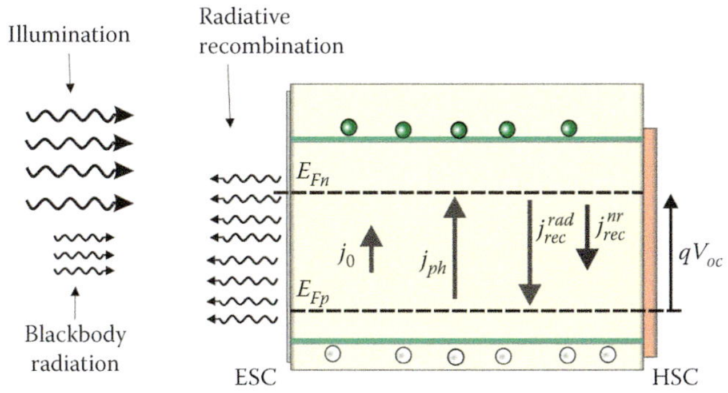

FIGURE 21.2 Scheme of a p-type semiconductor material, with selective contacts to electrons and holes, ESC and HSC. The voltage between the contacts corresponds to energy qV, associated with the splitting of the Fermi levels. The recombination rate is composed of both radiative j_{rec}^{rad} and nonradiative j_{rec}^{nr} mechanisms. The device can be operated as a solar cell or as an LED.

However, a solar cell may have additional nonradiative recombination pathways as shown in Figure 21.2, which invariably produce the loss of additional carriers and a degradation of the performance. There may exist a number of additional mechanisms, as discussed in Section 18.4, such as SRH and Auger recombination. Then, the rate of recombination can be increased with respect to Equation 21.3. These mechanisms can be studied from PL of the contactless films of absorbers as shown in Figure 18.8, and additional recombination channels are usually introduced by the addition of contacts. Actual recombination current in the dark equilibrium, j_0, is larger than the radiative parameter due to the limitation of compensation of incoming photons by two effects: the nonradiative recombination and the inefficiency of photon extraction from the device. The additional components of j_0 have been described by Cuevas (2014).

Following Equation 18.23, the relationship between ideal radiative current (the current $j_{0,rad}$ needed to cancel the absorbed photons) and the total recombination current j_0 is established by the external quantum efficiency of the diode operated as an LED:

$$j_0 = \frac{j_{0,rad}}{EQE_{LED}} \tag{21.5}$$

In summary, the central parameter determining the recombination properties in the solar cell, $j_{0,rad}$, can be obtained by optical and charge collection properties of the device. The specific methods to achieve this goal will be explained in Chapter 23. The nonradiative recombination modes included in j_0 are accounted for by the quantity EQE_{LED}.

21.2 RECOMBINATION CURRENT

The properties of recombination current play a determinant role in the characteristics of a solar cell. To quantify the current, we use the fundamental model in which the forward bias creates the separation of flat Fermi levels and induces the recombination current by injection of carriers (Figure 21.1b). Solving Equation 21.2 gives the result

$$j_{rec} = qdU_{rec} \tag{21.6}$$

We consider the bimolecular recombination model of Equation 21.3. Thus,

$$j_{rec} = qdBnp \tag{21.7}$$

For the analysis of the performance of solar cells, it is necessary to determine $j_{rec}(V)$, so the dependence of carrier density on voltage is needed. This is normally an exponential dependence as discussed below. We can express the current-voltage behavior using m, the diode quality factor explained in Equation 20.1. The recombination current is then written as

$$j_{rec} = j_0 e^{qV/mk_BT} \tag{21.8}$$

Usually, m takes values between 1 and 2. Note in Equation 21.8 that when the voltage is increased, the emission is simply enhanced by an exponential factor. According to the normal application of detailed balance described in Section 20.6.1, the parameter j_0 is established in dark equilibrium, zero voltage conditions, as indicated in Equation 21.4, and it is assumed to hold when the system is biased in one direction, that of voltage-enhanced recombination.

In general, the dependence $n(V)$ required in Equation 21.7 is a device property that can be obtained solving a full transport-recombination model, as will be described in Chapter 26. In our present simplified model with flat Fermi levels, we can establish the dependence of carrier density on voltage in two important types of situations.

1 If the generated carrier does not exceed majority carrier density, $p = p_0$, the minority carrier density n is connected to the voltage as

$$n = n_0 e^{qV/k_BT} \tag{21.9}$$

Hence, $k_{rec} = Bp_0$ and $m = 1$, with

$$j_0 = qdk_{rec}n_0 \tag{21.10}$$

2 When generated carrier number exceeds majority carrier density, we can express electron and hole densities as

$$n = n_0 e^{(E_{Fn}-E_{F0})/k_BT} = N_c e^{-(E_c-E_{Fn})/k_BT} \quad (21.11)$$

$$p = p_0 e^{-(E_{Fp}-E_{F0})/k_BT} = N_v e^{(E_v-E_{Fp})/k_BT} \quad (21.12)$$

Note the product

$$np = N_c N_v e^{-(E_g-qV)/k_BT} = n_0 p_0 e^{qV/k_BT} \quad (21.13)$$

The electroneutrality condition imposes equal concentration of oppositely charged carriers

$$n = p \quad (21.14)$$

Using Equation 21.7, we have

$$j_{rec} = qdBn^2 = j_0 e^{qV/k_BT} \quad (21.15)$$

The *saturation current* j_0 in this model has the value

$$\begin{aligned} j_0 &= qdBn_i^2 \\ &= qdBn_0 p_0 \\ &= qdBN_c N_v e^{-E_g/k_BT} \end{aligned} \quad (21.16)$$

The carrier density depends on voltage as

$$n = n_i e^{qV/2k_BT} \quad (21.17)$$

Hence, $k_{rec} = B$ and $m = 1$.

Recombination may include a number of nonidealities or complex combination of mechanisms, thus one can use an effective general recombination order β as indicated in Equation 18.42. Then, we have

$$j_{rec} = qdk_{rec}n^{\beta} \quad (21.18)$$

Thus, for model (a), $\beta = 1$ and for model (b), $\beta = 2$, as in Equation 21.15. The ideality factor may correspond to a variety of physical factors such as the voltage distribution at injection contacts, as was discussed in Chapter 20. This also will be discussed in Section 24.3 in connection with the nonideal behavior of the photovoltage.

The expression in Equation 21.16 assumes that all photons generated by recombination, Equation 21.2, are expelled out of the cell and cancel the incoming blackbody radiation. However, as discussed in Section 18.3, some of the photons may be reabsorbed and emitted again in the photon-recycling effect. Therefore, the derivation of the reverse saturation current requires a more detailed analysis that will be developed in Section 23.4. Equation 21.16 can be considered as an expression valid under certain restrictions. We will see in Chapter 23 that Equation 21.4 needs to be calculated from the incoming radiation flux.

21.3 DARK CHARACTERISTICS OF DIODE EQUATION

Let us consider the shape of the current density-voltage curve in the absence of external illumination. The recombination current has been described in Equation 21.8 as the parameter j_0 enhanced by the exponential term due to applied voltage. The current density-voltage characteristic, Equation 21.1, takes the form already given in Equation 20.1:

$$j = j_0\left(e^{qV/mk_BT} - 1\right) \quad (21.19)$$

A number of examples of current-voltage characteristics of diodes in the dark are shown in Figure 21.3. We remark upon two different regimes of behavior:

1 Forward bias is the region of positive potential in Equation 21.19. At forward bias, the injection process produces an increase of the recombination current that depends exponentially with respect to voltage. In Figure 21.3a, we note very large differences in the onset voltage of forward current due to differences in j_0. A large j_0 favors recombination and produces strong forward current at relatively small bias. Figure 21.3b shows the current density-potential characteristic in log vertical axis. The forward current branch can obtain different slopes according to the value of parameter m.
2 Reverse bias, shown in Figure 21.1c, inhibits the recombination current, so that total current tends to the constant value j_0, the reverse saturation current, which is a bulk property. In general, the reverse bias tends to extract carriers from the active layer. Due to the suppression of recombination, the carriers created by thermal generation are available for extraction. Therefore, the reverse saturation current is independent of the applied reverse voltage. However, in real diodes, in addition to the thermal generation of carriers, a leakage current occurs as well

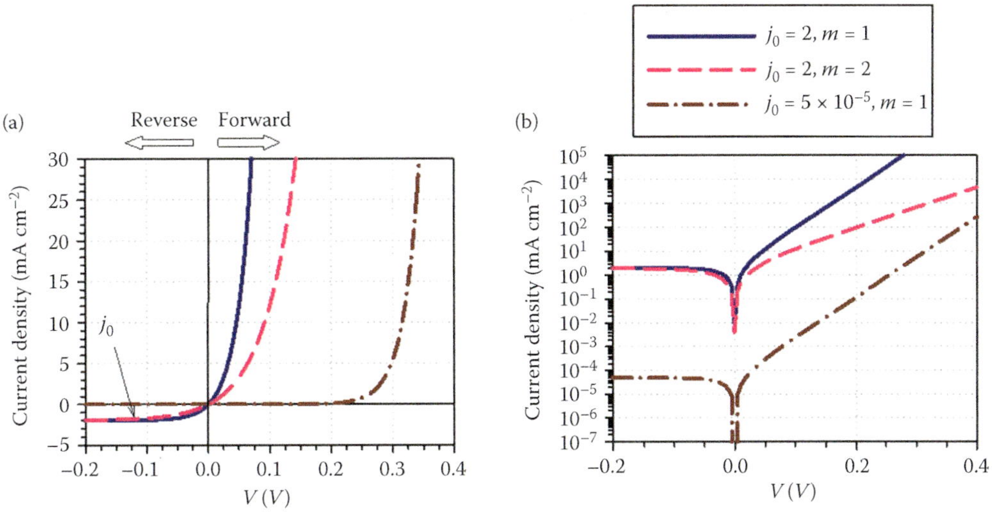

FIGURE 21.3 (a) Current–potential characteristics of a diode with dark saturation currents $j_0 = 2$ mA cm^{-2}, 5×10^{-5} mA cm^{-2}, and different diode quality factors. (b) Current-voltage characteristic in semilogarithmic plot.

in reverse bias. Ascribing the reverse dark current to the same fundamental process associated with recombination at forward bias requires a careful investigation. For instance, thin organic diodes are very sensitive to shunt paths and edge effects that may dominate the reverse current.

21.4 LIGHT-EMITTING DIODES

The LED is a device that generates light by electroluminescence. Charge carriers of both signs are injected at forward bias in a semiconductor with appropriate selective contacts. As shown in Figure 21.1b, when a diode is forward biased, an enhanced recombination current is produced that results in photon generation. The radiative recombination process converts the carriers to photons. The basic structure of an LED based on organic materials (OLED) containing a light-emitting polymer is shown in Figure 21.4, and a more realistic structure including relevant optical pathways is shown in Figure 18.1. The selectivity of contacts is realized using special transport layers at the cathode and anode as explained later in this section. These *injection* layers are very thin (10 nm) and optically inactive. They have the important role of confining the injected electrons and holes in the recombination layer where light is produced. Hence, efficient recombination occurs due to the overlapping distribution of electrons and holes (often facilitated through the formation of excitons). A very important feature of a practical LED is to avoid light trapping within the device by internal reflection, as indicated in Figure 18.1.

The first practical LED was invented by Holonyak and Bevacqua (1962) based on the inorganic semiconductor GaAsP and became commercially available in the late 1960s. Electroluminescence from organic crystals was first observed in 1963 (Pope et al., 1968), applying a large voltage of several hundred volts to 10 μm thick crystals of anthracene. In the late 1980s, OLEDs requiring low-operation voltage were first developed using a double layer structure including organic fluorescent dyes based on the small organic molecule tris(8-hydroxy quinoline)

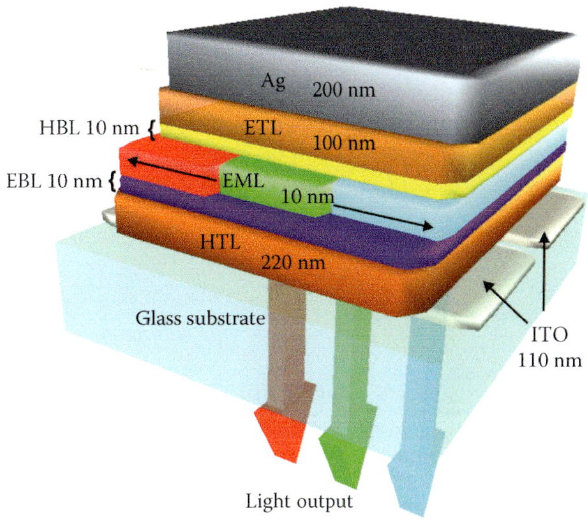

FIGURE 21.4 Basic structure of an OLED—a thin film stack consisting of indium-tin oxide (ITO) as transparent anode on a glass substrate, covered by a thermally evaporated stack of hole transport (HTL), electron blocking (EBL), emitting (EML), hole blocking (HBL), and electron transport layer (ETL), contacted by a silver (Ag) cathode. The EML consists of RGB layers prepared by co-evaporating matrix material and emitting phosphorescent dyes. (Adapted from Flämmich, M. et al. *Organic Electronics* 2011, 12, 1663–1668.)

Al (Alq_3) (Tang and VanSlyke, 1986). These results promoted extensive research on a wide variety of thin film OLED materials. Friend and coworkers were the first to develop a highly fluorescent conjugated polymer, poly(*p*-phenylene-vinylene) (PPV) as the active material in a single layer OLED (Burroughes et al., 1990), opening the way for the facile preparation of large area films by solution-processed deposition methods that has developed into a mature technology.

The materials that function as the hole-blocking layer in OLEDs should have both good electron-accepting and electron-transporting properties (Shirota, 2000). Wide bandgap electron-transporting materials with deep HOMO energies are used to block holes and wide bandgap hole-transporting materials with shallow LUMO energies block electrons (Sarasqueta et al., 2010). As an example, Figure 21.5 shows different configurations of contact layers in a diode formed with the emissive polymer F8T2 (poly(9,9-dioctylfluorene-alt-bithi ophene)) (Jin et al., 2009). The devices are shown with and without TFB interlayers, using either Al or Ca as the cathode material to achieve weak or strong electron injection, respectively. Poly(9,9-dioctylfluorene-alt-*N*-(4-butylphenyl)-diphenylamine) (TFB) is a semiconductor of relatively high hole mobility 2×10^{-3} cm^2 V^{-1} s^{-1}, a bandgap of 3 eV and low-electron affinity, as shown in Figure 21.5a, so that it effectively serves as an electron-blocking layer. Another approach is to use graded layers in which the p and n conductors are progressively mixed. This will have the effect of creating a gradient of donor–acceptor compositions that channels each carrier in the desired direction (Sullivan et al., 2004).

The use of high-work function transition metal oxides such as molybdenum (MoO_3), tungsten (WO_3), and vanadium (V_2O_5) oxide was first reported by Tokito et al. (1996) who demonstrated the use of evaporated MO_x (where M = vanadium, molybdenum, or ruthenium and x is the oxygen stoichiometry) to reduce the operating voltage of OLEDs. Many works have shown that transition metal oxide contacts such as MoO_3, WO_3, V_2O_3, and NiO form suitable hole selective contacts that can substitute PEDOT:PSS in organic solar cells and OLEDs. The photon emission spectroscopy measurements on interfaces between MoO_3 and organic hole transport materials indicate that MoO_3 has a very large work function and is strongly n-type, probably due to oxygen vacancies. As shown in Figure 21.6, the electron affinity is about 6.7 eV, therefore the electron-transport states (LUMO) of organic materials in contact with MoO_3 are unlikely to overlap with the 3 eV bandgap of MoO_3 and hence cannot

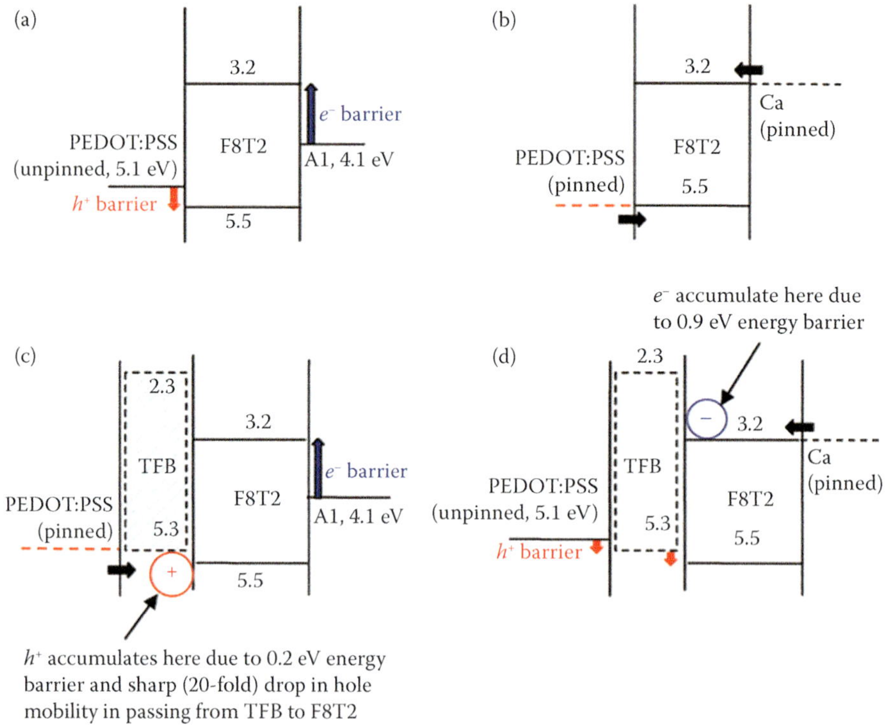

FIGURE 21.5 Energy level diagrams for four devices containing different types of contacts to F8T2. The Fermi level of the Ca cathode is pinned to the LUMO level of F8T2 in devices (b) and (d). The Fermi level of the PEDOT:PSS is pinned to the HOMO level of the adjacent organic layer in devices (b) and (c). (Reproduced with permission from Jin, R. et al. *Physical Chemistry Chemical Physics* 2009, 11, 3455–3462.)

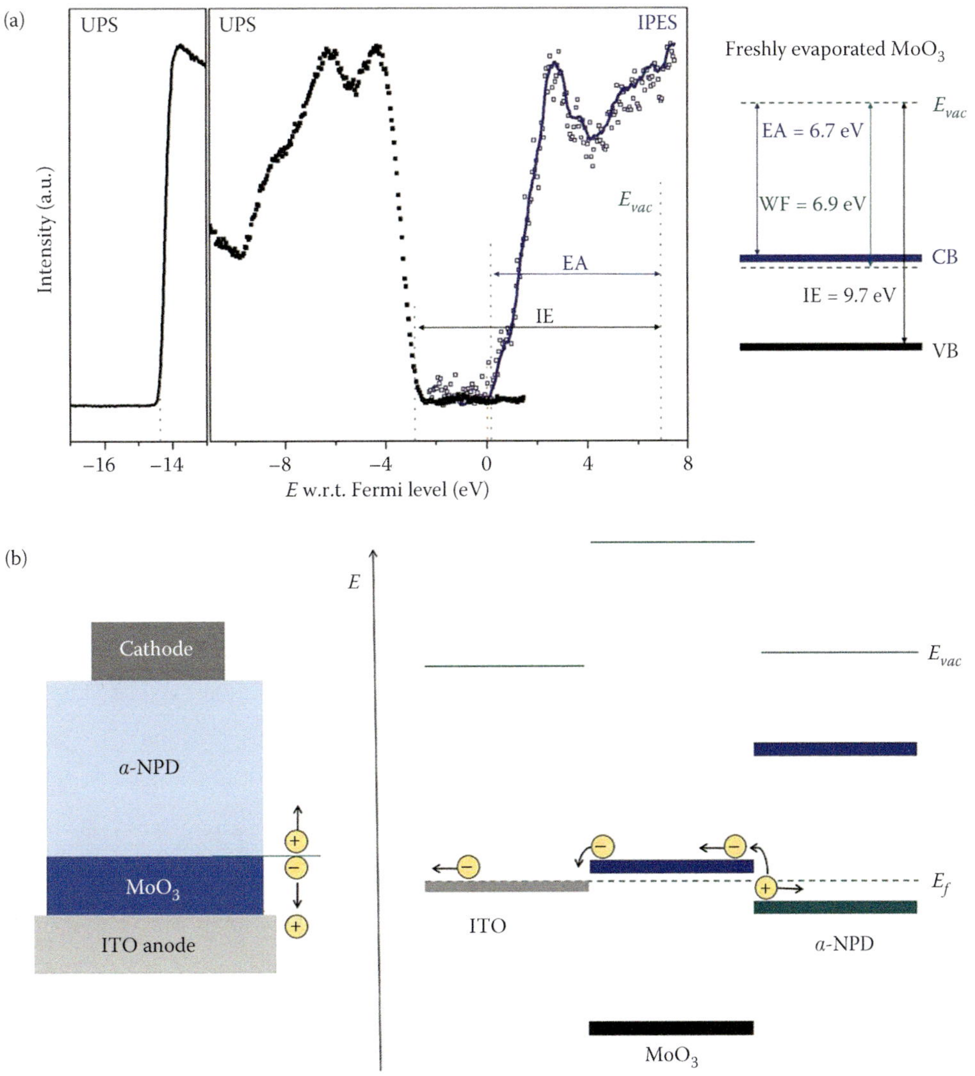

FIGURE 21.6 (a) (*Left*) Ultraviolet photoelectron spectroscopy (UPS) and inverse photoemission spectroscopy (IPES) spectra of clean MoO_3. (*Right*) Schematic energy diagram of the MoO_3 film. (b) Scheme of charge injection at a MoO_3/α-NPD interface. (Reproduced with permission from Meyer, J.; Kahn, A. *Journal of Photonics for Energy* 2011, 1, 011109.)

work as an electron-blocking layer. Despite the large difference between electron affinity of MoO_3 and ionization energy of the α-NPD, it is likely that the conduction band of MoO_3 aligns with the HOMO level of the organic layer, while the energy difference is taken by an interface dipole. The mechanism of hole extraction is indicated in Figure 21.6. Hole injection to the organic layer from the valence band of MoO_3 should not be facile, in view of the fact that the oxide valence band is separated several electron-volts (2.5–3.0 eV) below the indium-tin oxide (ITO) Fermi level. However, hole injection from ITO can occur via electron extraction through the conduction band of the oxide, which is situated only 0.6–0.7 eV above the HOMO of the organic film (Meyer and Kahn, 2011). This mechanism is observed in organic solar cells with MoO_3 hole contact layer (Yunlong and Russell, 2013).

Transport layers can readily modify not only the structure of the injection barrier but the full operation mode of the device under bias. An example of this effect is schematically presented in Figure 21.7, and it plays a double role on the device operation (Hoven et al., 2008, 2009). First, the injection of electrons is considerably improved. Second, the electrical field in the central emissive layer is nearly completely screened, as observed from the vanishing electroabsorption signal (Lane et al., 2003).

Electrons and holes injected electrically in a conjugated polymer get captured into excitons that form either the singlet or one of the three triplet states, with equal probability. Then, only 25% of the electron-hole pairs are useful for light production, so that $\chi_{op} = 0.25$ in Equation 18.26.

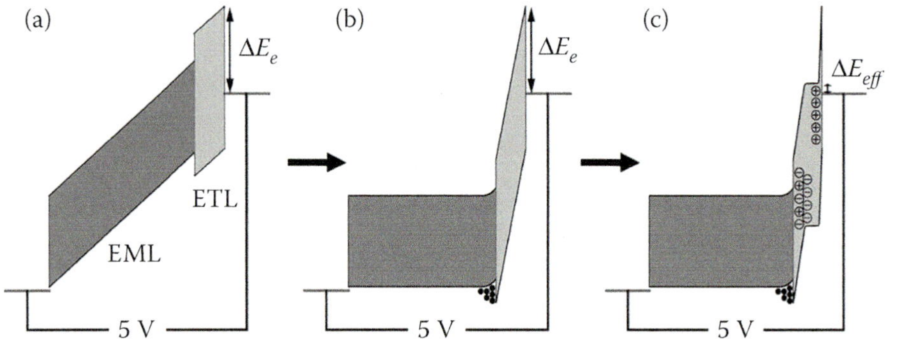

FIGURE 21.7 Schematic response of an anode/EML/ETL/cathode LED under 5 V applied bias. (a) The electric field (slope of the energy levels) is uniform across the device. (b) Injected holes (*black circles*) accumulate at the EML/ETL interface, screening the electric field to the ETL. (c) Ion charges within the conjugated polyelectrolyte ETL redistribute to screen the electric field to the two ETL interfaces. (Reproduced with permission from Hoven, C. V. et al. *Applied Physics Letters* 2009, 94, 033301–033303.)

21.5 DYE SENSITIZATION AND MOLECULAR DIODES

The fundamental structure of the recombination diode that has been discussed in Chapter 20 requires an asymmetry of contacts to a medium that realizes the recombination of injected electrons and holes. It is conceivable to shrink this structure to the level of one molecule as indicated in Figure 21.8. This model represents a two-state molecular system contacted by n- and p-type semiconductors. In the forward mode, holes are injected to the ground state and electrons to the LUMO, so that the carriers recombine by downward electron transition to the hole-occupied ground state. The LUMO is closely related to the first excited state of the molecule; hence, the diode can emit radiation forming an electroluminescent device, a model for the OLED.

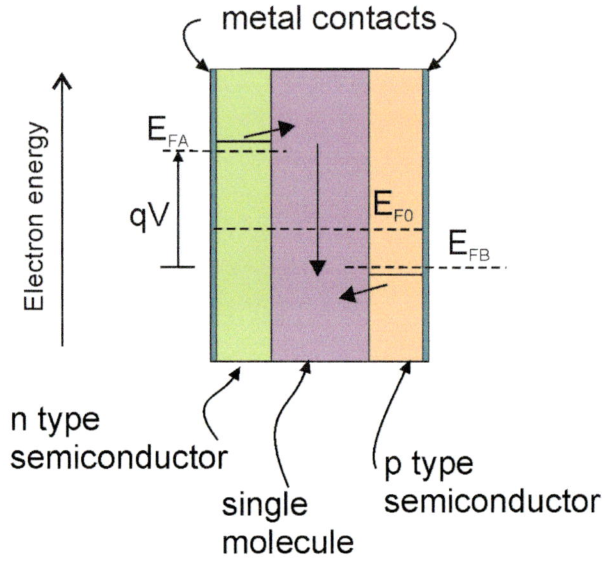

FIGURE 21.8 Basic energy structure of a single molecule diode formed by a molecule with a ground (E_1) and first excited state (E_2), n- and p-type semiconductors with the Fermi levels E_{FA} and E_{FB}, and metal contacts. The arrows indicate injection and recombination of electron and hole carriers.

The key element for using a molecule as the active material of a recombination diode is the existence of a *kinetic* preference for the electrons to be injected to the LUMO and holes to the ground state. This structure can be readily achieved using as one contact a metal or a semiconductor platform in which the organic molecule is grafted by appropriate linkers and an electrolytic contact at the other side. Electrochemical rectifiers formed by monolayers or thin molecular films grafted on electrodes have been widely studied (Abruña et al., 1981; Fujihira and Yamada, 1988). If the molecule is an efficient light-absorbing chromophore, then carriers can be separately extracted from the excited state to form a solar cell, as suggested by Gerischer et al. (1968). This structure uses an optically inactive wide bandgap metal oxide that electronically matches the excited state of a dye and also serves as electron conductor. On the other side, a very positive redox couple serves as selective contact for holes in the dye and realizes the regeneration of the oxidized chromophore, see Figure 21.9.

Photo-induced electron injection from a dye molecule anchored at the semiconductor surface constitutes a central model for heterogeneous electron transfer and has been extensively studied by experiment and simulation. Figure 21.10 shows the results of characterization of the excited states of the perylene molecule attached on the surface of TiO_2 by UPS and femtosecond time-resolved two-photon photoemission (2PPE). Due to limited extinction coefficient of organic dyes, a monolayer of dye on a flat electrode produces little effectiveness

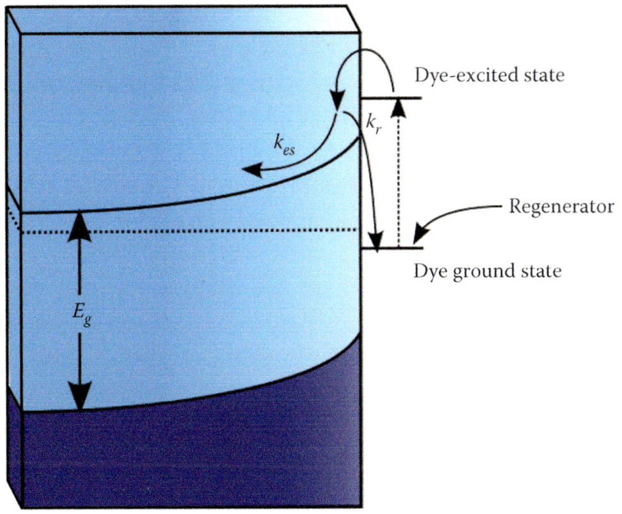

FIGURE 21.9 Schematic diagram of the competing rate processes when an electron is injected into the conduction band of an n-type semiconductor. The *dashed arrow* represents the photoexcitation process. k_{es} is the rate constant for escape from the dye into the semiconductor bulk to be collected as photocurrent and k_r is the rate constant for capture of a conduction band electron. (Reproduced with permission from Parkinson, B. A.; Spitler, M. T. *Electrochimica Acta* 1992, 37, 943–948.)

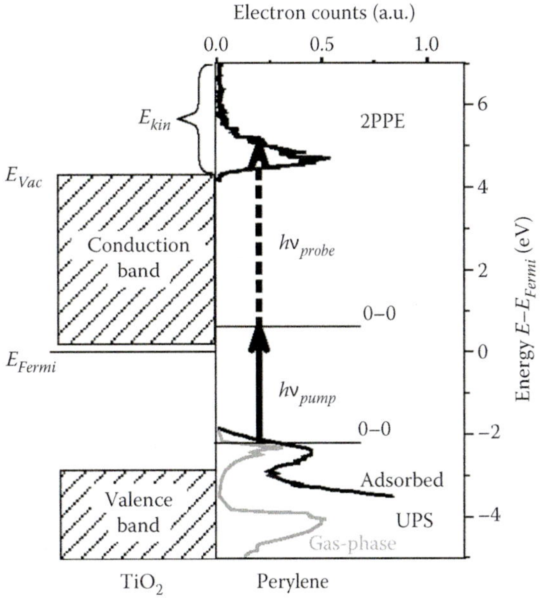

FIGURE 21.10 Alignment of the ground state and excited state of perylene with the tripod anchor/bridge group on the (110) rutile TiO_2 surface deduced from UPS and 2PPE measurements. The *gray curve* is a perylene gas-phase spectrum. (Reproduced with permission from Gundlach, L. et al. *Progress in Surface Science* 2007, 82, 355–377.)

for sunlight energy harvesting. A major breakthrough toward an efficient dye-sensitized solar cell (DSC) was realized by O'Regan and Grätzel (1991) using a nanostructured TiO_2 electrode about 10 μm thick in combination with I_3^-/I^- redox mediator. The operation of the dye with respect to selective contacts to extraction of electrons and holes forming a diode structure is indicated in Figure 8.12 and described in more detail later in Chapter 24. The nanostructured film greatly increases the internal area and facilitates charge shielding and transport. Figure 21.11 shows the kinetic timescales of the interfacial processes in a DSC using the paradigmatic dye termed N719, a polypyridyl-type ruthenium complex (*cis*-Ru(dcbpy)$_2$(NCS)$_2$) that yields overall solar-to-electric power conversion efficiencies close to 12%. This dye molecule has been designed to set an electron acceptor group close to the TiO_2 surface by two equivalent bipyridine ligands functionalized by carboxylic groups that ensure stable anchoring to the TiO_2 surface, allowing for the strong electronic coupling required for efficient excited-state charge injection. Under photoexcitation, there is a strong preference of the electron in the first excited state of the dye to be injected to TiO_2 due to an effective interaction of the dye with the titania levels, as shown by density functional theory calculations in Figure 21.12. In addition, the remaining hole stays away from the titania surface to facilitate regeneration

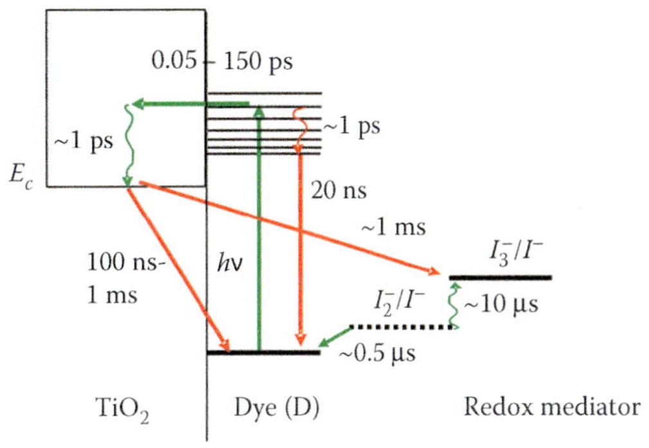

FIGURE 21.11 Kinetics of the N719 sensitized TiO_2 solar cell with I_3^-/I^- redox mediator. Typical time constants of the forward reactions (*green*) and recombination reactions (*red*) are indicated. (Reproduced with permission from Boschloo, G.; Hagfeldt, A. *Accounts of Chemical Research* 2009, 42, 1819–1826.)

by reduction from the redox electrolyte, and to avoid recombination of the photogenerated electron-hole pair. This simple view of the DSC neglects other significant recombination channels, but it effectively shows that the selectivity requirements of the diode structure of Figure 21.8 demand a specific control of intermolecular interactions at the interface. A combined design of the different elements that ensures rectification is required, by creating kinetic and energetic preference that directs carrier flows (Pastore and de Angelis, 2013).

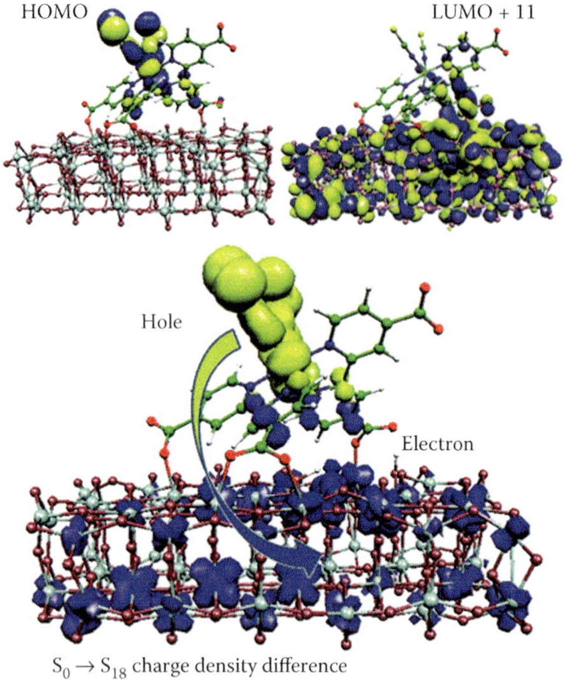

FIGURE 21.12 Isodensity plots of the HOMO and LUMO + 11 of the N719/TiO_2 system. *Bottom*: Charge density difference between the ground state (S_0) and the S_{18} excited state. A blue (*green*) color signifies an increase (decrease) of charge density upon electron excitation. (Reprinted with permission from De Angelis, F. et al. *The Journal of Physical Chemistry C* 2011, 115, 8825–8831.)

In the absence of external selective contacts, it is much more difficult to build a diode using a single molecular unit. Aviram and Ratner (1974) proposed that a single organic molecule of the type D–σ–A could be a rectifier, see Figure 21.13. Here, D is a good organic one-electron donor (but poor acceptor), σ the covalent saturated bridge, and A is a good organic one-electron acceptor. In this model, system selectivity must be built internally in the molecular assembly in contact with two metal electrodes and the process involves electron transfer between molecular orbitals, whose probability amplitudes are asymmetrically placed within the molecule (Metzger, 2003). The realization of this idea has not been conclusively demonstrated. The bias potential between the electrodes not only injects carriers in the required molecular levels but also creates an electrical field that distorts those levels (Krzeminski et al., 2001). It was suggested that the main impediment to realize D–σ–A molecular rectifier is that the system loses unidirectionality of charge transfer under the applied bias (Mujica et al., 2002). Other work showed that the bias voltage actually aligns the D and A levels creating a resonant state delocalized over the entire molecule (Stokbro et al., 2003). This state provokes a large rise of current but no rectification. It is, therefore, likely that experimental observation of rectifying behavior in unimolecular systems is mainly a result of asymmetric coupling between

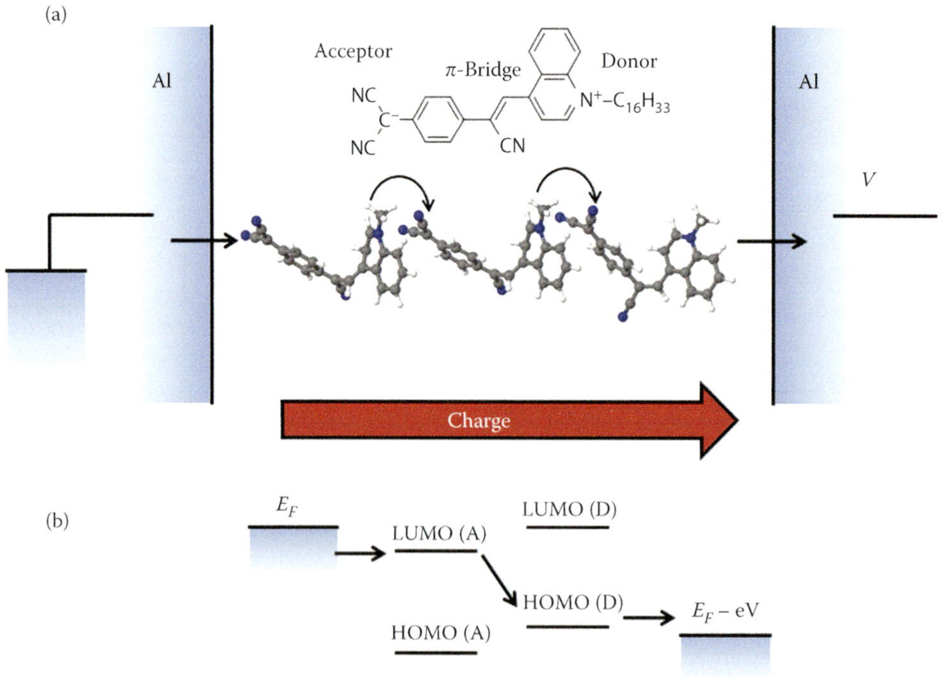

FIGURE 21.13 (a) Schematic representation of $C_{16}H_{33}Q$-3CNQ in contact with metal electrodes under applied bias voltage. (b) The Aviram–Ratner mechanism for molecular rectification. (Adapted from Krzeminski, C. et al. *Physical Review B* 2001, 64, 085405.)

the molecule and the electrodes, so that the rectification actually originates at the interface (Zhao et al., 2010).

GENERAL REFERENCES

Dye-sensitized solar cells: O'Regan and Grätzel (1991), Hagfeldt and Grätzel (1995), Hagfeldt and Grätzel (2000), Hagfeldt et al. (2010), Kalyanasundaram (2010), Barea and Bisquert (2013), and Bisquert and Marcus (2013).

Photo-induced electrons transfer from dye molecules: Rego and Batista (2003), Duncan et al. (2005), Duncan and Prezhdo (2007), and Gundlach et al. (2007).

REFERENCES

Abruña, H. D.; Denisevich, P.; Umana, M.; Meyer, T. J.; Murray, R. W. Rectifying interfaces using two-layer films of electrochemically polymerized vinylpyridine and vinylbipyridine complexes of ruthenium and iron on electrodes. *Journal of the American Chemical Society* 1981, *103*, 1–5.

Aviram, A.; Ratner, M. A. Molecular rectifiers. *Chemical Physics Letters* 1974, *29*, 277–283.

Barea, E. M.; Bisquert, J. Properties of chromophores determining recombination at TiO_2-dye-electrolyte interface. *Langmuir* 2013, *29*, 8773–8781.

Bisquert, J.; Marcus, R. A. Device modeling of dye-sensitized solar cells. *Topics in Current Chemistry* 2013. doi: 10.1007/1128_2013_1471.

Boschloo, G.; Hagfeldt, A. Characteristics of the iodide/triiodide redox mediator in dye-sensitized solar cells. *Accounts of Chemical Research* 2009, *42*, 1819–1826.

Burroughes, J. H.; Bradley, D. D. C.; Brown, A. R.; Marks, R. N.; MacKay, K.; Friend, R. H.; Burn, P. L.; et al. Light-emitting diodes based on conjugated polymers. *Nature* 1990, *347*, 539–541.

Cuevas, A. The recombination parameter J_0. *Energy Procedia* 2014, *55*, 53–62.

De Angelis, F.; Fantacci, S.; Mosconi, E.; Nazeeruddin, M. K.; Grätzel, M. Absorption spectra and excited state energy levels of the N719 Dye on TiO_2 in dye-sensitized solar cell models. *The Journal of Physical Chemistry C* 2011, *115*, 8825–8831.

Duncan, W. R.; Prezhdo, O. V. Theoretical studies of photo-induced electron transfer in dye-sensitized TiO_2. *Annual Review of Physical Chemistry* 2007, *58*, 143–184.

Duncan, W. R.; Stier, W. M. Prezhdo, O. V. Ab initio non-adiabatic molecular dynamics of the ultrafast electron injection across the alizarin-TiO_2 interface. *Journal of the American Chemical Society* 2005, *127*, 7941–7951.

Flämmich, M.; Frischeisen, J.; Setz, D. S.; Michaelis, D.; Krummacher, B. C.; Schmidt, T. D.; Brütting, W.; et al. Oriented phosphorescent emitters boost OLED efficiency. *Organic Electronics* 2011, *12*, 1663–1668.

Fujihira, M.; Yamada, H. Molecular photodiodes consisting of unidirectionally oriented amphipathic acceptor-sensitizer-donor triads. *Thin Solid Films* 1988, *160*, 125–132.

Gerischer, H.; Michel-Beyerle, M. E.; Rebentrost, F.; Tributsch, H. Sensitization of charge injection into semiconductors with large band gap. *Electrochimica Acta* 1968, *13*, 1509–1515.

Gundlach, L.; Ernstorfer, R.; Willig, F. Ultrafast interfacial electron transfer from the excited state of anchored molecules into a semiconductor. *Progress in Surface Science* 2007, *82*, 355–377.

Hagfeldt, A.; Boschloo, G.; Sun, L.; Kloo, L.; Pettersson, H. Dye-sensitized solar cells. *Chemical Reviews* 2010, *110*, 6595–6663.

Hagfeldt, A.; Grätzel, M. Light-induced redox reactions in nanocrystalline systems. *Chemical Reviews* 1995, *95*, 49–68.

Hagfeldt, A.; Grätzel, M. Molecular photovoltaics. *Accounts in Chemical Research* 2000, *33*, 269–277.

Holonyak, N.; Bevacqua, S. F. Coherent (visible) light emission from $Ga(As_{1-x}P_x)$ junctions. *Applied Physics Letters* 1962, *1*, 82–83.

Hoven, C. V.; Peet, J.; Mikhailovsky, A.; Nguyen, T.-Q. Direct measurement of electric field screening in light emitting diodes with conjugated polyelectrolyte electron injecting/transport layers. *Applied Physics Letters* 2009, *94*, 033301.

Hoven, C. V.; Yang, R.; Garcia, A.; Crockett, V.; Heeger, A. J.; Bazan, G. C.; Nguyen, T.-Q. Electron injection into organic semiconductor devices from high work function cathodes. *Proceedings of the National Academy of Sciences* 2008, *105*, 12730–12735.

Jin, R.; Levermore, P. A.; Huang, J.; Wang, X.; Bradley, D. D. C.; deMello, J. C. On the use and influence of electron-blocking interlayers in polymer light-emitting diodes. *Physical Chemistry Chemical Physics* 2009, *11*, 3455–3462.

Kalyanasundaram, K. *Dye-Sensitized Solar Cells*; CRC Press: Boca Raton, 2010.

Kennard, E. H. On the interaction of radiation with matter and on fluorescent exciting power. *Physical Review* 1926, *28*, 672–683.

Krzeminski, C.; Delerue, C.; Allan, G.; Vuillaume, D.; Metzger, R. M. Theory of electrical rectification in a molecular monolayer. *Physical Review B* 2001, *64*, 085405.

Lane, P. A.; deMello, J. C.; Fletcher, R. B.; Bernius, M. Electric field screening in polymer light-emitting diodes. *Applied Physics Letters* 2003, *83*, 3611–3613.

Metzger, R. M. Unimolecular electrical rectifiers. *Chemical Reviews* 2003, *103*, 3803–3834.

Meyer, J.; Kahn, A. Electronic structure of molybdenum-oxide films and associated charge injection mechanisms in organic devices. *Journal of Photonics for Energy* 2011, *1*, 011109.

Mujica, V.; Ratner, M. A.; Nitzan, A. Molecular rectification: Why is it so rare? *Chemical Physics* 2002, *281*, 147–150.

O'Regan, B.; Grätzel, M. A low-cost high-efficiency solar cell based on dye-sensitized colloidal TiO_2 films. *Nature* 1991, *353*, 737–740.

Parkinson, B. A.; Spitler, M. T. Recent advances in high quantum yield dye sensitization of semiconductor electrodes. *Electrochimica Acta* 1992, *37*, 943–948.

Pastore, M.; de Angelis, F. Intermolecular interactions in dye-sensitized solar cells: A computational modeling perspective. *Journal of Physical Chemistry Letters* 2013, *4*, 956–974.

Pope, M.; Kallmann, H. P.; Magnante, P. Electroluminescence in organic crystals. *Journal of Chemical Physics* 1968, *38*, 2042.

Rego, L. G. C.; Batista, V. S. Quantum dynamics simulations of interfacial electron transfer in sensitized TiO_2 semiconductors. *Journal of the American Chemical Society* 2003, *125*, 7989–7997.

Sarasqueta, G.; Choudhury, K. R.; Subbiah, J.; So, F. Organic and inorganic blocking layers for solution-processed colloidal PbSe nanocrystal infrared photodetectors. *Advanced Functional Materials* 2010, *21*, 167–171.

Shirota, Y. Organic materials for electronic and optoelectronic devices. *Journal Materials Chemical* 2000, *10*, 1–25.

Stokbro, K.; Taylor, J.; Brandbyge, M. Do Aviram and Ratner diodes rectify? *Journal of the American Chemical Society* 2003, *125*, 3674–3675.

Strickler, S. J.; Berg, R. A. Relationship between absorption intensity and fluorescence lifetime of molecules. *Journal of Chemical Physics* 1962, *37*, 814–820.

Sullivan, P.; Heutz, S.; Schultes, S. M.; Jones, T. S. Influence of codeposition on the performance of CuPc–C_{60} heterojunction photovoltaic devices. *Applied Physics Letters* 2004, *84*, 1210–1212.

Tang, C. W.; VanSlyke, A. Organic electroluminescent diodes. *Applied Physics Letters* 1986, *51*, 913–915.

Tokito, S.; Noda, K.; Taga, Y. Metal oxides as a hole-injecting layer for an organic electroluminescent device. *Journal of Physics D: Applied Physics* 1996, *29*, 2750–2759.

Yunlong, Z.; Russell, J. H. Influence of a MoO_x interlayer on the open-circuit voltage in organic photovoltaic cells. *Applied Physics Letters* 2013, *103*, 053302.

Zhao, J.; Yu, C.; Wang, N.; Liu, H. Molecular rectification based on asymmetrical molecular electrode contact. *The Journal of Physical Chemistry C* 2010, *114*, 4135–4141.

22 Radiative Equilibrium in a Semiconductor

The materials for the conversion of solar photons to electrical flux require to be spectrally matched for optimal energy harvesting. Using a model absorber material with a step-like bandgap, we discuss the central features of maximal photocurrent and electrical power that can be obtained from the incident sunlight. We then move to fundamental considerations that have a large impact on the properties of solar cells. Semiconductors that absorb light also emit radiation by radiative recombination. Detailed balance is established in thermal equilibrium (no applied voltage in the dark) between the absorption of incoming blackbody radiation and the emission of photons by recombination of excited carriers, which imposes fundamental constraints to the rate of recombination. Furthermore, when the semiconductor receives extra illumination, the quasi-Fermi levels of electrons and holes are split. A reciprocity relationship between the absorbed and emitted radiation allows us to accurately determine the internal photovoltage and the light absorption characteristics of the material from PL measurements.

22.1 UTILIZATION OF SOLAR PHOTONS

The light absorption characteristics of the materials that convert radiant energy to useful electrical energy must be adapted optimally to the spectral properties of solar radiation. The principal feature of the available radiation is the number of photons per energy (or frequency) interval dE denoted as $\phi_{ph}(E)dE$ in Section 17.7. As discussed in Section 17.8, sunlight shows a rather broad spectrum including photons in the energy range from ultraviolet to infrared (280–2500 nm, 0.5–4.4 eV). Figure 22.1a shows the spectral shape of terrestrial solar irradiation represented by blackbody radiation at T_S = 5800 K, and equivalent projected solid angle $\varepsilon_S^{1.5AM} = 4.8 \times 10^{-5}$ with a total power $\Phi_{E,tot}$ = 1000 W m^{-2}.

The photocurrent in a solar cell consists of the extraction of the photogenerated electrons and holes and is measured by the short-circuit photocurrent density, j_{ph}. Solar energy converters utilize the photons *one by one*, and each absorbed photon creates one electron-hole pair exactly. Therefore, the maximal value of the electrical current that can be extracted in photovoltaic cells is given by the number of photons that are absorbed, times the elementary charge. The primary limitations to the photocurrent are quantified by the spectral absorptivity or absorptance of the device, $a(E)$. In addition, in order to convert the incoming photons into an electron current in the external circuit, several steps that determine the internal quantum efficiency (IQE_{PV}) must function efficiently, as commented in Section 18.3; note the relationship in Equation 18.28, $EQE_{PV}(E) = a(E)IQE_{PV}(E)$. If the photovoltaic external quantum efficiency EQE_{PV} of the solar cell has been determined, the short-circuit photocurrent under arbitrary illumination corresponds to

$$j_{ph} = q\int_0^\infty EQE_{PV}(E)\phi_{ph}(E)dE \qquad (22.1)$$

Some materials have been proposed in which high-energy photons create more than one electron-hole pair, showing quantum efficiency larger than 1, although these have so far produced very low-power conversion efficiencies (Sambur et al., 2010; Semonin et al., 2011).

For an approximate evaluation of photovoltaic properties, it is useful to assume an ideal thick semiconductor that absorbs all photons of energy larger than the semiconductor band gap. The step absorptance takes the values

$$\begin{aligned} a(E) &= 0, \text{ for } E < E_g \\ a(E) &= 1, \text{ for } E \geq E_g \end{aligned} \qquad (22.2)$$

This model is shown in Figure 22.1a. It is a useful simplification for assessing the fundamental efficiency limit of solar cells described by just two properties, the semiconductor bandgap and the temperature of the absorber. Note that it is assumed that the optical pathway for above bandgap radiation is shorter than the film thickness, so that a = 1. Furthermore, we assume optimal charge collection at electrodes giving IQE_{PV} = 1. These two conditions are somewhat contradictory regarding the required film thickness, as total collection requires a short film, but nonetheless, we adopt this model as a hypothetical reference for optimal photovoltaic performance that will be qualified later on in Chapter 26. Obviously, the model works best for semiconductors that have a nearly step-like band gap like GaAs and $CH_3NH_3PbI_3$ perovskite rather than for indirect semiconductors (see Figure 19.4). However, the precise form of the onset of absorption has

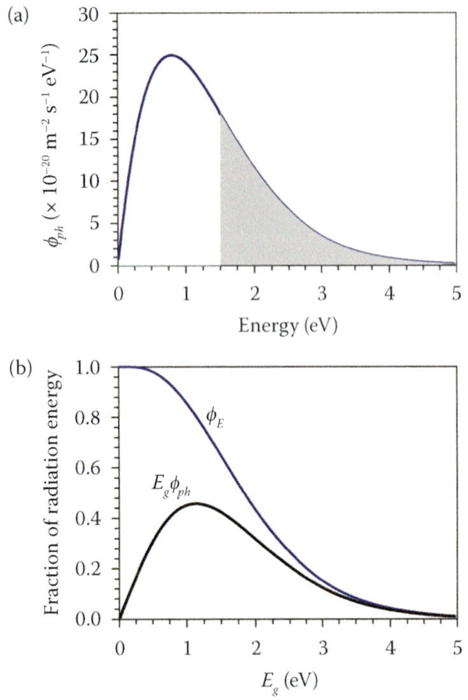

FIGURE 22.1 (a) Photon flux emitted by a blackbody at temperature 5800 K, with an *equivalent* projected solid angle $\varepsilon_S^{1.5AM} = D\pi = 4.8\times10^{-5}$ that gives a total power of 1000 W m^{-2}. The photons absorbed for a bandgap E_g = 1.5 eV are indicated. (b) Fraction of the incoming power absorbed as a function of the absorber gap. The lower line considers the thermalization to E_g of photon energy of each absorbed photon.

an important impact on photovoltaic performance as discussed in Section 25.5.

Based on Equation 22.2, the maximal current that can be obtained as a function of the bandgap of the absorber is

$$j_{ph} = q\Phi_{ph}\left(E_g,\infty\right) = q\int_{E_g}^{\infty}\phi_{ph}\left(E\right)dE \qquad (22.3)$$

The spectral flux for blackbody radiation with the étendue ε is indicated in Equation 17.32. The integral then takes the form

$$j_{ph} = q\frac{\varepsilon}{\pi}b_\pi\int_{E_g}^{\infty}\frac{E^2}{e^{E/k_BT}-1}dE \qquad (22.4)$$

where the prefactor for a hemisphere $\varepsilon = \pi$ is the constant $b_\pi = 2\pi/h^3c^2$. If $E_g \gg k_BT$, the integral is evaluated as follows:

$$\int_{E_g}^{\infty}E^2e^{-E/k_BT}dE = e^{-E_g/k_BT}k_BT\left(E_g^2+2E_gk_BT+2k_B^2T^2\right) \qquad (22.5)$$

We define the function

$$\Gamma_{ph}\left(E,T\right) = E^2+2Ek_BT+2k_B^2T^2 \qquad (22.6)$$

Hence, the photocurrent is approximately given by

$$j_{ph} = q\frac{\varepsilon}{\pi}b_\pi k_BT\Gamma_{ph}\left(E_g,T_S\right)e^{-E_g/k_BT_S} \qquad (22.7)$$

For sunlight modeled in terms of the étendue $\varepsilon_S^{AM1.5}$ in Equation 17.45, the photocurrent is

$$j_{ph} = q\frac{\varepsilon_S^{AM1.5}}{\pi}b_\pi k_BT\Gamma_{ph}\left(E_g,T_S\right)e^{-E_g/k_BT_S} \qquad (22.8)$$

The maximum current density that can be obtained from the blackbody radiation at T_S = 5800 K (k_BT_S = 0.5 eV) and $\varepsilon_S^{1.5AM}$, that is taking the integral from E_g = 0, is $j_{ph}^{max} = 73\,\text{mA}\,\text{cm}^{-2}$. The fraction of the incoming photon flux absorbed by a semiconductor of bandgap E_g can be observed in Figure 22.1a. The current as a function of the absorption edge, that is the semiconductor bandgap, is shown in Figure 22.2a both for the case of a 5800 K blackbody spectrum and for the *AM*1.5*G* spectrum. The actual current obtained in record inorganic cells of GaAs and Si and a lead-halide perovskite solar cell are also presented. We observe that the photocurrent in top quality devices is quite close to the value theoretically achievable from *AM*1.5*G*. The maximal efficiencies for more realistic absorption conditions are discussed in Figure 25.10.

The benchmark feature for a solar energy converter is the *power conversion efficiency* (PCE), that is the ratio of the electrical power P_{el} produced by the conversion device, with respect to the total incident radiant power $\Phi_{E,tot}$

$$\eta_{\text{PCE}} = \frac{P_{el}}{\Phi_{E,tot}} \qquad (22.9)$$

In principle, it is desirable to use a material that absorbs a large fraction of the spectrum, implying a small band gap, in order to generate a large photocurrent. But a second main aspect of spectral solar energy conversion relates to the utilization of the *energy* contained in each energy interval of the radiation field, $\phi_E(E)dE = E\phi_{ph}(E)dE$. The carriers initially obtain a separation of energy corresponding to that of the absorbed photon (see Figure 19.1), but they are rapidly thermalized to the edges of the bands corresponding to the quasiequilibrium energy configuration, E_g, that precedes recombination. Energy relaxation of the generated electrons and holes to the band gap has

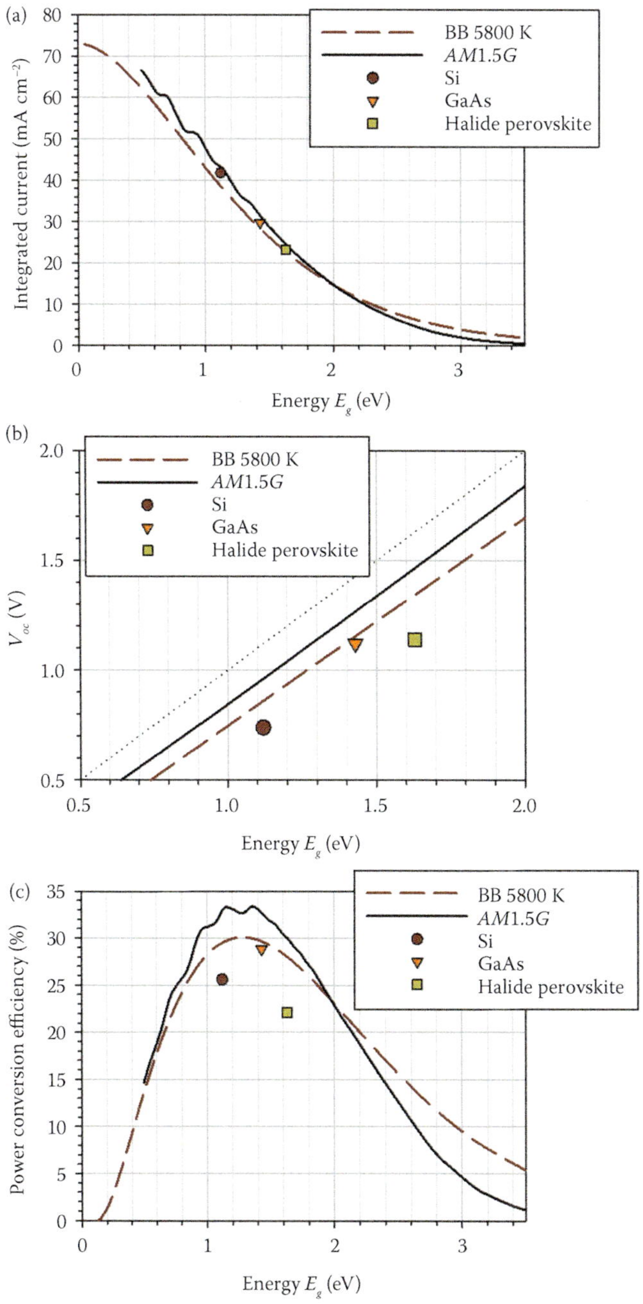

FIGURE 22.2 Photovoltaic parameters as a function of the bandgap with the assumption of sharp absorption edge and total charge collection. The quantities are shown for incident blackbody radiation at temperature T_S = 5800 K, with total energy flux $\Phi_{E,tot}$ = 1000 W m^{-2}, and for actual *AM*1.5*G* solar irradiance. (a) Integrated current density. (b) The radiative voltage. (c) The maximal photovoltaic Shockley–Queisser efficiencies. The points indicate the values for record cells.

been previously commented on in Section 5.6. After the fast thermalization in the absorber, the energy available is E_g for each absorbed photon (and electron-hole pair). Since the quantity of energy $h\nu - E_g$ per photon is lost, the PCE is severely reduced. Meanwhile, as taken into account in Equation 22.2, the photons with energy less than E_g are not absorbed.

Let us find the flux of energy contained in the incoming blackbody radiation that is absorbed by the semiconductor:

$$\Phi_E\left(E_g,\infty\right) = \int_{E_g}^{\infty} E\phi_{ph}\left(E\right)dE \tag{22.10}$$

$$\Phi_E\left(E_g,\infty\right) = \frac{\varepsilon}{\pi} b_\pi \int_{E_g}^{\infty} \frac{E^3}{e^{E/k_BT}-1} dE \tag{22.11}$$

If $E_g \gg k_BT$, the integral is evaluated as follows:

$$\Phi_E\left(E_g,\infty\right) = \frac{\varepsilon}{\pi} b_\pi e^{-E_g/k_BT} k_BT \chi_E\left(E_g,T\right) \tag{22.12}$$

using the function

$$\chi_E\left(E,T\right) = E^3 + 3k_BTE^2 + 6k_B^2T^2E + 6k_B^3T^3 \tag{22.13}$$

The lost power due to the transparent range is

$$\Phi_E\left(0,E_g\right) = \Phi_E\left(0,\infty\right) - \Phi_E\left(E_g,\infty\right) \tag{22.14}$$

The total flux is given in Equation 17.39

$$\Phi_{E,tot} = \frac{\varepsilon}{\pi} b_\pi \frac{\pi^4}{15}\left(k_BT\right)^4 \tag{22.15}$$

If the total flux is calculated from Equation 22.12, a prefactor 6 (instead of $\pi^4/15$) is obtained.

We can estimate the so-called Trivich-Flinn efficiency, which assumes that every electron-hole pair is extracted with a voltage equal to E_g/q (Green, 2012). It is

$$\eta = \frac{E_g\Phi_{ph}\left(E_g,\infty\right)}{\Phi_{E,tot}} \tag{22.16}$$

Taking the approximate expressions for blackbody radiation at the temperature T_S of the sun's surface, Equations 22.6 and 17.39, and the normalized energy $x_g = E_g/k_BT_S$, we obtain

$$\eta = \frac{15}{\pi^4} e^{-x_g} x_g\left(x_g^2 + 2x_g + 2\right) \tag{22.17}$$

This preliminary efficiency is shown in Figure 22.1b as a function of bandgap. It is *not* a well-defined conversion efficiency in the sense of Equation 22.9, since each electron-hole pair in a solar cell is extracted at a much smaller voltage than E_g, as discussed in Section 25.3. Nonetheless, Equation 22.17 interestingly reveals that maximal efficiency occurs at intermediate wavelength within the solar spectrum due to the two competing effects that determine solar energy conversion into electrical energy using a single semiconductor: (1) The transparency of the semiconductor to long wavelength photons and (2) the loss of a quantity of energy $E-E_g$ per each absorbed photon. These energy loss effects are severe. Even in the most favorable condition, which is $E_g \cong 1.1$ eV, more than 50% of the energy content of blackbody radiation at T_S is lost.

Additional considerations required for the physically correct theoretical efficiency of the ideal converter will be described in Section 25.4, and the practical photovoltaic efficiencies in Section 25.5. The complete approach is based on reciprocity principles that are derived in the following sections. We anticipate that the maximal efficiency is developed from a conservation argument. The number of photons per unit time that are absorbed from the incident spectrum minus the photons that are emitted by radiative recombination determines the electrical current, and the power is calculated at the voltage where the electrical output is optimal. The result of the calculation is the Shockley and Queisser (SQ) efficiency (1961) shown in Figure 22.2c for model blackbody radiation and $AM1.5G$ solar irradiance. The first step toward a determination of the maximal efficiency of a solar cell is to obtain the fundamental radiative lifetime of a semiconductor as we explain in the next section.

22.2 FUNDAMENTAL RADIATIVE CARRIER LIFETIME

The detailed balance of absorbed and emitted radiation in a semiconductor that has attained thermal equilibrium in the dark with the surrounding blackbody radiation enables one to establish a fundamental equation for the radiative lifetime, as first derived by van Roosbroeck and Shockley (1954); see also Pankove (1971). The lifetime is determined only by the absorption coefficient $\alpha(E)$ and the refraction index of the semiconductor n_r.

We consider a small fragment of a semiconductor, as shown in Figure 22.3. In equilibrium, the volume element exchanges radiation with the surroundings. The density of photons in the volume element is determined by Planck's law, established in Equation 17.19, as follows:

$$\frac{dn_{ph}(E)}{dE} = \frac{4b_\pi n_r^3}{c} \frac{E^2}{e^{E/k_BT}-1} \tag{22.18}$$

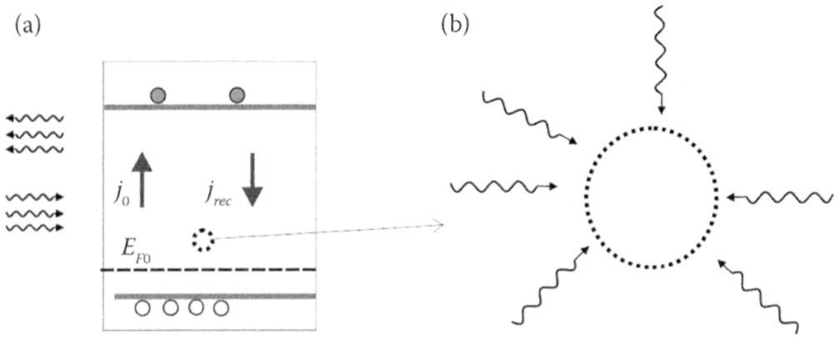

FIGURE 22.3 (a) Scheme of p-type semiconductor material in dark equilibrium, showing the balance of incoming and outgoing photons absorbed or produced in the material, and the equilibrium Fermi level E_{F0}. The generation by incoming thermal radiation produces a generation flux j_0/q and radiative recombination internal flux is j_{rec}/q. (b) In equilibrium, the volume element corresponding to a fragment of the semiconductor sees a black body radiation coming from its surroundings. The density of photons in the volume element results from two sources, the external radiation and the radiation from the other parts of the semiconductor.

The photon absorption rate in the volume $U_{abs}(E)$ is the product of the photon density and the absorption probability, which is given by the lifetime of the photons of energy E, $\tau_{ph}(E)$

$$U_{abs}(E)dE = \frac{1}{\tau_{ph}(E)} dn_{ph}(E) \tag{22.19}$$

The lifetime of the photons can be established by analyzing their mean free path in the medium according to the velocity of the photons:

$$\tau_{ph}(E) = \frac{1}{\alpha(E)c_\gamma} = \frac{n_r}{\alpha(E)c} \tag{22.20}$$

The recombination rate $U_{rec}(E)$ in the volume causes the emission of photons of energy E, as indicated in Equation 18.32. Using the intrinsic density, $n_i^2 = n_0 p_0$, we have the expression

$$U_{rec}(E)dE = B_{rad}(E) n_i^2 dE \tag{22.21}$$

Equating the rates, we obtain

$$B_{rad}(E) n_i^2 = 4b_\pi n_r^2 \alpha(E) \frac{E^2}{e^{E/k_BT}-1} \tag{22.22}$$

Integrating over the entire energy spectrum, we arrive at the radiative recombination constant

$$B_{rad} n_i^2 = 4b_\pi n_r^2 \int_0^\infty \alpha(E) E^2 e^{-E/k_BT} dE \tag{22.23}$$

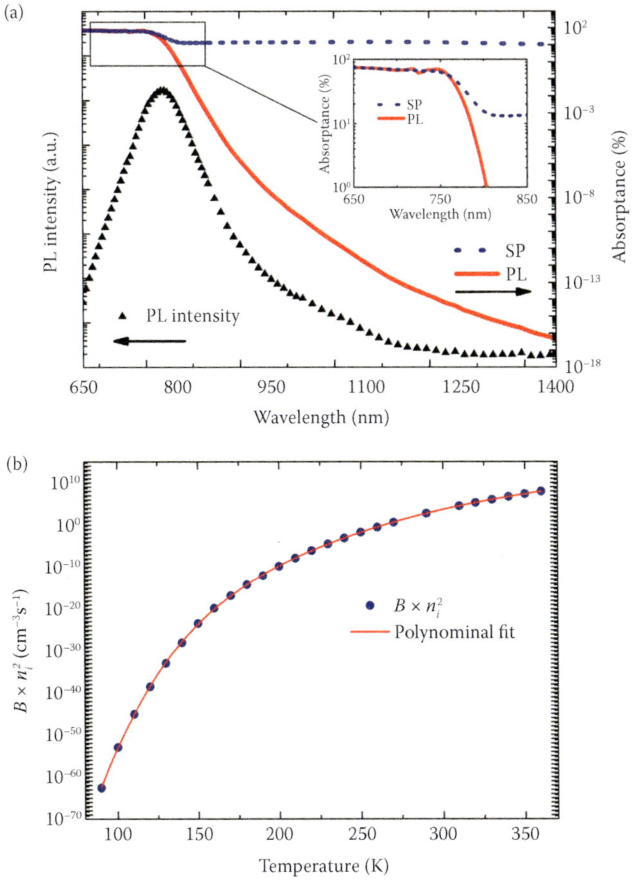

FIGURE 22.4 (a) PL spectra of 300 nm perovskite $CH_3NH_3PbI_3$ film (*left y*-axis). Spectrophotometer-measured absorptance and PL-extracted absorptance of $CH_3NH_3PbI_3$ as a function of wavelength (*right y*-axis). (b) Product of the radiative recombination coefficient and the square of the intrinsic carrier density $B_{rad} \times n_i^2$ plotted as a function of temperature. (Reproduced with permission from Barugkin, C. et al. *The Journal of Physical Chemistry Letters* 2015, 6, 767–772.)

Here, we have assumed that $E \gg k_B T$ for the parts of the spectrum where the absorption coefficient has non-negligible values. The determination of the recombination rate in Equation 22.23 is shown in Figure 22.4b for a perovskite $CH_3NH_3PbI_3$ film. This will be explained later in more detail.

We concluded in Equation 17.34 that the flux of photons of blackbody emission into a solid angle Ω of is

$$\phi_{ph}^{bb}(E) = \frac{\varepsilon}{\pi} b_\pi \frac{E^2}{e^{E/k_BT} - 1} \tag{22.24}$$

We may express Equation 22.22 more compactly:

$$B_{rad}(E) n_i^2 = \alpha(E) \phi_{ph}^{bb}(E, 4\pi n_r^2) \tag{22.25}$$

This form shows that the factor 4 in Equation 22.23 is associated to isotropic emission.

22.3 RADIATIVE EMISSION OF A SEMICONDUCTOR LAYER

The result of Equation 22.22 has been achieved for a small fragment of semiconductor in equilibrium with its surrounding. However, in actual measurements, we deal with the absorption and emission of an entire semiconductor film, as indicated in Figure 20.7a. We then need to extend the reciprocity reasoning to the light absorbed and emitted by the complete film or solar cell device. We recall that the quantity describing the amount of absorbed radiation is the absorptance $a(E)$, which depends not only on the absorption coefficient of the medium, but also on the macroscopic characteristics of the film such as the thickness and reflectivity of the surfaces, see Equation 18.13.

In Section 17.5, we have defined a blackbody as one that absorbs all incoming radiation such that the absorptivity is $a = 1$ at all photon energies. The radiation emitted by a blackbody is thermal radiation (at zero chemical potential) at the given temperature, Equation 22.24. In bodies that are not perfect absorbers, the rate of emission of radiant energy is reduced with respect to the radiation of a blackbody. The radiant power of a surface is described by the *emissivity* $\varepsilon_{em}(E)$. It is the spectral ratio of the energy or photon flux emitted by the body to that of a blackbody of a similar temperature:

$$\varepsilon_{em}(E) = \frac{\phi_E^{em}(E)}{\phi_E^{bb}(E)} = \frac{\phi_{ph}^{em}(E)}{\phi_{ph}^{bb}(E)} \tag{22.26}$$

Consider the solar cell to be in thermal equilibrium with its surroundings. Then, the fraction of the incident energy $\phi_E^{bb}(E)dE$ absorbed per unit area per unit time is given by

$$a(E)\phi_E^{bb}(E)dE \tag{22.27}$$

while the energy radiated per unit surface is

$$\phi_E^{em}(E)dE = \varepsilon_{em}(E)\phi_E^{bb}(E)dE \tag{22.28}$$

Therefore,

$$\varepsilon_{em}(E) = a(E) \tag{22.29}$$

This is Kirchoff's law, which states that the emissivity of a body equals its absorptivity. Thus, for a blackbody,

$\varepsilon_{em} = 1$, implying that it is a perfect radiator as well as a perfect absorber. A more general proof of Kirchhoff's law can be established including directionality in a projected solid angle (or étendue ε) for any type of surface (Greffet and Nieto-Vesperinas, 1998). Thus, we have the relationship

$$\phi_{ph}^{em}(E,\varepsilon) = a(E)\phi_{ph}^{bb}(E,\varepsilon) \quad (22.30)$$

In order to discuss a particular and relevant form of a light-absorbing film in the photovoltaic device, we consider a planar geometry with lateral dimensions that are much larger than the thickness. At the front surface, we assume a perfect antireflecting coating that causes the zero reflectivity and at the rear surface a perfect reflecting coating that gives unity reflectivity. Under these conditions, absorptive and emissive fluxes stem only from the front surface of the solar cell and the étendue for both absorption of the blackbody radiation and emission from recombination is $\varepsilon = \pi$. From Equations 17.35, 22.28, and 22.29, we obtain the expression of the spectral emission of the film

$$\phi_{ph}^{em}(E) = b_\pi a(E)\frac{E^2}{e^{E/k_BT} - 1} \quad (22.31)$$

This central result provides the basis for determining a number of photovoltaic properties as will be explained in Chapters 23 and 25. Shockley and Queisser (1961) established a radiative balance that enables to predict the maximal efficiency of a solar cell based on a semiconductor with a sharp bandgap, as commented before. In a seminal work, Ross (1967) showed that the equilibrium between the distribution of photons and the recombination allows one to predict a photovoltage in terms of the quantum yield of any photochemical converter. Smestad and Ries (1992) developed this idea for the photovoltage of a solar cell. Lasher and Stern (1964), Ruppel and Würfel (1980), and Würfel (1982) established the equilibrium for the external flux in a semiconductor layer. Rau (2007) and Kirchartz and Rau (2007) determined the influence of the internal properties of the solar cell such as charge collection efficiency on the reciprocity between LED and photovoltaic properties.

22.4 PHOTONS AT NONZERO CHEMICAL POTENTIAL

In Section 17.5, we discussed the thermalization of radiation in a *cavity*. This process leads to an equilibrium distribution determined by the temperature of the walls and the Planck spectrum, which yields the spectral photon flux shown in Equation 17.35. When a semiconductor film characterized by the absorptance $a(E)$ interacts with the radiation in a cavity, a stationary situation is reached by reciprocal absorption and emission. The radiation field corresponds to an equilibrium spectral distribution according to the generalized Planck's law based on the occupation function of Equation 17.13, consisting of photons at non zero chemical and electrochemical potential $\eta_{ph} = \mu_{ph}$. The spectral photon flux, shown in Figure 22.5, is

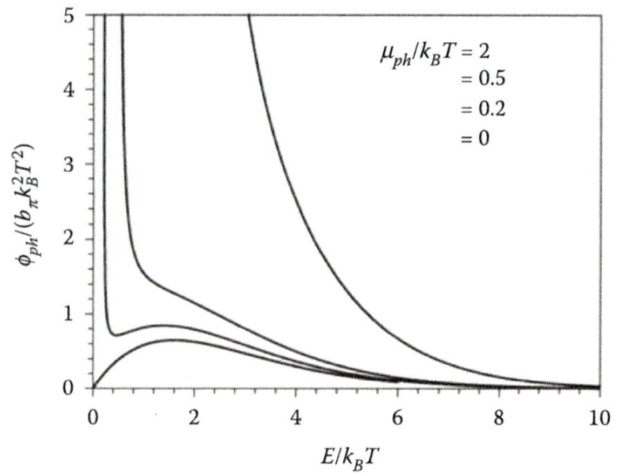

FIGURE 22.5 Dimensionless photon flux $\phi_{ph}^{bb}(E,\eta_{ph})/(b_\pi k_B^2T^2)$ as a function of dimensionless energy E/k_BT, for the indicated values of the dimensionless photon chemical potential.

$$\phi_{ph}^{bb}(E,\eta_{ph}) = b_\pi \frac{E^2}{e^{(E-\mu_{ph})/k_BT} - 1} \quad (22.32)$$

and the emitted photon flux has the form

$$\phi_{ph}^{em}(E) = b_\pi a(E)\frac{E^2}{e^{(E-\mu_{ph})/k_BT} - 1} \quad (22.33)$$

In the illuminated semiconductor, the electrochemical potential difference of electrons and holes is the net chemical potential μ_{np} as stated in Equation 20.6 and shown in Figure 20.7b. The interaction of the electron-hole gas and the photons can be viewed as a chemical reaction with a chemical equilibrium in which the chemical potential of the photons coincides with that of the electrons and holes, implying $\mu_{ph} = \mu_{np}$ (Smestad, 2002).

Under moderate illumination conditions, $\mu_{ph} = \mu_{np} < E_g$ is satisfied, therefore the Wien approximation can be used in Equation 22.33 and the result is

$$\phi_{ph}^{em}(E,V) = b_\pi a(E)E^2 e^{-E/k_BT} e^{\Delta E_F/k_BT} \quad (22.34)$$

where $\Delta E_F = \mu_{np}$ is the internal splitting of the Fermi levels. As an illustration to show the connection between light absorption and emission of a semiconductor film, we use the model of a sharp band gap, shown in Equation 22.2 and Figure 22.6a. The product with BB radiation shown in Figure 22.6b in Equation 22.34 implies the spectral shape of the emitted radiation field as shown in Figure 22.6c. We observe that there is no sub-band-gap emission while for energies above the band gap, the emission is given by the blackbody radiation. If a voltage enhances the splitting of Fermi levels as $qV = \Delta E_F$, then the photon density is increased as shown in Figure 17.8 and spectral emission is exponentially enhanced, as indicated in Figure 22.6c (Herrmann and Würfel, 2005).

The reciprocity between light absorption and emission in a semiconductor established in Equation 22.34 enables the determination of the absorption coefficient from photoluminescence (PL). The method was originally established for silicon (Daub and Würfel, 1995, Trupke et al., 1998). This approach exclusively provides the band to band absorption coefficient and achieves enormous sensitivity in the spectral region where α is rather small, as mentioned in Section 19.1. A model such as that indicated in Equation 18.13 is used to determine the absorption coefficient taking into account the reflection at interfaces. As it is often difficult to calibrate the exact amount of photon emission, the method only gives relative values of the absorption coefficient as a function of photon energy. The procedure to overcome this limitation is to match the results from PL with the absolute measurements obtained directly by light absorption at shorter wavelengths. An example for the lead-halide perovskite is shown in Figure 22.4a, where the absorption obtained from PL is compared with the absolute values determined by (De Wolf et al., 2014) and shown in Figure 19.4. From the absorption coefficient, the total radiative recombination rate of Equation 22.23 is calculated as shown in Figure 22.4b. An example for the

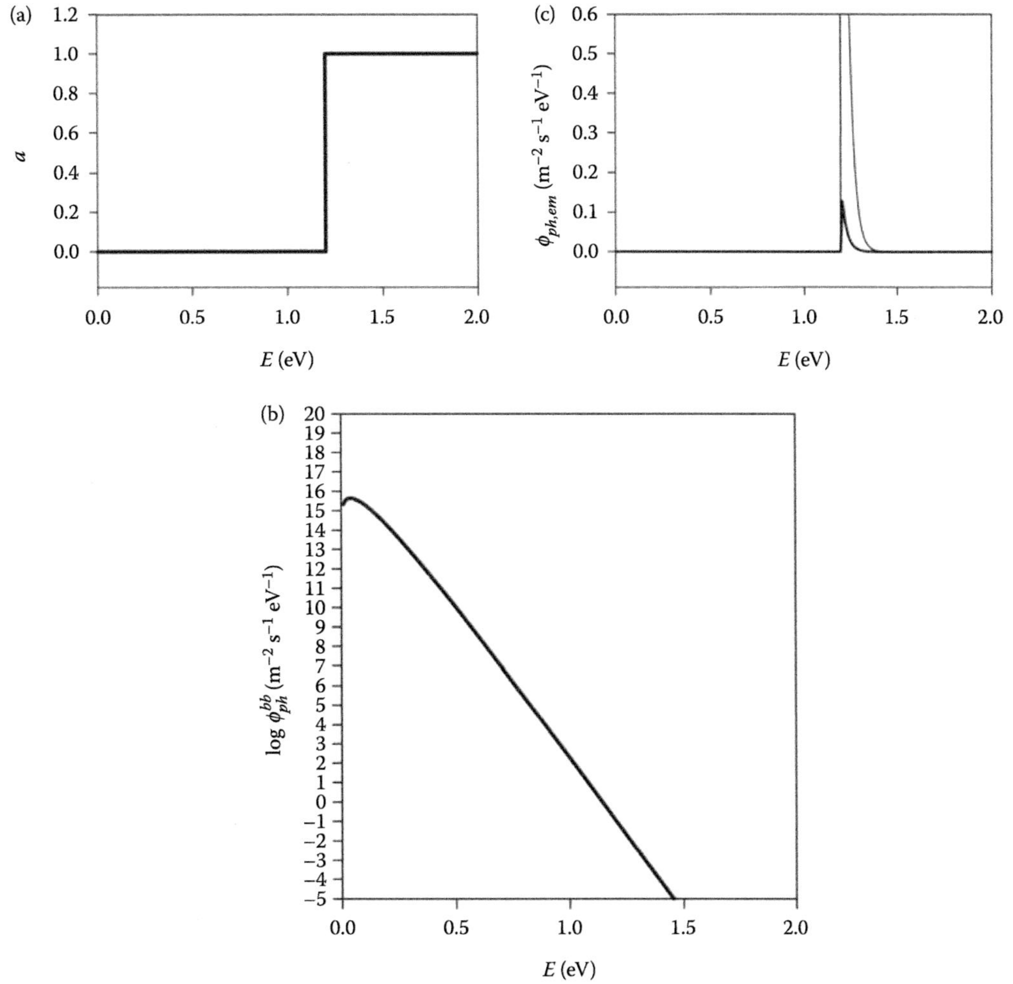

FIGURE 22.6 (a) Model semiconductor with sharp absorptance function at the bandgap value $E_g = 1.2$ eV. (b) The thermal radiation at $k_BT = 0.026$ eV received by the semiconductor. (c) The spectral light emission of the diode in thermal equilibrium (thick line) and at small applied voltage $qV = 3k_BT$ (thin line).

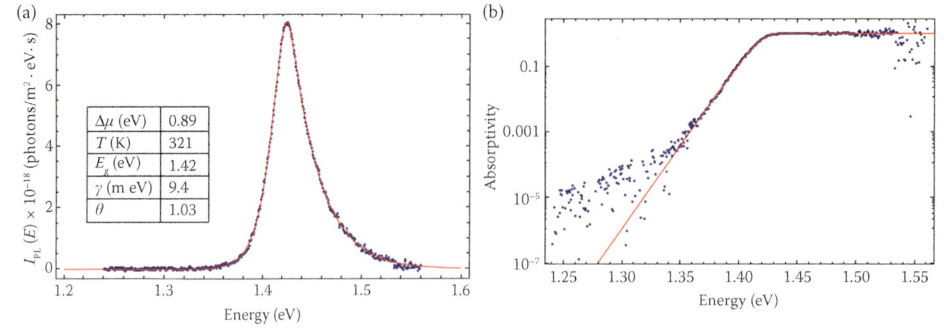

FIGURE 22.7 (a) PL (points) and full-spectrum fit (curve) of p-GaAs. (b) Absorptivity extracted from the PL data (points). (Reproduced with permission from Katahara and Hillhouse, 2014.)

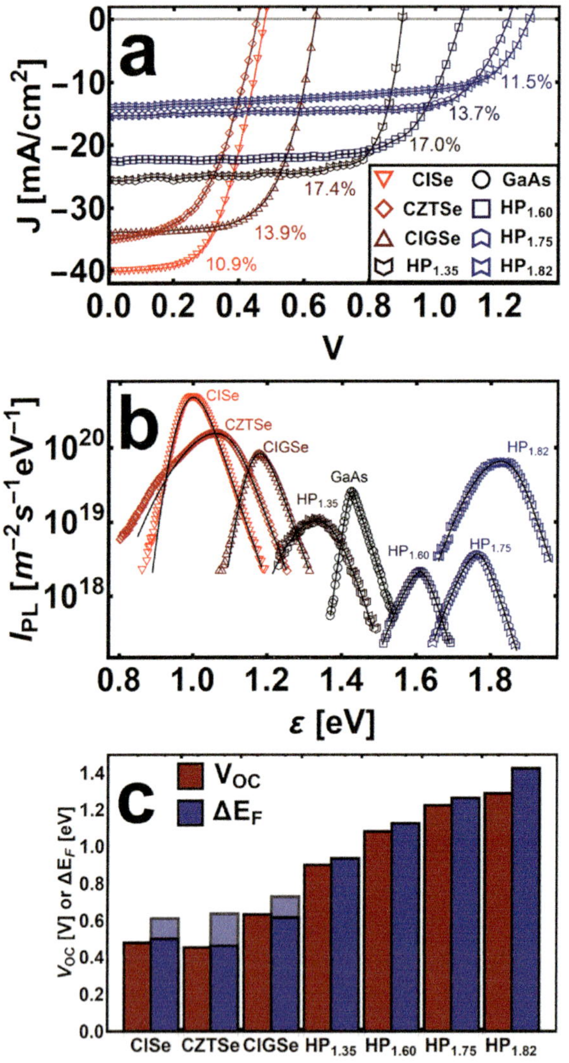

FIGURE 22.8 Comparison of photoluminescence results with device performance for four hybrid perovskite (HP) compositions of various band gaps (1.35–1.82 eV), CISe,CIGSe, and CZTSe. (a) Current density–voltage characteristics of $CuInSe_2$ (CISe), $Cu(In,Ga)Se_2$ (CIGSe), $Cu_2ZnSn\text{-}(S,Se)_4$ (CZTSe), $MA(Pb_{0.5},Sn_{0.5})(I_{0.8},Br_{0.2})_3(HP_{1.35})$, MAPbI3 ($HP_{1.6}$), $(GA_{0.10}FA_{0.58}Cs_{0.32})Pb(I_{0.73}Br_{0.27})_3$ ($HP_{1.75}$), and $PEA_2Pb\text{-}(I_{0.6},Br_{0.4})_4{\cdot}(MAPb(I_{0.6},Br_{0.4})_3)_{0.38}$ ($HP_{1.82}$) PV devices. (b) Absolute intensity photoluminescence spectra collected at 1 Sun equivalent light intensity from the same devices as those in (a), with the addition of GaAs (black circles). (c) Comparison of the device V_{oc} (red bars) and the splitting of Fermi levels from PL, ΔE_F (blue bars). The light blue extension for the chalcogenides shows the ΔE_F before intensity correction to 1 Sun, while all dark blue bars are 1 Sun ΔE_F values. (Reproduced with permission from Braly, I. L. et al., J. Phys. Chem. Lett. 9, 3779–3792, 2018.)

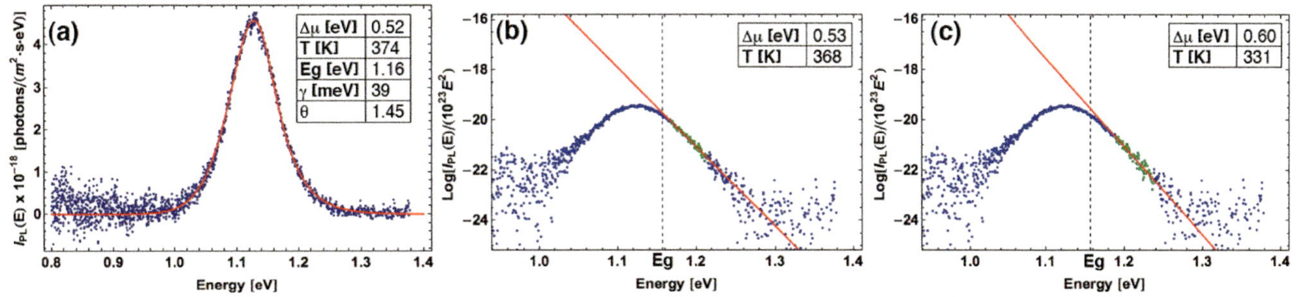

FIGURE 22.9 (a) Full-spectrum fitting of PL from CIGSe device. (b) High-energy-tail-only fitting (red line) over the region 1.17 to 1.21 eV (green points) assuming that a = 1 above band gap. (c) High-energy tail only fitting (red line) over the region 1.19 to 1.24 eV (green points) and assuming that a = 1 above band gap. (Reproduced with permission from Katahara and Hillhouse, 2014.)

radiative emission of a silicon diode with a textured surface is shown in Figure 18.2b. The light emission matches well with the blackbody spectrum for wavelengths shorter than the band gap (at 1102 nm), but rather than abruptly terminating it, the luminescence continues at longer wavelengths. An additional example is indicated in Figure 22.7. The transformation of PL data by Equation 22.34 provides values of absorptivity of GaAs down to more than four orders of magnitude below the maximum, compare Figure 19.3a.

The connection between PL and open circuit voltage is a powerful tool to evaluate the potential value of photovoltaic materials before the device engineering, avoiding the critical aspects of charge collection and optimized contacts. It is desirable to decouple absorber material quality evaluation from the engineering of the corresponding device architecture. The reciprocity relation shown in Equation 22.34 can be exploited to predict photovoltaic properties of a material, the internal splitting of the Fermi levels and the photovoltage in the semiconductor, even without the necessity to build a full device with contacts (Bauer and Gütay, 2007). This approach is particularly useful for new materials and devices.

Due to the fact that the term $e^{-E/kBT}$ of Equation 22.34 grows very rapidly as the energy decreases, the transition region of the absorptance function plays a dominant role for the accurate analysis of photovoltage. In particular, sub-band-gap states with significant absorption cross sections cause a large contribution to the emission spectrum (Katahara and Hillhouse, 2014).

By measurement of absolute intensity of PL emitted, Equation 22.34 provides the splitting of Fermi levels ΔE_F that can be achieved for an absorber material (Braly et al., 2018), which indicates a V_{oc} without effects of contacts. The results of this method for a variety of hybrid perovskites, CISe,CIGSe, and CZTSe solar cells are shown in Figure 22.8. A complementary method uses the measured EQE; this will be explained in Section 23.1.

A simple procedure to determine the photovoltage from Equation 22.34 is to fit only the high-frequency part of the spectrum, where it can be assumed that $a = 1$ so that

$$\phi_{ph}^{em}\left(E,V\right) = b_\pi E^2 e^{-E/k_BT} e^{\mu_{np}/k_BT} \tag{22.35}$$

Here, the PL is dominated by the exponential function commented previously in Figure 17.8. However, this method can be applied only if a full exponential dependence is experimentally observed, as shown in Figure 22.9.

GENERAL REFERENCES

Kirchoff's law: Landsberg (1978) and Howell et al. (2010).

Photons at nonzero chemical potential: Landsberg (1981), Würfel (1982), Ries and McEvoy (1991), and Herrmann and Würfel (2005).

Determination of Fermi level splitting from PL: Bauer and Gütay (2007), Unold and Gütay (2011), Katahara and Hillhouse (2014), and El-Hajje et al. (2016).

REFERENCES

Barugkin, C.; Cong, J.; Duong, T.; Rahman, S.; Nguyen, H. T.; Macdonald, D.; White, T. P.; et al. Ultralow absorption coefficient and temperature dependence of radiative recombination of $CH_3NH_3PbI_3$ perovskite from photoluminescence. *The Journal of Physical Chemistry Letters* 2015, *6*, 767–772.

Bauer, G. H.; Gütay, L. Analyses of local open circuit voltages in polycrystalline Cu(In,Ga)Se_2 thin film solar cell absorbers on the micrometer scale by confocal luminescence. *CHIMIA International Journal for Chemistry* 2007, *61*, 801–805.

Braly, I. L.; Stoddard, R. J.; Rajagopal, A.; Jen, A. K. Y.; Hillhouse, H. W. Photoluminescence and photoconductivity to assess maximum open-circuit voltage and carrier transport in hybrid perovskites and other photovoltaic materials. *Journal of Physical Chemistry Letters* 2018, *9*, 3779–3792.

Daub, E.; Würfel, P. Ultralow values of the absorption coefficient of Si obtained from luminescence. *Physical Review Letters* 1995, *74*, 1020–1023.

De Wolf, S.; Holovsky, J.; Moon, S.-J.; Löper, P.; Niesen, B.; Ledinsky, M.; Haug, F.-J.; et al. Organometallic halide perovskites: Sharp optical absorption edge and its relation to photovoltaic performance. *The Journal of Physical Chemistry Letters* 2014, *5*, 1035–1039.

El-Hajje, G.; Momblona, C.; Gil-Escrig, L.; Avila, J.; Guillemot, T.; Guillemoles, J.-F.; Sessolo, M.; et al. Quantification of spatial inhomogeneity in perovskite solar cells by hyperspectral luminescence imaging. *Energy & Environmental Science* 2016, *9*, 2286–2294.

Green, M. A. Analytical treatment of Trivich–Flinn and Shockley–Queisser photovoltaic efficiency limits using polylogarithms. *Progress in Photovoltaics: Research and Applications* 2012, *20*, 127–134.

Greffet, J.; Nieto-Vesperinas, M. Field theory for generalized bidirectional reflectivity: Derivation of Helmholtz's reciprocity principle and Kirchhoff's law. *Journal of the Optical Society of America* 1998, *15*, 2735–2744.

Herrmann, F.; Würfel, P. Light with nonzero chemical potential. *American Journal of Physics* 2005, *73*, 717–720.

Howell, J. R.; Siegel, R.; Menguc, M. P. *Thermal Radiation Heat Transfer*; CRC Press: New York, 2010.

Katahara, J. K.; Hillhouse, H. W. Quasi-Fermi level splitting and sub-bandgap absorptivity from semiconductor photoluminescence. *Journal of Applied Physics* 2014, *116*, 173504.

Kirchartz, T.; Rau, U. Electroluminescence analysis of high efficiency Cu(In,Ga)Se_2 solar cells. *Journal of Applied Physics* 2007, *102*, 104510.

Landsberg, P. T. *Thermodynamics and Statistical Mechanics*; Dover: New York, 1978.

Landsberg, P. T. Photons at non-zero chemical potential. *Journal of Physics C* 1981, *14*, L1025.

Lasher, G.; Stern, F. Spontaneous and stimulated recombination radiation in semiconductors. *Physical Review* 1964, *133*, A553–A563.

Pankove, J. I. *Optical Processes in Semiconductors*; Prentice-Hall: Englewood Cliffs, 1971.

Rau, U. Reciprocity relation between photovoltaic quantum efficiency and electroluminescent emission of solar cells. *Physical Review B* 2007, *76*, 085303.

Ries, H.; McEvoy, A. J. Chemical potential and temperature of light. *Journal of Photochemistry and Photobiology A: Chemistry* 1991, *59*, 11–18.

Ross, R. T. Some thermodynamics of photochemical systems. *The Journal of Chemical Physics* 1967, *46*, 4590–4593.

Ruppel, W.; Wurfel, P. Upper limit for the conversion of solar energy. *IEEE Transactions on Electron Devices* 1980, *27*, 877–882.

Sambur, J. B.; Novet, T.; Parkinson, B. A. Multiple exciton collection in a sensitized photovoltaic system. *Science* 2010, *330*, 63–66.

Semonin, O. E.; Luther, J. M.; Choi, S.; Chen, H.-Y.; Gao, J.; Nozik, A. J.; Beard, M. C. Peak external photocurrent quantum efficiency exceeding 100% via MEG in a quantum dot solar cell. *Science* 2011, *334*, 1530–1533.

Shockley, W.; Queisser, H. J. Detailed balance limit of efficiency of p-n junction solar cells. *Journal of Applied Physics* 1961, *32*, 510–520.

Smestad, G.; Ries, H. Luminescence and current-voltage characteristics of solar cells and optoelectronic devices. *Solar Energy Materials and Solar Cells* 1992, *25*, 51–71.

Smestad, G. P. *Optoelectronics of Solar Cells*; SPIE Publications, 2002.

Trupke, T.; Daub, E.; Würfel, P. Absorptivity of silicon solar cells obtained from luminescence. *Solar Energy Materials and Solar Cells* 1998, *53*, 103–114.

Unold, T.; Gütay, L. Photoluminescence analysis of thin-film solar cells. In *Advanced Characterization Techniques for Thin Film Solar Cells*; Abou-Ras, D., Kirchartz, T., Rau, U., Eds.; Wiley-VCH Verlag: Berlin, 2011; pp. 151–175.

van Roosbroeck, W.; Shockley, W. Photon-radiative recombination of electrons and holes in germanium. *Physical Review* 1954, *94*, 1558–1560.

Würfel, P. The chemical potential of radiation. *Journal of Physics C* 1982, *15*, 3967–3985.

23 Reciprocity Relations in Solar Cells and Fundamental Limits to the Photovoltage

The connection between incoming and outgoing radiation, complemented by quantum yields for carrier collection and light emission, provides important relationships between photovoltaic and LED operation modes of a semiconductor diode that can be checked experimentally. This approach introduces fundamental aspects of solar cell operation, which serve as a benchmark for the evaluation of the quality of a class of devices. We develop a detailed study of the photovoltage that can be obtained in a solar cell, with particular attention to the maximal photovoltage in the limit of pure radiative recombination, and we investigate the causes for a reduced photovoltage. For cells that operate close to the radiative recombination limit, the management of photons plays an important role in the performance, demanding control over the photon reabsorption and emission processes in the set of phenomena termed as photon recycling. We also describe the physical constraints of the phenomenon of luminescent refrigeration based on the fact that the radiative emission of a biased semiconductor can be applied to remove heat in the form of photons. Finally, we discuss the application of the reciprocity relation on an organic solar cell where light absorption is associated with the CT between different materials.

23.1 THE RECIPROCITY BETWEEN LED AND PHOTOVOLTAIC PERFORMANCE PARAMETERS

We have previously arrived at important results using arguments of reciprocity of light absorption and emission, and we now progress toward the determination of the radiative saturation current density $j_{0,\text{rad}}$ by the same type of reasoning, which will allow us to calculate central performance features, such as the maximal open-circuit voltage in the radiative limit.

In Chapter 22, we derived the rate of absorption and emission of radiation in a semiconductor film based on the measurement of the spectral absorptance $a(E)$, Equation 22.31. This result, that was obtained from the analysis of a semiconductor film as represented in Figure 20.7, does not yet take into account all material properties of a solar cell concerning the charge collection limitations. For the production of electricity, the solar cell contains selective contacts that make the conversion between a separation of Fermi levels caused by the radiation field, and the external voltage and electrical current. This property goes beyond the equilibrium of incoming and outgoing photons. There exists a connection between charge injection by the absorption of a photon and charge collection properties at the outer contacts, as shown in Figure 23.1. A photon, absorbed at a point x inside the semiconductor absorber, generates an electron-hole pair that requires suitable processes of transport and (sufficiently slow) recombination to produce a certain voltage at the contacts. We recall from Equation 18.28 that the photovoltaic external quantum efficiency (i.e., that of the diode operated as a solar cell) is $EQE_{PV}(E) = a(E)IQE_{PV}(E)$, where the internal quantum efficiency incorporates the limitations of separation and collection of photogenerated charge, that will be studied in detail in Chapter 26.

In the other direction, when the diode is operated as an LED, application of voltage to the selective contacts must split the Fermi levels so that the carrier concentration is increased at the internal point x. Physically, there is an equivalency between the two processes. It is thus necessary to extend Equation 22.31 to include the effect of IQE_{PV}, related to the formation of gradients of the Fermi level in Figure 23.1. The extension is given by the following expression that relates the spectral radiative emission of a solar cell to the blackbody radiation field via the photovoltaic EQE_{PV}

$$\phi_{ph}^{em}(E) = EQE_{PV}(E)\phi_{ph}^{bb}(E)e^{qV/k_BT} \tag{23.1}$$

This result has been shown by (Rau, 2007) under certain restrictive conditions. The exponential in Equation 23.1 takes into account the enhanced recombination by the applied voltage V as indicated in Equation 22.34.

The application of the reciprocity relation in Equation 23.1 in a solar cell to obtain the light emission of the device is not immediately obvious. The conditions in which the reciprocal relation between internal concentration and voltage at the contacts is strictly satisfied have been discussed by (Kirchartz and Rau, 2008; Kirchartz et al., 2016a). The reciprocity is directly justified by the Donolato theorem that

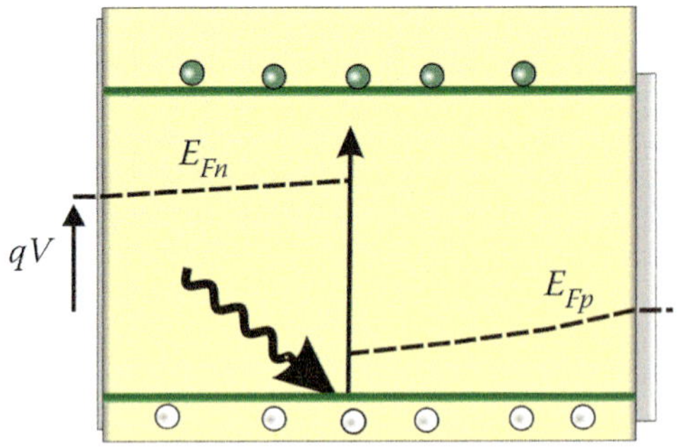

FIGURE 23.1 Scheme of p-type semiconductor material with selective contacts to electrons and holes, a model of a solar cell. The scheme shows the effect of absorption of a photon at a point x in creating a voltage in the external circuit. The Fermi levels are tilted due to limitations of charge collection that require a driving force for the extraction of carriers. Note that the diagram is illustrative of the reciprocity effect; however, the Fermi levels are ensemble properties not associated with a single electron-hole pair.

relates carrier collection and generation in a solar cell; see Section 26.5. However, the latter result only applies when recombination is linear in minority carrier density and when drift-currents can be neglected. This fails under high injection conditions (because recombination becomes non-linear in charge carrier density and minority carriers are no longer well defined), or when a substantial part of the absorber volume is depleted implying that the diffusion-only approximation for transport is violated. Therefore, the exact application of Rau's relationship needs to be carefully investigated for the measurement of external radiative efficiency (Wang and Lundstrom, 2013).

Since the light emission in a solar cell occurs in a restricted energy interval at $E \approx E_g \gg k_B T$, we can use Equation 17.34 to describe the thermal radiation field and Equation 23.1 can be simplified as

$$\phi_{ph}^{em}(E,V) = b_\pi EQE_{PV}(E)E^2 e^{-E/k_BT} e^{qV/k_BT} \quad (23.2)$$

with $b_\pi = 2\pi/h^3c^2$ as defined in Chapter 17. This formula is the generalization of Equation 22.34 for operating photovoltaic devices. A calculation of the radiative properties of different classes of solar cells using the EQE_{PV} based on Equation 23.2 is shown in Figure 23.2.

The radiative saturation current density $j_{0,\mathrm{rad}}$ introduced in Section 21.1 describes the minimum recombination situation and hence contains the main information about the optimal performance of a class of solar cells. We can obtain $j_{0,\mathrm{rad}}$ by integration of the emitted photon flux of the device in equilibrium, Equation 23.2, over all photon energies. The radiative saturation current for emission of the planar device to a hemisphere at the front surface is

$$j_{0,rad} = qb_\pi \int_0^\infty EQE_{PV}(E)E^2 e^{-E/k_BT} dE \quad (23.3)$$

This is the central parameter for evaluation of the materials and device performance.

We consider some simplified versions of Equation 23.3. For the ideal model of a semiconductor with a sharp absorption edge and $IQE_{PV} = 1$, we obtain from Equation 22.7 and Equation 23.3

$$j_{0,rad} = q\frac{\varepsilon_{out}}{\pi} b_\pi k_B T \Gamma_{ph}(E_g,T) e^{-E_g/k_BT} \quad (23.4)$$

For $E_g \gg k_BT$, we can simplify $\Gamma_{ph}(E,T) \approx E^2$, see Equation 22.6, and in the case $\varepsilon_{out} = \pi$, we have the value

$$j_{0,rad} = qb_\pi k_B T E_g^{\,2} e^{-E_g/k_BT} \quad (23.5)$$

Application of Equation 23.3 to calculate $j_{0,\mathrm{rad}}$ in experimental conditions requires an accurate measurement of the EQE_{PV} over a wide spectral band and especially at sub-band-gap values, whose contribution becomes dominant due to the rapid increase of the exponential in Equation 23.3 toward low energies, as remarked before. High sensitivity EQE_{PV} can be obtained with Fourier-transform based photocurrent spectroscopy (FTPS) technique (Vandewal et al., 2009; Tress et al., 2015; Müller and Kirchartz, 2016; Ledinsky et al. 2019; Roland et al. 2019).

Nevertheless, sensitivity of absorption techniques is rather limited. In order to obtain the EQE_{PV} at very low photon energies where absorption is rather weak, one may use the transformation of the LED emission in a similar procedure as that explained in Section 22.4. From EL data at voltage V, the quantum efficiency denoted $EQE_{PV\text{-}EL}$ can be obtained as follows

$$EQE_{PV-EL}(E) = f_g\, \phi_{ph}^{em}(E)E^{-2} e^{E/k_BT} e^{-qV/k_BT} \quad (23.6)$$

This last result provides the spectral shape but not the absolute values of $EQE_{PV}(E)$. However, the normalizing factor f_g may be adjusted from one point of the directly measured EQE_{PV}, in order to establish that the shapes of $EQE_{PV\text{-}EL}$ and EQE_{PV} are equal.

The complementary measurements of EQE_{PV} and $EQE_{PV\text{-}EL}$ is illustrated in Figure 23.3 for a dye-sensitized solar cell and in Figure 23.4 for a Cu(In,Ga)Se_2 solar cell. The combination of direct and indirect

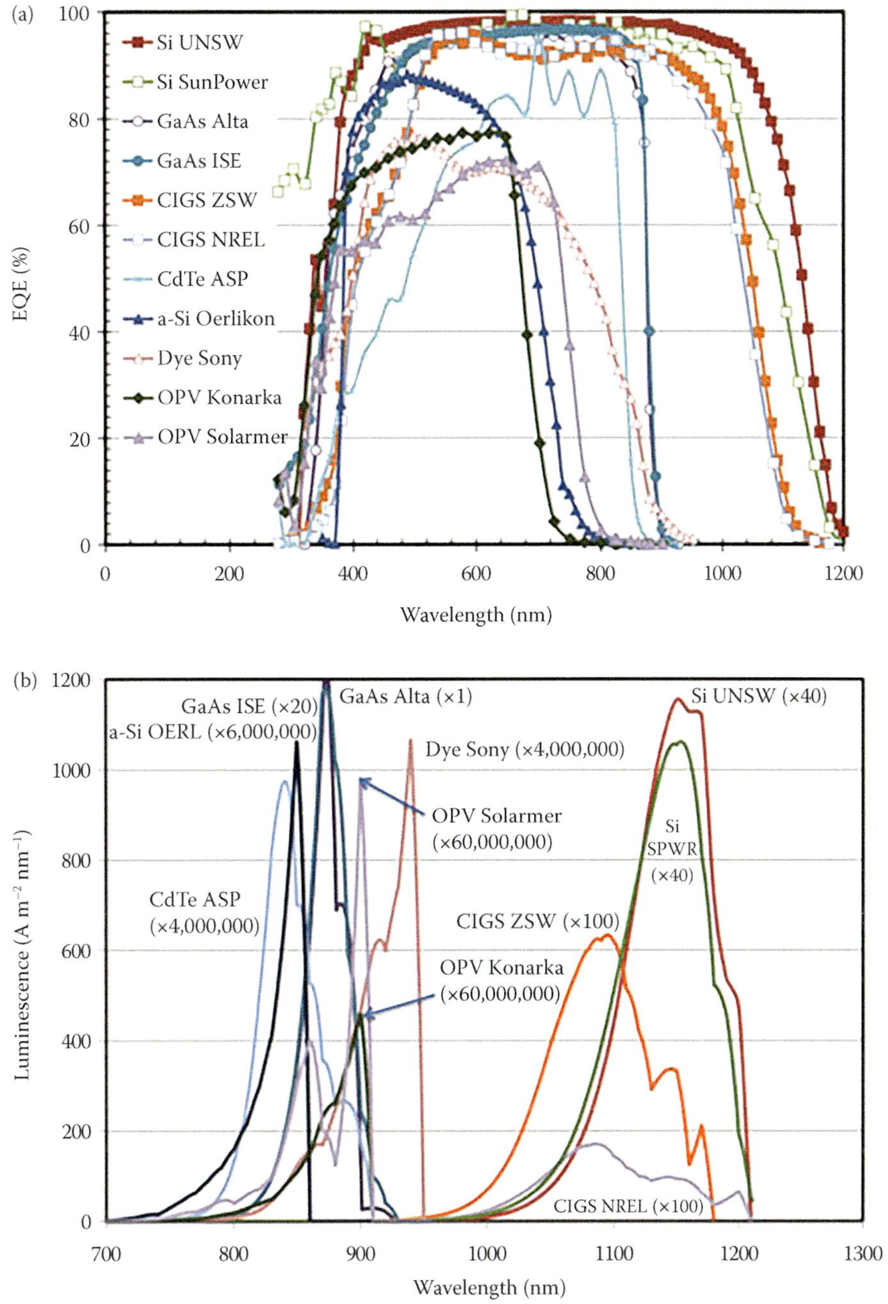

FIGURE 23.2 (a) Measured EQE_{PV} for different types of solar cells. (b) The calculated spectral luminescence by application of Equation 23.1., $\phi_{ph}^{em}(E) = EQE_{PV}(E)\phi_{ph}^{bb}(E)e^{\frac{qV}{k_BT}}$ (Reproduced with permission from Green, M. A. *Progress in Photovoltaics: Research and Applications* 2012, *20*, 472–476.)

methods therefore provides a full EQE_{PV} that enables very accurate determination of $j_{0,\mathrm{rad}}$ by Equation 23.3. This calculation forms a basis to determine the radiative limit to the photovoltage and it is explained later in Figure 23.18. A different method to achieve the same purpose avoiding the calibration factor uses the measurement of the absolute intensity PL, as indicated previously in Figure 22.9.

As remarked earlier, in contrast to the idealized model of sharp absorption onset in Figure 22.6, sub-band-gap states cause a large contribution to the emission spectrum in the onset region of photon energies, and hence determine to a large extent the operation of the radiative balance and the open-circuit voltage. This region of low energy absorptance is quantified by the shape of the Urbach tail and the tailing parameter E_U in Eq. 19.9. In

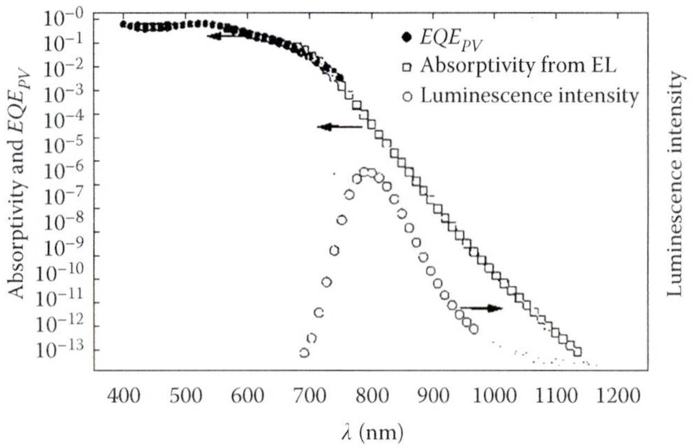

FIGURE 23.3 EL measured on a dye-sensitized solar cell and the absorptance that is calculated by its division by $E^2e^{-E/kBT}$ as in Equation 23.6. The obtained relative values for the absorptance have been fitted to the measured EQE_{PV} (*full circles*). (Reproduced with permission from Trupke, T. et al. *The Journal of Physical Chemistry B* 1999, *103*, 1905–1910.)

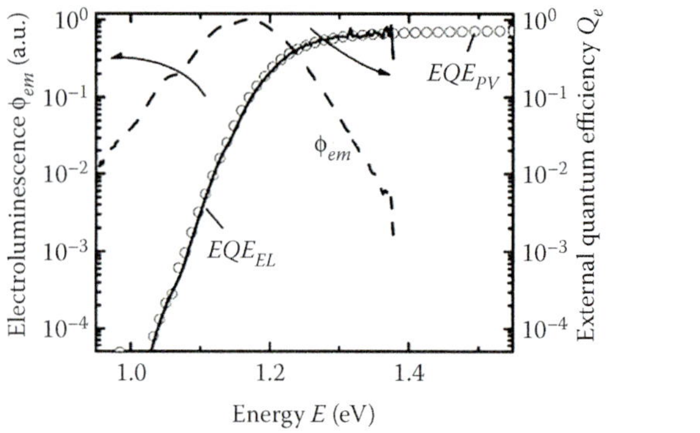

FIGURE 23.4 Measurement of an EL spectrum $\varphi_{ph}^{em}(E)$ of a Cu(In,Ga)Se_2 solar cell, which allows us to compute the solar cell quantum efficiency $EQE_{PV\text{-}EL}$. Thus, $EQE_{PV\text{-}EL}$ is derived from the EL measurement in relative units. A direct measurement of the quantum efficiency EQE_{PV} allows one to scale $EQE_{PV\text{-}EL}$ and to verify that the relative shape of $EQE_{PV\text{-}EL}$ and EQE_{PV} is equal. (Reproduced with permission from Kirchartz, T.; Rau, U. *Physica Status Solidi (A)* 2008, 205, 2737–2751.)

Figure 23.5 the absorptance of $CH_3NH_3PbI_3$ is found by the two different methods, the FTPS and the transformation of PL. The measurement as a function of temperature indicates the tendency of E_U in Figure 23.5b and the slope obtained in both techniques is nearly the same. The values obtained from PL spectra measurements can be considered more accurate, as they are unaffected by band-to-band absorption.

In the experimental analysis of photovoltaic devices, the recombination parameter must take into account all recombination pathways as outlined in Figure 21.2. By Equation 21.5, the actual diode saturation current takes the form

$$j_0 = \frac{1}{EQE_{LED}} qb_\pi \int_0^\infty EQE_{PV}(E)E^2 e^{-E/k_BT} dE, \quad (23.7)$$

where EQE_{LED} is the external quantum efficiency of a LED, defined in Equation 18.23. Note that EQE_{LED} is a scalar quantity unlike the solar cell EQE_{PV}, which is a function of wavelength or photon energy.

So far, it was naturally assumed that the environment provides a steady background of Planckian radiation at the same temperature of the solar cell, T_A. We now consider a different type of environment in which the solar cell device operates in an unconventional way.

From the property of equilibrium of outgoing and incoming radiation in dark equilibrium, we derived important relationships in Equations 21.4 and 22.31. However, by setting the thermal radiation coming from an emitter at a *lower* temperature than that of the solar cell device, $T_E < T_A$, an imbalance of incoming and outgoing radiation can be forced, as shown in Figure 23.6. Thus, the parameter of thermal generation is reduced and Equation 21.4 is displaced from equilibrium as $j_0^{th} < j_{rec}(V=0)$. Since the solar cell is still emitting radiation at T_A, in this situation of "negative radiation," we obtain at short circuit an observable negative current in the dark (Santhanam and Fan, 2016). The diode can be operated at reverse bias resulting in thermal-to-electrical energy conversion.

23.2 FACTORS DETERMINING THE PHOTOVOLTAGE

We now focus our attention on the open-circuit voltage V_{oc}, which is the voltage produced autonomously by the solar cell when it is irradiated with light. In Section 22.1, we discussed properties of the photocurrent. This is a straightforward quantity in that the absorbed photons are converted to electron-hole pairs that are measured as an electrical current j_{ph} at short circuit, as stated in Equation 22.1. The open-circuit voltage is rather easy to measure and it is a main performance parameter of a solar cell. However, in contrast to the interpretation of the photocurrent, the analysis of the photovoltage involves subtle aspects of the operation of the solar cell that we treat in the following sections.

In the present discussion, we continue to use the reference model of Figure 21.1 in which the solar cell is composed of a single homogeneous light-absorbing material with the property of very high carrier mobilities so that the application of either external illumination or bias voltage produces uniformly flat Fermi levels throughout

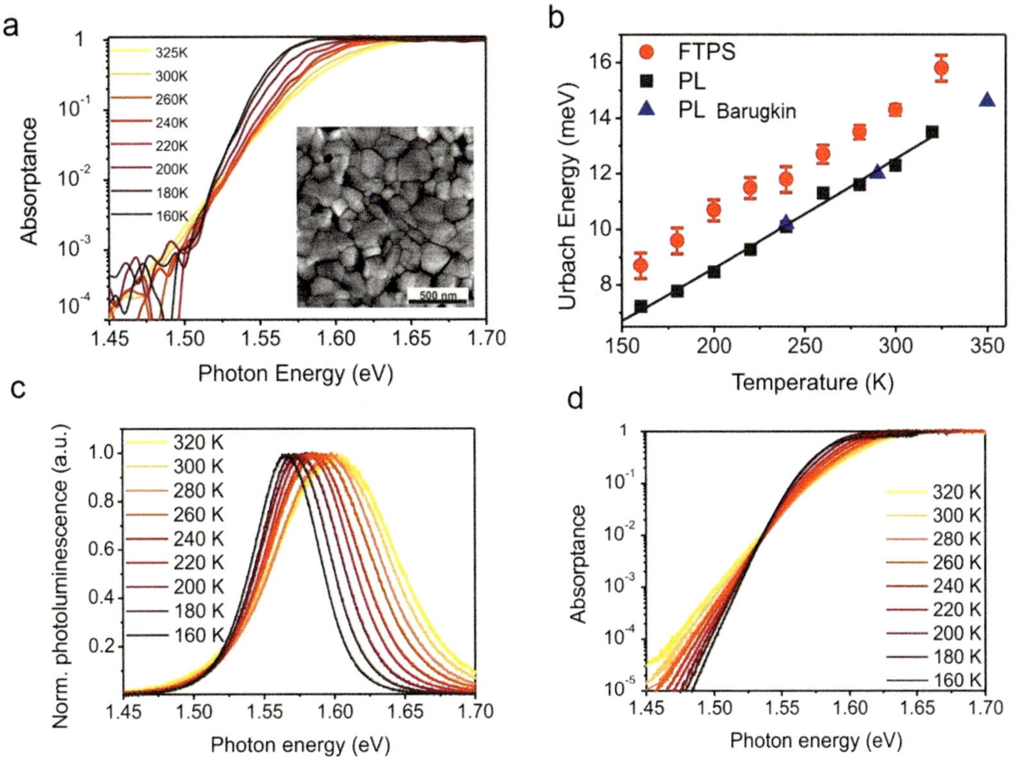

FIGURE 23.5 (a) Temperature dependence of the FTPS absorption spectra of $CH_3NH_3PbI_3$ films. When temperature decreases, and the band gap energy decreases, the absorption edge sharpens. The inset shows an SEM image of the $CH_3NH_3PbI_3$ film prepared on a glass substrate. (b) Urbach energy temperature dependence derived from FTPS (red dots) and PL (black squares) absorptance spectra. The black line is the Urbach energy dependence, the estimated error for Urbach energy calculated from PL spectra is 0.2 meV. Additionally, three blue triangles show E_U values calculated from PL absorption measurement published in (Barugkin et al., 2015). Lower row: Temperature dependence of the $CH_3NH_3PbI_3$ PL spectra (c) and PL based absorptance spectra (d). All absorptance spectra intersect at one point, the so-called Urbach focus. PL spectra, as well as the absorptance spectra, are normalized. (Reproduced by permission from Ledinsky et al., 2019).

the absorber. The solar cell possesses ideal selective contacts that transform the splitting of Fermi levels to the voltage as Equation 20.7. The physical and materials properties of the contacts can impose further limitations to the photovoltage in the solar cell that are ignored in this model. The band bending, surface recombination, and other usual features that occur during operation will be discussed in Chapter 26.

We consider the solar cell to be illuminated by a light source in open circuit conditions as shown in Figure 21.2. The optical radiation over the background thermal radiation creates excess carriers in the transport bands of the semiconductor. Since no carriers are extracted at the contacts, the balance equation reads

$$j_0 + j_{ph} = j_{rec} \tag{23.8}$$

On the left-hand side of Equation 23.8, we count the generation fluxes from the thermal and external sources, and at the right side, we have the internal recombination current, defined in Section 21.2, that removes photogenerated electrons and holes.

Under standard conditions, the solar cell receives solar AM1.5G radiation as explained in Section 17.8, denoted $\phi_{ph}^{AM1.5}(E)$, with an integrated photon flux $\Phi_{ph}^{AM1.5}$ given by Equation 17.48. In the following analysis, we will be mainly interested in the case in which a substantial illumination with a power flux of the order of 1 sun impacts the solar cell. Therefore, we will label the generation flux in terms of a photocurrent that clearly offsets the thermal generation, $j_{ph}^{sun} \gg j_0$, and the latter will be neglected.

When the charge separation steps that produce fresh electrons and holes in the respective energy levels have been realized, we obtain an ensemble of free charge carriers in the semiconductor, or in separate phases, where they are mobile. Since these are excess populations of electrons and holes with respect to the equilibrium situation, we have formed a situation of separation of Fermi levels and the recombination process starts immediately,

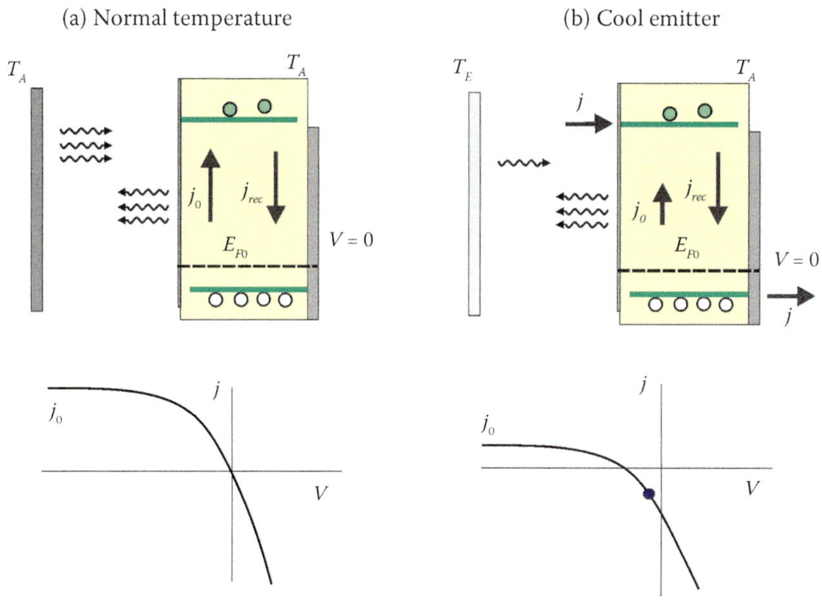

FIGURE 23.6 "Negative illumination" of a semiconductor diode. (a) The solar cell is at the same temperature T_A as the surrounding thermal radiation. We show the equilibrium in the dark of carrier generation by blackbody thermal radiation and recombination, which balances to net zero current at short circuit. At reverse voltage, recombination is suppressed and the diode saturation current j_0 corresponds to the extraction of the carriers generated by the thermal radiation. (b) The solar cell is at the ambient temperature T_A but now it receives thermal radiation from a colder source at T_E. The generation rate that determines j_0 decreases, while the recombination rate remains the same. Hence, a negative current occurs at short circuit, and power can be produced by operating the cell at negative voltage as shown by a point in the current density-voltage characteristic.

while the continuous absorption of photons produces a renewed supply of free electronic carriers. If the device is at open-circuit condition, the carriers accumulate inside causing a photovoltage V_{oc} but they cannot leave the device to produce electrical work. The recombination current at open circuit can be obtained from Equation 21.8, and the balance equation becomes

$$j_{ph}^{sun} = j_0 e^{qV_{oc}/mk_BT} \quad (23.9)$$

Therefore, we get the result

$$V_{oc} = \frac{mk_BT}{q} \ln\left(\frac{j_{ph}^{sun}}{j_0}\right) \quad (23.10)$$

Clearly, to obtain a high photovoltage, it is necessary to achieve a large separation of electron and hole Fermi levels, which means that the carrier densities have to be as large as possible, as suggested in Figure 21.2. Therefore, it is obvious that recombination is the main process limiting the photovoltage in a solar cell. According to Equation 23.10, the photovoltage is controlled only by the recombination parameter j_0, given the photogeneration rate determined by j_{ph}^{sun}.

For the evaluation of solar cell materials, it is very interesting to establish the largest V_{oc} that can be possibly obtained in a class of materials. The maximal open-circuit voltage is reached when the recombination parameter j_0 takes the minimal value, with all nonradiative pathways being suppressed. This situation is reached exactly if the saturation current density j_0 is equal to the radiative saturation current density $j_{0,\mathrm{rad}}$ defined in Equation 23.3. Assuming an ideality factor $m = 1$, the *radiative limit* to the photovoltage is

$$V_{oc}^{rad} = \frac{k_BT}{q} \ln\left(\frac{j_{ph}^{sun}}{j_{0,rad}}\right) \quad (23.11)$$

As mentioned earlier, Equation 23.3 is the critical parameter to establish the radiative voltage.

In preliminary arguments about the solar cell efficiency presented in Section 22.1, we regarded the output electrical power as the product of the photocurrent and the voltage corresponding to band gap energy. However a better estimation of the extracted electrical power is given by the voltage that carriers can produce when they are accumulated in the solar cell, which is the V_{oc}. Therefore, it is central to the discussion of power conversion efficiency to obtain the difference between qV_{oc} and E_g, whose magnitude causes a significant loss of extracted power. This important quantity is denoted the voltage deficit $W_{oc} = E_g/q - V_{oc}$, which will take the minimal value in the case of the radiative limit to the photovoltage. We explore now the main physical effects governing the radiative photovoltage of a solar cell.

We first analyze the value of the radiative voltage under the assumption of a planar cell with an opaque backside, sharp band gap, and excellent charge collection (IQE = 1). The sun is modelled as a blackbody at high temperature T_S. The current density has been previously described in Equation 22.7, and the saturation current at the cell temperature T_A is given in Equation 23.4. Inserting these conditions in Equation 23.11, we obtain the following equation, where we have expressed the radiative voltage in successive terms that reduce the carrier extraction energy with respect to the semiconductor band gap:

$$qV_{oc}^{rad} = E_g - \frac{T_A}{T_S}E_g - k_BT_A \ln\left(\frac{\varepsilon_{out}}{\varepsilon_{in}}\right) + k_BT_A \ln\left(\frac{T_S\Gamma(E_g,T_S)}{T_A\Gamma(E_g,T_A)}\right) \quad (23.12)$$

The first term of voltage reduction is the decrease of energy due to the Carnot factor $(1 - T_A/T_S)$, accounting for the equilibration of entropy necessary for extraction of work in a thermal machine as discussed in Section 5.9. The second loss term in Equation 23.12 represents entropy generation by the occupation of available states associated to expansion of the étendue of the incoming and outgoing photon fluxes. Note that the entropy loss can be eliminated by a concentrator that makes the solid angle of escape of the light to be the same as that of the light arriving to the device. The third term describes an *increase* of free energy per carrier as a result of the temperature of the absorbed and emitted photon distributions, see (Markvart, 2007; Hirst and Ekins-Daukes, 2011) for further details. These components of the voltage loss are shown in Figure 23.7a. The following approximation assuming $E_g \gg k_BT_S(\Gamma \approx E_g^{\ 2})$ is widely used (Ruppel and Würfel, 1980):

$$qV_{oc}^{rad} = E_g\left(1 - \frac{T_A}{T_S}\right) + k_BT_R \ln\left(\frac{\varepsilon_S}{\pi}\right) + k_BT_A \ln\left(\frac{T_S}{T_A}\right) \quad (23.13)$$

This approximation to the radiative losses is shown in Figure 23.7b and this is the expression used for the radiative photovoltage in Figure 22.2b. The average energy of absorbed photons is (Würfel, 2009)

$$\bar{E}(E_g,T_S) = \frac{\Phi_E(E_g,T_S)}{\Phi_{ph}(E_g,T_S)} = \frac{\chi_E(E_g,T_S)}{\Gamma_{ph}(E_g,T_S)} \quad (23.14)$$

In order to obtain more explicit expressions related to measured quantities, we write Equation 23.11 as

$$qV_{oc}^{rad} = k_BT \ln\left(j_{ph}^{sun}\right) - k_BT \ln\left(qb_\pi \int_0^\infty a(E)E^2 e^{-E/k_BT} dE\right) \quad (23.15)$$

For a solar cell with a sharp band gap and constant absorptivity a, we have from Equation 22.5

$$qV_{oc}^{rad} = E_g + k_BT \ln(j_{ph}^{sun}) - k_BT \ln\left(qb_\pi a k_BT E_g^{\ 2}\right) \quad (23.16)$$

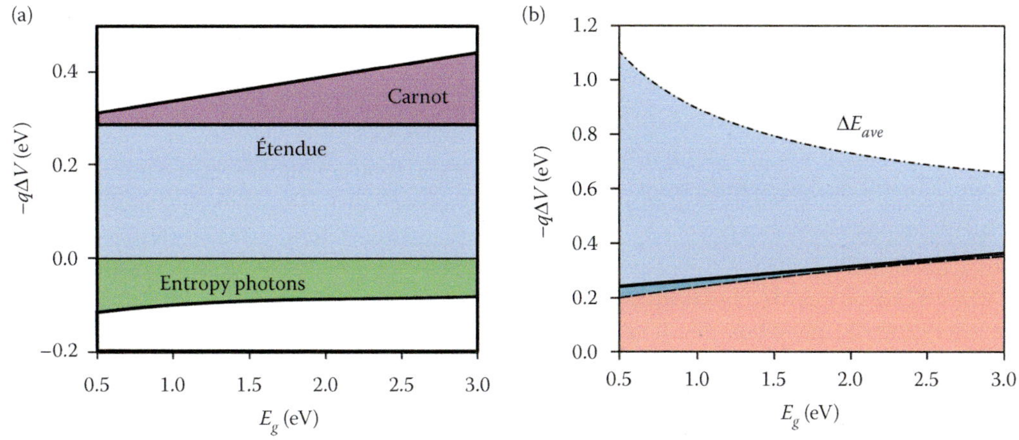

FIGURE 23.7 (a) Components of the thermodynamic loss in the radiative photovoltage for a solar cell that absorbs black body radiation at $T_S = 5800$ K, $T_A = 300$ K, according to Equation 23.13, using $\varepsilon_{in} = \varepsilon_S^{1.5AM} = 4.8\times10^{-5}$, $\varepsilon_{out} = \pi$. (b) The total voltage loss (thick line). The dashed line is the formula $qV_{oc}^{rad} = E_g\left(1 - T_A/T_S\right) + k_BT_R \ln\left(\varepsilon_S/\pi\right) - k_BT_A \ln\left(T_A/T_S\right)$. The dashed-dot line is the average energy of absorbed photons.

Alternatively, if the $EQE_{PV}(E)$ is known over a broad spectral range for a solar cell device, the radiative voltage of a device in standard calibration conditions can be obtained as follows

$$V_{oc}^{rad} = \frac{k_B T}{q} \ln\left(\frac{\int_0^\infty EQE_{PV}(E)\phi_{ph}^{AM1.5}(E)dE}{\int_0^\infty EQE_{PV}(E)\phi_{ph}^{bb}(E)dE} \right) \tag{23.17}$$

We remark that the V_{oc}^{rad} in Equation 23.16 shows the following general properties:

(1) It directly correlates with the value of the semiconductor band gap.
(2) It increases logarithmically with the photogeneration rate.
(3) Decreases logarithmically with the recombination rate, as indicated by the negative dependence on absorptance parameter a in Equation 23.16. The reason for this is that higher absorption implies a larger rate of radiative emission that causes the loss of carriers. However, a higher absorption will produce a larger j_{ph} that offsets by far the loss of V_{oc} in the calculation of the efficiency.

The voltage deficit $W_{oc} = E_g - V_{oc}$ is the energy price associated to the difference of the energy of thermalized photogenerated electrons and holes, and the actual separation of Fermi levels. The value E_g is considered an upper limit to V_{oc}. The reason for this is that the conduction and valence band levels contain a very large density of states and it is not likely that the Fermi levels overcome such energy levels by photogeneration. Thus even at very high light intensities we expect at most $W_{oc} = 0$. At reasonable light intensities the actual value $W_{oc} > 0$ is an important feature in determining solar cell efficiency.

As shown in Figure 22.2b there are two main components to the voltage deficit. The first is the loss associated to the radiative limit of the photovoltage, $E_g/q - V_{oc}^{rad}$, which can be calculated from basic principles as in Equation 23.16, and is about 300 mV for AM1.5G radiation at 1sun as indicated in the Figure 22.2b. Thus it is true that the photovoltage is correlated directly with the band gap value, but in normal operation conditions it is lower than E_g by reasons of fundamental balance of radiation, that requires an unavoidable amount of recombination.

The second quantity contributing to $W_{oc} > 0$ is the particular value of a solar cell device or class of materials, associated to additional loss processes as nonradiative recombination. In order to quantify the departure of the voltage from the radiative limit, we can use Equation 23.7 to express the actual recombination parameter j_0 in terms of the efficiency of the solar cell operated as a LED, EQE_{LED}. Thus, we have

$$V_{oc} = \frac{k_B T}{q} \ln\left(\frac{j_{ph}}{j_{0,rad}} EQE_{LED} \right) \tag{23.18}$$

Therefore, the actual open-circuit voltage relates to the ideal radiative limit as

$$V_{oc} - V_{oc}^{rad} = \frac{k_B T}{q} \ln\left(EQE_{LED} \right) \tag{23.19}$$

This is the result derived by (Ross, 1967). The conclusion is that when comparing cells of similar characteristics such as the band gap of the semiconductor, the solar cell with the highest radiative efficiency (large EQE_{LED}) will also show the highest photovoltage because it contains less nonradiative losses, and consequently, less recombination. Contrary to intuition, a solar cell must be a good radiator in order to approach the theoretical limits of energy conversion efficiency from sunlight to electricity (Miller et al., 2012).

Taking the typical form for the IQE_{LED} in Equation 18.24 and neglecting outcoupling effects, we can write Equation 23.19 as

$$V_{oc}^{rad} - V_{oc} = -\frac{k_B T}{q} \ln\left(\frac{j_{rad}}{j_{rad} + j_{nrad}} \right) \tag{23.20}$$

This equation indicates the fraction of recombination events that are radiative.

The correlation of the radiative voltage to the EQE_{LED} is shown in Figure 23.8 for a number of inorganic, perovskite and organic solar cells (Yao et al., 2015). In Figure 25.10, the value of the voltage deficit for optimal devices of a varied collection of PV technologies is presented. Normally, for solar cells that are not yet close to top quality performance, nonradiative recombination is the major problem to increase the obtained photovoltage and the factor j_{rad}/j_{nrad} can be a very small number. Nevertheless, every increase of a factor 10 in EQE_{LED} improves the photovoltage by 60 mV. On the other hand, when the recombination is mostly radiative, the efficient extraction of photons by the optical design of the cell becomes a dominant issue as will be analyzed in the next sections.

23.3 EXTERNAL RADIATIVE EFFICIENCY

So far, we have used information about the light absorption and recombination features to predict the performance of the photovoltage. As mentioned in Chapter 21,

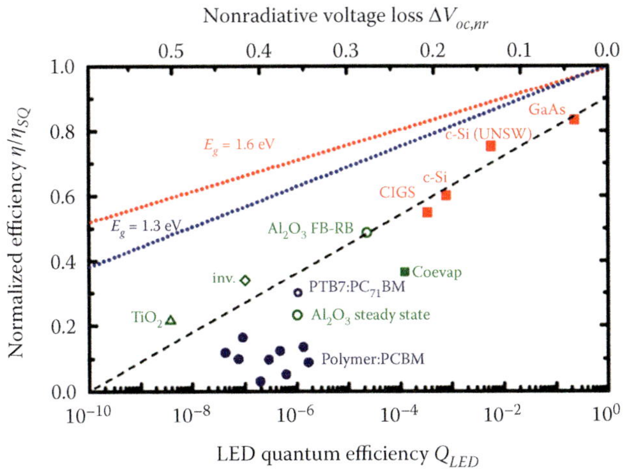

FIGURE 23.8 Power conversion efficiency for different solar cell technologies normalized to the Shockley-Queisser limit as a function of LED quantum efficiency EQE_{LED}. The *dotted lines* define the theoretical limits of various EQE_{LED} at two different optical band gaps: 1.6 and 1.3 eV. The top *x*-axis is the nonradiative voltage loss over the range of EQE_{LED}, referring to Equation 23.19. The data points for inorganic solar cells are shown in *red squares*. Different perovskite fabrication technologies are shown in *green*. *Open green* and *circle points* represent the solution-processed perovskite devices made on mesoporous TiO_2 and Al_2O_3 films, respectively. The open green diamond point is the solution-processed inverted device using PEDOT:PSS and PCBM interlayers. The solid square point is a coevaporated film. Different organic solar cells are shown in *blue*. PTB7:$PC_{71}BM$, the organic system with the highest PCE, is shown as the *open blue circle*. The *dashed line* is a guide to the eye representing the approximate experimental trend. (Reprinted with permission from Yao, J. et al. *Physical Review Applied* 2015, 4, 014020.)

it is also interesting to analyze the features of the radiative emission of the cell. Normally, the radiative saturation current density $j_{0,rad}$ is rather small, but one can operate the solar cell as an LED and obtain readily visible EL when a sufficient voltage is applied. We continue to assume that the applied voltage is translated into homogeneous quasi-Fermi levels, which requires high mobilities and other conditions that will be studied in the subsequent chapters. The total radiative flux at V_{oc} is

$$\Phi_{ph}^{em} = \frac{j_{0,rad}}{q} e^{qV_{oc}/k_BT} \tag{23.21}$$

The radiative emission in the radiative limit of the photovoltage would be

$$\Phi_{ph}^{em,rad} = \frac{j_{0,rad}}{q} e^{qV_{oc}^{rad}/k_BT} \tag{23.22}$$

Therefore, the LED quantum efficiency is

$$EQE_{LED} = \frac{\Phi_{ph}^{em}}{\Phi_{ph}^{em,rad}} = \frac{e^{qV_{oc}/k_BT}}{e^{qV_{oc}^{rad}/k_BT}} \tag{23.23}$$

Note that this result follows directly from Equation 23.19. Using Equation 23.11, we can write Equation 23.23 as

$$EQE_{LED} = \frac{j_{0,rad}e^{qV_{oc}/k_BT}}{j_{ph}^{sun}} \tag{23.24}$$

Equation 23.24 compares the theoretical radiative current at open circuit to the actual photocurrent. It can be written also as

$$EQE_{LED} = \frac{e^{qV_{oc}/k_BT}\int_0^\infty EQE_{PV}(E)\phi_{ph}^{bb}(E)dE}{\int_0^\infty EQE_{PV}(E)\phi_{ph}^{AM1.5}(E)dE} \tag{23.25}$$

As mentioned earlier, for high-efficiency solar cells, the radiative efficiency makes a major difference in the performance of otherwise similar devices. For example, two reported GaAs cells in Figure 23.2 (Green, 2012) show a substantial difference of radiative efficiency by a factor of 20, with EQE_{LED} = 22.5% for the best cell. This improvement produces a change of the respective cell efficiencies from 26.4% to 27.6%. Since these cells are rather close to the radiative limit, the gain of photonic quality is mainly responsible for the increase of efficiency.

The EQE_{LED} of different types of solar cells shown in Figure 23.8 has been tabulated by Yao et al. (2015). This figure shows that a good solar cell is a good LED too. Note that the solar cells based on organic materials have the lowest EQE_{LED}. This behavior is primarily due to the existence of several parallel recombination pathways in addition to recombination through the absorber, which, by itself, represents only 10^{-6} of the total recombination.

23.4 PHOTON RECYCLING

At this point, we wish to compute the actual rate of radiative emission from a solar cell, and we adopt the usual planar geometry for a recombination diode of thickness *d* with flat Fermi levels in which all recombination is radiative, that is $IQE_{LED} = 1$ and $EQE_{PV}(E) = a(E)$. Based on the previously obtained detailed balance considerations, the emitted photon flux at the semiconductor surface is

$$\Phi_{ph}^{em,ext} = \frac{j_0^{ext}}{q} = b_\pi \int_0^\infty a(E)E^2 e^{-E/k_BT} dE \tag{23.26}$$

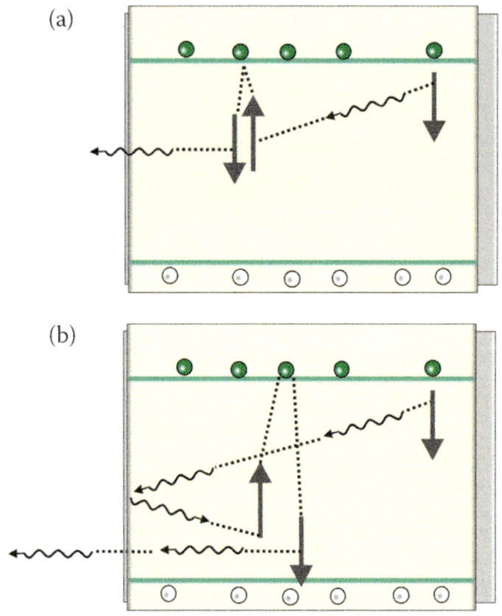

FIGURE 23.9 Illustration of photon recycling processes. (a) A photon is generated inside the absorber away from the transparent contact. It is reabsorbed and reemitted and then expelled out of the layer, thereby contributing to the luminescent flux. (b) The photon is reflected at the transparent contact, then reabsorbed and reemitted in a direction inside the escape cone of the semiconductor.

While Equation 23.26 is obtained viewing the cell from outside and is, therefore, labeled external emission, we may as well obtain the generated photons from within, starting from the diode model of Figure 21.1. The total photon production is determined by integration of the recombination rate U_{rec} over the whole diode thickness as previously shown in Equation 21.16. We obtained

$$j_0^{int} = qdB_{rad}n_i^2 \tag{23.27}$$

Here, n_i is the intrinsic carrier density.

The radiative lifetime has been calculated earlier in Equation 22.23 in terms of the absorption coefficient α and the refraction index n_i. The internally emitted flux has the expression

$$\Phi_{ph}^{em,int} = \frac{j_0^{int}}{q} = 4dn_r^2 b_\pi \int_0^\infty \alpha(E) E^2 e^{-E/k_BT} dE \tag{23.28}$$

Antonio Martí identified the Shockley paradox in that both formulae 23.26 and 23.28 do not match each other (Martí et al., 1997). It is remarked that using the van Roosbroeck–Shockley radiative lifetime in the fundamental diode equation, one cannot obtain the actual radiative emission of the solar cell, Equation 23.26. The latter equation is derived from the fundamental restriction of detailed balance of incoming and outgoing fluxes in equilibrium, as in the Shockley-Queisser approach. The failure of Equation 23.28 to produce the output emission is due to the fact that it assumes that all photons resulting from radiative recombination are automatically expelled out of the cell. But this is far from true, since luminescent photons can be reabsorbed producing a fresh electron-hole pair, in a process termed photon recycling (Asbeck, 1977), previously commented upon in Section 18.3. The significance of photon reabsorption and reemission in the operation of a solar cell is shown schematically in Figure 23.9. Equation 23.28 predicts that the emitted flux grows in proportion to the solar cell thickness but in reality, the radiative emission saturates when the absorptance of the cell becomes $a(E) = 1$ for a thick layer, as indicated in Equation 23.26.

Since the photons created by radiative recombination must be either expelled out of the cell or reabsorbed, an emission probability p_e can be defined as (Kirchartz et al., 2016)

$$p_e = \frac{\Phi_{ph}^{em,\text{ext}}}{\Phi_{ph}^{em,\text{int}}} = \frac{\int_0^\infty a(E)\phi_{ph}^{bb}(E,\varepsilon=\pi)dE}{\int_0^\infty dn_r^{\,2}\alpha(E)\phi_{ph}^{bb}(E,\varepsilon=4\pi n_r^{\,2})dE} = \frac{\int_0^\infty \pi a(E)E^2 e^{-E/k_BT}dE}{\int_0^\infty 4\pi dn_r^{\,2}\alpha(E)E^2 e^{-E/k_BT}dE} \tag{23.29}$$

Here $\phi_{ph}^{bb}(E,\varepsilon)$ is the blackbody emission with étendue ε established in Equation 17. 34. The integrand numerator is the external luminescence flux spectrum with a Lambertian distribution in solid angle π out of the front surface, into the vacuum with refractive index $n_r = 1$. And the integrand in the denominator is the spectrum of internal isotropic radiation with solid angle 4π, as explained in Equation 22.25, that should leave the cell of thickness d through the front face (Pazos-Outon et al. 2018).

The contribution of photon recycling to the power output of the solar cell is only significant when recombination is mainly radiative, which so far has been achieved by GaAs and lead halide perovskites. As shown in Figure 23.10, Thomas Kirchartz compared two cases of lead halide perovskite solar cells with different scattering properties of the front surface, where both cells have similar antireflective coating and total reflecting back surface, as in the usual model adopted in previous sections. In Figure 23.10a, the Lambert-Beer model allows only a double pass of the light ray, while in Figure 23.10b, the Lambertian diffuser scatters the rays randomly and increases the light trapping as commented in Section 18.1. Let us discuss the ratio

$$P(E) = \frac{a(E)}{4dn_r^2\alpha(E)} \tag{23.30}$$

which determines the emission probability according to Equation 23.29. The spectral shape of absorption coefficient $\alpha(E)$ of the lead iodide perovskite is shown in Figure 19.5.

For weak absorption of sub-band-gap wavelengths $(\alpha(E)d \ll 1)$, the photon will hit the surface many times before it has a chance to be internally reabsorbed. From Equation 18.12, the Lambert-Beer model gives the absorptance $a(E) = 2\alpha(E)d$, hence $P = 1/2n_r^2$. This is because the percentage of photons that have an angle that allows them to escape is just $(2n_r^2)^{-1}$, as discussed in Section 18.1. In the case of the light trapping model, Equation 18.18 gives the absorptance $a(E) = 4n_r^2 d\alpha(E)$ in the weak absorption range. Since the escape cone from a semiconductor is quite narrow, the external emission requires repeated attempts. The photon impacting the surface is randomized many times until it hits the escape cone and hence the probability of emission is $P(E) = 1$ in the limit of weak absorption.

However, for the domain of supra-band-gap photons with strong optical absorption, the behavior is rather different. In this range of photon energies, $\alpha(E)d > 1$ is satisfied. It is, therefore, not feasible to directly extract these photons (except those generated close to the transparent contact) as they have a very large chance of reabsorption, so that Equation 23.28 grossly overestimates the flux that can be extracted. The absorption coefficient grows steadily, as shown in Figure 19.5 while the absorptance stabilizes to $a(E) = 1$, so that $P(E) < 1$ makes a contribution to the integrals in Equation 23.26 that decreases p_e. The relevant terms in the integral and their ratios are depicted in Figure 23.11, see Kirchartz et al. (2016a, b) for a more detailed discussion. In conclusion, the smaller the p_e, the larger the need for efficient photon recycling. If $p_e = 1$, then photon recycling is nonexistent, but when

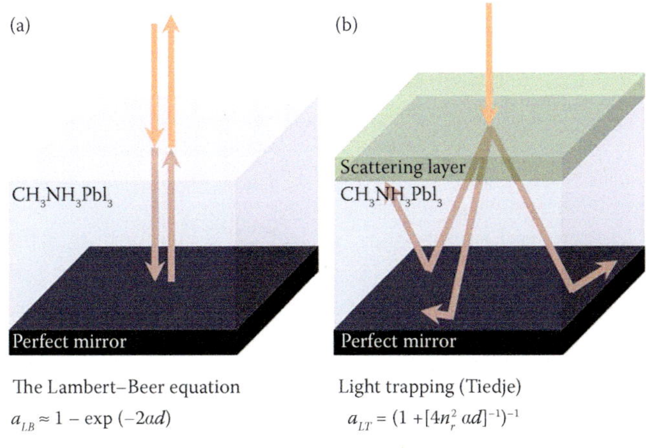

FIGURE 23.10 Schematic depiction of two optical models. Panel (a) describes perfect light incoupling and a perfect back reflector but no scattering. The absorptance is described for direct incidence by the Lambert-Beer equation. The integration over all angles of absorption and emission has very little effect on the total result. Panel (b) still considers perfect incoupling and a perfect back reflector but in addition includes a perfect Lambertian scattering layer that leads to a Lambertian distribution of angles. (Reproduced with permission from Kirchartz, T. et al. *ACS Energy Letters* 2016, 731–739.)

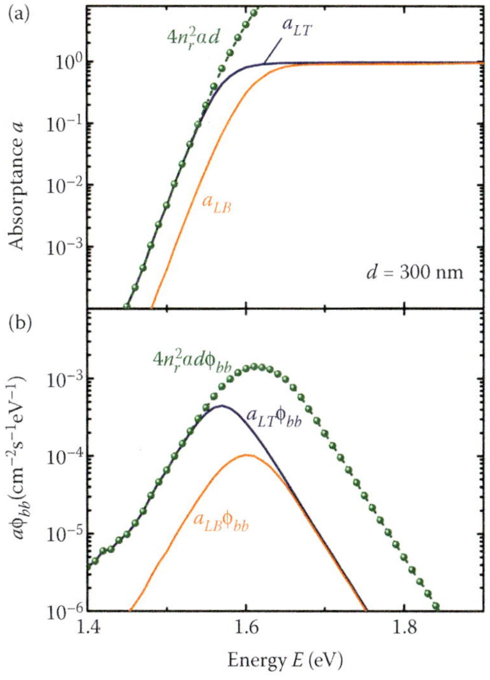

FIGURE 23.11 (a) Absorptance versus photon energy for different configurations (the Lambert-Beer or Lambertian light trapping) compared to the $4n_r^2\alpha d$ that defines the upper limit for the absorptance for $\alpha d \ll 1$. (b) Same as panel (a) but multiplied by the blackbody spectrum. The integral over the curves in panel (b) controls the ratio j_0^{int}/j_0^{ext} of the saturation current densities and thereby the probability of emission p_e. (Reproduced with permission from Kirchartz, T. et al. *ACS Energy Letters* 2016, 731–739.)

it is smaller than 1, then multiple reabsorption events are needed to maximize the external luminescence efficiency. This analysis shows that light trapping and light emission are correlated processes (Miller et al., 2012).

Consider the outcoupling efficiency defined in Section 18.3, which is the probability that a photon is emitted and not reabsorbed. In the absence of parasitic absorption, it is simply

$$\eta_{\text{outco}} = p_e \tag{23.31}$$

where p_e is given in Equation 23.29. Every cycle of reabsorbed photons increases the external quantum efficiency in the geometric series

$$\begin{aligned} EQE_{LED} &= IQE_{LED}\eta_{outco} + IQE_{LED}\left(1-\eta_{outco}\right)IQE_{LED}\eta_{outco} + \cdots \\ &= \frac{IQE_{LED}\eta_{outco}}{1-IQE_{LED}\left(1-\eta_{outco}\right)} \end{aligned} \tag{23.32}$$

An extension of the voltage losses including the photon recycling can be formulated using Equation 23.32 in 23.19 (Rau and Kirchartz, 2014; Rau et al., 2014).

In summary, for solar cells that are close to the radiative recombination limit, the optical features play an important role in increasing the photovoltage and the power conversion efficiency. Most of the photons may undergo total internal reflection upon reaching the semiconductor-air interface producing a much larger photon flux inside the semiconductor film than that emitted at the surface. In these conditions, with plenty of long-lived photons inside the cell, the output power of the solar cell becomes rather sensitive to a small nonradiative recombination current. Efficient external fluorescence is an indicator of low-internal optical losses, which will avoid the recycling of photons and successive loss by any remnant nonradiative pathways.

23.5 RADIATIVE COOLING IN EL AND PHOTOLUMINESCENCE

In Section 5.12, we remarked that a semiconductor diode can be operated as a Carnot heat engine in which thermal radiation and radiative emission play the role of heat baths (Rose, 1960; Berdahl, 1985). Emission of photons in EL takes away a quantity of heat (and entropy), which allows to operate an LED as a radiative refrigerator. Let us discuss the physical constraints to realize cooling by photon emission.

The energy that must be supplied to produce an electron-hole pair in a biased semiconductor diode is qV. As a result of the recombination, a photon of energy $h_\nu \approx E_g$ is emitted. The photon removes a quantity of heat from the semiconductor lattice

$$Q_{rh} = h\nu - qV \tag{23.33}$$

Therefore, cooling a semiconductor by EL photon emission requires that most of the photons be emitted at energy $h_\nu > qV$. The removal of one electron-hole pair by recombination decreases the entropy in the semiconductor by the quantity

$$\Delta s = -\frac{Q_{rh}}{T} \tag{23.34}$$

The production of the photon incurs a creation of the amount of entropy

$$\Delta s' = \frac{h\nu}{T*} \tag{23.35}$$

where $T*$ is the temperature of the radiation, which may be different from T. The process removes the heat from a source at temperature T to a sink at temperature $T*$ (Dousmanis et al., 1964). Since the total entropy production must be positive, we have

$$h\nu - qV < \frac{T}{T*}h\nu \tag{23.36}$$

Consider the case of the reversible process in which the entropy production is zero. It corresponds to the equality

$$\frac{h\nu - qV}{T} = \frac{h\nu}{T*} \tag{23.37}$$

Comparing Equations 22.32 and 23.37, we note that the latter corresponds to the equilibrium of emission and absorption when the radiation field obtains the temperature corresponding to a nonzero chemical potential $\mu_{ph} = qV$. This equilibrium will be discussed in Section 27.1.

We derive the rate of heat emission by radiative decay in the LED. If τ_{rad} is the radiative lifetime, after an initial excitation, the population of electrons and holes decreases by recombination as

$$\frac{dn}{dt} = \frac{dp}{dt} = -\frac{n}{\tau_{rad}} \tag{23.38}$$

By the definition of the chemical potential ζ in terms of the configurational entropy S of electrons and holes (see Equations 5.4 and 5.50), we have the rate of the

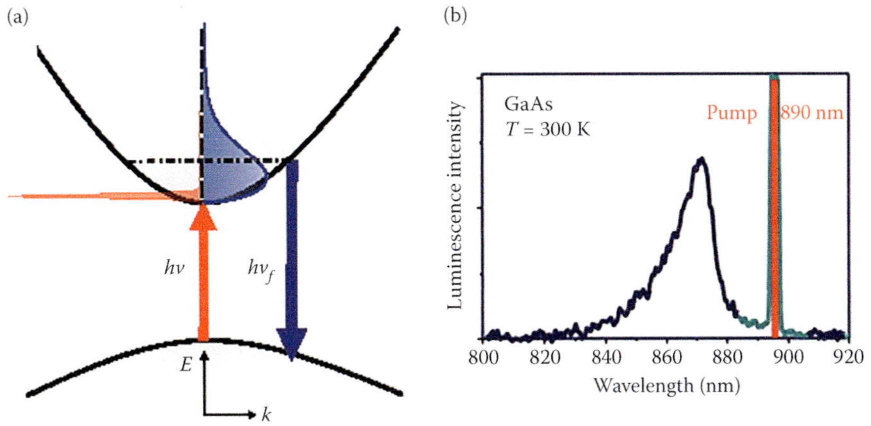

FIGURE 23.12 (a) Cooling cycle in laser refrigeration of a semiconductor in which absorption of laser photons with energy $h\nu$ creates a cold distribution of electron-hole carriers (only electron distribution is shown). The carriers then heat up by absorbing phonons followed by an up-converted luminescence at $h\nu_F$ (b). Typical anti-Stokes' luminescence observed in GaAs/GaInP double heterostructure. (Reproduced with permission from Sheik-Bahae, M.; Epstein, R. I. Laser cooling of solids. *Laser & Photonics Reviews* 2009, 3, 67–84.)

total entropy production by the removal of electrons and holes as

$$\frac{d(S_n + S_p)}{dt} = -\frac{\zeta_n + \zeta_p}{T}\frac{dn}{dt} \tag{23.39}$$

Applying the radiative decay process, we have (Weinstein, 1960)

$$\frac{d(S_n + S_p)}{dt} = \frac{\zeta_n + \zeta_p}{T}\frac{n}{\tau_{rad}} \tag{23.40}$$

From Equations 2.35 and 2.42, the sum of chemical potentials can be written as

$$\begin{aligned} \zeta_n + \zeta_p &= E_{Fn} - E_{Fp} - E_c + E_v \\ &= qV - E_g \end{aligned} \tag{23.41}$$

Finally,

$$\begin{aligned} \frac{dS}{dt} &= -\frac{E_g - qV}{T}\frac{n}{\tau_{rad}} \\ &\approx -\frac{Q_{rh}}{T}\frac{n}{\tau_{rad}} \end{aligned} \tag{23.42}$$

The entropy decreases when $qV < E_g$.

For the radiative cooling to be efficient, the EQE_{LED} must be large as otherwise, the cooling energy gain Q_{rh} per photon is lost by the fraction of photons that are reabsorbed or reflected at the edges and not extracted from the device. Photon recycling can enhance considerably the efficiency of the semiconductor EL cooling. Refrigeration can also be achieved with thermionic emission.

Refrigeration by photon emission in PL is called *laser cooling*. It requires anti-Stokes' emission so that the frequency of the pump source is lower than the mean luminescence frequency. A monochromatic source pumps photons in the low energy tail of the absorption spectrum of a material. The photogenerated electrons are thermalized in the conduction band so that they obtain an increase of energy of order k_BT from the semiconductor lattice. Subsequently, more energetic blue-shifted photons are emitted; see Figure 23.12. This emission process is very difficult to achieve in practice, since the cooling effects associated with nonradiative decay and multiphonon emission processes dominate over internal heating in the conduction band.

23.6 RECIPROCITY OF ABSORPTION AND EMISSION IN A CT BAND

The photo-induced transitions between organic donor (D) and acceptor (A) materials have been described in Figure 6.5. In Figure 19.26, we introduced the charge transfer (CT) transition across the interface between the D and the A. The combination of D-A and the resulting charge transfer and charge separation processes form the basis for the organic solar cells (Bisquert and Garcia-Belmonte, 2011). The energy diagram is shown in Figure 23.13. The purpose of using the interface to produce a photovoltage is to overcome the poor charge separation of the main absorber material, which would lead to recombination of most photogenerated carriers. The energy offset ΔE_{TOP} facilitates the transference of

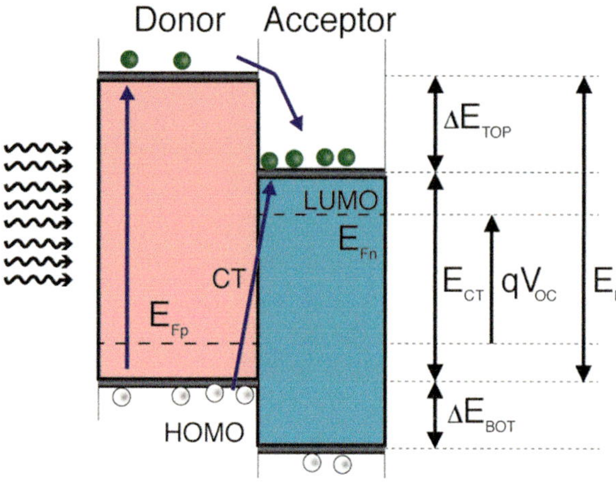

FIGURE 23.13 Schematic representation of a solar cell based on donor (D)-acceptor (A) system. Incoming light is absorbed both in the donor absorber and in the CT process across the interface. Excited electrons from the first type of absorber are transferred by charge separation to A. On the other hand, the CT process directly promotes an electron to the LUMO of the A, red shifted with respect to the intrinsic absorption. If the charge separation is efficient, most of the recombination occurs across the optical gap E_{CT}. The energy offsets ΔE_{TOP} and ΔE_{BOT} facilitate and maintain that separated charge remains confined in a single material until the collection at the respective selective contact.

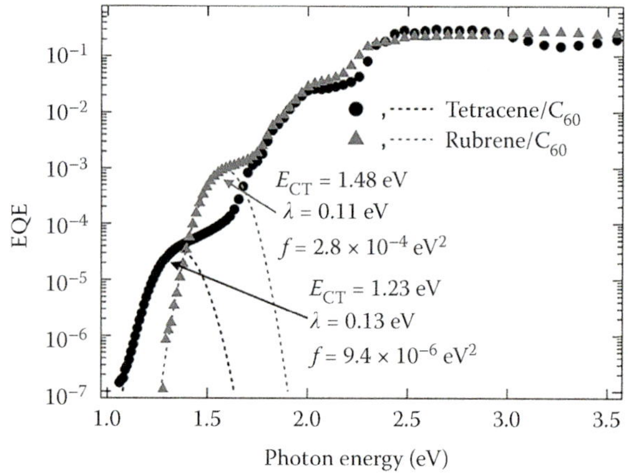

FIGURE 23.14 EQE_{PV} spectra for rubrene/C_{60} and tetracene/C_{60} bilayer organic solar cells. (Reproduced with permission from Graham, K. R. et al. *Advanced Materials* 2013, *25*, 6076–6082.)

electrons photogenerated in the donor material to the acceptor that functions as the electron transporting material until the collection of the electron charge at the respective selective contact. On its turn, the photogenerated holes remain in the donor side of the molecular heterojunction, confined by the energy barrier ΔE_{BOT}.

As indicated in Figure 23.13, there are two main pathways to create a photogenerated electron in A leaving the corresponding hole in D. First is a photon absorption across the intrinsic gap E_D of the main absorber material (here, the donor). This step is followed by fast injection down the energy gradient, the same process as that shown in Figure 6.5.

The second possibility is a CT transition across the optical gap that in this case is given by $E_{CT} = E_{LUMO(A)} - E_{HOMO(D)}$. More concretely the quantity $E_{LUMO(A)} - E_{HOMO(D)}$ is the *transport gap*, these are the dominant energy levels for transport of electron and holes in their respective material, up to the selective contacts. The *optical gap* associated to onset energy of absorption may actually be smaller than this difference due to excitonic effects, so that E_{CT} is defined as the energy difference between the CT complex ground state and the CT excited state. However, the specific energy value is of little relevance for the following considerations, and we will use E_{CT} to denote both the optical gap and the transport gap. We will remark the differences of these terms in the case of limitations to the photovoltage when required.

Here, we focus our attention on the characteristics of charge transfer process between molecular materials. In these materials, strong reorganization effects occur that have been described in Figure 6.5. We will analyze the impact of reorganization in the reciprocity relation, dominated by CT rather than the intrinsic transition, by the following reasoning. In Figure 23.14, we show the $EQE_{PV}(E)$ of bulk heterojunction polymer-fullerene devices. It is observed that the relevant values of $EQE_{PV}(E)$ for photovoltaic operation, i.e., between 0.1 and 1, occur at the larger spectral energies due to the fact that most of the photons are absorbed at energies larger than E_D in the interior of the polymer domains. But additional features are observed at lower photon energies, when the EQE_{PV} is reduced by several orders of magnitude, and these must correspond to the E_{CT} transition. While the effect of the latter is negligible in terms of photocurrent generation, it can play a dominant role in the radiative equilibrium, which is often dominated by sub-band-gap features as noticed previously. Therefore, the properties of this transition are relevant for the solar cell operation.

The thermal effects of reorganization are closely related to the maximum of absorbance and fluorescence, as previously suggested in Figure 6.9. According to Marcus theory, the spectral lineshape of the charge transfer absorption coefficient $\alpha_{CT}(E)$ at photon energy E is given by the following expression (Marcus, 1989,Gould et al., 1993,Vandewal et al., 2010)

$$\alpha_{CT}(E) = \frac{f_{abs} N_{CTC}}{E\sqrt{4\pi\lambda k_B T}} \exp\left(\frac{-(E_0 + \lambda - E)^2}{4\lambda k_B T}\right) \quad (23.43)$$

Here λ is the reorganization energy, N_{CTC} is the number of charge transfer complexes (CTC) per unit volume and f_{abs} is a constant proportional to the square of the electronic coupling element. Equation 23.43 can be derived from the picture of Figure 6.9 using the following assumptions: (a) Boltzmann population of vibronic states within the ground and excited states, (b) standard relation between Einstein A and B coefficients and (c) the reorganization energies for ground and excited states are the same (Koen Vandewal, private communication).

The external quantum efficiency for a photovoltaic cell based on the CT absorption in the weak absorption limit can be written

$$EQE_{PV}(E) = \frac{f_{CT}}{E\sqrt{4\pi\lambda k_B T}} \exp\left(\frac{-(E_0 + \lambda - E)^2}{4\lambda k_B T}\right) \quad (23.44)$$

where the prefactor is

$$f_{CT} = IQE_{PV}(E) N_{CTC} 2d f_{abs} \quad (23.45)$$

in terms of the internal quantum efficiency for charge separation and collection, and the thickness of the layer d, assuming a back reflector so that the optical pathway is $2d$ (Vandewal et al., 2010). Equation 23.44 is represented in Figure 23.15a. The experimental data in Figure 23.14 suggests the observation of the charge transfer band in measurements of EQE_{PV} of molecular organic solar cells.

In order to derive the relationship between light emission efficiency and absorbance, we now apply the reciprocity relationship 23.2. The spectral emission flux in dark equilibrium can be stated as

$$\phi_{ph}^{em}(E) = b_{\pi} \frac{f_{CT}}{\sqrt{4\pi\lambda k_B T}} E e^{-E/k_B T} \exp\left(\frac{-(E_0 + \lambda - E)^2}{4\lambda k_B T}\right) \quad (23.46)$$

This last equation can be rewritten as follows

$$\phi_{ph}^{em}(E) = b_{\pi} \frac{f_{CT}}{\sqrt{4\pi\lambda k_B T}} E e^{-E_0/k_B T} \exp\left(\frac{-(E_0 - \lambda - E)^2}{4\lambda k_B T}\right) \quad (23.47)$$

Note that the sign of λ is inverted in Equation 23.47 with respect to 23.46. Therefore, representation of a reduced normalized EQE, $\bar{\eta} \propto E \times EQE_{PV}$, and a reduced normalized photon flux that can be measured by EL, $\bar{\varphi} \propto \varphi_{ph}^{em}/E$, produces mirror images, as illustrated in Figure 23.15b. The intersection of the spectra is separated by a distance λ from the maxima of both absorption and emission. An observation of the mirror images for the reduced quantities for CT absorption and emission in molecular complexes is shown in Figure 23.16 (Gould et al., 1993). The measurements of organic solar cell devices for donor-acceptor materials have been presented in Figure 23.14. Further examples of the connection of EQE_{PV} and EL in BHJ organic solar cells based on PCPDTBT absorber are shown in Figure 23.17 (Kurpiers et al., 2018). The reduced quantities are shown more clearly in Figure 23.18 for a TQ1:PC_{71}BM blend.

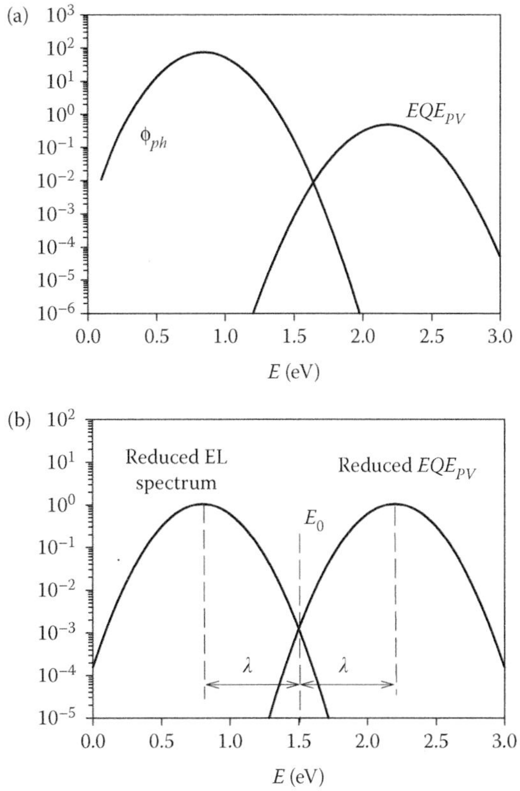

FIGURE 23.15 (a) EQE_{PV} and photon emission (m^{-2} s^{-1} eV^{-1}) at equilibrium with blackbody radiation of a CT complex that obeys the Marcus model with CT gap energy E_0 and reorganization energy λ. (b) Reduced EQE, $\bar{\eta} \propto E \times EQE_{PV}$, and photon emission spectra $\bar{\varphi} \propto \varphi_{ph}^{em}/E$.

These results and many others show that the reciprocity relationship between absorption and emission is well obeyed at the CT transition at the internal interface of D-A blend, as assumed in the original derivation (Marcus, 1989). In order to derive the radiative voltage limit associated to the solar cell structure of Figure 23.13, we now obtain the fundamental recombination parameter for radiative recombination. From Equation 23.47 we arrive at

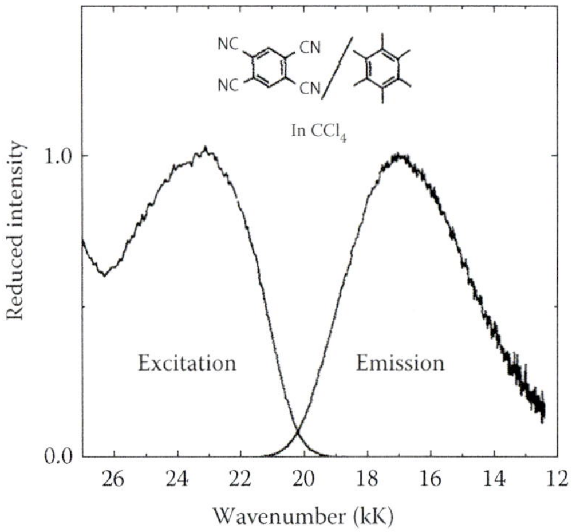

FIGURE 23.16 Reduced excitation and emission spectra for the excited CT complex of 1,2,4,5-tetracyanobenzene/hexamethylbenzene in carbon tetrachloride at room temperature. (Reproduced with permission from Gould, I. R. et al. *Chemical Physics* 1993, 176, 439–456.)

$$j_{0,rad} = qb_\pi \frac{f_{CT}}{\sqrt{4\pi\lambda k_B T}} e^{-E_0/k_BT} \int_0^\infty E \exp\left(\frac{-(E_{CT}-\lambda-E)^2}{4\lambda k_B T}\right) dE \quad (23.48)$$

It can be shown that $j_{0,\mathrm{rad}}$ can be calculated by the approximated expression (Vandewal et al., 2010)

$$j_{0,rad} = qb_\pi (E_{CT} - \lambda) f_{CT} e^{-E_0/k_BT} \quad (23.49)$$

Finally, the radiative limit to the photovoltage, associated to CT recombination, is given as

$$qV_{oc}^{rad} = E_0 + k_B T \ln(j_{ph}^{sun}) - k_B T \ln\left(qb_\pi (E_{CT} - \lambda) f_{CT}\right) \quad (23.50)$$

An analysis of radiative voltage in comparison to actual photovoltage for the organic solar cell shown in Figure 23.18a is presented by (Roland et al, 2019). It is based on the direct calculation of the recombination parameter in Equation 23.3 (and not by Equation 23.49). First, a complete EQE_{PV} spectrum is obtained in combination of the directly measured one, with EQE_{PV-EL}, as shown in Figure 23.18b. The result of the composition with the BB spectrum is shown in Figure 23.18c. Note that values in the lowest energies below 1.3 eV, obtained from *EL* in a region where EQE_{PV} cannot be directly measured, form an essential contribution for the final calculation of $j_{0,\mathrm{rad}}$. Then V_{oc}^{rad} is obtained from the Equation 23.11. From separate determination of EQE_{EL} being 2.6×10^{-6} at low injection conditions (corresponding to nonradiative V_{oc} loss of 0.325 V), the open-circuit voltage is predicted by Equation 23.19 to be 0.892 V, in excellent agreement to the measured V_{oc} at 1 sun. Note that qV_{oc} is about 0.57 eV below the CT energy ($E_{CT} \approx 1.46$ eV), which is a considerable voltage deficit.

In summary, the reciprocity relations connecting EL and light absorption properties provide significant insights about charge generation, charge transfer and recombination, in a spectral region where the absorption coefficient of the blend is extremely tenuous. The *EL* characteristic reveals mechanisms that cannot be obtained only from EQE_{PV}.

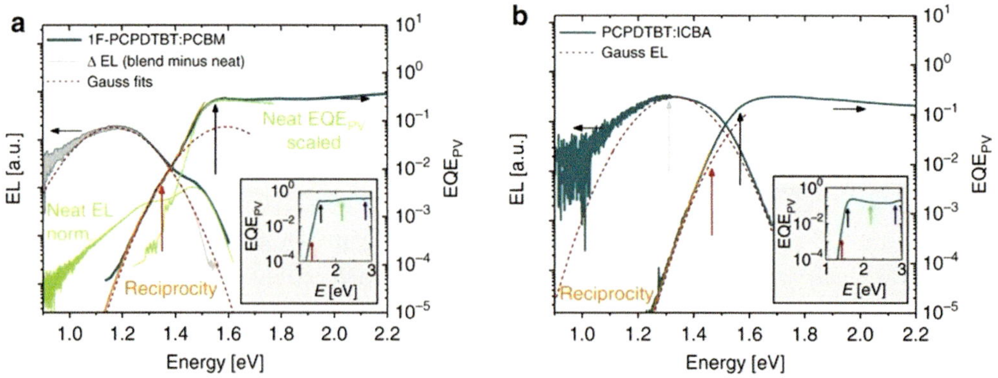

FIGURE 23.17 EQE_{PV} and *EL* spectra of (a) a fluorinated 1F-PCPDTBT blended with PCBM and (b) a non-fluorinated PCPDTBT blended with ICBA. Also shown are the EQE_{PV} and *EL* spectra of neat 1F-PCPDTBT, which are assigned to the polymer singlet exciton (light-green line). The gray lines show the blend *EL* corrected for the polymer emission. Red dashed lines show Gaussian fits to the low energy emission and EQE_{PV}, revealing CT energies of 1.38 eV (1F-PCPDTBT:PCBM) and 1.52 eV (PCPDTBT:ICBA). Arrows in the gray inset indicate the excitation energies. (Reproduced with permission from Kurpiers et al., 2018.)

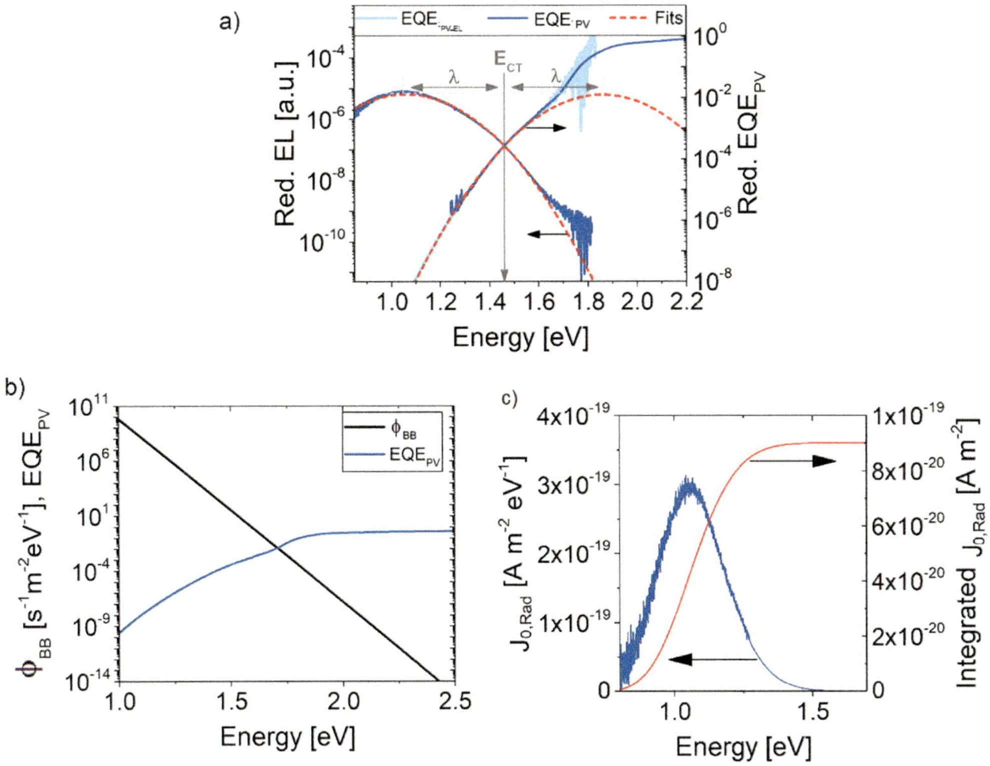

FIGURE 23.18 (a) Reduced electroluminescence and photovoltaic EQE_{PV} spectra measured for the TQ1:PC_{71}BM system. The EQE_{PV-EL} spectrum calculated via the reciprocity relation from the depicted *EL* spectrum is given in light blue. The CT state energy and the reorganization energy λ can be deduced from the scaled Gaussian fits as indicated by the gray arrows. (b) Black body photon flux spectrum and the extended EQE_{PV} spectrum obtained by the combination with EQE_{PV-EL} spectrum obtained from *EL*. (c) The convolution of both spectra shown in (b) and multiplication with the elementary charge gives the radiative dark current $j_{0,rad}$ density spectrum (blue). The integral over this spectrum (red line) gives the value of the absolute radiative dark current density. (Reproduced with permission from Roland et al., 2019.)

If we compare the reciprocity plot for CIGS in Figure 23.4, and for a BHJ in Figure 23.17a, we note an important difference. For CIGS, the whole EQE_{PV-EL} corresponds to the transformation of the measured EL emission. In contrast, the CT characteristics indicated in 23.17a is an effect restricted to the lowest energies of the spectrum. The absorption of 1F-PCPDTBT assigned to the polymer singlet exciton is indicated to the light-green line. Therefore the dominant component of EQE_{PV} does not correspond to the CT. This is due to the fact that the main light absorption channel in a BHJ is not the CT but rather the interior of the organic absorber, as indicated in Figure 23.13. Further studies about the impact of CT absorption in heterogenous solar cells require considering the full light absorption features and influence of the relative energetics of the two materials. While V_{oc}^{rad} is related to the dominant radiative recombination channel and can be predicted from the EL – EQE_{PV} analysis of the CT state spectral region, the prediction of the radiative limits to the solar cell efficiency requires consideration also of the light absorption characteristics, which will be based on a more general analysis presented in Section 25.2 (Figure 25.16). Further discussion on the correlation of V_{oc} with E_{CT} will be presented in Section 24.4 (Figure 24.13).

GENERAL REFERENCES

Reciprocity relationships: Ross (1967), Ruppel and Würfel (1980), Weber and Dignam (1984), Baruch (1985), Smestad and Ries (1992), Markvart and Landsberg (2002), Markvart (2007), Rau (2007), and Kirchartz and Rau (2008).

Radiative voltage in perovskite solar cells: Tress (2017).

Experimental relation of EQE to electroluminescence: Kirchartz and Rau (2008), Vandewal et al. (2009), Tvingstedt et al. (2014), Tress et al. (2015), and Yao et al. (2015).

Donolato theorem: Donolato (1985), Donolato (1989), and Green (1997).

Photon recycling in solar cell devices: Asbeck (1977), Martí et al. (1997), Miller et al. (2012), Kirchartz et al. (2016a, b), and Pazos-Outón et al. (2016).

Radiative cooling: Mahan (1994), Sheik-Bahae and Epstein (2004), Rakovich et al. (2009), Yen and Lee (2010), and Lee and Yen (2012).

REFERENCES

Asbeck, P. Self-absorption effects on the radiative lifetime in GaAs-GaAlAs double heterostructures. *Journal of Applied Physics* 1977, *48*, 820–822.

Baruch, P. A two level system as a model for a photovoltaic solar cell. *Journal of Applied Physics* 1985, *57*, 1347–1355.

Barugkin, C.; J. Cong; et al. T. Duong, S. Rahman, H. T. Nguyen, D. Macdonald T. P. White, K. R. Catchpole Ultralow absorption coefficient and temperature dependence of radiative recombination of $CH_3NH_3PbI_3$ perovskite from photoluminescence. *Journal of Physical Chemistry Letters* 2015, 6, 767–772.

Berdahl, P. Radiant refrigeration by semiconductor diodes. *Journal of Applied Physics* 1985, *58*, 1369–1374.

Bisquert, J.; Garcia-Belmonte, G. On voltage, photovoltage, and photocurrent in bulk heterojunction organic solar cells. *Journal of Physical Chemistry Letters* 2011, *2*, 1950–1964.

Donolato, C. A reciprocity theorem for charge collection. *Applied Physics Letters* 1985, *46*, 270–272.

Donolato, C. An alternative proof of the generalized reciprocity theorem for charge collection. *Journal of Applied Physics* 1989, *66*, 4524–4525.

Dousmanis, G. C.; Mueller, C. W.; Nelson, H.; Petzinger, K. G. Evidence of refrigerating action by means of photon emission in semiconductor diodes. *Physical Review* 1964, *133*, A316.

Gould, I. R.; Noukakis, D.; Gomez-Jahn, L.; Young, R. H.; Goodman, J. L.; Farid, S. Radiative and nonradiative electron transfer in contact radical-ion pairs. *Chemical Physics* 1993, *176*, 439–456.

Graham, K. R.; Erwin, P.; Nordlund, D.; Vandewal, K.; Li, R.; Ngongang Ndjawa, G. O.; Hoke, E. T. Re-evaluating the role of sterics and electronic coupling in determining the open-circuit voltage of organic solar cells. *Advanced Materials* 2013, *25*, 6076–6082.

Green, M. A. Generalized relationship between dark carrier distribution and photocarrier collection in solar cells. *Journal of Applied Physics* 1997, *81*, 268–271.

Green, M. A. Radiative efficiency of state-of-the-art photovoltaic cells. *Progress in Photovoltaics: Research and Applications* 2012, *20*, 472–476.

Hirst, L. C.; Ekins-Daukes, N. J. Fundamental losses in solar cells. *Progress in Photovoltaics: Research and Applications* 2011, *19*, 286–293.

Kirchartz, T.; Nelson, J.; Rau, U. Reciprocity between charge injection and extraction and its influence on the interpretation of electroluminescence spectra in organic solar cells. *Physical Review Applied* 2016a, *5*, 054003.

Kirchartz, T.; Rau, U. Detailed balance and reciprocity in solar cells. *Physica Status Solidi A* 2008, *205*, 2737–2751.

Kirchartz, T.; Staub, F.; Rau, U. Impact of photon recycling on the open-circuit voltage of metal halide perovskite solar cells. *ACS Energy Letters* 2016b, 731–739.

Kurpiers, J.; Ferron, T.; Roland, S.; Jakoby, M.; Thiede, T.; Jaiser, F.; Albrecht, S.; et al. Probing the pathways of free charge generation in organic bulk heterojunction solar cells. *Nature Communications* 2018, *9*, 2038.

Lee, K.-C.; Yen, S.-T. Photon recycling effect on electroluminescent refrigeration. *Journal of Applied Physics* 2012, *111*, 014511.

Ledinsky, M.; Schönfeldová, T.; Holovsky, J.; Aydin, E.; Hájková, Z.; Landová, L.; Neykova, N.; et al. Temperature dependence of the urbach energy in lead iodide perovskites. *Journal of Physical Chemistry Letters* 2019, *9*, 1368–1373.

Mahan, G. D. Thermionic refrigeration. *Journal of Applied Physics* 1994, *76*, 4362–4366.

Marcus, R. A. Relation between charge transfer absorption and fluorescence spectra and the inverted region. *The Journal of Physical Chemistry* 1989, *93*, 3078–3086.

Markvart, T. Thermodynamics of losses in photovoltaic conversion. *Applied Physics Letters* 2007, *91*, 064102.

Markvart, T.; Landsberg, P. T. Thermodynamics and reciprocity of solar energy conversion. *Physica E* 2002, *14*, 71–77.

Martí, A.; Balenzategui, J. L.; Reyna, R. F. Photon recycling and Shockley's diode equation. *Journal of Applied Physics* 1997, *82*, 4067.

Miller, O. D.; Yablonovitch, E.; Kurtz, S. R. Strong internal and external fluorescence as solar cells approach the Shockley-Queisser limit. *IEEE Journal of Photovoltaics* 2012, *2*, 303–311.

Müller, T. C. M.; Kirchartz, T. Absorption and photocurrent spectroscopy with high dynamic range. In *Advanced Characterization Techniques for Thin Film Solar Cells*; Abou-Ras, D., Kirchartz, T., Rau, U., Eds.; Wiley-VCH Verlag GmbH & Co. KGaA: Berlin, 2016; pp. 189–214.

Pazos-Outón, L. M.; Szumilo, M.; Lamboll, R.; Richter, J. M.; Crespo-Quesada, M.; Abdi-Jalebi, M.; Beeson, H. J. Photon recycling in lead iodide perovskite solar cells. *Science* 2016, *351*, 1430–1433.

Pazos-Outón, L. M.; Xiao, T. P.; Yablonovitch, E. Fundamental efficiency limit of lead iodide perovskite solar cells. *Journal of Physical Chemistry Letters* 2018, *9*, 1703–1711.

Rakovich, Y. P.; Donegan, J. F.; Vasilevskiy, M. I.; Rogach, A. L. Anti-Stokes cooling in semiconductor nanocrystal quantum dots: A feasibility study. *Physica Status Aolidi A* 2009, *206*, 2497–2509.

Rau, U. Reciprocity relation between photovoltaic quantum efficiency and electroluminescent emission of solar cells. *Physical Review B* 2007, *76*, 085303.

Rau, U.; Kirchartz, T. On the thermodynamics of light trapping in solar cells. *Nature Materials* 2014, *13*, 103–104.

Rau, U.; Paetzold, U. W.; Kirchartz, T. Thermodynamics of light management in photovoltaic devices. *Physical Review B* 2014, *90*, 035211.

Roland, S.; Kniepert, J.; Love, J. A.; Negi, V.; Liu, F.; Bobbert, P.; Melianas, A.; et al. Equilibrated charge carrier populations govern steady-state nongeminate recombination in disordered organic solar cells. *Journal of Physical Chemistry Letters* 2019, *10*, 1374–1381.

Rose, A. L. Photovoltaic effect derived from the Carnot cycle. *Journal of Applied Physics* 1960, *31*, 1640–1641.

Ross, R. T. Some thermodynamics of photochemical systems. *The Journal of Chemical Physics* 1967, *46*, 4590–4593.

Ruppel, W.; Wurfel, P. Upper limit for the conversion of solar energy. *IEEE Transactions on Electron Devices* 1980, *27*, 877–882.

Santhanam, P.; Fan, S. Thermal-to-electrical energy conversion by diodes under negative illumination. *Physical Review B* 2016, *93*, 161410.

Sheik-Bahae, M.; Epstein, R. I. Can laser light cool semiconductors? *Physical Review Letters* 2004, *92*, 247403.

Sheik-Bahae, M.; Epstein, R. I. Laser cooling of solids. *Laser & Photonics Reviews* 2009, *3*, 67–84.

Smestad, G.; Ries, H. Luminescence and current-voltage characteristics of solar cells and optoelectronic devices. *Solar Energy Materials and Solar Cells* 1992, *25*, 51–71.

Tress, W. Perovskite solar cells on the way to their radiative efficiency limit—Insights into a success story of high open-circuit voltage and low recombination. *Advanced Energy Materials* 2017, 1602358.

Tress, W.; Marinova, N.; Inganäs, O.; Nazeeruddin, M. K.; Zakeeruddin, S. M.; Graetzel, M. Predicting the open-circuit voltage of $CH_3NH_3PbI_3$ perovskite solar cells using electroluminescence and photovoltaic quantum efficiency spectra: The role of radiative and non-radiative recombination. *Advanced Energy Materials* 2015, *5*. doi: 10.1002/aenm.201400812.

Trupke, T.; Würfel, P.; Uhlendorf, I.; Lauermann, I. Electroluminescence of the dye-sensitized solar cell. *The Journal of Physical Chemistry B* 1999, *103*, 1905–1910.

Tvingstedt, K.; Malinkiewicz, O.; Baumann, A.; Deibel, C.; Snaith, H. J.; Dyakonov, V.; Bolink, H. J. Radiative efficiency of lead iodide based perovskite solar cells. *Scientific Reports* 2014, *4*, 6071.

Vandewal, K.; Tvingstedt, K.; Gadisa, A.; Inganäs, O.; Manca, J. V. On the origin of the open-circuit voltage of polymer–fullerene solar cells. *Nature Materials* 2009, *8*, 904–909.

Vandewal, K.; Tvingstedt, K.; Gadisa, A.; Inganas, O.; Manca, J. V. Relating the open-circuit voltage to interface molecular properties of donor:acceptor bulk heterojunction solar cells. *Physical Review B* 2010, *81*, 125204.

Wang, X.; Lundstrom, M. S. On the use of Rau reciprocity to deduce external radiative efficiency in solar cells. *IEEE Journal of Photovoltaics* 2013, *3*, 1348–1353.

Weber, M. F.; Dignam, M. J. Efficiency of splitting water with semiconducting electrodes. *Journal of the Electrochemical Society* 1984, *131*, 1258–1265.

Weinstein, M. A. Thermodynamics of radiative emission processes. *Physical Review* 1960, *119*, 499.

Würfel, P. Physics of solar cells. In *From Principles to New Concepts*, 2nd ed.; Wiley: Weinheim, 2009.

Yao, J.; Kirchartz, T.; Vezie, M. S.; Faist, M. A.; Gong, W.; He, Z.; Wu, H. Quantifying losses in open-circuit voltage in solution-processable solar cells. *Physical Review Applied* 2015, *4*, 014020.

Yen, S.-T.; Lee, K.-C. Analysis of heterostructures for electroluminescent refrigeration and light emitting without heat generation. *Journal of Applied Physics* 2010, *107*, 054513.

24 Charge Separation and Material Limits to the Photovoltage

In the photovoltaic conversion process, the excitation created by a photon in the light absorber is converted into separate electrons and holes, which produce current and voltage. In this chapter, the separation of the excitation into distinct carriers is analyzed, with particular emphasis on those devices composed of different nanostructured phases that realize the charge separation by energetic gradients close to the generation point, such as DSCs and organic bulk heterojunction cells. Realistic material limitations to the photovoltage beyond the radiative limit are summarized. In particular, the use of separate phases for electron and hole transport imposes constraints on the open-circuit voltage value because the Fermi level is limited by the density of states of the transport materials.

24.1 LIGHT ABSORPTION

The operation of a solar cell can be analyzed in terms of a sequential set of processes, which eventually cause the conversion of the free energy of the incoming photons into a voltage and current extracted at the outer contacts of the device. We will discuss in turn the processes of light absorption and charge separation. The final step of charge extraction is treated in Chapter 26.

The first process is the light absorption that converts the absorbed photon into an excitation in the semiconductor such as a bound electron-hole pair or exciton. The effective harvesting of solar photons is an important requirement for effective solar energy conversion, as discussed in Chapter 18. One primordial property is the thickness of the absorber, d. Since the low-energy photons usually have the longest pathway for absorption, the required layer thickness will be determined by the value of absorption coefficient in the photon frequency range of the absorption onset. The required range of values of d is also determined by light reflection/scattering properties of the device that increase the optical path length of weakly absorbed photons in the absorber. The thickness of the absorber layer is much smaller for strongly absorbing semiconductors, which typically necessitate only $d \approx 1\ \mu m$, and these are termed thin film solar cells. A thinner absorber produces a significant advantage from the point of view of manufacturing and cost.

24.2 CHARGE SEPARATION

When an electron-hole pair is created by the absorption of a photon, bound states like excitons may be formed. Charge separation means that initially formed electron-hole pairs after photogeneration dissociate to form two separate unbound charge carriers of opposite polarity. Effective local charge separation is an essential feature to avoid immediate geminate recombination of individually photogenerated electron-hole pairs in low permittivity absorbers and/or at low temperatures (that decrease the thermal energy available to overcome the binding state). After the initial separation of the geminate pair, the so-called nongeminate recombination may occur between electrons in the conduction and holes in the valence band.

Specific mechanisms of charge separation may differ broadly depending on the type of absorber and its relative permittivity. In homogeneous semiconductors, such as standard inorganic thin films or crystalline silicon (c-Si) with relatively high-dielectric permittivities and low-exciton binding energies, local charge separation is very efficient without the assistance of electrical fields, at least at room temperature. Thus, each photogenerated carrier rapidly forms part of the respective ensemble of free carriers in the conduction or valence band, after a carrier thermalization time; see Section 5.6. In lead halide perovskite solar cells, the exciton population at room temperature seems to be negligible (Manser and Kamat, 2014; Yamada et al., 2014). The photogenerated carriers are separated on ps time scale and the radiative recombination occurs between uncorrelated electron-hole carriers rather than geminate pairs (Chen et al., 2014).

Many types of solar cells rely on a homogeneous semiconductor absorber layer that realizes both the functions of charge separation and carrier transport for their collection at the selective contacts. However, several types of nanostructured solar cells that use organic materials as absorbers establish a combination of materials to realize effective charge separation and subsequently, the transport of carriers occurs in separated phases. The morphology of a DSC, based on a TiO_2 nanostructure to which the dye is attached and filled with redox electrolyte that actuates as the hole conductor, is shown in

Figure 24.1. The operation of the DSC has been commented upon in Chapters 8, 9 and 21, see Figures 8.12 and 21.9. The structure of a fully organic, bulk heterojunction organic solar cell (BHJ) is shown in Figure 24.2. Organic blends consist of a mixture of electron donor and electron acceptor, as discussed previously in Figures 6.5, 19.26 and 23.10. Light is absorbed at a polymer material, although small molecules or oligomers also give very good results as absorbers (Mishra and Bäuerle, 2012). The electron donor functions as the hole-transport material (HTM). Microstructure and morphology of the BHJ are key issues for the blend operational properties, and there is enormous variability of reported structures, but ideally, each phase should be formed by aggregates or crystallites a few nm thick, continuously connected along the film thickness.

All of these types of solar cells employ the structure of a blend of nanomaterials and/or molecules to achieve the efficient separation of photogenerated electrons and holes, as indicated in Figure 24.3. The junctions are arranged to facilitate charge separation immediately after absorption of a photon. The interfaces that exist very close to the generation point offer a downhill gradient of energy to at least one of the carriers. The relaxation of carriers takes place in different materials and the relevant energy gap for determining the photovoltage, the *transport gap* (Bredas, 2014), Figure 23.13, is formed by the difference between the electron energy level in the electron-transport material (ETM) and the hole energy level in the HTM.

The detailed steps for charge generation and charge separation in a DSC and BHJ are indicated in Figure 24.4a and b. In the DSC, the photoexcited electrons are preferentially injected to TiO_2 and the hole is thus transferred to the redox carrier or a solid hole conductor such as Spiro-OMeTAD, while the opposite pathway is kinetically and energetically forbidden for each carrier. The diode functionality is achieved by adequate energy level matching and kinetic selectivity, as discussed in Section 21.5 and suggested in Figure 24.3. The photovoltaic operation of the standard DSCs relies to a large extent on unique properties of I^-/I_3^- redox species, the redox hole carrier that ensures excellent kinetically induced rectification of electron transfer both at TiO_2/dye and liquid/counterelectrode contacts (O'Regan and Grätzel, 1991; Peter, 2007). Kinetic preference for charge extraction can switch the photovoltage sign, as shown in a layer of quantum dots in Figure 20.10b (Mora-Sero et al., 2013). In an organic BHJ, there is a rapid separation of the electron-hole pair by transfer of an electron to the nanoscaled ETM units, although it has been suggested that dissociation of an excitonic charge transfer state may become an important step for efficient conversion (Clarke and Durrant, 2010).

So far, we have obtained a picture of heterogeneous solar cells that is composed of a fine grained morphology that combines two or three material phases, as in Figures 24.1 and 24.2, which form a local charge-separation structure in the energy axis represented in Figure 24.4. The first reason for the nanostructured morphology in a BHJ is to realize charge separation in the polymer absorber phase in close vicinity of an interface. The active layer of the DSC is formed by a monolayer of dye that is locally sandwiched between nanostructured electron and hole conductors.

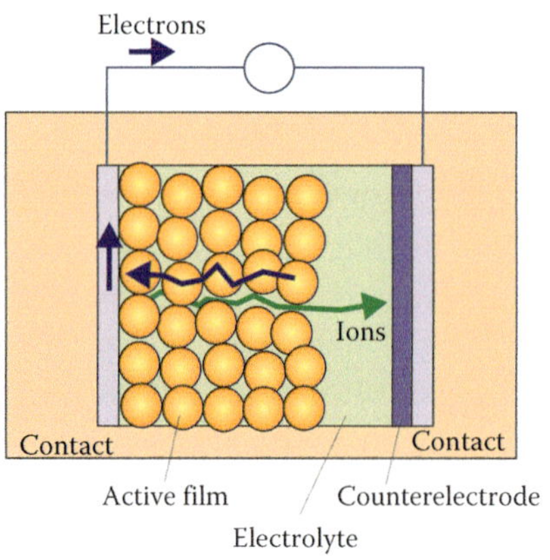

FIGURE 24.1 Scheme of a DSC indicating electron transport in TiO_2 metal-oxide nanoparticulate network and hole transport by ion carriers in the liquid electrolyte.

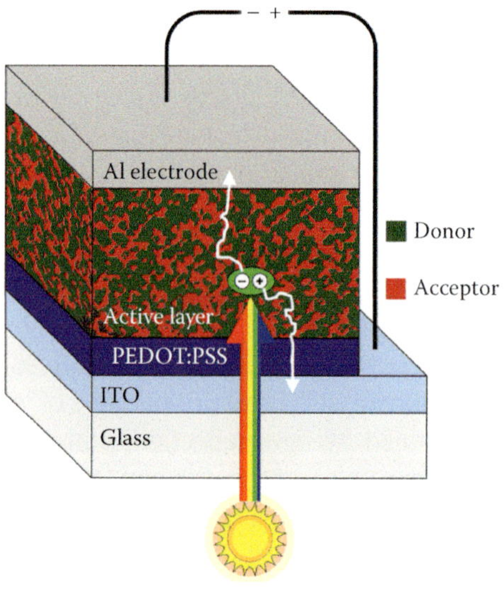

FIGURE 24.2 Structure of a BHJ organic solar cell. (Reproduced with permission from Verploegen, E. et al. *Advanced Functional Materials* 2010, 20, 3519–3529.)

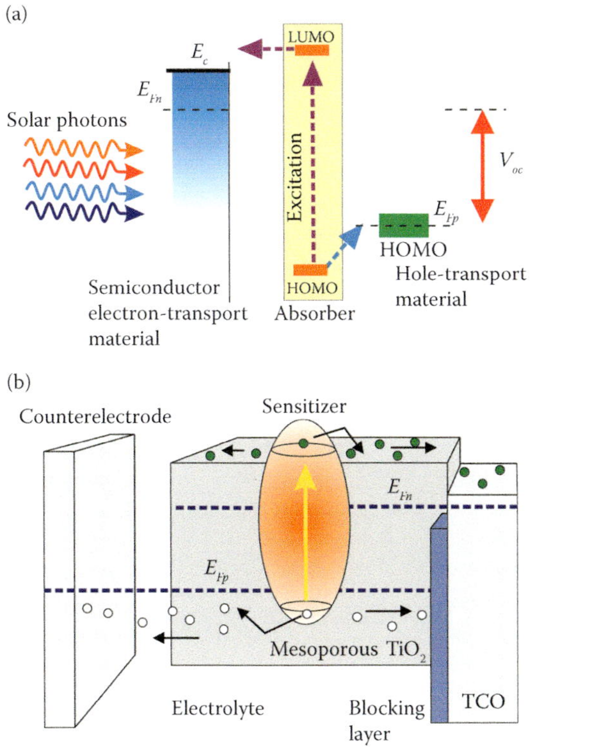

FIGURE 24.3 (a) Schematic steps of charge separation after the excitation of the dye in a DSC. (b) Charge separation and transport in a DSC. Electrons are injected from the excited state of a sensitizer to the TiO_2 framework, and they travel by diffusion across the semiconductor until collected at the transparent conducting oxide (TCO) substrate. Holes in the ground state of the absorber are ejected into a hole-transport medium. Holes are prevented to enter the electron collecting TCO by a blocking layer, and they are accepted at the counter-electrode (CE). The separation of the Fermi levels of electrons (E_{Fn}) and holes (E_{Fp}) produces a photovoltage. (Reproduced with permission from Kirchartz, T. et al. *Physical Chemistry Chemical Physics* 2015, 17, 4007–4014.)

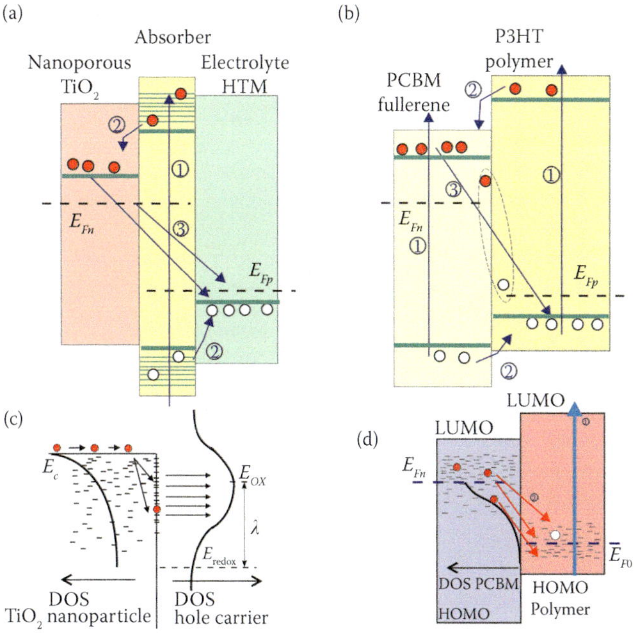

FIGURE 24.4 Energetic scheme of the different phases in (a) DSC and (b) organic BHJ indicating the different steps of (1) photogeneration of carriers, (2) charge separation from the absorber to the acceptor(s), and (3) recombination. (c, d) Represent the energy levels and charge transfer steps in the disordered media.

The small size of the constituents of the two transport phases facilitates effective electrical charge shielding, avoiding the space charge formation in long-range transport. If the sizes of the units of the phases are smaller than the Debye length, then the overall electroneutrality of positive and negative carriers can be maintained and transport can be realized by diffusion. These issues have been commented on in detail in Chapter 12. Charge shielding in a DSC is supported by the electrolyte that causes the effective compensation of electronic charge injected to the metal-oxide framework, as indicated in Figure 8.12b. Organic blends for BHJ contain a significant amount of doping, usually p-type, that also facilitate the existence of a minority carrier. For example, P3HT forms a defect complex with oxygen that promotes p-doping in P3HT-PCBM molecular blends, where the doping is in the range 10^{14}–10^{16} cm^{-3} (Abdou et al., 1997).

Another important issue for the organic nanostructured cells is the electronic disorder indicated in the bottom row of Figure 24.4. Facile low-temperature deposition techniques adopted for the preparation of nanostructured materials and blends often lead to components that contain a large amount of structural defects and irregular morphologies. Therefore, charge transport and charge transfer determining the recombination rates adopt the properties characteristic of disordered materials, which have been amply described in Chapter 13. By following the methods of chemical capacitance explained in Chapters 7 and 8, it is possible to establish the density of states (DOS) in the ETM and HTM that form organic nanostructured cells, as suggested in Figure 24.4. The exponential DOS usually found in nanostructured TiO_2 has been discussed in Figures 7.7, 7.8, and 9.34. The result of a quantitative determination of the TiO_2 energy levels in a DSC in reference to dye and redox electrolyte levels are shown in Figure 24.5. Figure 24.6 shows an example of the determination of the DOS of BHJ cells that use two different fullerene acceptors. Figure 24.7 shows a comparison of the dominant energy levels for photovoltaic operation in a DSC and in a BHJ.

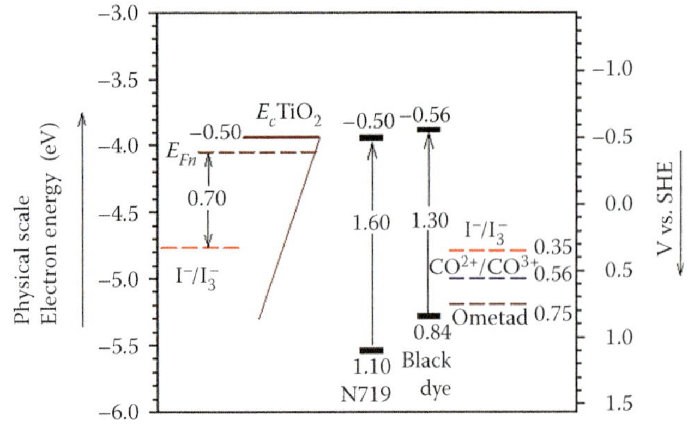

FIGURE 24.5 Energetic scheme of the components of a DSC. On the left is the DOS of TiO_2 and the position of the conduction band, indicating also the photovoltage by difference of the Fermi level of electrons and the redox potential in the electrolyte. On the right are shown the redox potentials of conventional hole conductors. In the center, the ground and excited state of standard dyes are shown.

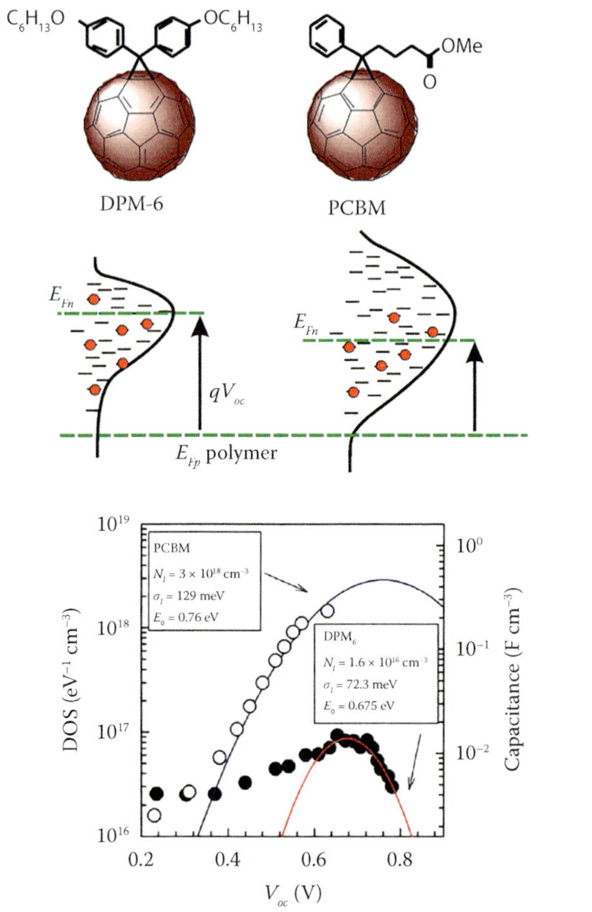

FIGURE 24.6 Capacitance values measured from low-frequency values of impedance spectroscopy as a function of V_{oc} reached under varying illumination levels. White dots correspond to PCBM-based solar cells and black dots to DPM_6-based solar cells. Gaussian DOS (*solid lines*) and distribution parameters resulting from fits. The top drawing shows the fullerene molecules and a pictorial representation of the DOS in each case. (Reproduced with permission from Garcia-Belmonte, G. et al. *Journal of Physical Chemistry Letters* 2010, 1, 2566–2571.)

24.3 MATERIALS LIMITS TO THE PHOTOVOLTAGE

The photovoltage is a central feature of the solar cell determining to a great extent the power conversion efficiency. In Section 22.2, we characterized the ideal properties and fundamental limitations of the photovoltage. We now examine different types of non-idealities and limitations imposed by materials properties on the V_{oc}.

The number of electrons and holes that determine the separation of Fermi levels is established by equilibrium of the processes of generation and recombination. From the expression that relates the Fermi level separation to the product densities of electrons and holes, Equation 5.37, we obtain the expression

$$qV_{oc} = E_g + k_B T \ln\left(\frac{np}{N_c N_v}\right) \tag{24.1}$$

Thus the photovoltage depends on the product of carrier densities np. Previously, we have stated two central models for recombination current under the assumptions of minority carrier and electroneutrality dominated recombination in Section 21.2. Investigation of the recombination process in the bulk absorber is an essential part of establishing the photovoltaic properties. Any deviation of the recombination from a fundamental product law, as noted in Equation 21.42, will have a signature on the V_{oc} of the solar cell. We now explore a more general expression of the bimolecular recombination rate as a function of local carrier densities as follows (Correa-Baena et al., 2017)

$$U_n = B_{rec}\left[(np)^{1/m_d} - n_i^{2/m_d}\right] \tag{24.2}$$

The parameter B_{rec} is a kinetic recombination rate and m_d is an index that accounts for the nonideal behaviour of recombination that departs from the strict bimolecular law, corresponding to Equation 21.42. Using Equation 24.2 and the product given in Equation 21.13, the diode Equation 21.19 is obtained directly. The ideal value for band to band recombination is $m_d = 1$ by Equation 21.15, but any source of trap-assisted recombination leads to $m_d > 1$ (Ansari-Rad et al., 2012).

The local generation rate by absorbed light flux Φ_{ph} is approximated as

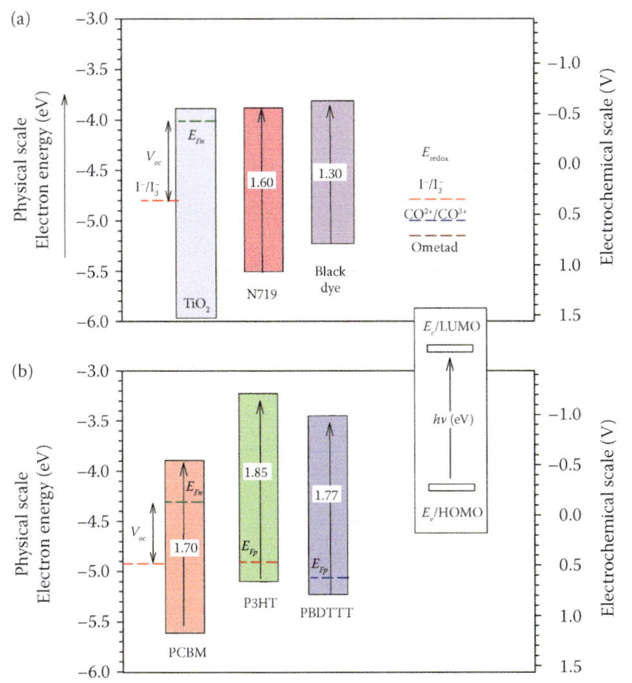

FIGURE 24.7 Schematic energy diagrams of the materials components of (a) dye-sensitized and (b) organic BHJ solar cells. Standard energy levels are given on the solid state, one electron energy scale, while the origin of the electrochemical scale is taken at −4.44 eV. E_{Fn} and E_{Fp} are the Fermi levels of electrons and holes (the I^-/I_3^- redox potential in most DS cells). Their difference relates to V_{oc} as indicated. TiO_2 and dye energy levels depend on the solution components; dye levels may also depend on criteria for the absorption onset. (Reproduced with permission from Nayak, P. K. et al. *Advanced Materials* 2011, 23, 2870–2876.)

$$G \propto \frac{1}{d}\Phi_{ph} \tag{24.3}$$

Under intense illumination, the number of photogenerated carriers is larger than the native doping density. Using electroneutrality condition 21.14 and neglecting the constant term in Equation 24.2, the recombination rate takes the form

$$U_n = B_{rec} n^{2/m_d} \tag{24.4}$$

The open-circuit condition $G = U_n$ implies that the carrier density depends on photon flux as

$$n = \left(\frac{1}{dB_{rec}}\Phi_{ph}\right)^{m_d/2} \tag{24.5}$$

The photovoltage dependence on illumination flux can be written

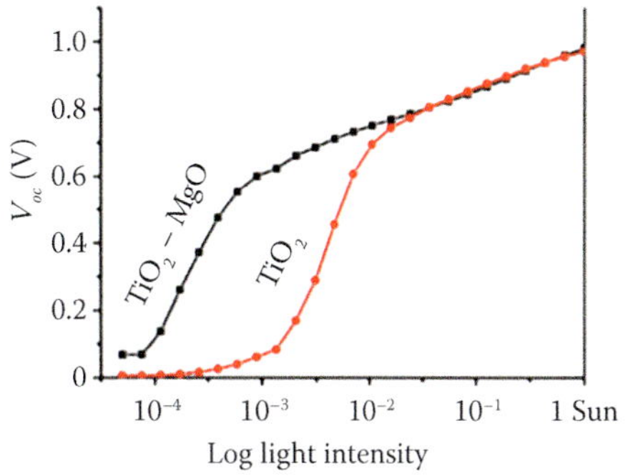

FIGURE 24.8 V_{oc} versus log I_0 (light intensity) curves. Comparison between a TiO_2 perovskite solar cell and a TiO_2–MgO perovskite solar cell. (Reproduced with permission from Gouda, L. et al. *The Journal of Physical Chemistry Letters* 2015, 6, 4640–4645.)

$$\frac{dV_{oc}}{d\ln\left(\Phi_{ph}\right)} = \frac{m_d k_B T}{q} \tag{24.6}$$

Therefore, the non-ideality of the recombination model is reflected in the diode parameter and produces a change of V_{oc} dependence on illumination flux in agreement with the Equations 21.19 and 23.10. Consequently, the measurement of the slope of the open-circuit voltage vs. light intensity is a useful method to determine recombination mechanisms in a solar cell device (Huang et al., 1997,Gouda et al., 2015). Figure 24.8 shows the characteristic curve of metal halide perovskite solar cell synthesized on metal-oxide nanostructured frameworks of different composition. The low light intensity region is sensitive to the contact properties, while at high generation rates, both cells show the same behaviour, which allows to identify bulk recombination properties (Correa-Baena et al., 2017).

As suggested by Figure 24.8, besides the details of bulk recombination, layers that are used as selective contacts may impose important restrictions to the V_{oc}. Recombination at defects in the contact layers is a major source of decrease of performance in crystalline solar cells. In crystalline silicon solar cells, electrons and holes are separated, accumulated and transported in the absorber medium itself. The selective contacts consist of extremely localized regions of strong electrical fields at the edges of a thick absorber as further discussed in the final paragraphs of this section. To prevent the carrier loss in silicon solar cells, a back surface field is formed that reflects the majority carrier, or a passivation layer is introduced that may be made up of materials like SiO_2, amorphous Si or Al_2O_3 (Taguchi et al., 2000,Schmidt et al., 2008). In lead halide perovskite solar cells, it has been observed that surface recombination at the outer

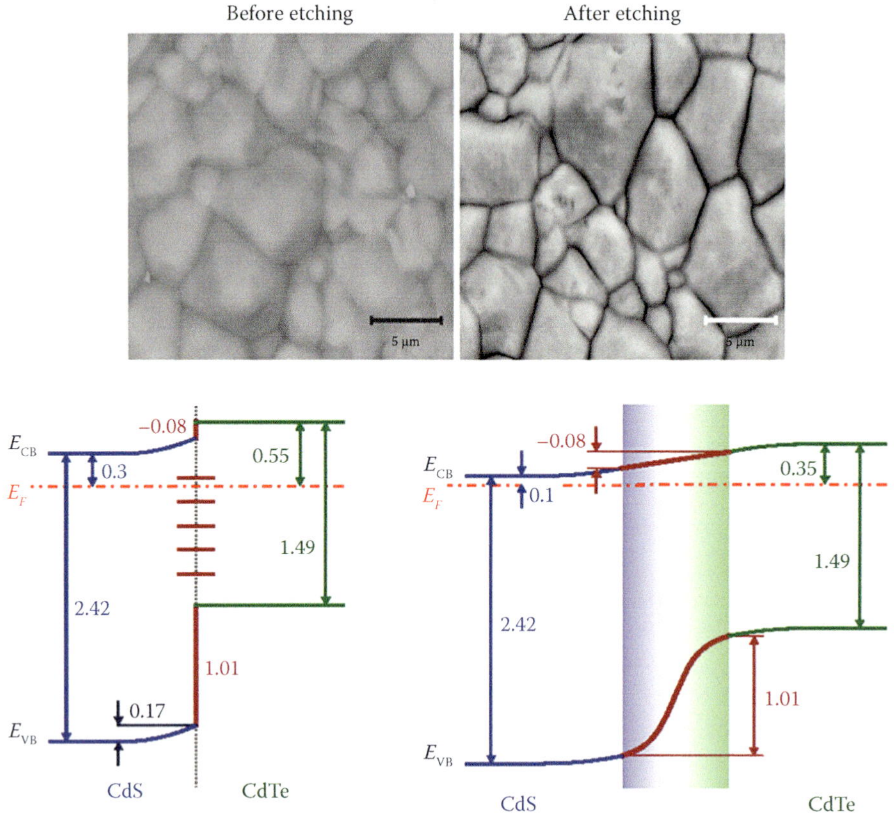

FIGURE 24.9 Grain structure and band energy diagram of CdS/CdTe interfaces prepared by PVD before activation (*left*) and after activation (*right*). (Reproduced with permission from Jaegermann, W. et al. *Advanced Materials* 2009, 21, 4196–4206.)

contact is a major source of loss of photogenerated carriers (Zarazua et al., 2016).

In thin film inorganic CdTe solar cells, a $CdCl_2$ "activation" treatment considerably enhances the performance by improving the properties of CdTe/CdS heterointerfaces (Jaegermann et al., 2009). As indicated in Figure 24.9, in the initial abrupt interface, a large density of interfacial electronic defects causes Fermi level pinning and large surface recombination rates. The $CdCl_2$ activation favors interdiffusion of CdS and CdTe at the phase boundary, which leads to a reduction of the interface density of states. In general, a proper passivation of the interfacial electronic defect levels is an essential precondition for the realization of efficient thin film solar cells.

We now consider the limitations to the photovoltage imposed by the materials properties of the ETM and HTM in nanostructured solar cells. In the DSC (composed by three materials: absorber, ETM and HTM), rapid charge extraction from the dye absorber requires a downhill driving force, as emphasized in Figure 24.3. Therefore, the conduction band in the ETM and the transport level in the HTM (such as the redox energy of the electrolyte) need to be lower and higher, respectively, than the energy level of the excited carriers in the dye. Since the V_{oc} is determined by the difference of Fermi levels in the ETM and HTM, the limits of Fermi level variation in these materials impose the constraint to the maximal V_{oc} that may be obtained. For the energetically favorable electron transfer, the energy of the excited state of the dye has to be about 0.3 eV higher than that of the conduction band (CB) of the electron transporter i.e. TiO_2. It is possible that E_{Fn} raises up to the CB edge, but here, a very large density of states prevents the Fermi level from further increase. On the hole transporter side, a driving force for the regeneration of the oxidized dye is required as well. This is up to 500 mV with the best carrier, an I_3^- /I^- redox couple in an organic solvent. Unfortunately, the redox energy of this vital element of the DSC lies high in the energy scale, at +0.35 V_{NHE}, see Figure 24.5. The V_{oc} will be higher if the HTM has a lower energy level in the energy scale, as indicated in several examples in Figure 24.5 and Figure 24.7, such as a Co-based redox couple. However, these alternative HTMs present lower kinetic selectivity than the I_3^- /I^- redox couple.

In summary, in a DSC, the primary available voltage is related to the separation of chemical potential of electrons and holes *in the dye*, but the photovoltage measured from outer contacts is produced by the difference of the Fermi level of electrons in TiO_2 and the redox level of the ionic carriers. The combination of the two energy steps for charge separation produces a limitation of the V_{oc}, as follows

$$qV_{oc,max} = E_{c,\text{ETM}} - E_{F,\text{HTM}} \tag{24.7}$$

Thus, maximal V_{oc} becomes unrelated to the ideal model of recombination in the absorber, which is manifested by the very low external luminescence, as already commented on in Section 23.3.

We have already indicated that the organic BHJ is a two-material solar cell with a distributed donor-acceptor interface. In Sections 19.6 and 23.5, we have remarked on the optical transitions that may occur in a donor-acceptor blend. The photon absorption across the intrinsic gap E_D is the main light absorption channel in an organic BHJ. For achieving effective charge separation in the blend, the injection state of the organic absorber needs to be about 0.3 eV higher than that of the electron transporter that corresponds to the LUMO level of the acceptor material (Veldmann et al., 2009). As discussed in Section 23.5 the transport gap is defined as $E_{CT} = E_{LUMO(A)} - E_{HOMO(D)}$. Similarly to Equation 24.7, the transport gap (and not the CT energy) becomes an important limitation to V_{oc} in BHJ solar cell, due to the fact that $E_{LUMO(A)}$ and $E_{HOMO(D)}$ poses a limitation to the separation of electron and holes Fermi levels in the blend.

Another closely related topic is the explanation of the relationship between measured CT energy and the photovoltage of a BHJ blend. It has been often remarked that the charge transfer state has a determining impact on the efficiency of organic solar cells. This is because the measured V_{oc} is often correlated to the measured energy of light emission of the CT state (Vandewal et al., 2010). If we look at this question from the point of view of the standard energy diagrams for solar cells, as those in Figure 24.4, we notice that a correlation of V_{oc} and the optical gap E_{CT} is rather natural. Figure 24.10 illustrates the properties of V_{oc} in cells that differ in the offset energy ΔE_{TOP}. If the recombination rate is similar in both, the V_{oc} is directly correlated to the transport gap. It follows that optimal match of energy levels of the separate components is essential to achieve a large V_{oc} with respect to the primary bandgap of the absorber. However, the reduction of energy offsets has a cost in the efficiency of kinetics of charge separation. In addition, the disposition of the interfaces often controls the rate of nonradiative recombination pathways, which always reduce the split of Fermi levels and reduce the photovoltage. The

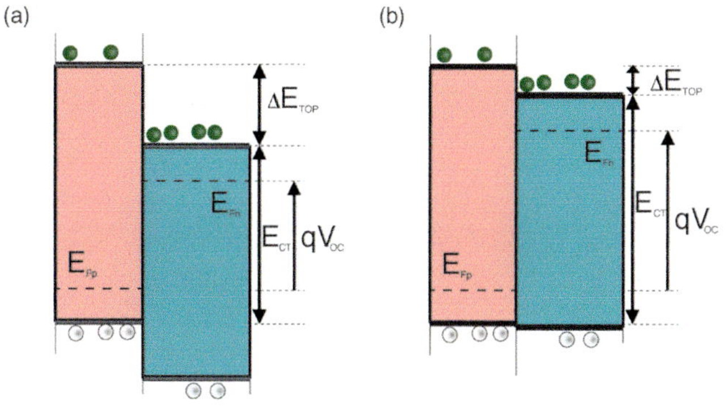

FIGURE 24.10 Schematic representation of a solar cell based on donor (D)-acceptor (A) system. In (b) the energy levels of the separate materials become better aligned than in (a) by materials design (though the intrinsic gaps are the same) and the energy offsets ΔE_{TOP} and ΔE_{BOT} are smaller. Under illumination, if the recombination rate is the same in both situations, the Fermi levels remain at the same distance of the respective energy levels. The V_{oc} becomes simply correlated with the optical (transport) gap, or CT energy, E_{CT}.

analysis of Figure 24.10 can be stated in very general terms to apply to different kinds of blend heterojunctions such as a DSC and organic BHJ. The correlation refers physically to energies of the separated electrons and holes, and hence to the transport gap, independently of the microscopic or spectroscopic significance of a CT state, if it has any for the given case.

The existence of a broad DOS in the electron and HTMs has important implications for the photovoltage, as already commented in Figure 8.11 (Nayak et al., 2012). The electron carriers generated will pile up at the bottom of the available states of the DOS, as shown in Figure 7.10. Hence, a material with a small density of states in the bandgap will favor a higher-electron Fermi level and consequently a larger V_{oc}, and a similar argument applies for holes. Similarly, a correlation between the DOS and the photovoltage can be observed in organic BHJ solar cells (Garcia-Belmonte and Bisquert, 2010). In the comparison shown in Figure 24.6, it is found that DPM_6 (4,4′-dihexyloxydiphenylmethano[60] fullerene) has a smaller DOS in the gap than PCBM, so that a full Gaussian DOS is observed in the former case. If the recombination rate is similar in both cases, then the Fermi level of electrons raises higher in the fullerene with the smaller DOS, producing a larger photovoltage as indicated in the scheme of Figure 24.6 and observed experimentally (Garcia-Belmonte et al., 2010).

The effect of accumulation of carriers in the Gaussian DOS upon obtaining a photovoltage can be quantitatively described using the properties of the distributions that have been discussed in Section 8.6 in terms of the disorder parameters σ_n, σ_p that give the width of

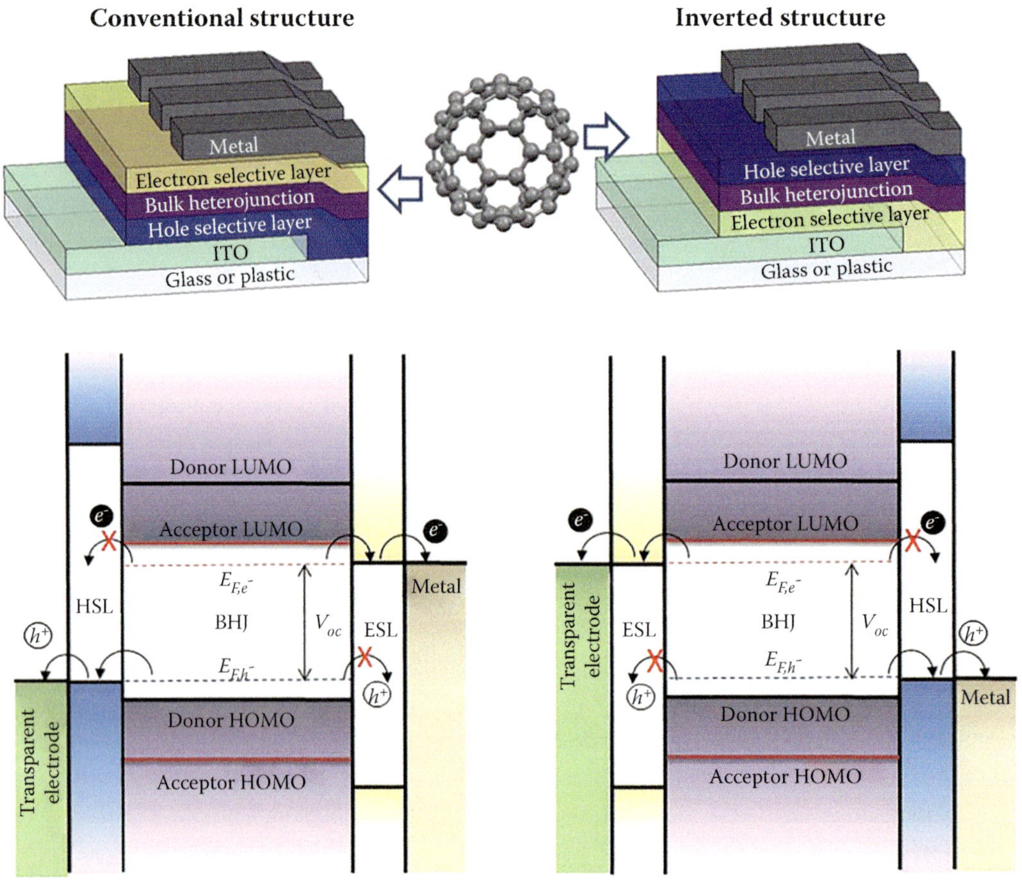

FIGURE 24.11 Schematic illustration of the structure of organic blend of fullerene/polymer layer and electron selective layer of conventional and inverted polymer solar cells, and their energy diagram. (Reproduced with permission from Li, C.-Z. et al. *Journal of Materials Chemistry* 2012, 22, 4161–4177.)

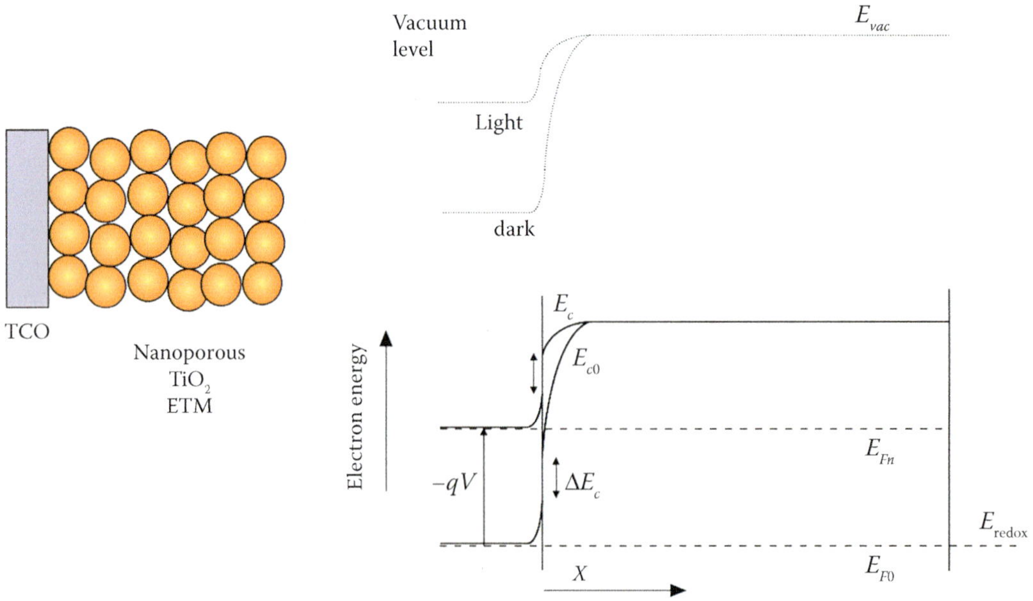

FIGURE 24.12 Changes of the position of the conduction band in a DSC when the voltage is increased, showing also the changes of the VL across the device. The left diagram shows the morphology of the active layer and the contact.

the distributions. For low-occupancy conditions, the carriers thermalize in the bottom of the DOS and the formula in Equation 24.1 is modified as follows (Garcia-Belmonte, 2010):

$$qV_{oc} = E_g - \frac{\sigma_n^2 + \sigma_p^2}{2k_BT} + k_BT \ln\left(\frac{np}{N_cN_v}\right) \quad (24.8)$$

Similar to inorganic thin film solar cells, in hybrid and organic nanostructured solar cells, the macroscopic contacts play an important role in selecting just one carrier from the mixture of phases upon arrival to the contact. This function is usually assisted by the use of blocking layers described in Section 20.6. Barrier layers at the contacts of a BHJ organic cell improve the selectivity of contacts by repelling one of the carriers, as shown in Figure 24.11. In solid DSCs, a thin TiO_2 blocking layer protecting the TCO contact has a crucial effect to prevent a shorting current from the hole transporting phase. The same type of blocking layer is usually adopted in lead halide perovskite solar cells.

The distribution of electric potential in a nanostructured cell in relation to the explanation of the origin of the photovoltage has been extensivelly discussed in the area of nanostructured cells (Pichot and Gregg, 2000). When the applied voltage is modified in a DSC device, a change of the vacuum level will necessarily occur, and similarly, when a photovoltage is generated under illumination. Since the Debye length in the electrolyte has the value of a few nm, establishing a macroscopic field in the direction normal to the contacts is unlikely, see Figure 9.26, which implies that the variation of local potential must be absorbed in a size of several nm, close to the outer interface with the TCO, as shown in Figure 24.12. To absorb the whole modification of the external potential difference in a region close to the interface, a change of band bending occurs in a highly doped medium, as in the TCO, or a change of voltage of the Helmholtz layer takes place at the substrate surface. Effectively, the interface dipole at the semiconductor-substrate interface is modified in equal magnitude to the change of the Fermi level, as remarked earlier in Figure 20.8c.

The operation of a p–n junction as a selective contact has been discussed in Figure 9.21. In Figure 24.13a, we indicate an operation of the p–n junction as a charge separating electrical field reminiscent of the model of Figure 20.11. This view is very typical in older texts on solar cells. It should be emphasized that silicon is an indirect semiconductor and poor light absorber; thus a thick absorber layer of about 200–300 μm (depending on light trapping schemes) is necessary to capture all the solar photons. For characteristic doping density of 10^{17} cm^{-3},

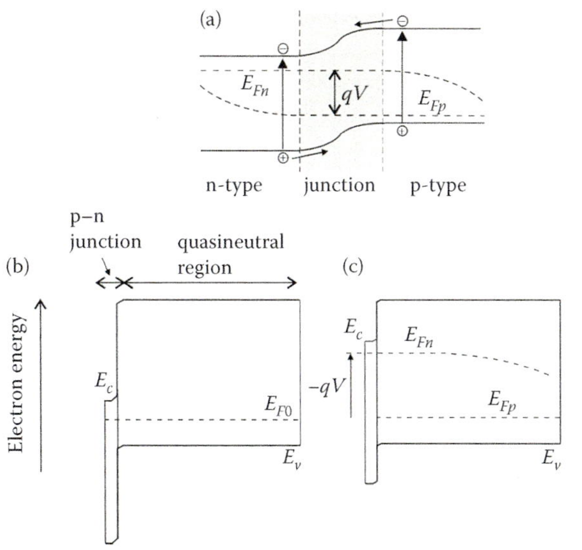

FIGURE 24.13 (a) Energy diagram of p–n junction. Depletion layers existing in the junction region are shown, as well as the Fermi levels of electrons and holes whose separation is the voltage across the junction. (b) Zero voltage and (c) forward voltage (or photovoltage under illumination) scheme of a p-type crystalline silicon solar cell. The p–n junction is a thin region at the left contact, where minority carrier electrons from the p-side are converted to majority carriers at the n-side of the junction.

the size of the depletion region in the p–n junction is about 50 nm (Figure 9.4). This means that in contrast to the drawing of Figure 24.13a, the p–n junction in a crystalline silicon solar cell occupies an extremely thin region and is located on the boundary.

In Figure 24.13b and c, we show the operation of the p–n junction as a very thin selective contact that converts the minority carrier Fermi level in the absorber side into a majority carrier Fermi level that is readily equilibrated with the metal. Charge separation of photogenerated electron-hole pairs *does not* occur mainly in the thin depletion region, but in the very large quasi-neutral region. From the conceptual point of view, the p–n junction is just another mechanism of selectivity that absorbs the variation of voltage in a rather thin interfacial layer, very similar to the TCO/TiO_2 contact in a DSC previously discussed in Figure 24.12.

GENERAL REFERENCES

Modification of potential at the contacts and origin of photovoltage: Cahen et al. (2000), Pichot and Gregg (2000), Kron et al. (2003), Turrión et al. (2003), Bisquert et al. (2004), and Bisquert and Garcia-Belmonte (2011).

Morphology of organic solar cells: Pfannmöller et al. (2011), Creddington and Durrant (2012), and Guerrero and Garcia-Belmonte (2016).

Effect of disorder on open-circuit voltage: Tiedje (1982), Yablonovitch et al. (1982), Garcia-Belmonte (2010), Garcia-Belmonte and Bisquert (2010), Nayak et al. (2011), Nayak et al. (2012), and Garcia-Belmonte (2013).

REFERENCES

Abdou, M. S. A.; Orfino, F. P.; Son, Y.; Holdcroft, S. Interaction of oxygen with conjugated polymers: Charge transfer complex formation with poly(3-alkylthiophenes). *Journal of the American Chemical Society* 1997, *119*, 4518–4524.

Ansari-Rad, M.; Abdi, Y.; Arzi, E. Reaction order and ideality factor in dye-sensitized nanocrystalline solar cells: A theoretical investigation. *The Journal of Physical Chemistry C* 2012, *116*, 10867.

Bisquert, J.; Cahen, D.; Rühle, S.; Hodes, G.; Zaban, A. Physical chemical principles of photovoltaic conversion with nanoparticulate, mesoporous dye-sensitized solar cells. *The Journal of Physical Chemistry B* 2004, *108*, 8106–8118.

Bisquert, J.; Garcia-Belmonte, G. On voltage, photovoltage, and photocurrent in bulk heterojunction organic solar cells. *Journal of Physical Chemistry Letters* 2011, *2*, 1950–1964.

Bredas, J.-L. Mind the gap! *Materials Horizons* 2014, *1*, 17–19.

Cahen, D.; Hodes, G.; Grätzel, M.; Guillemoles, J. F.; Riess, I. Nature of photovoltaic action in dye-sensitized solar cells. *The Journal of Physical Chemistry B* 2000, *104*, 2053–2059.

Chen, K.; Barker, A. J.; Morgan, F. L. C.; Halpert, J. E.; Hodgkiss, J. M. Effect of carrier thermalization dynamics on light emission and amplification in organometal halide perovskites. *The Journal of Physical Chemistry Letters* 2014, *6*, 153–158.

Clarke, T. M.; Durrant, J. R. Charge photogeneration in organic solar cells. *Chemical Reviews* 2010, *110*, 6736–6767.

Correa-Baena, J.-P.; Turren-Cruz, S.-H.; Tress, W.; Hagfeldt, A.; Aranda, C.; Shooshtari, L.; Bisquert, J.; et al. Changes from bulk to surface recombination mechanisms between pristine and cycled perovskite solar cells. *ACS Energy Letters* 2017, *2*, 681–688.

Creddington, D.; Durrant, J. R. Insights from transient optoelectronic analyses on the open-circuit voltage of organic solar cells. *Journal of Physical Chemistry Letters* 2012, *3*, 1465–1478.

Garcia-Belmonte, G. Temperature dependence of open-circuit voltage inorganic solar cells from generation–recombination kinetic balance. *Solar Energy Materials and Solar Cells* 2010, *94*, 2166–2169.

Garcia-Belmonte, G. Carrier recombination flux in bulk heterojunction polymer: Fullerene solar cells: Effect of energy disorder on ideality factor. *Solid-State Electronics* 2013, *79*, 201–205.

Garcia-Belmonte, G.; Bisquert, J. Open-circuit voltage limit caused by recombination through tail states in bulk heterojunction polymer-fullerene solar cells. *Applied Physics Letters* 2010, *96*, 113301.

Garcia-Belmonte, G.; Boix, P. P.; Bisquert, J.; Lenes, M.; Bolink, H. J.; La Rosa, A.; Filippone, S.; et al. Influence of the intermediate density-of-states occupancy on open-circuit voltage of bulk heterojunction solar cells with different fullerene acceptors. *Journal of Physical Chemistry Letters* 2010, *1*, 2566–2571.

Gouda, L.; Gottesman, R.; Ginsburg, A.; Keller, D. A.; Haltzi, E.; Hu, J.; Tirosh, S.; et al. Open circuit potential build-up in perovskite solar cells from dark conditions to 1 sun. *The Journal of Physical Chemistry Letters* 2015, *6*, 4640–4645.

Guerrero, A.; Garcia-Belmonte, G. Recent advances to understand morphology stability of organic photovoltaics. *Nano-Micro Letters* 2016, *9*, 10.

Huang, S. Y.; Schlichthörl, G.; Nozik, A. J.; Grätzel, M.; Frank, A. J. Charge recombination in dye-sensitized nanocrystallyne TiO_2 solar cells. *The Journal of Physical Chemistry B* 1997, *101*, 2576–2582.

Jaegermann, W.; Klein, A.; Mayer, T. Interface engineering of inorganic thin-film solar cells—Materials-science challenges for advanced physical concepts. *Advanced Materials* 2009, *21*, 4196–4206.

Kirchartz, T.; Bisquert, J.; Mora-Sero, I.; Garcia-Belmonte, G. Classification of solar cells according to mechanisms of charge separation and charge collection. *Physical Chemistry Chemical Physics* 2015, *17*, 4007–4014.

Kron, G.; Egerter, T.; Werner, J. H.; Rau, W. Electronic transport in dye-sensitized nanoporous TiO2 solar cells—Comparison of electrolyte and solid-state devices. *The Journal of Physical Chemistry B* 2003, *107*, 3556.

Li, C.-Z.; Yip, H.-L.; Jen, A. K. Y. Functional fullerenes for organic photovoltaics. *Journal of Materials Chemistry* 2012, *22*, 4161–4177.

Manser, J. S.; Kamat, P. V. Band filling with free charge carriers in organometal halide perovskites. *Nature Photonics* 2014, *8*, 737–743.

Mishra, A.; Bäuerle, P. Small molecule organic semiconductors on the move: Promises for future solar energy technology. *Angewandte Chemie International Edition* 2012, *51*, 2020–2067.

Mora-Sero, I.; Bertoluzzi, L.; Gonzalez-Pedro, V.; Gimenez, S.; Fabregat-Santiago, F.; Kemp, K. W.; Sargent, E. H.; et al. Selective contacts drive charge extraction in quantum dot solids via asymmetry in carrier transfer kinetics. *Nature Communications* 2013, *2*, 2272.

Nayak, P. K.; Bisquert, J.; Cahen, D. Assessing possibilities and limits for solar cells. *Advanced Materials* 2011, *23*, 2870–2876.

Nayak, P. K.; Garcia-Belmonte, G.; Kahn, A.; Bisquert, J.; Cahen, D. Photovoltaic efficiency limits and material disorder. *Energy & Environmental Science* 2012, *5*, 6022–6039.

O'Regan, B.; Grätzel, M. A low-cost high-efficiency solar cell based on dye-sensitized colloidal TiO_2 films. *Nature* 1991, *353*, 737–740.

Peter, L. M. Transport, trapping and interfacial transfer of electrons in dye-sensitized nanocrystalline solar cells. *Journal of Electroanalytical Chemistry* 2007, *599*, 233–240.

Pfannmöller, M.; Flügge, H.; Benner, G.; Wacker, I.; Sommer, C.; Hanselmann, M.; Schmale, S. Visualizing a homogeneous blend in bulk heterojunction polymer solar cells by analytical electron microscopy. *Nano Letters* 2011, *11*, 3099–3107.

Pichot, F.; Gregg, B. A. The photovoltage-determining mechanism in dye-sensitized solar cells. *The Journal of Physical Chemistry B* 2000, *104*, 6–10.

Schmidt, J.; Merkle, A.; Brendel, R.; Hoex, B.; de Sanden, M. C. M. v.; Kessels, W. M. M. Surface passivation of high-efficiency silicon solar cells by atomic-layer-deposited Al_2O_3. *Progress in Photovoltaics: Research and Applications* 2008, *16*, 461–466.

Taguchi, M.; Kawamoto, K.; Tsuge, S.; Baba, T.; Sakata, H.; Morizane, M.; Uchihashi, K.; et al. HITTM cells—High-efficiency crystalline Si cells with novel structure. *Progress in Photovoltaics: Research and Applications* 2000, *8*, 503–513.

Tiedje, T. Band tail recombination limit to the output voltage of amorphous silicon solar cells. *Applied Physics Letters* 1982, *40*, 627–629.

Turrión, M.; Bisquert, J.; Salvador, P. Flatband potential of $F:SnO_2$ in a TiO_2 dye-sensitized solar cell: An interference reflection study. *The Journal of Physical Chemistry B* 2003, *107*, 9397–9403.

Vandewal, K.; Tvingstedt, K.; Gadisa, A.; Inganas, O.; Manca, J. V. Relating the open-circuit voltage to interface molecular properties of donor:acceptor bulk heterojunction solar cells. *Physical Review B* 2010, *81*, 125204.

Veldman, D.; Meskers, S. C. J.; Janssen, R. A. J. The energy of charge-transfer states in electron donor-acceptor blends: Insights into the energy losses in organic solar cells. *Advanced Functional Materials* 2009, *19*, 1939–1948.

Verploegen, E.; Mondal, R.; Bettinger, C. J.; Sok, S.; Toney, M. F.; Bao, Z. Effects of thermal annealing upon the morphology of polymer–fullerene blends. *Advanced Functional Materials* 2010, *20*, 3519–3529.

Yablonovitch, E.; Tiedje, T.; Witzke, H. Meaning of the photovoltaic band gap for amorphous semiconductors. *Applied Physics Letters* 1982, *41*, 953–955.

Yamada, Y.; Nakamura, T.; Endo, M.; Wakamiya, A.; Kanemitsu, Y. Photocarrier recombination dynamics in perovskite $CH_3NH_3PbI_3$ for solar cell applications. *Journal of the American Chemical Society* 2014, *136*, 11610–11613.

Zarazua, I.; Han, G.; Boix, P. P.; Mhaisalkar, S.; Fabregat-Santiago, F.; Mora-Seró, I.; Bisquert, J.; et al. Surface recombination and collection efficiency in perovskite solar cells from impedance analysis. *The Journal of Physical Chemistry Letters* 2016, *7*, 5105–5113.

25 Operation of Solar Cells and Fundamental Limits to Their Performance

We describe the main characteristics of the operation of a solar cell with respect to voltage, electrical current, and illumination. We aim to explain the central feature of the solar cell performance in terms of the current density-voltage curve, considering the photovoltage, photocurrent, and the fill factor (FF), and how these are combined to produce the power conversion efficiency (PCE). In Section 25.4, we derive the limits to efficiency conversion based on the Shockley–Queisser approach and we present a detailed discussion of the different fundamental intrinsic losses of light to electrical power conversion that lead to the maximal theoretical efficiency as a function of the band gap energy of the absorber. The materials limits to solar energy conversion efficiencies by dynamic and static disorder in semiconductors are examined in the final section.

25.1 CURRENT-VOLTAGE CHARACTERISTICS

We analyze the operational properties of a solar cell under a substantial incident illumination using the basic model shown in Figure 21.1. We start with Figure 21.1d that shows the situation *of open circuit* in which no current is extracted. The excitation by light produces excess electrons and holes in the carrier bands, which causes the separation of the corresponding Fermi levels, as commented previously in Section 20.4. By controlling the voltage V that exists between the two contacts of the solar cell, the amount of recombination can be changed. In the case of Figure 21.1e, the voltage is $V < V_{oc}$. Here, the solar cell generates a current and acts as a battery. Another important operational variable is the current extracted from the solar cell, already indicated in Section 22.1. The nonrecombined carriers, by conservation, produce a photocurrent density, as indicated in Figure 21.1f.

In Equation 21.1, we established that the current injected to a diode in the dark is the difference between recombination current and the dark generation current j_0. The photocurrent, j_{ph}, arising from the excess radiation that impacts the solar cell, is added to the recombination and dark generation processes as represented in Figure 21.1. In summary, we have

$$j = j_{ph} - j_{rec} + j_0 \quad (25.1)$$

The balance between the three terms on the right-hand side of Equation 25.1 gives rise to the current that flows in the external circuit in terms of electrons leaving the excited state through their selective contact and entering the ground state through the hole selective contact. This process can be, therefore, viewed as a recombination through the external circuit and it is the current in which we are directly interested, as it is the one capable of doing useful work on external elements.

From Equation 21.8, the current density-voltage (j–V curve) characteristic of the solar cell is

$$j = j_{ph} - j_0\left(e^{qV/mk_BT} - 1\right) \quad \left(V_{\text{forward}} > 0\right) \quad (25.2)$$

Figure 25.1a shows the shape of the j–V curve corresponding to the ideal diode performance. We observe that under illumination, the dark characteristic is shifted up in the current density axis. At open circuit ($j = 0$), the recombination current balances the photogeneration rate from the external sources. The open-circuit voltage V_{oc} is given by the expression

$$V_{oc} = \frac{mk_BT}{q} \ln\left(\frac{j_{ph}}{j_0} + 1\right) \quad (25.3)$$

The second point of interest is the *short-circuit* condition, shown in Figure 21.1e. In this case, we have $j_{rec} = j_0$ as indicated in Equation 21.4. Therefore, all the excess generated carriers contribute to the external current, the *short-circuit photocurrent*,

$$j_{sc} = j_{ph} \quad (25.4)$$

Thus, we obtain an operational definition of the quantity j_{ph}: it is the current density at short circuit.

Between open circuit and short circuit, the voltage can have any value $0 \le V \le V_{oc}$, and this situation is indicated in Figure 21.1e. Finally, at strong forward bias, a large recombination current is induced.

Usually, the current in the solar cell grows very rapidly when V departs from V_{oc} toward lower voltages, see Figure. 25.1a, but when the voltage further decreases, the current becomes stabilized to $j \approx j_{ph}$. This is because at $V \approx V_{oc}$, we induce a large internal recombination

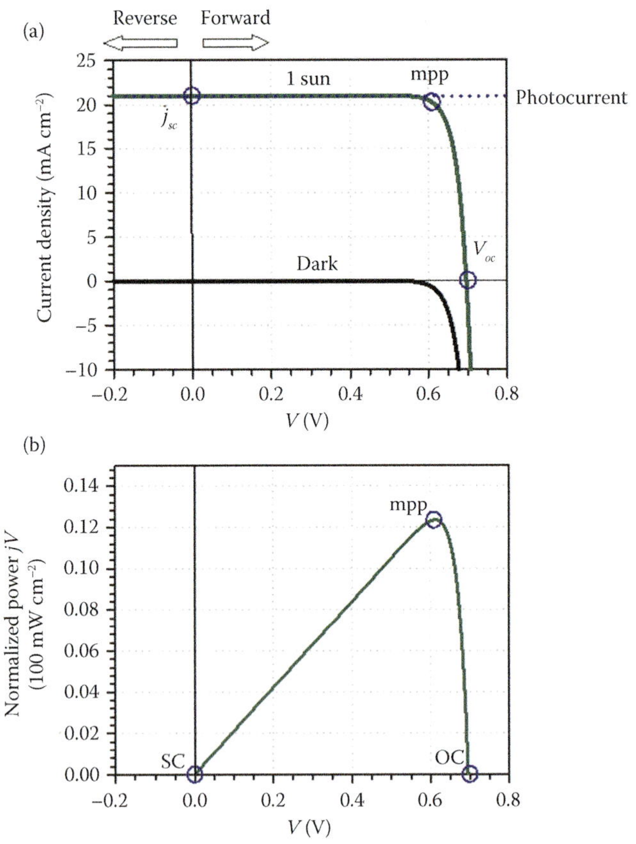

FIGURE 25.1 (a) Current density-potential characteristics of a solar cell that consists of a fundamental diode with dark saturation current $j_0 = 4.34 \times 10^{-11}$ mA cm^{-2} and diode quality factor $m = 1$. Shown are the short-circuit current and open-circuit voltage points. (b) The electrical power supplied by the solar cell as a function of voltage. The power is normalized to 100 mW cm^{-2} (corresponding to 1 sun light intensity). Shown is the maximum power point.

current that compensates the photogeneration, but when V decreases considerably, the recombination current is negligible with respect to j_{ph}. At strong reverse bias, the current in the ideal model is

$$j_{sat} = j(V \ll 0) = j_{ph} + j_0 \tag{25.5}$$

Figure 25.2 shows the j–V and EQE_{PV} characteristics of two lead halide perovskite solar cells, including one high-performance cell and another one with lower performance (Li et al., 2016).

25.2 POWER CONVERSION EFFICIENCY

The main application of a solar cell is to serve as a power supply unit that produces electricity from sunlight, or from ambient light in indoor applications. The central feature to assess the solar cell performance is the electrical power that can be extracted from the available radiation level. The electrical power at a given voltage operation point of the solar cell, P_{el}, has the value

$$P_{el} = jV \tag{25.6}$$

Figure 25.1b shows the power supplied by the solar cell as a function of the voltage. The power is zero at both open- and short-circuit conditions. In between the extreme cases of low power lies the *maximum power point* (mpp) at which voltage (V_{mp}) the solar cell should be operated for electricity production. The maximum power provided by the photovoltaic device is

$$P_{el,max} = j_{mp} V_{mp} \tag{25.7}$$

For the characterization of energy converter devices, the main figure of merit is the maximum conversion efficiency of the solar cell, that is, the PCE, that consists of the electrical power supplied at the mpp with respect to the incoming photon energy, as stated in Equation 22.9,

$$\eta_{\text{PCE}} = \frac{j_{mp} V_{mp}}{\Phi_{E,tot}^{\text{source}}} \tag{25.8}$$

The PCE depends on the operation conditions that need to be defined. The PCE of a solar cell is usually reported under simulated standard terrestrial spectrum $AM1.5G$, which bears an integrated power of $\Phi_E^{\text{AM1.5G}} = 1\,\text{kWm}^{-2} = 100\,\text{mW cm}^{-2}$. This type of illumination is usually denominated "1 sun," see Section 17.8.

The efficiency increases when the short-circuit photocurrent j_{ph} and open-circuit voltage V_{oc} increase, but η_{PCE} also depends critically on the mpp. The shape of the j–V curve determines at which voltage we extract the electrons as electric current. As observed in Figure 25.1, there is a trade-off between current and voltage. At low voltage, extraction is easy, and the current is determined by the quantum yield (conversion of photons to electron carriers) of the absorber. However, a high voltage produces a current that is contrary to the photocurrent, and eventually the power decreases. In general, if we extract the electrons at high useful energy (voltage), there is a price in the lowering of the number we can extract. If this point is close to V_{oc}, then the operational voltage and current are much larger than if V_{mpp} occurs at low voltage close to $V_{oc}/2$. The parameter that tracks this property is the *fill factor* (FF) defined as

$$\text{FF} = \frac{j_{mp} V_{mp}}{j_{ph} V_{oc}} \tag{25.9}$$

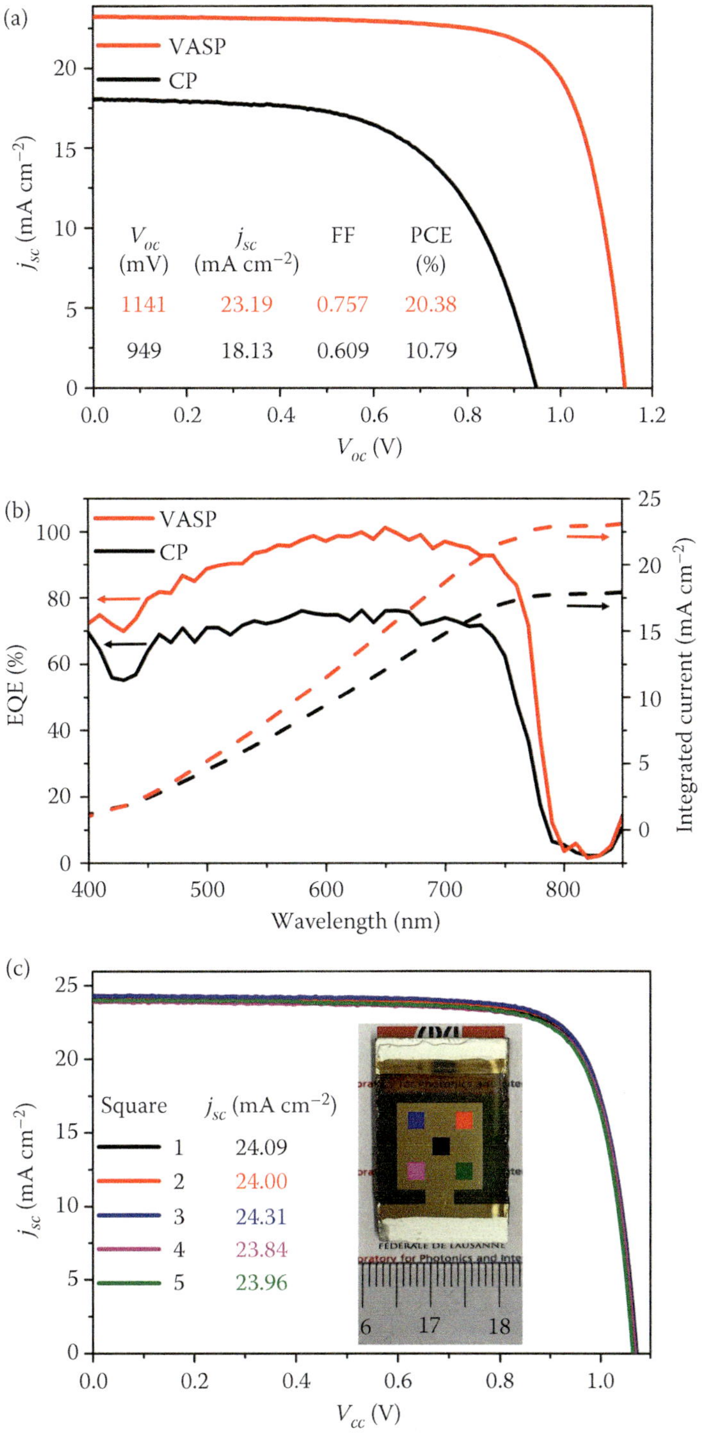

FIGURE 25.2 (a) Current-voltage curves for $FA_{0.81}MA_{0.15}PbI_{2.51}Br_{0.45}$ (FA: formamidinium, MA: methylammonium) perovskite solar cell devices using perovskite films prepared by the single-step solution deposition process (CP) and vacuum-flash solution processing (VASP) methods measured under standard *AM*1.5 solar radiation. (b) *Solid line*: EQE_{PV} curves of cells fabricated by the CP and VASP method. Measurements were taken with chopped monochromatic light under a white light bias corresponding to 5% solar intensity. *Dashed lines*: Short-circuit current density calculated from the overlap integral of the EQE_{PV} spectra with the standard *AM*1.5 solar emission. (c) Photovoltaic parameters for a representative $FA_{0.81}MA_{0.15}PbI_{2.51}Br_{0.45}$-based perovskite device fabricated by VASP measured from five different spots with an aperture area of 0.16 cm^2 selected from the total active area of 1.2 × 1.2 cm^2 under standard *AM*1.5*G* illumination. All *j–V* curves were recorded at a scanning rate of 50 mV s^{-1} in reverse direction. (Reproduced with permission from Li, X. et al. A vacuum flash-assisted solution process for high-efficiency large-area perovskite solar cells. *Science* 2016, 10.1126/science.aaf8060.)

We obtain the convenient expressions

$$P_{el,max} = j_{ph} \times \text{FF} \times V_{oc} \tag{25.10}$$

$$\eta_{\text{PCE}} = \frac{j_{ph} \times \text{FF} \times V_{oc}}{\Phi_{E,tot}^{\text{source}}} \tag{25.11}$$

25.3 ANALYSIS OF FF

The FF depends on the form of the j–V curve, determining the mpp and exerting a large influence on the PCE of the solar cell. If the FF is high, the drop in current at high voltage is delayed, and we can extract the electrons at high voltage while the current is still close to j_{ph}. For a good diode characteristic, the power P increases linearly at low voltage, Figure 25.3b, and decreases abruptly at $V > V_{mp}$. Figure 25.3a compares the j–V characteristic of an ideal diode described by Equation 25.3, which has FF = 0.85, with a j–V "ohmic" characteristic that consists of the straight line that represents a resistor in the j–V plane, displaced upward with the addition of a photocurrent. In the last case, the photocurrent at maximum power is $j_{mp} = j_{ph}/2$, $V_{mp} = V_{oc}/2$, thus FF = 0.25, and the efficiency is very low. A low FF is frequently found in solar cells made with a poorly conducting material (Ji et al., 2010; Lopez-Varo et al., 2016).

While the measurement of the open-circuit photovoltage and short-circuit photocurrent is straightforward, the determination of the mpp requires a tracking of the point of maximum energy conversion (Christians et al., 2015; Zimmermann et al., 2016). Assuming that a perfectly stable characteristic has been achieved, the voltage of maximum power V_{mp} is obtained by the solution of the equation $d(jV)/dV = 0$. For the ideal diode model, the equation can be expressed as

$$V_{mp} = V_{oc} - \frac{mk_BT}{q}\ln\left(1 + \frac{qV_{mp}}{mk_BT}\right) \tag{25.12}$$

Obtaining the mpp requires to solve numerically this transcendental equation for V_{mp}. Some useful approximated solutions have been developed. By an iteration of Equation 25.12, we find

$$V_{mp} = V_{oc} - \frac{mk_BT}{q}\ln\left(1 + \frac{qV_{oc}}{mk_BT} - \ln\left[1 + \frac{qV_{mp}}{mk_BT}\right]\right) \tag{25.13}$$

We still have the unknown at the right-hand side, but this term can be neglected with respect to the second term in the parenthesis. Thus, we obtain

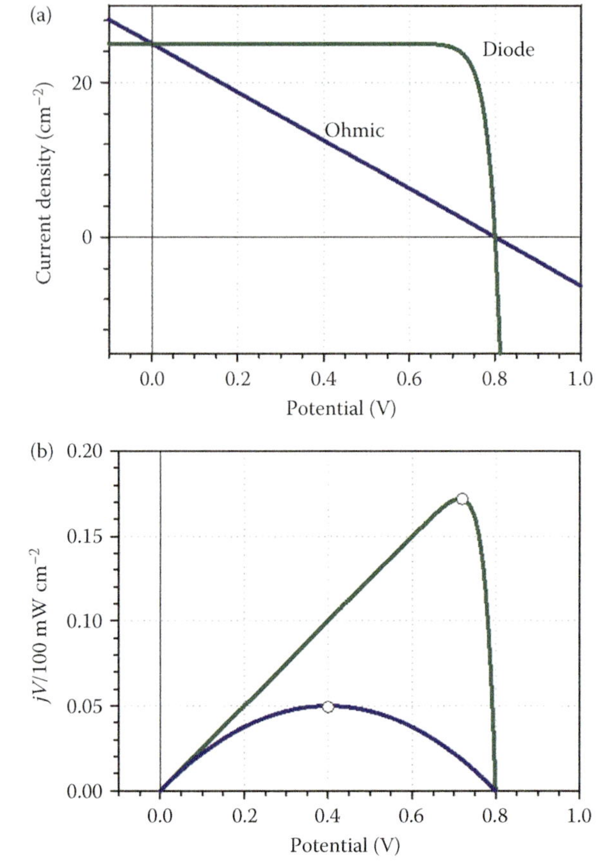

FIGURE 25.3 (a) Current density-potential characteristics of two solar cells: an ideal diode with quality factor m = 1 and a straight characteristic with added photocurrent denoted ohmic. (b) The electrical power supplied by the solar cells as a function of voltage. The power is normalized to 100 mW cm^{-2}. The mpp is indicated with a circle.

$$V_{mp} = V_{oc} - \frac{mk_BT}{q}\ln\left(1 + \frac{qV_{oc}}{mk_BT}\right) \tag{25.14}$$

We can calculate the current from Equations 25.2 and 25.3

$$j_{mp} = j_{ph}\frac{1}{1 + \dfrac{mk_BT}{qV_{oc}}} \tag{25.15}$$

Hence,

$$\text{FF} = \frac{\dfrac{qV_{oc}}{mk_BT} - \ln\left(1 + \dfrac{qV_{oc}}{mk_BT}\right)}{\dfrac{qV_{oc}}{mk_BT} + 1} \tag{25.16}$$

A variety of approximations of this type have been proposed, and the most accurate formula is the following (Green, 1982):

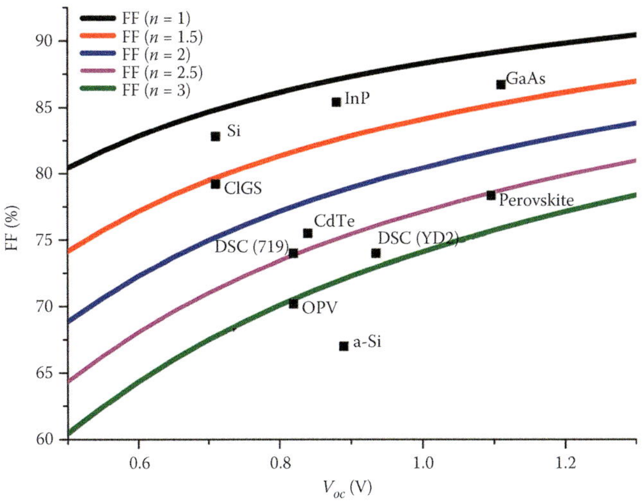

FIGURE 25.4 FFs of best laboratory cells of different categories. The solid lines represent the expected FF value as a function of V_{oc} for a given value of the diode ideality factor (see Equation 25.17). CIGS stands for Cu(In,Ga)Se_2; OPV stands for organic photovoltaic cell; DSC indicates dye-sensitized solar cell with the dye used in parentheses. (Adapted from Nayak, P. K. et al. *Energy & Environmental Science* 2012, 5, 6022–6039.)

$$\mathrm{FF} = \frac{\dfrac{qV_{oc}}{mk_BT} - \ln\left(0.72 + \dfrac{qV_{oc}}{mk_BT}\right)}{\dfrac{qV_{oc}}{mk_BT} + 1} \qquad (25.17)$$

In conclusion, in the diode model, the FF is completely determined by the values of V_{oc} and m. The lines in Figure 25.4 indicate the FF value depending on the diode quality factor, and the values for top performing cells in different photovoltaic technologies are also indicated. A more advanced analytical solution for j–V characteristics including the series resistance is described by Banwell and Jayakumar (2000).

Let us discuss some examples that illustrate the j–V characteristics in a solar cell. We compare in Figures 25.5 and 25.6 two different solar cells A and B. The photovoltaic parameters are indicated in Table 25.1. As observed in Figure 25.5, as the illumination power is reduced, the photocurrent decreases but the recombination characteristics (in this simple model) are unchanged, so the curve is displaced downward. As the collection efficiency of this exercise is $EQE_{PV} = 1$, the photocurrent is simply proportional to the illumination intensity. However, the photovoltage varies slowly with illumination as indicated in Equation 23.10, and the FF consequently changes very slowly as implied by Equation 25.17. Therefore, the PCE shows only a slight decrease as the light intensity

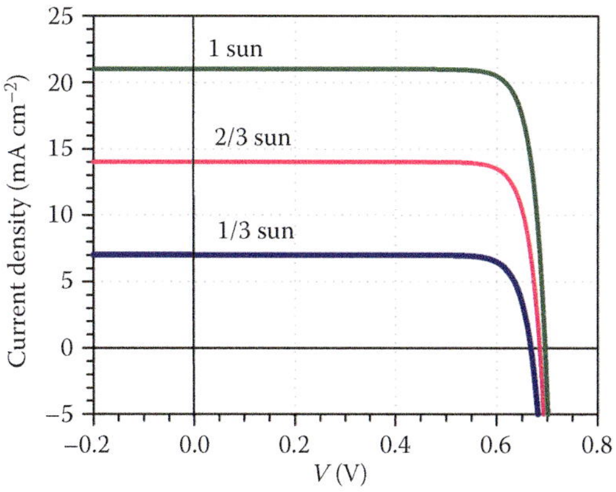

FIGURE 25.5 Current density-potential characteristics of a solar cell (A) that consists of a fundamental diode with dark saturation current $j_0 = 4.34 \times 10^{-11}$ mA cm^{-2} and diode quality factor $m = 1$, under different illumination intensities.

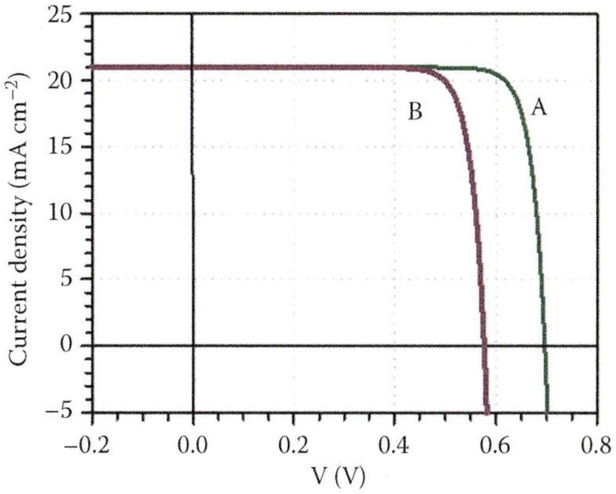

FIGURE 25.6 Current density-potential characteristics of two solar cells that consist of fundamental diodes with dark saturation current $j_0 = 4.34 \times 10^{-11}$ mA cm^{-2} (cell A), $j_0 = 4.34 \times 10^{-9}$ mA cm^{-2} (cell B) and diode quality factor $m = 1$ in both cases, under illumination of 1 sun.

decreases. Normally in experiments, the PCE increases as the light intensity decreases, and increases for cells with smaller area, as discussed in Section 26.10. The cell B in Figure 25.6 has larger recombination rate, as determined by the reverse saturation current density j_0, which causes the bending down of the curve at a lower forward voltage and consequently, the open-circuit voltage decreases.

One effect that produces a serious modification of the FF is the series resistance R_s. The series resistance may correspond to the external resistance of the selective contact layers or to internal transport effects. For simplicity, we now model a constant series resistance (per area) due to the ohmic transport in one contact layer that takes a potential drop jR_s. The external voltage is distributed as a combination of the ohmic drop and the "Fermi level

TABLE 25.1
Performance Parameters of Solar Cells (see Figures 25.5 and 25.6)

Cell	Illumination Intensity Φ_E (mW cm^{-2})	Short-Circuit Photocurrent j_{ph} (mA cm^{-2})	Open-Circuit Voltage V_{oc} (V)	Fill Factor *FF*	Efficiency η_{PCE}(%)
A	(1/3) × 100	7.0	0.671	0.840	11.84
A	(2/3) × 100	14.0	0.690	0.844	12.21
A	100	21.0	0.700	0.845	12.42
B	100	21.0	0.580	0.820	10.00

voltage" associated with the splitting of the Fermi levels at the outer edges of the absorber layer, V_F, as explained in Equation 25.18. Therefore,

$$V = V_F - jR_s \tag{25.18}$$

Since the solar cell continues to work internally as in Equation 25.3, the current density depends on voltage as

$$j = j_{ph} - j_0\left(e^{q(V+jR_s)/mk_BT} - 1\right) \tag{25.19}$$

The resulting current-voltage characteristics are shown in Figure 25.7 for different values of R_s. Note that all the curves have the same shape in terms of V_F, but when plotted against the external voltage, the FF becomes progressively degraded. The value $R_s = V_{oc}/j_{ph}$ causes the characteristic to be ohmic in the first quadrant as discussed earlier in Figure 25.3, which decreases the FF to a value 0.25.

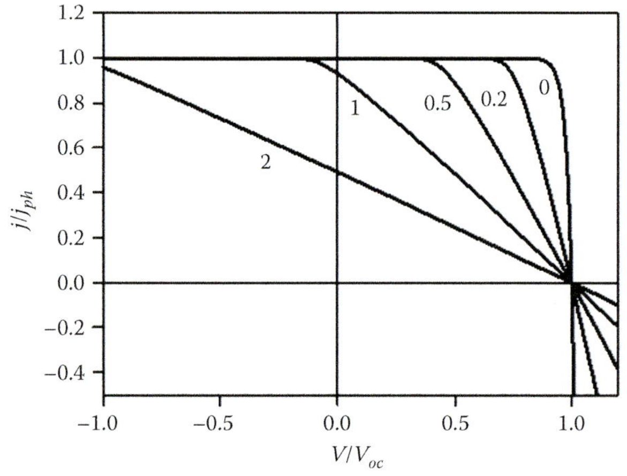

FIGURE 25.7 Current-voltage characteristics of the fundamental diode model including a series resistance. The value indicated in each curve is $R_s j_{ph}/V_{oc}$.

25.4 SHOCKLEY–QUEISSER EFFICIENCY LIMITS

The fundamental model of the solar cell has been studied in the previous chapters. Based on this model of the photovoltaic converter we can now analyze in detail the fundamental limitations to solar energy conversion. We aim to quantitatively establish the fraction of the incoming energy that can be transformed into electrical power by the use of a given absorber material. The central method to provide these results was derived by Shockley and Queisser (1961) (SQ). This method is universally acknowledged as the central figure of merit to characterize the target PCE of a given class of photovoltaic materials.

The rationale of the method is based on the following steps, in which the results corresponding to the radiative limit that have been obtained in previous chapters are used.

SQ1. Describe a solar cell in terms of a semiconductor absorber material with selective contacts, at local temperature T_A, surrounded by blackbody radiation at the same temperature. The solar cell can be operated electrically.

SQ2. The solar cell interacts optically with an external radiation source. The absorber material is a semiconductor with optically sharp band gap, characterized by a single parameter, the band gap E_g that defines the onset of light absorption.

SQ3. To obtain the minimal possible losses in power produced by the *j–V* curve, assume detailed balance in the radiative limit, in which the only recombination channel is radiative emission.

SQ4. Use the photocurrent in the radiative limit, Equation 22.8.

SQ5. Use the photovoltage in the radiative limit, Equation 23.11.

SQ6. Calculate the FF by Equation 25.17.

SQ7. Apply Equation 25.11 to obtain the PCE.

Having described the SQ method, we now consider the different factors θ associated with fundamental effects that reduce the energy conversion capabilities of a single absorber solar cell. To obtain the influence of each

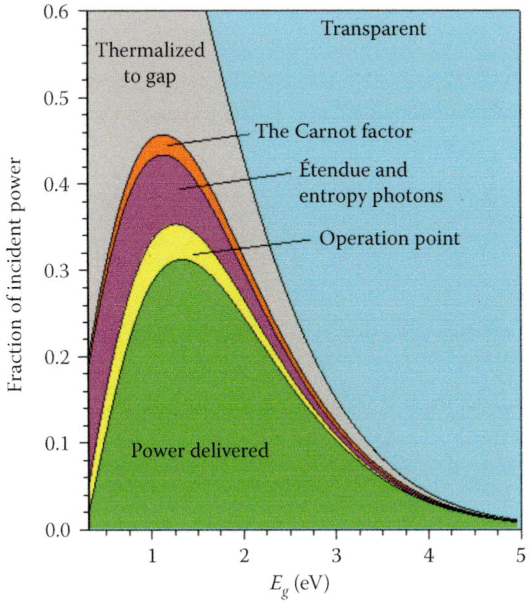

FIGURE 25.8 Different fundamental intrinsic losses of power with respect to that of incident radiation and the delivered electrical power dependence on E_g.

effect that limits the conversion of photons to electrical energy, the corresponding power loss with respect to the incoming solar power $\Phi_{E,tot}(T_S)$ (see also Hirst and Ekins-Daukes, 2011) is calculated. Sunlight is modeled by blackbody radiation at T_S, as discussed in Section 22.1, and the approximation $E_g \gg k_B T_S$ ($\Gamma \approx E_g^2$). Each of the factors of the following list are represented in a different color in Figure 25.8:

1. The transparency of the cell below E_g, with the expression

$$\theta_{E,trans}\left(E_g,T_S\right)=\frac{\Phi_E\left(0,E_g\right)}{\Phi_{E,tot}\left(T_S\right)} \tag{25.20}$$

where the quantity in the denominator is given in Equation 22.15. The effect of the factor $\theta_{E,trans}$ has already been described in Figure 22.1b. The fraction of power lost by the unabsorbed photons increases with the band gap energy.

2. The thermalization of photons. The fraction of power lost by the thermalization to the band gap is

$$\theta_{E,ther}\left(E_g,T_S\right)=\frac{\Phi_E\left(E_g,\infty\right)}{\Phi_{E,tot}}-\frac{E_g\Phi_{ph}\left(E_g,\infty\right)}{\Phi_{E,tot}} \tag{25.21}$$

This factor is also shown in Figure 22.1b, and together with transparency, they are the major losses of energy in photovoltaic conversion, giving rise to the Trivich-Flinn efficiency in Equation 22.16.

In the next set of losses, the reduction of electrical energy from E_g to V_{oc}, is discussed, using the fundamental expression of the radiative voltage in Equation 23.13. To each loss of voltage, there corresponds a loss of power given by the product of the voltage in units of energy and the absorbed photon flux, $qV_{oc}\Phi_{ph}$.

3. The Carnot losses

$$\theta_{E,C}\left(E_g,T_S\right)=\frac{T_A}{T_S}\frac{E_g\Phi_{ph}\left(E_g,\infty\right)}{\Phi_{E,tot}} \tag{25.22}$$

4. The étendue and photon entropy losses

$$\theta_{E,ee}\left(E_g,T_S\right)=\left[k_BT_A\ln\left(\frac{\varepsilon_S}{\pi}\right)+k_BT_A\ln\left(\frac{T_S}{T_A}\right)\right]\frac{\Phi_{ph}\left(E_g,\infty\right)}{\Phi_{E,tot}} \tag{25.23}$$

The previous factors complete the reduction of energy of electrical carriers so that the resulting efficiency is

$$\eta=\frac{qV_{oc}^{rad}\Phi_{ph}\left(E_g,\infty\right)}{\Phi_{E,tot}} \tag{25.24}$$

However, we have already noted that the carriers are extracted at current and voltage of the mpp, so that a further reduction given by the FF occurs in passing to Equation 25.11. The FF of this model has already been calculated in Equation 25.17. Therefore, SQ efficiency is

$$\eta=\frac{q\Phi_{ph}\left(E_g,\infty\right)\times FF\left(V_{oc}^{rad}\right)\times V_{oc}^{rad}}{\Phi_{E,tot}} \tag{25.25}$$

and it consists of the combination of the features shown in Figure 25.8.

Another way to represent the losses is shown in Figure 25.9. Here, a cost in power from the incident spectrum for each mechanism is associated to the product of current density and energy of the photons shown by a colored area. The final output power obtained by subtraction of all the losses is shown by the blue square.

25.5 PRACTICAL SOLAR CELLS EFFICIENCY LIMITS

The result of the SQ calculation when the incoming flux is described by blackbody radiation at the temperature of the sun is given by the dashed curve in Figure 22.2c. The more realistic theoretical efficiency based on the AM1.5G solar irradiance is shown in thin line. In Figure 25.10 we compare the maximal value of photocurrent, photovoltage, and the corresponding fill factor, according to the

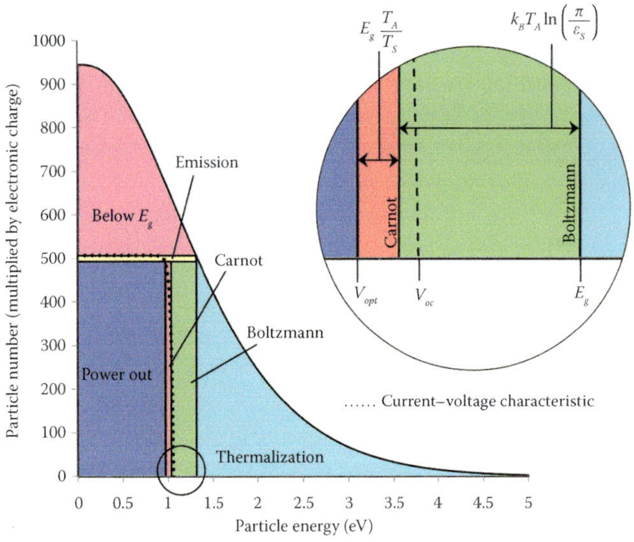

FIGURE 25.9 Intrinsic losses occurring in a device with optimal E_g (1.31 eV) under 1 sun illumination, indicated by the colored areas in the surface below the curve of incident-collected photons versus their energy. The Carnot, Boltzmann, and thermalization losses reduce the optimal operating voltage, while below-E_g radiation and emission losses reduce the current. The combination of both types of effects dictates the form of the current-voltage characteristic. Analytical descriptions of the Carnot and Boltzmann voltage drops are given. (Reproduced with permission from Hirst, L. C.; Ekins-Daukes, N. J. *Progress in Photovoltaics: Research and Applications* 2011, 19, 286–293.)

radiative limit obtained by detailed balance, to the values actually reached by the world record devices of a range of technologies in 2019. The maximal performance values have been calculated using the semiconductor band gap and assuming a sharp absorption edge, according to the SQ method. Figure 25.11 provides the SQ power conversion efficiencies and the maximal performance values obtained by record devices. Figure 25.12 shows several performance parameters with respect to SQ values of the different technologies: the photocurrent, the product $V_{oc} \times FF$, and the PCE.

The SQ method can be applied with different sources of illumination according to the required applications, for example using the spectral characteristic of warm while lamps used in ambient lighting for indoor photovoltaics (Ho, 2020; Venkateswararao, 2020). So far the calculation of efficiency has been presented on the assumption of a single band gap parameter that characterizes the light absorption properties. This idealized model associated to a sharp, stepwise absorptivity as suggested in Figure 22.6 is a very useful tool for general benchmarking and conclusions on the photovoltaic possibilities of different classes of materials. However, photovoltaic semiconductor materials present additional features that have a large

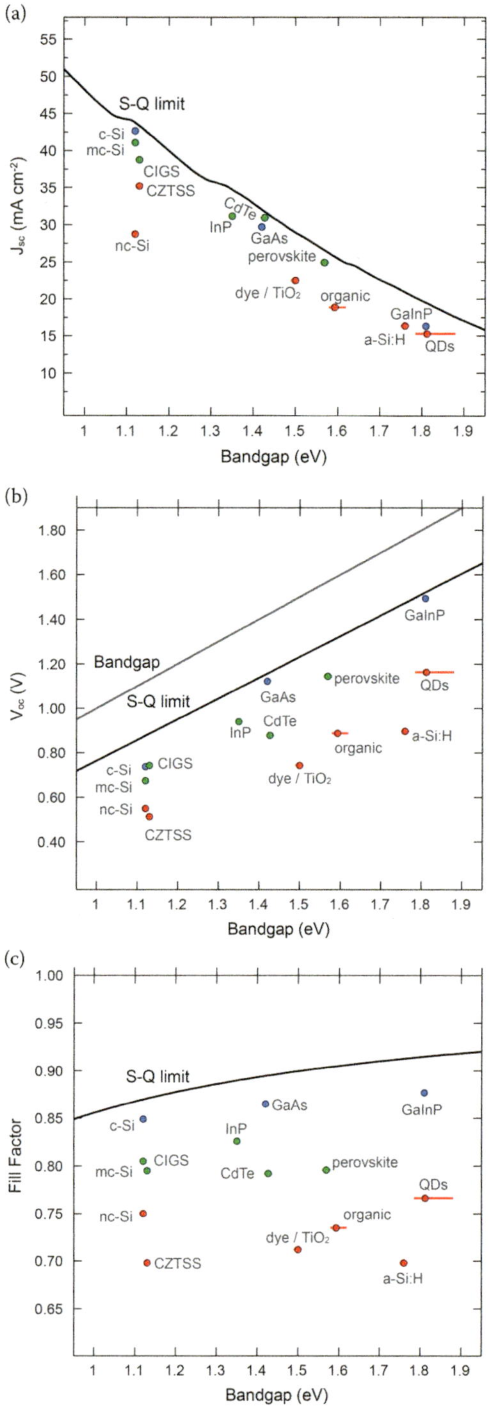

FIGURE 25.10 Record-efficiency cell parameters compared to the detailed-balance limit. Single-junction solar cell parameters are shown as a function of band gap energy according to the Shockley–Queisser limit (solid lines) and experimental values for record-efficiency cells. (A) Short-circuit current j_{sc}. (B) Open-circuit voltage V_{oc}. The voltage corresponding to the band gap is shown for reference, with the voltage deficit V_g-V_{SQ} indicated by the gray shaded region. (C) Fill factor FF. All data are for standard AM1.5 illumination at 1000 W/m². (Reproduced with permission from Polman, A. et al., *Science* 352, aad4424, 2016. Figures updated in 2019 by Tom Veeken from AMOLF Institute.).

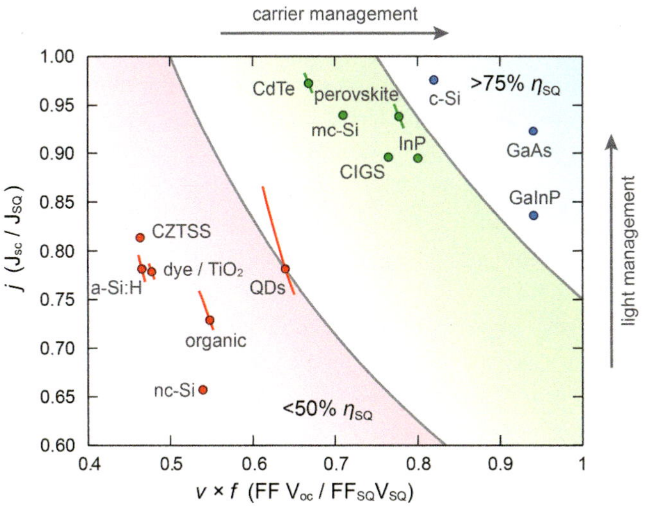

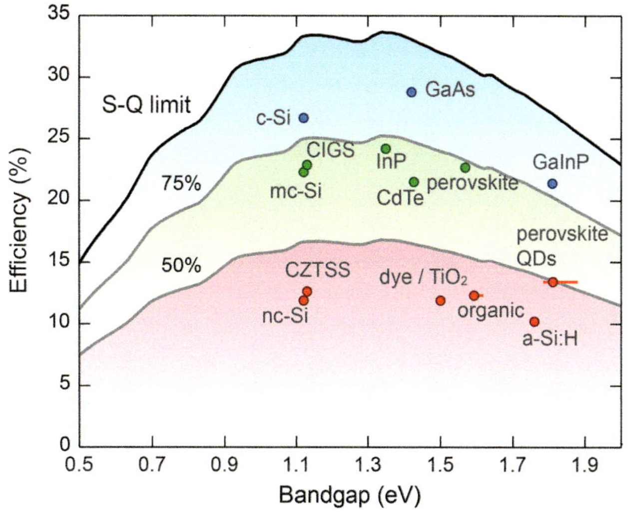

FIGURE 25.11 Theoretical Shockley–Queisser detailed-balance efficiency limit as a function of band gap (black line) and 75% and 50%of the limit (gray lines).The record efficiencies for different materials are plotted for the corresponding band gaps. (Reproduced with permission from Polman, A. et al., *Science* 352, aad4424, 2016.)

FIGURE 25.12 Limiting processes in photovoltaic materials. An efficient solar cell captures and traps all incident light ("light management") and converts it to electrical carriers that are efficiently collected ("carrier management"). The plot shows the short-circuit current and product of open-circuit voltage and fill factor relative to the maximum achievable values, based on the Shockley–Queisser detailed-balance limit, for the most efficient solar cell made with each photovoltaic material. The data indicate whether a particular material requires better light management, carrier management, or both. The lines around some data points correspond to a range of band gaps taken in the SQ calculations according to uncertainty in the band gap of the record cell. Arrows on top and right axes indicate how improved light management and charge carrier collection improve the cell efficiency. η_{SQ} denotes maximum achievable efficiency according to the SQ model. Colors correspond to cells achieving <50% of their SQ efficiency limit η_{SQ} (red), 50 to 75% (green), or >75% (blue). (Reproduced with permission from Polman, A. et al., *Science* 352, aad4424, 2016.)

impact on the actual performance and on the development of solar energy conversion technologies.

The main factor that causes a departure from the idealized characteristics is disorder. In Chapters 7 and 8, the effects of disorder were analyzed and associated with a broad density of states. This property causes the static localization of carriers mainly due to intrinsic structural disorder. In Chapter 13, we described the strong impact of static disorder on the carrier transport mechanisms. In addition to static disorder, dynamic disorder is usually associated to lattice fluctuations and it is often described in terms of interactions of carriers with phonons. Dynamic disorder becomes more relevant at higher temperatures and causes the broadening of characteristics, such as the chemical capacitance shown in Figure 7.5.

It has been amply discussed in previous chapters that the presence of static disorder in a material creates broad tails in the band gap that determine the position of Fermi levels and hence the photovoltage. In this respect we recall Figures 7.5, 7.8, 7.10, and the discussion of the Gaussian DOS in Figures 8.25, 8.26, and 8.27. The V_{oc} of amorphous silicon in relation to the exponential DOS is commented on in Figure 8.11, and the disorder in BHJ organic solar cells is discussed in Figures 24.4 and 24.6. Static disorder also has a strong effect on photovoltaic devices via the creation of defects that become nonradiative recombination centers.

The maximal PCE of a solar cell is linked by detailed balance to the light absorption characteristics. Therefore, the spectral shape of the light absorption coefficient for a certain material is a central property determining the performance. The actual physical features of a broad variety of PV materials can be observed in the EQE_{PV} characteristics shown in Figure 23.2.

The accuracy of the radiative limit efficiency for a specific device can be significantly improved by modifying step SQ2, describing the light absorption and emission properties of the materials by means of the actual absorptivity or EQE_{PV} of the device, as in Equation 22.1. More generally, however, we are interested in obtaining insights that affect a whole class of materials, by observing the dominant characteristics that influence the photovoltaic limits, in particular with respect to disorder and recombination. In any case, when considering more realistic features the radiative voltage will depend on the optical design, and it is not anymore a material constant as in the SQ classification (Pazos-Outón et al., 2018).

A very general feature of the onset of absorption is an Urbach tail, shown in Figure 8.9, which is quantified by the Urbach parameter E_U in Equation 19.9. As explained previously the origin of the exponential-linear dependence is varied, it can be of static nature, that is, impurity induced or a consequence of dislocations of the bulk lattice, or due to dynamic disorder related to thermal fluctuations, local coulombic interactions and polaronic effects. In semiconductors of high optical quality such as GaAs, CdTe and $CH_3NH_3PbI_3$ the absorption coefficient shows a relatively sharp transition, see Figure 19.5. A wide collection of absorption characteristics for the main PV classes of materials is shown in Figure 25.13a. By effectively considering the Urbach tail of a class of materials a more realistic efficiency limit can be obtained for the inorganic materials that give the highest practical efficiency. Alternatively, the case of a Gaussian shaped absorption band is explained in (Kirchartz, 2009).

As already remarked, and can be readily observed in part c of the Figure 22.6, the region of lowest light-absorbing energy levels forms the dominant range for light emission by the semiconductor. The significance of the lowest values of absorptance for the value of the radiative recombination parameter has been explained in the calculation outlined in Figure 23.18 for a BHJ organic cell. Even in highly crystalline materials, a modest amount of disorder can produce a significant impact of their photovoltaic properties. If the transition region becomes broader below the band gap, the recombination parameter $j_{0,\mathrm{rad}}$ is enhanced, with consequent large losses in V_{oc}. This is illustrated in Figure 25.14 that shows an enormous variation of the dark emission associated to the band tailing, that changes the $j_{0,\mathrm{rad}}$ by orders of magnitude. In addition to this, the disorder in the Urbach tail may induce nonradiative recombination. In summary, the lowest value of Urbach parameter is correlated with the highest PCE in a class of materials, as can be observed in Figure 25.13. Therefore, it is necessary to explore realistic spectral properties and the consequences they have on the theoretical limits of PV performance.

The temperature dependence of E_U is an important method to investigate the origin and significance of the Urbach tail in a class of semiconductors. In amorphous silicon a spontaneous decay of the weakest bonding orbitals forms nonbonding defects (Stutzmann, 1989). The results shown in Figure 23.5, provide information on the physical origin of the tailing parameter (Studenyak et al., 2014). There are two main additive factors governing the value: a static component that depends on the intrinsic disorder of the material, and a phonon contribution that describes how the vibrations of the material influence its electrical properties (an increasing phonon density should broaden the absorption edge). As the temperature increases, so does the Urbach parameter, and the voltage deficit increases, as shown in Figure 25.13c. A complete analysis of the radiative efficiency limits of the hybrid perovskite solar cells including Auger recombination and photon recycling is presented by (Pazos-Outón et al., 2018).

Based on the previous considerations, it appears that a realistic general exploration of materials for photovoltaic expectations needs to describe the shape of the absorption coefficient near the onset of absorption and nonradiative recombination channels. One advanced method of calculation for materials screening takes into account the band gap, the shape of absorption spectra and the material-dependent nonradiative recombination losses (Yu and Zunger, 2012). Figure 25.15 shows the screening of 256 materials belonging to the I-III-VI chalcopyrite group. The panel b shows the type of materials according to dipole allowed or dipole forbidden transitions, allowing to quantify nonradiative recombination which is dominant when the gap is smaller than the dipole allowed transition. The effect of different recombination pathways on the photovoltaic efficiency has been analyzed by Kirchartz and Rau (2017).

The realistic limiting efficiencies of organic bulk heterojunction solar cells open a different class of problems. It can be observed in Figure 23.8 and in Figure 25.10B that these devices often feature record V_{oc} values that lie considerably far below the radiative limit photovoltage associated to the band gap of the primary absorber, that is, the voltage deficit $W_{oc} = E_g/q - V_{oc}$ is characteristically high. One reason for the low performance of BHJ has been well identified as excess nonradiative recombination associated to the internal extended heterojunction (Vandewal et al., 2010), as confirmed by the analysis in Figure 23.18. However, there is a more profound limitation that occurs even if the deleterious nonradiative recombination channels can be suppressed. The energy offset between the primary gap and the optical gap $\Delta E_{\mathrm{TOP}} = E_g - E_{\mathrm{CT}}$ imposes a cost in photovoltage. In the BHJ, the semiconductor gap is not a unique quantity, with a step-like absorptivity as shown before in Figure 22.6, but contains two different features. This structure emerges from the diagram in Figure 23.13 and it is representative of a range of solar cells that are composed of a combination of materials that effect charge separation at the nanoscale. The general mechanism of operation of such classes of photovoltaic

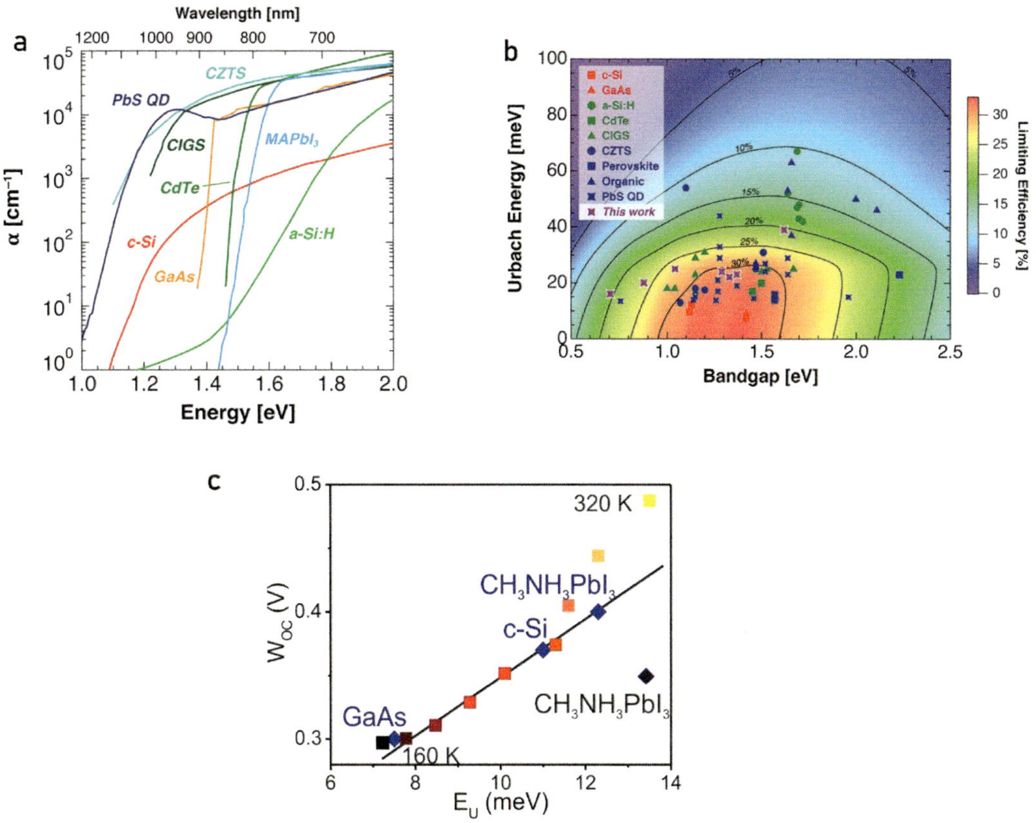

FIGURE 25.13 (a) Absorption spectra of commercial and emerging PV technologies. High-efficiency PV technologies including c-Si, GaAs, CdTe, CIGS, and $MAPbI_3$ have sharp band edges, while lower-efficiency technologies such as a-Si:H and CZTS have broader band edges. Because of its indirect band gap, c-Si appears to suffer from severe band tailing; however, at absorption coefficients lower than the range plotted, the tail drops off sharply. (b) Reported Urbach energies for commercial and emerging PV technologies at room temperature or below. Most commercial wafer-based (c-Si and GaAs) and thin-film (CdTe and CIGS) PV absorbers have low Urbach energies and radiative efficiency limits above 30%. In contrast, amorphous silicon and organic materials exhibit greater disorder and lower-efficiency limits. (Reproduced bypermission from Jean et al., 2017). (c) The voltage deficit $W_{oc} = E_g - V_{oc}$ and Urbach parameter E_U values, extracted from a $CH_3NH_3PbI_3$ solar cell (squares), data in 23.5b. Each data pair is measured at the same temperature, highest and lowest temperatures are marked in the graph, the temperature difference between neighboring points is 20 K. For comparison, room temperature data for GaAs, c-Si and $CH_3NH_3PbI_3$ minimal W_{oc} cells have been added (diamonds). (Reproduced by permission from Ledinsky et al., 2019.) Ledinsky, M.; Schönfeldová, T.; Holovsky, J.; Aydin, E.; Hájková, Z.; Landová, L.; Neykova, N.; Fejfar, A.; De Wolf, S. Temperature Dependence of the Urbach Energy in Lead Iodide Perovskites, *J. Phys. Chem. Lett.* **2019**, *10*, 1368-1373.

devices were more generally discussed in Section 24.2 (see Figure 24.4). The light absorption and emission features are shown in Figure 25.16. It is composed of a dominant absorptivity a_0 associated to bulk absorption in the donor D material, and a secondary absorption channel, due to CT excitation across the interface, a_{CT}. The latter corresponds to light absorber molecules at the interface with the acceptor, while the former applies to all D molecules, therefore a_{CT} is expected to be considerably weaker.

The radiative balance is outlined in Figure 25.16. While the main absorbance is at E_D, the E_{CT} level at lower energy is the prevalent one for radiative recombination. At open-circuit conditions the photon flux absorbed by the solar cell is approximately given by

$$a_0 \int_{E_D}^{\infty} \phi_{ph}^{bb,1\text{sun}}(E,5800)\,dE \tag{25.26}$$

The emitted flux close to the energy level E_{CT} corresponds to

$$e^{qV_{oc}/k_BT} a_{CT} \int_{E_{CT}}^{\infty} \phi_{ph}^{bb}(E,300)\,dE \tag{25.27}$$

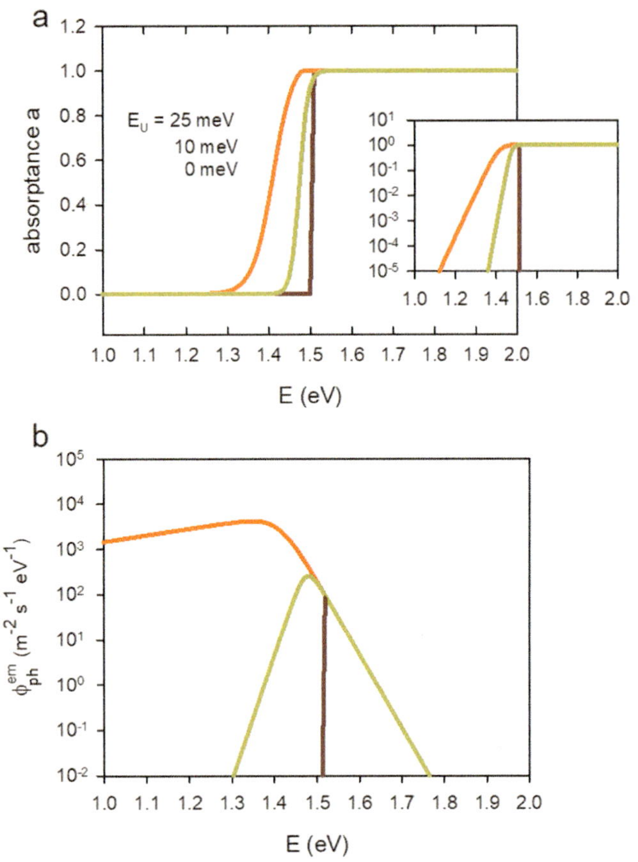

FIGURE 25.14 Radiative emission of semiconductors with Urbach band tailing. (a) Absorptance spectrum of an idealized solar cell with absorber band gap of $E_g = 1.5$ eV and absorption constant $\alpha_0 = 10^4$ cm^{-1} above the band gap, and band tailing characterized by the Urbach energy E_U values as indicated, as $\alpha(E) = \alpha_0\left[(E - E_g)/E_U\right]$. The absorptance is found assuming ideal light trapping (Lambertian front surface and perfect rear reflector), an absorber thickness of $d = 1$ μm, and refractive index of $n_r = 3$, with $a = (1 - t_{\text{cell}})/(1 - t_{\text{cell}}(1 - t_{\text{Lampert}}))$ where $t_{\text{Lampert}} = 1/n_r^2$ and $t_{\text{cell}} = (1 - 2\alpha d)/\exp(2\alpha d)$ as in (Jean et al., 2017). The inset shows the linear slope of the absorptance at low energies in log representation. (b) The spectral light emission of the solar cell in thermal equilibrium at $k_B T = 0.026$ eV, obtained by $\varphi_{ph}^{em}(E) = b_\pi a(E) E^2 e^{-E/k_B T}$ (Equation 22.31). Note the enhanced emission when the tailing becomes larger, which impacts the recombination parameter j_0 by Equation 23.3.

The quantities of absorbed and emitted photons, highlighted by colored shaded areas in Figure 25.16, represent the rates of excitation and de-excitation of electrons inside the BHJ blend, and both must coincide in equilibrium.

The fact that radiative emission emerges from $E_{CT} < E_D$ causes a decrease of the photovoltage with respect to a single band gap solar cell. The discussion of the photovoltage that can be obtained in a charge-separating blend of two materials has been already presented in Figure 24.10. It was shown that increasing the transport gap by better alignment of D-A energy levels to minimize the energy offset directly impacts in the increase of V_{oc}, and this is the main reason behind the best results obtained so far in practically improving the efficiency BHJ solar cells (Meng et al., 2018). If the energy offset ΔE decreases and the recombination rate is the same, then emission of the required photon number per unit time involves a larger separation of Fermi levels and hence larger V_{oc}. Thus, SQ photovoltage and efficiency are reduced in the presence of charge separation energy offset ΔE with respect to a solar cell with a single absorber and charge transporting material. This reduction is quantitatively described in Gruber et al. (2012).

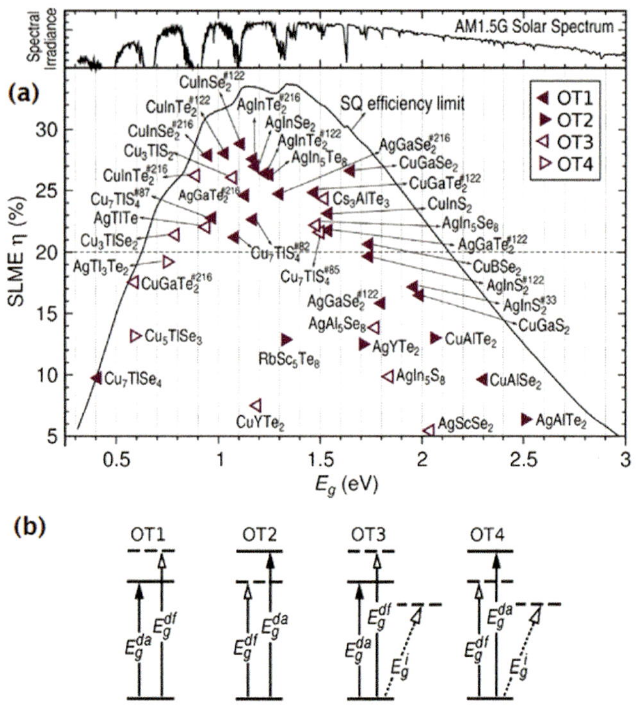

FIGURE 25.15 (a) Spectroscopic limited maximum efficiency (SLME) *vs.* the minimum gap E_g for generalized I-III-VI chalcopyrite materials at d = 0.5 μm. The compounds with SLME <5% are not shown. The space group number (superscript) is used to distinguish different materials with the same chemical formula. (b) Schematic diagrams of different types of semiconductors according to optical transitions. Electric-dipole-allowed (forbidden) direct optical transition is denoted by a line with an arrow pointing to solid (dashed) horizontal line. Indirect states are shown as laterally displaced dashed lines. (Reproduced with permission from Yu, L.; Zunger, A., Physical Review Letters 2012, 108, 068701.)

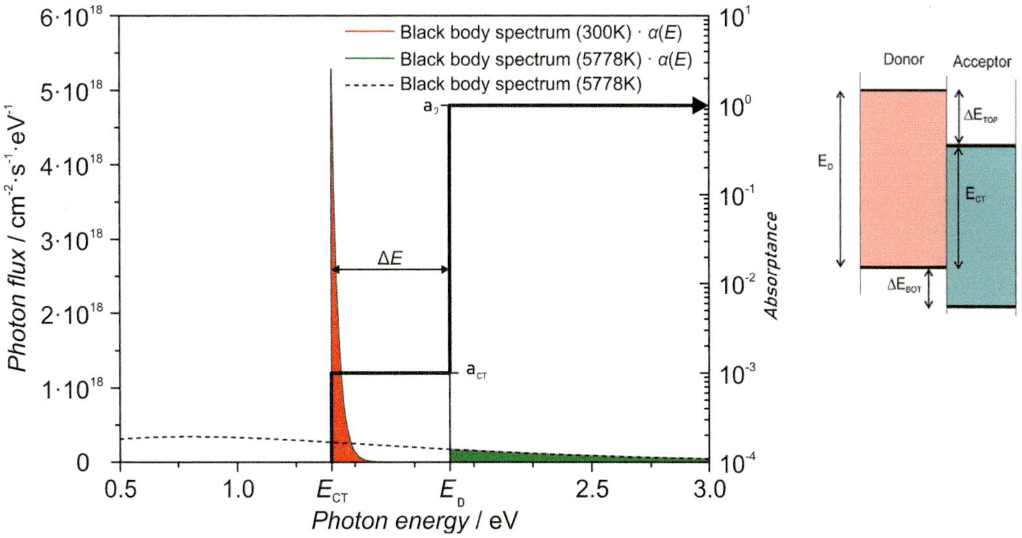

FIGURE 25.16 Spectral dependence of photon flux and step-like absorption in a donor/acceptor heterojunction solar cell. The dashed line shows the incident solar radiation, which after multiplication with the absorption spectrum yields the green-shaded area. The red-shaded area represents the emission of the solar cell under open-circuit conditions, which has to be equal to the incident photon flux. The scheme shows the relative positions of the energy levels and the resulting primary absorption energy gap E_D as well as the energy of the CT state, or transport gap, E_{CT}, and the energy offset $\Delta E = \Delta E_{TOP} + \Delta E_{BOT}$. (Reproduced with permission from Gruber, M. et al. Advanced Energy Materials 2012, 2, 1100–1108.)

GENERAL REFERENCES

Analysis of the fill factor: Green (1982), Banwell and Jayakumar (2000), Maa and Abdel-Motaleb (2002), De Soto et al. (2006), and Bashahu and Nkundabakura (2007).

Solar cell operation: Sze (1981), Partain (1995), Nelson (2003), Würfel (2009), Dittrich (2014), Fonash (2010), Tress (2014), and Schmidt-Mende and Weickert (2016).

Solar cell efficiency limits: Ross (1967), Henry (1980), Ruppel and Wurfel (1980), Araújo and Martí (1994), Baruch et al. (1995), Hanna and Nozik (2006), Markvart (2008), Würfel (2009), Hirst and Ekins-Daukes (2011), Yu and Zunger (2012), Rau et al. (2014), and Vossier et al. (2015).

Radiative efficiency limits, phonons and band tailing. (Kirchartz et al., 2009; De Wolf et al., 2014; Barugkin et al., 2015; Wright et al., 2016; Jean et al., 2017; Kirchartz et al., 2018)

REFERENCES

Araújo, G. L.; Martí, A. Absolute limiting efficiencies for photovoltaic energy conversion. *Solar Energy Materials and Solar Cells* 1994, *33*, 213–240.

Banwell, T. C.; Jayakumar, A. Exact analytical solution for current flow through diode with series resistance. *Electronics Letters* 2000, *36*, 291–292.

Baruch, P.; De Vos, A.; Landsberg, P. T.; Parrott, J. E. On some thermodynamic aspects of photovoltaic solar energy conversion. *Solar Energy Materials and Solar Cells* 1995, *36*, 201–222.

Barugkin, C.; Cong, J.; Duong, T.; Rahman, S.; Nguyen, H. T.; Macdonald, D.; White, T. P.; et al. Ultralow absorption coefficient and temperature dependence of radiative recombination of $CH_3NH_3PbI_3$ perovskite from photoluminescence. *The Journal of Physical Chemistry Letters* 2015, *6*, 767–772.

Bashahu, M.; Nkundabakura, P. Review and tests of methods for the determination of the solar cell junction ideality factors. *Solar Energy* 2007, *81*, 856–863.

Christians, J. A.; Manser, J. S.; Kamat, P. V. Best practices in perovskite solar cell efficiency measurements. Avoiding the error of making bad cells look good. *The Journal of Physical Chemistry Letters* 2015, *6*, 852–857.

De Soto, W.; Klein, S. A.; Beckman, W. A. Improvement and validation of a model for photovoltaic array performance. *Solar Energy* 2006, *80*, 78–88.

De Wolf, S.; Holovsky, J.; Moon, S.-J.; Löper, P.; Niesen, B.; Ledinsky, M.; Haug, F.-J.; et al. Organometallic halide perovskites: Sharp optical absorption edge and its relation to photovoltaic performance. *The Journal of Physical Chemistry Letters* 2014, *5*, 1035–1039.

Dittrich, T. *Materials Concepts for Solar Cells*; Imperial College Press: London, 2014.

Fonash, S. J. *Solar Cell Device Physics*, 2nd ed.; Academic Press: New York, 2010.

Green, M. A. Accuracy of analytical expressions for solar cell fill factors. *Solar Cells* 1982, *7*, 337–340.

Gruber, M.; Wagner, J.; Klein, K.; Hörmann, U.; Opitz, A.; Stutzmann, M.; Brütting, W. Thermodynamic efficiency limit of molecular donor-acceptor solar cells and its application to diindenoperylene/C_{60}-based planar heterojunction devices. *Advanced Energy Materials* 2012, *2*, 1100–1108.

Hanna, M. C.; Nozik, A. J. Solar conversion efficiency of photovoltaic and photoelectrolysis cells with carrier multiplication absorbers. *Journal of Applied Physics* 2006, *100*, 074510.

Henry, C. H. Limiting efficiencies of ideal single and multiple energy gap terrestrial solar cells. *Journal of Applied Physics* 1980, *51*, 4494.

Hirst, L. C.; Ekins-Daukes, N. J. Fundamental losses in solar cells. *Progress in Photovoltaics: Research and Applications* 2011, *19*, 286–293.

Ho J. K. W., Yin H., So S. K. From 33% to 57% – an elevated potential of efficiency limit for indoor photovoltaics. *Journal of Materials Chemistry A*. 2020, 8, 1717–23.

Jean, J., Mahony, T. S.; Bozyigit, D.; Sponseller, M.; Holovský, J.; Bawendi, M. G.; Bulović, V. Radiative efficiency limit with band tailing exceeds 30% for quantum dot solar cells. *ACS Energy Letters* 2017, *2*, 2616–2624.

Ji, W.; Yao, K.; Liang, Y. C. Bulk photovoltaic effect at visible wavelength in epitaxial ferroelectric $BiFeO_3$ thin films. *Advanced Materials* 2010, *22*, 1763–1766.

Kirchartz, T.; Markvart, T.; Rau, U.; Egger, D. A. Impact of small phonon energies on the charge-carrier lifetimes in metal-halide perovskites. *Journal of Physical Chemistry Letters* 2018, *9*, 939–946.

Kirchartz, T.; Rau, U. Decreasing radiative recombination coefficients via an indirect band gap in lead halide perovskites. *The Journal of Physical Chemistry Letters* 2017, *8*, 1265–1271.

Kirchartz, T.; Taretto, K.; Rau, U. Efficiency limits of organic bulk heterojunction solar cells. *Journal of Physical Chemistry C* 2009, *113*, 17958–17966.

Ledinsky, M.; Schönfeldová, T.; Holovsky, J.; Aydin, E.; Hájková, Z.; Landová, L.; Neykova, N.; Fejfar, A.; De Wolf, S. Temperature dependence of the urbach energy in lead iodide perovskites. *The Journal of Physical Chemistry Letters* 2019, *10*, 1368–1373.

Li, X.; Bi, D.; Yi, C.; Décoppet, J.-D.; Luo, J.; Zakeeruddin, S. M.; Hagfeldt, A.; et al. A vacuum flash-assisted solution process for high-efficiency large-area perovskite solar cells. *Science* 2016. doi: 10.1126/science.aaf8060.

Lopez-Varo, P.; Bertoluzzi, L.; Bisquert, J.; Alexe, M.; Coll, M.; Huang, J.; Jimenez-Tejada, J. A. Physical aspects of ferroelectric semiconductors for photovoltaic solar energy conversion. *Physics Reports* 2016, *653*, 1–40.

Maa, Y. J.; Abdel-Motaleb, I. M. Analysis of the diode characteristics using the thermodynamic theories. *Solid-State Electronics* 2002, *46*, 735–742.

Markvart, T. Solar cell as a heat engine: Energy–entropy analysis of photovoltaic conversion. *Physica Status Solidi A* 2008, *205*, 2752–2756.

Meng, L.; Zhang, Y.; Wan, X.; Li, C.; Zhang, X.; Wang, Y.; Ke, X.; et al. Organic and solution-processed tandem solar cells with 17.3% efficiency. *Science* 2018, *361*, 1094–1098.

Nayak, P. K.; Garcia-Belmonte, G.; Kahn, A.; Bisquert, J.; Cahen, D. Photovoltaic efficiency limits and material disorder. *Energy & Environmental Science* 2012, *5*, 6022–6039.

Nelson, J. *The Physics of Solar Cells*; Imperial College Press: London, 2003.

Partain, L. D. *Solar Cells and Their Applications*; Wiley: Weinheim, 1995.

Pazos-Outón, L. M.; Xiao, T. P.; Yablonovitch, E. Fundamental efficiency limit of lead iodide perovskite solar cells. *Journal of Physical Chemistry Letters* 2018, *9*, 1703–1711.

Rau, U.; Paetzold, U. W.; Kirchartz, T. Thermodynamics of light management in photovoltaic devices. *Physical Review B* 2014, *90*, 035211.

Ross, R. T. Some thermodynamics of photochemical systems. *The Journal of Chemical Physics* 1967, *46*, 4590–4593.

Ruppel, W.; Wurfel, P. Upper limit for the conversion of solar energy. *IEEE Transactions on Electron Devices* 1980, *27*, 877–882.

Schmidt-Mende, L.; Weickert, J. *Organic and Hybrid Solar Cells. An Introduction*; de Gruyter: Berlin, 2016.

Shockley, W.; Queisser, H. J. Detailed balance limit of efficiency of p-n junction solar cells. *Journal of Applied Physics* 1961, *32*, 510–520.

Studenyak, I.; Kranjec, M.; Kurik, M. Urbach rule in solid state physics. *International Journal Optical Applications* 2014, *4*, 76–83.

Stutzmann, M. The defect density in amorphous silicon. *Philosophical Magazine B* 1989, *60*, 531–546.

Sze, S. M. *Physics of Semiconductor Devizes*, 2nd ed.; John Wiley and Sons: New York, 1981.

Tress, W. *Organic Solar Cells. Theory, Experiment, and Device Simulation*; Springer: Switzerland, 2014.

Vandewal, K.; Tvingstedt, K.; Gadisa, A.; Inganas, O.; Manca, J. V. Relating the open-circuit voltage to interface molecular properties of donor:acceptor bulk heterojunction solar cells. *Physical Review B* 2010, *81*, 125204.

Venkateswararao A, Ho JKW, So SK, Liu S-W, Wong K-T. Device characteristics and material developments of indoor photovoltaic devices. *Materials Science and Engineering: R: Reports*. 2020; 139, 100517.

Vossier, A.; Gualdi, F.; Dollet, A.; Ares, R.; Aimez, V. Approaching the Shockley-Queisser limit: General assessment of the main limiting mechanisms in photovoltaic cells. *Journal of Applied Physics* 2015, *117*, 015102.

Wright, A. D.; Verdi, C.; Milot, R. L.; Eperon, G. E.; Pérez-Osorio, M. A.; Snaith, H. J.; Giustino, F.; et al. Electron–phonon coupling in hybrid lead halide perovskites. *Nature Communications* 2016, *7*, 11755.

Würfel, P. *Physics of Solar Cells. From Principles to New Concepts*, 2nd ed.; Wiley: Weinheim, 2009.

Yu, L.; Zunger, A. Identification of potential photovoltaic absorbers based on first-principles spectroscopic screening of materials. *Physical Review Letters* 2012, *108*, 068701.

Zimmermann, E.; Wong, K. K.; Müller, M.; Hu, H.; Ehrenreich, P.; Kohlstädt, M.; Würfel, U. Characterization of perovskite solar cells: Toward a reliable measurement protocol. *APL Materials* 2016, *4*, 091901.

26 Charge Collection in Solar Cells

This chapter completes the analysis of the operation of solar cells by describing the mechanisms of charge collection, composed of the combination of carrier transport, recombination, and extraction at the selective contacts. We establish the main intuitive ideas concerning the charge collection by either diffusion, or by drift in strong electrical fields associated with space-charge regions. The modeling of the solar cell by transport–recombination equations, as well as the boundary conditions, is analyzed in detail. Increasingly complex analytical models are reviewed, and then we discuss the numerical simulation that allows to treat any type of complex morphology of charge and current distribution. We form a basic classification of the main operation modes for charge collection, depending on the size of the depletion region with respect to the thickness of the absorber. We finish with some practical considerations for the measurement and reporting of solar cell performance.

26.1 INTRODUCTION TO CHARGE COLLECTION PROPERTIES

The photovoltaic operation leading to a photocurrent can be analyzed in terms of the functions of charge generation, charge separation, and charge collection, as commented in Chapter 24. Once the initial charge separation step is achieved, which in many inorganic and hybrid perovskite solar cells is an extremely fast process, it is necessary to establish a flux of both types of carriers toward separate contacts. The carrier collection across the active layer in competition with the recombination losses that take place either in the bulk or at interfaces of the material will be analyzed in this chapter.

In Chapter 20, we have presented and analyzed two archetype models for the operation of a solar cell with respect to charge collection. First is the model shown in Figures 20.8 and 21.1, where the bands are flat in all circumstances. Here, the problem of transport is suppressed by declaring that the mobilities are infinite, in other words, an internal quantum efficiency (IQE_{PV}) of 1 is assumed. This model represents an "ideal" solar cell where charge collection is optimal without the need for fields that produce drift transport, and the current-voltage characteristic is completely determined by recombination, which modulates the internal loss current depending on the applied voltage.

In Section 20.5 and Figure 20.11, however, we discussed a solar cell formed by an intrinsic absorber semiconductor with slanted bands, in which the performance is limited by charge collection determined by the drift current in the electrical field.

We now start a broader investigation of the mechanisms of generation, recombination, and transport in solar cells, based on the extensive discussion of the carrier transport properties examined in Part II of this book. Chapter 12 describes the drift-diffusion and conservation equations and is especially relevant for the general problem of solar cell operation. We reached the conclusion that the transport is governed by the gradient of each Fermi level, so that calculating the distribution of electron and hole Fermi levels is one important goal of the simulation and modeling of solar cells. We will treat this problem with increasing degrees of complexity, showing a set of representative and important models, and applying simulation tools to obtain a general description of the Fermi levels in a solar cell device under illumination. For most applications, it can be assumed that the excess carrier transport occurs by carriers thermalized to the band edges, so that the relationships $n(E_{Fn})$ and $p(E_{Fp})$ in Equations 21.11 and 21.12 can be used to change between carrier density and the Fermi level representation of the charge distribution.

In organic and nanostructured solar cells, we find the features of disorder that were discussed in Chapter 24. The presence of energy or structural disorder affects the photovoltage and implies that transport characteristics require the specific methods and concepts of disordered systems such as the hopping conductivity. These tools were reviewed in Chapter 13 and will not be considered in depth here.

The combination of light harvesting, charge generation, and charge collection phenomena, including recombination effects, determine the photovoltaic EQE, which has been described in Equation 18.28. The EQE_{PV} is the central function that establishes the practical value of the photocurrent. It also determines the fundamental operation properties of the solar cell, discussed in Chapter 23. The total photocurrent j_{ph} can be directly measured (Equation 25.4), but additional insight is obtained by the composition of EQE_{PV} and the spectral photon number, as can be seen in Equation 22.1.

To illustrate real characteristics of charge collection, we show the typical behavior of EQE_{PV} and total

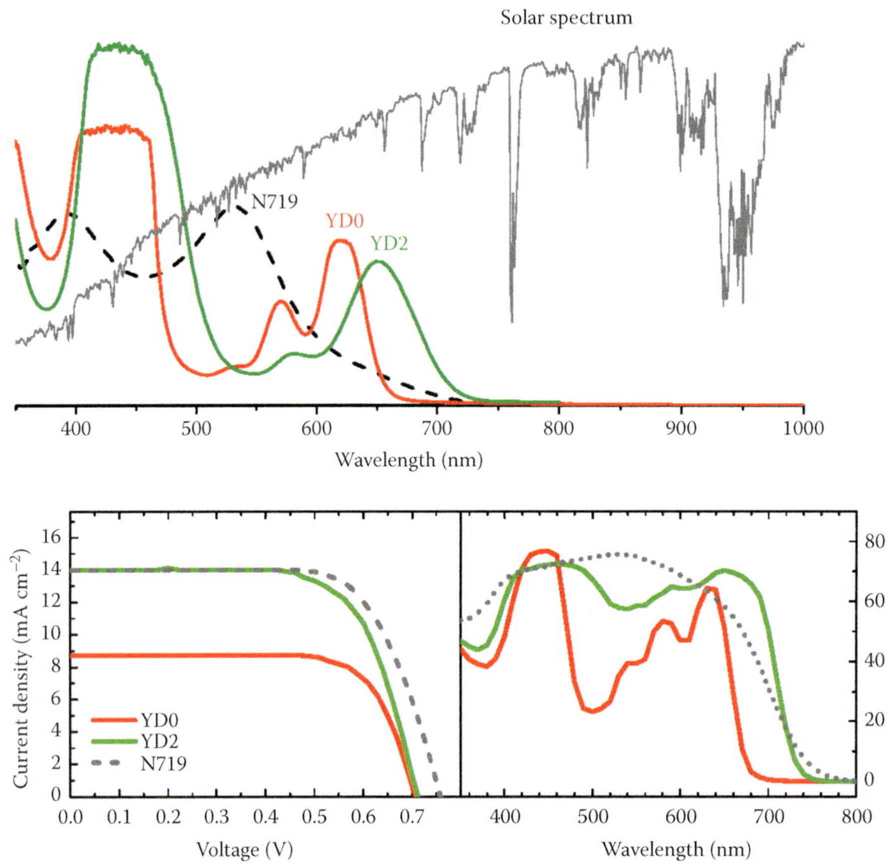

FIGURE 26.1 Absorption coefficient of N719 dye and two porphyrin dyes, in comparison with the spectral photon flux, and the corresponding photovoltaic performance and EQE_{PV}. (Adapted from Li, L.-L.; Diau, E. W.-G. *Chemical Society Reviews* 2012, 42, 291–304.)

photocurrent in a number of solar cell technologies. Figure 26.1 shows the features of the organic absorber molecules in a DSC. The figure compares the performance of the reference ruthenium dye N719 discussed in Figure 21.12 with two porphyrin dyes coded YD0 and YD2 (Barea et al., 2011). Porphyrins are molecules that contain a heterocyclic macrocycle with a π-aromatic core, showing an intense Soret band at 400–450 nm and moderate *Q* bands at 500–650 nm. The shift of the *Q* band toward longer wavelength by ligand modification in the case of YD2 results in a much better match of the absorption coefficient with the solar spectrum region, where the photon density is large. This implies a much larger photocurrent, as can be seen in Figure 26.1. However, since N719 has larger light absorption and EQE_{PV} in the region around 550 nm, the current resulting from both YD2 and N719 is similar.

A range of characteristics for different solar cell technologies are shown in Figure 26.2, with the normalized EQE_{PV}, and in Figure 26.3, with current-voltage characteristics. Figure 25.2 showed the match of the photocurrent with the integrated EQE_{PV} in perovskite solar cells.

26.2 CHARGE COLLECTION DISTANCE

In a solar cell under intense photogeneration, there occurs a competition of transport and recombination that determines whether the carriers generated at an internal point will reach the external contacts, which is a necessary condition for the efficient operation of the device under sunlight. Chapters 10 to 12 discussed the transport in semiconductors can often be separated into two distinct mechanisms, provided that interactions between carriers can be neglected. The diffusion transport is governed by the gradient of the carrier concentration, while the drift transport of the carriers occurs in an electrical field. In a cell with flat and horizontal bands, no electrical fields assist the transport, which occurs entirely by diffusion. The band bending or band slanting indicates the presence of electrical fields that assist charge separation and transport by directing oppositely charged carriers in contrary directions. This distinction is useful for establishing simple models of solar cell operation. It is often assumed that the recombination of carriers can be drastically reduced by intense drift in space-charge

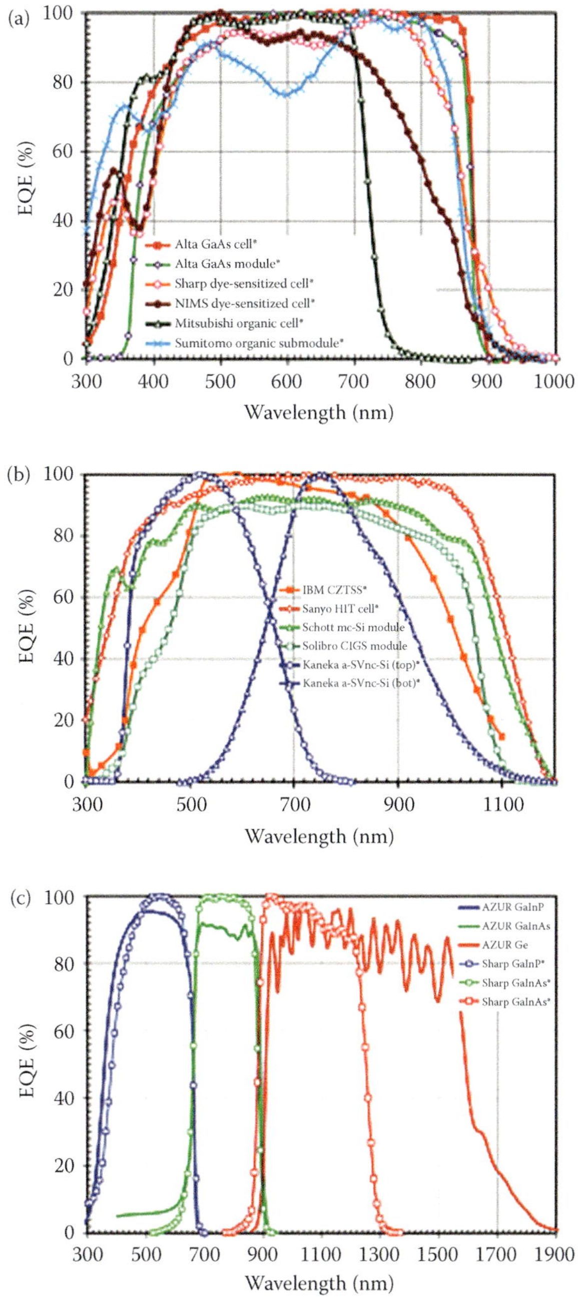

FIGURE 26.2 Normalized external quantum efficiency (EQE_{PV}) for solar cells and modules. (a) GaAs cell and module, dye-sensitized cell, organic cell, and module; (b) Silicon cell and module and copper zinc tin sulfide-selenide (CZTSe) cell; (c) Triple junction multijunction cells. (Reproduced with permission from Green, M. A. et al. Solar cell efficiency tables, version 39. *Progress in Photovoltaics: Research and Applications* 2012, 20, 12–20.)

regions. More generally, diffusion and drift can cooperate in a device, as explained in Section 26.6 in the analysis of the Gärtner model.

Let us assume that no electrical fields exist along the absorber thickness. The central parameter that establishes the charge collection features of the solar cell device is the diffusion length, explained in Section 11.3. We recall that the diffusion length is the most likely distance that a carrier can travel before recombination. For electrons, it is denoted

$$L_n = \sqrt{D_n \tau_n} \tag{26.1}$$

in terms of the electron diffusion coefficient D_n and recombination lifetime τ_n. By Equation 18.36, we also have

$$L_n = \left(\frac{D_n}{B_{rec} p_0} \right)^{1/2} \tag{26.2}$$

We consider first diffusion transport in the dark in a semiconductor layer of thickness d. Minority carrier electrons can be injected in the layer by forward bias applied to the electron selective contact (ESC). The resulting carrier distribution dramatically depends on the ratio L_n/d, as shown in Figure 26.4 (the graph is plotted with expressions described in Section 26.5). When the diffusion length is short, the electrons do not penetrate all the way to the back contact. When the diffusion length is long, the concentration of carriers is homogeneous in the layer. This last case corresponds to our previous assumptions about flat Fermi levels and homogeneous minority carrier density in Figure 21.1.

In the case of a short diffusion length, when the diode operates as a solar cell, only the carriers generated close to the contact are collected, as indicated in Figure 26.5, which is a schematic picture that will be developed more fully in Sections 26.5 and 26.8. Nevertheless, we obtain a valid and informative expression simply observing that the saturation current density in Equation 23.27 is modified as follows:

$$j_0 = q \int_0^{L_n} U_{rec} dx = q L_n B_{rec} n_0 p_0 \tag{26.3}$$

That is, only the carriers generated in dark at $x \leq L_n$ can be extracted and contribute to the saturation current. Using Equation 26.2, we obtain the recombination parameter

$$j_0 = q \frac{D_n}{L_n} n_0 \tag{26.4}$$

The current density-voltage curve of the solar cell is given by Equation 25.2, where j_0 in Equation 26.4 is

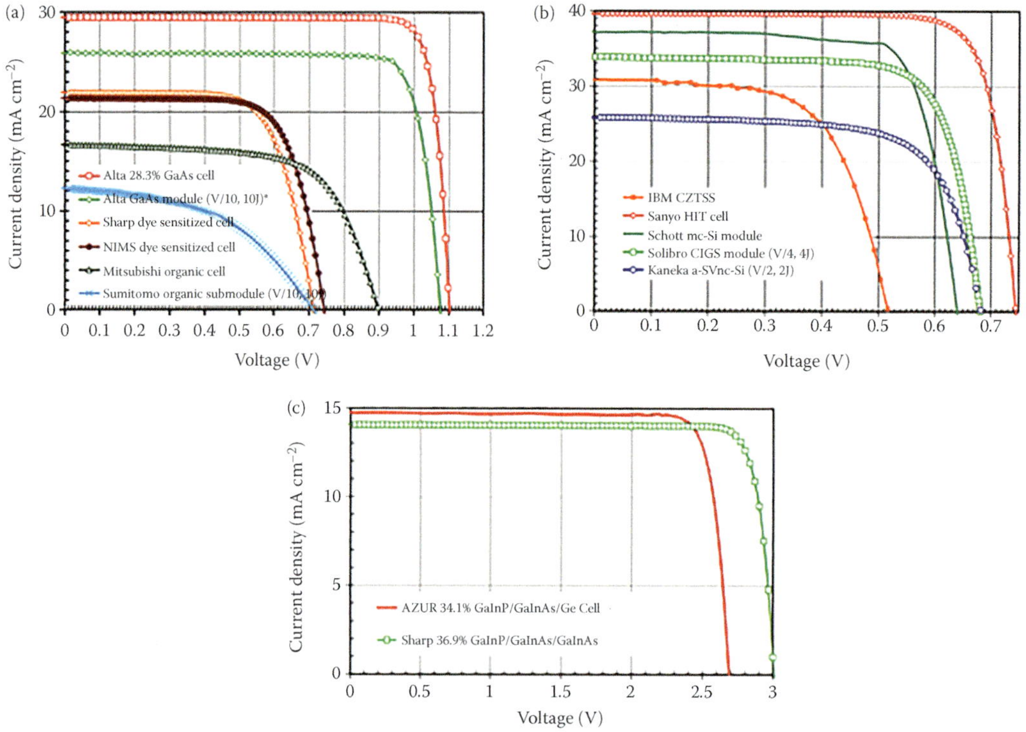

FIGURE 26.3 Current density–voltage (j–V) curves for solar cells and modules. (a) GaAs cell and module, dye-sensitized cell, organic cell, and module; (b) silicon cell and module and CZTSe cell; (c) triple junction multijunction cells. (Reproduced with permission from Green, M. A. et al. Solar cell efficiency tables, version 39. *Progress in Photovoltaics: Research and Applications* 2012, 20, 12–20.)

independent of the thickness of the sample. Using the density of acceptors N_A as described in Section 5.4, we obtain with respect to the measured photocurrent j_{ph}

$$V_{oc} = \frac{k_B T}{q} \ln\left(\frac{j_{ph} L_n N_A}{q D_n n_i^2}\right) \tag{26.5}$$

This relationship is nearly quantitatively obeyed for silicon solar cells with excellent interface contacts (Maldonado et al., 2008).

If the current is governed by drift in an electrical field, as in Figure 20.11, then the mean distance covered by a carrier before recombination is obtained by the product of the drift velocity and the recombination lifetime

$$L_{drift} = v\tau_n \tag{26.6}$$

Using Equation 10.9, we find the expression

$$L_{drift} = u_n \tau_n \mathbf{E} \tag{26.7}$$

in terms of the mobility and the electrical field. At given field strength, the mobility-lifetime product is the quantity determining the drift length.

26.3 GENERAL MODELING EQUATIONS

In order to analyze the dominant processes of operation of a solar cell for charge transport and charge extraction, we will adopt a model in which the carriers obey the general drift-diffusion equations for electrons and holes in a layer $0 \leq x \leq d$, combined with the continuity and the Poisson equations that were described in Section 12.4. We have the equations

$$j_n = -qJ_n = qnu_n E + qD_n \frac{\partial n}{\partial x} \tag{26.8}$$

$$j_p = +qJ_p = qpu_p E - qD_p \frac{\partial p}{\partial x} \tag{26.9}$$

$$j = j_n + j_p \tag{26.10}$$

$$\frac{1}{q}\frac{\partial j_n}{\partial x} + G_\Phi - U_{rec} = 0 \tag{26.11}$$

$$-\frac{1}{q}\frac{\partial j_p}{\partial x} + G_\Phi - U_{rec} = 0 \tag{26.12}$$

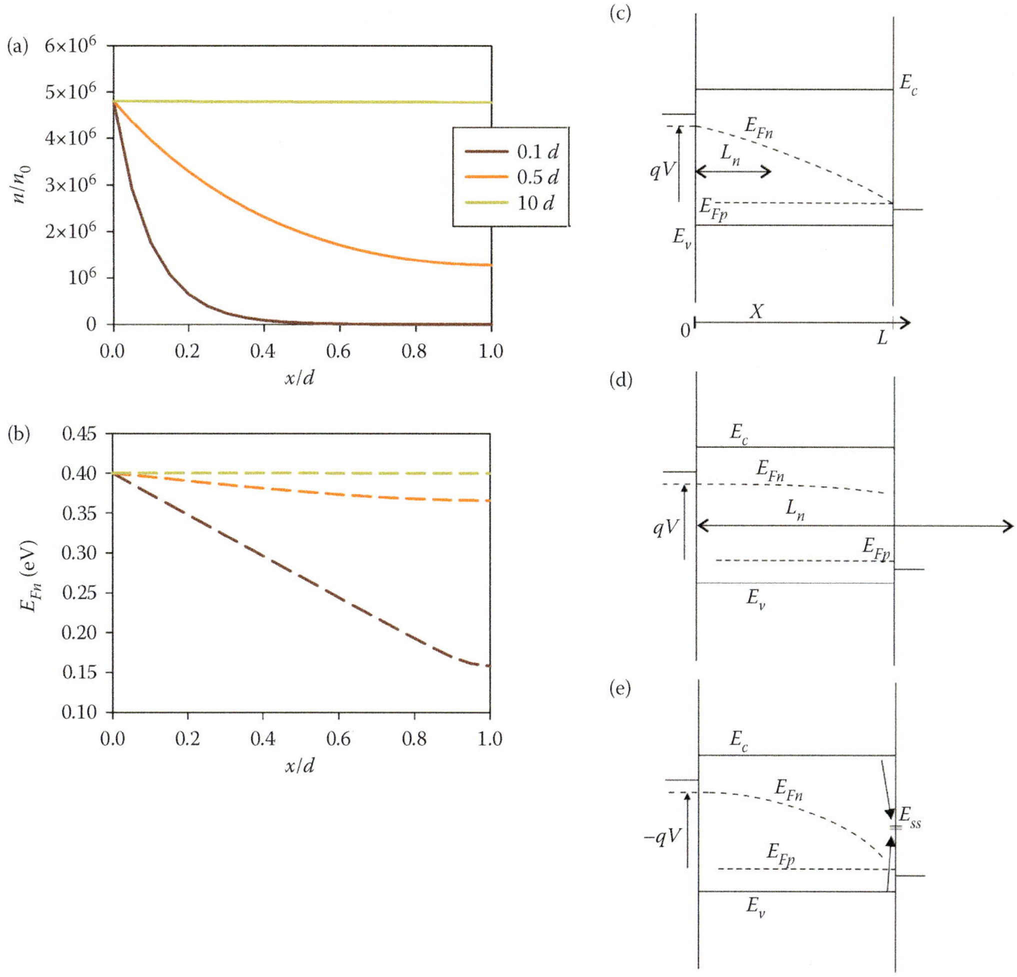

FIGURE 26.4 Electron distribution for injection of minority carrier electrons in the dark for a forward bias of 0.4 V, by diffusion and recombination, in a layer of thickness d for different values of the diffusion length as indicated. (a) electron density and (b) electron Fermi level. The right column shows the schemes of electrons and holes Fermi level under forward bias for the following cases: (c) Short diffusion length, (d) long diffusion length and reflecting boundary condition, and (e) long diffusion length and absorbing boundary condition by strong SRH recombination at the contact.

$$\frac{\partial E}{\partial x} = \frac{q}{\varepsilon_0 \varepsilon_r}\left(p - n - N_A\right) \tag{26.13}$$

$$E = -\frac{\partial \varphi}{\partial x} \tag{26.14}$$

If not otherwise stated, we adopt the bimolecular model for band-to-band recombination as defined in Equation 18.34.

$$U_{rec} = B_{rec}\left(np - n_i^2\right) \tag{26.15}$$

The carrier generation rate at the position x is given by the attenuation of the photon number, $G_\Phi(x) = d\Phi_{ph}(x)/dx$. Using the Beer-Lambert law, and integrating over the wavelength of the incident radiation λ,

$$G_\Phi(x) = \int (1 - R)\alpha(\lambda)\Phi_{ph}(0,\lambda)e^{-\alpha(\lambda)x}d\lambda \tag{26.16}$$

where $\Phi_{ph}(0,\lambda)$ is the incident photon number and R is the reflection coefficient. In addition, in the presence of photon recycling inside the semiconductor layer, an additional generation term corresponding to the secondary photons is added to the continuity equations (Ansari-Rad and Bisquert, 2018). For monochromatic radiation it is $G_\Phi(x) = \alpha(\lambda)\Phi_{ph}(0)e^{-\alpha(\lambda)x}$.

26.4 THE BOUNDARY CONDITIONS

We have remarked the central significance of selective contacts in the basic solar cell structure in Chapter 20. For a one-dimensional solar cell model, we use the differential equations summarized in Section 26.3 to establish

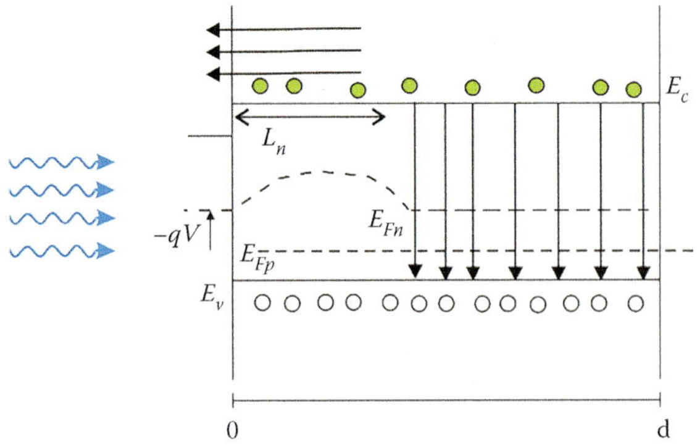

FIGURE 26.5 Schematic model indicating the photogeneration of electrons in a solar cell with flat bands where the diffusion length L_n is shorter than layer thickness. The minority carriers situated at $x \leq L_n$ are collected at the contact, while those created away from the contact recombine with majority carrier holes.

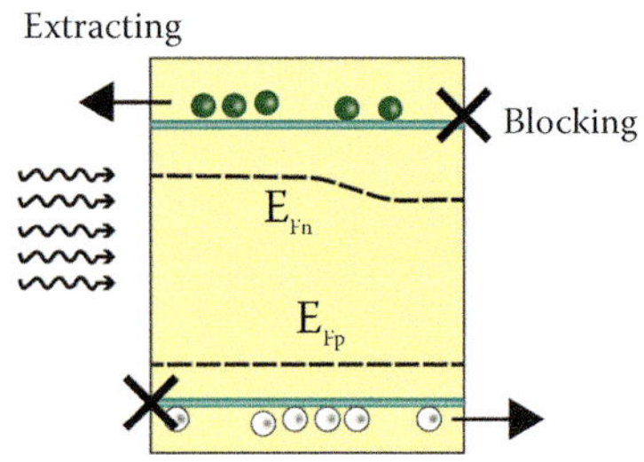

FIGURE 26.6 Fundamental solar cell model indicating the extracting and blocking properties of the contacts, which determine the carrier selectivity.

the interdependence of the charge distribution, electrical field, and carrier currents in the active layer $0 \leq x \leq d$. To complete the problem mathematically, it is necessary to establish a set of boundary conditions at the edges that express the operation of contacts. These conditions often play a determinant role in the physical properties of the model and require a judicious selection. The selective contacts must have the property of extracting one carrier and blocking the oppositely charged carrier, as emphasized in Figure 26.6. In real cases, the property of charge blocking occurs to a total or partial extent, combined with some amount of surface recombination. We will now examine the main types of boundary conditions for extraction and blocking of the carriers.

26.4.1 Charge Extraction Boundary Condition

Let us assume that the Fermi level of the majority carrier is generally fixed in the absorber layer while an increase of the minority carrier occurs under photogeneration, that is, we remain under the constraint $n \ll p \approx p_0$ or $p \ll n \approx n_0$. We can impose boundary conditions on the carrier density. One common assumption is to fix the carrier density at a constant value, the equilibrium value n_0 or p_0, as explained in Section 10.3. In Figures 9.2, 9.3 and 20.5 we have shown an n-doped semiconductor with a contact to the majority carrier electrons at the right side. Correspondingly, we write the condition for the number density at the surface as

$$n_s = n_0 \tag{26.17}$$

In mathematics, this is called the Dirichlet boundary condition, the absorbing boundary condition in the theory of diffusion. For a majority carrier in low-injection conditions, it is a natural condition implying that the carrier density is maintained at the contact with zero impedance for transport across the contact. It is called an "ohmic" boundary condition. However, for the minority carrier, the implication of a constant density is different, as discussed later in this chapter.

The most important contact of the solar cell structure is that of the extraction of minority carriers. In the simple model of Figure 20.8, the variation caused by the voltage is located completely at the minority carrier extraction interface. The change of the VL occurs at the surface dipole layer. At this contact, the surface concentration in the semiconductor relates directly to the voltage in the device as

$$n_s = n_0 e^{qV/k_BT} \tag{26.18}$$

This model boundary is realized in the cases of Figures 24.11 and 24.12.

However, as discussed in Figure 10.9, the original built-in voltage may be distributed either partly or fully inside the absorber layer. This situation was analyzed in the model SB diode discussed in Chapter 20. The changes of the SB are indicated in detail in Figure 26.7. The height of the barrier, which corresponds to the built-in potential, is given by the difference of the Fermi level of the semiconductor, $\Phi_s = E_{F0}$ and the cathode work function

$$V_{bi} = \left(E_{F0} - \Phi_c\right)/q \tag{26.19}$$

Polarization of the junction first causes the decrease of the space-charge region and size of the barrier, keeping the density of minority carrier constant as the Fermi level raises upward, as indicated in Figure 26.7c. At further forward bias, the density of electrons rises at the surface according to the expression

$$n_s = n_0 e^{q(V - V_{bi})/k_BT} \tag{26.20}$$

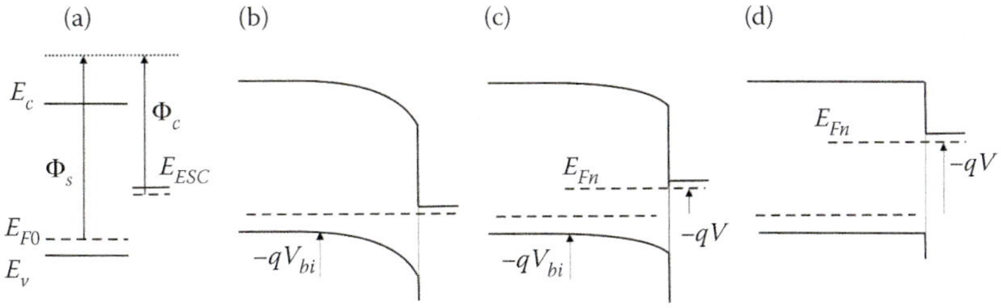

FIGURE 26.7 Schematic representative energy diagram of a p-type organic layer and ESC as contact for the device. (a) The separate materials. (b) In contact, the original difference of work functions forms a depletion zone with barrier qV_{bi}. (c) At forward bias, the SB decreases. (d) When the built-in potential is overcome by negative potential or by photogeneration of electrons, the electron density at the surface increases.

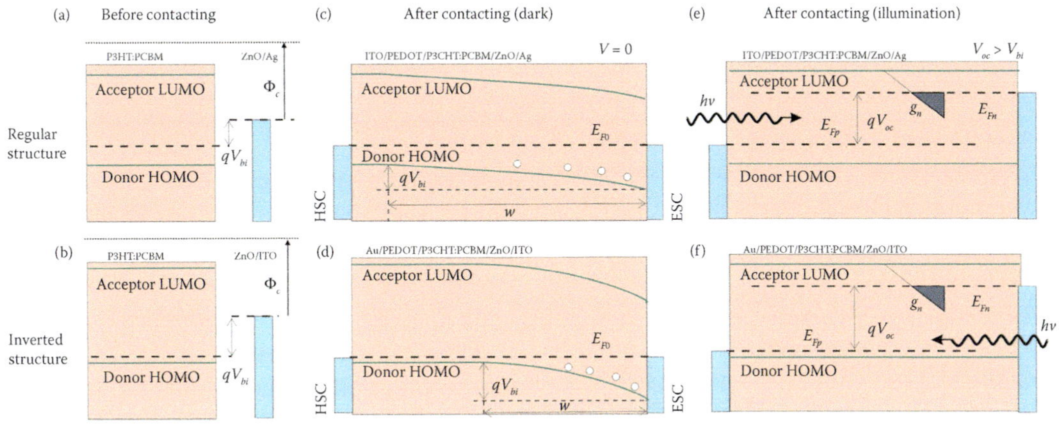

FIGURE 26.8 Energy diagram of a P3HT:PCBM organic bulk-heterojunction with a contact for holes in the left side (PEDOT) and a contact to electrons at the right side, in regular (ZnO/Ag) or inverted (ZnO/ITO) configurations. (a, b) Separate representation of the blend and the cathode. (c, d) Equilibrium after contact ($V_{app} = 0$). Band bending appears near the cathode. (e, f) Forward voltage larger than flat band condition at the cathode. The Fermi level of minority electrons scans the density of states g_n in the band gap associated with disorder. (Reproduced with permission from Boix, P. P. et al. *Journal of Physical Chemistry Letters* 2011, 2, 407–411.)

The formation and operation of the contact including depletion in the absorber layer is illustrated in Figure 26.8 for organic BHJ solar cells. In "regular" configuration, the hole selective contact (HSC) is the transparent combination of ITO/PEDOT, and in the "inverted" cell, the transparent contact is ZnO/ITO. A difference of size of the depletion region occurs due to the fact that the blend is spin-coated on top of the anode for regular cells, whereas, for inverted cells, it is deposited on top of the cathode, forming different degrees of doping in the two situations.

In the initial contact formation, the original difference of work functions indicated in Equation 26.19 can be distributed in two parts, a depletion region of barrier height corresponding to the built-in voltage V_{bi}, and the interfacial dipole Δ_i that accommodates part of the potential drop, as discussed in Section 4.6. Then Equation 26.19 is modified to the more general expression

$$V_{bi} = \left(E_{F0} - \Phi_c - \Delta_i\right)/q \tag{26.21}$$

26.4.2 Blocking Boundary Condition

At the blocking contact, we can find different situations depending on the assumed properties of the interface, as shown in Figure 26.4. The best case for applications is a reflecting boundary condition for the minority carrier at the majority extraction contact. The diffusion flux is zero:

$$\frac{\partial n}{\partial x}\Big|_{x=d} = 0 \tag{26.22}$$

This condition is called the von Neumann boundary condition, or a reflecting boundary condition. If the electrical field is intense at the boundary, the flux in Equation 26.22 must include the drift component (Beaumont and Jacobs, 1967).

We have remarked in the fundamental solar cell model of Figure 21.1 that the minority carrier density must increase under illumination, causing the splitting of the Fermi levels and hence generating the photovoltage. For enhanced solar cell performance, the minority carrier density should increase everywhere in the absorber including at the majority extraction contact, as seen in Equation 26.22, which imposes a horizontal Fermi level at the blocking boundary. In Figure 20.5b, the minority carrier density increases at the majority extraction contact. In contrast to this, in Figure 20.6, the minority carrier density is fixed at the dark equilibrium level, which has the consequence that its density cannot increase:

$$p_s = p_0 \tag{26.23}$$

This condition was remarked in Equation 26.17 for the extraction of majorities, but for the blocking of minorities, the interpretation is different. If the rate of surface recombination is very large, then the population of minority carrier at the contact cannot rise (Figure 26.4e). Thus, according to Equation 26.23, the Fermi levels of electrons and holes may separate in the bulk but are forced to coincide at this contact under the assumption of constant density at the boundary. A similar result is obtained with another approach that consists in fixing the energy barriers $\Phi_{B,n}$ and $\Phi_{B,p}$ at the contacts.

The use of the boundary conditions that fix the charge is a common procedure for p–i–n-type solar cells discussed in Figure 20.11, in which transport layers form the contacts to an insulator layer. This assumption has the effect that the charge density of electrons and holes at either edge of the active layer becomes fixed, and it is very large at the respective selective contact (due to the initial match of energy levels), as remarked in Equations 26.22 and 26.23. The resulting characteristic energy diagrams are shown in Figure 26.9, which correspond to short-circuit condition as in Figure 20.11a. The Fermi level of the carrier cannot change at the contact and the variation of the potential is necessarily absorbed across the active layer, with similar results to those of the MIM model of Figure 10.10. In order to allow a variation of the carrier density at the edge of the active layer, one can introduce additional contact layers that serve as a buffer from the highly doped transport layer, and then variations across the contact are permitted, as shown in Figure 26.10.

26.4.3 Generalized Boundary Conditions

A more general condition relates the carrier flux to the excess concentration at the contact, similar to Equation 20.3. This condition is stated in terms of a "surface recombination velocity" parameter S_n

$$J_n(0) = S_n(n_s - n_0) \tag{26.24}$$

Depending on the strength of S_n, either Equation 26.22 or 26.23 can be obtained in extreme cases of Equation 26.24, respectively $S_n = 0$ and $S_n \to \infty$. The evolution of the shape of the Fermi level from blocking to absorbing at the blocking contact is shown in Figure 26.11.

The condition of recombination at the surface can be more realistically described by SRH recombination, with the rate given in Equation 6.35. For a blocking boundary

$$J_n(0) = -N_t \frac{\beta_n\beta_p\left(n(0)p(0) - n_0p_0\right)}{\beta_n n(0) + \varepsilon_n + \beta_p p(0) + \varepsilon_p}\Delta d \tag{26.25}$$

Δ_d is the thickness close to the contact where the SRH recombination takes place. For an extracting boundary

$$J_p(0) = N_t \frac{\beta_n\beta_p\left(n(0)p(0) - n_0p_0\right)}{\beta_n n(0) + \varepsilon_n + \beta_p p(0) + \varepsilon_p}\Delta d + S_p\left(p(0) - p_0\right) \tag{26.26}$$

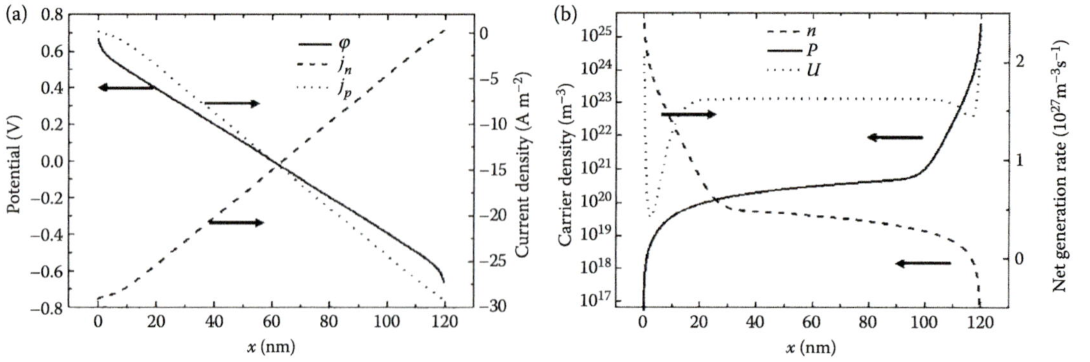

FIGURE 26.9 Device model with the assumption of constant density boundary conditions at short circuit: (a) shows the potential and current densities, (b) shows the carrier densities and the net recombination rate. (Reproduced with permission from Koster, L. J. A. et al. *Physical Review B* 2005, 72, 085205.)

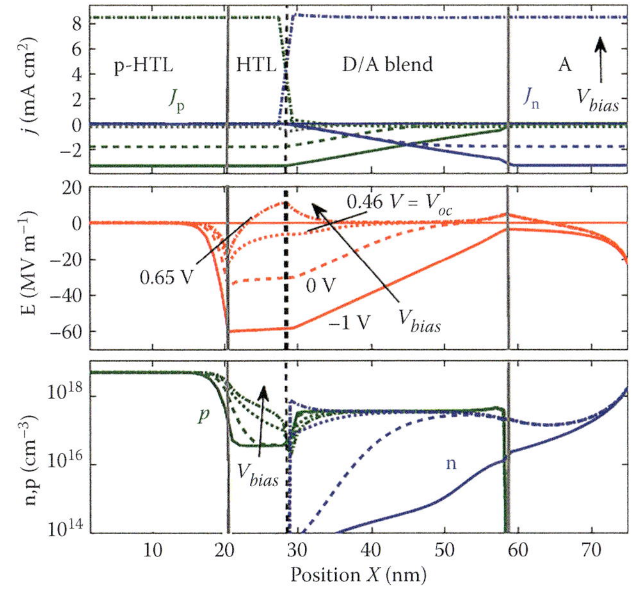

FIGURE 26.10 Model of a bulk-heterojunction solar cell with contact layers. Profiles of current density (j), electric field (**E**), and charge-carrier densities (n,p) with applied bias (V_{bias}) as parameter, green lines (high values on p-side) show hole currents and densities, blue lines (high values in acceptor [A]) show values for electrons. Field and charge-carrier density profiles indicate the applied boundary conditions: constant doping concentration on p-side, thermal injection on n-side with a work function of the cathode of 4.2 eV. $V_{bias} = -1$, 0, 0.46, 0.65 V. (Reproduced with permission from Tress, W.; Leo, K.; Riede, M. *Advanced Functional Materials* 2011, 21, 2140–2149.)

The diagrams in Figure 26.12 (discussed in more detail in Section 26.8) indicate the presence or absence of surface states according to the passivation of the back contact of the solar cell. Further discussion of charge transfer models at metal-semiconductor contact, which can actuate as boundary conditions, is presented in Section 6.7.

Charge transfer at semiconductor-electrolyte interface is a fundamental step in solar fuel production that will be described in Section 27.3. As discussed previously, in Figures 6.24 and 16.3, the electron or hole transfer may originate either from extended states in the band edges, or by an intermediate step of capturing at a surface state. The capture of an electron at the surface states increases the probability of recombination with a photogenerated hole in the valence band. The competition between charge transfer and recombination at the surface states is a major factor in the operation of metal oxide photoelectrodes for water splitting reactions (Klahr et al., 2012), and the boundary condition at the extraction contact needs to incorporate the mechanism of surface states capture (Gimenez and Bisquert, 2016).

26.5 A PHOTOVOLTAIC MODEL WITH DIFFUSION AND RECOMBINATION

We will now establish a complete photovoltaic model taking into account the detailed features of generation, transport, and recombination in a solar cell with perfect selective contacts. Our main objective is to obtain the current dependence on voltage and illumination intensity, and to discuss the features of IQE_{PV} and EQE_{PV} for monochromatic illumination. The situation we consider is shown in the scheme of Figure 26.5. We assume a large majority carrier density and nearly flat bands. We discuss the case of illumination from the contact of minority carrier extraction at $x = 0$. In Section 26.6, this model will be extended with the collection at the space-charge region, which as of now we assume is negligible.

The total electrical current density produced by the transport of carriers (see Equation 10.3) is

$$j = \sum_i j_i \tag{26.27}$$

The sum includes electrons, holes, and also other carriers, such as ions in a DSC. If we consider a semiconductor where ions are immobile, then we have the drift-diffusion problem indicated in Section 26.3, where the current corresponds to Equation 26.10.

Let us discuss the implications of the different conductivities of the two species, electrons and holes, which lead to the conclusion that diffusion of minority carriers is the dominant effect we need to describe analytically in our model.

We note that the Poisson equation is not needed in this model, as the compensation of charge already present in dark equilibrium is not disturbed by the presence of additional generated minority carrier electrons. The majority carrier concentration is orders of magnitude larger, $p_0 \gg n$. According to the dominant electrical conductivity in the bulk semiconductor layer, nearly all the electrical current is taken by the majority carrier. An imperceptible band slant, associated with a small electrical field, is enough to drive the required electrical current inside the device. However, this is not true at the boundary with the electron selective contact (ESC), as here the majority species holes cannot conduct the current. The problem of finding the outgoing current can then be reduced to a calculation of the minority carrier current just at this point. We have mentioned that the electrical field is negligible. Then the carrier transport is driven by diffusion.

Based on the previous remarks, the electrons operate in a medium where charge compensation is granted. Thus, the solar cell operation is controlled by diffusion

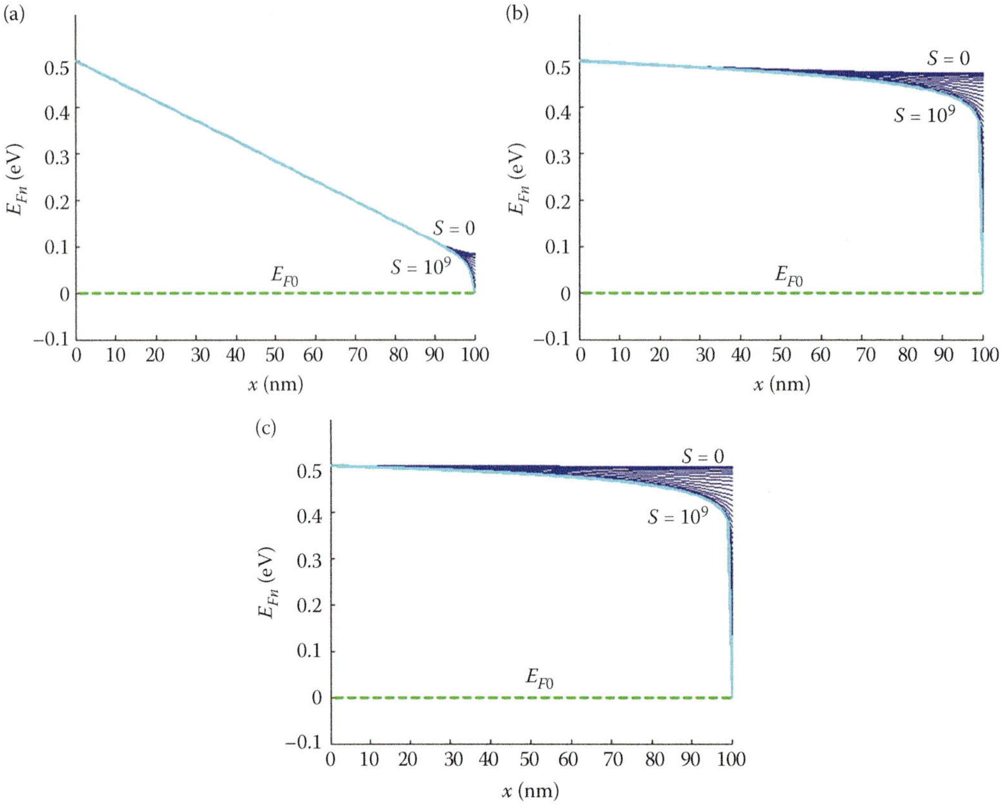

FIGURE 26.11 Injection of electrons in the dark in a semiconductor layer of thickness 100 nm. Electron Fermi level is shown for applied voltage at electron contact of 0.5 V, and diffusion lengths of (a) 6, (b) 60, and (c) 600 nm. At the right contact, surface recombination velocity S ranging from 0 (blocking boundary) to 10^9 cm s^{-1} (absorbing boundary condition), $n(d) = n_0$.

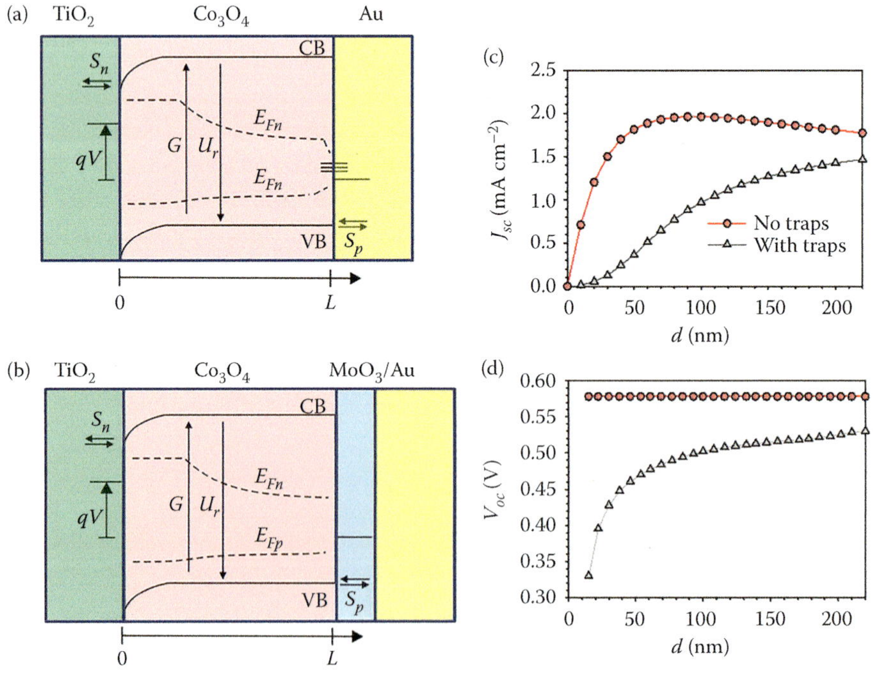

FIGURE 26.12 (a, b) Schemes of all-oxide solar cells with different structure of recombination at the back contact. (c, d) Model calculation of photovoltage and photocurrent as function of absorber thickness. (Reproduced with permission from Majhi, K. et al. *Advanced Materials Interfaces* 2016, 3, 10.1002/admi.201500405.)

and recombination, as described in Section 12.6, which produces the total diffusion current at the extraction ESC point. We require the solution of generation-diffusion-recombination in a quasi-neutral region, Equation 11.25, which is obtained by combining of Equations 26.8 and 26.12:

$$D_n \frac{\partial^2 n}{\partial x^2} - U_n + \alpha \Phi_{ph} e^{-\alpha x} = 0 \quad (26.28)$$

Using the linear recombination model of Equation 18.35 and the diffusion length of Equation 26.1, the conservation equation can be expressed as

$$\frac{\partial^2 n}{\partial x^2} - \frac{n - n_0}{L_n^2} + \frac{\alpha \Phi_{ph}}{D_n} e^{-\alpha x} = 0 \quad (26.29)$$

This equation is an extension of Equation 11.31, and it can be solved analytically with the solution

$$n = n_0 + A_1 \cosh\left(\frac{x}{L_n}\right) + A_2 \sinh\left(\frac{x}{L_n}\right) + \gamma_n e^{-\alpha x} \quad (26.30)$$

where

$$\gamma_n = \frac{L_n^2 \alpha \Phi_{ph}}{D_n\left(1 - L_n^2 \alpha^2\right)} \quad (26.31)$$

To determine the constants A_1 and A_2, we fix the boundary conditions. In the extraction contact, we take Equation 26.18, and we select the most favorable Equation 26.22, which will impose a flat Fermi level at the blocking contact. We obtain the following results:

$$A_1 = n_0\left(e^{qV/k_BT} - 1\right) - \gamma_n \quad (26.32)$$

$$A_2 = \frac{\alpha L_n \gamma_n e^{-\alpha d}}{\cosh(d/L_n)} - A_1 \tanh(d/L_n) \quad (26.33)$$

The carrier distribution in the dark has been shown earlier in Figure 26.4, and for the more general boundary condition (26.24), the solution is shown in Figure 26.11.

The diffusion flux at the extraction point is

$$J_n(0) = -D_n \frac{\partial n}{\partial x}\Big|_{x=0}$$
$$= \frac{D_n}{L_n}\left(-\frac{\alpha L_n \gamma_n e^{-\alpha d}}{\cosh(d/L_n)} + A_1 \tanh(d/L_n)\right) + \alpha D_n \gamma_n \quad (26.34)$$

The flux is the combination of the photocurrent and recombination flux

$$J_n(0,V) = -J_{ph} + J_{rec}(V) \quad (26.35)$$

The photocurrent has the expression (Södergren et al., 1994; Jennings et al., 2010)

$$j_{ph} = j_n(V=0) = qJ_{ph} \quad (26.36)$$

$$J_{ph} = \frac{L_n^2 \alpha^2}{\left(1 - L_n^2 \alpha^2\right)\cosh(d/L_n)}\left(e^{-\alpha d} + \frac{1}{L_n \alpha}\sinh(d/L_n) - \cosh(d/L_n)\right)\Phi_{ph} \quad (26.37)$$

and the recombination current is given by

$$j_{rec} = qJ_{rec}(V) = j_0\left(e^{qV/k_BT} - 1\right) \quad (26.38)$$

$$j_0 = \frac{qD_n n_0}{L_n} \tanh(d/L_n) \quad (26.39)$$

The ratio D_n/L_n has units of cm s^{-1} and this quantity is often referred to as the diffusion velocity v_d. The current-voltage characteristic is given by the equation

$$j = j_{ph} - j_0\left(e^{qV/k_BT} - 1\right) \quad (26.40)$$

The quantum efficiencies have the form

$$\mathrm{EQE_{PV}} = \frac{J_{ph}}{\Phi_{ph}} \quad (26.41)$$

$$\mathrm{IQE_{PV}} = \frac{J_{ph}}{\Phi_{ph}\left(1 - e^{-\alpha d}\right)} \quad (26.42)$$

If the diffusion length is long with respect to the film thickness, then

$$J_{ph} = \frac{L_n^2 \alpha^2 \Phi_{ph}}{L_n^2 \alpha^2 - 1}\left(1 - e^{-\alpha d}\right) \quad (26.43)$$

and we obtain

$$EQE_{PV} = \frac{1 - e^{-\alpha d}}{1 - (L_n \alpha)^{-2}} \quad (26.44)$$

$$IQE_{PV} = \frac{1}{1 - (L_n \alpha)^{-2}} \quad (26.45)$$

If $L_n \gg \alpha^{-1}$, we have total collection by diffusion and $IQE_{PV} \approx 1$. For $L_n \gg d$, the recombination parameter in Equation 26.39 reduces to that of Equation 23.27 corresponding to homogeneous (dark) recombination in the whole semiconductor layer

$$j_0 \approx \frac{qD_n dn_0}{L_n^2} = qdB_{rec} n_0 p_0 \tag{26.46}$$

However, if the diffusion length is short, then Equation 26.39 reduces to Equation 26.4. Furthermore, we get

$$EQE_{PV} = \frac{L_n \alpha}{1 + L_n \alpha} \tag{26.47}$$

Figure 26.13 shows the distribution of electrons (indicated by the Fermi level) in the diffusion-recombination model under illumination. A number of examples covering different cases of the absorption depth and diffusion length are shown, with parameters indicated in Table 26.1. When $\alpha^{-1} > d$, the electrons are generated homogeneously across the layer, and $EQE_{PV} \ll IQE_{PV}$ due to the fact that many photons leave the layer through the back contact. Thus, in panel (a) and (e), it is observed that the distribution of electrons is rather homogeneous. The gradient increases close to the extraction contact as here, the total current is carried exclusively by diffusion of the minority carrier, while in the rest of the active layer, the current is sustained by the majority carrier, as explained earlier. The diffusion currents for a more general situation of combined electronic and ionic transport in a DSC are shown in Figure 12.5. In the case of strong absorption $\alpha^{-1} \ll d$, electrons are generated only close to the transparent contact. Nevertheless, if the

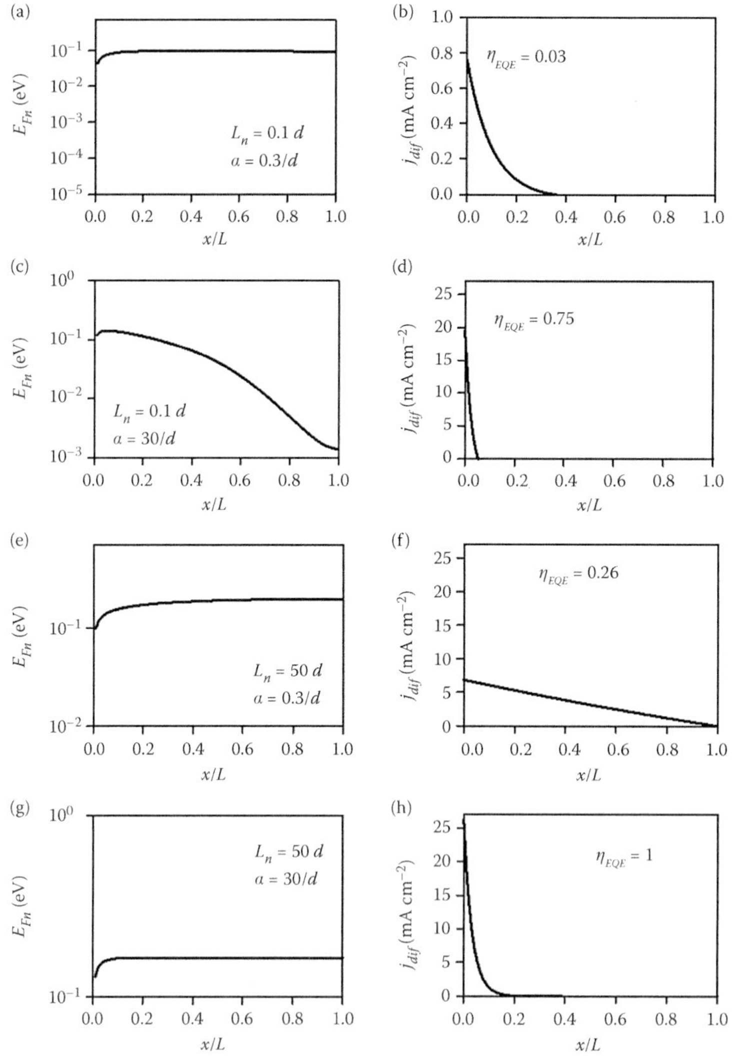

FIGURE 26.13 Electron Fermi level (*left column*) and local diffusion current density (*right column*) in a solar cell controlled by the diffusion-recombination model, covering different cases of the absorption depth and diffusion length, as indicated in Table 26.1. The simulations correspond to short-circuit situation. (Courtesy of Luca Bertoluzzi.)

TABLE 26.1
Characteristics of Simulated Solar Cells in Figure 26.13

	(a)	(c)	(e)	(g)
L_n/d	0.10	0.10	50	50
αd	0.30	30	0.30	30
j_{ph} (A cm^{-2})	0.763	19.6	6.79	26.2
V_{oc} (V)	0.101	0.185	0.420	0.456
EQE_{PV}	0.0291	0.750	0.259	1.00
IQE_{PV}	0.112	0.750	1.00	1.00
j_0 (A cm^{-2})	0.0160	0.0160	6.40×10^{-7}	6.40×10^{-7}
$k_{rec} = B_{red}p_0$ (s^{-1})	1000	1000	4.00×10^{-3}	4.00×10^{-3}

Parameters: $d = 10$ μm, $T = 300$ K, $n_0 = 10^{15}$ cm^{-3}, $D_n = 10^{-5}$ cm^2 s^{-1}, $\Phi_{ph} = 10^{16}$ cm^{-2} s^{-1}.

diffusion length is long, $L_n \gg d$, panel e, the electrons do not remain at the generation point but spread across the layer again creating a nearly homogeneous distribution. However, when both $\alpha^{-1} \ll d$ and $L_n \ll d$, the electrons are localized close to the generation point and a rather inhomogeneous profile is obtained as shown in panel (c).

Based on these observations, we consider the EQE_{PV} dependence on wavelength of a solar cell driven by electron diffusion. These characteristics are shown in Figure 26.14 from the landmark article by Södergren et al. (1994), which showed quantitatively the diffusive transport of electrons in a DSC. If the solar cell is illuminated from the TCO substrate side, it will always collect the electrons, even if the diffusion length is short. The EQE_{PV} will be unity at the shorter, strongly absorbed wavelengths and will decrease at longer wavelength due to the transparency of the device. But for illumination from the electrolyte side, the behavior of $EQE_{PV}(\lambda)$ is different. For strongly absorbed wavelengths and short diffusion lengths, the electrons cannot reach the collecting contact. At longer wavelengths, the behavior from illumination of either side must be rather similar. By fitting the spectra to the diffusion-recombination model, the diffusion length can be obtained. However, this methodology can be applied only to cells that show a low transport rate (Halme et al., 2008). As we remarked in Figure 26.13, if the diffusion length in the device becomes sufficiently long, such analysis will not be possible, as $IQE_{PV} = 1$ and all the electrons will be collected wherever they are generated.

Another way to calculate the external quantum efficiency is to apply a reciprocity theorem derived by Donolato (1985, 1989). It states that the probability of collection of a carrier generated at position x, $f_c(x)$, is related to the dark carrier distribution value, $n(x)$, divided by its thermal equilibrium value $n_0(x)$,

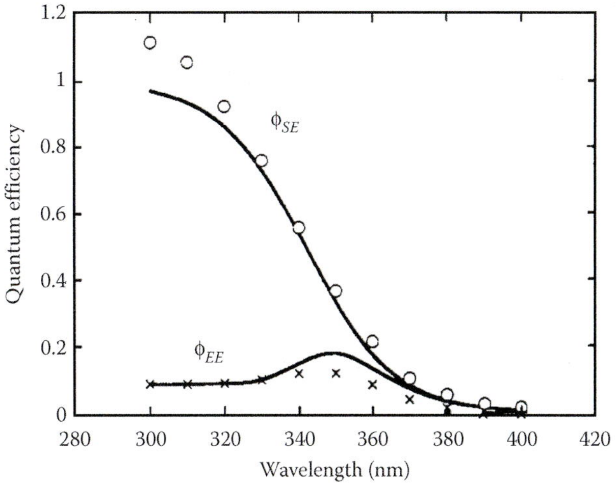

FIGURE 26.14 Experimental data for the EQE_{PV} of a 2.5 μm thick colloidal TiO_2 film electrode in 0.1 M KSCN in ethanol with a diffusion length of $L_n = 0.8$ μm, and fit to the diffusion-recombination model. φ_{SE} corresponds to the illumination from the substrate side and φ_{EE} to the illumination from the side of the electrolyte. The absorption coefficient as a function of wavelength was approximated by $\ln\alpha(\mu m) = 29-85\lambda$ (μm). (Reproduced with permission from Södergren, S. et al. *The Journal of Physical Chemistry* 1994, 98, 5552–5556.)

$$f_c(x) = \frac{n(x)}{n_0} \tag{26.48}$$

For a generation profile $G(x)$, the collected carrier flux at $x = 0$ is

$$J_{ph} = \int_0^d G(x) f_c(x) \tag{26.49}$$

Expressions for EQE_{PV} under general boundary conditions for surface recombination are given in Green (2009).

26.6 THE GÄRTNER MODEL

The solar cell with the band-bending region at one contact associated with an SB has been analyzed in Figures 9.3, 20.1, and 20.5. When the diffusion process is insufficient to extract the charge efficiently, the formation of an SB creates a strong electrical field that assists charge separation and facilitates charge collection in situations of low mobility and strong recombination, as in the application of some metal oxide electrodes to solar water splitting (Klahr and Hamann, 2011). In Figure 26.15a is shown the operation of this model under photogeneration, which is usually associated with the work of Gärtner (1959). The original model assumed a

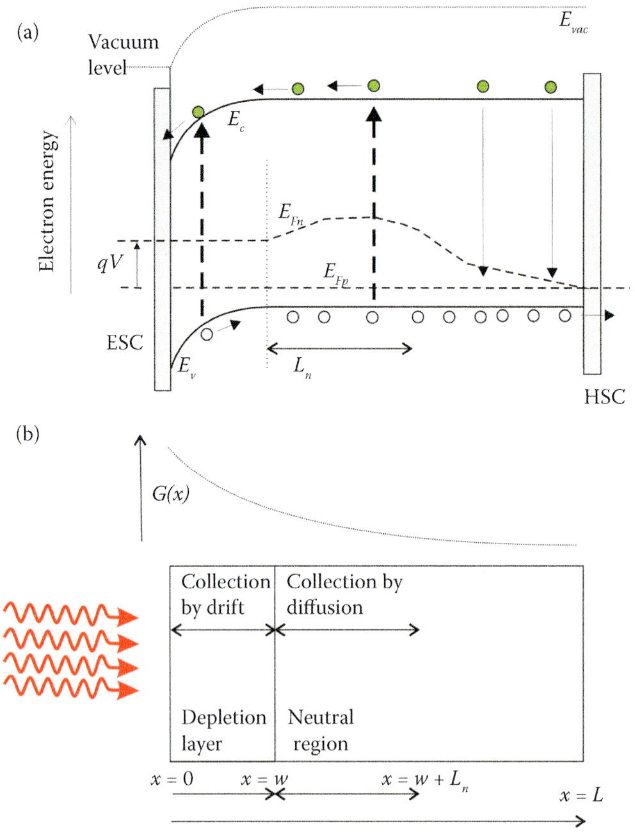

FIGURE 26.15 (a) Model for a p-type semiconductor forming an SB solar cell. Electronic processes under illumination are indicated. (b) Photogeneration gradient G in the two parts of the n-type semiconductor material, the depletion layer and the neutral region, indicating the collection in the region of a diffusion length.

short diffusion length with respect to film thickness. It is also assumed that the layer is sufficiently thick to absorb all incoming photon flux, $\alpha^{-1} \ll d$.

The charge collection process is composed of two elements, namely the space-charge zone and the diffusion length zone, as indicated in Figure 26.15b. In the depletion layer formed by the SB, there are two concerted factors that reduce recombination: first, the majority carrier population is depleted, which enhances the lifetime of minorities; and, second, the electrical field moves the carriers in opposite directions favoring rapid charge separation. In consequence, in a first approximation recombination can be neglected. The current of minority carriers generated in the depletion region of thickness w is

$$j_{ph,dl} = q\int_0^w G_\Phi dx = q\Phi_{ph}\left(1 - e^{-\alpha w}\right) \quad (26.50)$$

The second part of the collection process is that of carriers that are generated in the neutral region. These do not have a preferred direction for the displacement, as suggested in Figure 26.15a. Only the minority carriers generated at a distance less than L_n from the edge of the depletion region are collected, as discussed earlier in this chapter. The diffusion current arriving at $x = w$ is obtained by the solution of the diffusion-recombination model, Equation 26.29, with the boundary conditions correspondent to Figure 26.15a. The result derived in Equation 26.47 gives

$$j_{ph,diff} = q\Phi_{ph}e^{-\alpha w}\frac{\alpha L_n}{1+\alpha L_n} \quad (26.51)$$

Summing up the two terms, we obtain the photocurrent

$$j_{ph} = q\Phi_{ph}\left[1 - \frac{1}{1+\alpha L_n}e^{-\alpha w}\right] \quad (26.52)$$

From Equation 26.52, the quantum efficiency is given by

$$IQE_{PV}(\lambda) = \frac{j_{ph}(\lambda)}{q\Phi_{ph}(\lambda)} = \left[1 - \frac{1}{1+\alpha(\lambda)L_n}e^{-\alpha(\lambda)w}\right] \quad (26.53)$$

Due to the reflectivity R of the front surface, the EQE_{PV} relates to IQE_{PV} as

$$EQE_{PV} = (1-R)\eta_{IQE} \quad (26.54)$$

Equation 26.53 can be used to determine the minority carrier diffusion length, if the absorption coefficient $\alpha(\lambda)$ is known and depletion layer thickness is calculated independently by capacitance measurements (Ritenour et al., 2012). If $w \ll \alpha^{-1}$, then the approximate relation holds (Werner et al., 1993):

$$IQE_{PV}^{-1} = 1 + \frac{1}{\alpha L_n} \quad (26.55)$$

The Gärtner model has been amply used in the field of water splitting with metal oxide semiconductors, since these photoelectrodes normally rely on the space-charge region for the efficient transfer of holes in the water decomposition reaction (Gimenez and Bisquert, 2016). The assumption of full collection in the space-charge region is usually not well satisfied due to recombination in the case of slow transport, such as in typical hematite electrodes (Klahr and Hamann, 2011). This physical effect is suitably described by the model of Reichman (1980). Figure 26.16 shows the exact numerical calculation of the current-voltage curve using the methods

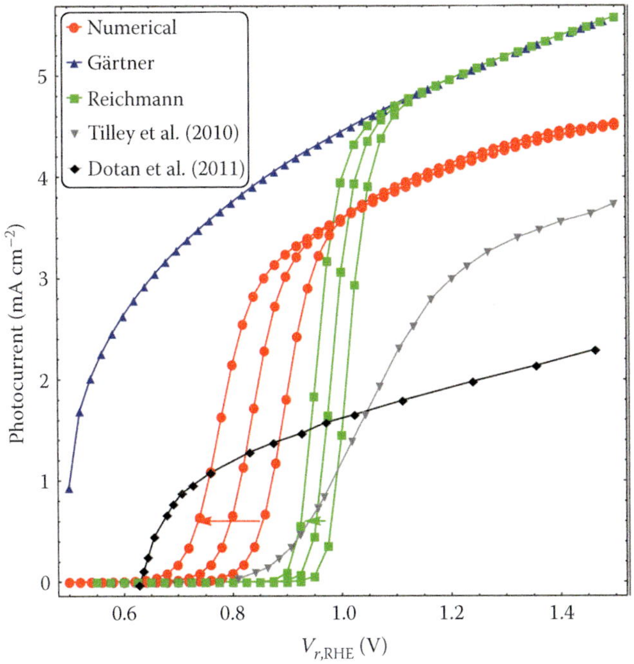

FIGURE 26.16 Comparison of photocurrent-voltage curves from: a numerical model, the Gärtner model, the Reichmann model, and the measured data from Tilley et al. (2010) and Dotan et al. (2011) for n-doped hematite photoelectrodes. The different simulated curves in each case are for different values of charge transfer constant. (Reproduced with permission from Cendula, P.; Tilley, S. D. et al. *The Journal of Physical Chemistry C* 2014, 118, 29599–29607.)

indicated in Section 26.3, compared with the models of Gärtner and Reichmann and showing also characteristic experimental results (Cendula et al., 2014).

26.7 DIFFUSION-RECOMBINATION AND COLLECTION IN THE SPACE-CHARGE REGION

We derive the general analysis of the model of Figure 26.15, consisting of the combination of unity collection in the space-charge region of width w and the diffusion-recombination in the neutral region with the reflecting boundary condition at $x = d$. The general solution (without a restriction to the case of short diffusion length) is

$$n = n_0 + R\cosh\left(\frac{x-w}{L_n}\right) + S\sinh\left(\frac{x-w}{L_n}\right) + \gamma_n e^{-\alpha x} \tag{26.56}$$

where

$$\gamma_n = \frac{L_n^2 \alpha \Phi_{ph}}{D_n\left(1 - L_n^2\alpha^2\right)} \tag{26.57}$$

$$R = n(w) - n_0 - \gamma_n e^{-\alpha w} \tag{26.58}$$

$$S = \frac{\alpha L_n \gamma_n e^{-\alpha d}}{\cosh y_d} - R\tanh y_d \tag{26.59}$$

$$y_d = \frac{d-w}{L_n} \tag{26.60}$$

The electron flux at the edge of the SCR is

$$\begin{aligned} J_n(w) &= -D_n \frac{\partial n}{\partial x}\Big|_{x=w} \\ &= \frac{D_n}{L_n}\left(-\frac{\alpha L_n \gamma_n e^{-\alpha d}}{\cosh y_d} + R\tanh y_d\right) + \alpha D_n \gamma_n e^{-\alpha w} \end{aligned} \tag{26.61}$$

In the SCR ($0 \le x \le w$), we have just the generation term

$$\frac{\partial J_n}{\partial x} = \alpha \Phi_{ph} e^{-\alpha x} \tag{26.62}$$

Integrating Equation 26.62, we obtain

$$J_n(0) = J_n(w) + \Phi_{ph}\left(e^{-\alpha w} - 1\right) \tag{26.63}$$

The concentration of electrons at both edges of the space-charge region is related by

$$n(w) = n(0) e^{-qV_{sc}/k_B T} \tag{26.64}$$

where V_{sc} is the voltage across the SCR. Inserting Equation 26.64 in Equation 26.58, we obtain the relation between electron flux and concentration at the edge of the semiconductor layer:

$$J_n(0) = -J_g + \frac{D_n}{L_n}\tanh y_d\left(n(0)e^{-qV_{sc}/k_B T} - n_0\right) \tag{26.65}$$

where

$$J_g = \Phi_{ph}\left[1 - e^{-\alpha w} + \frac{L_n^2\alpha^2}{\left(1 - L_n^2\alpha^2\right)\cosh y_d}\left(e^{-\alpha d} + \frac{e^{-\alpha w}}{L_n \alpha}\sinh(y_d) - e^{-\alpha w}\cosh(y_d)\right)\right] \tag{26.66}$$

The extracted flux in Equation 26.65 is a function of the electron density at the edge of the semiconductor layer. This allows for more complex boundary conditions to be added, such as the effect of a dipole layer, as shown in Figure 20.6 (Bisquert et al., 2008). If the electron concentration is simply controlled by the external voltage as indicated in Equation 26.18, then Equation 26.65 gives the diode model

$$j = qJ_g - \frac{qD_n n_0}{L_n} \tanh y_d \left(e^{qV/k_BT} - 1 \right) \quad (26.67)$$

In the particular case in which $d \gg L_n$, we obtain

$$j = j_{ph} - \frac{qD_n n_0}{L_n} \left(e^{qV/k_BT} - 1 \right) \quad (26.68)$$

where j_{ph} is the value of the Gärtner model given in Equation 26.52.

26.8 SOLAR CELL SIMULATION

We describe the photovoltaic operation of a generic solar cell with selective contacts of different extraction properties. Instead of using a simplified approach as in the previous sections, here we adopt the general drift-diffusion equations, combined with the continuity and the Poisson equations for both electrons and holes that have been described in Sections 12.4 and 26.3. We can add the specific recombination model and boundary conditions that apply to the given problem according to the interface properties and then run a numerical simulation, which will provide the band-bending, electron and hole concentrations, the Fermi levels, and currents.

We use a p-doped semiconductor with the ESC at $x = 0$, where an SB is formed as shown in Figure 26.17. The boundary conditions are

$$J_n(0) = qS_n \left(n(0) - n_0 e^{qV_{bi}/kT} \right) \quad (26.69)$$

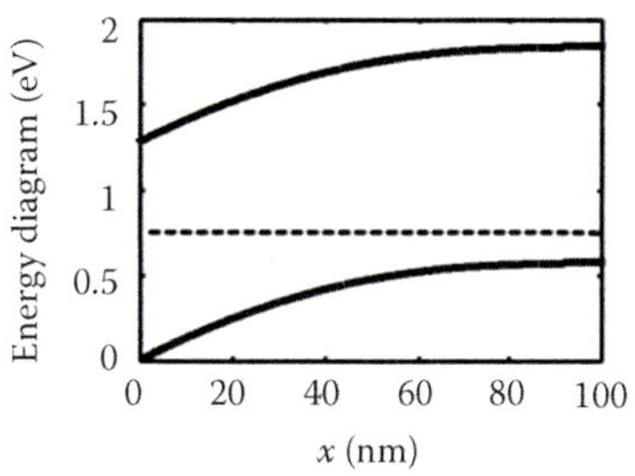

FIGURE 26.17 Band diagram of an SB solar cell in dark. The ESC is at $x = 0$ and the HSC is at $x = d$. The physical parameters used in the simulation are: $u_{n,p}$= 0.1 cm^2 V^{-1} s^{-1}, ε_r = 10, d = 100 nm, E_g = 1.2 eV, N_C, N_V = 10^{19} cm^{-3}, energy barrier for holes at the HSC Φ_a = 0.18 eV, energy barrier for electrons at the ESC Φ_b = 0.45 eV, αd = 1, $\tau_n = (5d)^2/D_n$, and N_A = 10^{17} cm^{-3}.

$$J_p(0) = 0 \quad (26.70)$$

The origin of the electrostatic potential is taken at the ESC at $x = 0$:

$$\varphi(0) = 0 \quad (26.71)$$

$$\varphi(d) = V - V_{bi} \quad (26.72)$$

At the HSC ($x = d$), we consider different cases of the properties of the interface:

1 A reflecting contact for electron and hole transfer limited by surface recombination of holes:

$$J_p(d) = qS_p \left(p(d) - p_0(d) \right) \quad (26.73)$$

$$J_n(d) = 0 \quad (26.74)$$

2 SRH recombination and hole transfer limited by surface recombination, Equations 26.25 and 26.26:

$$J_n(d) = -qN_t \frac{\beta_n \beta_p \left(n(d) p(d) - n_0 p_0 \right)}{\beta_n n(d) + \varepsilon_n + \beta_p p(d) + \varepsilon_p} \Delta d \quad (26.75)$$

$$J_p(d) = qN_t \frac{\beta_n \beta_p \left(n(d) p(d) - n_0 p_0 \right)}{\beta_n n(d) + \varepsilon_n + \beta_p p(d) + \varepsilon_p} \Delta d + qS_p \left(p(d) - p_0 \right) \quad (26.76)$$

3 Infinite recombination of electrons:

$$J_p(d) = qS_p \left(p(d) - p_0(d) \right) \quad (26.77)$$

$$n(d) = n_0 \quad (26.78)$$

In Figure 26.18 are shown the characteristics of the solar cell at open circuit in the case of generation across the whole active layer. When the diffusion length is higher than the device length ($L_n > d$), the concentration of electrons remains almost constant along the device, as already discussed in Section 26.5. In some cases as shown in Figure 26.18, holes accumulate at the ESC. At open-circuit voltage, the bands are almost flat along the device and the carriers can flow to any direction. The holes (majority carriers in this semiconductor) are blocked at the ESC and, in addition, they cannot recombine due to

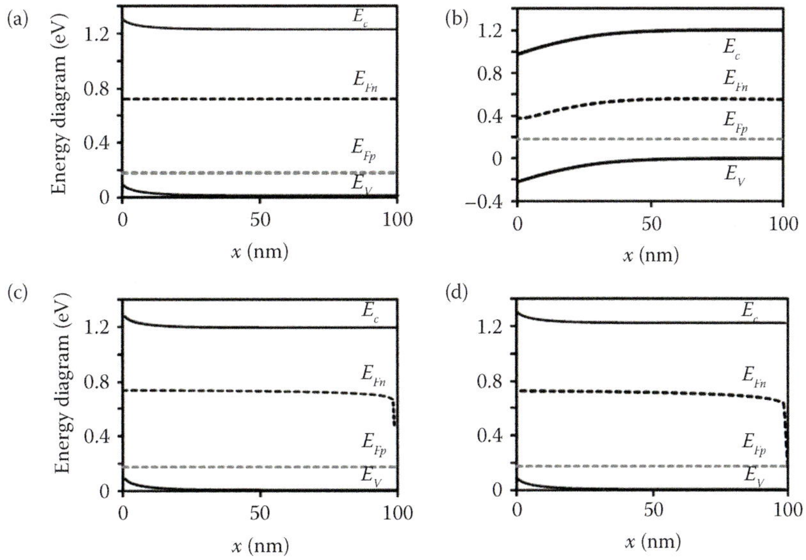

FIGURE 26.18 Drift-diffusion simulations of a p-type solar cell at open-circuit voltage for different cases in which the boundary condition for electrons at the back ($x = d$) and the diffusion length, L_n, are modified. (a) HSC is a reflecting contact and the semiconductor has a long diffusion length; (b) HSC is a reflecting contact and the semiconductors have a short diffusion length. (c) HSC with SRH recombination and semiconductor with long diffusion length. (d) Ideal absorbing contact and semiconductor with long diffusion length. The ESC is always a reflecting contact. The physical parameters used in the simulation are $u_{n,p}$= 1 cm^2 V^{-1} s^{-1}, $\varepsilon_r = 10$, $N_A = 10^{17}$ cm^{-3}, $d = 100$ nm, $E_g = 1.2$ eV, N_C, $N_V = 10^{20}$ cm^{-3}, energy barrier at the hole selective contact for holes $\Phi_{B,a} = 0.2$ eV, energy barrier at the ESC for electrons $\Phi_{B,b} = 0.6$ eV, $\Phi_{ph} = 10^{16}$ cm^{-1} s^{-1}, and $\alpha d = 1$. The particular parameters are in (a, c, d): $\tau_n = (5d)^2/D_n = 3.5 \times 10^{-12}$ s and in (b) $\tau_n = (0.03d)^2/D_n = 10^{-7}$ s. For SRH: $\beta_n = \beta_p = 1$ cm^3 s^{-1}, $N_c = N_v = 10^{21}$ cm^{-3}, $\varepsilon_n = 10^{-6}$, and $\varepsilon_p = 10^{26}$. (Courtesy of Pilar Lopez-Varo.)

the low recombination, thus, they accumulate at the ESC interface. This accumulation can be suppressed for different values of the parameters.

In Figure 26.19, a shorter penetration of the incident light occurs, which is reflected in the carrier distribution when the diffusion length is short, panel (b).

Figure 26.20 shows the characteristics of the p-type solar cell under different assumptions of light penetration and diffusion length for the parameters and results shown in Table 26.2. The diagrams obtained from the general simulation indicate the carrier distributions and the components of electron current, and they complement the previous calculation for the case of flat bands presented in Figure 26.13. The internal current components in the second column indicate internal flues due to generation close to the left contact.

The effect of the surface states on solar cell characteristics is discussed with the example in Figure 26.12. The scheme and calculations represent an all-oxide solar cell based on Co_3O_4 absorber (Majhi et al., 2016), with TiO_2 ESC at the transparent side and Au layer as HSC. To improve the quality of the back contact, an additional MoO_3 layer is added as shown in Figure 10.12b, which reduces the density of surface states. The interface extraction rate of each carrier at the respective contact is modeled by a surface velocity constant S_n, S_p. On the right side, it is observed that surface traps reduce substantially the voltage and photocurrent for short thickness, while in the case of thicker absorber layer, the recombination at the back contact is less influential in the solar cell performance.

26.9 CLASSIFICATION OF SOLAR CELLS

Having analyzed in full detail the different steps that compose the photovoltaic operation, we now establish a comparison of the mechanisms that lead to charge collection (Kirchartz et al., 2015).

As mentioned earlier, charge extraction in a solar cell consists of a competition between charge transport and recombination. On their way to the contacts, electrons and holes encounter each other in the central layer, which can lead to recombination events that are quantified in a generic sense by a recombination lifetime τ. This lifetime may depend on impurities, dislocations and grain boundaries in inorganic solar cells, or on the local microstructure of the donor–acceptor network in organic solar cells. In polycrystalline solar cells such as CdTe or $CH_3NH_3PbI_3$ perovskite, one may take care and control the special properties of the grain boundaries, which

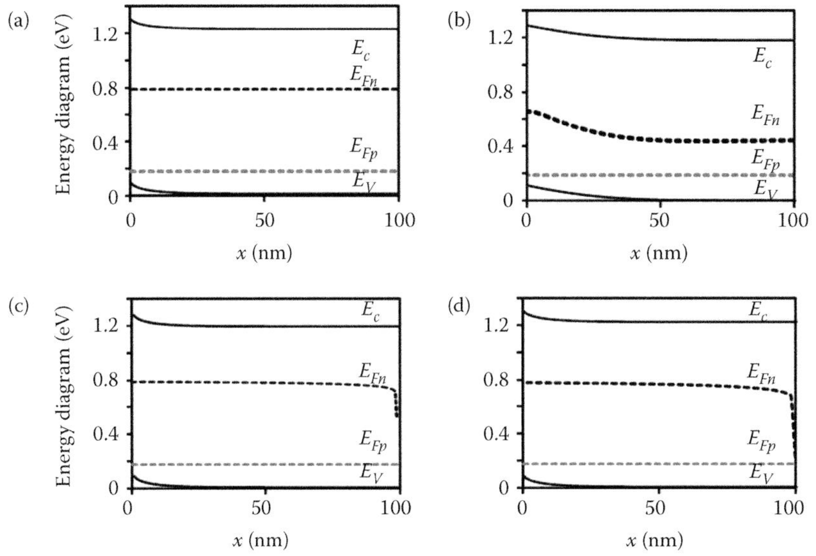

FIGURE 26.19 Drift-diffusion simulations for the same cases of 26.18. The light penetration is the only parameter with a different value, $\alpha d = 10$. (Courtesy of Pilar Lopez-Varo.)

may have electronic properties deleterious for the overall operation, as mentioned previously in Section 24.3 and shown in Figure 26.21, where a large grain structure (left) forms conduction pathways that provoke a short circuit. The adequate recrystallization, annealing, and control of grain size procedures will provide a more compact film with optimized photovoltaic properties, as shown in the right column of Figure 26.21, and the passivation of grain boundaries reduces the deleterious recombination surface states as indicated in Figure 24.9 (Jaegermann et al., 2009). Controlling crystal growth and morphologies by solvent engineering has been a determinant step for the development of high-efficiency perovskite solar cells (Jeon et al., 2014; Ahn et al., 2015).

In Section 20.5, we remarked on the construction of a built-in voltage in solar cell devices. In general, the selective contact to electrons must be ohmic and form a small injection barrier for electrons. Thus, materials with a low-work function are required. These will align well to the CB minimum energy level of the absorber layer. Oppositely, materials with a large work function may form an efficient hole extraction contact. In consequence, solar cell devices contain an intrinsic built-in potential between the outer contacts.

In some cases, the built-in field does not help much with charge extraction at short circuit. This is typically the case when the built-in field is confined to a dipole layer or to a narrow space-charge region, as in a DSC or crystalline Si (c-Si) solar cell, as discussed in Section 24.3 and shown in Figures 24.11 and 24.12. Nevertheless, the built-in voltage V_{bi} ensures that the applied forward bias V (e.g., at the maximum power point of an illuminated solar cell) can drop somewhere. As long as $V_{bi} - V$ is still positive, no barriers for extraction will form. Thus, the built-in voltage serves an important role even if, like in c-Si solar cells, charge collection is nearly entirely driven by diffusion. Alternatively, the built-in field, obtained after initial device formation, may be extended over the absorber thickness and take a leading role in charge extraction. In fact, the device might be even fully depleted at short circuit in which case the built-in electric field extends over the whole absorber, as in amorphous silicon (a-Si) solar cells (Schiff, 2003), see the model in Figure 20.11.

The ratio between the width w of the space-charge region and the absorber thickness d is therefore a useful criterion to classify solar cell types with regard to their way of separating charge carriers. Figure 26.22a compares important solar cell technologies in terms of their ratio w/d. Figure 26.22b shows the band diagram of a p-type semiconductor at short circuit in the dark. The doping is sufficiently high relative to the absorber thickness so that $w/d \ll 1$. c-Si as well as DSCs are typical examples of solar cells that have a tiny space-charge region relative to the total absorber width. Here, electron transport through the device will be mostly by diffusion, as explained earlier.

For typical inorganic thin film solar cells, the ratio w/d varies from around 1/10 in $Cu(In,Ga)Se_2$ (CIGS) to 1 in fully depleted devices like amorphous or microcrystalline Si (μc-Si). Figures 26.21c and d show the schematic band diagrams for devices with intermediate

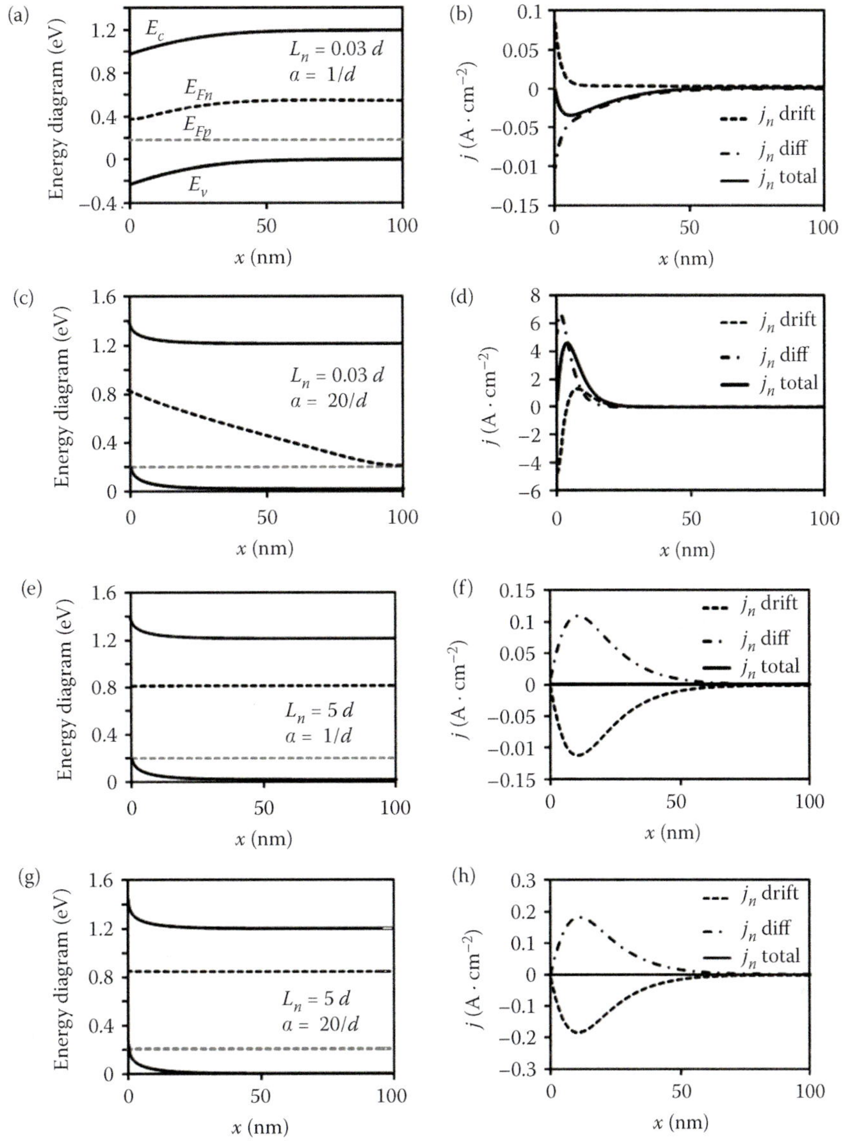

FIGURE 26.20 (a, c) Energy band diagrams including the quasi-Fermi levels in Schottky p-type solar cells at open circuit with different values of the absorption depth and the diffusion length. (b, d) Respective calculations of the drift, diffusion (diff), and total electron current densities. The simulation parameters are in Table 26.2. (Courtesy of Pilar Lopez-Varo.)

(c) and high (d) ratios of *w*/*d*. When going from small to large values of *w*/*d*, the way charge separation occurs in the solar cell changes. While devices with low *w*/*d* are controlled by diffusion, the larger the space-charge region becomes relative to the absorber thickness, the more drift will affect charge-carrier collection. Alternate situations can exist, as in the case of organic bulk heterojunctions where both band bending in the absorber and contact dipole account for the total V_{bi} (Guerrero et al., 2012). Figure 26.23 shows the distribution of the Fermi levels for the three cases of solar cells regarding the size of *w*/*d*, under illumination, from short- to open-circuit condition.

26.10 MEASURING AND REPORTING SOLAR CELL EFFICIENCIES

Investigating new types of solar cell materials requires adopting careful protocols of measurement to avoid communicating biased and unjustified claims, with devastating negative impact on the research field. The top performing cell of a given technology is routinely reported in efficiency tables and in the popular NREL chart (http://www.nrel.gov/ncpv). To become official, the device must have a minimum area of 1 cm^2 and the result must be certified by specialist labs. For solar cells still in their initial stages of development, the current published

TABLE 26.2
Characteristics of Simulated Solar Cells of Figure 26.20

	(a)	(c)	(e)	(g)
L_n/d	0.03	0.03	5	5
αd	1	20	1	20
j_{ph} (mA cm^{-2})				
$q\Phi_{ph0}(e^{-\alpha d} - 1)$ (mA cm^{-2})	10	16	10	16
j_{ph} (mA cm^{-2})	0.05	0.7	1	1.6
V_{oc} (V)	0.28	0.5	0.65	0.675
EQE_{PV}	0.03	0.4	0.66	1
IQE_{PV}	0.05	0.4	1	1

Parameters: d = 100 nm, T = 300 K, Φ_{ph} = 10^{16} cm^{-2} s^{-1}, N_A = 10^{17} cm^{-3}, and $u_{n,p}$= 1 cm^2 V^{-1}s^{-1}.

record may have been made with special materials that are not available to everyone, and with very unique skills of the staff or equipment in the lab, sometimes using sophisticated light management schemes. Although not widely recognized, some reported records have been achieved for very small area devices, which is not representative of practical solar cells.

Reporting the current density-voltage curve under standard conditions is obviously necessary when investigating new types of solar cells. Papers should also ideally report the EQE_{PV}, which provides an important test of calculation of the photocurrent as commented in Sections 22.1 and 26.1. If the photocurrent is larger than the integral of EQE_{PV}, a major flaw has occurred. Figure 25.2b shows the result of the integration that presents an excellent match with the photocurrent measured under 1 sun condition indicated in Figure 25.2a. Note that such good match requires some precautions to avoid the difference between the solar cell conditions in dark and light (Jennings et al., 2010). Li et al. (2016, Figure 25.8) remark that the measurements of EQE_{PV} were taken with chopped monochromatic light under a white light bias corresponding to 5% solar intensity. In solar cells with a large low frequency capacitance, the measured EQE_{PV} may show frequency dependence, which complicates the confirmation of the measured photocurrent. It is important to obtain reliable information about EQE_{PV} using IMPS technique (Ravishankar et al., 2018).

Larger cells always lower the efficiency to a significant extent with respect to small ones. The best module efficiency is typically about 60% the best small cell

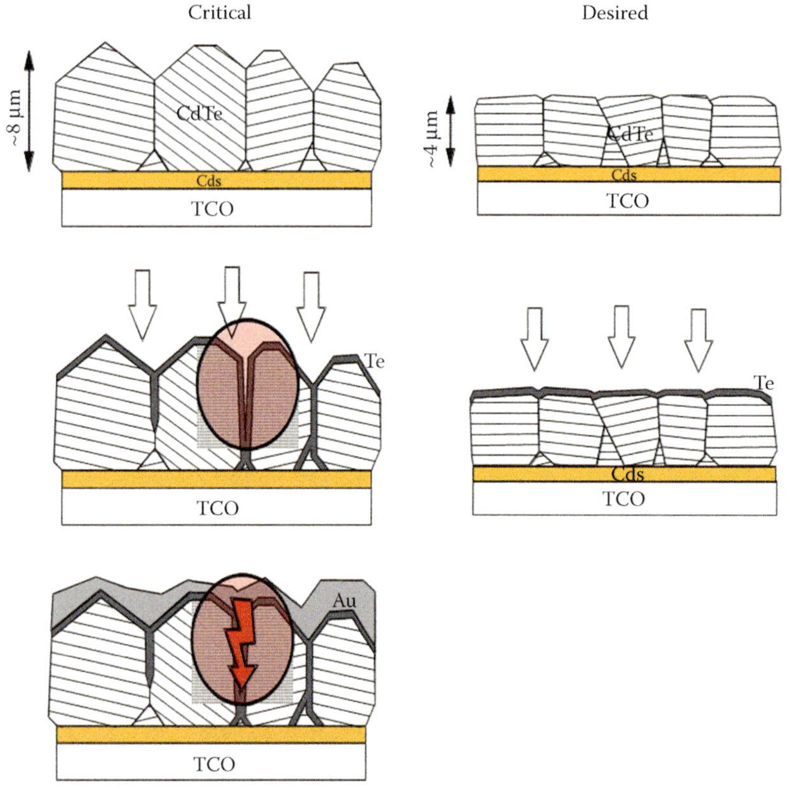

FIGURE 26.21 Schematic representation of recombination pathways producing low-shunt resistances along grain boundaries of three-dimensional, statistically oriented films (*left*), and the improved morphology of compact CdTe layers expected by better control of texture and nucleation (*right*). (Reproduced with permission from Jaegermann, W. et al. *Advanced Materials* 2009, 21, 4196–4206.)

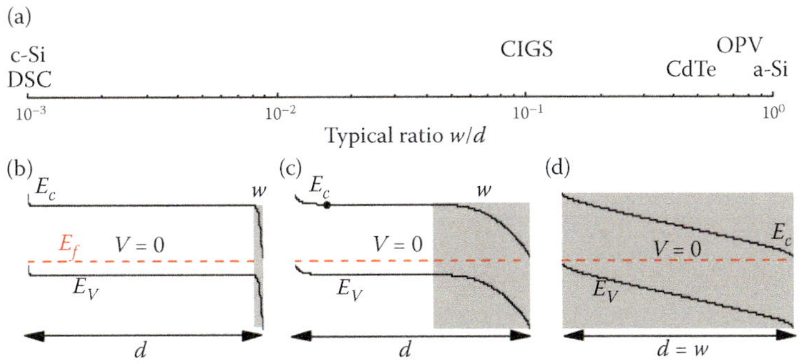

FIGURE 26.22 (a) The typical ratio of space-charge region width w and thickness d for different solar cell technologies. Band diagrams of solar cells with small w/d (b), intermediate w/d (c), and $w = d$ (d) at short circuit in the dark. The gray areas represent the main space-charge regions in the different diagrams at short circuit. (Reproduced with permission from Kirchartz, T. et al. *Physical Chemistry Chemical Physics* 2015, 17, 4007–4014.)

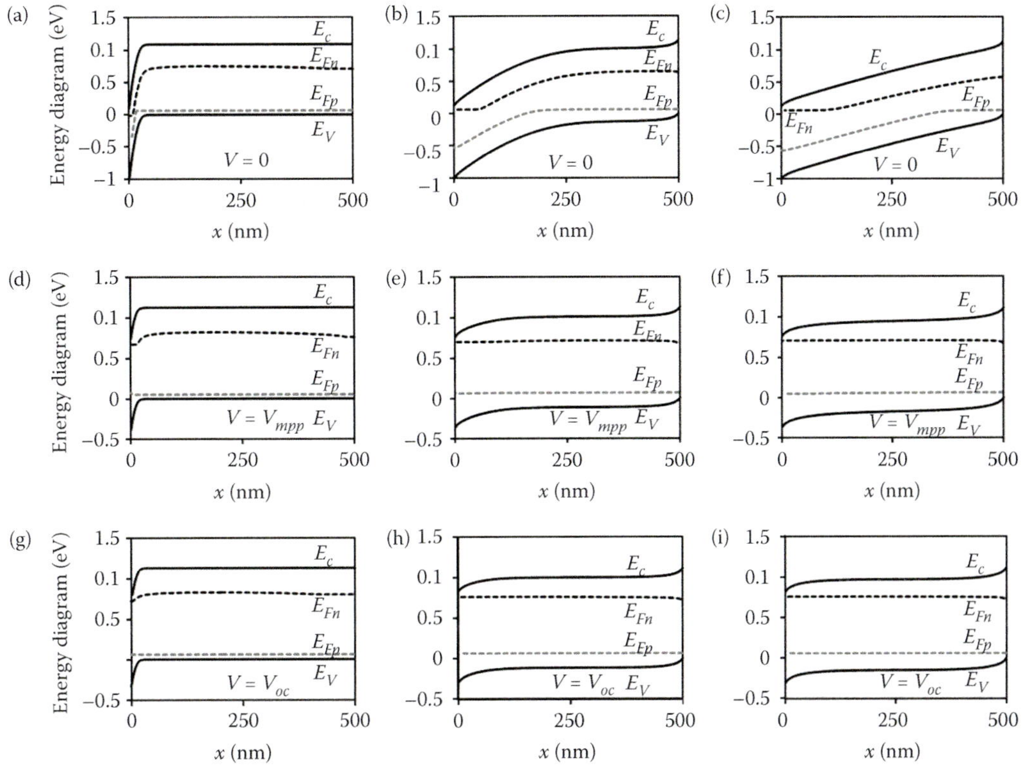

FIGURE 26.23 Solar cell simulations with different ratio between space charge region width w and device length d due to different doping ion acceptor concentrations. Band diagrams of the solar cells with small w/d (a), intermediate w/d (b), and $w = d$ (c) at short circuit, at maximum power point (d–f), and at open circuit (g–i). Parameters: $u_{n,p}$= 0.1 cm^2 V^{-1} s^{-1}, ε_r = 10, d = 500 nm, E_g = 1.2 eV, N_c, N_v = 10^{19} cm^{-3}, barrier at hole selective contact for holes $\Phi_{B,a}$ = 0.05 eV, barrier at ESC for electrons $\Phi_{B,b}$ = 0.05 eV, αd = 1, $\tau_n = (5d)^2/D_n$ and Φ_{ph} = 10^{16} cm^{-1} s^{-1}. The particular parameters are (a, d, g) N_A = 10^{18} cm^{-3}, (b, e, h), N_A = 10^{16} cm^{-3}, and (c, f, i) N_A = 1 × 10^{15} cm^{-3}. (Courtesy of Pilar Lopez-Varo.)

efficiency (Nayak et al., 2011). Therefore, the active area of the device and the illumination procedure should be clearly stated. Standard procedures of characterization require avoiding the extra incident light from lateral reflection and also contributions from the material indirectly illuminated through light scattering. When reporting efficiencies, masks with (preferably) the same size as the active area of the device should be used. The calibration of the setup with a solar cell that has similar spectral characteristics to the one measured is important to avoid mistakes due to the spectral features of a given light source (Doscher et al., 2016).

In this chapter, we have emphasized the power conversion efficiency under normal incident light, which is the standard for calibration. In real applications, there are other figures of merit such as the annual energy yield that depends on the light absorption at different incident angles and indirect light absorption across the year (Reale et al., 2014).

Perovskite solar cells and many other solution-processed photovoltaic devices, at their early stage of development, demonstrated an enormous variability of efficiency when manually prepared in the lab. The dispersion was huge until robust procedures and consistent protocols of materials and device making were established. It has happened often that a lab produces a large batch of samples and picks the best one for preparing a publication. Obviously, such a device may not be considered to be state-of-art. The authors should always present histograms of the performance parameters, and average as well as best values. The stability of the reported devices under prolonged simulated sunlight should also be stated.

GENERAL REFERENCES

Simulation of organic and thin film solar cells: Fonash (1981), Schiff (2003), Kirchartz and Rau (2011), Kirchartz et al. (2012), Tress et al. (2012), and Tress (2014).

Measuring and reporting solar cells: Smestad et al. (2008), Snaith (2012), Zimmermann et al. (2014), and Christians et al. (2015).

Stabilized measurement of power conversion efficiency: Christians et al. (2015), Zimmermann et al. (2016), and Bardizza et al. (2017).

Measuring the diffusion length: Werner et al. (1993), Södergren et al. (1994), Brendel and Rau (1999), Halme et al. (2008), Jennings et al. (2010), and Pala et al. (2014).

Distribution of electrons in a DSC: Lobato and Peter (2006), Lobato et al. (2006), and Jennings and Peter (2007).

Diffusion-recombination model: Gärtner (1959), Reichman (1980), Brendel et al. (1995), and Bae et al. (2015).

Effect of built-in voltage: Rau et al. (2003), Turrión et al. (2003), and Kirchartz and Rau (2011).

Perovskite Solar Cells: Como et al. (2016) and Park et al. (2016).

REFERENCES

Ahn, N.; Son, D.-Y.; Jang, I.-H.; Kang, S. M.; Choi, M.; Park, N.-G. Highly reproducible perovskite solar cells with average efficiency of 18.3% and best efficiency of 19.7% fabricated via Lewis base adduct of lead(II) iodide. *Journal of the American Chemical Society* 2015, *137*, 8696–8699.

Ansari-Rad, M.; Bisquert, J. Insight into photon recycling in perovskite semiconductors from the concept of photon diffusion. *Physical Review Applied* 2018, *10*, 034062.

Bae, D.; Pedersen, T.; Seger, B.; Malizia, M.; Kuznetsov, A.; Hansen, O.; Chorkendorff, I.; et al. Back-illuminated Si photocathode: A combined experimental and theoretical study for photocatalytic hydrogen evolution. *Energy & Environmental Science* 2015, *8*, 650–660.

Bardizza, G.; Pavanello, D.; Galleano, R.; Sample, T.; Müllejans, H. Calibration procedure for Solar Cells exhibiting slow response and application to a dye-sensitized photovoltaic device. *Solar Energy Materials and Solar Cells* 2017, *160*, 418–424.

Barea, E. M.; Gonzalez-Pedro, V.; Ripolles-Sanchis, T.; Wu, H.-P.; Li, L.-L.; Yeh, C.-Y.; Diau, E. W.-G.; et al. Porphyrin dyes with high injection and low recombination for highly efficient mesoscopic dye-sensitized solar cells. *The Journal of Physical Chemistry C* 2011, *115*, 10898–10902.

Beaumont, J. H.; Jacobs, P. W. M. Polarization in potassium chloride crystals. *Journal of Physical Chemistry of Solids* 1967, *28*, 657.

Bisquert, J.; Garcia-Belmonte, G.; Munar, A.; Sessolo, M.; Soriano, A.; Bolink, H. J. Band unpinning and photovoltaic model for P3HT:PCBM organic bulk heterojunctions under illumination. *Chemical Physics Letters* 2008, *465*, 57–62.

Boix, P. P.; Ajuria, J.; Etxebarria, I.; Pacios, R.; Garcia-Belmonte, G.; Bisquert, J. Role of ZnO electron-selective layers in regular and inverted bulk heterojunction solar cells. *Journal of Physical Chemistry Letters* 2011, *2*, 407–411.

Brendel, R.; Hirsch, M.; Stemmer, M.; Rau, U.; Werner, J. H. Internal quantum efficiency of thin epitaxial silicon solar cells. *Applied Physics Letters* 1995, *66*, 1261–1263.

Brendel, R.; Rau, U. Effective diffusion lengths for minority carriers in solar cells as determined from internal quantum efficiency analysis. *Journal of Applied Physics* 1999, *85*, 3634–3637.

Cendula, P.; Tilley, S. D.; Gimenez, S.; Bisquert, J.; Schmid, M.; Grätzel, M.; Schumacher, J. O. Calculation of the energy band diagram of a photoelectrochemical water splitting cell. *The Journal of Physical Chemistry C* 2014, *118*, 29599–29607.

Christians, J. A.; Manser, J. S.; Kamat, P. V. Best practices in perovskite solar cell efficiency measurements. Avoiding the error of making bad cells look good. *The Journal of Physical Chemistry Letters* 2015, *6*, 852–857.

Como, E. D.; Angelis, F. D.; Snaith, H.; Walker, A. (Eds.). *Unconventional Thin Film Photovoltaics*; Royal Society of Chemistry: London, 2016.

Donolato, C. A reciprocity theorem for charge collection. *Applied Physics Letters* 1985, *46*, 270–272.

Donolato, C. An alternative proof of the generalized reciprocity theorem for charge collection. *Journal of Applied Physics* 1989, *66*, 4524–4525.

Doscher, H.; Young, J. L.; Geisz, J. F.; Turner, J. A.; Deutsch, T. G. Solar-to-hydrogen efficiency: Shining light on photoelectrochemical device performance. *Energy & Environmental Science* 2016, *9*, 74–80.

Dotan, H.; Sivula, K.; Gratzel, M.; Rothschild, A.; Warren, S. C. Probing the photoelectrochemical properties of hematite (a-Fe_2O_3) electrodes using hydrogen peroxide as a hole scavenger. *Energy & Environmental Science* 2011, *4*, 958–964.

Fonash, S. J. *Solar Cell Device Physics*; Academic Press: New York, 1981.

Gärtner, W. Depletion-layer photoeffects in semiconductors. *Physical Review* 1959, *116*, 84–87.

Gimenez, S.; Bisquert, J. *Photoelectrochemical Solar Fuel Production. From Basic Principles to Advanced Devices*; Springer: Switzerland, 2016.

Green, M. A. Do built-in fields improve solar cell performance? *Progress in Photovoltaics: Research and Applications* 2009, *17*, 57–66.

Green, M. A.; Emery, K.; Hishikawa, Y.; Warta, W.; Dunlop, E. D. Solar cell efficiency tables (version 39). *Progress in Photovoltaics: Research and Applications* 2012, *20*, 12–20.

Guerrero, A.; Marchesi, L. F.; Boix, P. P.; Ruiz-Raga, S.; Ripolles-Sanchis, T.; Garcia-Belmonte, G.; Bisquert, J. How the charge-neutrality level of interface states controls energy level alignment in cathode contacts of organic bulk-heterojunction solar cells. *ACS Nano* 2012, *6*, 3453–3460.

Halme, J.; Boschloo, G.; Hagfeldt, A.; Lund, P. Spectral characteristics of light harvesting, electron injection, and steady-state charge collection in pressed TiO_2 dye solar cells. *The Journal of Physical Chemistry C* 2008, *112*, 5623–5637.

Jaegermann, W.; Klein, A.; Mayer, T. Interface engineering of inorganic thin-film solar cells—Materials-science challenges for advanced physical concepts. *Advanced Materials* 2009, *21*, 4196–4206.

Jennings, J. R.; Li, F.; Wang, Q. Reliable determination of electron diffusion length and charge separation efficiency in dye-sensitized solar cells. *The Journal of Physical Chemistry C* 2010, *114*, 14665–14674.

Jennings, J. R.; Peter, L. M. A reappraisal of the electron diffusion length in solid-state dye-sensitized solar cells. *The Journal of Physical Chemistry C* 2007, *111*, 16100–16104.

Jeon, N. J.; Noh, J. H.; Kim, Y. C.; Yang, W. S.; Ryu, S.; Seok, S. I. Solvent engineering for high-performance inorganic-organic hybrid perovskite solar cells. *Nature Materials* 2014, *13*, 897–903.

Kirchartz, T.; Agostinelli, T.; Campoy-Quiles, M.; Gong, W.; Nelson, J. Understanding the thickness-dependent performance of organic bulk heterojunction solar cells: The influence of mobility, lifetime and space charge. *Journal of Physical Chemistry Letters* 2012, *3*, 3470–3475.

Kirchartz, T.; Bisquert, J.; Mora-Sero, I.; Garcia-Belmonte, G. Classification of solar cells according to mechanisms of charge separation and charge collection. *Physical Chemistry Chemical Physics* 2015, *17*, 4007–4014.

Kirchartz, T.; Rau, U. *Advanced Characterization Techniques for Thin Film Solar Cells*; Abou-Ras, D., Kirchartz, T., Rau, U., Eds.; Wiley: Berlin, 2011; p. 14.

Klahr, B.; Gimenez, S.; Fabregat-Santiago, F.; Hamann, T.; Bisquert, J. Water oxidation at hematite photoelectrodes: The role of surface states. *Journal of the American Chemical Society* 2012, *134*, 4294–4302.

Klahr, B. M.; Hamann, T. W. Current and voltage limiting processes in thin film hematite electrodes. *The Journal of Physical Chemistry C* 2011, *115*, 8393–8399.

Koster, L. J. A.; Smits, E. C. P.; Mihailetchi, V. D.; Blom, P. W. M. Device model for the operation of polymer/fullerene bulk heterojunction solar cells. *Physical Review B* 2005, *72*, 085205.

Li, L.-L.; Diau, E. W.-G. Porphyrin-sensitized solar cells. *Chemical Society Reviews* 2012, *42*, 291–304.

Li, X.; Bi, D.; Yi, C.; Décoppet, J.-D.; Luo, J.; Zakeeruddin, S. M.; Hagfeldt, A.; et al. A vacuum flash-assisted solution process for high-efficiency large-area perovskite solar cells. *Science* 2016. doi: 10.1126/science.aaf8060.

Lobato, K.; Peter, L. M. Direct measurement of the temperature coefficient of the electron quasi-fermi level in dye-sensitized nanocrystalline solar cells using a titanium sensor electrode. *The Journal of Physical Chemistry B* 2006, *110*, 21920–21923.

Lobato, K.; Peter, L. M.; Würfel, U. Direct measurement of the internal electron quasi-fermi level in dye sensitized solar cells using a titanium secondary electrode. *The Journal of Physical Chemistry B* 2006, *110*, 16201–16204.

Majhi, K.; Bertoluzzi, L.; Rietwyk, K. J.; Ginsburg, A.; Keller, D. A.; Lopez-Varo, P.; Anderson, A. Y.; et al. Combinatorial investigation and modelling of MoO_3 hole-selective contact in $TiO_2|Co_3O_4|MoO_3$ all-oxide solar cells. *Advanced Materials Interfaces* 2016, *3*. doi: 10.1002/admi.201500405.

Maldonado, S.; Knapp, D.; Lewis, N. S. Near-ideal photodiodes from sintered gold nanoparticle films on methyl-terminated Si(111) surfaces. *Journal of the American Chemical Society* 2008, *130*, 3300–3301.

Nayak, P. K.; Bisquert, J.; Cahen, D. Assessing possibilities and limits for solar cells. *Advanced Materials* 2011, *23*, 2870–2876.

Pala, R. A.; Leenheer, A. J.; Lichterman, M.; Atwater, H. A.; Lewis, N. S. Measurement of minority-carrier diffusion lengths using wedge-shaped semiconductor photoelectrodes. *Energy & Environmental Science* 2014, *7*, 3424–3430.

Park, N.-G.; Grätzel, M.; Miyasaka, T. (Eds.). *Organic-Inorganic Halide Perovskite Photovoltaics: From Fundamentals to Device Architectures*; Springer: Switzerland, 2016.

Rau, W.; Kron, G.; Werner, J. H. Reply to comments on "Electronic transport in dye-sensitized nanoporous TiO_2 solar cells—Comparison of electrolyte and solid-state devices." *The Journal of Physical Chemistry B* 2003, *107*, 13547.

Ravishankar, S.; Aranda, C.; Boix, P. P.; Anta, J. A.; Bisquert, J.; Garcia-Belmonte, G. Effects of frequency dependence of the external quantum efficiency of perovskite solar cells. *The Journal of Physical Chemistry Letters* 2018, *9*, 3099–3104.

Reale, A.; Cinà, L.; Malatesta, A.; De Marco, R.; Brown, T. M.; Di Carlo, A. Estimation of energy production of dye-sensitized solar cell modules for building-integrated photovoltaic applications. *Energy Technology* 2014, *2*, 531–541.

Reichman, J. The current-voltage characteristics of semiconductor-electrolyte junction photovoltaic cells. *Applied Physics Letters* 1980, *36*, 574–577.

Ritenour, A. J.; Cramer, R. C.; Levinrad, S.; Boettcher, S. W. Efficient n-GaAs photoelectrodes grown by close-spaced vapor transport from a solid source. *ACS Applied Materials & Interfaces* 2012, *4*, 69–73.

Schiff, E. A. Low-mobility solar cells: A device physics primer with applications to amorphous silicon. *Solar Energy Materials and Solar Cells* 2003, *78*, 567–595.

Smestad, G. P.; Krebs, F. C.; Lampert, C. M.; Granqvist, C.-G.; Chopra, K. L.; Mathew, X.; Takakura, H. Editorial: Reporting solar cell efficiencies in solar energy materials and solar cells. *Solar Energy Materials and Solar Cells* 2008, *92*, 371–373.

Snaith, H. J. The perils of solar cell efficiency measurements. *Nature Photonics* 2012, *6*, 337–340.

Södergren, S.; Hagfeldt, A.; Olsson, J.; Lindquist, S. E. Theoretical models for the action spectrum and the current-voltage characteristics of microporous semiconductor films in photoelectrochemical cells. *The Journal of Physical Chemistry* 1994, *98*, 5552–5556.

Tilley, S. D.; Cornuz, M.; Sivula, K.; Grätzel, M. Light-induced water splitting with hematite: Improved nanostructure and iridium oxide catalysis. *Angewandte Chemie International Edition* 2010, *49*, 6405–6408.

Tress, W. *Organic Solar Cells. Theory, Experiment, and Device Simulation*; Springer: Switzerland, 2014.

Tress, W.; Leo, K.; Riede, M. Influence of hole-transport layers and donor materials on open-circuit voltage and shape of I–V curves of organic solar Cells. *Advanced Functional Materials* 2011, *21*, 2140–2149.

Tress, W.; Leo, K.; Riede, M. Optimum mobility, contact properties, and open-circuit voltage of organic solar cells: A drift-diffusion simulation study. *Physical Review B* 2012, *85*, 155201.

Turrión, M.; Bisquert, J.; Salvador, P. Flatband potential of F:SnO_2 in a TiO_2 dye-sensitized solar cell: An interference reflection study. *The Journal of Physical Chemistry B* 2003, *107*, 9397–9403.

Werner, J. H.; Kolodinski, S.; Rau, U.; Arch, J. K.; Bauser, E. Silicon solar cell of 16.8 μm thickness and 14.7% efficiency. *Applied Physics Letters* 1993, *62*, 2998–3000.

Zimmermann, E.; Ehrenreich, P.; Pfadler, T.; Dorman, J. A.; Weickert, J.; Schmidt-Mende, L. Erroneous efficiency reports harm organic solar cell research. *Nature Photonics* 2014, *8*, 669–672.

Zimmermann, E.; Wong, K. K.; Müller, M.; Hu, H.; Ehrenreich, P.; Kohlstädt, M.; Würfel, U. Characterization of perovskite solar cells: Toward a reliable measurement protocol. *APL Mater* 2016, *4*, 091901.

27 Spectral Harvesting and Photoelectrochemical Conversion

The main feature that hinders high efficiency of energy collection in solar cells is the poor match of the optical absorption of semiconductors to the solar spectrum. These effects cause large energy losses by thermalization of carriers and sub-band-gap transparency. Here we describe different schemes to overcome the limitations of solar energy harvesting. We first discuss the conversion of the spectral characteristics by use of luminescent layers and molecules that perform different processes of thermalization, splitting, and fusion of photons to significantly change the wavelength of the emission. These processes result in the spectral conversion methods of down-shifting using fluorescent collectors (FCs), and upconversion of long-wavelength photons via excited state absorption, energy transfer, or triplet–triplet annihilation. We discuss the tandem solar cell that combines different semiconductor layers of complementary band gaps to absorb different portions of the solar spectrum for more efficient photon utilization. Finally, we review the main methods of conversion of sunlight to chemical fuel by application of photoelectrochemical cells and photovoltaic cells combined with catalyst layers.

27.1 CONVERSION OF PHOTON FREQUENCIES FOR SOLAR ENERGY HARVESTING

From the studies of the physical limitations to photovoltaic energy conversion presented in Chapters 22 through 24, it emerges that one central condition for efficient solar energy harvesting is the spectral match of the semiconductor optical properties with the solar radiation. As the latter is distributed over a very wide range of energies, any single semiconductor device is doomed to substantial losses, either by reduction of the photon energy or by the failure to collect the photons in the transparent range of the semiconductor. In order to improve the harvesting of the energy of the solar photons, we can either transform the incoming spectrum to adapt it to the semiconductor features, as discussed in the following paragraphs, or we can combine several semiconductors toward a better spectral match. The second strategy consists of the tandem solar cell, which will be discussed in Section 27.2.

Solar photon conversion schemes consist of luminescent layers that produce a spectral change in a relevant part of the incoming spectrum. Three different luminescent processes can be used to increase the efficiency of single band gap solar cells: *down-shifting*, *quantum-cutting*, and *upconversion*. The conversion processes are indicated in Figure 27.1. The potential gain of these photon conversion methods is summarized in Figure 27.2, which shows the part of the solar spectrum absorbed by a typical silicon-based solar cell and the spectral regions that can be utilized through quantum-cutting and upconversion processes (Huang et al., 2013).

Down-shifting is a single photon conversion process, which consists of the transformation of a high-energy photon to a low-energy photon. Downconversion is usually investigated in the application of luminescent solar concentrators (LSC), also called fluorescent collectors (FCs). The aim of the collector is to capture light over a large area, which is then delivered to a smaller area solar cell reducing the cost of the overall energy-conversion system. FCs are also useful to complement solar cells that show poor spectral response to short-wavelength electromagnetic radiation.

The FC consists of a transparent plastic doped with luminescent species, normally either organic dye molecules or inorganic quantum dots (Erickson et al., 2014). The incident sunlight is absorbed by the dyes that perform isotropical reemission with high-quantum efficiency. By total internal reflection, a large part of the light is transmitted to the side of the plastic where it is collected by a solar cell, as shown in Figure 27.3. The emission is tuned to be just above the band gap of the solar cell, as indicated in Figure 27.4, to ensure near-unity conversion efficiency of the radiation that is trapped and waveguided to the edges of the slab. An optimum choice of the Stokes' shift of the dye is an important factor for solar energy harvesting. If the Stokes' shift is small, then a large part of the emitted light is reabsorbed producing photon transport losses. If, on the other hand, the Stokes' shift is large, then a substantial part of the photons in the spectral range between the absorption and fluorescence maxima are not collected.

The FC provides a practical example of radiation at nonzero chemical potential that we discussed in

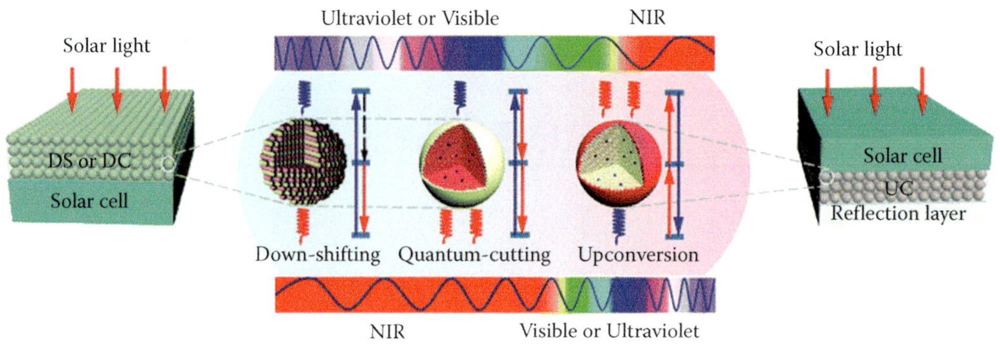

FIGURE 27.1 Spectral conversion design for PV applications involving down-shifting, quantum-cutting, and upconversion luminescent materials. In a typical down-shifting process, upon excitation with a high-energy photon, nonradiative relaxation takes place followed by radiative relaxation, thereby resulting in the emission of a lower-energy photon. In contrast, two-step radiative relaxation occurs in the quantum-cutting process upon excitation with a high-energy photon, leading to the emission of two (or more) lower-energy photons. The upconversion process can convert two (or more) incident low-energy photons into a single higher-energy photon. Note that the down-shifting and quantum-cutting materials are generally placed on the front surface of a monofacial solar cell, allowing the downconverted photons to be absorbed by the solar cell. The upconversion material is typically placed in between a bifacial solar cell and a light-reflection layer to harvest the sub-band gap spectrum of sunlight. (Reproduced with permission from Huang, X. et al. *Chemical Society Reviews* 2013, 42, 173–201.)

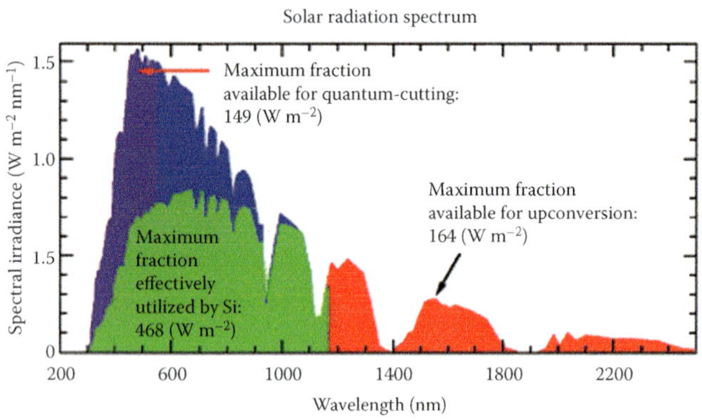

FIGURE 27.2 *AM*1.5*G* spectrum showing the fraction (highlighted in *green*) absorbed by a typical silicon-based PV cell and the spectral regions that can be utilized through quantum-cutting and upconversion processes (highlighted in *purple* and *red*, respectively). (Reproduced with permission from Huang, X. et al. *Chemical Society Reviews* 2013, 42, 173–201.)

Section 22.4. The excitation by the surrounding radiation promotes the dye to the excited state and produces a separation of the Fermi levels in the dye molecules, which has the value μ_{dye} when the radiation and the molecules come to equilibrium by absorption and reemission. The spectral flux emitted by the FC achieves the form of Equation 22.32, corresponding to a flux of photons at chemical potential $\mu_{flux} = \mu_{dye} > 0$.

Figure 27.5 shows the scheme of an optimized FC used to funnel the collected light to a solar cell, which covers the surface of the FC. The window open to light capture poses a problem in that light generated in the escape cone will be lost, decreasing the efficiency of effective light harvesting for the production of electricity in the solar cell. This is prevented with the use of a photonic crystal that suppresses reemission, as shown in Figure 27.5. Note the ideal characteristics of absorption and emission used in this model, as in Figure 22.6. A real example of radiation produced in the FC is shown in Figure 27.6. The dye does not have a sharp absorption edge as in the model of Figure 27.5, but nonetheless at energies larger than the emission peak Equation 22.35 is well satisfied. The radiation in Figure 27.6 is characterized by a chemical potential $\mu_{flux} \approx 1.7$ eV > 0.

Quantum-cutting is a form of downconversion in which one incident high-energy photon is split into two or more low-energy photons with a quantum yield larger than 100%. Long-wavelength photons of the solar spectrum are not suitable for direct photoelectric conversion since they are transmitted through usual semiconductors, as shown in Figure 27.2. Upconversion involves the fusion of photons to significantly change the longer wavelength photons of the sunlight spectrum toward shorter wavelengths. In the scheme of photon upconversion or anti-Stokes' shift, two photons of low energy are combined to produce one photon of higher energy. Upconversion is a considerable challenge as it requires the simultaneous or sequential absorption of two or more photons with lower energy than that of the emitted photon.

There are two main methods for the combination of two excitations. Excited state absorption (ESA) consists of sequential absorption of two photons in a single ion

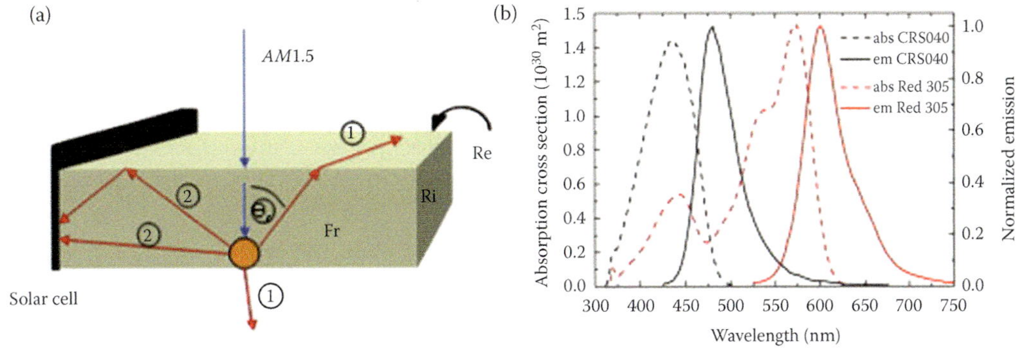

FIGURE 27.3 (a) Schematic three-dimensional view of a luminescent concentrator. *AM*1.5 light is incident on the top. The light is absorbed by a luminescent particle, and its luminescence is randomly emitted. Part of the emission falls within the escape cone and is lost from the luminescent concentrator at the surfaces (1). The other part (2) is guided to the solar cell by total internal reflection. (b) Absorption cross section and normalized emission spectra of two fluorescent dyes, illustrating Stokes' shift. (Reproduced with permission from van Sark, W. G. J. H. M. *Renewable Energy* 2013, 49, 207–210.)

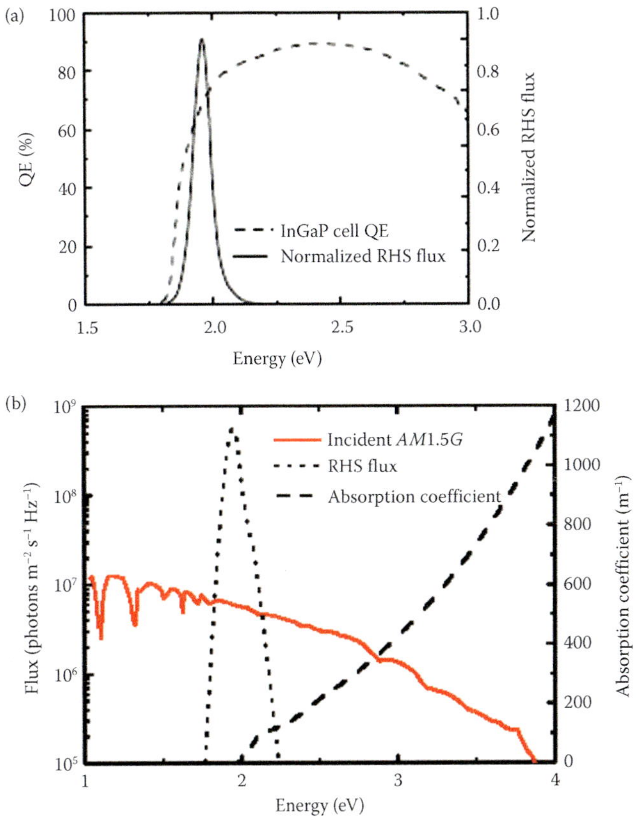

FIGURE 27.4 (a) Quantum efficiency (spectral response) of a GaInP cell used in modeling the idealized LSC together with the modeled luminescence escaping the right-hand surface (RHS) of the LSC that would be coupled into the cell. (b) Absorption of the LSC material used in the calculations for the idealized system together with the flux incident on the top surface and the predicted concentrated average luminescent flux escaping the right-hand surface of the idealized LSC. (Reproduced with permission from Van Sark, W. G. et al. *Optics Express* 2008, 16, 21773–21792.)

using an intermediate level and then causing the promotion of the excited state to a higher level, see Figure 27.7c. In energy transfer upconversion (ETU), two neighboring ions are excited by pump photons. Then, one ion passes the excitation by energy transfer to the neighbor, which becomes promoted to a higher-excited state (Figure 27.7d). These upconversion processes are possible with lanthanide ions that were mentioned in Section 18.2 as phosphor materials. These ions possess metastable intermediate levels that are able to store the first excitation until the arrival of a second excitation, which provides sufficient energy for fluorescence at shorter wavelength. Actual energy transfer schemes for the couple Er^{3+} – Yb^{3+} are shown in Figure 27.8. Here Er^{3+} emits in the green and red after upconversion (de Wild et al., 2011).

Another approach to upconversion of low-energy photons is based on triplet–triplet annihilation (TTA). In this method, the triplet state of the sensitizer is excited by singlet excitation, Figure 27.9, or metal-to-ligand charge transfer (MLCT) excitation (Figure 27.10). The excitation and emission wavelengths of TTA upconversion can be selected by independent choice of the triplet sensitizer and triplet acceptor. Heavy metal containing porphyrins and phthalocyanines as well as metallated sensitizers are the compounds most studied as sensitizers since the presence of the π-conjugated aromatic rings place their absorption and emission maxima in the red and NIR (near-infrared) region of the spectrum (Singh-Rachford and Castellano, 2010). The first process in the chain after light absorption is the intersystem crossing (ISC), which produces the metastable triplet state in the sensitizer. The excitation is passed to the annihilator by triplet–triplet energy transfer. The efficient ISC within the sensitizer molecules ensures a large population of the sensitizer triplet level

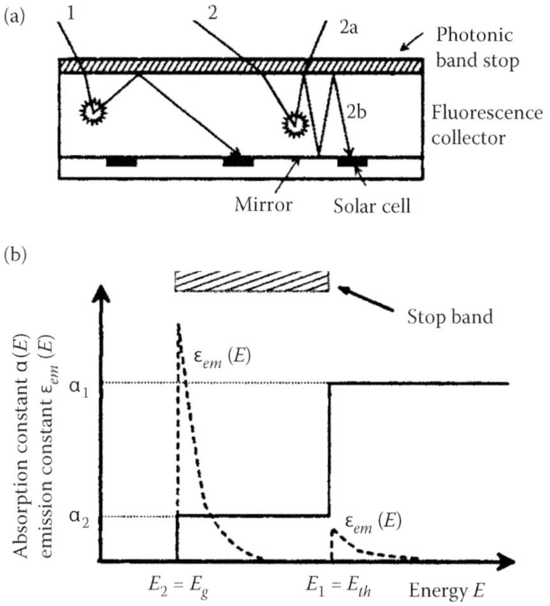

FIGURE 27.5 (a) Schematic drawing of an FC with solar cells at its bottom. The fluorescent layer absorbs incoming photons (1) and emits them at lower photon energy. Without photonic band stop some of these photons leave the collector (2a). The photonic band reflects these photons back into the collector (2b). (b) Spectral dependence of the absorption and emission coefficients, α and e, of the ideal fluorescent dye with $\alpha = \alpha_2$ for photon energies, E, in the range $E_2 \leq E \leq E_1$ and $\alpha = \alpha_1$ for $E_1 \leq E$. The photonic band has an omnidirectional reflectance $R = 1$ for $E_2 \leq E \leq E_1$. (Reproduced with permission from Rau, U.; Einsele, F.; Glaeser, G. C. *Applied Physics Letters* 2005, 87, 171101.)

after single photon absorption. On the other hand, the depopulation of the excited emitter triplet states via phosphorescence is impeded by the very weak ISC of the emitter molecules. The created triplet population of the emitter is, therefore, preserved for the process of TTA. Then, two triplets in two sensitizers recombine to form one singlet, which emits by fluorescence emission that is blueshifted to shorter wavelengths with respect to the incoming light.

27.2 TANDEM SOLAR CELLS

The tandem solar cell is another solution to the problem of optimal match of the light absorber system to the spectral characteristics of sunlight. It consists of a stack of different semiconductor absorber layers with complementary spectral characteristics that are suitably connected electrically to combine their energy production rates. A device that uses a single absorber is termed a single-junction solar cell, and then one can use two junctions, three junctions, and so on. The two-junction device is composed of a wide band gap material as top absorber that produces a high voltage, followed by a smaller band gap bottom absorber material that utilizes the photons that pass through the top absorber, see Figure 27.11. The power output gain is mainly obtained by a reduction of the thermal losses, as the more energetic photons in the solar spectrum are thermalized in the high band gap material and contribute a higher voltage than by thermalization in a lower band gap material. An example of the power output gain for a perovskite-on-silicon stack is shown in Figure 27.12. The theoretical reduction of energy losses is better observed in Figure 27.13, cf. Figure 24.9.

The simplest possibility of a tandem device is a mechanically integrated device also called a four-terminal (4T) device, in which individual cells are operated separately by independent electrical contacts, Figure 27.14a. In the monolithically integrated or two-terminal (2T) device shown in Figure 27.14b, two solar cells are connected in series so that the current must be the same and the voltages of the individual cells are added, as in the scheme of Figure 27.11. In the 2T device, it is required to build a stack of two, very high-efficiency solar cells including an intermediate, highly transparent recombination contact that aligns the Fermi levels. Figure 27.15 shows the use of organic polymers of complementary absorption ranges. When using inorganic semiconductors, the short-wavelength absorption region of the bottom cell is obscured by the wide band gap top cell. The advantage of converting the photons to a larger voltage at the front cell must compensate the darkening of the bottom cell, Figure 27.12b. The optical management of the incoming photons is crucial for an effective operation of the device, as indicated by the improved charge collection obtained with the use of an antireflective coating in Figure 1.11c. The technological aspects of the fabrication of these cells are rather demanding (Bush et al., 2017) due to the need for matching the spectral range with thin film layers, avoiding reflection losses, constructing intermediate recombination layers, and extracting the photocurrent with the appropriate contacts.

By combining absorbers of approximately 1 and 1.6 eV, an optimal PCE of 44%–45% can be obtained under SQ maximal energy-conversion conditions, as shown in Figure 27.16 (Henry, 1980). The morphologic and photovoltaic characteristics of a perovskite-on-silicon tandem cell of 25% PCE are shown in Figures 27.17 and 27.18. A large scale prototype of the perovskite-on-silicon cell is shown in Figure 27.19. The features of a four-junction record device achieving 44% PCE under concentrated sunlight are shown in Figure 27.20.

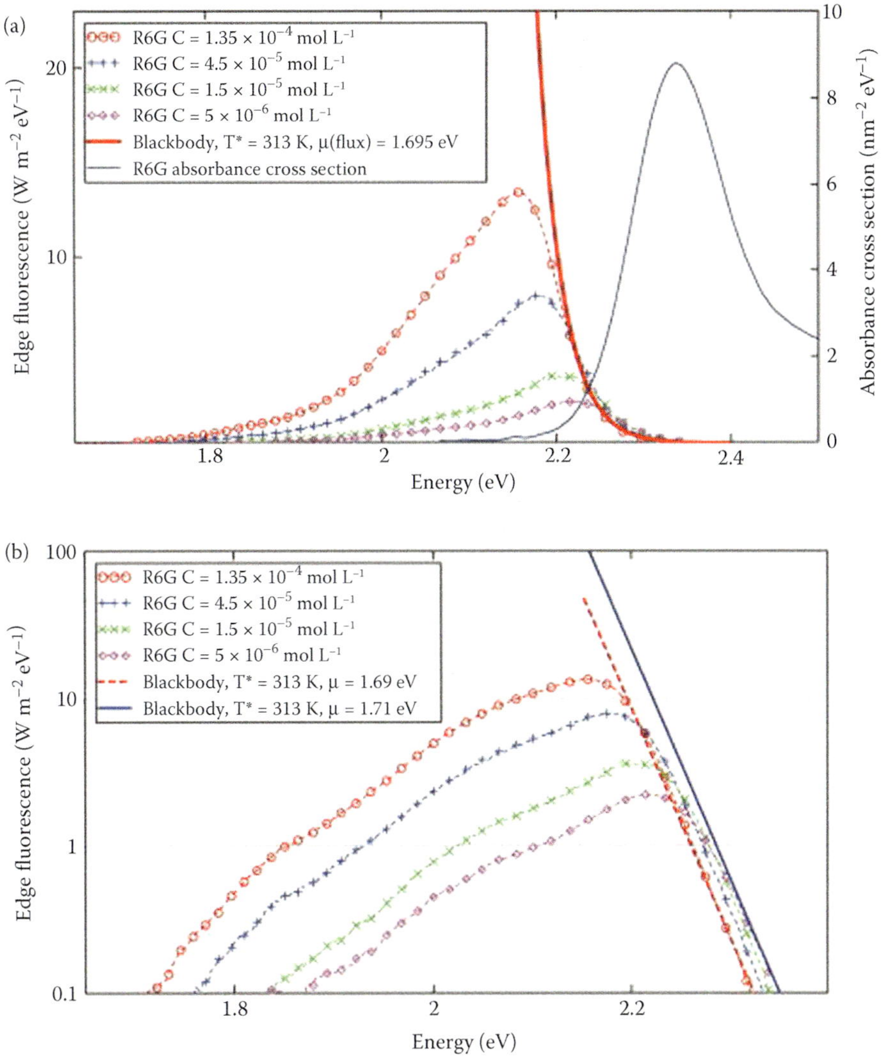

FIGURE 27.6 (a) Edge fluorescence spectra of Rhodamine 6G (R6G) compared with a blackbody function at $T = 313$ K and $\mu_{flux} = 1.695$ eV. (b) Log plot of (a), showing the fit lines to Equation 22.35. The numbers in the legend give the dye concentration in the solvent. (From Meyer, T. J. J.; Markvart, T. *Journal of Applied Physics* 2009, 105, 063110.)

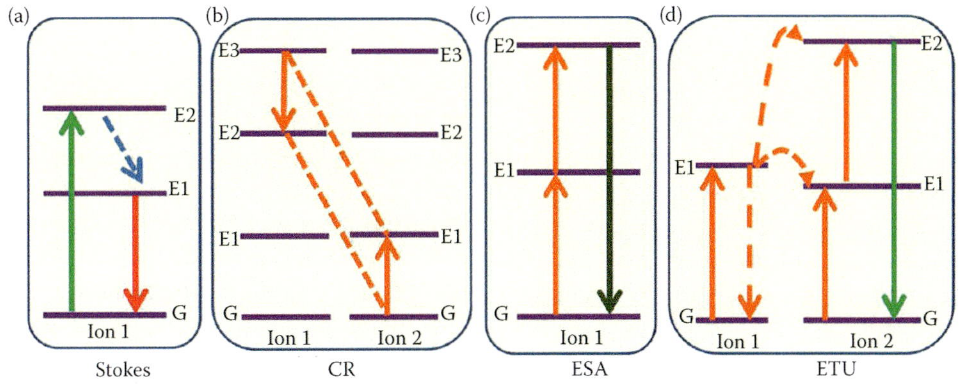

FIGURE 27.7 Mechanisms of (a) downconversion, (b) cross relaxation, (c) excited state absorption, and (d) energy transfer upconversion. (Reproduced with permission from Chen, G. et al. *Accounts of Chemical Research* 2012, 46, 1474–1486.)

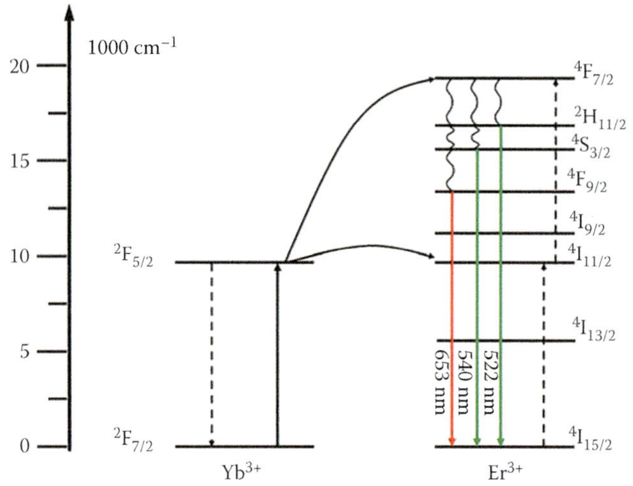

FIGURE 27.8 Schematic energy level scheme for the Yb/Er couple. The Yb_{2+} ion absorbs around 980 nm and transfers the energy from the $^2F_{5/2}$ level to the $^4I_{11/2}$ level of Er^{3+}. Subsequent energy transfer from a second excited Yb^{3+} ion to Er^{3+} ($^4I_{11/2}$) excites Er^{3+} ion to the $^4F_{7/2}$ excited state. After multiphonon relaxation to the lower lying $^4S_{3/2}$ and $^2F_{9/2}$ states, green and red emissions are observed, as indicated. (Reproduced with permission from de Wild, J. et al. *Energy & Environmental Science* 2011, 4, 4835–4848.)

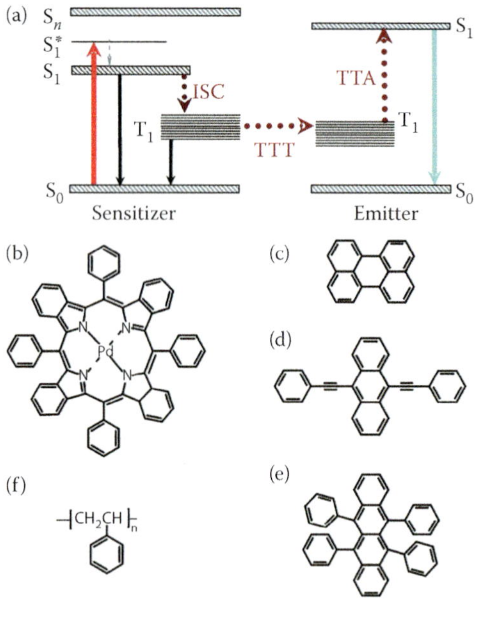

FIGURE 27.9 (a) Energetic scheme of the TTA-supported upconversion process. The structures of sensitizer PdPh4TBP (b), and the emitters: perylene (c), BPEA (d), rubrene (e), and the matrix-styrene oligomers (f). (Reprinted with permission from Miteva, T. et al. *New Journal of Physics* 2008, 10, 103002.)

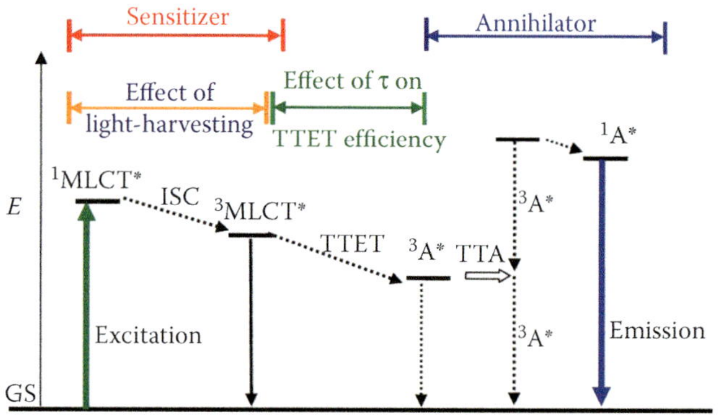

FIGURE 27.10 Qualitative diagram illustrating the sensitized TTA upconversion process between triplet sensitizer and acceptor (annihilator/emitter). The effect of the light-harvesting ability and the excited state lifetime of the sensitizer on the efficiency of the TTA upconversion is also shown. *E* is energy. GS is ground state (S_0). ^{3}MLCT* is the metal-to-ligand-charge-transfer triplet-excited state. TTET is triplet–triplet energy transfer. ^{3}A* is the triplet-excited state of annihilator. ^{1}A* is the singlet-excited state of annihilator. (Reproduced with permission from Zhao, J. et al. *RSC Advances* 2011, 1, 937–950.)

27.3 SOLAR FUEL GENERATION

The generation of fuels with semiconductor materials offers a versatile strategy to efficiently capture and store the solar energy incident on the Earth's crust. One of the most interesting approaches involves the utilization of solar photons to realize the reduction of water to H_2 or CO_2 to carbon-based molecules. In order to efficiently carry out these processes, an energy-conversion device based on visible light-absorbing semiconductors that will capture solar photons and use them to carry out the required electrochemical reactions must be developed, for example, the oxygen evolution reaction (OER) for water oxidation and hydrogen evolution reaction (HER) for H^+ reduction to gas molecules. This topic is treated in a monograph by Gimenez and Bisquert (2016).

There are aspects of contrast in the application of semiconductor light absorbers for either photovoltaic or solar fuel conversion devices. For a photovoltaic material, a larger band gap increases the open-circuit voltage, but decreases the absorption range and hence the photocurrent, as discussed in Section 22.1. The benchmark method to calculate the optimal semiconductor property for maximal electrical power output is the SQ approach that provides a value of about 1.1 eV, as shown in Section 25.4. However, a solar fuel generator must deliver a voltage to exceed the difference of free energies of the proposed reaction scheme, including the overvoltage required by the surface catalysts for the specific redox reactions, as shown in Figures 6.29 and 27.21. For water splitting resulting in hydrogen production, a potential of 1.23 V is needed thermodynamically, Figure 3.11, plus 0.2 V for each overvoltage, which adds up to about 1.8 V depending on the catalyst quality. This voltage must occur at the operating point of the device

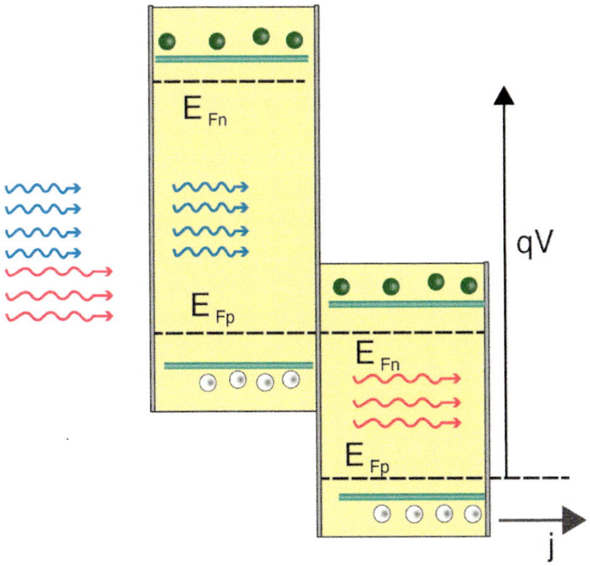

FIGURE 27.11 Scheme of a monolithically integrated tandem solar cell. The first large band gap material absorbs short-wavelength photons in the solar spectrum, while the long-wavelength photons are not absorbed. These are collected in a second absorber layer consisting of a low band gap material.

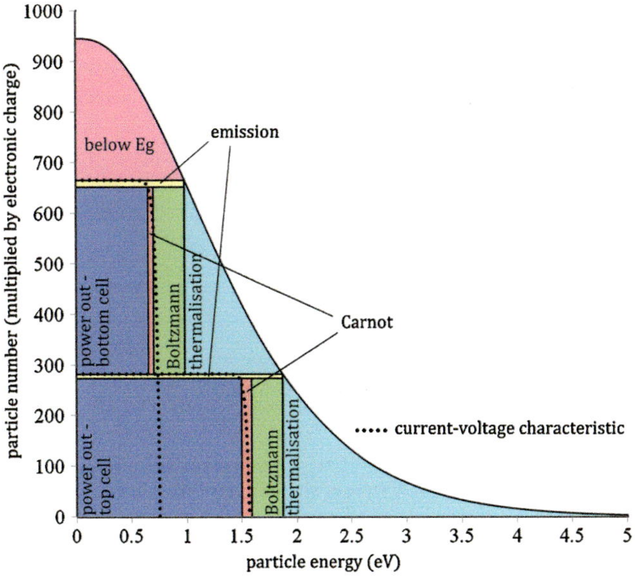

FIGURE 27.13 Intrinsic losses occurring in an unconstrained double junction device with optimal band gaps (0.98 and 1.87 eV) under 1 sun illumination. The Carnot, Boltzmann, and thermalization losses reduce the optimal-operating voltage in each absorber. (Reproduced with permission from Hirst, L. C. et al. *Progress in Photovoltaics: Research and Applications* 2011, 19, 28–93.)

so that, even with an excellent fill factor, the photovoltage should be larger.

The second essential requirement that marks a contrast with electricity-producing photovoltaics is the necessary contact of the system with the active electrolyte in which effective catalysts must carry out the desired reactions. This feature poses the need for stability at the reaction site. This is a particularly critical aspect that determines the chosen approach among a range of alternatives that we discuss in the following (Guerrero and Bisquert, 2017).

The photoelectrochemical cell (PEC), already discussed in Section 6.8 and shown in Figure 27.21a, is among the simplest arrangements to produce the water-splitting reaction and generation of hydrogen as fuel. Here, a semiconductor electrode realizes the functions of light absorption, charge separation in the space charge region, and catalytic OER. The complementary reaction occurs at a counterelectrode. Suitable semiconductor materials must satisfy stringent conditions in terms of light absorption in the visible range, adequate alignment

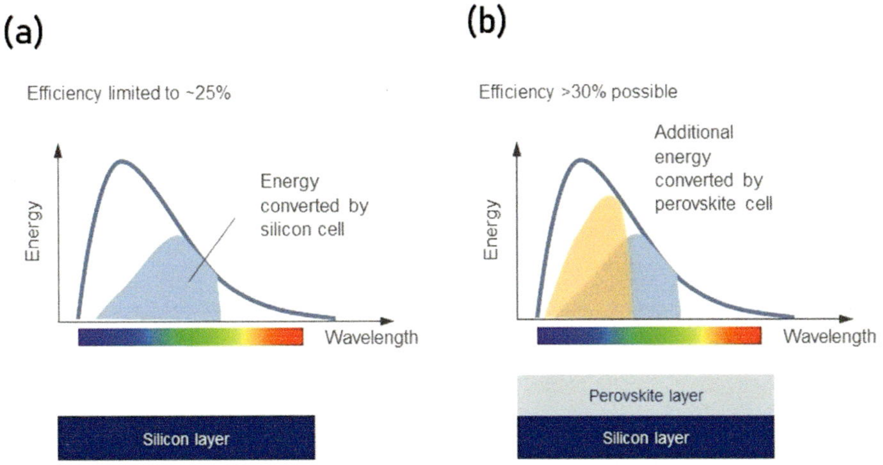

FIGURE 27.12 The comparison of solar energy converted into electricity with a conventional silicon solar cell (a) and a perovskite-on-silicon tandem solar cell (b). (Courtesy of Chris Case from Oxford PV™.)

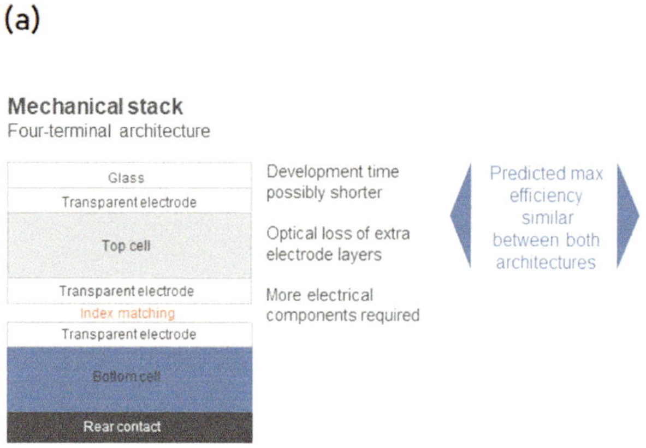

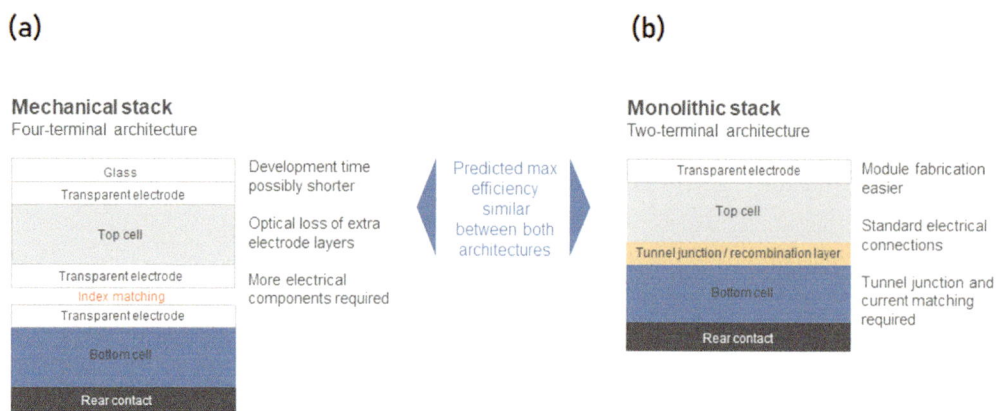

FIGURE 27.14 Two-terminal and four-terminal architecture perovskite-on-silicon tandem solar cells. (Courtesy of Chris Case from Oxford PV™.)

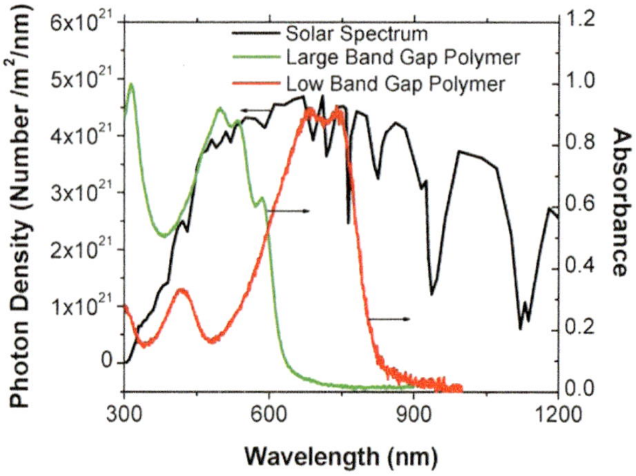

FIGURE 27.15 Complementary absorbing organic materials: absorbance of P3HT, a large band gap polymer *(green)* and PSBTBT (Poly[(4,4′-bis(2-ethylhexyl)dithieno[3,2-b:2′,3′-d]silole–,6-diyl-alt-(2,1,3- benzothiadiazole–,7-diyl]), a low band gap polymer *(red)*, compared to the solar spectrum. (Reproduced with permission from Sista, S. et al. *Energy & Environmental Science* 2011, 4, 160–620.)

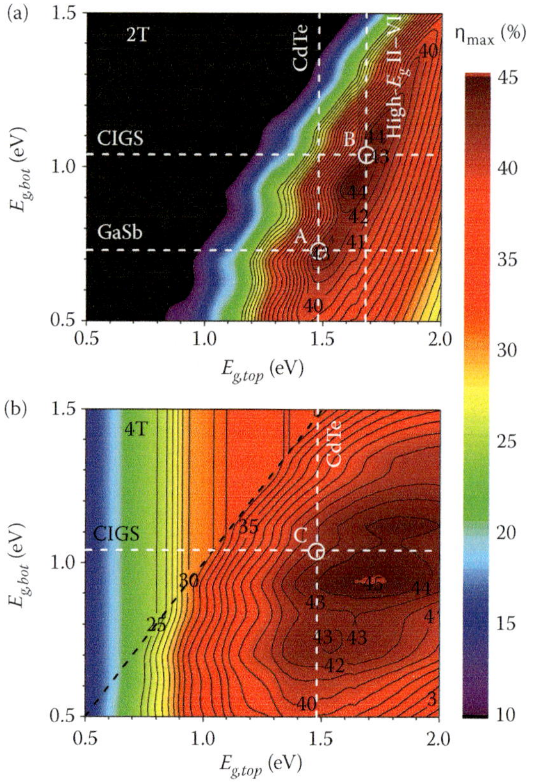

FIGURE 27.16 Two-junction radiatively limited efficiency for (a) series-connected monolithic 2T tandem and (b) mechanically stacked 4T tandem solar cells. In the 4T architecture, for $E_{g,top} < E_{g,bot}$, there is no contribution from the bottom cell. The band gap of CdTe, GaSb, CIGS, and high-E_g II–VI are marked with white dashed lines while different architectures A, B, and C are marked using the white circle marks. (Reproduced with permission from Mailoa, J. P. et al. *Energy & Environmental Science* 2016, 9, 264–653.)

of band edges with the relevant redox potentials, overall efficiency, cost, and stability under operating conditions. The main visible-light-driven junction water-splitting photo(electro)catalysts reported include $BiVO_4$, Fe_2O_3, Cu_2O, and C_3N_4, but they show important shortcomings, which result in relatively low efficiencies. For example, oxide materials such as WO_3, $BiVO_4$, or Fe_2O_3 are suitable for OER generation but cannot produce hydrogen on the surface even after absorbing visible light because their conduction band edges are more positive (in the electrochemical scale; see Figure 2.7) than the hydrogen evolution potential. Furthermore, traditional oxide materials for water splitting based on a surface SB show very poor fill factor characteristics due to low mobilities and large recombination close to flatband conditions; see Figure 26.16. Alternatively, chalcogenides containing CuI ions, such as $CuGaSe_2$, Cu_2ZnSnS_4, and Cu(Ga,In)$(S,Se)_2$, are suitable for HER but not for water oxidation

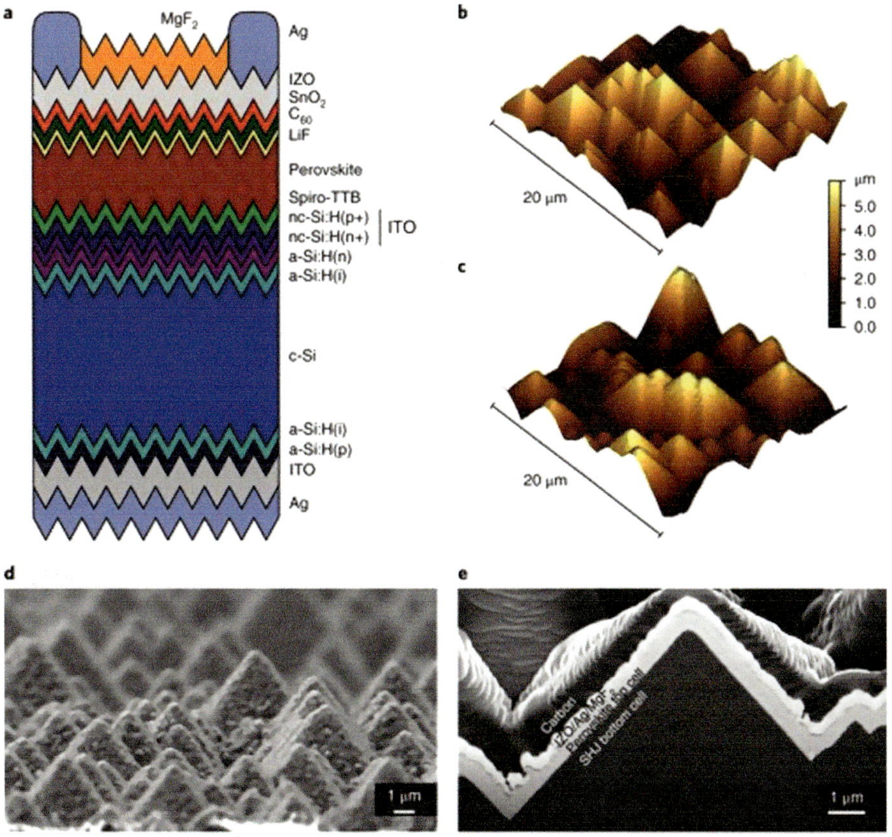

FIGURE 27.17 Cell design and microstructure of the perovskite top cell on a textured silicon bottom cell. a Schematic view of a fully textured monolithic perovskite/silicon tandem. AFM surface morphology 3D views of b, bare c-Si pyramids and c, c-Si pyramids covered with the perovskite layer. d,e, Secondary electron SEM image of the perovskite layer (d) and a cross-section of the full perovskite top cell deposited on the SHJ bottom cell (e). (Reproduced with permission from Sahli, F.; Werner, J.; Kamino, B. A.; Bräuninger, M.; Monnard, R.; Paviet-Salomon, B.; Barraud, L.; Ding, L.; Diaz Leon, J. J.; Sacchetto, D.; Cattaneo, G.; Despeisse, M.; Boccard, M.; Nicolay, S.; Jeangros, Q.; Niesen, B.; Ballif, C. Fully textured monolithic perovskite/silicon tandem solar cells with 25.2% power conversion efficiency, *Nat. Mater.* **2018**, *17*, 820-826.

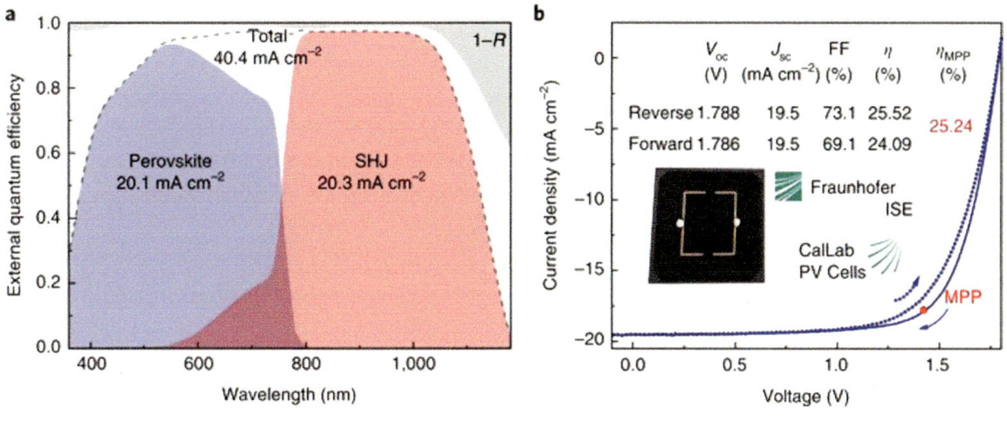

FIGURE 27.18 a) EQE spectra of a current-matched fully textured monolithic perovskite/SHJ tandem cell featuring a 1.6 eV perovskite absorber with a thickness of 440 nm alongside the 1−R curve, both excluding losses due to the front side metal grid. b) Corresponding certified j–V data (1.42-cm^2 aperture area, device shown in the inset), which was measured with a scan rate of 100 mV s^{-1}. (Reproduced with permission from Sahli, F.; Werner, J.; Kamino, B. A.; Bräuninger, M.; Monnard, R.; Paviet-Salomon, B.; Barraud, L.; Ding, L.; Diaz Leon, J. J.; Sacchetto, D.; Cattaneo, G.; Despeisse, M.; Boccard, M.; Nicolay, S.; Jeangros, Q.; Niesen, B.; Ballif, C. Fully textured monolithic perovskite/silicon tandem solar cells with 25.2% power conversion efficiency, *Nat. Mater.* **2018**, *17*, 820-826.)

FIGURE 27.19 Oxford PV™ full sized, two-terminal perovskite-on-silicon tandem solar cell. (Courtesy of Oxford PV™.)

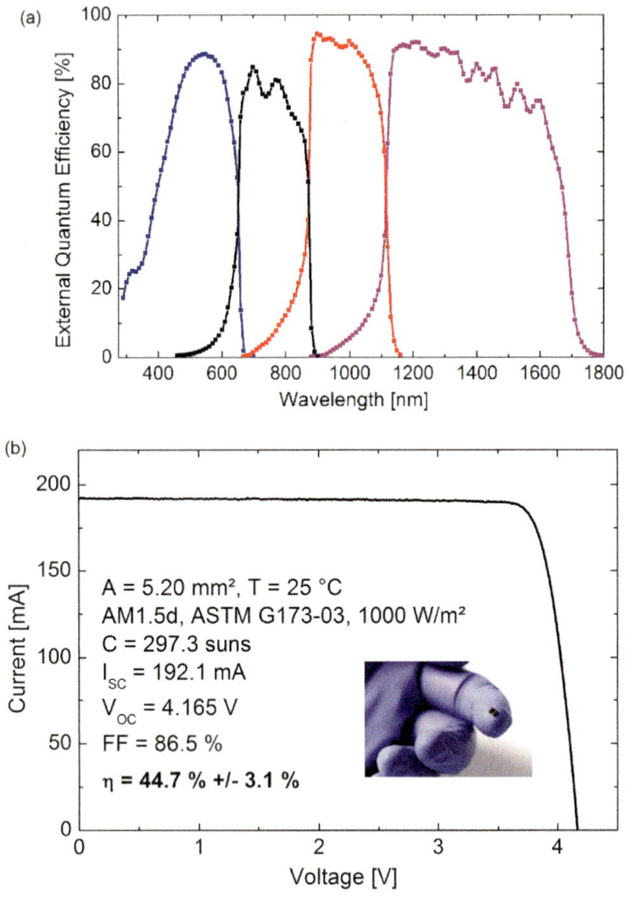

FIGURE 27.20 (a) EQE of the four-junction solar cell prepared jointly by Fraunhofer Institute for Solar Energy Systems ISE, Soitec, CEA-Leti and the Helmholtz Center Berlin. (b) Current density-voltage characteristic under AM1.5d ASTM G17-3 spectrum at a concentration of 297 suns. The inset shows an image of the cell.

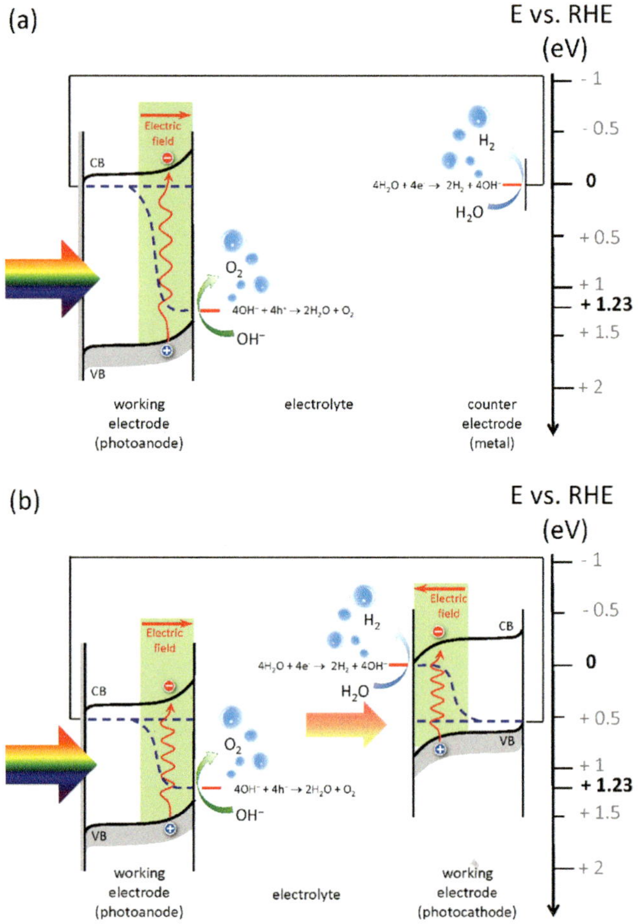

FIGURE 27.21 Schematic illustration of (a) a single absorber PEC with a photoanode and a metal cathode, and (b) a dual absorber system consisting of a photoanode and a photocathode in a tandem configuration. *Blue dashed lines* represent Fermi levels under illumination. CB and VB indicate the conduction and valence bands, respectively. (Reproduced with permission from Smith, W. A. et al. *Energy & Environmental Science* 2015, 8, 2851–2862.)

due to photocorrosion and an insufficient band gap. For this reason, in order to accomplish water splitting using these oxides or chalcogenides, a combination of semiconducting materials and an electric power supply are needed, similar to the tandem solar cell, in PEC water-splitting arrangement, shown in Figure 27.21b. A tandem configuration of semiconductor–electrolyte junctions using the earth's abundant elements that realize the complementary redox reactions is a suitable option for the competitive conversion of water to hydrogen (competing with US\$2–3 kg^{-1} for the steam reforming of natural gas) (Prévot and Sivula, 2013). Optimized nanostructured catalysts are often deposited in order to promote the desired reactions at the surface.

In order to relax some of these highly demanding material specifications, other configurations of solar fuel

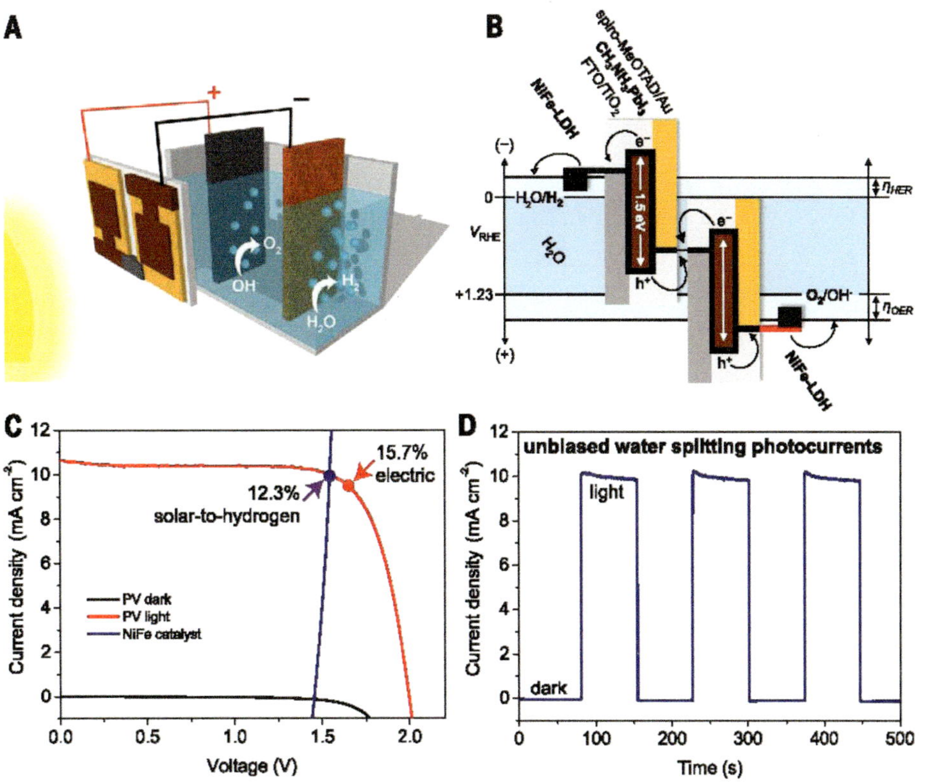

FIGURE 27.22 (a) Schematic diagram of the water-splitting device. (b) A generalized energy schematic of the perovskite tandem cell for water splitting. (c) *j–V* curves of the perovskite tandem cell under dark and simulated *AM*1.5*G* 100 mW cm^{-2} illumination, and the NiFe/Ni foam electrodes in a two-electrode configuration. The illuminated surface area of the perovskite cell was 0.318 cm^2 and the catalyst electrode areas (geometric) were ~5 cm^2 each. (d) Current density–time curve of the integrated water-splitting device without external bias under chopped simulated *AM*1.5*G* 100 mW cm^{-2} illumination. (Reproduced with permission from Luo, J. et al. *Science* 2014, 345, 1593–1596.)

converters have been developed. The photovoltaic part can be placed outside the electrolytic solution while the catalytic fragment will be in direct contact with water, both portions being connected by a wire. By using this approach, the photovoltaic part does not need to be water resistant and the catalytic fragment of the system does not need to offer good optical properties, that is, fulfill all the band gap requirements. In Figure 27.22, we show the use of a tandem solar cell configuration achieving stable photocurrents of ≈10 mA cm^{-2} for the production of H_2/O_2 (Luo et al., 2014) with solar-to-hydrogen (STH) efficiency of 12.3%. In this setup, two hybrid perovskite ($CH_3NH_3PbI_3$) solar cells are connected in series in a tandem configuration. An electrolyzer is connected by an external wire to an electrode containing earth-abundant catalysts such as nickel–iron-layered double hydroxide NiFe-LDH, which carries out the desired electrochemical reaction. Similarly, monolithic tandem solar cells based on lead halide perovskites and $BiVO_4$ have also been described with photocurrents of ≈2 mA cm^{-2} for the production of H_2/O_2 (Chen et al., 2015). The main disadvantage of this approach is the high complexity of the devices, which obviously will pose a negative impact by increasing the final cost of the device.

A third approach aims at converting a photovoltaic device into a PEC by using the protecting layers to avoid the chemical reaction of the liquid solution with the materials used in the photovoltaic device. Khaselev and Turner (1998) converted GaAs/GaInAs tandem photovoltaic devices into PECs for the production O_2/H_2 in the absence of protecting layers. Although highly efficient, this type of configuration lacks long-term stability and the photocurrent decreased by 15% over the initial 20 h. Similarly, photovoltaic devices based on CIGS have been turned into PEC and stability in the liquid solution has been gained by the use of a Ti protective layer (Kumagai et al., 2015). Indeed, it was observed that a configuration such as CIGS/Ti/Mo/Pt could improve the stability over a period of 10 days with gradual decrease in efficiency. With current low-voltage photovoltaic technologies such as silicon, a multiple tandem is required to attain the required voltage (Nocera, 2012; Cox et al., 2014). STH efficiency of 10%–14% has been demonstrated with extremely expensive III–V PV materials (May et al.,

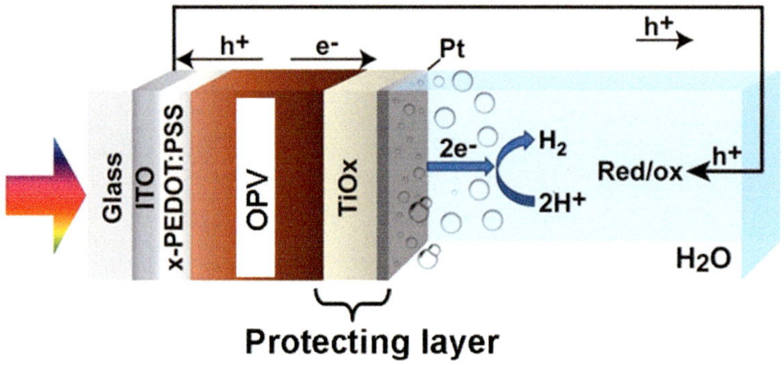

FIGURE 27.23 Organic PEC. (Reproduced with permission from Haro, M. et al. *The Journal of Physical Chemistry C* 2015, 119, 6488–6494.)

2015; Verlage et al., 2015). Similarly, the conversion of organic photovoltaic devices into PECs has been conferred by the use of protective TiO_2 layers that enhance protection against the aqueous solution and electrically communicates the organic layer with a Pt catalyst, see Figure 27.23 (Haro et al., 2015).

GENERAL REFERENCES

Luminescent solar concentrators: Yablonovitch (1980), Rau et al. (2005), Markvart et al. (2012), van Sark (2013), and Klimov et al. (2016).

Upconversion: Trupke et al. (2002), Auzel (2003), and Huang et al. (2013).

Triplet–triplet annihilation: Miteva et al. (2008), Singh-Rachford and Castellano (2010), Ji et al. (2011), and Zhao et al. (2011).

Solar fuel production and photoelectrochemistry: Morrison (1980), Nozik and Memming (1996), Memming (2001), van de Krol et al. (2008), Walter et al. (2010), Prévot and Sivula (2013), Hisatomi et al. (2014), Nielander et al. (2015), and Gimenez and Bisquert (2016).

Solar fuel conversion efficiency: Weber and Dignam (1984), Chen et al. (2010), Ager et al. (2015), Coridan et al. (2015), McCrory et al. (2015), Smith et al. (2015), Doscher et al. (2016), and Guerrero and Bisquert (2017).

REFERENCES

Ager, J. W.; Shaner, M. R.; Walczak, K. A.; Sharp, I. D.; Ardo, S. Experimental demonstrations of spontaneous, solar-driven photoelectrochemical water splitting. *Energy & Environmental Science* 2015, *8*, 2811–2824.

Auzel, F. Upconversion and anti-Stokes processes with f and d ions in solids. *Chemical Reviews* 2003, *104*, 139–174.

Bush, K. A.; Palmstrom, A. F.; Yu, Z. J.; Boccard, M.; Cheacharoen, R.; Mailoa, J. P.; McMeekin, D. P. 23.6%-efficient monolithic perovskite/silicon tandem solar cells with improved stability. *Nature Energy* 2017, *2*, 17009–17014.

Chen, G.; Yang, C.; Prasad, P. N. Nanophotonics and nanochemistry: Controlling the excitation dynamics for frequency up- and down-conversion in lanthanide-doped nanoparticles. *Accounts of Chemical Research* 2012, *46*, 1474–1486.

Chen, Y.-S.; Manser, J. S.; Kamat, P. V. Allsolution-processed lead halide perovskite-$BiVO_4$ tandem assembly for photolytic solar fuels production. *Journal of the American Chemical Society* 2015, *137*, 974–981.

Chen, Z.; Jaramillo, T.; Deutsch, T.; Kleiman-Shwarsctein, A.; Forman, A.; Gaillard, N.; Garland, R.; et al. Accelerating materials development for photoelectrochemical hydrogen production: Standards for methods, definitions, and reporting protocol. *Journal of Materials Research* 2010, *25*, 3–16.

Coridan, R. H.; Nielander, A. C.; Francis, S. A.; McDowell, M. T.; Dix, V.; Chatman, S. M.; Lewis, N. S. Methods for comparing the performance of energy-conversion systems for use in solar fuels and solar electricity generation. *Energy & Environmental Science* 2015, *8*, 2886–2901.

Cox, C. R.; Lee, J. Z.; Nocera, D. G.; Buonassisi, T. Ten-percent solar-to-fuel conversion with nonprecious materials. *Proceedings of the National Academy of Sciences* 2014, *111*, 14057–14061.

de Wild, J.; Meijerink, A.; Rath, J. K.; van Sark, W. G. J. H. M.; Schropp, R. E. I. Upconverter solar cells: Materials and applications. *Energy & Environmental Science* 2011, *4*, 4835–4848.

Doscher, H.; Young, J. L.; Geisz, J. F.; Turner, J. A.; Deutsch, T. G. Solar-to-hydrogen efficiency: Shining light on photoelectrochemical device performance. *Energy & Environmental Science* 2016, *9*, 74–80.

Erickson, C. S.; Bradshaw, L. R.; McDowall, S.; Gilbertson, J. D.; Gamelin, D. R.; Patrick, D. L. Zero-reabsorptiondoped-nanocrystal luminescent solar concentrators. *ACS Nano* 2014, *8*, 3461–3467.

Gimenez, S.; Bisquert, J. *Photoelectrochemical Solar Fuel Production. From Basic Principles to Advanced Devices*; Springer: Switzerland, 2016.

Guerrero, A.; Bisquert, J. Perovskite semiconductors for photoelectrochemical water splitting applications. *Current Opinion in Electrochemistry* 2017. doi: 10.1016/j.coelec.2017.04.003.

Haro, M.; Solis, C.; Molina, G.; Otero, L.; Bisquert, J.; Gimenez, S.; Guerrero, A. Toward stable solar hydrogen generation using organic photoelectrochemical cells. *The Journal of Physical Chemistry C* 2015, *119*, 6488–6494.

Henry, C. H. Limiting efficiencies of ideal single and multiple energy gap terrestrial solar cells. *Journal of Applied Physics* 1980, *51*, 4494–4500.

Hisatomi, T.; Kubota, J.; Domen, K. Recent advances in semiconductors for photocatalytic and photoelectrochemical water splitting. *Chemical Society Reviews* 2014, *43*, 7520–7535.

Huang, X.; Han, S.; Huang, W.; Liu, X. Enhancing solar cell efficiency: The search for luminescent materials as spectral converters. *Chemical Society Reviews* 2013, *42*, 173–201.

Ji, S.; Guo, H.; Wu, W.; Wu, W.; Zhao, J. Ruthenium(II) polyimine–coumarin dyad with non-emissive 3IL excited state as sensitizer for triplet–triplet annihilation based upconversion. *Angewandte Chemie International Edition* 2011, *50*, 8283–8286.

Khaselev, O.; Turner, J. A. A monolithic photovoltaic-photoelectrochemical device for hydrogen production via water splitting. *Science* 1998, *280*, 425–427.

Klimov, V. I.; Baker, T. A.; Lim, J.; Velizhanin, K. A.; McDaniel, H. Quality factor of luminescent solar concentrators and practical concentration limits attainable with semiconductor quantum dots. *ACS Photonics* 2016, *3*, 1138–1148.

Kumagai, H.; Minegishi, T.; Sato, N.; Yamada, T.; Kubota, J.; Domen, K. Efficient solar hydrogen production from neutral electrolytes using surface-modified Cu(In,Ga)Se_2 photocathodes. *Journal of Materials Chemistry A* 2015, *3*, 8300–8307.

Luo, J.; Im, J.-H.; Mayer, M. T.; Schreier, M.; Nazeeruddin, M. K.; Park, N.-G.; Tilley, S. D.; et al. Water photolysis at 12.3% efficiency via perovskite photovoltaics and Earth-abundant catalysts. *Science* 2014, *345*, 1593–1596.

Mailoa, J. P.; Lee, M.; Peters, I. M.; Buonassisi, T.; Panchula, A.; Weiss, D. N. Energy-yield prediction for II-VI-based thin-film tandem solar cells. *Energy & Environmental Science* 2016, *9*, 2644–2653.

Markvart, T.; Danos, L.; Fang, L.; Parel, T.; Soleimani, N. Photon frequency management for trapping & concentration of sunlight. *RSC Advances* 2012, *2*, 3173–3179.

May, M. M.; Lewerenz, H.-J.; Lackner, D.; Dimroth, F.; Hannappel, T. Efficient direct solar-to-hydrogen conversion by *in situ* interface transformation of a tandem structure. *Nature Communications* 2015, *6*, 8286.

McCrory, C. C. L.; Jung, S.; Ferrer, I. M.; Chatman, S. M.; Peters, J. C.; Jaramillo, T. F. Benchmarking hydrogen evolving reaction and oxygen evolving reaction electrocatalysts for solar water splitting devices. *Journal of the American Chemical Society* 2015, *137*, 4347–4357.

Memming, R. *Semiconductor Electrochemistry*; Wiley-VCH: Weinheim, 2001.

Meyer, T. J. J.; Markvart, T. The chemical potential of light in fluorescent solar collectors. *Journal of Applied Physics* 2009, *105*, 063110.

Miteva, T.; Yakutkin, V.; Nelles, G.; Baluschev, S. Annihilation assisted upconversion: All-organic, flexible and transparent multicolour display. *New Journal of Physics* 2008, *10*, 103002.

Morrison, S. R. *Electrochemistry at Semiconductor and Oxidized Metal Electrodes*; Plenum Press: New York, 1980.

Nielander, A. C.; Shaner, M. R.; Papadantonakis, K. M.; Francis, S. A.; Lewis, N. S. A taxonomy for solar fuels generators. *Energy & Environmental Science* 2015, *8*, 16–25.

Nocera, D. G. The artificial leaf. *Accounts of Chemical Research* 2012, *45*, 767–776.

Nozik, A. J.; Memming, R. Physical chemistry of semiconductor-liquid interfaces. *The Journal of Physical Chemistry* 1996, *100*, 13061–13078.

Prévot, M. S.; Sivula, K. Photoelectrochemical tandem cells for solar water splitting. *The Journal of Physical Chemistry C* 2013, *117*, 17879–17893.

Rau, U.; Einsele, F.; Glaeser, G. C. Efficiency limits of photovoltaic fluorescent collectors. *Applied Physics Letters* 2005, *87*, 171101.

Singh-Rachford, T. N.; Castellano, F. N. Photon upconversion based on sensitized triplet-triplet annihilation. *Coordination Chemistry Reviews* 2010, *254*, 2560–2573.

Sista, S.; Hong, Z.; Chen, L.-M.; Yang, Y. Tandem polymer photovoltaic cells-current status, challenges and future outlook. *Energy & Environmental Science* 2011, *4*, 1606–1620.

Smith, W. A.; Sharp, I. D.; Strandwitz, N. C.; Bisquert, J. Interfacial band-edge energetics for solar fuels production. *Energy & Environmental Science* 2015, *8*, 2851–2862.

Trupke, T.; Green, M. A.; Wurfel, P. Improving solar cell efficiencies by up-conversion of sub-band-gap light. *Journal of Applied Physics* 2002, *92*, 4117–4122.

van de Krol, R.; Liang, Y. Q.; Schoonman, J. Solar hydrogen production with nanostructured metal oxides. *Journal of Materials Chemistry* 2008, *18*, 2311–2320.

van Sark, W. G.; Barnham, K. W.; Slooff, L. H.; Chatten, A. J.; Büchtemann, A.; Meyer, A.; Mc.Cormack, S. J.; et al. Luminescent solar concentrators. A review of recent results. *Optics Express* 2008, *16*, 21773–21792.

van Sark, W. G. J. H. M. Luminescent solar concentrators. A low cost photovoltaics alternative. *Renewable Energy* 2013, *49*, 207–210.

Verlage, E.; Hu, S.; Liu, R.; Jones, R. J. R.; Sun, K.; Xiang, C.; Lewis, N. S.; et al. A monolithically integrated, intrinsically safe, 10% efficient, solar-driven water-splitting system based on active, stable earth-abundant electrocatalysts in conjunction with tandem III-V light absorbers protected by amorphous TiO_2 films. *Energy & Environmental Science* 2015, *8*, 3166–3172.

Walter, M. G.; Warren, E. L.; McKone, J. R.; Boettcher, S. W.; Mi, Q.; Santori, E. A.; Lewis, N. S. Solar water splitting cells. *Chemical Reviews* 2010, *110*, 6446–6473.

Weber, M. F.; Dignam, M. J. Efficiency of splitting water with semiconducting electrodes. *Journal of the Electrochemical Society* 1984, *131*, 1258–1265.

Yablonovitch, E. Thermodynamics of the fluorescent planar concentrator. *Journal of the Optical Society of America* 1980, *70*, 1362–1363.

Zhao, J.; Ji, S.; Guo, H. Triplet-triplet annihilation based upconversion: From triplet sensitizers and triplet acceptors to upconversion quantum yields. *RSC Advances* 2011, *1*, 937–950.

Appendix

A.1 PHYSICAL CONSTANTS AND CONVERSION FACTORS

Quantity	Symbol, equation	Value*
speed of light in vacuum	c	2.998×10^{8} m s^{-1}
electron charge magnitude	q	1.602×10^{-19} C
Planck constant	h	6.626×10^{-34} J s = 4.136×10^{-15} eV s
Planck constant, reduced	$\hbar = h / 2\pi$	1.0546×10^{-34} J s
permittivity of free space	ε_0	8.85×10^{-12} F m^{-1} = 8.85×10^{-14} F cm^{-1}
permeability of free space	μ_0	$4\pi \times 10^{-7}$ N A^{-2}
electron mass	m_0	9.11×10^{-31} kg
proton mass	m_p	1.67×10^{-27} kg
unified atomic mass unit (u)	(mass ^{12}C atom)/12 = (1 g)/(N_A mol)	1.66×10^{-27} kg
(e^- Compton wavelength)/2π	$\lambdabar_e = \hbar / m_e c$	3.86×10^{-13} m
Bohr radius ($m_{\text{nucleus}} = \infty$)	$a_\infty = 4\pi\varepsilon_0 \hbar^2 / m_e e^2$	0.529×10^{-10} m
Rydberg energy	$hcR_\infty = m_e e^4 / 2(4\pi\varepsilon_0)^2 \hbar^2$	13.6 eV
Bohr magneton	$\mu_B = e\hbar / 2m_e$	9.274×10^{-24} J T^{-1}
Nuclear magneton	$\mu_N = e\hbar / 2m_p$	5.051×10^{-27} J T^{-1}
Avogadro constant	N_A	6.02×10^{23} mol^{-1}
Boltzmann constant	k_B	1.3806×10^{-23} J K^{-1} = 8.617×10^{-5} eV K^{-1}
Gas constant	$R = N_A k$	8.314 J K^{-1} mol^{-1}
Thermal energy at 300 K	$k_B T$	0.0259 eV
Wien displacement law constant	$b = \lambda_{\max} T$	2.87×10^{-3} m K
Faraday constant	$F = N_A q$	96 490 C mol^{-1}
Étendue of solar disc	ε_S	6.8×10^{-5}
Constant Blackbody flux for radiation to the hemisphere	$b_\pi = \dfrac{2\pi}{h^3 c^2}$	9.883×10^{26} eV^{-3} m^{-2} s^{-1}
Stefan-Boltzmann constant	$\sigma = \pi^2 k^4 / 60\hbar^3 c^2$	5.67×10^{-8} W m^{-2} K^{-4}

*Complete values with uncertainties at www.physics.nist.gov/constants

0° C = 273.15 K 1 dina = 10^{-5} N 1 eV = 1.602×10^{-19} J

1 Å = 10^{-10} m 1 erg = 10^{-7} J 1 Wh = 3600 J

1 eV corresponds to 96.48 kJ mol^{-1} = 23.05 kcal mol^{-1} or to the energy of a quantum of wavelength 1240 nm.

$\hbar\omega\lambda = hc_{vac} = 1240$ eV nm

1 debye = 3.33564×10^{-30} C m

A.2 GENERAL LIST OF ACRONYMS AND ABBREVIATIONS

4T	Four terminal device
A	Acceptor (of charge)
AFM	Atomic Force Microscopy
APCE	Absorbed-photon-to-collected-electron-efficiency
ARF	Anti-reflective foil
BE	Binding energy
BHJ	Bulk heterojunction
CB	Conduction band
CCT	Correlated color temperature
CE	Counterelectrode
CNL	Charge neutrality level
CPD	contact potential difference
CRI	color rendering index
CT	Charge transfer
CTC	Charge transfer complex
CV	Cyclic voltammogram
D	Donor (of charge)
DFT	Density Functional Theory
DOS	Density of states
DSC	Dye-sensitized solar cells
EA	Electron affinity
EC	equivalent circuit
EL	Electroluminescence
emf	Electromotive force
EQE	External quantum efficiency
ESA	Excited state absorption
ESC	Electron selecting contact
ETU	Energy transfer upconversion
FC	Fluorescent collector
FET	Field-effect transistor
FRET	Förster resonant energy transfer
FTO	Fluor-doped tin oxide
FTPS	Fourier transform photocurrent spectroscopy
HER	hydrogen evolution reaction
HOMO	Highest occupied molecular orbital
HSC	Hole selecting contact
IE	Ionization energy
IMPS	Intensity modulated photocurrent spectroscopy
IPCE	incident-photon-to-current-collected-electron-efficiency
IPES	Inverse Photoemission Spectroscopy
IQE	Internal quantum efficiency
IR	Infrared
IS	Impedance Spectroscopy
ISC	Intersystem crossing
ITO	Indium-doped tin oxide
IUPAC	International Union of Pure and Applied Chemistry
KMC	Kinetic Monte Carlo
KP	Kelvin probe
KPFM	Kelvin probe force microscopy
LED	Light emitting diode
LHE	Light harvesting efficiency
LIB	Lithium ion battery
LSC	Luminescent solar concentrators
LUMO	Lowest unoccupied molecular orbital
LVL	Local vacuum level
MA	Millar-Abrahams hopping rate
MIM	Metal-insulator-metal
MIS	Metal-insulator-semiconductor
MISFET	metal-insulator-semiconductor field-effect
MLCT	Metal-to-ligand charge transfer
MOS	metal-oxide-semiconductor
mpp	Maximum Power Point
MS	Mott-Schottky
NAS	Sodium-sulfur battery
NHE	Normal Hydrogen Electrode (same as SHE)
NIR	Near infrared
OCV	Open circuit voltage
OER	Oxygen evolution reaction
OET	Optoelectronics and Transport (this volume)
OFET	Organic field-effect transistor
ohp	Outer Helmholtz plane
OLED	Organic light emitting diode
OPV	Organic photovoltaics
ORR	Oxygen reduction reaction
OSC	Organic solar cell
PA	Pair approximation
PA	Pair approximation
PCE	Power conversion efficiency
PEC	Photoelectrochemical cell
PES	Photoelectron Spectroscopy
PL	Photoluminescence
PSC	Physics of Solar Cells (accompanying volume of this book).
pzc	Point of zero charge
QD	Quantum dot
QE	Quantum efficiency
QY	Quantum yield
RE	Reference electrode
RHE	Reversible hydrogen electrode
RNN	Random nanoparticulate network
SAM	Self-assembled monolayer
SB	Schottky barrier
SCE	Saturated calomel electrode
SCLC	Space-charge limited current
SCR	Space charge region
SEI	Solid electrolyte interface
SHE	Standard hydrogen electrode
SPV	Surface photovoltage
STC	Standard test conditions
STH	Solar to hydrogen
SVL	Surface vacuum level
TAS	Transient Absorption Spectroscopy
TCO	Transparent conducting oxide
TFT	Thin film Transistor
TOF	Time-of-flight

TRMC	Time-resolved microwave conductivity
TRTS	Time-resolved Terahertz spectroscopy
TTA	Triplet-triplet annihilation
UHV	Ultra high vacuum
UPS	Ultraviolet Photoelectron Spectroscopy
UV	Ultraviolet Photoelectron Spectroscopy
VB	Valence band
VL	Vacuum level
VLA	Vacuum level alignment
VRH	variable-range hopping
WE	Working electrode

A.3 NOMENCLATURE

Alq	8-hydroxyquinoline aluminium
Alq3	Tris(8-hydroxyquinolino)aluminium
a-Si	Amorphous silicon
BAlq	Bis(8-hydroxy-2-methylquinoline)-(4-phenylphenoxy)aluminum; Aluminium 4-biphenylolate 2-methyl-8-quinolinolate
Biphen	1,10-Phenanthroline
BPEA	9,10-Bis(phenylethynyl)anthracene
CBP	(4,4'-N, N'-dicarbazolyl-biphenyl)
CIGS	$Cu(In,Ga)Se_2$
c-Si	Crystalline silicon
CZTSe	$Cu_2ZnSnSe_4$
DCV4T	Dicyanovinyl end-substituted quaterthiophene
DEC	Diethyl carbonate
DMC	Dimethyl carbonate
DPM6	4,4'-dihexyloxydiphenylmethano[60]fullerene
DPNTCI	N,N'-diphenyl-1,4,5,8-naphthyltetracarboxilicimide
DSB	Distyrylbenzene
EC	Ethylene carbonate
EMC	Ethyl methyl carbonate
F8	Poly(9,90-dioctylfluorene)
F8T2	poly(9,9-dioctylfluorene-alt-bithiophene)
FA	Formamidinium
FTO	Fluor-doped tin oxide
H2 TPP	5,10,15,20-Tetra(phenyl)porphyrin
H2 TPyP	5,10,15,20-Tetra(4-pyridyl)porphyrin
HATNA	Hexaaza-triaaphthylene
Ir(MDQ)2	Bis(2-methyldibenzo[f,h]quinoxaline)(acetylacetonate)iridium(III)
ITO	Indium-doped tin oxide
MA	Methylammonium
MEH-PPV	Poly(2-methoxy-5-(2 ′ -ethylhexyloxy)-1,4-phenylene vinylene
MTDATA	(4,4,4''-Tris[3-methyl-phenyl)phenyl)amino]-triphenylamine
NTCDA	1,4,5,8-Naphthalene-tetracarboxylic-dianhydride
OF	Oligofluorenes
OMeTAD	2,2'7,7'-Tetrakis(N,N-di-p-methoxyphenyl-amine)-9,9'-spiro-bifluorene)
OTiPc	Titanyl phthalocyanine
P3HT	Poly(3-hexylthiophene)
PBDTTT	Poly[4,8-bis(5-(2-ethylhexyl)thiophen-2-yl)benzo[1,2-b;4,5-b']dithiophene-2,6-diyl-alt-(4-(2-ethylhexyl)-3-fluorothieno[3,4-b]thiophene-)-2-carboxylate-2-6-diyl)]
$PC_{60}BM$	[6,6]- phenyl-C60-butyric acidmethyl ester
PCDTBT	Poly[[9-(1-octylnonyl)-9H-carbazole-2,7-diyl]-2,5-thiophenediyl-2,1,3-benzothiadiazole-4,7-diyl-2,5-thiophenediyl]
PCPDTBT	Poly[2,6-(4,4-bis-(2-ethylhexyl)-4H-cyclopenta[2,1-b;3,4-b']-dithiophene)-alt-4,7-(2,1,3-benzothiadiazole)]
$PdPh_4TBP$	Meso-tetraphenyl-tetrabenzoporphyrin palladium
PEDOT	Poly(3,4-ethylenedioxythiophene)
PPV	4-phenylene vinylene
Ppy	Polypyrrole
PSBTBT	Poly(4,4-dioctyldithieno(3,2-b:2',3'-d)silole)-2,6-diyl-alt-(2,1,3-benzothiadiazole)-4,7-diyl)
PTB7	Poly[[4,8-bis[(2-ethylhexyl)oxy]benzo[1,2-b:4,5-b']dithiophene-2,6-diyl][3-fluoro-2-[(2-ethylhexyl)carbonyl]thieno[3,4-b]thiophenediyl]]
PTB7	Poly[[4,8-bis[(2-ethylhexyl)oxy]benzo[1,2-b:4,5-b']dithiophene-2,6-diyl][3-fluoro-2-[(2-ethylhexyl)carbonyl]thieno[3,4-b]thiophenediyl]]
PTCDA	Perylene-3,4,9,10-tetracarboxylic dianhydride
TCNQ	Tetracyanoquinodimethane
TCO	Transparent conducting oxide
TFB	poly(9,9-dioctylfluorene-alt-N-(4-butylphenyl)-diphenylamine)
TPBi	2,2′,2''-(1,3,5-Benzinetriyl)-tris(1-phenyl-1-H-benzimidazole)
TPD	N,N'-diphenyl-N,N'-(3-methylphenyl)-1,1'-biphenyl-4,4'diamine
TQ1	Poly[2,3-bis-(3-octyloxyphenyl)quinoxaline-5,8-diyl-alt-thiophene-2,5-diyl]
TTC	Tetratetracontane
TTN	Tetrathianaphthacene
ZnPc	Zinc phthalocyanine
ZnTPP	5,10,15,20-Tetraphenylporphynatozinc(II)
α-NPD	N,N'-diphenyl-N,N'-bis(1-naphthyl)-1,1'-biphenyl-4,4'-diamine
μc-Si	Microcrystalline silicon

Index

Absorptance, 306–308, 365, 369–373
Accumulation region, 204–205
Activity, 44–46
Admittance, capacitor, 40
AFM, *see* Atomic force microscopy
Air mass (AM), 299
Ambipolar diffusion transport, 221–222
Annihilator/emitter, 309
Anode, 34–35
Applied voltage, 35–37
Atomic force microscopy (AFM), 62
Attempt frequency, 102–103
Auger recombination, 312–313, 355, 416
Avogadro constant, 13

Bacteria, in energy conversion, 2
Band bending, 17
Band edge movement, 190–191
Band tail, 134
Barriers
    energy, 59, 68, 105–106, 109–112, 167, 170, 388
    injection, 66–69, 200
    Schottky, 14, 87, 167–170, 343
Batteries, 39
    capacity, 50
    lead-acid, 47
    materials, 48
    primary, 47
    secondary, 47
Bethe thermionic-emission theory, 344
Bimolecular recombination, 313, 355, 398
Blackbody radiation, 293–294
    cavity, 294
    photons in
        energy density of distribution of, 295–297
        and energy fluxes, 297–299
    physical variables for, 297
    spectrum of, 293–295
    thermal, 293, 298
Bloch functions, 14
Boltzmann distribution, 77
Boltzmann factor, 15
Bose-Einstein distribution, 295
Bragg–Williams approximation, 96
Built-in potential, 28
Bulk material, 13
Burstein-Moss shift, 324
Butler-Volmer equation, 115–116, 118, 146
    charge transfer, kinetic constant for, 117–120
    electronic species, availability of, 116
    redox species, availability of, 116–117

Cahn–Hilliard theory, 96
Candela, 293
Capacitance, 37–38, 271–272
    admittance, 40
    chemical, 131–142
    conduction band, 135–136
    defect levels, response of, 172–173
    dielectric, 39, 131–132
    differential, 40, 131–132
    diffusion, 135–136
    of DSC, 155–156
    energy storage in, 40
    in equivalent circuit models, 274–279
    Helmholtz, 42
    measurement of, 38–40
    quantum, 140–142
    supercapacitors, 39
    surface states, response of, 172–173
    time domain, relaxation in, 279–281
    unit of, 38
Capacity, 50–51
Carnot efficiency factor, 88, 381
Carriers, 86
    diffusion of, 213
    flux density, 198
    hot, 82–83
    majority, 80
        devices, 344–345
        injection mechanisms, 343–344
    minority, 80
        devices, 345–346
Cathode, 34–35
CCT, *see* Correlated color temperature
Cell
    DSC, 5–6
    electrochemical, 43–45
    electrolytic, 47
    galvanic, 47
    inorganic quantum-dot sensitized, 5–6
    organic solar, 5–6
    perovskite solar cells, 6–8, 277, 279, 280, 313, 314, 366, 385, 395, 399, 403, 408, 409, 416, 421, 422, 438, 442
    photoelectrochemical, 173–176
    photovoltaic, 429–433
    reversible, 50
    solar, 2–3, 5–9, 200, 231
Charge
    screening, 132, 222–223
    separation, 395–398
Charge collection in solar cells
    boundary conditions, 425–429
        blocking, 427–428
        charge extraction, 426–427
    distance, 422–424
    general modeling equations, 424–425
    properties, 421–422
Charge-neutrality level, pinning of, 71–73
Charge-transfer complex (CTC), 335
Charge transfer (CT), 25
    absorption and emission, reciprocity of, 387–391
    complex, 335
    electrochemical reactions and, 41
    kinetic constant for, 117–120
    phenomenological rates, 102
    resistance, 119, 146, 274
    state, 396, 401
Chemical capacitance, 131–142
    amorphous/disordered semiconductors, localized electronic states in, 133–135
    carrier accumulation in, 131–133
    of DOS, 136–138
        conductivity, 138–139
        voltage, 138–139
    energy storage in, 131–133
    of graphene, 140–142
    Li intercalation materials, 139–140
    of single state, 135–136
Chemical diffusion coefficient, 209, 212–215, 217–220, 232–239
Chemical energy, 1, 47
Chemical fuel, 2, 5, 125, 173, 295, 445
Child's law, 263
Coefficient of reflection, 305
Cole-Cole model, 274, 280
Color Rendering Index (CRI), 293
Colors, 291–293
    hue, 291
    human eye and, 292
    saturation, 291
Conduction
    band capacitance, 135–136
    band states, equilibrium of, 76–79
    current, 197
Conductivity
    DOS, chemical capacitance of, 138–139
    frequency-dependent, 281–283
    hopping transport, 241–242
    TRMC, 324–325
Contact
    blocking layer, 33–34
    CPD, 28–29, 35, 61, 168
    injection layer, 198–202
    ohmic, 61
    selective, 348–351
Contact potential difference (CPD), 28–29, 35, 61, 168
Conversion reaction, 56–57
Correlated color temperature (CCT), 292
Coulomb repulsion, 76
CPD, *see* Contact potential difference
CRI, *see* Color Rendering Index
Crystalline semiconductors, 16, 75

CTC, *see* Charge-transfer complex
Curie-von Scweidler law, 274, 280
Current
collector, 33
conduction, 197
diffusion, 132, 217–220, 431–434
displacement, 197
electrical, 197
equilibrium, 60
injection-limited, 199
overpotential equation, 118
photocurrent, 407
recombination, 355–356
reverse saturation, 85–87, 343–344, 354–356
transient, 223–224
Cyclic voltammetry (CV), 145
capacitive current in, 145–149
kinetic effects in, 149–150
reactive current in, 145–149

Debye–Huckel theory, 327
Debye relaxation, 273–274
Debye screening length, 84–85
Degenerate semiconductor, 81
Density of states (DOS), 15–16, 80
in conduction band states, 76–79, 319
exponential
in amorphous semiconductors, 150–152
multiple trapping in, 237
in nanocrystalline metal oxides, 152–156
Gaussian, 160–162, 201–202
activated transport in, 237–239
in TFTs, 252–255
in valence band states, 76–79, 319
Depletion approximation, 170
Depletion region, 14, 167
Detailed balance, principle of, 101–104
Devices, 1; *see also specific* devices
Dexter energy transfer, 310
Dielectric
capacitor, 39, 131–132
constant, 13, 24–25, 42, 84, 112, 123, 131, 171–172, 223, 265, 272
loss, 272–274
relaxation functions, 272–274
Differential capacitance, 40, 131–132
Diffusion
capacitance, 135–136
current, 132, 217–220, 431–434
length, 212–213
theory, 344
transport, 209
chemical diffusion coefficient, 213–215
length, 212–213
macroscopic equation, 211–212
parameters, 214
in random walk model, 209–211
thermodynamic factor, 213–215
Diffusion coefficient, 139
ambipolar, 222
chemical, 209, 212–215, 217–220, 232–239
jump, 209–210, 213, 217, 219, 232–233, 236–237, 239
Tracer, 209–210
Diode law, *see* Shockley ideal diode equation
Diodes, 85–86
dye sensitization and molecular, 360–363
equation, dark characteristics of, 356–357
ideality factor, 343
LEDs, 6, 59, 231, 357–360
majority carriers, 80
devices, 344–345
injection mechanisms, 343–344
minority carriers, 80, 87
devices, 345–346
recombination, 75, 347
current, 355–356
reverse saturation current, 86–87
unipolar, 347
Dipole
layer, 23–24, 29
moment, 24
surface, 13, 18, 24–25
Dirichlet boundary condition, 426
Discharge lamps, 292–293
Displacement current, 197
Distribution
Boltzmann, 77
Bose-Einstein, 295
Fermi–Dirac, 15, 76–77, 135, 148, 241
Gaussian, 122–123, 149, 160–162, 229, 231, 237–240, 252–253
Maxwell–Boltzmann, 82–83
Donor-acceptor system, 108, 335, 358, 389, 401, 437
DOS, *see* Density of states
Down-shifting, 445
Drift-diffusion transport, 217
ambipolar, 221–222
Einstein relation, 219–220
electrochemical potential, equation in terms of, 217
equations, 220–221
injected charge, relaxation of, 222–223
modeling problems, 224–227
resistance, 217–219
transient current in insulator layers, 223–224
Drift transport
in electrical field, 197–198
flux density with, 198
DSC, *see* Dye-sensitized solar cells
Dye sensitization diodes, 360–363
Dye-sensitized solar cells (DSC), 5–6, 152, 360–363, 396–397, 400–403
Efficiency
Carnot factor, 88, 381
collection, 311, 370, 411
external quantum, 310
external radiative, 382–383
internal quantum, 310
light harvesting, 310
power conversion, 366–367, 408–410
quantum, 310–311
Shockley–Queisser limits, 412–413
solar cells
limits, 413–419
measuring, 439–442
reporting, 439–442
Trivich-Flinn, 367, 413
Einstein relation, 219–220
Electrical current, 197
Electrical double layer, 41
Electrical force, 197
Electric energies, 1, 13
Electric potential difference, 13, 17, 27, 35–37
Electroabsorption, 359
Electrochemical batteries
charge capacity, 50
energy content, 50–51
Li-ion, 53–57
materials, 48
principles of, 47–50
sodium-sulfur, 52–53
zinc-silver, 51–52
Electrochemical cells
electrode potentials in, 43
redox potential in, 44–45
Electrochemical potentials, 21–23, 75
of electrons, 75–76
Electrochemical reactions, 41
Electrochemical systems, 40–42
Electrochemical transistor, 258–259
Electrochemistry, 40
Electrodes, 3, 33, 40
Butler-Volmer equation, 115–116
charge transfer, kinetic constant for, 117–120
electronic species, availability of, 116
redox species, availability of, 116–117
equilibrium capacitance of, 147
metal oxide mesoporous, 3–4
n-type semiconductor, 175
polarization, 283–284
potential, 42–44
reference, 42–44
Electroluminescence (EL), 307
radiative cooling in, 386–387
Electrolyte levels with pH, 46–47
Electrolytes, 2
Electrolytic cell, 47
Electromagnetic radiation, 289
Electromotive force (emf), 33, 35, 49
Electrons, 2, 13

affinity rule, 182
chemical potential of, 21–23
detailed balance, principle of, 101–104
effective mass, 15
electrochemical potential, equilibration of, 75–76
entropy of, 76
Fermi level of, 20–21
free, 14–17
hot, 82–83
lifetime, 213, 233–235, 312–313
momentum, 15
selective contact for, 87
in semiconductors
crystal, 14–17
potential energy, 17
in thermal equilibrium, 15
transfer, 101–104
adiabatic, 112
frequency of, 101
Marcus's theory, 107–112
metal semiconductor contact, 120–121
nonadiabatic, 111–112
semiconductor/electrolyte interface, 121–126
vacuum level of, 17–20
work function of, 20–21
Electron-transport material (ETM), 396–397, 400–402
Electrostatic potential, 13–14
depletion region, 14
neutral region, 14
Electrostatic potential difference, 35–37
Electrostatics, 13
Energies
band, 6, 14, 17, 19, 22, 122, 157, 186, 190, 255, 439
barriers, 59, 68, 105–106, 109–112, 167, 170, 388
chemical, 1, 47
content, 50–51
conversion process, 3–6
definition, 1
delivery, 1
demarcation, 240
electric, 1
free, 17
free electrons, 14–17
holes, 14–17
ionization, 17–20
kinetic, 1
photon, 63–64, 83, 114, 126, 151, 289, 295–298
production, 1–2
renewable, 1
reorganization, 110–113
solar, 1–2
into chemical fuel, 2
conversion, 6–9
harvesting, photon frequencies conversion for, 445–448
storage, 1
thermal, 1
transfer, 309–310
transport, 242–243
utilization, 1
Energies devices, 1–9, 231, 275, 289, 305
archetypical, 3
functions, 2–3
nanostructured, 33, 87
operation of, 2
production, 13
storage, 13
use of, 1
Energy transfer upconversion (ETU), 447
Entropy
configurational, 76
electrons, 75
mixing, 76
EQE, *see* External quantum efficiency
Equilibrium, 2
of conduction band states, 76–79
of contact potential difference, 28–29
current, 60
of emission of radiation, 353–355
Fermi level, 78
of metal junctions, 28–29
of radiation absorption, 353–355
semiconductor junction, across, 29–30
of valence band states, 76–79
Equivalent circuit models, 274–279
ESA, *see* Excited state absorption
Escape cone, 307, 384–385
Étendue ε, 290
ETM, *see* Electron-transport material
ETU, *see* Energy transfer upconversion
Exchange current density, 118
Excited state absorption (ESA), 446–447
Excitons, 325–328
binding energy, 327–328
Frenkel, 327
Wannier–Mott, 327
External quantum efficiency (EQE), 310

Faradaic pathway, 42
Farad (F), 38
FCs, *see* Fluorescent collectors
Fermi–Dirac distribution function, 15, 76–77, 135, 148, 241
Fermi level, 20–21, 42–43, 75
pinning of, 69–71, 200
in semiconductors, 78–81
of two electronic conductors in contact, 27–28
Fermi's rule, 319
FETs, *see* Field effect transistors
FF, *see* Fill factor
Field effect transistors (FETs), 231, 249
Field emission, 120
Fill factor (FF), 408–412
Flatband potential, 167
Fluor-doped tin oxide (FTO), 81
Fluorescence, 307, 331
Fluorescent collectors (FCs), 445–446
Flux density, drift transport, 198
Forster resonant energy transfer (FRET), 309–310
Forward voltage, 87
Fossil fuels, 1
Franck–Condon principle, 101, 107–108, 331
Free carriers, 15
absorption, 323–325
lifetime, 234
Free electrons, 14–17
Free energy, 17, 76
Frenkel exciton, 327
Frequency-dependent conductivity, 281–283
Frequency dispersion, 274
FRET, *see* Forster resonant energy transfer
FTO, *see* Fluor-doped tin oxide

Galvanic cell, 47
Galvani potential difference, 27
Gärtner model, 433–435
Gaussian distribution, 122–123, 149, 160–162, 229, 231, 237–240, 252–253
Gaussian DOS, 160–162, 201–202
Geminate recombination, 312, 355
Graphene, 250

Half-cell potential, 42
Half reaction, 34
Helmholtz capacitance, 42
Helmholtz equation, 24
Helmholtz layer, 41–42
HER, *see* Hydrogen evolution reaction
Heterojunctions
CT band in, 333–336
semiconductors, 181–183
Highest occupied molecular orbital (HOMO), 19–20, 63, 157, 159–160
Hole affinity, 17
Holes energies, 14–17
Hole-transport material (HTM), 7, 396–398, 400–401
HOMO, *see* Highest occupied molecular orbital
Homojunction, semiconductors, 181
Hopping models, *see* Single jump models
Hopping transport, 229–231
conductivity, 241–242
Miller-Abrahams rate, 103
in single level, 231–233
VRH, 243–245
Hot electrons, 82–83
HTM, *see* Hole-transport material
Hydrogen evolution reaction (HER), 450

Ideal solution, 44
IE, *see* Ionization energy
Illuminance, 293

Impedance
definition, 271
frequency domain measurements, 271–272
time domain, relaxation in, 279–281
transfer function, 271
Impedance spectroscopy (IS), 265, 271
measurement, 271–272
IMPS, *see* Intensity-modulated photocurrent spectroscopy
Incandescence, 292
Incandescent lamps, 292
Incident radiation, 305–307
coefficient of reflection, 305
quantum efficiency, 310–311
scattering, 307
transmittance, 305
Indium-doped tin oxide (ITO), 81, 201
Injection
barriers, 66–69, 200
at contacts, 198–202
Injection-limited current, 199
Inner-sphere electron transfer, 108
Inorganic quantum-dot sensitized cells, 5–6
Inorganic semiconductors, light absorption process in, 319–323
direct transition, 319–320
indirect transition, 320
Inorganic solid material, light absorption process in, 319–323
Inorganic transistors, 249
Insertion, 53, 55, 94–98
Intensity-modulated photocurrent spectroscopy (IMPS), 271, 277
Intensity of light, 293
Intercalation, 139–140
Internal quantum efficiency (IQE), 310
Intersystem crossing (ISC), 332, 447–448
Inverse Photoemission Spectroscopy (IPES), 63
Inversion region, 181
Ionization energy (IE), 17–20
IPES, *see* Inverse Photoemission Spectroscopy
IQE, *see* Internal quantum efficiency
IS, *see* Impedance spectroscopy
ISC, *see* Intersystem crossing
ITO, *see* Indium-doped tin oxide

Jump rate, 103
Junctions
heterojunctions
CT band in, 333–336
semiconductors, 181–183
homojunction, 181
metal/semiconductors, 28–30
energy diagram of, 180
equilibrium across, 29–30
p–n, 181

Kelvin probe force microscopy (KPFM), 62, 255
Kelvin Probe (KP) method, 61–63
Kinetic energy, 1, 14, 63–65
Kinetics, definition of, 101
Kinetics of localized states, 106–107
Kirchhoff's law, 370
Kohlrausch-Williams-Watts (KWW) function, 280–281
KPFM, *see* Kelvin probe force microscopy
Kramers-Kronig relationship, 273

Lambert's law, 291
Lamps for lighting, 292
discharge, 292–293
incandescent, 292
solid state, 293
Lattice gas model, 94–98
Layer
dipole, 23–24, 29
electrical double, 41
Helmholtz, 41–42
Nernst diffusion, 116–117
organic, properties of, 156–160
semiconductors accumulation, 180–181
transport, 349
Lead-acid batteries, 47
LEDs, *see* Light emitting diodes
LHE, *see* Light harvesting efficiency
LIB, *see* Lithium-ion battery
Lifetime
electron, 213, 233–235, 312–313
free carriers, 234
radiative carrier, 368–369
recombination, 314–316
Light, 289, 305
absorption, 395
direction, 289–291
energy transfer, 309–310
étendue of, 290
intensity of, 293
luminescence, 307–309
spread of, 289–291
Light emitting diodes (LEDs), 6, 59, 231, 357–360
quantum efficiencies for, 310, 383
reciprocity relation in, 375–378
Light harvesting efficiency (LHE), 310
Light-modulated emission spectroscopy, 277
Linear sweep voltammetry, 145
Lithium battery, 2
Lithium-ion battery (LIB), 41, 53–57
cell potential in, 90–94
Load, 33
Local vacuum level (LVL), 18
Lower energy, 17
Lowest unoccupied molecular orbital (LUMO), 19–20, 63, 157, 159–160
LSC, *see* Luminescent solar concentrators
Luminescence, 293, 307–309
Luminescent solar concentrators (LSC), 445–446
LUMO, *see* Lowest unoccupied molecular orbital
LVL, *see* Local vacuum level

Macroscopic diffusion equation, 211–212
Majority carriers, 80, 343–345
Marcus inverted region, 113, 123
Marcus model, 101, 103, 107–112
Mass action law, 80
Master equation, 103–104
Material systems, electron energy in, 45–46
Maximum power point (mpp), 408, 438, 441
Maxwell–Boltzmann distribution, 82–83
Metal-insulator-metal (MIM) model, 202–205, 349–350
Metal-insulator-semiconductor field effect transistor (MISFET), 249
Metal-insulator-semiconductor (MIS), 180–181, 183
Metal junctions, equilibration of, 28–29
Metal-to-ligand charge transfer (MLCT), 332
Metropolis rate, 105–106
Miller-Abrahams jump rate, 103
Minority carriers, 80, 345–346
MIS, *see* Metal-insulator-semiconductor
MISFET, *see* Metal-insulator-semiconductor field effect transistor
MLCT, *see* Metal-to-ligand charge transfer
Mobility, hopping transport, 198
Mobility edge, 134
Molecular assemblies, 157
Molecular crystals, 157
Molecular diodes, 360–363
Molecular extinction coefficient, 306
Molybdenum oxide (MoO_3), 201
Monomers, 156
Monte Carlo simulation, 105
Mott-Gourney square law, 263
Mott–Schottky (MS) plots, 66, 171–172, 179, 283
Mpp, *see* Maximum power point
Multiple trapping transport, 229–231
in exponential DOS, 237
in time domain, 239–241

Nanostructured materials, 3
Nanotechnology, 2
Nearest neighbor hopping (NNH), 241–242
Nernst diffusion layer, 116–117
NHE, *see* Normal hydrogen electrode
NNH, *see* Nearest neighbor hopping
Nonfaradaic pathway, 42
Nonradiative recombination, 312
Normal hydrogen electrode (NHE), 43

OER, *see* Oxygen evolution reaction
"Ohmic" boundary condition, 426
Ohmic contact, 61, 199
Ohmic region, 119

OLEDs, *see* Organic light-emitting diodes
Oligomers, 156
Optical density, 306
Optical region, 289
Organic layers, properties of, 156–160
Organic light-emitting diodes (OLEDs), 200, 231, 264, 305–306, 350–351
Organic materials/molecules
  CT band in, 333–336
  light absorption in, 330–333
Organic solar cells, 5–6, 396
  bulk heterojunction, 396
  DSC, 5–6, 360–363, 396–397
  types, 396
Organic TFTs, 249–251
Outcoupling, 310, 382, 386
Outer-sphere electron transfer, 105, 108
Oxidation, 34
Oxygen evolution reaction (OER), 450

Pauli principle, 15
PCE, *see* Power conversion efficiency
PEC, *see* Photoelectrochemical cell
PEDOT:PSS, 201
Perovskite solar cell, 6–8, 277, 279, 280, 313, 314, 366, 385, 395, 399, 403, 408, 409, 416, 421, 422, 438, 442
PES, *see* Photoelectron emission spectroscopy
Phonon, 75, 83, 103, 106–107, 113–115, 241, 309
Phosphorescence, 307, 332
Phosphors, 309
Photocurrent, 407
Photoelectrochemical cell (PEC), 173–176, 451–452
Photoelectron emission spectroscopy (PES), 63–66
Photoluminescence (PL), 307
  radiative cooling in, 386–387
Photometry, 291–293
Photons, 289
  in blackbody radiation
    energy density of distribution of, 295–297
    and energy fluxes, 297–299
  energy, 63–64, 83, 114, 126, 151, 289, 295–298
  at nonzero chemical potential, 370–373
  reabsorption, 310–311, 375, 384–386
  recycling, 310, 375, 383–386
Photopic response, 293
Photosynthesis, 1–2
Photovoltage
  devices, 5–6
  effect, 347
  external radiative efficiency, 382–383
  factors determining, 378–382
  materials limits to, 398–403
  model with diffusion and recombination, 429–433
  performance parameters, reciprocity relation in, 375–378
Pinning
  band, 176
  of charge-neutrality level, 71–73
  of Fermi level, 69–71, 200
Planar depletion layer, 170–171
Planck's constant, 14
Planck's law, 296, 368
Planck spectrum (blackbody spectrum), 294–295
Plants, in energy conversion, 2
Plasma, 323
Plasmons, 323
Poisson equation, 13–14, 17, 60, 84, 170, 186, 217, 221–222, 225, 264, 268, 424, 429, 436
Polarization, electrodes, 283–284
Polaron hopping, 101, 112–115
Polymer films, 157
Polymers, 156
Poole-Frenkel mechanism, 229
Potential
  built-in, 28–29, 33, 38
  cell, 90–94
  chemical, 13–14, 21–23, 29, 44, 370–373
  contact, 27–28
  CPD, 28–29, 35, 61, 168
  electrochemical, 20, 27, 217
  Galvani, 27
  half-cell, 42
  ionization, 19, 181
  redox, 44
  Volta, 25–27
  of zero charge, 42
Potentiostat, 33
Power conversion efficiency (PCE), 366–367, 408–410
Primary batteries, 47
P-type semiconductor, 85–86

QDs, *see* Quantum dots
Quantum capacitance, 140–142
Quantum-cutting, 446–447
Quantum dots (QDs), 19, 200, 328–330, 349
  colloidal, 244–245
  films, 349
Quantum yield (QY), 305, 309–311
Quasineutral region, 80
Quasi-static approximation, 234
QY, *see* Quantum yield

Radiance, 293
Radiant intensity, 293
Radiation, 289
  absorption, dark equilibrium of, 353–355
  blackbody, 293–294
    cavity, 294
    energy density of distribution of photons in, 295–297
    Lambertian, 353
    photons and energy fluxes in, 297–299
    physical variables for, 297
    spectrum of, 293–295
    thermal, 293, 298
  cooling, 386–387
  direction, 289–291
  electromagnetic, 289
  emission, dark equilibrium of, 353–355
  incident, absorption of, 305–307
  Planck's law, 296
  quantum efficiencies for, 310–311
  spread, 289–291
  thermal, 293, 296, 298
Radiative recombination, 312
Radiative transfer mechanism, 309
Random walk model, diffusion transport in, 209–211
RE, *see* Reference electrode
Reactions
  conversion, 56–57
  coordinate, 103
  electrochemical, 41
  half, 34
  hydrogen evolution, 450
  oxygen evolution, 450
  redox, 34
Reciprocity relationships, 387–391
Recombination
  Auger, 312–313, 355, 416
  bimolecular, 313, 355, 398
  center, 106
  diode model, 75, 347
  energetic disorder in, 5
  geminate, 312, 355
  lifetime, 314–316
  nonradiative, 312
  radiative, 312
  of semiconductors carriers, 311–314
  Shockley-Read-Hall (SRH) model, 105–107, 312–313
  valence band in, 16
Rectification, 86
Rectifiers, 85–88
Redox potential, 44
Redox reaction, 34
Reduction, 34
Reference electrode (RE), 42–44
Refraction index, 307–308, 368, 384
Region
  accumulation, 204–205
  depletion, 14, 167
  neutral region, 14
  quasineutral, 80
Renewable energy, 1
Reorganization energy, 110–113
Resistance
  in equivalent circuit models, 274–279
  total, 276
Resistor, 33
Resonant energy transfer, 309

Reverse saturation current, 85–87, 343–344, 354–356
Reverse voltage, 87
Reversible hydrogen electrode (RHE), 46–47
Richardson constant, 59, 61, 120
Richardson–Dushmann equation, 59–61
Rydberg constant, 326

SB, *see* Schottky barrier
Scattering, 307
Schottky barrier (SB), 14, 87, 343
  applied voltage, changes by, 168–170
  charge distributions in SCR, 180
  semiconductors contacts, structure at, 167–168
Schottky diode, *see* Majority carriers
SCLC, *see* Space-charge-limited current
Screening, 84–85
Secondary batteries, 47
Secondary electrons cutoff, 63
Selective contacts
  for electrons, 87
  hole, 85–87
  solar cells, physical properties of, 348–351
Semiconductors, 13–17
  accumulation layer, 180–181
  amorphous, exponential DOS in, 150–152
  band unpinning, changes of, 176–180
  carrier number in, 79–81
  carriers, recombination of, 311–314
  crystalline, 16, 75
  degenerate, 81
  direct, 321–322
  disordered, localized electronic states in, 133–135
  electrodes, 173–176
  electronic carrier mobility in, 198
  electrons in
    potential energy of, 17
    transfer of, 121–126
  energy diagram of, 19
  Fermi level in, 15, 78–81
  flatband potential, 173–176
  free carrier absorption, 323–325
  heterojunctions, 181–183
  highly doped nanocrystalline, effect of voltage on, 183–188
  homojunction, 181
  indirect, 335, 365, 403
  inorganic, light absorption process in, 319–323
  inversion, 180–181
  junction
    energy diagram of, 180
    equilibrium across, 29–30
    p–*n*, 181
  layers, characteristics of, 172
  low doped nanocrystalline, homogeneous carrier accumulation in, 188–192
  Mott–Schottky (MS) plots of, 171–172, 179
  n-type, 175
  optical gap, 324
  organic
    CT band in, 333–336
    light absorption in, 330–333
  p-type, 85–86
  quantum dots, 6, 328–330
  radiative carrier lifetime, 368–369
  redox level, changes of, 176–180
  Schottky barrier
    changes by applied voltage, 168–170
    structure, 167–168
  solar photons, utilization of, 365–368
  as thermal machines, 88–90
SHE, *see* Standard hydrogen electrode
Shift of conduction band, 190
Shockley ideal diode equation, 343
Shockley–Queisser efficiency limits, 412–413
Shockley-Read-Hall (SRH) recombination model, 105–107, 312–313
Short-circuit photocurrent, 407
Single jump models, 102–103
Singlet state, 330
Sodium–sulfur (NAS) batteries, 52–53
Solar fuels, 1–2, 450–456
Solar cells, 2–3, 5–9, 200, 231, 344–345, 354
  charge collection in
    boundary conditions, 425–429
    distance, 422–424
    general modeling equations, 424–425
    properties, 421–422
  classification of, 437–439
  current-voltage characteristics, 407–408
  DSC, 5–6, 152, 360–363
  efficiency
    limits, 413–419
    measuring, 439–442
    reporting, 439–442
  fill factor, analysis of, 408–412
  fundamental properties, 346–348
  Gärtner model, 433–435
  operation, 375
  performance parameters of, 412
  perovskite, 6–8, 277, 279, 280, 313, 314, 366, 385, 395, 399, 403, 408, 409, 416, 421, 422, 438, 442
  photovoltaic model, 429–433
  physical properties of selective contacts in, 348–351
  power conversion efficiency, 408–410
  reciprocity relation in, 375–378
  recombination current in, 355–356
  Shockley–Queisser efficiency limits, 412–413
  simulation, 436–437
  space-charge region, diffusion-recombination and collection in, 435–436
  tandem, 448–450
Solar energy, 1–2
  into chemical fuel, 2
  conversion, 6–9
  harvesting, photon frequencies conversion for, 445–448
Solar photons, utilization of, 365–368
Solar spectrum, 299–302
Solid angle, 289–291, 293, 296
Solid state lamps, 293
Space-charge-limited current (SCLC), 60–61, 263–265
  in double injection, 267–268
  injected carrier capacitance in, 265–267
  regime, 263
  for trap-free layer, 265–267
Spectral flux, 297
Spectral irradiance, 297
Spectroscopy
  impedance, 265, 271–272
  IMPS, 271, 277
  IPES, 63
  light-modulated emission, 277
  PES, 63–66
  TRTS, 325
  UPS, 63–66, 69
  XPS, 63
Spectrum
  of blackbody radiation, 293–295
  Planck, 294–295
  solar, 299–302
Standard chemical potential, 44
Standard hydrogen electrode (SHE), 43, 45–46
States
  charge transfer, 396, 401
  chemical capacitance of single, 135–136
  kinetics of localized, 106–107
  singlet, 330
  surface, 134
  triplet, 330–331
  valence band, equilibrium of, 76–79
Stefan–Boltzmann law, 299
Steradian (sr), 289–291
Stirling approximation, 76
Stokes' law, 308
Stokes' shift, 111, 308–309
Supercapacitors, 39
Surface dipoles, 24–25
Surface states, 134
Surface vacuum level (SVL), 18
SVL, *see* Surface vacuum level

Tandem solar cells, 448–450
TAS, *see* Transient absorption spectra
TCO, *see* Transparent conducting oxide
TFTs, *see* Thin film transistors
Thermal energy, 1
Thermal engine, 2
Thermal radiation, 293, 296, 298
Thermionic emission, injection to vacuum in, 59–60

Thermionic emission–diffusion theory, 344
Thin film transistors (TFTs), 249
  channel, carrier density in, 250–252
  current-voltage characteristics, 255–257
  disordered semiconductors, mobility in, 257–258
  DOS in, 252–255
  energy scheme of, 251–255
  operating principles, 249
  organic, 249–251
  schematic geometry and voltages of, 250
  structure, 249
  threshold voltage, 249
Time-of-flight (TOF) method, 205–206
Time-resolved microwave conductivity (TRMC), 324–325
Time-resolved Terahertz spectroscopy (TRTS), 325
TiO_2 nanoparticles, 3–4
Tracer diffusion coefficient, 209–210
Transfer coefficient, 117
Transfer function, 271
Transient absorption spectra (TAS), 324
Transistors
  electrochemical, 258–259
  space-charge-limited, 263–268
  TFTs, 249
    channel, carrier density in, 250–252
    current-voltage characteristics, 255–257
    disordered semiconductors, mobility in, 257–258
    DOS in, 252–255
    inorganic, 249
    operating principles, 249
    organic, 249–251
    structure, 249
Transition
  band-to-band, 311, 320
  direct, 319–320
  indirect, 320
  rates, form of, 104–106
Transparent conducting oxide (TCO), 81–82, 152–153, 201
Transport
  diffusion, 209
    chemical diffusion coefficient, 213–215
    length, 212–213
    macroscopic equation, 211–212
    in random walk model, 209–211
    thermodynamic factor, 213–215
  drift
    in electrical field, 197–198
    flux density with, 198
  drift-diffusion, 217
    ambipolar, 221–222
    Einstein relation, 219–220
    equation in terms of electrochemical potential, 217
    equations, 220–221
    injected charge, relaxation of, 222–223
    modeling problems, 224–227
    resistance, 217–219
    transient current in insulator layers, 223–224
  energy, 242–243
  in Gaussian DOS, 237–239
  hopping, 229–231
    conductivity, 241–242
    in single level, 231–233
    VRH, 243–245
  multiple trapping, 229–231
    in exponential DOS, 237
    in time domain, 239–241
  trapping
    kinetic constants, factors in, 233–235
    two-level (single-trap) model, 235–236
Transport layer, 349
Trapping transport
  kinetic constants, factors in, 233–235
  multiple, 229–231
    in exponential DOS, 237
    in time domain, 239–241
  two-level (single-trap) model, 235–236
Traps, 106
Triplet states, 330–331
Triplet–triplet annihilation (TTA), 447
Trivich-Flinn efficiency, 367, 413
TRMC, *see* Time-resolved microwave conductivity
TRTS, *see* Time-resolved Terahertz spectroscopy
TTA, *see* Triplet-triplet annihilation
Tungsten oxide (WO_3), 201
Two-level (single-trap) model, 235–236
Ultraviolet photoelectron spectroscopy (UPS), 63–66, 69, 200
UPS, *see* Ultraviolet photoelectron spectroscopy
Urbach tail, 134, 151, 321–322, 377, 416

Vacuum level alignment (VLA), 66, 202
Vacuum level (VL), 17–20
  away from surface, 18
  local, 18
  surface, 18
Valence band, 14
Valence band states, equilibrium of, 76–79
Vanadium oxide (V_2O_5), 201
Variable range hopping (VRH), 243–245
VL, *see* Vacuum level
VLA, *see* Vacuum level alignment
Voltage, in device, 33–34
  applied, 35–37
  forward, 87
  reverse, 87
  threshold, 249–251
Volta potential, 25–27
Von Neumann boundary condition, 427
VRH, *see* Variable range hopping

Wannier–Mott excitons, 327
Water, 1–2
Wien's displacement law, 294
Work function, metals/organic/inorganic semiconductor, 20–21, 59
  charge-neutrality level, pinning of, 69–73
  Fermi level, pinning of, 69–71
  injection barriers, 66–69
  Kelvin Probe (KP) method, 61–63
  PES, 63–66
  Richardson–Dushmann equation, 59–61
  thermionic emission, injection to vacuum in, 59–60

XPS, *see* X-ray photoelectron spectroscopy
X-ray detectors, 6
X-ray photoelectron spectroscopy (XPS), 63

Yablonovitch limit, 307

Zinc-silver batteries, 51–52